REVIEWS IN MINERALOGY
AND GEOCHEMISTRY

Volume 91A 2025

Surface Complexation Models

EDITORS

Piotr Zarzycki
Lawrence Berkeley National Laboratory, CA, USA

Johannes Lützenkirchen
Karlsruhe Institute of Technology, Germany

Benjamin Gilbert
Lawrence Berkeley National Laboratory, CA, USA
University of California, Berkeley, CA, USA

Series Editor: ***Ian Swainson***

MINERALOGICAL SOCIETY OF AMERICA
GEOCHEMICAL SOCIETY

COVER ILLUSTRATION

The photo in the background Diego de Almagro, Chile by Alberto E. Regazzoni, Gerencia Química, Centro Atómico Constituyentes, Comisión Nacional de Energía Atómica, Villa Maipú, Argentina, schematic drawings in the foreground by Piotr Zarzycki, Lawrence Berkeley National Laboratory, Berkeley, CA, USA.

Reviews in Mineralogy and Geochemistry, Volume 91A
Surface Complexation Models

ISSN 1529-6466 (print)
ISSN 1943-2666 (online)
ISBN 978-1-946850-16-4

THE MINERALOGICAL SOCIETY OF AMERICA
14200 PARK MEADOW DRIVE
SUITE 310-S
CHANTILLY, VA 20151
HTTPS://MSAWEB.ORG/

Surface Complexation Models

91A *Reviews in Mineralogy and Geochemistry* **91A**

PREFACE

Chemical reactions at aqueous interfaces play major roles in Earth terrestrial and subsurface cycles of most reactive elements, trace elements, heavy/radioactive ions, and environmental pollutants. Understanding these interfacial geochemical reactions requires identifying the products that may occur and form. Predicting them requires the reactions to be quantitatively defined through the reaction stoichiometries and associated equilibrium constants. These descriptions are the basis of thermodynamic models, known as Surface Complexation Models (SCMs), which aim to predict the chemical equilibrium state of a given (charged) interface in contact with solution. SCMs appear to be one of the most successful models in geochemistry. SCMs have provided the ability to understand, predict, and explain the sorption of protons, aqueous ions, and molecules to hydrated mineral surfaces using physical models for interfacial processes and energies.

SCMs are, however, often challenging to construct and apply realistically. First, the identity and the stoichiometries of the reactants and products are not easily determined. They must be inferred from bulk measurements, partially glimpsed through molecular probes, or predicted using state-of-the-art simulation methods. Second, the equilibrium constants are influenced by local mesoscale chemical phenomena that alter interfacial free energies while they are, in turn, coupled to changes in speciation at the interface and in the solution. In particular, the distribution of the charges close to the interface—the electric double layer—and the structuring of interfacial water directly influence energetics in a dynamical system. While some aspects of the molecular and mesoscale interactions and energetics remain enigmatic at the fundamental level, SCMs can deliver self-consistent thermodynamic descriptions that can predict interfacial effects in natural systems at scale.

The short course is devoted to the foundational science, illustrates the state-of-the-art experimental and molecular modeling methods for determining structural and energetic data, and demonstrates how these thermodynamic models may be developed for increasingly complex systems such as nanoscale confinement. The volume is intended to provide instruction and guidance to new scientists, suggest new problems and methods for experienced practitioners, and identify opportunities for new collaborations between fields.

The last decades have seen spectacular advances in our ability to experimentally test the mineral/electrolyte interfaces. This volume also describes the progress in surface diffraction, scattering, and absorption techniqwues and how the new experimental insights are incorporated in the SCM construction.

Finally, most systems in the natural world, including most systems we are interested in, are not at thermodynamic equilibrium. Mineral particles frequently undergo interfacial reactions that challenge the assumptions of surface complexation modeling. Mineral surfaces can be chemically and topographically remodeled by small changes in saturation state, slight shifts in redox potential, variation in stress state, and chemical reactions with aqueous ions. Understanding the pathways through which geochemical systems evolve towards equilibrium is a considerable challenge. In some settings with large-scale gradients in chemical parameters, mineral-fluid systems may be well approximated, assuming local equilibrium.

This volume is intended to present the past, present, and future of Surface Complexation Modeling and to nucleate the concepts for the next generation of more comprehensive SCMs that account for more complex chemistries and non-equilibrium phenomena.

Piotr Zarzycki (Lawrence Berkeley National Laboratory, CA, USA)
Johannes Lützenkirchen (Karlsruhe Institute of Technology, Germany)
Benjamin Gilbert (Lawrence Berkeley National Laboratory, CA, USA,
University of California, Berkeley, CA, USA)

1529-6466/25/0091A-0000$00.00 (print)
1943-2666/25/0091A-0000$00.00 (online)
http://dx.doi.org/10.2138/rmg.2025.91A.00

PREFACE

Surface Complexation Models

91A *Reviews in Mineralogy and Geochemistry* **91A**

TABLE OF CONTENTS

1 Solution and Surface Complexation: The European Perspective

Staffan Sjöberg

2 Development and Modus Operandi relating Surface Structure and Ion Complexation Modeling for Important Metal (Hydr)oxides

Tjisse Hiemstra, Johannes Lützenkirchen

3 Impedance Spectroscopy of the Mineral–Electrolyte Interfaces

Youzheng Qi, Yuxin Wu

4 Molecular Controls on Complexation Reactions and Electrostatic Potential Development at Mineral Surfaces

Jean-François Boily

5 Surface Complexation at Charged Organic Surfaces

Maryam Salehi

6 Surface Complexation and Reactivity of Ferrihydrite in Relation to its Surface and Mineral Structure, with Applications to Natural Systems

Tjisse Hiemstra, Annette Hofmann, Juan C. Mendez, Yilina Bai

7 Ligand and Charge Distribution Modeling of Natural Organic Matter Adsorption on Metal (Hydr)oxides: State-of-the-art

Yun Xu, Yilina Bai, Tjisse Hiemstra, Liping Weng

8 Ion-Dependent Calcium Carbonate Cohesion: Insights from Surface Forces Measured between Calcite Surfaces

Joanna Dziadkowiec, Anja Røyne

9 Measurements of the Electrostatic Potential at the Mineral/Electrolyte Interface

Tin Klačić, Jozefina Katić, Davor Kovačević, Danijel Namjesnik, Ahmed Abdelmonem, Tajana Begović

10 Surface Complexation Reactions in Oxide Nanopores

Anastasia G. Ilgen

11 Transport and Surface Complexation in Subsurface Flow-through Systems

Massimo Rolle, Lucien Stolze, Jacopo Cogorno, Muhammad Muniruzzaman

12 History, Algorithms, Model Uncertainty, and Common Pitfalls of Traditional SCM Fitting Procedures

Norbert Jordan, Frank Heberling, Jeffrey Kelling, Johannes Lützenkirchen

13 Practical Application of Surface Complexation Models: Evolution, Approaches, and Examples

David A. Dzombak, Jerry D. Allison, Ted P. Lillys, Jason Mills

Reviews in Mineralogy & Geochemistry
Vol. 91A pp. 1–12, 2025

Solution and Surface Complexation: The European Perspective

Staffan Sjöberg

*Department of Chemistry, University of Umeå,
S-90187 Umeå, Sweden*

Staffan.Sjoberg@umu.se

OBJECTIVES

I am a solution chemist and in 2024 I became 80 years old. I tend to forget what I did yesterday, but I remember the "good old days" quite well. I will share some memories with you. Some solution/surface chemists I have met and never will forget. It is not intended to be an exhaustive overview and I apologize for the omission of pioneering work by someone older as well as younger than me.

I have for more than 50 years now been working with the characterization of the chemical speciation in different aquatic systems involving, in particular, aluminium, and silicon in combination with a great number of inorganic or organic ligands. Since 1990 my main research interest has shifted to studies of surface complexation, especially in different Fe-and Al(hydr)oxide systems.

Based upon automatic, precise potentiometric titrations, equilibrium constants for formed complexes were determined. Furthermore, spectroscopic techniques (e.g., NMR, FTIR, EXAFS, XPS) have given molecular level information of different species.

I will try to give an overview of strong European groups in surface complexation studies. The objective of my paper is also to highlight what I found important in characterizing solution complexation. Could this knowledge be useful when doing surface complexation studies? I will expand in some detail on this below.

Finally, my opinion on the future of research in solution and surface complexation modeling will be presented.

SOLUTION COMPLEXATION

I like to share with you what I have learnt from the "Sillén school" at the Royal Institute of Technology (RIT) in Stockholm and implemented at the Department of Chemistry, Umeå University. **Lars Gunnar Sillén (1916–1970)** was appointed professor in Inorganic Chemistry at the Royal Institute of Technology, Stockholm in 1950.

At that time solution chemists involved in equilibrium studies in general regarded the hydrolysis of metal ions as disturbing side reactions in their measurements. Therefore, experimental conditions were chosen to be able to suppress the hydrolysis of the metal ion to avoid the formation of complicated speciation schemes involving mixtures of mono- and polynuclear binary or ternary hydroxy species. Long equilibration times, often found in

1529-6466/25/0091A-0001$05.00 (print)
1943-2666/25/0091A-0001$05.00 (online)
http://dx.doi.org/10.2138/rmg.2025.91A.01

systems with polynuclear complexes, was also a problem for persons with little (or insufficient) patience that manually performed the titrations. It is difficult to be patient when it's time for lunch or time to go home. To circumvent such problems related to human nature, Sillén now systematically developed the experimental titration technique and introduced automated computer assisted titrators. This meant that systems with slow kinetics were allowed to reach equilibrium even if the time needed amounted to hours, or even days. This also meant that titrations could be performed day and night and during week-ends.

In parallel to this, comprehensive work was devoted to developing graphical methods to find the composition of formed polynuclear hydrolytic species as well as of different iso- and heteropoly anions.

Sillén and coworkers also developed the least squares computer program LETAGROPVRID (Sillén et al. 1962; Sillén and Warnqvist 1964; Ingri and Sillen 1965) The program made it possible to search for composition and stability of "best" fitting complex(es) of the system. In addition, another program HALTAFALL (Ingri et al. 1967) was developed, which made it possible to use known formation constants to construct speciation diagrams in multicomponent systems involving an aqueous phase, several solid phases as well as a gas phase.

The" Sillén School" in Stockholm now became regarded as a leading center in solution chemistry attracting a great number of visiting chemists from all over the world. This is understandable thinking of the facilities put at your disposal: automated titrators, computer programs (probably the best available at that time) and a group of very qualified supervisors. Amongst the many visitors I remember Alberto Vacca (University of Florence), Diego Ferri and Liberato Ciavacca, both from the University of Naples as well as Paul W. Schindler, the University of Bern.

In 1968, I had the privilege of meeting Sillén, for the first and only time. in Umeå. He gave a lecture on "The ocean as a chemical system" (Sillén 1967) and I remember when he was ready to start his presentation, he locked the door leaving late arriving students outside. He did not like people arriving late.

He presented an equilibrium model for seawater. It was assumed that silicate phases of the sediments regulated the pH and the chemical composition of the aqueous phase. However, he made a reservation stating that he did not suggest there would be true equilibrium in the real system. Instead, the model should be regarded as a first approximation to the real system. Later, McDuff and Morel (McDuff and Morel 1980) instead pointed out that the current view of how seawater got is chemical composition is due to the balance between inputs and outputs.

After returning from a visit to Africa, Sillén became very ill and passed away only 54 years old. I was told that to the very end (at the hospital) he was working on the compilation of thermodynamic data for the inorganic part of "Stability Constants" (Sillén et al. 1964).

Figure 1. Lars-Gunnar Sillén Nils Ingri (**left**) and SS (**right**) (1972)

One of Sillén´s students was **Nils Ingri (1929–2014).** In 1965, he was appointed the first professor in Inorganic chemistry at Umeå University and I became a graduate student of his solution chemistry group. Unfortunately, Nils met with an accident in the laboratory in 1966 that made him almost blind for the rest of his life. Nils was of course hampered by his handicap but the door to his office always stood open and he was very eager to help you in solving problems related to your experiment, LETAGROP calculations, and in writing manuscripts.

Good experimental data

Below follow some guidelines regarding potentiometric titrations in equilibrium studies of homogeneous aquatic systems. A more detailed approach is presented in "The experimental determination of thermodynamic properties for aqueous aluminium complexes" (Öhman and Sjöberg 1996). This paper also presents strategies for finding the "best" composition of formed complexes $A_pB_qC_r$ - a so called p,q,r, analysis.

i) A description of electrode calibration should be given. Also, the characteristics of measured electromotive force (e.m.f.) values typical for *stable* potentials at equilibrium should be presented.

ii) Whenever possible the proposed equilibrium model should be complemented by some spectroscopic method to give information about the coordination characteristics of formed complexes.

iii) It is recommended that a constant ionic medium is applied.

iv) Reproducibility and reversibility criteria of the titrations must be demonstrated. This is obvious to everyone but is frequently overlooked.

These guidelines are also very much relevant for surface complexation studies (see below).

Literature equilibrium constants

Choosing the most reliable constants is a serious challenge even for experienced solution chemists. Constants will have been determined by a range of experimental methods, some less appropriate than others, and by laboratories with diverse levels of expertise. Even when the experimental measurements are on a simple binary system, the choice of different experimental pH ranges or total metal ion concentrations or ionic strengths or the nature of the ionic medium can lead to different speciation models and numerically different results.

Noncritical compilations of metal complex formation constants and related equilibrium constants are found in the two extensive large volumes of *Stability Constants*. The first volume, edited by L. G. Sillén (inorganic ligands) and A. E. Martell (organic ligands) was published in 1964 (Sillén et al. 1964). The second volume was published in 1971 (Sillén and Martell 1971). The compilations frequently contained ten to twenty values for a single equilibrium constant. In many cases a stability constant would vary up to two orders of magnitude, which of course made it difficult to identify the most reliable value. The need for *critical* compilations of equilibrium constants became obvious.

Critical compilations. Arthur E. Martell and Robert M. Smith accepted the challenge to critically evaluate published thermodynamic data for binary metal-ligand systems (Martell and Smith 1974). Six volumes (Amino acids, Amines, Other organic ligands, Inorganic complexes, First Supplement and Second Supplement) cover the literature up to 1985. References of these volumes and the ensuing collections of stability constants can be found in a recent review article (Hummel et al. 2019).

Here, I also like to highlight the book *The Hydrolysis of Cations* by Charles F. Baes Jr. and Robert E. Mesmer (Baes and Mesmer 1976). It contains critically evaluated equilibrium constants for mono- as well as polynuclear hydrolytic species of most of the metal ions in the Periodic Table. Furthermore, solubility products of solid metal(hydr)oxides are evaluated. Besides speciation schemes, formation constants and their ionic strength dependency are tabulated. The book also contains a great number of distributions, solubility and predominance area diagrams, very useful to solution and surface complexation chemists. A more recent compilation is available as well (Brown and Ekberg 2016).

Where can I find recommended equilibrium constants? Today, there are several large compilations of equilibrium constants for metal complexes and solids, each having their own limitations, availability, and coverage. In a the recent paper already mentioned above (Hummel et al. 2019), eleven major compilations including those from the authoritative groups, the *International Union of Pure and Applied Chemistry (IUPAC),* the *National Institute of Standards and Technology (NIST)* and the *Nuclear Energy Agency (NEA),* were surveyed. The paper contains a lot of useful references to different data bases and compilations. It is recommended that solution/surface chemists take a closer look at this paper to find the most reliably constants available for their modelling purposes.

The role of IUPAC. One responsibility of IUPAC to the international community is the critical evaluation of published experimental data. For equilibrium constants this includes the identification and recommendation of the most reliable values for each metal-ligand system. It is a very time-consuming process which initially involves the critical evaluation of experimental methods and numerical treatment of data in each publication and at all ionic strengths reported, applying clearly stated criteria to identify reliable publications and reject others, testing the reliability of data by establishing correlations between reliable ("accepted") data for different ionic strengths and rejecting statistical outliers.

This process of critical evaluation of metal + ligand equilibrium (stability) constants has been applied to the complexes formed between the natural-water-dominant inorganic ligands OH^-, Cl^-, CO_3^{2-}, SO_4^{2-} and PO_4^{3-}, the proton H^+, and the environmentally significant heavy-metal ions Hg^{2+}, Cu^{2+}, Pb^{2+}, Cd^{2+} and Zn^{2+} which all occur at trace levels in fresh water and sea water. I was chairing a group of solution chemists involved in an IUPAC project entitled: *Chemical speciation of environmentally significant heavy metals with inorganic ligands.* The results are published in a completed series of five IUPAC Technical Reports. References are included in the recent review paper (Hummel et al. 2019).

In addition to the experimental equilibrium constants, values of the standard state (infinite dilution, zero ionic strength) constants, K_n^{o} or $\beta_{p,q,r}^{o}$, are also reported. These standard state constants are obtained by extrapolation of the accepted experimental values at finite ionic strengths using the Brønsted–Guggenheim–Scatchard Specific Ion-Interaction Theory, SIT. The underlying rationale to use the SIT for the correlation of stability constant data is given in the first Technical Report. In essence, SIT represents a reasonable trade-off between theoretical rigor and numerical simplicity.

An unsought outcome from this critical evaluation is the realisation that there is still a lack of detailed high-quality data for several metal + inorganic ligand systems scrutinized. This is particularly true for the different $M^{2+} + PO_4^{3-}$ systems for which no values could be recommended at $I_m = 0$ mol dm^{-3}. Reliable data in several of the systems $M^{2+} + CO_3^{2-}$ (e.g., M = Cd, Pb, Zn) are also missing at finite ionic strengths, as is the case for the systems $M^{2+} + SO_4^{2-}$. Very few enthalpy values have been possible to evaluate, let alone recommend, which means that a rigorous thermodynamic modelling of the chemical speciation at temperatures different from 25 °C is still not possible.

A challenge to the modelling of surface complexation is the absence of reliable thermodynamic data for metal complexes with the relevant inorganic/organic ligand in solution. The worst-case scenario is that data are missing (or so bad that they cannot be used at all), which means that a determination of the solution speciation must be studied as well. It is recommended that the speciation in solution is controlled by doing some potentiometric titrations to be able to compare the obtained titration curves with the corresponding ones calculated from recommended formation constants.

THE FUTURE

The critical evaluations also showed a trend in the number of published papers per year. Covering a period of more than a century (1900–2010) 900 papers were critically evaluated. The number of papers was highest during 1970–1980, with numbers amounting to 194, and the interest (or the funding) subsequently continued to decrease, with numbers dropping to 78 early this century (2000–2010).

This clear trend shows the field to diminish with time and groups today doing solution complexation studies involving precise potentiometry combined with spectroscopic studies today, are few. In the future I think spectroscopic studies will survive but precise potentiometric titrations like those performed in the "good old days" will be very rare. Consequently, the future for solution complexation modelling, in the absence of accurate pH measurements, is not bright.

In some domains, the study of aqueous complexation is still ongoing though. The nuclear waste community mostly via solubility measurements or spectroscopic studies (Neck et al. 2009; Altmaier et al. 2013) needs stability constants for radionuclides for the safety analysis that is mandatory for nuclear waste repositories. But very few laboratories are still able to do these experiments (also due to the required infrastructure) or receive the funding for long-term studies, that besides the required patience for obtaining results also does not fit into the current pressure to publish a lot.

SURFACE COMPLEXATION

The importance of the mineral/water interface has been emphasized by one of the giants of environmental chemistry (Stumm 1987): "*Almost all the problems associated with understanding the processes that control the composition of our environment concern interfaces, above all, the interfaces of water with naturally occurring solids.*" With this statement in mind, it seems important to generate models that will increase our knowledge about speciation and equilibria at the particle/water interface.

To me the European groups that had the greatest impact on surface complexation modelling (SCM) are the Swiss groups with Paul Schindler and Werner Stumm and the Dutch group in Wageningen with Gerhard Bolt, Willem van Riemsdijk, Tjisse Hiemstra, and Luuk Koopal.

The Swiss giants Schindler and Stumm

In 1988, I met **Paul W. Schindler** in Stockholm. We were both members of an examining board for a Ph.D. defense at RIT in Stockholm. Paul told me that he had spent a year in Stockholm working as a visiting professor with Lars Gunnar Sillén. Paul was known for being a very careful experimentalist and I remember his comment "*Models come and go but good experimental data always survive*". I fully agreed with him on this point. We had a discussion on the future of solution/surface chemistry. I was aware of the pioneering works done by Paul regarding his surface complexation studies in different mineral systems. I indicated that I had an interest in surface complexation and asked him if he was interested in visiting me and my group in Umeå to teach us about the solution/surface interface. I was happy that he gladly accepted my invitation and a collaboration extending over many years was initiated.

Paul has done some pioneering studies on SCM, the first paper published in 1968 (Schindler and Kamber 1968). A 2-pK approach, which involves two consecutive deprotonation steps on one (generic) surface group to give the species $\equiv SOH_2^+$, $\equiv SOH$ and $\equiv SO^-$. Such a mechanism had been earlier proposed by Parks for example (Parks 1965). Based on the work in solution chemistry these interfacial studies were also done at constant, "high" salt levels and the required mean-field electrostatic models were used for a given electrolyte concentration. The CCM (constant capacitance model) that was later used by Schindler's group can best be compared with solution speciation studies using the constant ionic medium method (to keep activity coefficients as constant as possible) at high ionic strengths greater than approximately 0.1 M.

Paul was a gourmet with an interest in cooking and a wine connoisseur. One of his favorite wines was the Rioja wine Faustino 1 (Grand Reserva), which he enjoyed in norther Sweden with a piece of reindeer filet (easy to find here in Umeå), vegetables, and a potato gratin. He owned a cottage in the Swiss Alpes (Col le Mosses) where we did some downhill skiing together before attending a conference in Les Diablerets. He had rented a couple of downhill skies for me, and we both managed to survive the steep downhill slopes! In the evening he prepared a Raclette with melted Gruyere cheese, potatoes, vegetables, and a bottle of chilled white wine.

Paul was a very kind and humble person. He was a heavy smoker, which became a problem when visiting Sweden. Here, it was forbidden to smoke in the office, at work in general and in restaurants. I remember an occasion when Paul was visiting Luleå Technical University because he had been invited to act as opponent at a dissertation. His host was Professor Willis Forsling, who had reserved an office for Paul at the department. When Willis told him that it is not allowed to smoke in the room, Paul became very upset and said: "If you cannot solve this problem, I will go back to Bern!" Paul was allowed to stay in his room and the following day he acted as an opponent to the respondent's satisfaction.

Our collaboration was quite fruitful and successful. Paul became honorary doctor at Umeå University in 1993. He passed away 2006 leaving many fond memories.

My feeling is that there was a kind of rivalry between Paul and **Werner Stumm** (1924–1999). They were two different personalities. Stumm was outgoing and dominating the scene (extrovert), whereas Schindler was more standing behind him with less "sharp elbows". However, I know that Stumm showed Schindler great respect and in a letter to me he wrote that he regarded Paul as a "giant" in the fields of solution and surface complexation modelling.

Figure 2. Paul in a coffee bar. Paul becoming honorary doctor in 1993. Lars Lövgren (**left**) and S. S. (**right**)

Werner Stumm had his PhD in inorganic chemistry at the University of Zürich in 1952. Then he moved to the U.S. where he worked as a professor at Harvard University. From 1970 until 1992, he was head of the Swiss Federal Water Resources Centre EAWAG. Many scientists interested in environmental, aqueous interfaces spent time at EAWAG during that period.

Early in his academic career Stumm was influenced by the ideas of Lars-Gunnar Sillén and Robert Garrels regarding chemical equilibria and kinetics in aqueous geochemistry. He made significant contributions to our understanding of the reactions at the mineral/water interface. Besides studies of surface chemical equilibria, he showed that the rate by which a mineral weathers depends on the surface charge, which is turn varies with pH and the chemical composition of the aqueous phase.

Stumm has written several books. The most influential one is *Aquatic Chemistry* written with James J. Morgan (Stumm and Morgan 1970).

I have met Werner Stumm several times at EAWAG, conferences, and at a workshop outside Stockholm. I still remember him commenting on potentiometric titrations: "Potentiometry has belonged to a poor man´s laboratory. However, today more automated, accurate and slightly more expensive potentiometric titrations can be performed, running day and night."

The Dutch group in Wageningen

I consider the **group in Wageningen**, the Netherlands, to be the most innovative one within the SCM community. Principal Investigators like **Gerhard Bolt, Willem van Riemsdijk, Tjisse Hiemstra, or Luuk Koopal,** have made significant contributions to the development of surface complexation models and tuned them more mechanistically realistic with a view of applying them to a very complex reality. Maybe, one important ingredient was the knowledge on electric double layer implanted in the colloid chemistry school of Wageningen. This and the realization of surface heterogeneity allowed major advances in SCM. Their 1-pK approach exemplified by the reaction:

$$\equiv FeOH_2^{0.5+} \rightleftharpoons FeOH^{0.5-} + H^+$$

defines a pK value equal to the experimentally determined PPZC (pristine point of zero charge) value, resulting in fewer adjustable parameters in comparison with the 2-pK models. This concept was introduced as early as 1982 (Bolt and Van Riemsdijk 1982). Later, it became the backbone of what is now known as the Multi Site Complexation (MUSIC) model (Hiemstra et al. 1989a,b). The possibility to distribute a charge in the compact part of the interface resulted in the Charge Distribution (CD) MUSIC model (Hiemstra et al. 1996; Hiemstra and Van Riemsdijk 2006; van Riemsdijk and Hiemstra 2006).

In 2010, I was invited to be a member of the committee scrutinizing the thesis by Tjisse Hiemstra for the doctorate degree. As expected, the thesis (Hiemstra 2010) was of very high scientific quality, comprising 9 chapters on 383 pages! I enjoyed reading it. The party after the ceremony was nearly as nice as the corresponding parties we have in Umeå. The major difference is that we do more singing!

The group in Umeå, Sweden

From 1988 and onwards, the research within the solution chemistry group in Umeå became more oriented towards surface complexation studies. The finances became better, and the group grew in number. It was obvious from our solution and early surface complexation studies that we were lacking competence within surface sensitive methods like FTIR, EXAFS and XPS. We were very lucky to be able to recruit a young, very promising spectroscopist named **Per Persson.** After having finished a post-doctoral stay at Stanford University under the supervision of Gordon Brown, he joined us in 1992 and stayed in the north for almost twenty years. He was appointed professor at our university, but later moved to his region of origin Skåne, and now works at Lund University, Sweden.

One of my colleagues, Lars Lövgren, met **Jean-Francois Boily** at the 5th Goldschmidt conference (Penn State University) in 1995. At that time "Frasse" had just started his graduate studies at McGill University, Montreal, Canada. He asked Lars about the possibility to visit us in Umeå as a post-doctoral position after finishing his ongoing PhD studies. However, it was decided that he could come for a 3 month stay already the next summer (1996). He turned out to be a very smart and hardworking guy, so I asked him to stay with us to finish his PhD studies. Three years later in 1999 he defended his PhD thesis with Jim Davis as the opponent. After a post-doctoral stay at ETH-Zürich, Switzerland, he received a position as Scientist at Pacific Northwest National Laboratory, USA. He is since 2008 now back in Umeå as professor in molecular geochemistry with a research group of his own. I regard Frasse as one of the very best Ph. D. students I have had.

Another young surface chemist we have benefitted a lot from is **Johannes Lützenkirchen.** He is now working at the Karlsruhe Institute of Technology, and there at the Institute for Nuklear Waste Disposal, in Germany. He also arrived at our department in 1996 and stayed for a bit more than 4 years. Our collaboration has been very fruitful. I regard Johannes as a leading expert in SCM.

Figure 3. Picture taken during an internal mini-workshop in the Swedish mountains in February 1997. The average outdoor temperature was –22 °C. **Back row:** Johannes Lützenkirchen, Johan Lindgren, Per Persson, Jean-Francois Boily, and Lars Lövgren. **Middle row:** Daniela (Johannes's wife), Laurence Bochatay, Julia Sheals and Lars-Olof Öhman. **Front row:** A hare, SS, and a fox.

Good experimental data

SCM requires the determination of additional experimental parameters in comparison with those in solution complexation modelling. A determination of the specific surface area (SSA), the solid concentration, as well as a characterization of the different surface sites (singly, doubly and triply coordinated oxygens) with the corresponding surface site densities will be required.

Guidelines regarding experimental approaches and SCM of reactions at the mineral/water interface have been presented by Lützenkirchen (2002) or Payne et al. (2013).

Specific Surface Area (SSA)

The most popular method of determining the SSA for small particles is the gas sorption BET method (Brunauer et al. 1938), which involves the drying of the particles followed by N_2-gas sorption at constant T and variable P. The uncertainty in this method is estimated to

10% (Hackerman and Wade 1964; Hackley and Stefaniak 2013). The drawback with the BET method is related to the drying step, which can change the structure and surface characteristics of particles that interact strongly with water like metal(hydr)oxides. It is therefore not surprising that much larger differences have been recently reported for goethite particles between BET-method measurements and geometrically determined averaged values (Livi et al. 2017).

Site densities (SD)

It is obvious that to generate a SCM in agreement with molecular level characterizations of the surface coordination of metal ions, metal complexes and inorganic/organic ligands, proton active as well as non- proton active surface sites must be quantified. This is possible for microcrystalline particles of known surface area and with known cleavage plans of the crystals provided the crystallographic parameters are known. Furthermore, surface heterogeneity must be of minor importance. Basic studies of such well-defined particle surfaces are well motivated by the buildup of new knowledge of their coordination chemistry.

We are aware of the limitations of the potentiometric titration technique in determining proton active site densities using proton excess data. It has been shown (Lützenkirchen et al. 2002) that the "active" proton site concentration of goethite accessible down to pH about 1 is just 70% of the crystallographic value. However, from chromate adsorption on Goethite data, Villalobos and Pérez-Gallegos (2008) have shown that the macroscopic adsorption of chromate agrees with the crystallographic values for goethite. This observation is very interesting as it opens the possibility to quantify the surface sites of less well-defined particles (e.g., amorphous) for which crystallographic parameters are missing. This is a challenge to solution/surface chemist in the future.

Equilibration times

The characteristics of measured potentials at equilibrium in solution (in the absence of a solid phase) is a stability often of the magnitude of ±0.2 mV. In general, equilibrium is reached within minutes, but longer times are often needed when polynuclear complexes are formed or when measurements are performed close to precipitation boundaries. In the presence of metal(hydr)oxide particles the reaction kinetics become different. The proton uptake/release by the surface seems to occur in two steps. The first step is claimed to be complete within a few minutes. The second step may continue for days and can be attributed to structural changes at the surface. In the potentiometric titrations at the Umeå laboratory, our strategy has always been to accept the suspension to equilibrate with an accepted *drift* in the measured emf amounting to 0.1–0.6 $mV \cdot h^{-1}$ (0.002–0.01 pH $units \cdot h^{-1}$).

Fe(III) (Hydr)Oxides. **Hydrous Ferric Oxide (HFO) ($Fe(OH)_3$).** A compilation and interpretation of experimental data for the sorption of protons and inorganic ions onto HFO is given in the book *Surface Complexation Modeling. Hydrous Ferric Oxide* (Dzombak and Morel 1990). They have determined surface complexation constants from analysis of "rapid" titration data. A corresponding critical evaluation of published (up to 1995) experimental data for **Goethite (α–FeOOH)** was presented by Mathur and Dzombak (2006). Even, here "rapid" titration data were analyzed. Contrary to this approach we have, in our studies on different goethite system, accepted a drift in the measured emf values of less than 0.6 $mV \cdot h^{-1}$. According to the study by Lövgren et al. this was achieved within ca 15 minutes for pH < 4.5 and took 80 to 100 minutes for 4.5 < pH < 9.5 (Lövgren et al. 1990). When adsorbing Al(III) the waiting time increases up to six hours within the adsorption edge. Kozin et al. (2013) studied the electrolyte ion binding at two goethite samples with different surface areas, 69 and 122 $m^2 \cdot g^{-1}$, and in two different ionic media, 10 mM NaCl and $NaClO_4$. In their potentiometric titrations the drift criteria was set to 0.6 $mV \cdot h^{-1}$. It is interesting to note that the "near" equilibration times were considerably longer for the smaller particles than for the bigger ones. Furthermore, the NaCl solutions required longer equilibrium times than $NaClO_4$

maybe due to the higher charge to radius ratio for Cl^-. The smaller goethite particles exibit a larger level of microporosity than the bigger ones, Obviously, microporosity as well as size of the medium anion affects the "near "equilibration times.

Al(III)(Hydr)oxides. **Gibbsite α-$Al(OH)_3$.** In the study by Rosenqvist et al. (2002), a maximum drift of 0.2mV·h^{-1} was allowed as the "equilibrium" criterion. This value was reached within 8–14 hours after each addition of acid or base (often coulometrically generated). Rapid titrations were also performed. In these, pH was registered 2 min after the addition of acid or base. The uptake/release of protons was significantly lower in comparison with the "slower" titrations and the drift in measured emf values amounted to several mV per hour. In the case of **Boehmite (γ-AlOOH)** the maximum drift in the emf values was set to 0.2 mV·h^{-1}. This criterion was always achieved within 24 h but the time to achieve it showed a considerable variation dependent on pH and the presence/absence of the ligand. Similar reaction times were observed for **Aged Y-Alumina (γ-Al_2O_3)** in our laboratory (Laiti et al. 1995).

Reproducibility and reversibility

Reversibility and reproducibility in equilibrium studies should be demonstrated. Our studies on different Al(III)(Hydr)Oxides as well as goethite have been checked with respect to these criteria to ascertain that the experimental data are at "near" equilibrium. In the **gibbsite** system (Rosenqvist et al. 2002), reproducibility and reversibility were found to be good. This was also observed for **γ-Alumina (γ-Al_2O_3)** and **boehmite (γ-AlOOH)** by Laiti et al. (1995, 1996) and Laiti and Öhman (1996). It was found that γ Al_2O_3 suspensions had to be aged for about one month in aqueous solution before it attained reversible acid/base properties of surface sites. This was related to a surface phase transformation in which **bayerite (β-$Al(OH)_3$** formed at the surfaces of the original particles (Laiti et al. 1998).

Lövgren et.al. found that reproducibility was obtained in the **goethite** system after allowing the goethite batch to age for about 6 months (Lövgren et al. 1990). They also found the acid/ base titrations of goethite to be reversible. However, when Al(III) was added to the suspension, reversibility was poor, probably due to a replacement of Fe for Al in the surface layer(s) within the adsorption/desorption edge ($3.5 < pH < 4.5$). Obviously, surface transformations may lead to longer equilibration times.

THE FUTURE

Today, adsorption studies are much more popular than solution complexation studies. Various levels of sophistication exist, from very simple uptake studies that ignore the need to characterize the acid–base properties of adsorbents that are required for obtaining surface complex stability constants to comprehensive studies that even include a comprehensive spectroscopic characterization of surface complexes. The latter are time consuming not only due to the various experimental approaches involved but also because the associated surface complexation modelling requires time and patience.

So far, we have learned much from surface complexation studies in systems with well characterized microcrystalline particles. We know more about the surface coordination chemistry (inner/outer sphere complexation, mono-, and bidentate surface binding) of adsorbed metal ions and inorganic/organic ligands as well as metal–ligand complexes. We also know more about the characteristics of the electrical double layer (charge distribution). In my opinion, this knowledge will continue to increase with the developments of more sensitive surface spectroscopic techniques.

This opens possibilities for a better understanding of surface complexation in less crystalline, amorphous particle systems. However, it must be emphasized that sophisticated computer programs for surface complexation modelling are available. However, it remains unclear how we should collect good experimental data in particle systems where crystallographic information is lacking. Such systems are well represented in natural waters as well as aquatic systems in industry. In the future, solution/surface chemists should be recruited to help design equilibrium models that describe the chemistry of particle–water interfaces.

ACKNOWLEDGMENTS

I want to thank Jean-Francois Boily and Johannes Lützenkirchen for valuable suggestions, comments, and corrections to the manuscript.

REFERENCES

Altmaier M, Gaona X, Fanghänel T (2013) Recent advances in aqueous actinide chemistry and thermodynamics. Chem Rev 113:901–943

Baes CF, Mesmer RE (1976) The Hydrolysis of Cations. Wiley

Bolt G, Van Riemsdijk W (1982) Ion adsorption on inorganic variable charge constituents, Chapter 13. *In:* Soil Chemistry B. Physico-chemical Models. GH Bolt (ed), Elsevier, Amsterdam, p 459–504

Brown PL, Ekberg C (2016) Hydrolysis of Metal Ions. John Wiley & Sons

Brunauer S, Emmett PH, Teller E (1938) Adsorption of gases in multimolecular layers. J Am Chem Soc 60:309–319, https://doi.org/10.1021/ja01269a023

Dzombak DA, Morel FMM (1990) Surface Complexation Modeling: Hydrous Ferric Oxide. Wiley

Hackerman N, Wade W (1964) Certain aspects of the interpretation of immersional heats of gels. J Phys Chem 68:1592–1594

Hackley VA, Stefaniak AB (2013) "Real-world" precision, bias, and between-laboratory variation for surface area measurement of a titanium dioxide nanomaterial in powder form. J Nanopart Res 15:1742, https://doi.org/10.1007/s11051-013-1742-y

Hiemstra T (2010) Surface Complexation at Mineral Interfaces: Multisite and Charge Distribution Approach. PhD dissertation, Wageningen University

Hiemstra T, Van Riemsdijk WH (2006) On the relationship between charge distribution, surface hydration, and the structure of the interface of metal hydroxides. J Colloid Interface Sci 301:1–18, https://doi.org/10.1016/j.jcis.2006.05.008

Hiemstra T, Van Riemsdijk WH, Bolt GH (1989a) Multisite proton adsorption modeling at the solid/solution interface of (hydr) oxides: A new approach: I. Model description and evaluation of intrinsic reaction constants. J Colloid Interface Sci 133:91–104

Hiemstra T, De Wit JCM, Van Riemsdijk WH (1989b) Multisite proton adsorption modeling at the solid/solution interface of (hydr)oxides: A new approach: II. Application to various important (hydr)oxides. J Colloid Interface Sci 133:105–117, https://doi.org/10.1016/0021-9797(89)90285-3

Hiemstra T, Venema P, Van Riemsdijk WH (1996) Intrinsic proton affinity of reactive surface groups of metal (hydr) oxides: The bond valence principle. J Colloid Interface Sci 184:680–692

Hummel W, Filella M, Rowland D (2019) Where to find equilibrium constants? Sci Total Environ 692:49–59

Ingri N, Sillen LG (1965) High-speed computers as a supplement to graphical methods 4. An ALGOL version of LETAGROP VRID. Arkiv for Kemi 23:97

Ingri N, Kakolowicz W, Sillén LG, Warnqvist B (1967) High-speed computers as a supplement to graphical methods—V: Haltafall, a general program for calculating the composition of equilibrium mixtures. Talanta 14:1261–1286

Kozin PA, Shchukarev A, Boily J-F (2013) Electrolyte ion binding at iron oxyhydroxide mineral surfaces. Langmuir 29:12129–12137

Laiti E, Öhman L-O (1996) Acid/base properties and phenylphosphonic acid complexation at the boehmite/water interface. J Colloid Interface Sci 183:441–452

Laiti E, Öhman L-O, Nordin J, Sjöberg S (1995) Acid/base properties and phenylphosphonic acid complexation at the aged γ-Al_2O_3/water interface. J Colloid Interface Sci 175:230–238

Laiti E, Persson P, Öhman L-O (1996) Surface complexation and precipitation at the H^+–orthophosphate–aged γ-Al_2O_3/water interface. Langmuir 12:2969–2975

Laiti E, Persson P, Öhman L-O (1998) Balance between surface complexation and surface phase transformation at the alumina/water interface. Langmuir 14:825–831

Livi KJT, Villalobos M, Leary R, Varela M, Barnard J, Villacís-García M, Zanella R, Goodridge A, Midgley P (2017) Crystal face distributions and surface site densities of two synthetic goethites: Implications for adsorption capacities as a function of particle size. Langmuir 33:8924–8932, https://doi.org/10.1021/acs.langmuir.7b01814

Lövgren L, Sjöberg S, Schindler PW (1990) Acid/base reactions and Al(III) complexation at the surface of goethite. Geochim Cosmochim Acta 54:1301–1306, https://doi.org/10.1016/0016-7037(90)90154-D

Lützenkirchen J (2002) Surface complexation models of adsorption: a critical survey in the context of experimental data. *In*: Adsorption : Theory, Modeling, and Analysis, Toth J (ed), Surfanctant Science Series, Vol. 107, Marcel Dekker, p 631–710

Lützenkirchen J, Boily J-F, Lövgren L, Sjöberg S (2002) Limitations of the potentiometric titration technique in determining the proton active site density of goethite surfaces. Geochim Cosmochim Acta 66:3389–3396, https://doi.org/10.1016/S0016-7037(02)00948-1

Martell AE, Smith RM (1974) Critical Stability Constants. Springer

Mathur SS, Dzombak DA (2006) Chapter 16—Surface complexation modeling: Goethite. *In*: Interface Science and Technology. Vol 11. Lützenkirchen J (ed) Elsevier, p 443–468

McDuff RE, Morel FM (1980) The geochemical control of seawater (Sillen revisited). Environ Sci Technol 14:1182–1186

Neck V, Altmaier M, Rabung T, Lützenkirchen J, Fanghänel T (2009) Thermodynamics of trivalent actinides and neodymium in NaCl, $MgCl_2$, and $CaCl_2$ solutions: Solubility, hydrolysis, and ternary Ca-M(III)–OH complexes. Pure Appl Chem 81:1555–1568

Öhman L-O, Sjöberg S (1996) The experimental determination of thermodynamic properties for aqueous aluminium complexes. Coord Chem Rev 149:33–57

Parks GA (1965) The isoelectric points of solid oxides, solid hydroxides, and aqueous hydroxo complex systems. Chem Rev 65:177–198

Payne TE, Brendler V, Ochs M, Baeyens B, Brown PL, Davis JA, Ekberg C, Kulik DA, Lutzenkirchen J, Missana T (2013) Guidelines for thermodynamic sorption modelling in the context of radioactive waste disposal. Environ Modell Software 42:143–156

Rosenqvist J, Persson P, Sjöberg S (2002) Protonation and charging of nanosized gibbsite (α-$Al(OH)_3$) particles in aqueous suspension. Langmuir 18:4598–4604, https://doi.org/10.1021/la015753t

Schindler P, Kamber HR (1968) Die Acidität von Silanolgruppen. Vorläufige Mitteillung. Helvetica Chimica Acta 51:1781–1786, https://doi.org/10.1002/hlca.19680510738

Sillén LG (1967) The ocean as a chemical system. Science 156:1189–1197

Sillén LG, Martell AE (1971) Stability Constants of Metal-ion Complexes. Spec Publ 25, (Suppl 1 to Spec Publ 17), Chem Soc (London)

Sillén L, Warnqvist B (1964) High-speed computers as a supplement to a graphical methods. Acta Chem Scand 18:1085–1098

Sillén LG, Timm D, Motzfeldt K, Theander O, Flood H (1962) High-speed computers as supplement to graphical methods. I. Functional behavior of the error square sum. Acta Chemica Scand 16:159–172

Sillén LG, Martell AE, Bjerrum J (1964) Stability constants of metal-ion complexes. Chemical Society, London

Stumm W (1987) Aquatic surface chemistry: Chemical processes at the particle–water interface. John Wiley & Sons

Stumm W, Morgan JJ (1970) Aquatic Chemistry: An Introduction Emphasizing Chemical Equilibria in Natural Waters. Wiley-Interscience

van Riemsdijk WH, Hiemstra T (2006) Chapter 8—The CD-MUSIC model as a framework for interpreting ion adsorption on metal (hydr) oxide surfaces. *In*: Interface Science and Technology. Vol 11. Lützenkirchen J, (ed) Elsevier, p 251–268

Villalobos M, Pérez-Gallegos A (2008) Goethite surface reactivity: A macroscopic investigation unifying proton, chromate, carbonate, and lead(II) adsorption. J Colloid Interface Sci 326:307–323, https://doi.org/10.1016/j.jcis.2008.06.026

Reviews in Mineralogy & Geochemistry
Vol. 91A pp. 13–84, 2025

2

Development and Modus Operandi relating Surface Structure and Ion Complexation Modeling for Important Metal (Hydr)oxides[1]

Tjisse Hiemstra

Wageningen University, Department of Soil Chemistry, Droevendaalsesteeg 3a, 6708 PB Wageningen, The Netherlands

tjisse.hiemstra@wur.nl

Johannes Lützenkirchen

Institute for Nuclear Waste Disposal (INE), Karlsruhe Institute of Technology (KIT), Hermann-von-Helmholtz-Platz 1, 76344 Eggenstein-Leopoldshafen, Germany

johannes.luetzenkirchen@kit.edu

INTRODUCTION AND HISTORICAL NOTE

It is said that Wolfgang Pauli (1900–1958) stated that if God made materials, surfaces were the work of the Devil. The complexity of surfaces remains elusive despite all the tools that surface scientists, geochemists, environmental scientists, and colloid chemists have at their disposal. Simultaneously, it is even more amazing how scientists, who did not have the possibilities of today, have been drawing valid conclusions by developing theories and models that still make sense. One might wonder what they would be capable of nowadays if given time and patience to dig deep and focus.

The range of disciplines interested in surfaces is vast. Surfaces are important since no solid or liquid material exists without surfaces. After all, surfaces are the border between two phases, like the solid and solution phase which is the main focus of this volume. As such, surfaces are interfacial elements, where part of the characteristics originates from the underlying materials. Yet, their behavior is also determined by the bordering phase, the aqueous solution. The solid-solution interfaces can be studied from different perspectives with different aims: theoretically via thermodynamics or state-or-the-art quantum chemistry, and experimentally via the numerous techniques originating from colloid chemistry, crystallography and mineralogy, inorganic and physical chemistry, encompassing wet chemistry, electrokinetics, and spectroscopy. Comprehensive studies involving multiple experimental techniques and model approaches are preferred for deeper understanding, covering the microscopic and macroscopic scales. Mechanistic surface complexation models have the potential to bridge both scales.

[1] This chapter was developed based on the Introduction of the PhD thesis of Tjisse Hiemstra (2010), which was an extended and updated version of earlier work in Encyclopedia of Surface and Colloid Science (p. 3773–3799, Marcel Dekker, Inc. 2002), written by Tjisse Hiemstra and Willem H. van Riemsdijk, on invitation. Substantial additions and updates by Johannes Lützenkirchen are based on courses on surface complexation modeling at Paris Technical Institute of Chemistry. The literature search (we will never claim that all relevant papers were found, since there are far more than we expected) on the very early origins of the models revealed various possible pathways in the development of this kind of modeling. It is impossible to say, to what extent the literature retrieved here was available, and known to the pioneers, but for sure, we found references that surprised us in various ways.

1529-6466/25/0091A-0002$10.00 (print)
1943-2666/25/0091A-0002$10.00 (online)
http://dx.doi.org/10.2138/rmg.2025.91A.02

An historical note on the development of interface chemistry may start in 1895 with the discovery of an unknown kind of radiation, X-rays, by Wilhelm Conrad Röntgen (1898), which was going to play a vital role in inorganic and interface chemistry in general. In 1912, Max von Laue discovered that X-rays can be diffracted. In the same year, the diffraction pattern was interpreted by Laurence Bragg, and used, together with his father, William Bragg, to identify the structure of minerals (Bragg 1912). Atoms were proven to be ordered in regular patterns. It was the birth of crystallography (Thomas 2012). In the following years, structure was found to be the key to understanding materials.

In 1929, the rules that lead to a stable atomic arrangement in crystals were established in an influential paper by Linus Pauling (1929). More than a century after Röntgen's discovery, X-rays still play an essential role in unraveling structures, particularly those at interfaces. For instance, for the first time, the surface structure of adsorbed ions (selenite and selenate on goethite particles) was measured in 1987 (Hayes et al. 1987), and the use of X-rays in the study of water interactions with mineral surfaces started in 2000 for alumina (Eng et al. 2000), followed by a survey of the interfacial ordering of water on mica in 2001 (Cheng et al. 2001) and on quartz in 2002 (Schlegel et al. 2002). The enigmatic structure of ferrihydrite was finally resolved in 2007 (Michel et al. 2007). Grazing-incidence XAS, X-ray standing wave, and crystal truncation rod techniques (Bargar et al. 1996; Trainor et al. 2002; Catalano et al. 2005) are other examples in which X-rays were used. Like in the early years of Lawrence Bragg, knowledge of structures remains a landmark, currently in understanding the chemistry of interfaces with staggering developments and advances in X-ray techniques. A recent review and perspectives paper gives more details on these and other related techniques that have the potential to experimentally elucidate solid/liquid interfaces, even time-resolved (Bañuelos et al. 2023). All these developments in experimental methods contribute to the structural cornerstone of understanding surfaces from an atomistic level.

Another fundamental ingredient for Surface Complexation Models (SCMs), also with a long tradition, comes from the Electrical Double Layer (EDL) theory, a mean-field approach. The early ideas have in the meantime been shown to hold remarkably. In the following, we will first introduce the development of the EDL theory before turning to surface chemistry in SCMs as an analog to solution chemistry, focusing on structural aspects of state-of-the-art SCMs. During the chapter, various examples will be discussed and additional (needs for) developments in EDL approaches will be addressed as well as open questions. Part of the current work includes original, yet unpublished results, calculations, and interpretations

ELECTRICAL DOUBLE LAYER

Concepts

One of the most eye-catching properties of interfaces is the ability to carry charge. Its influence on the adsorption behavior of ions can sometimes be quite counter-intuitive. A good insight can be of help. Here, we describe the basic elements of the double layer.

Charge separation. The presence of surface charge, due to the separation of charge, was conceptually already described by Hermann von Helmholtz in 1879 (von Helmholtz 1879), as visualized in the left upper panel of Figure 1 showing original figures from the publication of Otto Stern (Stern 1924). The charge separation, typically quantified in terms of charge densities ($+\eta_0 - \eta_0$), expressed in Coulombs per surface area (C/m^2), results in an electrostatic potential difference ψ_0 (V), a measure for the corresponding energy E in Joule per unit charge expressed in Coulomb (V = J / C).

Diffuse double layer. The surface charge of colloids in suspension is balanced by counter and co-ions in the aqueous phase. The combination is called the "electrical double layer", EDL. The counter and co-ions are diffusely distributed with respectively decreasing and increasing concentrations towards the solution as experienced by colloid and interface chemists when

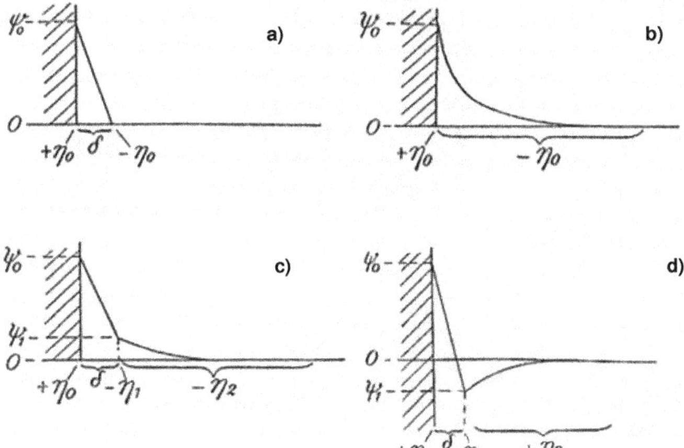

Figure 1. Schematic EDL structures, published by Stern (1924) visualizing the basic concepts of the potential-distance relationship. The symbol η represents charge density. The sign of charge is explicitly given and, therefore, it is different from the symbol σ, in which the sign is implicitly part of the definition of charge density. **a)** depicts the charging concept of von Helmholtz (1879), **b)** shows the diffuse double layer picture of Gouy (1910) and Chapman (1913). **c)** and **d)** give the model of Stern himself (Stern 1924), showing some neutralizing charge ($-\eta_1$) in the 1-plane, present at a distance (δ) from the surface, which leads to a break in the slope of the potential-distance curve. If more counterion charge is present at this location, the net particle charge ($\sigma_0 + \sigma_1$) may reverse, as shown in d). The location of the charge (η) is indicated at the horizontal axis, being at the surface η_0, in the Helmholtz plane $-\eta_1$, or the diffuse double layer $-\eta_2$. Note $\eta_0 - \eta_1 - \eta_2 = 0$ but $\sigma_0 + \sigma_1 + \sigma_2 = 0$. [©2021 John Wiley & Sons, Inc.]

studying colloidal properties. In 1910 and 1913, Louis Georges Gouy (1910) and David Leonard Chapman (1913) developed independently of one another a theory, known as the diffuse double layer (DDL) theory. It describes the diffuse distribution of charged counter- and co-ions as a function of distance from a planar surface. The corresponding potential–distance relationship is shown in Figure 1b.

Stern layer. In the decade after the formulation of the DDL concept, Otto Stern extended the double layer theory in 1924, explaining the relationship between the charge of a mercury electrode and the interface potential between the electrode surface and an electrolyte solution (Stern 1924). He concluded that the charge in the DDL was separated from the surface charge of the mercury electrode by a layer, empty of charge, nowadays called the Stern layer (Figs. 1c,d). The physical explanation of the Stern layer was the finite size of the counter ions with a corresponding minimum distance of approach to the surface, i.e., the structure of ions was involved. Stern recognized that ions should not be considered as point charges at the scale of the compact part of the EDL. Stern also recognized that electrolyte ions might be specifically adsorbed at the Helmholtz plane (Figs. 1c,d). David Grahame (1947) later refined this concept. The adsorption of electrolyte ions that retain their primary hydration shell and therefore are placed at some distance from the surface plane is called nowadays outer-sphere (OS) complex formation (Sposito 1984).

The sketches in Figure 1 have been the ingredients for surface complexation models (SCMs). They appear in the terminology used by Westall and Hohl (1980) as the constant capacitance (CC) model (Fig. 1a), the diffuse layer model (DLM) given in Figure 1b, and the basic Stern (BS) model represented in Figures1c,d. From these, Dzombak has chosen the DLM model as a pragmatic approach to developing a thermodynamic database for surface complexation modeling (Dzombak and Morel 1990; Mathur and Dzombak 2006; Karamalidis and Dzombak 2011), frequently used by practitioners.

Generalization. The generalized representation of the EDL model is known as the Gouy–Chapman–Stern-model or Grahame model. In this model (Fig. 2a), the Stern layer involves two distinct layers, bounded by the surface plane, the inner Helmholtz plane, and the outer Helmholtz plane. This picture is the most comprehensive EDL model that is frequently used in surface complexation modeling. Still, the charge allocation of adsorbed ions in the electrostatic planes leads to different versions. It has been used since the late 1970s (Yates et al. 1974; Davis et al. 1978), and the various versions are named the triple layer model, the three-plane model (Hiemstra and Van Riemsdijk 1996), and the Extended Stern layer model (Hiemstra and Van Riemsdijk 2006).

A four-layer model can be found in the paper by Bousse et al. (1983). Compared to the three-plane model, the inner layer is subdivided to accommodate ions of different degrees of hydration (Fig. 2b). In the inner layers a linear drop of the interfacial potential is assumed. The derivation of the resulting electrostatic equations can be found elsewhere (Ohshima 2006). Noteworthy, that derivation implies the absence of a potential gradient within the solid (Ohshima 2006). If the counter-ions are treated as point charges, the capacitance values in the multilayer models may have to vary when trying to simulate the relative position of the various ions treated as point charges. Another approach for treating electrolyte ions is a charge distribution option (Rahnemaie et al. 2006). It results in a comprehensive model framework that can in principle cover any solution composition (i.e., also mixtures) with a unique set of parameters.

Application in surface complexation models. Figures 1 and 2 depict the principles concerning the structure of the EDL. Starting in the late sixties, these double layer representations have been implemented in a large variety of SCMs that also comprise different formulations of surface chemical reactions. Various proposed models are the Constant Capacitance (CC) model (Atkinson et al. 1967; Schindler and Kamber 1968; Kummert and Stumm 1980;

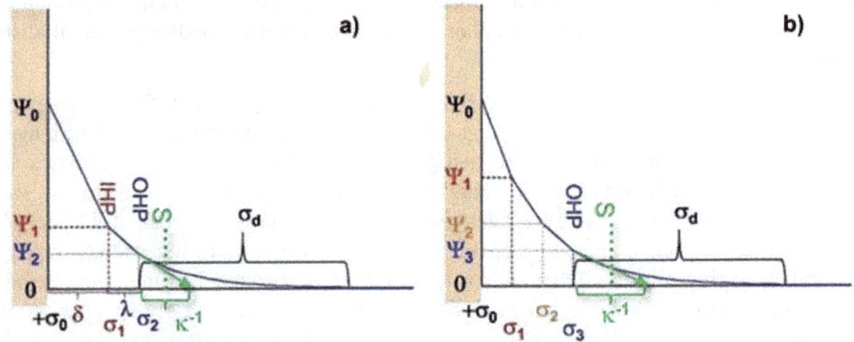

Figure 2. **a)** Schematic EDL structure according to Grahame (1947). Compared to the Stern–Gouy–Chapman model, the Stern layer is split into two layers, involving three planes, i.e., the surface plane, the inner Helmholtz plane (IHP), and the outer Helmholtz plane (OHP). Any change in the slope of the potential–distance curve occurs if the strength of the electrical field, radiating from the surface charge, is changed by the presence of a net amount of counter- and co-ion charge. For instance, the absence of discrete charges in the OHP will result in equal slopes of the potential–distance curve on both sides of the OHP. This is assumed in the traditional Triple Layer model (TLM). **b)** Four-layer (FL) model according to Bousse et al. (1983), in which the IH layer from a) is split into two layers with specific planes where cations and anions are accommodated. At electroneutrality, $\Sigma \sigma_i = 0$. The potential-distance relation $\psi(x)$ at the head end of the DDL can be extrapolated linearly (**green arrow**) to a zero potential using the slope of the potential at the head-end of the diffuse double layer, yielding the Debye length (κ^{-1}) (given in **green**) at the x-axis, which is a measure for the thickness of the DDL. Interpretation of zeta-potential becomes possible with surface complexation modeling by positioning the slip-plane away from the head end of the DDL, at distances between 0.17 and 0.5 times the Debye length (κ^{-1}), as will be discussed later in this contribution.

Sigg and Stumm 1981; Goldberg and Sposito 1984), the extended CC (ECC) model (Nilsson et al. 1996), the Diffuse Double Layer (DDL) model (Dzombak and Morel 1990); (Stumm et al. 1970), the Basic Stern (BS) model (Westall and Hohl 1980; Hiemstra et al. 1987; Borkovec 1997; Schudel et al. 1997; Felmy and Rustad 1998; Lützenkirchen 1998; Machesky et al. 1998; Christl and Kretzschmar 1999), the Triple Layer (TL) model, (Yates et al. 1974; Davis et al. 1978; Hayes and Leckie 1987), the extended triple layer (ETL) model (Sverjensky 2005), the Variable Charge–Variable Potential (VC–VP) model (Bowden 1973; Bowden et al. 1977, 1980; Barrow et al. 1980; Barrow 1987; Barrow and Bowden 1987) or the Four Layer (FL) models (Bousse et al. 1983).

Physical–chemical representations of the EDL

The above representations of the EDL can be translated into physical-chemical pictures being schematic or molecular as illustrated in Figure 3 (Bockris and Reddy 1976a). Figure 3a shows an interface with a defect-free metal oxide and oriented water with cations and anions. Depending on water dipole orientation, different dielectric constants are attributed to the various assumed layers of the EDL. Figure 3b gives a more detailed molecular picture of the interface that involves surface defects, chemical heterogeneity, and constraints like confinement or processes such as the flow of water. In this representation, the focus is solely on a microscopic view without a link to the conceptual potential profiles of Figure 2.

At some point, it will be required to link such microscopic information to the macroscopic level at which many laboratory and environmental observations are made. There is little doubt that interfacial charge is a major ingredient in the overall picture and that the microscopic mechanisms must be interpreted in light of the electrostatics. One major difference between both views is that in the older one, the dipole arrangement is such that one might speak of dielectric saturation, where the bare surface charge can align the water dipoles. In the current view on the other side, the H-bonding network is another factor in the water structure at these charged interfaces. The subtle interplay between both will determine how the water structure organizes itself. To what extent the resulting water structure may also cause feedback on the bare charging or other phenomena is an interesting problem. Ultimately, at equilibrium, the overall energy must be minimized for a given solution composition.

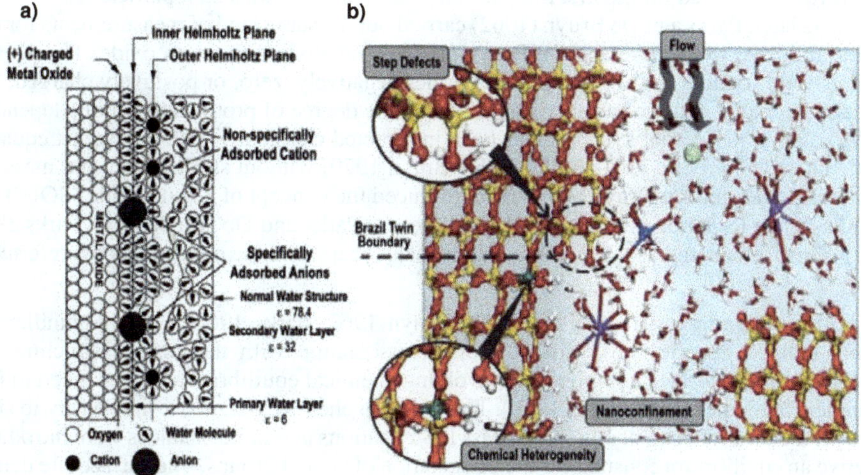

Figure 3. Interfacial representations from a recent review article (Bañuelos et al. 2023). The left figure is inspired by the corollaries of Bockris and Reddy (1976a). [Reprinted with permission from Bañuelos et al. (2023) Chem Rev 123(10):6413–6544. Copyright 2023 American Chemical Society.]

SURFACE CHARGE

Charge separation is the basis of the EDL, as emphasized above. In the following, the origin of surface charge will be described. It will also be discussed, how it can be related in SCMs to the solution conditions using chemical reactions implementing the concept of the electrostatic potential and energy.

Origin of surface charge

The primary charge of colloids and minerals may originate from two different sources. In the lattice of some minerals, ions with a higher valence have replaced ions of approximately the same size but with a lower valence, known as isomorphic substitution. The resulting charge deficiency is compensated for by ions outside the primary structure. Part of these ions may form a DDL in contact with water. Since the charge deficit is located in the primary structure, the mineral has a permanent charge, which is independent of the conditions in the solution. In environmental systems, clays and clay minerals are known to exhibit permanent charge (Sposito 1984).

Metal (hydr)oxides form a different group of charged minerals. These materials generally have a pH-dependent variable charge. There is common agreement that the mechanism that causes variable charge, is adsorption or desorption of protons. This is also widely accepted for the edge faces of the clay and clay minerals that were mentioned in the previous section, making them a complex kind of particle with overlapping permanent and variable charge, resulting in regulated situations, which are more complex than simple variable charge minerals. In the case of metal (hydr)oxides, the development of primary charge handles the potential profiles given in Figures 1 and 2 and is among the most frequently studied adsorption phenomena, particularly with wet chemistry in the beginning, but also by spectroscopies and increasingly by advanced quantum chemistry. An EDL may also develop in the presence of dipoles (Levine and Bell 1980). Hunter states that in the interfacial region, the effects of dipole orientation and polarization effects are expected to be small and restricts his discussion of zeta-potential to charges, but states that closer to the surface the two other contributions become more important and should be considered (Hunter et al. 2013).

Variable charge

Strong interest in variable charge materials started in the late 50s and early 60s. Gerard Bolt in 1957 measured for the first time the variable charge of silica nanoparticles (Bolt 1957). Five years later, Parks and De Bruyn (1962) carried out the same kind of measurements for iron oxides, followed by Bérubé and De Bruyn (1968) who studied titanium oxide. These oxide surfaces were found to be amphoteric, i.e., either negatively, zero, or positively charged. The variation in charge at these surfaces was related to the degree of proton adsorption, depending on the solution pH. Variable charge has been interpreted classically with the Nernst equation (Lyklema and Overbeek 1961; Blok and De Bruyn 1970) without surface sites and mass law equations. Parks and De Bruyn and Parks introduced the concept of a neutral site (SOH^o) that may dissociate (SO^-) or associate (SOH_2^+) a proton (Parks and De Bruyn 1962; Parks 1965, 1967). This formalism has been embraced for many years, and the concept is now referred to as the "2-pK model".

Following earlier work (Parks and De Bruyn 1962; Parks 1965, 1967), Schindler and Stumm defined generic sites without a direct association with the surface structure. For amorphous silica, surface complexation involving chemical equilibria can be assigned to Paul Schindler (Schindler and Kamber 1968). This solution chemist assigned a site density to silica based on the titration data of silica gel. Mass law equations and mole balances were introduced to derive an equilibrium constant for the dissociation of silanol groups. The surface site density was assessed from the titration to a maximum dissociation at high pH. The value was found to be electrolyte-concentration-dependent due to the development of an electrostatic field. Therefore, this group later used the CCM (Schindler and Gamsjäger 1972). At approximately the same time, another Swiss, Werner Stumm, used a similar approach for Al_2O_3 (Stumm et al. 1970; Huang and Stumm 1973), invoking the DLM to account for electrostatic corrections, that date back to the work of Atkinson et al. (1967), who used equilibria in terms of adsorbed amounts, without directly resorting to surface sites.

Electrostatic potential

Overall, and in agreement with what one of the authors of this chapter retained from discussions in the early 1990s[2], the idea of using electrostatic corrections came from the treatment of polyelectrolyte titration data. The paper by Atkinson et al. (1967) refers to a work from 1944, where chemical species (states) were used (Gilbert and Rideal 1944), building on even earlier work of Steinhardt et al. (1940), who covered electrostatic contributions with a Donnan-type potential.

Several groups can be identified that involved chemical equilibria when treating the adsorption of various ions on hydrous ferric oxide or silica (Kurbatov et al. 1951a; Ahrland et al. 1960b; Stanton and Maatman 1963; Dugger et al. 1964). The work on HFO involved mass law equations with non-integer coefficients for the proton (Kurbatov et al. 1945), required due to the absence of an electrostatic term. This work is built on experimental information collected by scientists like Kurbatov, Kolthoff, and others (Kurbatov 1931, 1949; Kolthoff and Stenger 1932, 1933, 1934; Kolthoff 1935; Kurbatov et al. 1945, 1951a,b; Kurbatov and Kurbatov 1947), concerning the effect of pH and background electrolyte concentration on the adsorption of ions. Work in the early 1960ties already proposed an equation that resembles the formation of bidentate uranyl surface complexes or related adsorption constants for various cations to their first hydrolysis constant (Ahrland et al. 1960a,b; Dugger et al. 1964). All the work mentioned above has contributed to a better understanding of ion adsorption and the development of concepts implemented in current state-of-the-art SCMs.

Surface complexation modeling has largely benefited from the development of computer programs facilitating the use of electrostatic models. The first charge regulation code is from Ninham and Parsegian (1971) for the surfaces of biological cells having ionizable groups. In their approach, the proton concentration at the surface (c_s) was related to that in the bulk solution (c_o) via an electrostatic interaction factor, known as the Boltzmann factor, according to

$$c_s = c_o\, e^{-\frac{z\mathrm{F}\psi(x)}{\mathrm{R}T}} \tag{1}$$

where z is the proton charge (+1), F is the Faraday constant (96485 C·mol^{-1}), $\psi(x)$ is the local electrostatic potential (V) at location x, R is the gas constant (8.31 J·mol^{-1}·K^{-1}) and T is the absolute temperature (K). The use of the Boltzmann factor can be traced back to the work of Weiss (1963), McLaren and Babcock (1959), and Hartley and Roe (1940). The charge regulation model is nowadays used to simulate self-consistently the electrostatic contributions to force–distance measurements (Zhmud and House 1997; Zhmud et al. 2000; Trefalt et al. 2016)[3].

Electrostatic energy corrections

In the context of environmental chemistry, the development of computer programs was initiated by Morel and Morgan (1972), subsequently pursued by Westall (Westall et al. 1976; Westall 1979; Westall and Hohl 1980) for forward modelling, and later extended to inverse modeling (Westall 1982). It replaced for instance graphical extrapolations for deriving surface hydroxyl acidity constants (Schindler and Gamsjäger 1972; Davis et al. 1978).[4]

---

[2] When talking to Lars-Olof Öhman during his post-doc at Umeå University, JL understood that Paul Schindler during visits at the same university had referred to the origins of the ideas in polyelectrolyte chemistry when developing the idea of applying solution chemical concepts to oxide surfaces.

[3] Measurement and interpretation of such data are the topic of a separate chapter in this volume.

[4] Algorithms and codes are discussed in a separate chapter of this volume.

Understanding surface charge behavior is not feasible without considering electrostatics. It is possible to describe surface charge development without formulating a surface reaction. In that case, one assumes that the surface behaves as a Nernstian surface. The Nernstian approach can also be used to describe the proton co-adsorption of metal ions to oxides (Fokkink et al. 1987). However, if a protonation reaction is formulated, the surface will react *by definition* non-Nernstian, which means that at the change of the pH by 1 unit, the surface potential changes by less than 2.3R*T*/F or ~59.2 mV at 25 °C. Another suggestion made is to measure *in-situ* the surface and zeta potential and introduce these in the electrostatic expression for a capacitor, assuming that the zeta potential equals the Stern layer potential (Brown et al. 2016b). The validity of the measurements and interpretation will be discussed later in a case study.

In the mass law equations for the formation of surface complexes, activity coefficients for solution species are typically used to represent the electrostatic interaction energy of ions in the solvent (E_{sol}) that can be calculated with the Debye–Hückel spherical double layer theory. The lateral electrostatic interaction energy of the ions in the adsorption phase at the surface (E_{surf}) is calculated with one of the available EDL models. Both activity corrections are usually present in the expression in the mass law expression energy (log *K*). This can be exemplified by the adsorption reaction of species A^z to $n = 1$ sites S with $S + A^z \rightleftharpoons SA^z$ (monodentate adsorption reaction) for which an equilibrium expression $Q = K$ can be formulated:

$$Q = \frac{\theta_A}{(1-\theta_A)^{n=1}} \frac{1}{\left[A^z\right]} = K_{ads} \equiv e^{-E_{sol}/(RT)} \times e^{-E_{surf}/(RT)} \times e^{-E_{intr}/(RT)} \quad (2)$$

in which the ionic product, or reaction quotient *Q* is thermodynamically defined using mole fractions (θ) of occupied sites. For the solution species, the mole fraction is replaced by a concentration, denoted here with [ ][5]. In the case of a multiple (*n*) dentate reaction, the denominator can be defined as $(1-\theta_A)^n$. In that definition, the use of the mole fraction is essential to make the quotient of Equation (2) dimensionless, avoiding the *K* value becoming system-size-dependent.

The overall adsorption energy E_{ads} of Equation (2) has three energy terms, namely the energy E_{sol} for ions in solution, the electrostatic interaction energy E_{surf} for ions bound to the surface, and an intrinsic adsorption energy E_{intr} that comprises all other energy contributions. It seems that the paper by Stumm et al. (1970) introduced the terminology of an intrinsic stability constant devoid of mean-field electrostatic contributions. In some cases, E_{sol} has been discarded (Hayes et al. 1991). In the expression, no additional energy contributions are assumed for ions in the adsorption phase, meaning no additional activity corrections (Yates et al. 1974). In the case of the formation of a co-precipitate, a mole fraction of the ions in the adsorption phase can be additionally introduced (Farley et al. 1985). So far, this has been mainly restricted to the formation of ideal solid solutions.

Point of zero charge or PZC

To explain the PZC values of (hydr)oxides, partial charges on surface oxygens have been introduced (Ray and Sen 1974; Yoon et al. 1979). The work of Yoon et al. (1979) involved the Pauling bond valence (v) and they introduced a single chemical equilibrium between a fully deprotonated and fully protonated site according to $\equiv MO^{-2+v} + 2H^+ \rightleftharpoons \equiv MOH_2^{+v}$. However, the surface species in this equilibrium are not expected to co-exist as the difference in proton

[5] [i] is often used to denote the molar concentration of species i. In most applications the activity of species i is used, so that the same way of extrapolating to infinite dilution for the aqueous species is used as for the concomitant treatment for pure solution systems. The rigorous notation of activity involves either a mole fraction multiplied by the associated activity coefficient or a given concentration divided by the corresponding standard state concentration. Concentration can be for example molar (m) or molal (M). Since the standard states are typically at 1 m (or M) for solution species, it is often omitted, and only the concentration in molar units is given. For many environmental conditions molar and molal concentrations hardly differ. For high salt concentrations, there can be significant differences in the numerical values of a certain concentration. In addition, the molar concentration changes with temperature, while the molal concentration does not. For surface species, activity coefficients are not available and are omitted. It has been stated that the ratio of the two activity coefficients for the surface species involved in Equation (2) cancels or that the activity coefficient related to the non-ideality is expected to be much smaller in magnitude compared to the electrostatic factor discussed below.

affinity of the consecutive pronation steps $\equiv MO^{-2+v} + 2H^+ \rightleftharpoons \equiv MOH^{-1+v} + H^+ \rightleftharpoons \equiv MOH_2^{+v}$ is extremely large as argued by Bolt and Van Riemsdijk (1982) and later shown by Hiemstra et al. (1989a). It implies that only one of both protonation steps determines the charging behavior within the limitations of the experimental pH window. Various other approaches to relate the PZC of oxide minerals are described in more detail elsewhere (Lützenkirchen et al. 2002). Published values for the PZC of minerals have been summarized by Kosmulski (2001, 2016) and are being regularly updated by that author (Kosmulski 2004, 2006, 2009, 2011, 2014, 2018, 2020, 2021, 2023). Note that there are various notions relating to PZCs. It is necessary to explicitly define which kind of measurement a reported PZC refers to. The above refers to the pristine PZC as being defined by the reactions. More detailed discussions of PZCs can be found elsewhere (Sposito 1998; Lützenkirchen et al. 2002).

Physical surface sites

The first attempt to identify the imaginary site in terms of a physical structure came from Hingston et al. (1968), who proposed that for sesquioxides (sesqui ~ 1½) with octahedral metal coordination, the site is a combination of two surface groups, i.e., an $\equiv OH$ and an $\equiv OH_2$. Implicitly, this can be interpreted as an application of the Pauling bond valence concept (Pauling 1929) when attributing to each ligand an average charge of +1/2 valence unit (v.u.), leading to $(\equiv OH^{-0.5})_2]^{-1}$, $(\equiv OH^{-0.5})(\equiv OH_2^{+0.5})]^0$, and $(\equiv OH_2^{+0.5})_2]^{+1}$ for the surface species SO^{-1}, SOH^o, and SOH_2^{+1}. Bolt and Van Riemsdijk (1982) applied the Pauling bond valence to the surface groups of gibbsite. They formulated the protonation reaction of the singly coordinated surface groups as $\equiv MOH^{-1/2} + H^+ \rightleftharpoons \equiv MOH_2^{+1/2}$.

Spectroscopy has played an important role in directing the development of theoretical concepts. In the early years, infrared (IR) spectroscopy showed the existence of different types of surface groups for example on silica (Mcdonald 1958). Surface groups can be conveniently defined based on the number of metal ions coordinating with the surface oxygen ions using a crystallographic analysis as illustrated in work from about 50 years ago (Peri and Hensley 1968; Jones and Hockey 1971; Marshall et al. 1974; Yates 1975; Rochester and Topham 1979). The importance of metal ion coordination on the reactivity of surface groups became increasingly clear in the seventies through the work of Parfitt et al. (Parfitt and Russell 1977; Parfitt et al. 1975, 1976, 1977a,b,c), and others[6]. Yet, no attempts were made to include such information in SCMs, prior to the publication of the Multi-Site Complexation (MUSIC) model in 1989 (Hiemstra et al. 1989a,b).

Proton affinities

Proton affinity of individual types of surface groups is essential in any multiple-site SCM. The first *a priori* assessment of the affinities of individual surface groups for various metal (hydr)oxides was published by Hiemstra et al. (1989a). The proton (H) affinity of the surface groups was related to the Pauling bond valence charge of the coordinating metal ion (M) of the solid and combined with an H–M distance parameter, inspired by a similar electrostatic interpretation of the PZC of Parks (1967) and Yoon et al. (1979). While the Pauling concept is only adequate for a symmetrical coordination environment of metal ions, in many minerals such as goethite, hematite, corundum, anatase and rutile, the actual coordination sphere is asymmetric[7]. After realizing that the bulk of goethite consists of only triply coordinated groups and that these differ very strongly in proton affinity ($\equiv Fe_3O_IH$ and $\equiv Fe_3O_{II}$)[8], the concept had

[6] It is noted here that the authors will not claim that all relevant publications are mentioned. This is virtually impossible. A more detailed discussion of the nature of surface sites can be found in Boily (2025, this volume)

[7] Some of these minerals are discussed in separate chapters in this volume.

[8] This notion started with the CD modeling of the PO_4 adsorption to goethite showing that the effective density of triply coordinated surface groups was much lower than can be understood from treating all triply coordinated groups at the goethite surface equally. Asking Alain Manceau at a meeting in Paris whether he believed that triply coordinated surface groups of goethite might differ in proton affinity, he answered, without any hesitation, with "Of course", adding, "Look to the bulk structure, one O has a proton, the other not".

to be further developed. In the subsequent attempt to estimate the proton affinities (Hiemstra et al. 1996), asymmetry in the coordination environment was introduced, via the bond valence-bond length relation developed by I. David Brown (1978). For applying the full MUSIC model knowledge of the morphology of the target particles is a prerequisite.

More recently, many attempts have been made to find the proton affinities by molecular modeling techniques (Rustad et al. 1996a, 1998; Vlcek et al. 2007; Aquino et al. 2008; Machesky et al. 2008; Cheng and Sprik 2010; Sulpizi et al. 2012; Liu et al. 2013; Cheng and Sprik 2014; Cheng et al. 2015; Pfeiffer-Laplaud et al. 2015; Zhang et al. 2021; Smith et al. 2022; Li et al. 2023; Wen et al. 2023; Yuan et al. 2024). Such studies are valuable in the sense that the results can be used in the MUSIC framework for surface complexation modeling. Some of these approaches can be qualified as very successful when examined and evaluated with surface complexation modeling, as will be discussed and illustrated in later sections for SiO_2, gibbsite, goethite, and rutile. In other cases, the outcome might require a closer look.

An important aspect of assessing proton affinities of surface groups is validation. Most frequently, the results of modeling are compared with experimental values for the PZC. Ideally, the experimental PZC value results for measurements done for well-defined crystal faces with little defects and irregularities. This can be exemplified for α-TiO_2(110) surfaces. For this material, AFM (Bullard and Cima 2006), electrokinetic potentials (Preočanin et al. 2017), or second harmonic generation (Fitts et al. 2005) data have been collected, pointing to a PZC/ IEP of 5.2±0.3, 5.1, and 4.8±0.1, respectively that can be compared to computational values. Another relevant experimental feature is the difference in proton affinity constants. For rutile, initially (Cheng and Sprik 2010) the calculated proton affinity of the $\equiv TiOH^{-1/3}$ site was pK_a = 9 or 8, and for $\equiv Ti_2O^{-2/3}$, it was pK_a = −2 or −1. These values are too high and too low respectively, leading to a too-large ΔpK_a (~10), strongly suppressing the development of surface charge, particularly below pH 7, as shown in Figure 4 with dotted lines. This conflicts with titration data for TiO_2 (Bourikas et al. 2001; Ridley et al. 2009) and does not agree with published spectroscopic results (Connor et al. 1999). In a later update (Cheng et al. 2015), the pK_a values were adapted to pK_a = 6.5 for $\equiv TiOH^{-1/3}$ and pK_a = 2.4 for $\equiv Ti_2O^{-2/3}$, resulting in a small ΔpK_a of 4. More recently, an even smaller ΔpK_a of 1 has been calculated (Wen et al. 2023).

A novel experimental approach to proton affinity has been recently developed using a non-contact atomic force microscope (nc-AFM) in an ultra-high vacuum. The approach measures the force between a probing tip and protonated surface group, which is then translated by correlation to the log K_H under aqueous conditions (Wagner et al. 2021). The proton affinities for anhydrous TiO_2 (110) and In_2O_3 (111) surfaces are in good agreement with MUSIC model predictions covering a very wide range of pK_a values (~3–20), as shown in another contribution to this volume[9]. The large difference in pK_a can be understood from the differences in metal ion valence of the oxide, the coordination of the surface oxygens as well as the M–O bond length. The highest and lowest pK_a values in the measured range are both for doubly coordinated groups, i.e., $\equiv Ti_2O^{-2/3}$ (log K_H ~3±1) and $\equiv In_2O^{-1}$ (log K_H ~20±1). Part of this difference (ΔpK~7) is due to the difference in the valence of the cation of the oxides (In^{3+}, Ti^{4+}) and the remaining part (ΔpK~10) can be explained by the difference in surface relaxation of In_2O_3 and TiO_2. For the surface of the latter material, the Ti–O bond lengths decrease strongly from 1.95 Å in the bulk to 1.83 Å at the α-TiO_2 (110) surface in UHV, while for the In_2O_3 surface, contractions are small.

There are still many open questions and challenges that remain in the interpretation of charging phenomena on oxidic surfaces in aqueous electrolyte solutions. Not only has it become obvious that many particle surfaces will not be ideal and involve defects (Klaassen et al. 2017; Livi et al. 2017, 2023), but there are also discrepancies between *a priori* estimates of proton affinity constants resulting in very different surface speciation schemes, as discussed later for

[9] In a separate chapter included in this volume, this method is detailed.

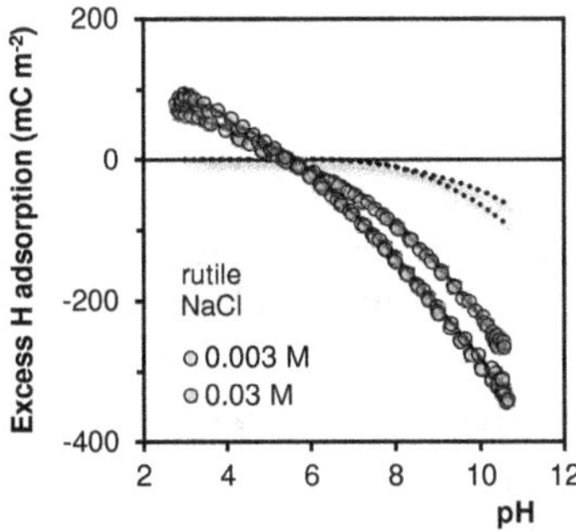

Figure 4. Excess proton adsorption on rutile particles dominated by (110) surfaces in NaCl solutions measured (**spheres**) and previously modeled (**full lines**) (Ridley et al. 2009) using the Basic Stern model ($C = 0.64$ F·m^{-2}), implementing published proton affinity constants (Machesky et al. 2008). In the modeling, the specific adsorption of Na^+ and ion pair formation with Cl^- in agreement with MD simulations (Předota et al. 2007) are included. The dotted lines represent the predicted excess proton charge according to the MO/DFT modeling of Cheng and Sprik (2010), illustrating that their log K_H values derived do not allow the development of a significant amount of surface charge. This not only conflicts with all experimental titration data collected for rutile particles (Bourikas et al. 2001), but also with second harmonic generation (Fitts et al. 2005) and force-distance measurements (Bullard and Cima 2006) that suggest a sharp point of zero charge (i.e., no plateauing) in 1 mM NaCl and KCl at pH around 5. The failure to describe proton titration data is related to the large difference in the proton affinity constants for the two sites ($\Delta pK = 10$). In a later update (Cheng et al. 2015), the ΔpK value decreased to 4, making the surface chargeable.

goethite. We will then also discuss the performance of the MUSIC model in predicting the proton affinity of the hydrous α-TiO_2 (110) surface, as it has been challenged by Cheng and Sprik (2010).

SURFACE COMPLEXATION MODELS

Thermodynamic nature of surface complexation models

The various available thermodynamic models differ in the formulations of adsorption reaction(s) and the calculated electrostatic contribution to the overall Gibbs free energy change. Over the years, it became clear that the same adsorption phenomena could be described with very different sets of surface species using different models. From a thermodynamic point of view, it is relatively easy to describe the adsorption with SCMs, provided that one can freely choose the type and number of surface species and surface sites. It is even possible to describe ion adsorption phenomena thermodynamically correctly without electrostatics using for example a large series of sites, that differ in chemical affinities (Cerník and Borkovec 1996; Borkovec et al. 1998) or a restricted number of surface sites (Bradbury and Baeyens 1997). The non-electrostatic models are called "mechanistic" and indeed capable of simulating for instance specific spectroscopic results (Bradbury and Baeyens 2005) or even predicting uptake data that require for the aqueous species an independent ion activity model for high salt levels, where the typically used approaches are no longer valid (Schnurr et al. 2015), showing the strength and weaknesses of a *purely* thermodynamic approach. The recently found applicability to high salt concentrations is relevant to nuclear waste storage (Schnurr et al. 2015; García et al. 2019, 2021; Morelová et al. 2020) and application even extends to interfacial redox reactions in the case of clay (Banik et al. 2017; Marsac et al. 2017) in such brines.

Importantly, the calculated surface speciation of most models has often been and still can be hypothetical and model-dependent. Yet, SCMs, even if thermodynamically correct with hypothetical species on generic sites, may have severe limitations. It is *a priori* not obvious and from a thermodynamic point of view not necessary that models, which correctly describe the pH- and concentration-dependent binding of one component in a simple system, will accurately predict the ion interactions in multicomponent systems. A major reason is that the interaction of ions is very strongly determined by the electrostatic contributions to the Gibbs free energy change. If not well represented, the models cannot be linked well to the microscopic surface speciation and used to gain mechanistic insights at that level or predict changes when macroscopic conditions alter. The challenge is to develop a mechanistic SCM that has these characteristics as it can then be a great help to guide the interpretation of microscopic data and can be used to design optimal conditions for microscopic measurements.

Towards a mechanistic approach in surface complexation modeling

Locating ion charge in the interface. The electrostatic interaction energy strongly depends on the location of the ion charge in the EDL profile, which became clear in the work of Fokkink et al. (1987). They showed that the location of the charge of an ion is a key factor in the co-adsorption or co-desorption of protons upon the adsorption of an ion. Two years earlier, Perona and Leckie (1985) had shown, on a thermodynamic basis, that the co-adsorption or release of protons is related to the pH dependency of adsorption. A combination of both findings leads to the conclusion that *the location of the ion charge in the electrostatic double-layer profile and the pH dependency are intimately related.* For the adsorption of anions, the co-adsorption is a positive number, for the adsorption of cations, it is negative, i.e., in the latter case, protons are released. This difference explains the well-known opposite pH dependency of the adsorption of many anions and cations, creating adsorption edges. In specific cases, e.g., H_4SiO_4 or $As(OH)_3$, adsorption envelopes are found in the pH dependency. These envelopes show that proton co-adsorption changes with pH, which is primarily due to a shift of the protonation of the species in solution. We will discuss the thermodynamic consistency principle later in detail, as it is a crucial concept in elucidating the role of the structure of surface complexes in the pH dependency of ion adsorption, apart from the role of the (de)protonation of solutes and surface groups (representing e.g., ligands for metal cations).

Point charge. Most SCMs traditionally located innersphere (IS) complexes entirely as a point charge at a single electrostatic position in the EDL profile of Figures 1 and 2. In the CC and the DL models, the charge of specifically adsorb cations and anions is *by definition* present on the surface. In the TL model, Davis and Leckie (1978, 1980) located these specifically bound ions at the same position in the EDL as the electrolyte ions that form outersphere (OS) complexes. However, it was argued that in principle the charge of a species, e.g., a $ZnOH^+$ ion, may be located in two different positions, e.g., Zn^{2+} at one position and OH^- at another one (Davis and Leckie 1978), i.e., charge can be distributed.

Charge distribution. Interfacial charge distribution of ions is key to understanding ion adsorption with surface complexation modeling. The recognition of the importance of charge distribution in ion adsorption modeling dates back to the famous publication of Davis and Leckie (1978) on the modeling of ion adsorption to hydrous ferric oxide (HFO). In their approach, this was done by allocating the charge of IS-complexes in the interface as point charges, placing the proton charge at the surface and the charge of the adsorbing ion in the β-plane of the 2p*K*-TL model. However, it is evident from the structure of IS complexes that part of the IS is shared with the surface, while the remaining part must be at some distance. The corresponding distribution of charge is rooted in the structure of the surface complexes and the position of the ligands in the EDL, collectively forming a *structure–charge–potential* relationship.

The formation of IS surface complexes involves *by definition* ligand exchange (Sigg and Stumm 1981), but the interfacial charge distribution of the ligands can be very different (Fig. 5). Rietra et al. (1999) showed that the difference in structure is reflected in the co-adsorption of protons and the corresponding pH dependency of the adsorption, as discussed later.

For a long time, the notion of ligand exchange for IS complexes, observed by spectroscopy was considered an obstacle in relating structure to surface complexation modeling. The ion adsorption with ligand exchange was treated in a rather simplistic way, in the sense that the adsorbing ions were considered as point charges, located near the surface.

In a new model approach (Hiemstra and Van Riemsdijk 1996), ligand exchange was structurally rationalized via the bond valence concept resulting in a charge distribution (CD) of an IS complex in the compact part of the interface. The charge of an adsorbed ion is distributed over its ligands, which in turn are distributed over the interfacial locations, as illustrated in Figure 5a for two divalent anions. From the perspective of using bond valences, this view can be considered a logical extension of the MUSIC approach. The combination of the CD model and the MUSIC model integrates the knowledge of the structure of crystals and surfaces, the structure of surface complexes, and EDL structure. Essential to the approach is the relation between electrostatics and structure which yields a powerful framework for ion adsorption modeling needed if the aim is using physically realistic species as observed with state-of-the-art *in-situ* spectroscopic methods and/or characterized by computational techniques. The bond valence approach has been used by an independent group later to rationalize the stability of IS Pb surface complexes by calculating the degree of charge saturation of the oxygens ligands common with the surface (Bargar et al. 1997a,b).

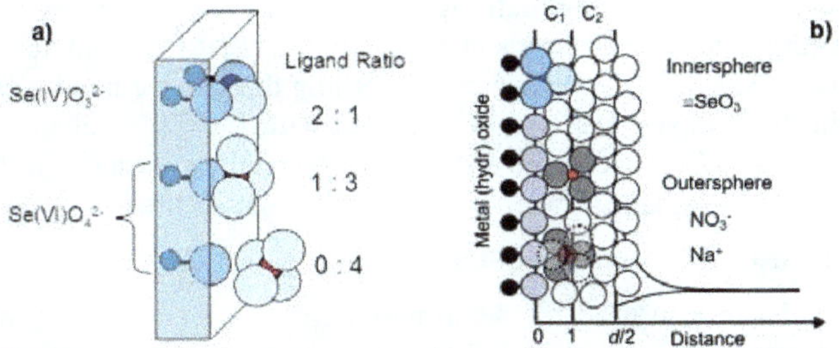

Figure 5. **a)** Ligand distribution of two divalent anions between surface and solution for surface complexes. The bidentate IS complex of SeO_3^{2-} shares two ligands with the surface while SeO_4^{2-} (which may form a monodentate IS or an OS complex) shares one or no ligand with the surface. The different ligand distributions lead to different interfacial distributions of charge. **b)** Schematic representation of IS and OS complexes in the interface with three electrostatic planes. In the diffuse double layer, ions can take nearly every position, but closer to the surface water gets increasingly ordered. The compact part of the interface is conceptually separated from the DDL by the d-plane. The minimum distance of approach of ions that keep their primary ligands (OS complexes) coincides with the electrostatic 1-plane. IS complexes have one or more ligands in common with the surface. The charge in the IS surface complexes is distributed according to the ligand ratio (Fig. 5a), in a first approximation.

Adsorption energy and electrostatic potentials

The concept of charge distribution is crucial for properly calculating the electrostatic contribution (ΔG_{elec}) to the overall Gibbs free energy of adsorption ($\Delta G_{adsorption}$) according to:

$$\Delta G_{adsorption} \equiv \Delta G_{chem} + \Delta G_{elec} = \Delta G_{chem} + \Delta z_0 F\psi_0 + \Delta z_1 F\psi_1 \quad (3)$$

in which ΔG_{chem} is the intrinsic Gibbs free energy due to chemical bond formation. The change in electrostatic energy (ΔG_{elec}) depends on the location of the interfacial charge (Δz_0 and Δz_1). In the CD model, the total change of charge (Δz) due to the adsorption of an ion is distributed between the ligands common with the surface (Δz_0) and the ligands outside the surface, present on the solution side of the Stern layer (Δz_1) as illustrated in Figure 5. In this formalism, $\Delta z = \Delta z_0 + \Delta z_1$, and the "Δ" indicates a change compared to the charge of the reference group in the reaction, which must be distinguished from absolute charges.

The CD model allows a distribution of the charge of an adsorbed ion over two (or even more) electrostatic planes, leading in the present example (Fig. 5) to two electrostatic energy contributions ($\Delta z_0 F\psi_0$ and $\Delta z_1 F\psi_1$). These contributions ($\Delta z_i F\psi_i$) may have opposite signs, caused by the respective potential value (ψ), so that part of the ion may experience attraction while the other part may experience repulsion. This occurs for SeO_3^{2-} binding to goethite at relatively high surface loading and low pH. In the absence of SeO_3 adsorption, the surface is positively charged (pH<PZC), but due to significant adsorption of SeO_3^{2-}, the particle as a whole becomes negatively charged. Both O-ligands of adsorbed SeO_3, common with the surface, experience attraction ($\psi_0>0$). In contrast, the remaining ligand of the adsorbed SeO_3 experiences repulsion because the overall charge of the particle is negative, making the Stern layer potential negative ($\psi_1<0$). This occurs at high SeO_3 loading. Overcharging or charge reversal can be explained with the concept.

Since the Gibbs free energy change is linked to the $\log K$ of the reaction, i.e., $\Delta G_r^o = -RT\ln K$, Equation (3) can be rewritten to:

$$K_{\text{adsorption}} = K_{\text{chem}} \cdot e^{\frac{-\Delta z_0 F\psi_0 - \Delta z_1 F\psi_1}{RT}} \tag{4}$$

The concomitant values of Δz_0 and Δz_1, expressing the charge distribution, depend on the structure of the complex formed, highlighting the link between a surface complex structure and the overall affinity. With the development of spectroscopy, surface complex structures have become accessible[10]. Details of the geometry concerning the relative bond length and charge distribution within the complex can also be obtained with MO/DFT calculations (Hiemstra and Van Riemsdijk 2006). This makes the results of CD modeling verifiable if the independent (e.g., spectroscopic) methods are sufficiently quantitative and accurate.

Determining electrostatic interfacial potentials

Single crystals have been used for determining streaming potentials (Smit and Holten 1980), and surface potentials (Kallay et al. 2008; Boily et al. 2011; Chatman et al. 2013). For surface potential data[11], collected for various solids in the absence of typically specifically adsorbing solute ions, it was concluded that the results look suspiciously similar for all kinds of surfaces (Preočanin et al. 2017), but that they often disagree with MUSIC model predictions. For concomitant zeta-potential measurements on identical surfaces, the zeta-potential values are higher in magnitude than the concomitant surface potential, at odds with the conventional double layer model (Fig. 2). On the other hand, for rather unspecific flat surfaces, surface potential measurements, obtained with the ISFET technique (Bousse et al. 1983), combined with zeta-potentials, give support to the classical EDL picture (Bousse et al. 1992).

The isoelectric points (IEP), derived from zeta potential data for single crystals specifically for the (0001) face of α-Al_2O_3, which in the meantime have been reported in comparatively many studies are often low (around pH 4) and have confirmed the early measurements (Smit and Holten 1980; Higgins et al. 1998). This low value has caused suspicion about contamination (Kosmulski 2003). Such contamination may easily occur due to the near absence of chemical buffering caused by the extremely low solid-to-solution ratios used in single-crystal experiments. For example, exposure of clean single crystals to air already leads to contamination by carbonic and organic acids (Balajka et al. 2018). The low values for the IEPs challenged others (Kershner et al. 2004; Franks and Gan 2007; Wang et al. 2016), including one of the present authors (Lützenkirchen et al., 2010), to do additional measurements. All this work confirmed previous results.

[10] Note that this was quite different in the early 1990s, when EXAFS measurements for example had just started.

[11] See the chapter by Klačić et al. (2025, this volume).

Charging data (d-plane potentials) for single crystals can also be obtained from AFM (atomic force microscope) and SFA (surface force apparatus) experiments, but will involve interpretation of the force curves via some DLVO model variant (Israelachvili and Adams 1976; Bullard and Cima 2006). This is also possible for distinct facets of sufficiently well-defined particles (Siretanu et al. 2014; Su et al. 2021).

From the measurement with single crystals, one might conclude that the collected data for surface potentials, zeta-potentials, and force measurement do not yield a consistent picture and show a discrepancy with the standard model. A comprehensive non-standard explanation is possible but involves a lot of assumptions (Lützenkirchen et al. 2010). It has been used to design a SCM for the isostructural surface of hematite (Lützenkirchen et al. 2015), which is equally able to explain the data available for the (0001) plane of α-Al_2O_3.[12]

Interfacial water

Close to the surface, the properties of water may change as visualized in Figure 6a. As evidenced by different experimental approaches such as force measurements (Pashley and Israelachvili 1984; Israelachvili and Wennerstrom 1996), X-ray reflectivity (Toney et al. 1995; Fenter and Sturchio 2004; Catalano et al. 2006), and Sum Frequency Spectroscopy (Yeganeh et al. 1999; Kataoka et al. 2004; Ostroverkhov et al. 2005; Shen and Ostroverkhov 2006), water near the surface is increasingly ordered over a distance of 0.3–0.7 nm, depending on the type of face, which is equivalent to about 1–3 layers of water molecules (Hiemstra and Van Riemsdijk 2006).

Interfacial water may differ from liquid water in density. Water as well as the major minerals of the earth's crust (silicates and oxides) can be considered as a collection of oxygen ions predominantly neutralized by respectively protons and relatively small cations in between. A major difference between water and minerals is the packing density of the oxygen ions, about 0.10 mole O per cm^3 for minerals and 0.056 $mol\,O{\cdot}cm^{-3}$ for water (Hiemstra and Van Riemsdijk 2009; Hiemstra et al. 2009). At the interface, both phases meet, leading to adjustments that create interfacial water.

Although a water molecule as a whole is neutral, the orientation of water may contribute to the interfacial charge distribution, since water molecules have a dipole that may orientate in an electrostatic field, contributing to the field strength. The relatively small effect can be incorporated into the CD framework (Hiemstra and Van Riemsdijk 2006). Potential reorientation of interfacial water is thought to be a consequence of the electric field, but the optimization of the H-bonding environment between surface hydroxyls and interfacial water and among the interfacial water molecules is another ingredient. The subtle interplay between both will lead to the equilibrium interfacial water structure. To what extent the H-bonding can have repercussions on the electrostatics (and the associated protonation/deprotonation reactions) has rarely if ever been discussed. The situation becomes even more complex in the presence of interfacial ions, which have to be accommodated by the interfacial water partially (IS) or entirely (OS). The hope for the future is that non-linear optical or other surface-specific methods and simulations might shed light on these subtle interplays, ideally in a quantifiable way.

To gain more insight into the structure of the interface in contact with water, single crystals have been used. Measurement of the force–distance relation for two charged crystal surfaces with double layer overlap in the muscovite mica system is the classical example (Pashley and Israelachvili 1984; Israelachvili and Wennerstrom 1996). The force increases gradually (Fig. 6a) as expected from the theories for the diffuse part of the double layer. At short distances of plate separation (~1.2–1.6 nm), the force oscillates with a discreet distance frequency of 0.25 nm, corresponding to the packing distance of water (Fig. 6a). Per crystal, about 2–3 layers of water molecules in the EDL in contact with the surface are structured (Fig. 6a). Single crystals of sapphire have been studied similarly (Horn et al. 1988).

[12] Unpublished results by JL.

X-ray reflectivity/CTR can also be used to study water layering at solid–water interfaces. Figure 6b shows results for the Al_2O_3 (021)–water interface (Catalano et al. 2006). Water ordering is found for the first 3 layers. As pointed out above, this ordering can be disturbed by the presence of electrolyte ions, as demonstrated by MD simulations (Zarzycki 2023). Moreover, in the CTR experiment, there is no EDL overlap.

For the (001) faces of α-Al_2O_3 and α-Fe_2O_3 less water ordering is found (Catalano 2011), which can be explained by the low surface charge according to the MUSIC model. However, for the (001) faces, MD simulations suggest a bi-layer of water (Hass et al. 2000, Du et al. 2024), which may involve oxygen layers that are too close to each other to be resolved by CTR. In the case of (100) and (101) of quartz (natural and annealed at 400 °C), the water ordering is restricted to a single layer (Schlegel et al. 2002). This is also found for the (110) face of rutile (Zhang et al. 2007). Less water ordering for these quartz and rutile specimens concurs with the absence of a second Stern layer in modeling approaches.

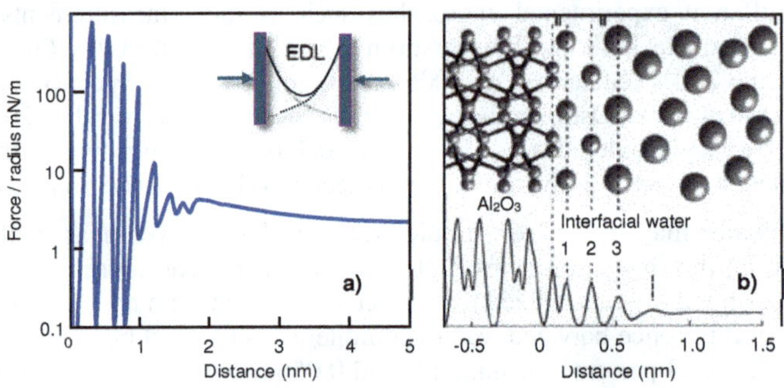

Figure 6. a) Force as a function of the distance needed to compress the EDL between two parallel mica plates. At large distances, the repulsive force gradually increases, as predicted by the diffuse double-layer theory. At short plate distances of about 1.2–1.6 nm, the force oscillates with a distance frequency of about 0.25 nm, close to the diameter of one water molecule. About 2–3 layers of water molecules in the EDL at each plate side are structured. Data from Pashley and Israelachvili (1984). **b)** Structural model of the Al_2O_3(012)–water interface in relation to the electron density profile determined by X-ray reflectivity (Catalano et al. 2006). The first three interfacial water layers are ordered, aligning with an extension of the basic Stern model. [Reprinted with permission from Langmuir 22:4668–4673 Copyright 2006 American Chemical Society.]

Challenging systems

Specific systems may have specific charging and ion speciation mechanisms that differ from metal (hydr)oxides but nevertheless can be treated along similar lines, although with a twist.

Some materials are seemingly inert and do not have functional groups on their surfaces, for instance, oil, air, diamond, and various others (Healy and Fuerstenau 2007). Nevertheless, these interfaces, unexpectedly, have variable charge characteristics (Lützenkirchen et al. 2008) that can be attributed to interfacial water molecules acting as surface sites. These surfaces are still subject to discussion when it comes to the origin of their variable charging properties. In many cases, the observations rely on zeta-potential measurements. If the no-slip-boundary condition in an electrokinetic experiment is violated (as may be expected for hydrophobic surfaces), the zeta-potentials as usually obtained with commercial set-ups might be shaky according to the chapter on electrokinetics in this book. Others have developed a theory according to which the slip amplifies zeta-potentials (Joly et al. 2004; Audry et al. 2010). However, the low isoelectric point of these materials and the pH-dependent charging have not only been established for many hydrophobic surfaces (Healy and Fuerstenau 2007) but also using non-electrokinetic methods like acid–base titration of pure oil droplets (Beattie and Djerdjev 2004), contact angle titration of PTFE surfaces (Hamadi et al. 2009), sum frequency generation studies on hydrophobic surfaces (Tian and Shen 2009), and others. Complementary measurements involving titrations and electrokinetics (Preočanin et al. 2017) or contact angle titration and electrokinetics (Lützenkirchen et al. 2010) yield consistent values for the points of zero charge/isoelectric points for PTFE and silanated silica, respectively, which also agree with the bulk of reported electrokinetic results. The origin of the pH-dependent charge remains ambiguous.

The charging behavior of a very classical model colloid material (AgI) is highly asymmetrical around the PZC (Fig. 7b) and cannot be modeled with traditional SCM approaches. However, the strong asymmetry can be understood from the simultaneous depletion of Ag^+ ions from the surface, creating negative charge on the surface, and the specific adsorption of Ag^+ ions to I^- ions of the surface, creating positive charge outside the surface (Hiemstra 2012). In the respective model, the Ag^+ ion is the only potential-determining ion, deviating from the classical view.

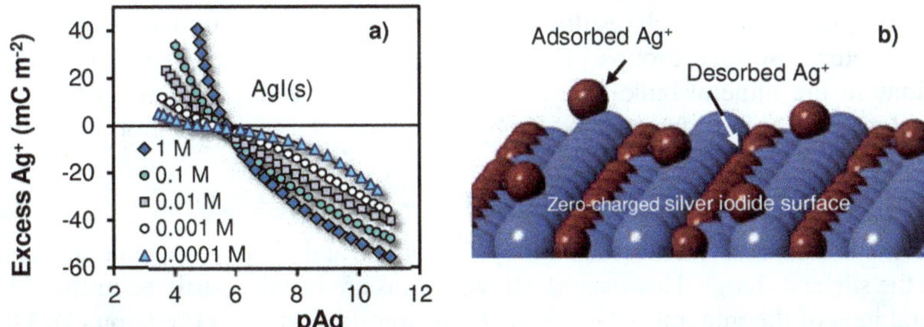

Figure 7. a) Excess silver ion adsorption, $\Gamma_{Ag}-\Gamma_{I}$ as a function of the negative logarithm of the Ag^+ activity in the solution (pAg) for AgI (s) colloids at different KNO_3 electrolyte concentrations (Bijsterbosch and Lyklema 1978). The **lines** have been calculated with the CD model for AgI (Hiemstra 2012). **b)** Visualization of the adsorption of silver ions at the PZC, with the simultaneous depletion of Ag^+ from the surface and complexion by I^- ions of the surface, creating charge separation. [Reprinted with permission from Langmuir 28:15614–15623. Copyright 2012 American Chemical Society.]

Concepts of surface complexation can also be used to reveal the oxidative stabilization of metallic Ag nanoparticles (AgNP) by the formation of a semi-metallic surface layer with sub-valent $\equiv Ag_3OH^0$ surface groups (Molleman and Hiemstra 2015). At high pH, these groups protect AgNP kinetically against oxidation. With pH lowering, surface reconstruction occurs, temporarily releasing Ag^+ ions, before a new metastable equilibrium with Ag^+ in solution is reached through the formation of a new sub-valent oxidation surface species, e.g., $\equiv Ag_5(OH)_2^0$ (Molleman and Hiemstra 2017). The kinetics and partial equilibrium can be described well with the formulated model, but the structure of the new sub-valent surface structure remains unclear.

Another challenging material is ferrihydrite[13], because the nanoparticles of this material are extremely small. The ion adsorption behavior has been studied for almost a century, but due to a lack of understanding of the mineral structure, mechanistic surface complexation modeling, constrained by surface structural information, was missing. With unraveling the mineral (Michel et al. 2010) and its surface structure (Hiemstra 2013), this has changed as outlined in a separate contribution to this volume. Similarly, advances have been made for carbonate, sulfides, and sulfates, to which SCMs have been applied.

Scope and further outline

The above survey intended to reveal the origin and development of concepts that have become relevant for current state-of-the-art surface complexation modeling. It is beyond the scope to review or even mention all the developments of a plethora of approaches and ideas that can in principle contribute to improving surface complexation modeling.

[13] See chapter by Hiemstra et al. (2025, this volume).

In the following parts, we will elaborate on pivotal factors for a structure-based surface complexation modeling of well-defined systems. We intend to exemplify the approaches with sometimes new results or interpretations that give a deeper insight into the development of charge on the surfaces of key minerals with an increasing complexity of the surface structure, and how the charge-potential-structure relationship influences ion adsorption and its modeling. Moreover, we will discuss the measurement and interpretation of EDL features including electrokinetics, and what can be learned from it. In the last part, we will explore approaches for less- or ill-defined systems and suggest a modus operandi for those who are not familiar with the field.

ATTRIBUTION OF CHARGE TO SITES

Minerals and solutions can be considered as three-dimensional networks of bonds. The interface is a discontinuity in the network or some kind of transition from the mineral phase to the aqueous one, which results in the presence of mineral surface groups and changes in water structure. Surface groups of oxides may have lower metal ion coordination than oxygen ions in the mineral bulk. The charge of the surface oxygen ions is then less well-neutralized, which can be compensated for by the uptake of one or two protons, written as:

$$\equiv O^{?} + 2H^{+}(aq) \rightleftharpoons \equiv OH^{?+1} + H^{+}(aq) \rightleftharpoons \equiv OH_2^{?+2} \quad (5)$$

Counting the number of the various groups involved and their corresponding charge can establish the surface charge. However, the oxygen ions are usually partly neutralized by one or more metal ions of the mineral, which leads to the question mark in Equation (5). The charge attribution can be calculated via the Pauling bond valence concept as a bookkeeping tool. The Pauling bond valence ν represents the mean available charge per bond, i.e., the charge z of the metal ion divided by the coordination number CN,

$$\nu = \frac{z}{CN} \quad (6)$$

The lowest metal coordination number in oxides and hydroxides is two. The oxide quartz (SiO_2) and the hydroxide gibbsite ($Al(OH)_3$) are important representatives and will be used for illustration in the following. Only two types of surface groups are found for these minerals in the sense of ion coordination, i.e., singly and doubly coordinated surface groups making these minerals relatively simple. Their structure and charging will be discussed first.

Quartz and silica[14]

Figure 8 shows a two-dimensional representation of the quartz structure with linked SiO_4 tetrahedra. In the bulk, all oxygen ions are doubly coordinated. Each Si–O bond represents 1 valence unit (v.u.), as is found by applying Equation (6) with $z = 4$ and CN = 4, resulting in $\nu = +1$ v.u. Two Si ions in the bulk neutralize the charge (–2 v.u.) of the oxygen, as calculated with the bond valence sum rule, $\sum\nu - 2 = 0$. Doubly coordinated $\equiv Si_2O$ groups are also present at the surface, not only for quartz, but also at the (001) faces of phyllosilicates such as kaolinite, mica, and montmorillonite. These $\equiv Si_2O$ groups have a very low proton affinity, i.e., the reaction $\equiv Si_2O^0 + H^+ \rightleftharpoons \equiv Si_2OH^+$ has an extremely low log K (Hiemstra et al. 1989a), implying a chemical inertness of $\equiv Si_2O^0$ concerning proton binding.

[14] At this point, it is worthwhile to discuss the terms 1-pK and 2-pK. For modeling acid–base titration of silica, in many cases, it has been sufficient to use a single deprotonation step with a charge transition 0 to –1, i.e., a model with one pK value. The charge transition step from 0 to -1 is the very same as used in the classical 2-pK model that has the charge transitions +1, 0, –1. In contrast, the term "1-pK model" is often used when charges of –0.5 and +0.5 are involved for a single (generic) type of site so that the pK value is equal to the PZC. So, in this sense, silica is different and not a 1-pK surface in that sense. To complicate things, some data sets for silica/quartz point to two sites each with a 0 to –1 charge transition and corresponding pK value (Ong et al. 1992b; Higgins et. al. 1998; García et al. 2019), and sometimes, this is also referred to as a 2-pK model. For describing the PZC of silica/quartz simultaneously, additional protonation reactions are required, leading to e.g., a 2×2-pK model. In general, the MUSIC model suggests a maximum of two protonation steps per type of surface site. However, both pK values of each type of site are so different (ΔpK ~ 12) that only one is seen in the normal pH window, apart from some exceptions (e.g., doubly coordinated groups on the hematite or corundum basal planes).

The second type of surface group at the solution–water interface of quartz is the singly coordinated group that arises from the absence of one coordinating Si^{4+} ion. The negative charge (–1 v.u.) can be found by applying the bond valence sum $\sum v - 2 = +1 - 2 = -1$ v.u. The charge can be compensated by the uptake of one proton, according to the reaction:

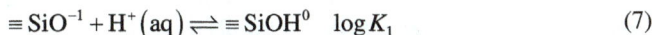

$$\equiv SiO^{-1} + H^{+}(aq) \rightleftharpoons \equiv SiOH^{0} \quad \log K_1 \tag{7}$$

The $\equiv SiOH^0$ groups at the surfaces of quartz and silica are the only proton-reactive entities, but they may be present in isolated, $\equiv(SiO)_3SiOH$ or Q^3, and germinal, $\equiv(SiO)_2(SiOH)_2$ or Q^2, forms. At SiO_2 faces with a relatively low site density, only isolated $\equiv$SiOH groups are found. Examples (Fig. 8b) are the pyramidal (101) and (011) faces of α-quartz with 5.3 $\equiv$SiOH nm^{-2} or the (001) face of tridymite (4.6 nm^{-2}) and the (001) face of α-cristobalite (5.8 nm^{-2}). At SiO_2 faces with a high site density, geminal groups are found, for example at the prismatic (100) faces (7.5 nm^{-2}) and top-end (001) faces (8.5 nm^{-2}) of quartz (Fig. 8b). By constructing SiO_2 spherical nanoparticles (Fig. 8b), we derive a $\equiv$SiOH site density is 5.0 ± 0.4 nm^{-2}, very well aligning with experimental results for many SiO_2 samples (Zhuravlev 2000).

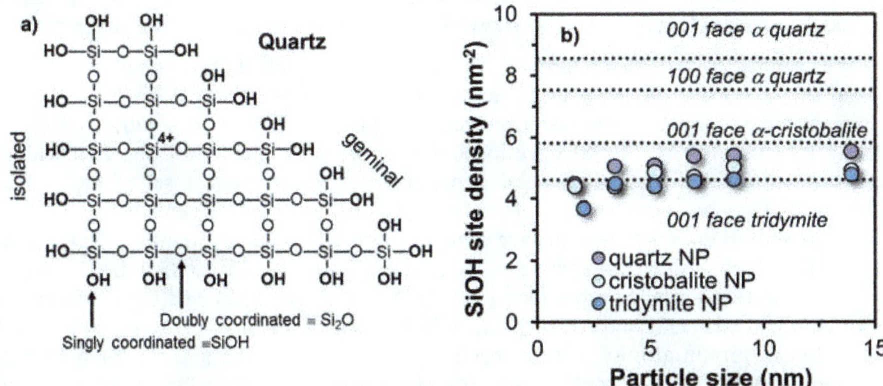

Figure 8. a) Schematic mineral and surface structure of SiO_2. The oxygen ions in the bulk are doubly coordinated. This group ($\equiv Si_2O$) is also present at the surface. Singly coordinated surface groups result from the absence of a coordinating Si. The binding of a proton can compensate for the missing charge. **b)** Site densities derived by constructing nanoparticles from which non-coordinated oxygen ions are removed (**spheres**). The excess of oxygen ions relative to Si are the $\equiv SiOH^o$ surface groups. For spherical SiO_2 nanoparticles with various SiO_2 cores (quartz, cristobalite, tridymite), the mean site density is 5.0 ± 0.4 nm^{-2}, in full agreement with the arithmetical mean of the site density (4.9 ± 0.5 nm^{-2}) of 100 amorphous silica samples analyzed (Zhuravlev 2000). The site density of individual crystal faces of the various SiO_2 materials varies from ~4.6 to 8.6 nm^{-2}. Isolated $\equiv$SiOH groups are found for faces with a rather low site density (5.2 ± 0.6 nm^{-2}) and geminal $\equiv Si(OH)_2$ groups are present on faces with a relatively high site density (8 ± 0.5 nm^{-2}).

Proton affinity of the first protonation step. The $\log K_1$ of Equation (7) can be evaluated by modeling experimental charging curves. Modeling the data of Bolt (1957) for silica with the Basic Stern option yielded a $\log K_1$ value of 7.5 (Hiemstra et al. 1989b). Schindler and Kamber (1968) found $\log K_1 = 6.8 \pm 0.2$ for amorphous silica. Sonnefeld (1995) showed that the fitted $\log K_1$ varied with the applied site density (N_s), and obtained for spherical SiO_2 nanoparticles $\log K_1 = 7.0$ for $N_s = 5$ nm^{-2}. A mean value of $\log K_1 = 7.3 \pm 0.3$ was obtained for a series of SiO_2 materials published in the literature (Sahai and Sverjensky 1997). From IR spectroscopy, a value of $\log K_1 = 7.2$ was inferred (Hair and Hertl 1970; Marshall et al. 1974). These findings suggest that the intrinsic proton affinity of silica is about 2.5 $\log K$ units lower than the related one in solution ($\log K = 9.7$).

Some studies (Michael and Williams 1984; Seigel 1995) for ground quartz point to a single $\log K_1$ of ~7. On the other hand, the presence of two types of ≡SiOH groups with quite different $\log K_1$ is also supported by experimental work, ranging from XPS investigations for quartz crystals (Duval et al. 2002), second harmonic generation (SHG) (Ong et al. 1992; Higgins 1998; Azam et al. 2012) for fused silica prisms, as well as attempts to model column experiments with 1 mm quartz (Stolze et al. 2020), and the pH dependency of the specific cation adsorption and proton titrations of ground 5 μm quartz (García et al. 2019) with a surface area of 6 $m^2 \cdot g^{-1}$. The latter authors found by fitting a site density of about 1 nm^{-2} for groups with a low proton affinity ($\log K_H = 4$). The obtained density aligns well with the experimental and theoretical site density (0.8–1.1 nm^{-2}) of the acidic ≡SiOH groups of ≡Si(OH)$_2$ on fractured quartz (Murashov and Demchuk 2005). With AFM, a bimodal behavior is also found for large (5 μm) spherical silica particles with A = 0.5 $m^2 \cdot g^{-1}$ (Morag et al. 2013), showing an opposite pH-dependent trend in the double layer potential for various types of electrolyte ions above and below pH 6.

For silica nanoparticles (NP), bimodal behavior is sometimes reported for Binzil 40/220 and 15/500 via electrokinetics (Blute et al. 2007) and LUDOX HS via a titration technique (Scheinost et al. 2001). Other experimental data also suggest the two-site scenario (Abbas et al. 2013). However, in most cases, the charging behavior of silica nanoparticles can be described using a single type of site, as shown in a case study at the end of this chapter.

The affinity of the ≡SiO$^-$ groups can also be approached theoretically. A Pauling bond valence approach for the ≡SiOH^o group (Hiemstra et al. 1996) yielded $\log K$ = 7.9, assuming interactions with three water molecules, similar to H_4SiO_4 in solution. Molecular modeling by Rustad et al. (1998) gave $\log K$ = 8.5. Leung et al. (2009) obtained with the combination of MO/DFT and molecular dynamic (MD) simulations a set of $\log K$ values for the (100) face of β-cristobalite of around 7.6±0.3 for geminal and isolated ≡SiOH groups. For ≡SiOH in a strained surface structure, $\log K_H$ values as low as 4.5±0.5 could be found (Leung et al. 2009). A few years later, Sulpizi et al. (2012) studied the (0001) face of quartz. This (high site density) face (Fig. 8b) has only geminal ≡Si(OH)$_2$ groups that form in-plane SiO–H···OSi and out-of-plane SiO–H···OH$_2$ bonds in a 1:1 ratio. Both ≡SiOH groups differ in proton affinity, with respectively $\log K_{1a}$ = 8.5±0.6 and $\log K_{1b}$ = 5.6±0.6. For various particles of SiO_2 polymorphs (quartz, tridymite, cristobalite, coesite, vitreous silica), the average total ≡SiOH site density is 5.9±1.4 nm^{-2}. Isolated ≡SiOH groups dominate and those that may form out-of-plane ≡SiOH groups have an average density in the range of ~0.3 to 1.0 nm^{-2} (Murashov and Demchuk 2005).

In more recent studies involving a model surface for amorphous silica, MO/DFT–MD (Pfeiffer-Laplaud et al. 2015) yielded very low values for some geminal ≡Si(OH)$_2$ (Q^2) groups ($\log K_H$=2.9) as well vicinal (2×) ≡SiOH (Q^3) groups ($\log K_H$=2.1). Additionally, high values were reported for isolated SiOH groups ($\log K$=10) and germinal groups with another local structure ($\log K$=8.9). *The authors concluded that the local Si environment, as well as the local H-bond network, are key factors in determining the acidity of these surface groups.* The very low $\log K_H$ values of 2 to 3 would lead to a strong charging of amorphous silica while the high affinities would prevent any contribution of those groups to surface charging. Both are contradictory to the experimental results for most silica nanoparticles, though some have been reported to involve two pK values (Scheinost et al. 2001) or show two-step charging in electrokinetic experiments (Abbas et al. 2013).

Proton affinity second protonation step. According to Equation (7), quartz and silica are neutral or negatively charged, which contrasts with electrophoresis and shows positive particles at very low pH. The iso-electric point (IEP) is (still) typically reported as close to pH 2 (Sposito 1984), even in more recent reviews. At high salt concentrations, the electrolyte ion adsorption may lead to an increase in the IEP (Kosmulski 1998). Concerning zeta-potential measurements, the recent observation that CO_2 can decrease the zeta potentials of SiO_2 may be due to the allocation of CO_2 molecules near the surface changing the water structure and suppressing the development of surface charge (Vogel et al. 2022; Vogel and Palberg 2024). This may have been overlooked in other electrokinetic measurements because it is simply assumed that anions do not adsorb to SiO_2 surfaces.

The positive charge at very low pH suggests the formation of $\equiv SiOH_2^+$ according to:

$$\equiv \mathrm{SiOH}^0 + \mathrm{H}^+(\mathrm{aq}) \rightleftharpoons \equiv \mathrm{SiOH}_2^+ \qquad \log K_2 \tag{8}$$

with $\log K_2 = -3.9$ (Hiemstra et al. 1996), i.e., close to $\log K_2 = -5$ found by Sulpizi et al. (2012). When relating this to the value for $\log K_1$ from the same surfaces, it turns out that the difference between $\log K_1$ and $\log K_2$ is very large, typically $\Delta \log K_H \sim 11$. The important message of the results from both modeling titration data and AIMD simulation is that *the difference in log K value between the first and second protonation step on one and the same surface group is huge.* This large ΔpK is supposed to be intrinsic to surface groups on all oxides, which is fully supported by first-principles MD simulations (Smith et al. 2022). It implies that in most cases only one protonation step can be observed in the experimental pH window (Hiemstra et al. 1989a).

Outlook. The above shows that SiO_2 materials and surfaces may differ in behavior, likely due to differences in the surface structure. The bimodal behavior is quite well established for well-crystalline SiO_2, whereas silica NPs most often have a behavior that can be represented by a single type of site with a $\log K_H$ that differs from MD predictions, possibly due to surface relaxation and/or reconstruction. One puzzle for the future is to further elucidate the origin of the different behaviors of this seemingly simple surface. Interactions with anions like chloride should also be considered (Pfeiffer-Laplaud et al. 2016) as well CO_2 (Vuceta and Morgan 1978; Vogel et al. 2022; Vogel and Palberg 2024), particularly in electrokinetic studies. Presently, a full picture and straightforward conclusions are hampered by the absence of consistent data sets collected with multiple techniques using the same sample or material at well-defined conditions.

Gibbsite α-Al(OH)$_3$

Basal plane with doubly coordinated surface groups. As for quartz, the structure is formed by the coordination of two cations to the oxygen ion, but in addition, a proton is bound. The Al^{3+} ions are present in octahedra. The hexa-coordination leads to a mean bond valence of $\nu = +3/6 = 0.5$ v.u. (Eqn. 6). Two Al–O bonds are needed to neutralize the OH^- in the bulk. Doubly coordinated hydroxyls are also present on crystal faces. The dominant (001) face of the flat hexagonal gibbsite crystals (Fig. 9) has, ignoring imperfections, only $\equiv Al_2OH^o$ groups (Fig. 9). This group is uncharged but may dissociate or associate a proton:

$$\equiv \mathrm{Al_2O^{-1}} + 2\mathrm{H}^+(\mathrm{aq}) \rightleftharpoons \equiv \mathrm{Al_2OH^0} + \mathrm{H}^+(\mathrm{aq}) \rightleftharpoons \equiv \mathrm{Al_2OH_2^+} \tag{9}$$

The variation in the (001) face contribution in synthetic gibbsite preparations has shown (Hiemstra et al. 1999) that the (001) face remains uncharged ($\equiv Al_2OH^o$) between pH 4–10 (Fig. 10). The fast titrations did not show hysteresis (Hiemstra et al. 1987, 1999) implying equilibrium for the process studied, i.e., protonation of singly coordinated surface groups. The experimental observations suggest a very high $\log K_1$ (Eqn. 9, left-hand side) and a very low $\log K_2$ (Eqn. 9, right-hand side) i.e., the large ΔpK is retrieved and leads to the dominance of $\equiv Al_2OH^o$, in agreement with first principles molecular dynamics simulations for the uncharged (001) face of gibbsite (Liu et al. 2013) or the basal plane of boehmite (γ-AlOOH) (Smith et al. 2022). Moreover, sum frequency generation (SFG) data for the hydrated (0001) faces of sapphire (α-Al_2O_3) (Zhang et al. 2008)—a surface that also is dominated by doubly-coordinated hydroxyls—indeed shows a behavior that is predicted by the MUSIC model (Hiemstra et al. 1996), i.e., $\log K_1 \approx 12$ and $\log K_2 \approx 0$, although Zhang et al. report $\log K_1 \approx 9.7$ and $\log K_2 \approx 3$ for the doubly coordinated surface groups of that material.

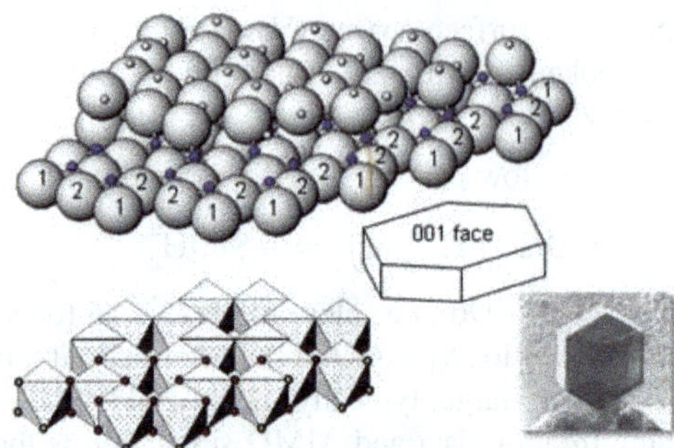

Figure 9. Structure of an $Al(OH)_3$ layer of a hexagonal gibbsite crystal. In the **upper figure**, half of the doubly coordinated OH ligands on top are removed to show the coordination with Al^{3+} ions. The coordination of the OH at the edges is indicated with a number, i.e., 1 for singly, and 2 for doubly coordinated groups. Only coordinated protons of the top layer of oxygen ions are shown. The polyhedra representation of the structure is shown on the lower left. By coating gibbsite crystals with C-sputtering at a certain known angle, one may calculate from the TEM micrograph (**right-hand side**) the height of the hexagonal crystals (Hiemstra et al. 1999b) and thereby, the edge surface area. TEM observations (Hiemstra et al. 1999b) further suggest that the total surface areas of their gibbsite samples are usually significantly larger than measured with BET, which can be attributed to face-to-face contact upon drying.

Proton excess charge at gibbsite basal faces. Although no significant charge development on the basal plane was found by Hiemstra et al. (Fig. 10), for other preparations, charge development has been reported (Kavanagh et al. 1975). Rosenqvist et al. (2002) found that an aged sample may consume much more acid/base, which is remarkable since the suspensions of Hiemstra et al. (1987, 1999), after exhausting dialysis at neutral pH and storage for a very long time (years), did not suggest such behavior.

A difference between both results is the speed of titration. The strong charging was attributed to the slow kinetics of protonation of surface groups, a phenomenon unobserved for (hydr)oxides, but on hematite single crystals, the kinetics of surface potential development (Chatman et al. 2013) has been explained similarly, including the isostructural (0001) surfaces. Remarkably, for non-aged gibbsite, the total proton adsorption was considerably less (Rosenqvist et al. 2002), while also using slow titrations. It suggests that some factors in the aging process may have caused the unusual development of charge.

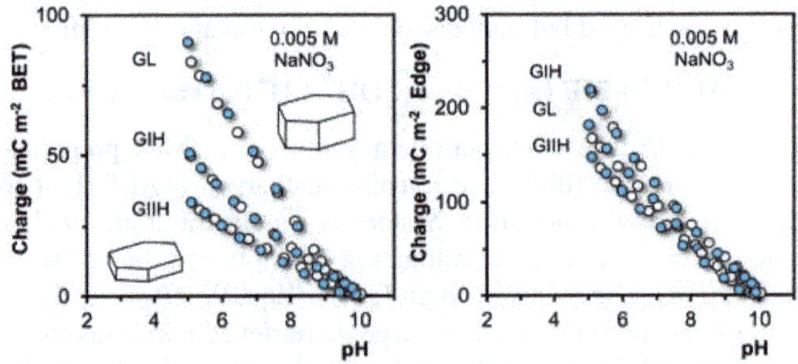

Figure 10. Charging of hexagonal gibbsite crystals, differently prepared to vary the ratio of edge to planar faces. The charge has been scaled to the total BET surface area (**left**), and the edge face area (**right**). When the surface charge is scaled to the BET surface area, a large variation is found, and moreover, the surface charge is unusually low compared to other metal(hydr)oxides, including bayerite. When scaled to only the edge surface area obtained from BET measurements and the known morphologies, the charge is higher and becomes more realistic in relation to other metal(hydr)oxides. The reactivity of the gibbsite can predominantly be attributed to the presence of edge faces rather than the planar (001) faces. [Reprinted (adapted) with permission from Langmuir 15:5942–5955. Copyright 1999 American Chemical Society.]

Adeloka et al. (2011) showed that, with the very same equipment that had been used by Rosenqvist, slow titrations do not lead to high charging of a "Hiemstra-type" gibbsite (Liu et al. 2013). It is noted that in the experiments of Hiemstra et al. (1999), one of the gibbsite materials used was prepared at a relatively low temperature of dialysis (20 °C). That gibbsite did show a strong charging. After a second dialysis at 70 °C for 2–3 weeks, most of this excess charge had disappeared, while the suspension was self-acidifying during this procedure, suggesting a transformation of singly into doubly coordinated groups ($2 \equiv AlOH_2^{1/2+} \rightleftharpoons \equiv Al_2OH^o + H_3O^+$). After this, the samples behaved as expected.

Ion pair formation. Experimentally, a large quantity of Cl^- has been found in dialyzed material by Rosenquist (pH ~ 4.6), i.e., ~3 μmol·m^{-2} which is similar to the amount of H^+ expected from slow titrations. This contrasts with the work of others (Hingston et al. 1972; Wendelbow 1987). Wendelbow (1987) did not report large amounts of adsorbed Cl^- in his experiments. He studied ^{36}Cl and ^{22}Na counter and co-ion adsorption as a function of pH for a well-crystallized gibbsite. The equilibrium time in these experiments was about 18–20 hours. At pH 5, Wendelbow observed about 0.25 μmol $^{36}Cl \cdot m^{-2}$ BET for 0.001 M NaCl. This adsorbed amount can be understood assuming only charge development at the edge faces of the gibbsite crystals, which contribute 8 m^2·g^{-1} (Hiemstra et al. 1999). The behavior of this gibbsite preparation is quantitatively in agreement with the behavior of the gibbsite samples presented in Figure 10. It is noteworthy to mention that after visiting the lab in Wageningen, Wendelbow used the same method for his gibbsite preparation, yielding consistent results.

Surface complexation modeling revealed for Al (hydr)oxides in general ion pair formation affinities of $\log K_{Na} = +0.2$ and $\log K_{Cl} = -0.2$ (Hiemstra et al. 1999), which are higher than e.g., for Fe (hydr)oxides. Rosenqvist et al. (2002) modeled the surface charge of the (001) faces of their aged gibbsite by the formation of strong ion pairs, i.e., $\equiv Al_2OH_2^+{-}Cl^-$ and $\equiv Al_2O^-{-}Na^+$, which suppresses the formation of $\equiv Al_2OH_2^{+1}$ and $\equiv Al_2O^{-1}$. Consequently, no DDL charge develops on the (001) face *by definition*, which may have repercussions for AFM measurements.

AFM data. Atomic force measurements (AFM), support the absence of a significantly developed DL for the (001) faces of gibbsite (Gan and Franks 2006). Interpretation of the AFM data leads to a very small, but pH-dependent diffuse double layer potential (~ 0–10 mV) that develops in the pH range from pH 9 to pH 4 (Fig. 11) in 1 mM NaCl for a large gibbsite crystal. The corresponding diffuse double layer charge (σ_d) is extremely low σ_d ~10^{-3} μmol·m^{-2} or lower. More recently, higher values for σ_d have been reported, varying between 0.05 and 0.16 μmol·m^{-2} for respectively 1 and 100 mM NaCl at pH 5.8 (Siretanu et al. 2014) for 250 nm platelets.

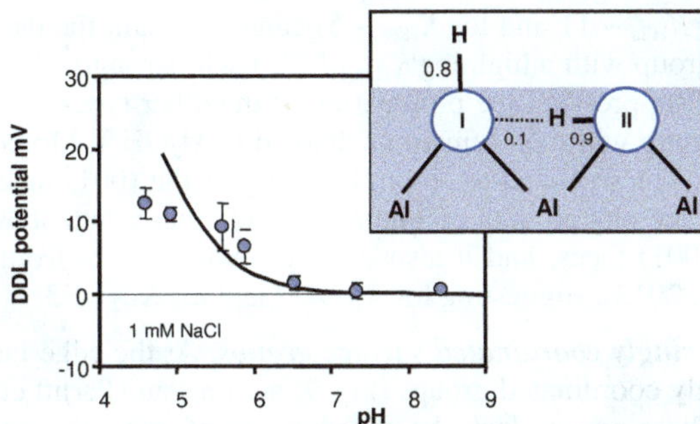

Figure 11. Diffuse double layer potential (DDL) measured with AFM on the (001) face of gibbsite (**data points**) as a function of pH in 0.001 M NaCl (Gan and Franks 2006). The **line** has been calculated assuming affinity constants for successive protonation steps (Eqn. 9) that are approximately 2 units higher than predicted previously, i.e., $\log K_{H1} = 11.9 + 2 = 13.9$ and $\log K_{H2} = 0 + 2 = 2$. Two types of doubly coordinated surface groups can be found at the (001) face. Site I can dissociate a proton and site II may adsorb one.

The same group later showed that their basal plane was heterogeneous and that in particular, the rims showed a different behavior compared to the flat parts (Klaassen et al. 2017). The degree of heterogeneity as well as the surface charge density appears to increase with the salt level and pH (Klaassen et al. 2017), which supports the measurements of Rosenqvist et al. (2002).

The origin of a gibbsite sample and history with pretreatments may generate basal planes with a different degree of defects. At the basal plane of gibbsite with few defects, Curium (Cm^{3+}) forms a surface species (Rabung et al. 2004; Kupcik 2011) that is unique for the isostructural (0001) faces of α-Al_2O_3, as determined with laser-fluorescence spectroscopy for some gibbsite samples (personal communication with Kupcik, 2015). Such a gibbsite had been equilibrated for a long time at the solubility minimum before the measurements started at pH 4. The work suggested that the presence of this specific Cm species strongly depends on the sample and pre-treatment. It underscores that the origin of the gibbsite samples and its treatment may generate basal planes with few defects, but that in other cases, the surfaces may have steps, kinks, and other defects that create charge. If locally present on the surface, a non-linear electrical field will develop that is much less repulsive, intensifying the protonation of the surface groups present. This all supports the maybe not-so-surprising view that gibbsite properties may strongly depend on the synthesis and the treatment after synthesis. Similar as for silica, the seemingly simple solid may often be far more complex than for example samples from well-controlled laboratory synthesis.

Internal H-bridging. Even on ideal surfaces, the molecular scale situation is more complicated than the picture of a simple surface with one surface site reacting with the solution (Jodin et al. 2005; Du et al. 2024). On the surface of gibbsite, $\equiv AlO_{II}H$ may form OH···O bonds with an adjacent $\equiv Al_2O_IH$ group (Jodin et al. 2005), as visualized in Figure 11. Such asymmetric charge distribution leads to more charge neutralization for $\equiv Al_2O_{II}H$ and less for $\equiv Al_2O_IH$. For type I, this may lead to an increase of the affinity constants for the protonation, by 2 units (Jodin et al. 2005) according to the MUSIC model, i.e., $\log K_1(\equiv Al_2O_{II}^{-1}) = 11.9 + 2 = 13.9$ and $\log K_2(\equiv Al_2O_IH^0) = 0 + 2 = 2$. The increase in the latter value of $\log K$ is sufficient to explain the AFM data of Gan and Franks (2006). An increase of $\log K_2$ creates a DDL potential of ψ_d ~15 mV at pH = 5 (Fig. 11) while the DDL potential remains near zero at the highest pH. The σ_d values reported by Siretanu et al. (2014) from their AFM measurements at the (001) face of their gibbsite at pH 6 can be translated to the ψ_d potentials, resulting in a range of 55 to 110 mV, i.e., substantially higher than those reported by Gan and Franks (2006). This substantial charging conflicts with the experimental acid–base titration results of Figure 10 since these point to little or no charge development at the (001) faces.

Others (Bickmore et al. 2004) have proposed the presence of two types of $\equiv Al_2OH^0$ surface groups with relatively high $\log K_{H2}$ values or the protonation reaction $\equiv Al_2OH^0 + H^+ \rightleftharpoons \equiv Al_2OH_2^+$. The $\log K$ values ($\log K_{H2a}$ ~ 11 and $\log K_{H2b}$ ~ 5) cannot explain the data of Gan and Franks (2006), because the group with a high $\log K_{H2a}$ of ~ 11 will strongly charge the (001) faces at a decrease of the pH, suppressing the protonation of the other types (Hiemstra et al. 1989b). That type of acidic group will then remain uncharged ($\equiv Al_2OH^0$). Moreover, the presence of a group with $\log K_{H2a}$ ~ 11 will lead to strong charging of the (001) face. This conflicts with the experimental surface charge data of Figure10, since these data show little or no charge development at the (001) faces, and it also disagrees with results from first principles MD simulations (Liu et al. 2013), suggesting $\log K_{H1}$ =22 and $\log K_{H2}$ =1.3.

Edge faces with singly coordinated surface groups. At the edge faces of the hexagonal gibbsite crystals, singly coordinated groups (Fig. 9) are present (Parfitt et al. 1977b; Hiemstra et al. 1987, 1999). These groups lack the coordination of one Al^{3+} ion with corresponding neutralization. Therefore, the charge deficit will be on average +0.5 v.u. based on the bond valence sum applying the Pauling bond valence concept (Eqn. 6), i.e., $\sum \nu - 2 = +0.5 + 1 - 2 = -0.5$ v.u. meaning that the groups can be represented as $\equiv AlOH^{-0.5}$. In principle, this group may accept or donate a proton. The formal adsorption reactions can be written as follows:

$$\equiv AlO^{-3/2} + 2H^+(aq) \rightleftharpoons \equiv AlOH^{-1/2} + H^+(aq) \rightleftharpoons \equiv AlOH_2^{+1/2} \tag{10}$$

The acid–base titration experiments of Figure 10 imply that gibbsite is positively charged in the pH range of 4–10, illustrating the dominance of $\equiv AlOH_2^{+0.5}$ over $\equiv AlOH^{-0.5}$. Since a large Δp*K* exists between the first and second step on the same group, $\equiv AlO^{-3/2}$ will practically not exist in the common pH range, which fully agrees with results of first-principles molecular dynamics showing a typical value of $\Delta \log K = 12 \pm 1.5$ for boehmite (Smith et al. 2022). In contact with an aqueous solution, the large undersaturation of charge of any $\equiv AlO^{-3/2}$ immediately leads to protonation, forming $\equiv AlOH^{-1/2}$. This implies that the charge on gibbsite must be determined by the protonation of $\equiv AlOH^{-1/2}$. The $\equiv Al_2OH^o$ groups at the edge faces are chemically inert according to the $\log K_H$ values found by first-principles molecular dynamics (Liu et al. 2013) and due to the electrostatic suppression of charge development caused by the presence of the reactive singly coordinated groups (Hiemstra et al. 1989b).

At the edge face, the singly coordinated groups are present in pairs, but both have a very similar proton affinity (Liu et al. 2013), which justifies the use of a 1-p*K* approach for these groups. For this reason, it is possible to experimentally find the log *K* value from the surface charge curves. In the absence of other variable charge groups, at zero charge, equal numbers of $\equiv AlOH^{-0.5}$ and $\equiv AlOH_2^{+0.5}$ must be present. This leads to log *K* = PZC (Bolt and Van Riemsdijk 1982) with a value of 10 ± 0.5 (Hiemstra et al. 1987, 1999), which in turn equals the $\log K_2$ value predicted with the bond valence approach (Hiemstra et al. 1996), and was found via first principles molecular dynamics simulations for gibbsite (Liu et al. 2013) and boehmite (Smith et al. 2022).

If the titration data for gibbsite are scaled to the overall BET surface area, the charge density stays (too) low (in comparison to other oxide minerals). In 0.1 M electrolyte solution, the mean slope of the charging curve is ~6–18 mC/m^2BET/pH, but it increases to ~30–45 mC/m^2/pH if scaled to the edge surface area. This latter value is in line with those for goethite, ~30–60 mC/m^2/pH (Hiemstra et al. 1989b), or rutile and anatase, ~30–50 mC/m^2/pH (Bourikas et al. 2001), all at 0.1 M electrolyte solution.

With a decrease of the pH from 10 to 4, the fraction of the double protonated ($\equiv AlOH_2^{0.5}$) surface groups gradually increases, but the transition remains rather limited due to the strong feedback of the electrostatic field that is formed. This can be calculated from the experimental surface charge σ_o ($C \cdot m^{-2}$) using the expression $\sigma_o \equiv F([AlOH_2^{+1/2}] - \frac{1}{2} N_s)$ in which N_s is the site density of the edge face (13.5×10^{-6} $mol \cdot m^{-2}$). For $\sigma_o = 0.2$ $C \cdot m^{-2}$ at pH 4 (Fig. 10), 65% of the sites are present as $AlOH_2^{+1/2}$ while in the PZC, it is 50%. The gain is just ~15%, while a factor of 3 more is needed to approach the theoretical maximum. Even at the low pH, *protonation is* not yet limited by a lack of sites but *inhibited by the electrostatic field.* Complete deprotonation or protonation as sometimes claimed (Gao and Mucci 2001) is not observable in potentiometric titrations (Lützenkirchen et al. 2002). At the extreme pH ranges, which would need to be probed, dissolution typically becomes a problem as well as technical obstacles such as liquid junction potentials or changes in ionic strength. Plateauing zeta-potential curves are not related to site saturation either.

CHARGE–POTENTIAL RELATIONSHIP

Although the proton reactivity of silica, as well as ideal gibbsite, are determined by essentially one type of surface group, i.e., the singly coordinated one, both surfaces exhibit clearly different reactivity in terms of charging behavior. The mathematical formalism of both one-step protonation reactions ($S + H \rightleftharpoons SH$) is equivalent, but the charge attribution differs. This results in a quite different relationship between surface charge and surface potential, due to a very different ratio [SH]/[S] (mixing entropy) at the PZC, resulting in completely differently shaped charging and potential curves. The charge development is shown in Figure 12, illustrating the fundamental difference in the charging of both types of (hydr)oxides that is rooted in the structure, which is one of the important achievements of the MUSIC approach.

Figure 12 illustrates that two materials can have very different charging curves at the *same* proton affinity ($\log K$), the *same* site density N_s, and the *same* Stern layer capacitance C. This is due to the different charge attribution to the sites and corresponding transitions. For both curves, the same chemical equilibrium is defined as $S^z + H^+ \rightleftharpoons SH^{z+1}$. Only the values of z differ, which is related to the differences in bond valence. In other words, *the different behavior has a structural origin.*

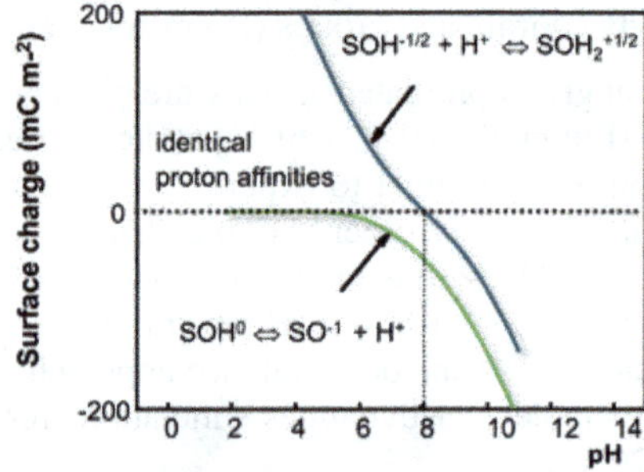

Figure 12. Charging curves of two metal (hydr)oxides with one reactive type of surface group ($N_s = 6\ nm^{-2}$). Both surfaces have a single-step protonation and the same proton affinity ($\log K = 8$), but with different charge attribution (0/–1 and –0.5/+0.5), which can be related to the mineral and surface structure as described in the text. The PZC of the surface with $\equiv SOH^0$ groups is very low. Note the discrepancy between the $\log K$ and PZC for this oxide. A result corresponding to the **green curve** was also obtained in Figure 4 (**dotted lines**) but for very different reasons.

From a more generalizing point of view, the shape of the experimental charging curves of iron, aluminum, and titanium oxide particles correspond to a –0.5/+0.5 type of transition, while the charging of quartz and silica particles, ideal basal planes of gibbsite, and the (0001) faces of oxygen terminated α-Al_2O_3 and α-Fe_2O_3 correspond to the –1/0 type of transition. These different pH-potential relationships have been found experimentally (Siu and Cobbold 1979; Diot et al. 1985; Van Hal et al. 1996) using electrolyte-oxide-semiconductor (EOS) and Ion-Sensitive Field-Effect Transistors (ISFET) devices (Fig. 13a), as well as using single crystal electrodes (Preočanin et al. 2017) of TiO_2 and α-Al_2O_3 (Fig. 13b). Around the PZC, SiO_2 and (0001) faces of α-Al_2O_3 react strongly non-Nernstian, while Al_2O_3 ISFETS (probably polycrystalline) and TiO_2 behave near-Nernstian.

The full lines of Fig. 13b have been modeled with the MUSIC approach. For sapphire (0001) faces, the $\log K_H$ values for doubly coordinated oxygens were fitted ($\log K_1 = 13.6$ and $\log K_2 = 0.5$) and can very well describe the potential behavior. For rutile (100) faces, the model line was shifted to the experimental values that are defined relatively to an arbitrarily chosen reference value. The data illustrate the fundamental difference in the development of the surface potential of both materials. The fitted $\log K_2$ for the sapphire (0001) faces is in good agreement with the value of 1.9 obtained with AIMD simulations (Zhang et al. 2024). However, for $\log K_1$, a value of 18 is suggested with AIMD simulations. Higher $\log K_1$ values have been earlier reported for the corresponding hematite surface (Gittus et al. 2018). The mismatch may be solved in future studies.

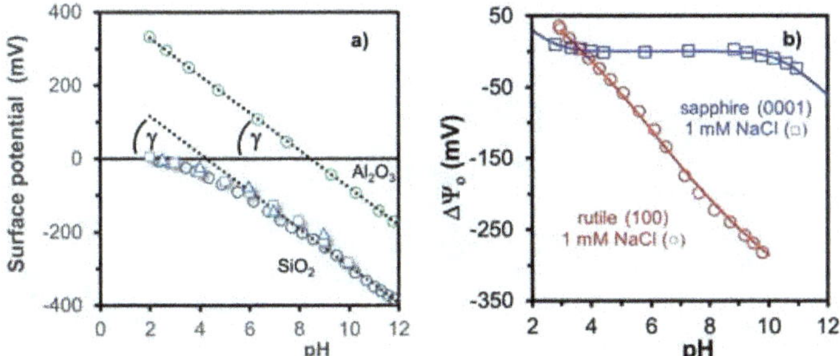

Figure 13. a) Experimental surface potentials measured for a SiO_2 and Al_2O_3 -ISFET (**spheres**) in 1 mM TBACl (Van Hal et al. 1996) and measured for SiO_2 (**squares, triangles**) with electrolyte-oxide-semiconductor (EOS) devices in 0.05–1 M NaCl (Siu and Cobbold 1979; Diot et al. 1985). Silica reacts strongly non-Nernstian, in particular around the PZC. The Al_2O_3 ISFET reacts near-Nernstian and has a slope (γ) of approximately 51 mV/pH. **b)** Comparison of experimental data (**symbols**) (Preočanin et al. 2017) and MUSIC model predictions (**full lines**) for rutile (100) and sapphire (0001) faces in 1 mM NaCl solutions.

SURFACE SPECIATION OF SITES

In the preceding section, two materials with only one kind of proton active surface sites in the normal pH range, but with a very different pH dependency of charge, were compared to illustrate the role of the charge attribution to surface sites in the pH-dependency of surface charge. By using the elementary Pauling bond valence concept, the difference in the charging behavior of both materials can be understood. Heterogeneity of samples and particles hamper straightforward generalization as has been shown for the SiO_2 and gibbsite samples. The results were also discussed in relation to AIMD simulations, showing for both relatively uncomplicated materials a mixture of success and failure. The situation will not become simpler for other (hydr) oxides, as the coordination sphere of the metal ions in (hydr)oxides can be largely asymmetrical. This gives rise to another surface speciation of sites, as will be illustrated and discussed in detail for a well-known metal (hydr)oxide with a relatively large asymmetry in the coordination sphere.

Brown bond valence concept

In the Pauling bond valence concept, the charge is equally distributed over all coordinating bonds. This is equivalent to equal bond lengths in the coordination sphere. However, for a given metal (hydr)oxide, bond lengths may differ considerably. In goethite (α-FeOOH) as a typical example, each oxygen ion is coordinated to three Fe^{3+} ions, saturating oxygen charges with contributions of an asymmetrical H bond, leading to the $Fe_3OH–OFe_3$ moiety in Figure 14 (left-hand side).

Due to the higher proton affinity of O_I over O_{II}, with both oxygen ions being triply coordinated to Fe, the proton attributes more charge to the short $H–O_I$ bond than to the long $H\cdots O_{II}$ bond. Consequently, the Fe ions coordinated to O_{II} of the non-protonated part ($\equiv Fe_3O_{II}$) contribute more to the neutralization of this oxygen while the Fe ions of $\equiv Fe_3O_IH$ contribute less than expected based on the Pauling bond valence. This difference in neutralization concurs with larger $Fe–O_I$ bond lengths in $\equiv Fe_3O_IH$ and smaller distances in $O_{II}Fe_3\equiv$, implying that bond valence, charge distribution, and bond length are related. For a bond length R (om), the actual bond valence (s) can be calculated (Brown and Altermatt 1985; Brown 2009) with:

$$s = e^{-(R-R_o)/B} \tag{11}$$

in which R_0 is an element-specific reference bond length and B is a constant (usually B = 37 pm). In the case of protons in an $O–H\cdots O$ bridge, one may use for the bond valence ($s_{OH\cdots O}$) at the long

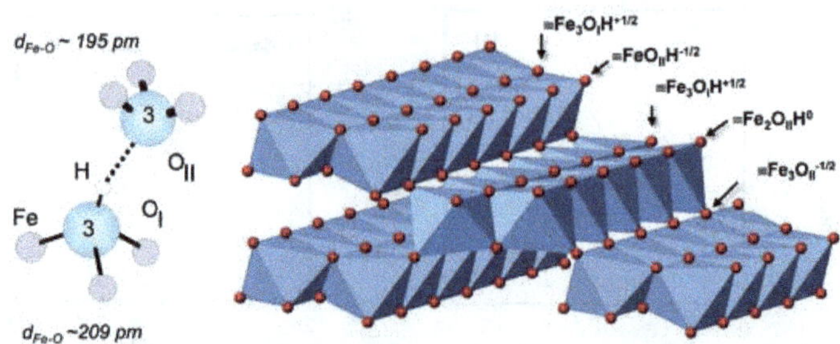

Figure 14. Left-hand-side, two types of oxygen ions in the bulk of α-FeOOH (goethite) and their O–H bonds. Both oxygen ions (**red spheres**) are triply coordinated with respect to Fe(III). The Fe–O_IH distance (in pm) is larger than the Fe–O_{II} distance, implying a lower charge attribution of the coordinating Fe ions to oxygen I than to II. This leads to a higher proton affinity of oxygen I, making it a hydroxyl ion and a hydrogen bond donor, while oxygen II is a hydrogen bond acceptor. **Right-hand-side**, the octahedral structure of the (110), also indexed as (101), face of goethite with **red** oxygens at the corners showing 5 alternating rows of triply, singly, triply, doubly, and triply coordinated surface oxygens. Protons are not shown.

side of the bridge the same equation with $R_o = 73$ pm and $B = 73$ pm, while the complementary part ($s_{O\text{-}H}$) with the short H-bond follows from the charge balance $s_{O\text{-}H} = 1 - s_{O\cdots HO}$ (Hiemstra 2018). Application to the H bridge of Figure 14 yields $s_{O\cdots HO} = 0.2$ v.u. and $s_{O\text{-}H} = 0.8$ v.u. A similar distribution is found in water.

The actual bond valence concept has also been used to assess the proton affinities of aqueous species. Brown (1978) applied it to aqueous organic acids, but did not include the formation of H-bonds. Its importance was highlighted later (Bleam 1993). In the refined bond valence approach, charge neutralization was calculated using donating and accepting H bonds. An unstressed H bond distributes its charge asymmetrically with ~0.8 v.u. for the donating O–H bond and ~0.2 v.u. for the accepting O····H bond (Brown 1992). This distribution is typical for the network of neutral water, which is used as a thermodynamic reference for the proton affinity of aqueous species. By using bond valences for metal ions and H-bonds, the charge saturation of surface groups can be calculated ($\Sigma s_i - 2$), which usually deviates from zero, i.e., $\Sigma s_i - 2 \neq 0$. *The calculated degree of charge saturation of the oxygen can be considered as a measure for the tendency to change H bonds and accept proton charge*. Therefore, the value will correlate with the proton affinity, which can be shown for species in aqueous solutions for which proton affinity can be measured (Hiemstra et al. 1996).

The foremost factor in the above bond valence sum ($\Sigma s_i - 2$) of an oxygen ion on the surface is the number of coordinating ions of the solid. In addition, it is important to know the number of accepting (n) H-bonds and the corresponding strength. Donating (m) H-bonds will contribute too. Without having microscopic information, differences and trends in $\log K$ values have to be rationalized making (disputable) assumptions in the calibration of the relation between ($\Sigma s_i - 2$) and $\log K_H$ using solution species and in its application to surfaces (Hiemstra et al. 1996). In the approach, the non-protonated state of the ligands is used as the reference, while in MD-MO/DFT approaches, the protonated state of a species is used to infer the proton affinity.

Formal charge. Various concepts of formal charge are used in chemistry. Although the elementary charge of a single electron or atom may be rather well-defined, the charge in a collection of atoms in a solution or solid is less so. In quantum chemistry, the atomic charge of molecules atoms can be calculated from the electron densities, but even at this level, arbitrary choices have to be made in the attribution of electron charge to individual atoms, and the compromise is a certain formal charge. As an example, the quantum chemical Mulliken charge of PO_4^z in the moiety $PO_4^{3-}(H_2O)_{12}$ is approximately −1, i.e., very different from the −3 v.u. obtained from a classical valence analysis, illustrating that formal charge is definition-dependent. Despite these complications, bond valence can be considered a simple and adequate tool to define formal charges that can be used in the bookkeeping of charges for surfaces.

For surface complexation modeling, a formal surface charge is defined based on the excess of adsorbed ions in the interface with the corresponding valence charge. For instance, the addition of an acid such as HCl to an oxide suspension in the PZC will lead to simultaneous adsorption of H^+ and Cl^-. Overall, the adsorption phase is and remains uncharged, but locally, excess of charge can be defined in the interface. In a physical-chemical picture, the protons are bound at the surface to oxygen ions, and the chloride ions are present in a diffuse pattern in the solution close to the surface. Conceptually, this results in charge separation (Fig. 1) Thermodynamically, electrostatic energy can be related to the concentration profile. MD simulations suggest that the sum valence rule is not obeyed for surface groups (Machesky et al. 2008), while it usually is approximately in mineral lattices.

In SCMs, the formal proton charge is (in an idealizing fashion) attributed to a defined location, a so-called electrostatic plane (Figs. 1 and 2). The accumulated proton charge is smeared out in that plane standing for a mean electrostatic field emitted by the surface of the plane. The mean field is gradually attenuated by the accumulated counterions at the solution side of the interface. If the aim is to calculate the electrostatic energy, any bond valence sum rule is to be neglected deliberately. To relate the local proton excess to surface charge, one needs a charge bookkeeping tool for the reference state. The Pauling bond valence concept is very useful for this (Hiemstra et al. 1989a), but other choices can be made too (Hiemstra et al. 1996; Machesky et al. 2008; Ridley et al. 2009). With the proton affinities and formal charge assignments, surface speciation can be evaluated, i.e., the state and abundance of different surface species including their charges.

Surface site speciation of goethite (α-FeOOH)

Goethite particles expose different crystal faces, (110) faces typically being the predominant[15] (Fig. 14). AFM studies reveal the presence of (100) faces in some preparations, which are found to be less stable than the (110) faces (Weidler et al. 1999). At the top-end of the needle-shaped prismatic crystals, (021) and (001) faces can be found (Weidler et al. 1996). The contributions of the various crystal faces are crucial when trying to unify SCMs between preparations (Gaboriaud and Ehrhardt 2003; Lützenkirchen et al. 2008; Han and Katz 2019) yet for ideal surfaces. Even more details of the goethite surfaces and sites have been established in the meantime (Livi et al. 2017, 2023; Martinez et al. 2023).

(110) face. The faces of goethite differ in types of sites and densities (Barrón and Torrent 1996). For goethite produced by slow neutralization of Fe(III) solution, the (110) faces contributed 85–90% (Hiemstra et al. 1989b; Martinez et al. 2023). The predominant (110) face of goethite has 4 different types of surface groups, one singly-, one doubly-, and two types of triply-coordinated oxygen ions (Fig. 14). The predicted $\log K$ values for the different proton reactive groups of the main crystal face (110) are given in Table 1. The affinity constants in Table 1 were obtained from different approaches. A comparison shows that three of the approaches appear to lead to quite comparable results. Only the static ab initio molecular orbital (MO) approach using density functional theory (DFT) (Aquino et al. 2008) yields an important difference.

Figure 15 shows the calculated surface speciation for the (110) faces of goethite using the affinity constants from Table 1 as derived with a) the MUSIC model (Hiemstra et al. 1996), b) a static ab initio MO/DFT approach (Aquino et al. 2008), and c) first principles molecular dynamic (FPMD) simulations (Zhang et al. 2021). The results based on the first and latter (Fig. 15a and c) are in close agreement, from which the results in Figure 15b clearly differ.

[15] The (110), (021), and (001) can also be indexed as (101), (210), and (010) faces, respectively.

The pH-dependency of surface charge in Figures 15a,c is mainly due to the proton reactivity of two types of surface groups, i.e., $\equiv FeO_{II}H^{-1/2}$ and $\equiv Fe_3O_IH^{+1/2}$ (I and II are defined in Fig. 14). The charge of half the sites of $\equiv Fe_3O_IH^{+1/2}$ is compensated by the charge of $\equiv Fe_3O_{II}^{-1/2}$, leading to a reduced apparent site density of 3 nm^{-2} for triply coordinated surface groups. As shown in Figure 15a,c, the change in the speciation of $\equiv FeO_{II}H^{-1/2}$ overall dominates the change in surface charge below the PZC. In the normal pH range (pH 4–10), the triply coordinated groups act mainly via permanent negative charge $\equiv Fe_3O_{II}^{-1/2}$, or positive charge $\equiv Fe_3O_IH^{+1/2}$, each with behavior similar to the corresponding groups in the bulk crystal where O_I is protonated in contrast to O_{II} (Fig. 14).

In Figure 15a), the singly coordinated surface group (green diamond) is $\equiv FeO_{II}H^{-1/2}$ at high pH and it is gradually protonated at lowering pH. This group is mainly responsible for the pH-dependent charge development. Both types of triply coordinated surface groups (yellow and purple triangles) behave very similarly as in the bulk mineral, i.e., $\equiv Fe_3O_IH$ is mainly protonated and $\equiv Fe_3O_{II}$ not. In Figure 15b, the singly coordinated groups are always protonated $\equiv FeO_{II}H_2^{+1/2}$, which strongly contrasts with the speciation in Figures 15a,c, and is problematic in light of interpreting the IR spectrum, see text. The doubly coordinated groups (squares) in Figure 15b protonate at a pH decrease. The speciation of Figures 15a c is very similar since the log K values in both models closely match (Table 1). For the (110) face, the calculated PZC values are respectively 9.8, 9.8, and 9.1.

The calculated surface speciation of Figure 15b strongly differs from Figure 15a,c: The singly coordinated surface groups are mainly present as water ligands ($\equiv OH_2$), i.e., aquo groups, while in Figure 15a,c, the groups are predominantly OH ligands, which start to accept a proton at low pH. According to the speciation in Figure 15b, charge development is attributed to $\equiv Fe_3O_I(H)$ and $\equiv Fe_2O(H)$ due to the combination of a relatively (too) high $\log K_H$ for $\equiv FeO_{II}H^{-1/2}$ and a relatively (too) low $\log K_H$ for $\equiv Fe_2O_{II}H^o$ (Table 1), promoting proton release from $\equiv Fe_2O_{II}H^o$ forming $\equiv Fe_2O_{II}^{-1}$. The curves in Figure 15 also highlight that the intrinsic stability constants differ from the corresponding "observable ones". In solution chemistry, the pK_a for e.g., a monoprotic acid can be determined via spectroscopic methods. It corresponds to the pH at which

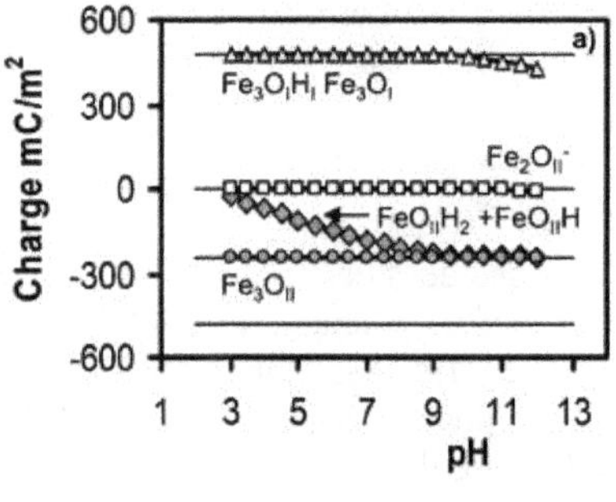

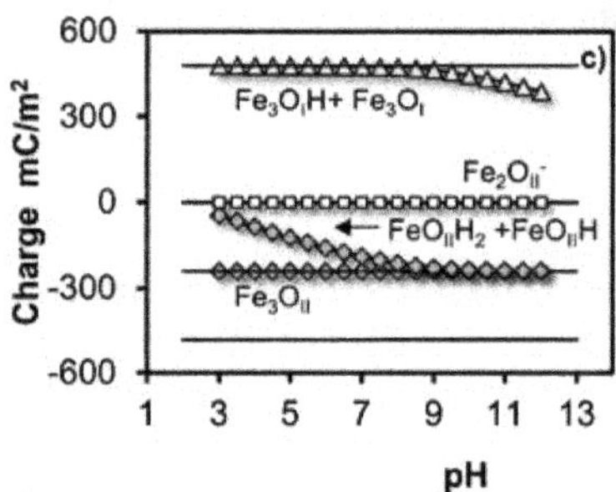

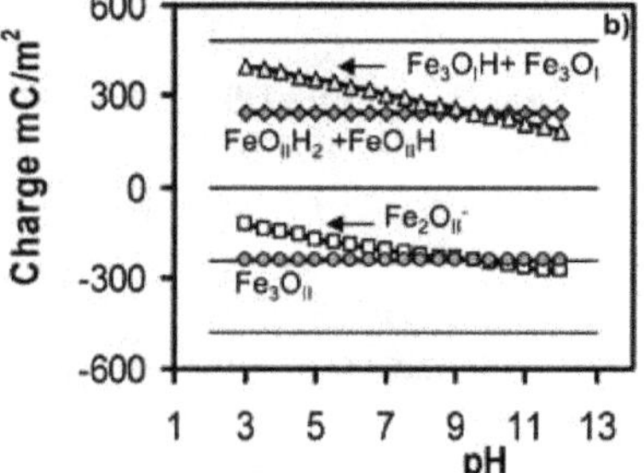

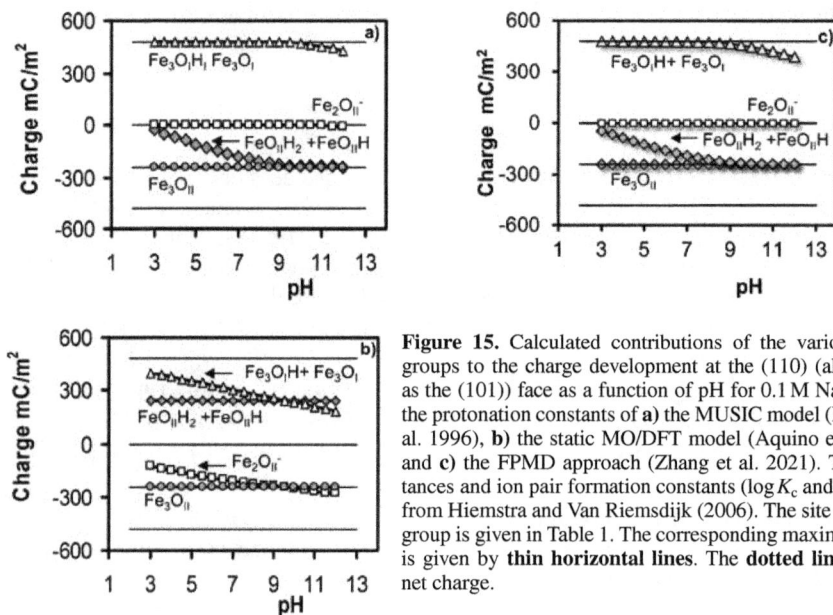

Figure 15. Calculated contributions of the various surface groups to the charge development at the (110) (also indexed as the (101)) face as a function of pH for 0.1 M $NaNO_3$ using the protonation constants of **a)** the MUSIC model (Hiemstra et al. 1996), **b)** the static MO/DFT model (Aquino et al. 2008), and **c)** the FPMD approach (Zhang et al. 2021). The capacitances and ion pair formation constants ($\log K_c$ and $\log K_a$) are from Hiemstra and Van Riemsdijk (2006). The site density per group is given in Table 1. The corresponding maximum charge is given by **thin horizontal lines**. The **dotted lines** give the net charge.

the concentrations of protonated and deprotonated species equal each other. This would for the singly coordinated goethite hydroxyls be the pH at which the net charge of the two relevant species equals zero. In Figure 15a, this would be a bit below pH 3, but the corresponding value in Table 1 is 7.7. This difference is caused by the electrostatic effects through the charges on all the sites present on the face making the situation very different from that on gibbsite.

Table 1. Surface groups of the goethite (110) (also indexed as (101)) face with corresponding site density (N_s), potential undersaturation of charge (Σs_i-2) and resulting proton affinity ($\log K$). The indexes I and II indicate the type of oxygen in the lattice (Fig. 14). The $\log K_{H2}$ values (*) for the protonation of $\equiv Fe_2O_{II}H^0$ reveal a large ΔpK.

Species [#)]	N_s nm^{-2}	Σs_i-2	$\log K$ [a)]	$\log K$ [b)]	$\log K$ [c)]	$\log K$ [d)]
$\equiv FeO_{II}H^{-1/2}$ (-OH)	3	−0.39	+7.7	+12.1	+7.3	+7.0
$\equiv Fe_2O_{II}^{-1}$ (μ-O$^-$)	3	−0.60	+12.3*	+9.5*	+13.9*	
$\equiv Fe_3O_I^{-1/2}$ (μ_3-O_I^-)	6	−0.60	+11.9	+10.0	+10.7	
$\equiv Fe_3O_{II}^{-1/2}$ (μ_3-O_{II}^-)	3	+0.02	+0.4	low	−0.2	
Calculated PZC			9.8	9.8	9.1	-

Notes:
#) Between brackets, the alternative chemical name for the species is given.
a) From Hiemstra et al. (1996) with $\log K_{H2} = -0.5$*
b) Calculated with static MO/DFT (Aquino et al. 2008) with $\log K_{H2} = -4$*
c) Calculated with FPMD (Zhang et al. 2021) with $\log K_{H2} = -0.5$*
d) Calculated with MO/DFT-MD (Leung and Criscenti 2012)

Testing $\log K_H$ ***models.*** The differences in speciation presented in Figure 15 should require a test with surface complexation modeling of the surface speciation. This is typically omitted by the scientists who derive the proton affinity constants *ab initio*. A comparison of calculated affinity constants of individual surface groups with data is in general problematic, because the protonation of the surface is the result of the interplay of various types of surface groups and is strongly masked by the electrostatic field. Moreover, experimental results often have contributions from more than one crystal face.

The model inherent value of the PZC can be used as a first test. The calculated PZC value is 9.8 when using the MUSIC model and the static MO/DFT approach, and it is 9.1 when using the FPMD data. For well-crystallized goethite, the experimental PZC is 9.1 ± 0.2. At first glance, all models are in fair agreement with the measured PZC values of goethite. However, goethite may also have a considerable contribution from the top-end faces (Hiemstra et al. 1996; Gaboriaud and Ehrhardt 2003; Prelot et al. 2003; Livi et al. 2017, 2023; Han and Katz 2019). Surface roughness and crystal face ratios scales with the total surface area leading to higher contributions of ≡FeOH(H) sites (Livi et al. 2023), well aligned with the earlier observed higher charge density (Hiemstra et al. 1989b) of goethites with relatively low BET surface areas and higher adsorption of oxyanions (Hiemstra and Van Riemsdijk 1996; Villalobos et al. 2009; Martinez et al. 2023) and heavy metal ions (Venema et al. 1996b). With their relatively low proton affinity (Table 1), the experimental PZC for goethite as a whole will be lower than the PZC calculated for the (110) face. Therefore, the PZC of the (110) face predicted with the FPMD data (PZC = 9.1) might be slightly too low for a goethite particle, due to a too-low value for the $\log K_H$ of the $\equiv Fe_3O_I$ group.

To discuss the validity of the approaches in terms of the differences in surface speciation, spectroscopic results are the gold standard. The very different surface speciation according to the static MO/DFT approach (Aquino et al. 2008), suggests the predominance of $\equiv FeOH_2^{+1/2}$, whereas the other approaches suggest a predominance of the $\equiv FeOH^{-1/2}$ surface species with a small fraction becoming protonated at low pH. The goethite surface has two main O–H stretching bands, i.e., ~3660 cm^{-1} and 3490 cm^{-1} (Boily and Felmy 2008;

Kanematsu et al. 2018), comparable to the stretching found in D_2O. According to that interpretation of the spectroscopic results, $\equiv FeOH^{-1/2}$ dominates in the FTIR spectra of dried goethite, but changes intensity when lowering the pH from 7 to 4, due to the protonation of $\equiv FeOH^{-1/2}$ to $\equiv FeOH_2^{+1/2}$. This corroborates the speciation according to Figures 15a,c. Furthermore, at high pH, FTIR data of dried goethite (Boily and Felmy 2008) show that the intensity of the lower frequency band decreases with increasing negative surface charge above the PZC, which also agrees with the deprotonation of $\equiv Fe_3O_IH$ (Figs. 15a,c).

(021) face. On goethite, the relative contribution of the (021) (also indexed as (210)) face varies with the rate of Fe(III) neutralization at preparation and correlates with the specific surface area of the product formed (Hiemstra et al. 1989b; Martinez et al. 2023). At instant neutralization, the BET SSA may be as low as 30 $m^2 \cdot g^{-1}$ yielding an estimated (021)-face contribution of about 50 % while at very slow neutralization of 10 % per hour, the resulting goethite may have a BET SSA close to 100 $m^2 \cdot g^{-1}$ and a (021) face contribution of about 10%.

At the (021) face, four types of surface oxygens can be distinguished (Venema et al. 1998), two types of singly coordinated $\equiv FeO_{I/II}H$ groups (both with fixed $\log K_H$ of 8), and two types of doubly coordinated $\equiv Fe_2O_{I/II}$ (with selected $\log K_H$ values of 12/0 and 20/8). These MUSIC values rely strongly on the number and strength of the H-bonds as well as on the persistence of internal H-bonds (Venema et al. 1998). With the preselected set, a PZC of 8.3 can be calculated. The pH-dependency of the charge is predominantly due to the protonation of both types of singly-coordinated $\equiv FeO_{I/II}H$ groups and one type of doubly coordinated group ($\equiv Fe_2O_I$).

The calculation of the proton affinity constants of the various groups at the (021) face with the FPMD approach (Zhang et al. 2021) yields a very large difference in proton affinity for both types of $\equiv FeOH$ groups. The site with a high proton affinity ($\log K_H = 10$) occurs predominantly as $\equiv FeOH_2^{+1/2}$ and the other one with a low affinity ($\log K_H = 2.3$) as $\equiv FeOH^{-1/2}$. Both groups are approximately present in a 1:1 ratio, which implies that their net contribution to the surface charge is low. One of the two $\equiv Fe_2OH^o$ groups contributes to the variable surface charge by dissociating a proton ($\log K_H = 5.2$) in the pH range 4–10, forming $\equiv Fe_2O$, while the other one is only present as $\equiv Fe_2OH^o$. For the (021) faces, the PZC according to FPMD is very low (PZC ~3.9), which would imply the development of a significant negative surface charge (–150 $mC \cdot m^{-2}$ in 0.1 M $NaNO_3$ at pH 9.1). Such a low PZC value would also imply the absence of significant binding of oxyanions such as phosphate, arsenate, or chromate on the (021) faces while experiments do not suggest this according to tests for a series of goethites with a range of (021) contributions (Han and Katz 2019). Goethites with a lower BET surface area develop more proton charge (Hiemstra et al. 1989b) and have a higher oxyanion adsorption capacity (Hiemstra and Van Riemsdijk 1996; Villalobos et al. 2009; Martinez et al. 2023), which can has been attributed to differences in crystallinity and more (021) face contributions.

Proton affinity of sites at the α-TiO_2 (110) surface

In the MUSIC model, the proton affinity depends on the bond valence saturation of the surface oxygen which is determined by the number of coordinating metal ions, adsorbed protons (m), and weakly interacting protons (n). For the latter, MD simulations (Machesky et al. 2008) (Cheng and Sprik 2010) showed that surface groups may have a different number of weak H-bond interactions. This will affect the bond valence saturation of the surface oxygen and thereby the $\log K_H$ value. For calculating the $\log K_H$ with the MUSIC model, the average number (n_{mean}) was used (Machesky et al. 2008; Cheng and Sprik 2010). However, an average performs differently from a combination of discrete physical sites each with one of two possible states of interaction with either $n = 1$ or 2. Defining both states of the reactants separately leads to two protonation constants that differ by 4 log units, as given in Table 2. In the calculations, the applied Ti-O bond lengths were derived with molecular modeling. For TiO_2, these values are substantially shorter than for the bulk as used initially in the MUSIC model approach, due to a lack of this information.

Table 2. Log K_H predictions using the bond valence (BV) approach of the MUSIC model in which the neutralization of surface oxygen charge is determined by the number of coordinating metal ions (Ti), the number of protons bound (m), and the number of weak H-bond interactions (n). For the latter, discrete surface species with a different state of interaction (n = 1 or n = 2) are defined, rather than assuming a single hypothetical mean site with an average number, derived with MD simulations (see text). The applied Ti–O distances have been derived with classical MD, DFT, and DFTMD modeling (Machesky et al. 2008; Cheng and Sprik 2010).

Site	Ti–O	s_{Ti}	m	n	$\log K_H$
$\equiv TiOH^{-1/3}$	193±5 pm	1	1	1	5.6±2.1
$\equiv TiOH^{-1/3}$	193±5 pm	1	1	2	1.6±2.1
$\equiv Ti_2O^{-2/3}$	189±2 pm	2	0	1	3.9±1.6
$\equiv Ti_2O^{-2/3}$	189±2 pm	2	0	2	0.0±1.6

The mean n-values obtained for the reactants with molecular dynamic modeling (n_{mean} = 1.7±0.2 for $\equiv TiOH^{-1/3}$ and n_{mean} = 1.2±0.1 for $\equiv Ti_2O^{-2/3}$) can be used as a constraint to reveal the surface speciation. Modeling with the equilibria of Table 2 discloses that the singly- and doubly-coordinated groups with a relatively high proton affinity are by far the most important species (~85%). The calculated surface densities is N_s = 4.48 and 4.46 nm^{-2}, respectively. The surface groups with a relatively low proton affinity are minor and present in low densities (N_s = 0.72 and 0.84 nm^{-2}, respectively).

The rather high mean n-value (n_{mean} = 1.7±0.2) for the terminal $\equiv TiOH^{-1/3}$ groups, obtained with MD simulations (Machesky et al. 2008; Cheng and Sprik 2010), can be understood from the dominance of the reactant $\equiv TiOH^{-1/3}$ with n = 2 over $\equiv TiOH^{-1/3}$ with n = 1 at equilibrium. Most of the latter reactant has been transformed into $\equiv TiOH_2^{+2/3}$ by protonation in contrast to the former. For the bridging $\equiv Ti_2O^{-2/3}$ sites, the situation is the opposite because the protonation of these sites is quite limited. The mean value of n for the reactants of $\equiv Ti_2O^{-2/3}$ is rather low (n_{mean} = 1.2±0.1) according to MD simulations. This value can be rationalized from the relatively large contribution of the $\equiv Ti_2O^{-2/3}$ site with n = 1 that remains largely unprotonated and the relatively small contribution of the $\equiv Ti_2O^{-2/3}$ site with n = 2 that also remains unprotonated at equilibrium.

The MUSIC model (Table 2), made consistent with the MD results, reveals a predicted PZC of 4.4±0.2. Within the uncertainly of ±1, due to variation in the reported Ti–O distances, the calculated PZC agrees well with the PZC value (4.4±2) obtained with DFTMD simulations (Cheng et al. 2015) and the experimental PZC values (~5.0±0.5) measured with different techniques for the hydrous α-TiO_2(110) surface of single crystals (Fitts et al. 2005; Bullard and Cima 2006; Preočanin et al. 2017). These values are also close to the results of the extensive work done by Machesky et al. (2008) for well-crystallized rutile with predominantly (110) faces (PZC = 5.4±0.2), but also some surface roughness (Livi et al. 2013).

If in the bond valence approach of the MUSIC model, a single proton interaction state for each site is assumed using the corresponding n_{mean} value, the $\log K_H$ for $\equiv TiOH^{-1/3}$ decreases to 2.9 and for $\equiv Ti_2O^{-2/3}$ to 3.1, resulting in a PZC value of 3.0±1 which is evidently too low. If the (110) surface is represented by only the dominant surface sites (~85%) both with a high $\log K_H$, the PZC will be 4.8±1, which is close to the PZC value calculated using two interaction states (n =1, n = 2). Synthesis of the above suggests that for a practical SCM, the reactivity of TiO_2 surfaces can be still rather well represented by only one state per type of site.

It is interesting to compare the above $\log K_H$ of $\equiv Ti_2O^{-2/3}$ in Table 2 with the $\log K_H$ value earlier derived using the bulk distance (d_{Ti-O} = 1.96 Å) and assuming two weak OH bonds (n = 2). This early value of 4.4 is nearly the same as the $\log K_H$ value using the DFT-derived distance (d_{Ti-O} = 1.89 Å) and $n = 1$ (Table 2). The use of a higher Ti–O bond length (bulk versus surface) leads to a higher value for the $\log K_H$ ($\Delta = +4.4$). In the early approach, this increase is compensated by a higher number of accepting H bonds (n = 2) leading to a corresponding change in $\log K_H$ ($\Delta = -4.0$), ending with a net difference of only 0.4 units. This example demystifies the apparent success of the original MUSIC model for rutile while using a discussible water–surface interaction. It shows that it is crucial to use appropriate surface M–O distances in the bond valence calculations.

THERMODYNAMIC CONSISTENCY FOR ION ADSORPTION

Before discussing ion adsorption modeling, our focus will be on a key thermodynamic principle that relates the pH dependency of ion adsorption to its proton-co-ad(de)sorption. This key concept is intimately related to the location of the charge of the adsorbed ion in the interface, where it interacts with the protons at the surface. The concept self-consistently relates the microscopic structure of surface complexes to macroscopic data about charging behavior and proton exchange in the presence of specifically adsorbing ions. It also highlights the relevance of measuring proton ad(de)sorption and not relying on pH-edge/envelope and isotherm data only.

Thermodynamic consistency rule

Ion-surface interactions are strongly influenced by interfacial electrostatics (Hiemstra and Van Riemsdijk 1999). Cations are generally absorbed relatively weakly at low pH, where adsorbed protons on many oxidic minerals generate a repulsive electrostatic potential, which in turn promotes the adsorption of anions. With increasing pH, the repulsive electrostatic action on cation adsorption is reduced or may even become attractive, resulting in higher cation adsorption. For anions, the opposite trend is observed.

From a thermodynamic perspective, the pH dependency of ion adsorption is related to the co-ad(de)sorption of protons (Perona and Leckie 1985) via the following thermodynamic relation:

$$\chi = \left(\frac{\partial \Gamma_H}{\partial \Gamma_i}\right)_{pH} = \left(\frac{\partial \log a_i}{\partial pH}\right)_{\Gamma_i} \tag{12}$$

i.e., the change in H adsorption (Γ_H) due to a change in the uptake of ion i at the surface (Γ_i) for a given pH equals the change in the logarithm of the ion activity a_i in solution with the pH at a given level of ion adsorption (Γ_i).

The slope ($\partial\Gamma_H/\partial\Gamma_i$) is negative for cations (i.e., proton co-desorption) and positive for anions (i.e., proton co-adsorption), so their pH dependency is opposite. The proton co-adsorption is illustrated in Figure 16a for the adsorption of fluoride ions on goethite. The activity of F^- (a_F) in the solution can be measured experimentally using an ion-selective electrode. Figure16a displays proton co-adsorption as a function of the F^- uptake. The slope of the line represents $\partial\Gamma_H/\partial\Gamma_F$ or χ (left-hand side of Eqn. 12). In this specific case, χ changes little with F^- loading, i.e., ≈0.7 at 1.5 μmol $F^-\cdot m^{-2}$ (Fig. 16a). According to the thermodynamic consistency, this leads, at a given F^- loading, to ($\partial\log a_F/\partial pH$) ≈ +0.7. If $\partial\Gamma_H/\partial\Gamma_F$ is pH-independent, a similar value should then be obtained for $\Delta\log a_F/\Delta pH$ and in Figure 16b indeed $\Delta\log a_F \approx +0.7$ at $\Delta pH = +1$. The same approach has been used by Girvin et al. (1991) for NpO_2^+ adsorption on ferrihydrite.

The proton co-adsorption can also be derived for the plot in Figure16c ($\Delta pH \approx +1.5$ at $-\Delta pF = \Delta\log a_F = +1$). The analysis shows that the experimental pH dependency can be used to predict proton-ion stoichiometry χ, and *vice versa*.

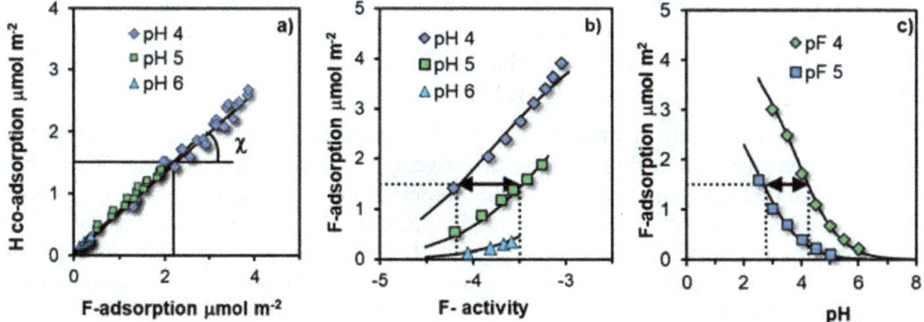

Figure 16. Fluoride adsorption behavior (μmol·m^{-2}) on well-crystallized goethite in 0.1 M $NaNO_3$ (Hiemstra and Van Riemsdijk 2000). **a)** Proton co-adsorption (μmol·m^{-2}) for three pH values. **b)** Fluoride adsorption isotherms at constant pH. **c)** Fluoride adsorption (μmol·m^{-2}) at a constant activity (pF). The differential proton co-adsorption ratio $\partial\Gamma_H / \partial\Gamma_F = \chi$ in Figure 16a is approximately 0.7. According to the thermodynamic consistency principle (Eqn. 12), the value of χ leads at a given F^- loading to an equal differential change of the F^- activity with pH ($\partial\log a_F / \partial$pH). If $\partial\Gamma_H / \partial\Gamma_F$ is pH-independent, $\Delta\log a_F / \Delta$pH in Figure 16b and $-\Delta$pF/ΔpH in Figure 16c equal the value in Figure 16a.

Role of solution speciation in the proton co-adsorption of ions

The above thermodynamic consistency (Eqn. 12) has been derived for an ion with a fixed protonation state. However, the protonation state of an ion may significantly vary with pH, for example for $Hg(OH)_m^{+2-m}$ or $H_nPO_4^{-3+n}$. Thus, the activity a_i of an ion (as a species) in solution is pH-dependent even if total ion concentration C_t in solution remains constant. This can be included in the thermodynamic principle. Equation (12) can be generalized (Rietra et al. 2000) and the proton co-adsorption (χ_H), corrected for the excess protonation state of the species in solution (n_H), is then related to the overall pH dependency of the ion adsorption ($\partial\log C_t / \partial$pH), according to:

$$\left(\frac{\partial \log C_t}{\partial \text{pH}}\right)_{\Gamma_i} = \left(\frac{\partial \Gamma_H}{\partial \Gamma_i} - n_H\right)_{\text{pH}} \equiv \left(\chi_H - n_H\right)_{\text{pH}} \tag{13}$$

where $\chi_H \equiv \partial\Gamma_H/\partial\Gamma_i$ is the proton co-adsorption relative to a chosen reference state of component *i*, and *n* is the change of the mean excess number of protons per ion *i*, defined as a proton excess relative to the same reference state of *i*. This number can be calculated with traditional solution speciation approaches.

An important step in using the equation is the choice of the reference state from which protons are counted. Typically, unprotonated /non-hydrolysed species, like Hg^{2+} and PO_4^{3-} are selected as references. With PO_4^{3-} as a reference on goethite, one measures $\chi_H \approx 2.3$ (Hiemstra and Van Riemsdijk 1996) for pH 9 and $\Gamma = 1$ μmol·m^{-2}. Using the same reference (PO_4^{3-}), one obtains for the mean proton excess of the species in solution a value of $n_H \approx 1.0$, because the main solution species is HPO_4^{2-} (aq) at that pH. This yields a difference of $\chi_H - n_H = 2.3 - 1.0 = 1.3 = \partial\log C / \partial$pH. When selecting HPO_4^{2-} as a reference to do the counting, the excess number of protons in the solution, relative to HPO_4^{2-}, will decrease to $n_H \approx 0$. Similarly, the co-adsorption will be less when counted relative to the adsorption of HPO_4^{2-} ($\chi_H \approx 1.3$). While the individual values of n_H and χ_H decrease by the same integer when the reference state is changed, the difference, $\chi_H - n_H$, remains the same and is independent of the choice of the reference if the same reference species is used for both χ_H and n_H. Negative values of *n* may also occur if species in solution have fewer protons than the selected reference.[16]

[16] Equation (13) is exemplified in another contribution (Hiemstra et al. 2025, this volume)

Proton co-adsorption ratio χ. Equation (13) shows that the overall change in adsorption with pH involves an intrinsic surface property (χ_H) and a chemical property of the component in solution (n), meaning for example that changes in solution speciation affect adsorption. This had been realized early on (Hingston et al. 1967).

Figures 16b and c demonstrate how χ_H can be measured experimentally with batch ion adsorption experiments. It is also possible to derive χ_H from acid–base titrations with a fixed amount of an added ion, sampling the solution to determine the fraction adsorbed. Moreover, it can be assessed by only modeling the classical acid–base titrations, as shown by Sjöberg and co-workers in several examples (Lövgren et al. 1990; Gunneriusson and Sjöberg 1993; Gunneriusson 1994). Such high-end titrations are laborious and require patience and therefore hardly performed anymore.

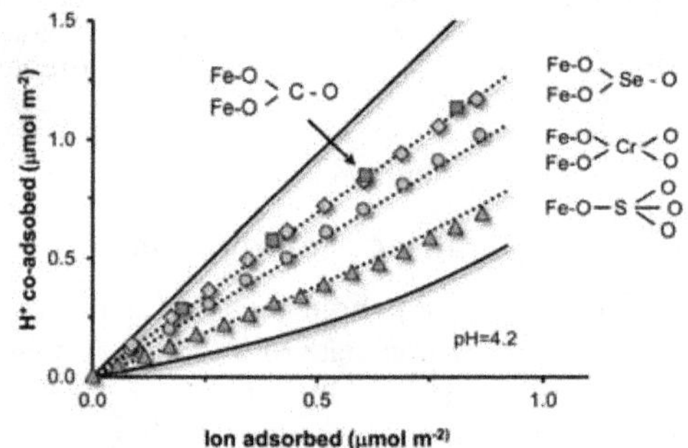

Figure 17. Co-adsorption of protons on goethite in 0.01 M $NaNO_3$ at pH=4.2 due to the binding of CO_3^{2-} (Villalobos and Leckie 2000), SeO_3^{2-}, CrO_4^{2-}, and SO_4^{2-} (Rietra et al. 2000). **Full lines**: expected co-adsorption if ion charge (–2) is fully present at the surface or in the 1-plane (BS approach), yielding a very high and a low co-adsorption, respectively. **Dotted lines** are calculated using a Pauling distribution for the charge over the coordinating ligands, leading to a surface charge attribution of respectively ⅔, ½, and ¼ of the anion charge (–2). [Reproduced from Hiemstra et al. (2004) with permission of Elsevier]

An elegant and simple approach to measure proton co-adsorption has been proposed by Rietra et al. (2000). In that approach (Fig. 17), pH-stat ion titrations at high solid concentrations ($m^2 \cdot L^{-1}$) are used, allowing direct measurement of χ_H without measuring the total concentration C_t in solution as long as the vast majority of the added ions is adsorbed. The same method also gives information about the pH dependency of the ion adsorption at extremely low solution concentrations that are experimentally not or very difficult to access. This widens the conditions for evaluating a SCM.

Proton co-adsorption and surface complex structure. Proton co-adsorption and pH dependency can be related to the structure of surface complexes. The interfacial ligand distribution is a leading factor. When a negative charge moves towards a positively charged metal (hydr)oxide surface, the surface reacts by adsorbing an increasing number of protons. Maximum interaction with a corresponding number of co-adsorbed protons is achieved when the negative test charge reaches the surface where the proton charge is located. If the surface is Nernstian, the number of protons adsorbed equals the valence of the test charge (Fokkink et al. 1987; Venema et al. 1996a). Thus, χ *depends on the location of the charge of an ion in the electrostatic field.* In the limiting case of a Nernstian surface, χ becomes independent of the formulated reaction including the number of involved protons, i.e., independent of the intrinsic stoichiometry of the reaction! Even for Near-Nernstian surfaces, electrostatics are often the most important explanatory factor in the pH dependency of adsorption, apart from the behavior of the species in solution, (i.e., n_H in Eqn. 13).

Measurements of the co-adsorption for a series of divalent anions (Fig. 17) on goethite at low pH using the pH-state ion adsorption technique (Rietra et al. 1999) involves a series of different innersphere surface complexes (Hayes et al. 1987; Fendorf et al. 1997; Hug 1997; Peak et al. 1999; Wijnja and Schulthess 2000). The measured co-adsorption can be explained by a different location of the ligands in the interface. The average distance of the ion charge from the surface follows the order $SeO_3^{2-} \approx CO_3^{2-} < CrO_4^{2-} < SO_4^{2-}$ and can be related to differences in the ligand distribution of the surface complexes.

The SeO_3^{2-} ion forms a bidentate surface complex (Fig. 17) having only one non-coordinated ligand (Hayes et al. 1987). At low pH, the dominant surface complex of SO_4^{2-} is a monodentate innersphere with three of the four ligands not coordinated with the surface (Hug 1997; Peak et al. 1999; Wijnja and Schulthess 2000). In addition, some SO_4 outer-sphere complexation may occur, and the charge of sulfate is then expected at some larger distance from the surface. Chromate, for which a bidentate surface complex was reported (Fendorf et al. 1997), has an intermediate position (Fig. 17). Experimentally, the slopes of the lines in Figure 17 ($\partial\Gamma_H/\partial\Gamma_{ion}$) differ with the ion loading.

Rationalizing proton co-adsorption with electrostatics. If all anion charge is attributed to the surface plane, where the protons are adsorbed, and the surface is Nernstian, the same amount of charge is co-adsorbed as protons, i.e., 2 H^+ per divalent anion, such that the surface charge σ_o and surface potential ψ_o do not change, which approaches a surface with a near-Nernstian behavior (upper full line in Fig. 17).

If instead all divalent anion charge is placed in the non-Nernstian environment of the Helmholtz/Stern plane, the charge (σ_1), and correspondingly the interfacial potential (ψ_1) in that plane (1-plane), change with anion loading, which in turn affects σ_o, while the surface potential (ψ_o) remains nearly constant (Near-Nernstian). The surface charge σ_o increases, as can be rationalized with the electrostatic relation for a capacitor:

$$\sigma_0 = C(\psi_0 - \psi_1) \tag{14}$$

At a constant (positive) surface potential ψ_o but decreasing ψ_1 due to OS complexation, σ_o will increase (Eqn. 14). This increase in σ_o reflects in that case the number of protons co-adsorbed.[17]

The decrease in ψ_1 is non-linear with the net charge of the particle ($\sigma_o + \sigma_1$) which follows from the Gouy–Chapman equation in combination with the charge balance:

$$\sigma_0 + \sigma_1 = -\sigma_d = \pm\sqrt{2000\,\varepsilon_o\varepsilon_d RT\,\Sigma_i c_i\left(\exp-\frac{z_i F\Psi_d}{RT} - 1\right)} \tag{15}$$

where $\psi_1 = \psi_d$ in the Basic Stern model. The nonlinearity results in a curve with a slope that changes with loading (full lower line in Fig. 17). In applications, the relative dielectric constant of the DDL (ε_d) is generally set to the value in pure water or that of the electrolyte solution. Due to ion crowding, it can be less (Hiemstra and Van Riemsdijk 2006).

In summary, the pH dependency of ion adsorption is governed by two factors: solution chemistry (n) and co-adsorption or co-desorption of protons (χ). The latter is a true surface property, and very strongly determined by the electrostatics caused by introducing charge into the electrostatic field at the metal (hydr)oxide interface. The distribution of this charge depends on the ligand distribution, i.e., the structure of the surface complex formed. For the pH dependency, the intrinsic stoichiometry of a surface chemical reaction equation is usually secondary in contrast to electrostatics. This electrostatic concept may not be very intuitive for scientists who are not familiar with it.

[17] In general, σ_o may have different contributions. It can be due to the adsorption of protons but may also have a contribution from IS surface complex formation. The proton contribution cannot be specified experimentally by doing potentiometric titrations, because the ligands of IS species may also adsorb or desorb protons, apart from aqueous solution contributions. Only a model can separate the contributions.

Allocation of charge in relation to adsorption phenomena. The electrostatic potentials of the surface and first Stern plane (Eqns. 14 and 15) determine two major ion adsorption features that have to be described simultaneously in a correct manner. One of these is the proton co-adsorption, i.e., pH dependence of adsorption, which is related to the charge distribution over the surface and Stern plane, as discussed above (Fig. 17). The second feature to be described simultaneously and correctly is the shape of the adsorption isotherm. This feature is particularly sensitive to the allocation of charge in the 1-plane. The shape of the isotherm is pH- and concentration-dependent, regulating ion loading, as illustrated in Figure 18 for the adsorption of Cd^{2+} to goethite.

For a given intrinsic affinity, $\log K_{chem}$ (Eqn. 4), adsorption isotherms are predominantly determined by the allocation of charge in the first Stern or 1-plane, changing the corresponding potentials and adsorption energy contributions ($\Delta z_1 F\psi_1$). At a given pH, the electrostatic potential of the 1-plane will change strongly with the amount of charge that is located there by an adsorbed species, as illustrated in Figures 1c and 1d for an anion, while the surface potential is hardly affected due to a proton buffering by the free sites, as long as these are available. The amount of charge located in the Stern plane also contributes to the proton co-adsorption and pH dependency and thereby, to the pH dependency of the adsorption isotherms, as discussed in the context of Figure 17.

Ion adsorption modeling, constrained by the allocation of charge. For well-defined systems, surface complexation modeling has revealed that, for a given solution chemistry, the pH-dependency of the ion adsorption is strongly influenced by the interaction of ionic charge with the surface, where adsorbed protons reside.

The allocation of charge to the Stern plane is not only determined by the structure of the surface complex (Figs. 5, 17) but may also be influenced by the protonation or deprotonation of one or more of the outer ligands of the species. In surface complexation modeling, the challenge

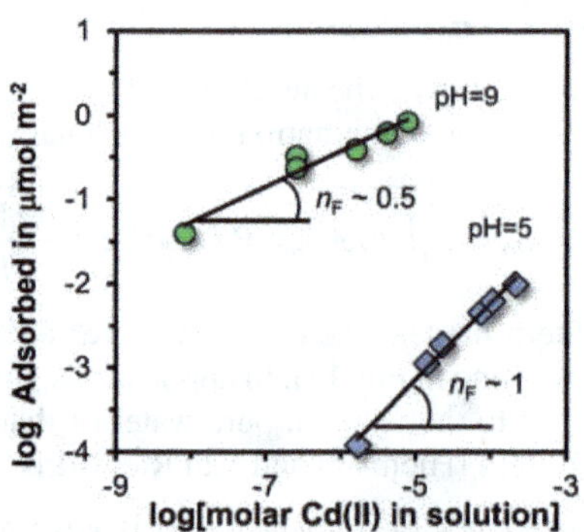

Figure 18. Isotherms of the Cd^{2+} adsorption to goethite in 0.1 M $NaNO_3$ for two pH values, in a double logarithmic plot. At low pH, the Cd loading is low. The slope of the isotherm is then nearly n_F ~1, which is equivalent to a linear adsorption isotherm on the linear scale. The reason for this linearity is the low adsorption of Cd^{2+} (nanomole range) and correspondingly a low contribution of Cd^{2+} charge to the Stern plane. Consequently, the corresponding electrostatic contributions, i.e., $\Delta z_0 F\psi_0 + \Delta z_1 F\psi_1$ (Eqn. 3), will be nearly constant at a constant pH and the overall Gibbs free energy of adsorption will not vary. At higher pH, the repulsive potential ψ_0 will decrease, leading to higher adsorption. With the increase of the concentration at a given pH, the Cd^{2+} adsorption (micromole range) will increasingly attribute charge to the Stern plane. This will increase the corresponding repulsive potential ψ_1 and $\Delta z_1 F\psi_1$. It implies that the overall $\log K_{ads}$ will no longer be constant (Eqn. 4), but will decrease with loading, leading to $n < 1$ in a log–log plot and a Freundlich type of adsorption isotherm at the linear scale. Data from Venema et al. (1996b).

is to disentangle the contribution of various factors. The starting point is the structure of the complex, determining the ligand distribution with the corresponding charge, which is followed by considering protonation or deprotonation as required. Ignoring the first makes a SCM purely thermodynamic, weakening its capability to predict. By measuring adsorption phenomena over a large range of pH, ionic strength, and concentration conditions, preferably augmented with ion competition experiments and proton balances, the contribution of protonation or hydrolysis of the outer ligands can be disentangled, as will be discussed in the next section.

SURFACE COMPLEXATION OF CATIONS AND ANIONS

Outersphere complex formation

Monovalent electrolyte ions are usually adsorbed without the exchange of primary ligands. These ions form OS surface complexes (Sposito 1984). Monovalent electrolyte ions like Na^+, K^+, Cl^-, and NO_3^- are typical examples, but also Ca^{2+} (Rietra et al. 2001a), Sr^{2+} (Axe et al. 1998), SO_4^{2-}, SeO_4^{2-} (Rietra et al. 2001b) and some groups of organic acids under certain pH conditions (Filius et al. 1997, 1999, 2000).

The concept of OS adsorption of electrolyte ions has been used to relate surface charge and potential (Stern 1924; Grahame 1947). Experimentally, simultaneous adsorption of electrolyte cations and anions in the PZC has been found (Smit and Holten 1980; Shiao and Meyer 1981; Sprycha 1984, 1989b), which depended on the electrolyte concentration. The interaction is weak and has a large electrostatic contribution. The OS complexes have no common ligands with the metal ions of the solid (Fig. 5) and remain at some minimum distance of approach. From an electrostatic point of view, these complexes may be treated with less structural detail. They are often considered as point charges, located at the head end of the diffuse double layer. In recent work on single crystals, several observations clearly differ from those made on the corresponding particles. In particular OS adsorption of e.g., oxyanions like arsenate on alumina or hematite (Xu et al. 2019; Catalano et al. 2008) has never been reported in numerous studies involving the equivalent particles.

In the simplest model approach for ion adsorption, the same affinity is assumed for the outersphere complex formation of electrolyte cations and anions. However, it has been shown that surface charge varies with the type of electrolyte cations and anions (Dumont and Watillon 1971; Breeuwsma and Lyklema 1973; Sprycha 1984; Kallay et al. 1994; Sahai and Sverjensky 1997; Bourikas et al. 2001), indicating differences in affinity and/or location.

Rahnemaie et al. (2006) have revisited ion pair formation using acid–base titrations that are internally consistent by starting from a common salt-free suspension stock, acidified to a known pH of 5–6, that is continuously purged with moist N_2 gas. The connected acid–base titrations were done using a series of solutions containing different combinations of electrolyte cations and anions. In the experimental approach, it is important to scale the data of each type of salt not individually by using a common intersection point (CIP), but rather by referring to the charge of a stock suspension (Rietra et al. 2000; Hiemstra and Van Riemsdijk 2006; Rahnemaie et al. 2006). In this way, the potentiometric titrations are connected, and all data become consistent. A simultaneous interpretation of all data points leads to a consistent set of parameters (Hiemstra and Van Riemsdijk 2006).

Others used electrokinetic experiments to unambiguously determine the isoelectric point of a given system. Combining this information (plus reliable zeta-potentials[18]) with titration results in a consistent way ideally also leads to such a set of parameters.

Innersphere complex formation

Many ions form IS surface complexes on oxide surfaces in aqueous solutions. Primary ligands of ions, O or OH ligands are shared with one or more metal ion(s) of the solid. Typical examples of solutes that form IS surface complexes, and have been proven as such by spectroscopy, on oxide surfaces are SeO_3^{2-} (Hayes et al. 1987), PO_4^{3-}, (Tejedor-Tejedor and Anderson 1990), Cr^{3+} (Charlet and Manceau 1992), UO_2^{2+} (Manceau et al. 1992; Waite et al. 1994), NpO_2^{+} (Combes et al. 1992), AsO_4^{3-} (Waychunas et al. 1993), Cd^{2+} (Spadini et al. 1994; Collins et al. 1999b; Randall et al. 1999), Pb^{2+} (Bargar et al. 1997a,b, 1998), CrO_4^{2-} (Fendorf et al. 1997), $As(OH)_3^0$ (Manning et al. 1998), Hg(II) (Collins et al. 1999a), Cu^{2+} (Parkman et al. 1999), Eu^{3+} and Cm^{3+} (Rabung et al. 2000, 2004), Sr^{2+} (Fenter et al. 2000) and other alkaline earth ions (Mendez and Hiemstra 2020b). However, in some cases, monovalent electrolyte ions such as Na^+, K^+, and Rb^+ ions may also form IS complexes as found spectroscopically for rutile (α-TiO_2) for example (Fenter et al. 2000; Zhang et al. 2004, 2007) on single crystal samples and by SCM Ridley et al. (2009).

IS surface complexation of monovalent cations only occurs at conditions of a highly negative charge and, therefore, it is easier observable for oxides with a relatively low PZC, such as rutile (α-TiO_2). For this oxide, IS complexation was inferred via modeling of high-precision proton titration data (Ridley et al. 2009). The obtained charge distribution can be well understood from the Pauling distribution of charge in these complexes (Fig. 19a).

Some divalent ions may form OS as well as IS surface complexes. The pH-dependent partitioning of an ion over OS and IS complexes is related to the structure of the complexes, thereby determining the location of the charge in the electrostatic field. The ion charge of the OS complexes is present at some distance from the surface in the 1-plane, while the IS complex will also attribute some charge to the surface (0-plane). This leads to a different proton co-adsorption and a different pH dependency, according to the thermodynamic consistency rule. Consequently, IS complexation will become relatively more important for cations with increasing pH and anions with decreasing pH (Fig. 19b). IS species introduce a significant amount of charge in the 1-plane which affects the competitive adsorption of the OS complexes. It leads to suppression of OS complexes as shown in Figure 19b.

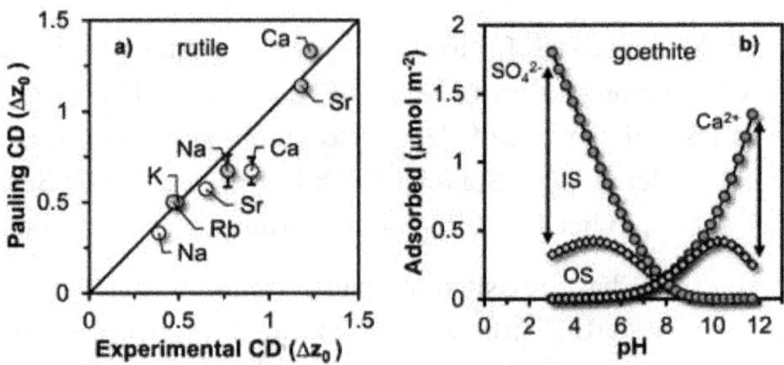

Figure 19. **a)** Relationship between the surface charge attribution (Δz_0) of IS complexes of electrolyte cations for rutile, resolved by CD modeling versus the expected values applying the Pauling bond valence concept to the various coordination spheres. Bidentate (Na, Ca, **white**) and tetradentate (Na, K Rb, Ca, Sr, **green**) complex formation is involved (Ridley et al. 2009). **b)** Calculated total adsorption of SO_4^{2-} and Ca^{2+} at a solution concentration of 0.001 M in a goethite system with 0.1 M $NaNO_3$ (**blue spheres**). OS complexation is given with **orange symbols**. The difference (**vertical arrow**) gives the contribution of the IS complexes. OS and IS formation have a different pH dependency of adsorption, leading to a relative change in surface speciation with pH. Parameters are from Rietra et al. (2001a,b).

[18] This might be more of a problem than is commonly thought in the SCM community, as can be concluded from the chapter on electrokinetics in this volume.

Factors influencing the interfacial distribution of charge

Spectroscopy can resolve the structure of surface complexes formed on various surfaces. First-principles calculations increasingly contribute by producing relevant details of the structure of IS complexes. For surface complexation modeling, these insights require translation into parameters that can be used. Exploration of a wide range of ion adsorption phenomena with surface complexation modeling, constrained by insights into the structure of IS complexes, has elucidated major factors that determine in various interfaces the allocation of charge. As discussed below, that picture is rich.

A) Distribution of ligands. The different electrostatic locations of the ligands of specifically adsorbed ions imply that charge is spatially distributed in the surface complex formed. As illustrated in Figure 5a, the charge distribution is in the first place determined by the interfacial ligand distribution. The charge distribution depends on the overall charge of the ion adsorbed, the coordination number of the central ion, and the denticity of the complex. As an example, for adsorbed SeO_3^{2-}, two of the three ligands are common with the surface, and 1/3 is not, as illustrated in Figures 5 and 17. The total change of charge ($\Delta z = -2$) is distributed accordingly, resulting in an attribution to the 0-plane of $\Delta z_0 = -1.33$ v.u. and to the 1-plane of $\Delta z_1 = -0.67$ v.u.

The role of the coordination number in the charge attribution is evident when we compare two ions equal in charge ($z = -2$) and denticity (bidentate), namely SeO_3^{2-} and CrO_4^{2-} in Figure 17. For the former, the charge is distributed over 3 oxygen ligands and for the latter over 4 ligands, implying less charge per bond. Consequently, for CrO_4^{2-} less charge is attributed to the surface ($\Delta z_0 = -1$), while more is attributed to the 1-plane ($\Delta z_1 = -1$). In the above calculation, we assumed a symmetrical distribution of charge over the ligands, i.e., the Pauling bond valence concept was applied.

B) Relative bond length. The charge distribution in surface complexes of oxyanions like PO_4^{3-}, SeO_3^{2-}, and SO_4^{2-} can often be approximated using, as a first approach, the Pauling bond valence concept (Hiemstra and Van Riemsdijk 1996, 1999). This may also apply to the Cd surface complex formed on goethite $\equiv(FeOH)_2Cd(OH_2)_4$ (Venema et al. 1996b). However, for many important ions, the coordination environment is highly asymmetrical. This can be exemplified by a few divalent cations.

Very strong asymmetry has been revealed by EXAFS for the Hg(II) innersphere complex at the surface of goethite (Collins et al. 1999a), in contrast to Hg^{2+} in solution, where it is hexacoordinated with water with a mean $d_{Hg\text{–}O}$ of 2.22 Å (Partana et al. 2019). The detected distance of Hg(II) with the two surface oxygen ions of $\equiv(FeO)_2Hg(OH_2)_n$ is very short ($d_{Hg\text{–}O} = 2.04$ Å) compared to the Hg–O distance with water molecules at the solution side (Fig. 20a). The large asymmetry leads to deprotonation of the surface ligands while preventing hydrolysis of the solution-oriented ligands (Fig. 20a). This asymmetry is reflected in the surface charge attribution ($\Delta z_0 \approx 1.7$ v.u.) found by describing the literature data of the Hg(II) adsorption to goethite (Barrow and Cox 1992; Gunneriusson and Sjöberg 1993) with the CD model.

On goethite, Pb(II) ions are bound to singly and triply coordinated groups (Ostergren et al. 1999b), and the coordination environment is also distorted (Ostergren et al. 1999b). The common surface ligands are almost fully neutralized (Bargar et al. 1997a,b). The adsorbed Pb ion does not remove the proton from the singly coordinated ≡FeOH (Fig. 20b), illustrating that distortion is weaker than for Hg(II). This aligns with the lower surface charge attribution ($\Delta z_0 \approx 1.2$ v.u.), found by CD-modeling of published Pb adsorption data (Hayes and Leckie 1987; Kooner 1993; Gunneriusson et al. 1994). No hydrolysis of the Pb-OH_2 ligands is found below pH 7 with CD modeling.

At high Cd loading, the main Cd complex on goethite is a bidentate, involving singly coordinated surface groups (Spadini et al. 1994; Randall et al. 1999). Boily has measured the

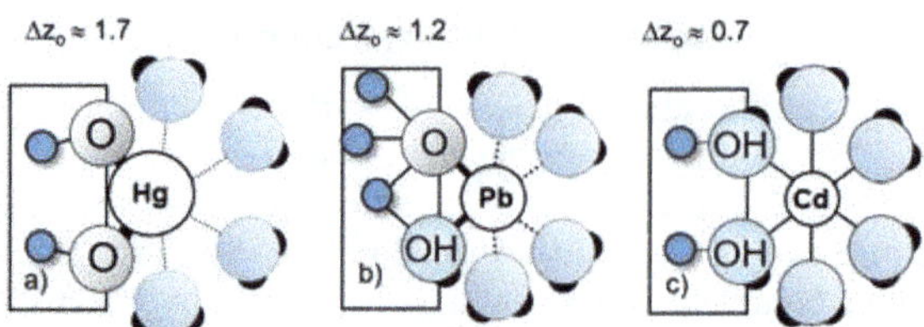

Figure 20. Comparing surface complexes of polarizable Hg(II), Pb(II), and Cd(II) on goethite with contrasting structures, resulting in a different surface charge attribution (Δz_o), fitted with the CD model. The metal ions strongly differ in size. The coordination environment of the Hg(II) ion is highly distorted, resulting in a rather linear O–Hg–O structure. The relatively short bonds with the surface ligands attribute so much charge that no protons can be retained on the common ligands. The same asymmetry prevents hydrolysis of the loosely bound water molecules due to the low charge attribution to the ligands. Asymmetry is weaker in the Pb surface structure. The common ligand of the singly coordinated ≡FeOH group keeps its proton. The asymmetry causes longer Pb–OH_2 bonds, suppressing the hydrolysis of those ligands. The cadmium (Cd) environment is rather symmetrical. The bond valence to each ligand is still relatively low, preventing hydrolysis of the coordinated –OH_2 ligands, as found with the CD modeling.

Cd adsorption for these high-loading conditions (Boily 1999). Modeling in combination with the data of Venema et al. (1996b) leads to a fitted charge attribution coefficient of $\Delta z_0 \approx 0.7$ v.u. This value suggests a rather symmetrical distribution of charge over the six ligands of the complex (Fig. 20c).

The examples of Fig. 20 show that the interfacial charge distribution of three divalent cations, differing in the asymmetry of the coordination sphere, is reflected in the CD coefficients (Δz_0, Δz_1) found by fitting the adsorption data, guided by information about the complex structure. In general, fitting of the CD value requires high-precision data, collected over a wide range of conditions for well-defined surfaces. This approach can be successful, but if multiple species are present, resulting from protonation or deprotonation, discrimination between contributions of various species can be rather difficult, and a reliable determination of the CD values is hampered. In that case, a good practice can be to rely on the CD values that have been derived from MO/DFT optimized geometries using the Brown bond valence approach.

C) Protonation or hydrolysis of ligands. The protonation state of the ligands of an IS surface complex may be quite different from the situation in the solution. A major factor is the presence of an electrostatic field that induces or suppresses protonation. The structure of the solvent is also different for dissolved and adsorbed species. Another important factor is the difference in the above-discussed (a)symmetry of the coordination sphere. In the examples of Figure 20, no hydrolysis is found for the –OH_2 ligands of the metal ion surface species, while this occurs in solution for Hg(II) and Pb(II) ions. The suppression of interfacial hydrolysis is due to the asymmetry in the coordination sphere of the IS complex. Interestingly (enhanced) hydrolysis of adsorbed cations has been advocated in many studies both in combination with electrostatic (Lövgren et al. 1990)(Gunneriusson and Sjöberg 1993; Gunneriusson et al. 1994; Huittinen et al. 2009; Kupcik et al. 2016; Neumann et al. 2021) and even more so with non-electrostatic surface complexation models (Bradbury and Baeyens 1997, 2005).

Evaluation of pH-dependent time-resolved laser fluorescence has been frequently interpreted in terms of surface species with different degrees of hydrolysis (Rabung et al. 2000, 2004; Stumpf et al. 2004; Kupcik et al. 2016). This is based both on peak deconvolution and evaluation of fluorescence lifetimes. The outcome of such a treatment for Cm(III) adsorption on montmorillonite was in very good agreement with the above-mentioned non-electrostatic SCM that had been independently developed (Bradbury and Baeyens 2005). This result would contradict the above statement derived from the goethite CD-MUSIC models. The major relevance of the fluorescence lifetimes is in the number of water molecules bound to Eu(III) or Cm(III)[19]. It is estimated based on an equation calibrated for solutions (Horrocks and Sudnick 1979; Kimura and Choppin 1994). The application of this equation to interfacial species with a far more distorted geometry and exposed to significant electrostatic fields should be reviewed, and more studies involving both TRLFS and EXAFS (as well as other suitable techniques) on the same kind of system would lead to advances.

Unlike cation adsorption, oxyanion adsorption has frequently required different protonation states of surface species in CD-SCM approaches. However, for phosphate, for example, this difference in protonation state relates to different surface species. Thus, in agreement with EXAFS (Abdala et al. 2015), phosphate binds to well-crystallized goethite predominately as a bidentate complex according to CD modeling (Rahnemaie et al. 2007), showing CD values of $\Delta z_0 = +0.48$ v.u and $\Delta z_1 = -1.48$ v.u. for the adsorption reaction 2 $\equiv FeOH^{-1/2} + 2H^{+} + PO_4^{3-}$. At the (110) (or (101)) faces, this may be a binuclear bidentate complex (Fig. 21a), but at the (021) (or (210)) faces, mononuclear bidentate complexes may be formed too (Abdala et al. 2015). At low pH, the fitted CD values for an additionally adsorbed PO_4 species point to the formation of a singly protonated PO_4 surface species, i.e., $\equiv FeOPO_2OH^{-1.5}$ (Rahnemaie et al. 2007). The charge distribution of the species is not only changed by the denticity from bidentate to monodentate, but also by the allocation of a proton charge at one of the outer ligands ($\Delta z_0 = +0.32$ v.u. and $\Delta z_1 = -1.32$ v.u. for the adsorption reaction 2 $\equiv FeOH^{-1/2} + 2H^{+} + PO_4^{3-}$. Between the two structures, the differences in charge distribution coefficients and reaction stoichiometries are relatively small, complicating an unambiguous resolution of both species through the modeling of the ion adsorption data. By using CD values derived from MO/DFT optimized structures, this can be improved. It reduces the number of adjustable parameters to the affinity constants ($\log K$) only.

Another approach to get insight into the surface speciation due to protonation is to asses the proton affinity with a high level of MO/DFT theory, as recently done for glyphosate (Geysels et al. 2024). This organic herbicide with three types of functional groups (-PO_3, -NH, and -COO) is bound to goethite by ligand exchange of the -PO_3 group, forming monodentate (M) and bidentate (B) surface complexes that bind additional protons. BH and MH are dominant around neutral pH. According to CD modeling, MH_2 is formed at a low pH and BH deprotonates at a high pH. The intrinsic $\log K_H$ constants found by CD modeling compared to MO/DFT values calculated using functionals of different levels of theory (Table 3) show good agreement. This result is promising, considering future developments of MO/DFT.

Table 3. Intrinsic protonation constants ($\log K_H$) of glyphosate adsorbed to goethite (Geysels et al. 2024), experimentally and calculated with MO/DFT using functionals from an increasing level of theory. The ωB97X-D and ωB97M-V functional are range-separated hybrids. Realistic assessment of absolute values for the Gibbs free energy of protonation is notoriously difficult, particularly the G-value of H^+ (aq) (Aquino et al.2008; Ridley et al. 2019; Malloum et al. 2021). However, from the proton transfer between equimolar surface species ($B + MH_2 \rightleftharpoons BH + MH$ and $BH + M \rightleftharpoons B + MH$), $\Delta \log K_H$ values can be calculated and translated into absolute values using the experimental value of $B + H^+(aq) \rightleftharpoons BH$ as a reference.

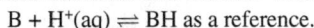

Equilibrium	Experimental*	B3LYP	ωB97X-D	ωB97M-V
$M + H^+(aq) \rightleftharpoons MH$	high**	10.8	11.4	11.1
$B + H^+(aq) \rightleftharpoons BH$	8.2±0.5	8.2***	8.2***	8.2***
$MH + H^+(aq) \rightleftharpoons MH_2$	5.1±1.1	4.1	4.4	5.1

Notes:
* CD – modeling of a large data set over a wide pH range
** No contribution of M is found, which is due to a high $\log K_H$
*** Experimental value used as a reference to derive both others

[19] Terbium has been much less used in studies on interfaces.

D) Ternary complex formation / polymerization. The outer ligands of an adsorbed oxyanion may additionally interact with a cation, forming a ternary surface complex. In the example of PO_4 and Ca, the dominant ternary complexation involves a PO_4 monodentate surface complex with an attached Ca ion (Fig. 21b). This species was resolved with the fitting of an extended set of data in which the PO_4^{3-} and Ca^{2+} loading varied strongly with pH (Mendez and Hiemstra 2020a). Modeling indicated that part of the charge of the Ca^{2+} ion, attached to the adsorbed PO_4, is located further away from the surface and 1-plane. The adsorption can be described by a distribution of charge over three electrostatic planes of the extended Stern layer model option (Fig. 2a). For the species of Figure 21b, the formation reaction is $\equiv FeOH^{-1/2} + H^+ + Ca^{2+} + PO_4^{3-}$, with CD coefficients $\Delta z_o = +1\ -0.76$ v.u., $\Delta z_1\ -2.24\ +0.94$ v.u., and $\Delta z_2 = +1.06$ v.u. These are built from $\Delta z_H = +1$, $\Delta z_{PO_4} = -0.76 - 2.24 = -3$, and $\Delta z_{Ca} = +0.94 + 1.06 = +2$ ($\Sigma\Delta z = +1$). The charge distribution in the PO_4 moiety was derived from MO/DFT geometry optimization of the $\equiv FeOPO_3Ca$ surface complexes. The charge distribution of Ca^{2+} can be derived by fitting showing that about half of the Ca^{2+} charge was allocated to the 2-plane. For the surface complex considered, a monodentate Ca–PO_4 interaction was assumed, but recent molecular modeling (Koca Fındık et al. 2024) suggests that Ca^{2+} in solution forms a bidentate complex with $H_2PO_4^{1-}$ and a tridentate complex with HPO_4^{-2} and PO_4^{3-}. In the latter case, the sum bond valence of three Ca–OP bonds is ~0.5–0.7 v.u. The Ca–PO_4 denticity for the ternary surface complex cannot be revealed from our modeled charge distribution as the interfacial ligand distribution is not sufficiently known.

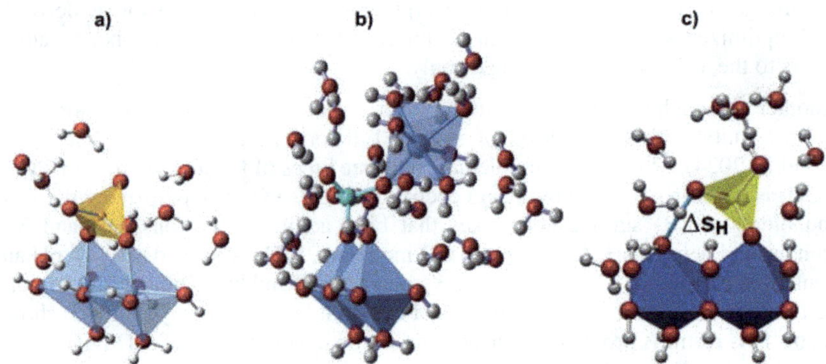

Figure 21. a) Calculated geometry of a hydrated FeOOH cluster with a partly fixed geometry to mimic the mineral structure with PO_4 bound as a bidentate (Rahnemaie et al. 2007). These bond lengths accessible via EXAFS agree with MO/DFT calculations (Rose et al. 1997; Abdala et al. 2015). **b)** Ternary complex of PO_4–Ca, leading to distribution of charge over three electrostatic planes (Mendez and Hiemstra 2020a). **c)** Binding of a $Si(OH)_4$ as a monodentate complex where one of the OH ligands of Si strongly interacts with an adjacent FeOH group, transferring charge ($\Delta s_H \sim 0.2$ v.u.) (Hiemstra 2018; Wang et al. 2018). [Reproduced (adapted) from Hiemstra (2018) and Mendez and Hiemstra (2020a) with permission of Elsevier.]

E) H-bond formation of ligands with an adjacent surface group. Another concern for the relation between structure and interfacial charge distribution is the formation of (very) strong O···H–O bonds between free ligands of a surface complex and an adjacent surface group, as reported for $\equiv FeOSi(OH)_3$ (Wang et al. 2018) and modeled subsequently (Hiemstra 2018). MO/DFT geometry optimization suggests significant transfer ($\Delta s_H \sim 0.20$ v.u.) of bond valence charge, nearly doubling the value of Δz_o to +0.48 v.u. and Δz_1 to -0.48 v.u. The formation of this H bond is supported by a free fit of the CD values ($\Delta z_o = +0.55 \pm 0.03$ v.u. and $\Delta z_1 = -0.55 \pm 0.03$ v.u.). Due to the formation of the strong H-bond with an adjacent group, the resulting complex could be considered a bidentate complex in the sense that two $\equiv FeOH$ groups are involved. Only one forms a monodentate innersphere complex with a $\equiv FeOH$ surface group and the other contributes with an H-bridge.

Exploring the formation of a similar complex for a monodentate $\equiv FeOPO_2OH$ interacting with an adjacent $\equiv FeOH$ surface group shows that a strong H-bond can be formed with a bond valence change of $\Delta s_H \sim 0.17$, resulting in $\Delta z_o = +1 - 0.65 + 0.17 = +0.52$ v.u. and $\Delta z_1 = -2.35 + 1 - 0.17 = -1.52$ v.u. for the formation reaction $2{\equiv}FeOH^{-1/2} + 2H^+ + PO_4^{3} \rightleftharpoons {\equiv}FeOPO_2OH{\cdots}O(H)Fe{\equiv} + H_2O$. Note that one proton of the reaction is located at the 0-plane and the second one at the 1-plane. From the perspective of the reaction stoichiometry, the reaction is identical to the formation of an ordinary binuclear bidentate complex $\equiv(FeO)_2PO_2$. Our MO/DFT-calculated Fe–P distance for the $\equiv FeOPO_2OH{\cdots}O(H)Fe{\equiv}$ complex is 331 pm, close to the EXAFS value of 328 ± 0.05 pm (Abdala et al. 2015). The latter has been attributed to a binuclear bidentate complex $\equiv(FeO)_2PO_2$, but the alternative cannot be excluded. In addition, the CD values of both complexes are nearly identical since the CD for the ordinary bidentate complex is $\Delta z_o = +2 - 1.54 = +0.46$ v.u. and $\Delta z_1 = -1.46$ v.u. according to MO/DFT geometry optimization. This illustrates the challenges in identifying the structure of a surface complex via both EXAFS and CD modeling.

Finally, we mention that additional factors may influence the interfacial distribution of charge. Electron transfer of adsorbed Fe(II) (Hiemstra and Van Riemsdijk 2007) and induced water dipole orientation (Hiemstra and Van Riemsdijk 2006) can also contribute, but are excluded here from detailed discussion.

ELECTRICAL DOUBLE LAYER FEATURES

Capacitances of the compact part of the EDL

For metal (hydr)oxides, the primary surface charge density depends on pH, ionic strength, and background electrolyte. Charging curves can be interpreted with an electrostatic model, involving an interfacial capacitance (Eqn. 14). If the value of C is interpreted with the macroscopic capacitor model, a value for the relative dielectric constant (ε_r) can be generated assuming a distance (δ) of charge separation (Figs. 1a,c,d). For an average size of the counter ions as a measure of the minimum distance of approach, the equivalent relative dielectric constant (ε_r) can be calculated with the following equation:

$$C = \frac{\varepsilon_o \varepsilon_r}{\delta} \tag{16}$$

in which ε_o is the permittivity of vacuum (8.854×10^{-12} V·m^{-1}).

Capacitances of inner Stern layers. For well-crystallized goethite, the capacitance value of the first Stern layer is typically $C = 0.9$–1.0 F·m^{-2}, resulting in $\varepsilon_r \approx 35$ using $\delta \sim 0.35$ nm (Hiemstra and Van Riemsdijk 2006). For rutile (TiO_2), the lower capacitance of $C = 0.64$ F·m^{-2} (Ridley et al. 2009) is equivalent to $\varepsilon_r \approx 25$. The first number is about halfway between the dielectric constant of free water ($\varepsilon_r = 78$) and the solid (for goethite $\varepsilon_r = 11$), but for rutile ($\varepsilon_r = 120$), the value is not halfway. Both rutile and water have a high dielectric constant. Nevertheless, the mean ε_r is estimated to be lower than for goethite. This indicates that the interfacial dielectric constant is not necessarily a property that follows from the solid. This is supported by the similar behavior of TiO_2 (rutile) and isostructural SnO_2 (cassiterite). The latter has a low relative dielectric constant of $\varepsilon_r \sim 20$ (Borges et al. 2010).

Modeling of the surface charge for a large set of silica nanoparticles (NP) suggests a capacitance value of $C = 1.0 \pm 0.1$ F·m^{-2} when described using ion pair formation, as will be shown later. Applying Equation (16) with $\delta \sim 0.35$ nm leads to $\varepsilon_r \sim 40$. This value is similar to the value for goethite and slightly higher than for rutile. It is evident that the model-inherent value is very different from the relative dielectric constant of the solid ($\varepsilon_r = 3$–4). It implies that the Born solvation energy will not be exceptionally low, as suggested earlier (Sverjensky 1994; Sahai 2000).

The dielectric constant of interfacial water is related to the freedom of the orientation of the dipoles of the water molecules. This can be illustrated by comparing the currently available capacitance values for metal (hydr)oxides with the values resulting from modeling AgI (Fig. 7a). The capacitance value of AgI is very different from the values for metal (hydr) oxides, which has been attributed to another interface structure (Hiemstra and Van Riemsdijk 1991). The charging curves of AgI can be described well (Fig. 7a) when using the extended Stern layer model (Fig. 2a). Assuming a typical Stern layer space of $\delta \sim 0.35$ nm, the fitted capacitance value for the first Stern layer ($C_1 = 0.15 \pm 0.01$ F·m^{-2}) results in a relative dielectric constant of $\varepsilon_r \sim 6$. This very low value can be attributed to a strong orientation of the water molecules that is due to the binding of H_2O molecules as primary hydration water of the Ag^+ ions of the solid. This molecular picture has been confirmed by MD simulations (Zarzycki and Rosso 2010).

MD simulations for the calcite–water interface show that near the surface, the dielectric constant increases (Zarzycki 2023). This can be attributed to an enhanced density of water (1–2 g·cm^{-3}) near the surface, which is caused in general by the mismatch of the oxygen density of minerals and water, being respectively 0.10 and 0.0556 mol O^{2-}·cm^{-3} (Hiemstra and Van Riemsdijk 2009). The higher interfacial water density leads to a higher dielectric constant than for free water ($\varepsilon_r \sim$ 80–200), as shown by MD simulations (Zarzycki 2023). However, the ordering of water is largely disturbed by the presence of electrolyte ions, leading to a value for the relative dielectric constant below 80. This is in line with the dielectric behavior of concentrated salt solutions. In a 2 M NaCl solution, the typical value is $\varepsilon_r \sim 60$ (Bockris and Reddy 1976b). At charged interfaces, the local ion concentration near the head end of the DDL will be generally very high (> 1–2 M). In addition, ion pair formation occurs. This will lower values for ε_r relative to pure water. The formation of IS surface complexes will also cause a change in the electrical field potentially changing water structure and dipole orientation, which in turn will affect the interfacial charge distribution (Hiemstra and Van Riemsdijk 2006), but may also affect the relative dielectric constant[20].

Capacitances of outer Stern layers. For SiO_2 as well as TiO_2, modeling does not suggest the presence of a second Stern layer, in line with the observation of less water ordering (Fenter et al. 2000; Schlegel et al. 2002). In other cases, modeling points to the presence of a second Stern layer. For the AgI–water interface, the fitted number for the capacitance of the second Stern layer is $C_2 = 0.57 \pm 0.01$ F·m^{-2}. This capacitance value is somewhat lower than the second capacitance of the goethite–water interface ($C_2 = 0.74 \pm 0.10$ F·m^{-2}). The latter was derived by measuring the charging behavior of goethite with a consistent methodology for a large series of different types of electrolytes (Hiemstra and Van Riemsdijk 2006; Rahnemaie et al. 2006). The experimental approach involved the use of an electrolyte-free suspension as the common reference to which all the titrations were scaled. Using individual sets of titrations with their respective CIPs as reference(s) can lead to different results. Alternatively, the isoelectric point can be used as a reference, and its shift with salt concentration and composition can be a stringent test. For $\delta \sim 0.35$ nm, the dielectric constant for the second Stern layer of AgI would be $\varepsilon_r \sim 23$. It is substantially higher than for the first layer having primary hydration water ($\varepsilon_r \sim 6$). For goethite, the relative dielectric constant for the second Stern layer is also relatively low ($\varepsilon_r \sim 30$). These relative dielectric constant values may point to some structuring of water between the head end of the DDL (2-plane) and the minimum distance of approach of electrolyte ions (1-plane). However, this may also be explained by ion crowding, decreasing the value of ε_r.

Overall capacitance. The overall capacitance C_T of an interface with two Stern layers in series can be calculated with $1/C_T = 1/C_1 + 1/C_2$. Applying this relation to the compact part of the EDL of AgI leads to $C_T = 0.12$ F·m^{-2}. In the past, this low number for AgI(s) has been a motivation to set, by definition, the capacitance of the second Stern layer of the traditional Triple Layer model to a very low value ($C_2 \equiv 0.2$ F·m^{-2}). Combined with an intermediate value

[20] Additional points of view on interfacial dielectric constants can be found in the chapters by Boily (2025, this volume), Zarzycki (2025, next volume), and Heberling (2025, next volume)

for the relative dielectric constant ($\varepsilon_r \sim 40$), it would lead to a picture in which the head end of the DDL and the minimum distance of approach of electrolyte ions would be 3 nm. This huge distance is equivalent to a layer thickness of 10 or more water molecules, illustrating the misconception that the overall capacitance of the compact part of the EDL would be very low for metal (hydr)oxides (Hiemstra and Van Riemsdijk 1991).

Finally, we note that in modeling, permittivity is generally considered constant. However, in a homogeneous continuum, the dielectric constant may depend on the field strength. A potential fall of 0, 100, or 200 mV over the Stern layer will lead to respectively $\varepsilon_r = 78$, $\varepsilon_r \approx 40$, and $\varepsilon_r \approx 20$ (Hiemstra and Van Riemsdijk 2006). However, the actual situation in the Stern layer can be more complicated since surfaces are crowded by hydrated ions while the capacitor model assumes a structureless dielectric agent (Zarzycki 2023). Not only this sets limits to a detailed interpretation. To achieve the same molecular-based level for the dielectric constant as can be achieved for the interface structure of the bare surfaces or the structures of surface complexes, additional input of new experimental data and techniques is welcome. Presently, modeling of macroscopic data is mostly done while ignoring such issues. Examples include the validity of zeta-potentials (see Ahuali et al. 2025, next volume) or the rare use of electrochemical methods (see Boily 2025, this volume) and Ahuali et al.(2025, next volume). As in other contexts of this chapter, the application of various methods to the (very) same system can help solve inconsistencies.

Surface curvature

The interfacial double layer models with corresponding equations for the capacitance are typically implemented in codes for planar geometries without double layer overlap. For flat surfaces, an exact analytical solution exists for the Poisson–Boltzmann equation, yielding the well-known charge-potential relationship between σ_d and ψ_d (Eqn. 15). However, for small particles, it becomes important to consider the effects of geometry. Ohshima et al. (1982) derived an accurate charge-potential relationship for purely diffuse layer models. Ohshima (2006) also described the treatment for cases including Stern layers.

Currently, there does not seem to be a widely available code that allows for consistent modeling of homogeneous spherical or cylindric particles or the edges of crystal faces, where the surface curvature plays a role. Although much of the chemistry and electrostatics is probably captured in the Stern layer and accounted for by the respective equations for spherical capacitors (Hiemstra and Van Riemsdijk 2009), it has to be realized that in the traditionally formulated electroneutrality condition and double layer equations for the Basic Stern model (Eqns.14 and 15), the charge in the various electrostatic planes is expressed per unit oxide surface area ($C \cdot m^{-2}{}_{ox}$), which is appropriate for flat surfaces. However, if a surface is curved, the surface areas required in the scaling are not equal but differ considerably depending on the position of the electrostatic plane in the interface. It implies that in the electroneutrality equation, absolute values for charge are required rather than charge densities (Ohshima 2006). For the extended Stern layer model option, the appropriate electroneutrality condition can be given $A_o\sigma_o + A_1\sigma_1 + A_2\sigma_2 + A_{DDL}\sigma_d = 0$.

For a given specific oxide surface area (A_o) of a spherical particle, the value for A_1 and A_2 or A_{DDL} will depend on a choice of the distance between the electrostatic Stern layer planes[21]. The surface charge (σ_0) follows from the potential gradient in the first Stern layer, applying Equation (14), yielding $A_o\sigma_o = A_oC_1(\psi_0 - \psi_1)$. The charge in the 1-plane ($A_1\sigma_1$) follows the potential gradient over the second Stern plane applying $A_o\sigma_o + A_1\sigma_1 = A_1C_2(\psi_1 - \psi_2)$, and the charge in the spherical DDL ($A_{DDL}\sigma_d$) is calculated numerically by applying spherical diffuse layer theory or using an approximation (Ohshima 2006). Introducing these three terms in the electroneutrality equation finally leads to $A_2\sigma_2$. After scaling the calculated charges ($A_i\sigma_i$) to the surface area A_o of the oxide particle, the values can be introduced in the traditional Tableau approach for solving chemical equilibria.

[21] Spherical double layers and double layer overlap are evaluated for ferrihydrite in the contribution of Hiemstra et al. (2025, this volume)

Double layer interactions

Where different types of faces meet at an edge, the electrostatic fields of these faces may interact. For clay particles, some attempts have been made to include the "spillover" from the permanent charge on the basal planes to the variable charge on the edges (Tournassat et al. 2018). The relevance of this probably depends on the overall size of the particle and the relative contributions of the relevant faces.

Even on ideal particles of gibbsite (Fig. 9), locally, at regions where different facets meet, a transition from the potentials of the basal and edge-face plane must occur, resulting in atypical charge-regulated situations[22]. Where a charged edge meets an uncharged basal face, the electrostatic field will be weaker, due to a flaring of the field. The same occurs at charged steps on uncharged basal planes as well as at localized defects. It implies that at these locations, the field will be less strong, which will increase locally the surface charge density and change the corresponding surface speciation, including the protonation or deprotonation of $\equiv Al_2OH^o$ of gibbsite. This has consequences for the rate of dissolution of gibbsite at these steps as the latter process is considered rate-limiting (Hiemstra and Van Riemsdijk 1990a). Indeed, these locations are most vulnerable to acidic attack as observed with *in-situ* AFM (Peskleway et al. 2003).

Force measurements on well-defined particles with morphologies that allow the different faces to be studied individually (Su et al. 2021, 2024), combined with electrokinetic or potentiometric measurements, could help to design reliable SCMs including charge regulation. With the available techniques and the possibilities in particle synthesis, great progress is to be expected, including sophisticated analysis of the experimental results.

For goethite, the particle morphology has been studied using atomic-resolution scanning transmission electron microscopy (Livi et al. 2023), showing relatively large contributions of (021) faces surface roughness for goethites with a low specific surface area. This will largely affect the overall site density of $\equiv FeOH$ and $\equiv Fe_3O$ surface groups as the (021) faces have much higher site densities for $\equiv FeOH$ than the (110) faces and have no $\equiv Fe_3O$ surface groups. This will lead to a much higher oxyanion adsorption capacity (Hiemstra and Van Riemsdijk 1996). As both types of faces differ not much in PZC, the effect of double layer interactions is probably rather low, which then justifies, as an approximation, the use of a single electrostatic model for the entire surface.

Zeta-potential from electrokinetics

Previously, we have discussed measurements of the surface potential (Fig. 13). Here we will focus on the potentials in the diffuse double layer[23]. The diffuse layer potential can be derived by measuring force-distance curves and the zeta potential can be obtained with various electrokinetic methods. The electromobility measurements are probably one of the most easily accessible quantities for researchers. Measured particle mobilities are translated into zeta potentials (ζ) with theory requiring assumptions about e.g., the particle geometry, surface conductance, and two states of water mobility in the interface, namely mobile and immobile water.[24] Despite these limitations, comparison among many different data and methods suggests that on average the values are not unreasonable. The obtained zeta potentials are a function of pH and electrolyte ions, as illustrated for gibbsite (Rowlands et al. 1997) in Figure 22a.

Zeta potentials cannot *a priori* be assigned to a single location in the EDL. MD simulations (Předota et al. 2016) show that the mobility of interfacial water gradually decreases towards the surface. There is no clear difference between mobile and immobile water. When trying to

[22] We distinguish the terms "variable charge" (meaning the surface charge is variable and the potentials are changing according to charge/potential relationships) and "charge regulation" (referring to overlapping double layers).

[23] The reader is strongly encouraged to read Ahualli et al. (2025, next volume) in this context.

[24] More details can be found in Ahualli et al. (2025, next volume)

simulate zeta-potentials with surface complexation modeling for gibbsite and aluminum oxides, it turned out that the slip plane had to be located at some distance away from the head end of the DDL (Hiemstra et al. 1999). Without this assumption, the potential is significantly higher (factor 2 or more). The calculated distance (S) varies with the electrolyte ion concentration as shown in Figure 22b. The values range over about 2 orders of magnitude from about 10 to 0.1 nm. This has been found for a range of metal (hydr)oxide materials as well as air bubbles (Lützenkirchen et al. 2008) as shown in Figure 22b. On the other hand, force measurements (Israelachvili and Adams 1976; Pashley and Israelachvili 1984), surface diffraction studies (Eng et al. 2000; Catalano 2011), or molecular dynamics simulations (Brkljača et al. 2018) all suggest that ordered water is restricted to the compact part of EDL (Fig. 6), which in turn was found to have a more or less constant thickness, independent of electrolyte concentration conditions. It suggests that interfacial water with low mobility may be still non-ordered, in line with the MD simulations of Předota et al. (2016).

The distance of the model-inherent slip-plane from the head-end of the DDL can be compared with the thickness of the DDL. The equivalent distance of the DDL can be expressed in the Debye length (κ^{-1}), which is a function of the 1:1 electrolyte concentration c_o (mol·L^{-1}), according to:

$$\kappa^{-1} = \sqrt{\frac{\varepsilon_r \varepsilon_o R T}{2000\, F^2 c_o}} \tag{17}$$

In the EDL visualizations of Figure 2, the Debye length or κ^{-1} can be found from the potential–distance relation, $\psi(x)$, using the slope at the head-end of the DDL, $(\partial\psi/\partial x)_d$, for linear extrapolation to a zero potential (green arrow in Fig. 2).

The upper line in Figure 22a represents 0.5 times the Debye length of the DDL ($0.5\,\kappa^{-1}$) as a function of the ionic strength. The lower line is for $0.17\ \kappa^{-1}$. For $0.5\,\kappa^{-1}$, the slip-plane location is in the range of 0.5–15 nm from the head end of the DDL in 1:1 electrolyte solution with an ionic strength of 10^{-1}–10^{-4} M (Fig. 22b)[25]. Translated to an equivalent number of water molecule layers, it ranges from 2–60. These substantial numbers, particularly at low

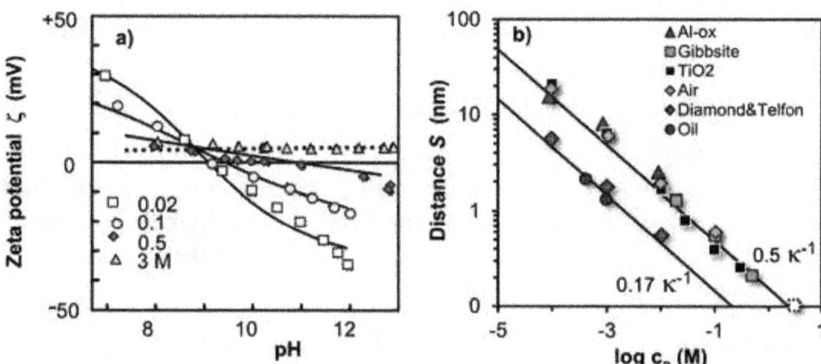

Figure 22. a) Zeta potentials as a function of the pH and 1:1 electrolyte level for gibbsite, using electro-acoustophoretic measurements (Rowlands et al. 1997). The lines have been calculated using the MUSIC model, adapting the location of the ζ- plane in the DDL. Because of uncertainty in the activity coefficient for 3 M, the corresponding **line and symbol** for distance S have been **dotted**. **b)** Distance S of the plane of shear from the head-end of the DDL as a function of the monovalent 1:1 electrolyte concentration for Al_2O_3 and $Al(OH)_3$ (Hiemstra et al. 1999b), Ti-oxides (Bourikas et al. 2001), and air (**green diamonds**), as well as inert hydrophobic surfaces (Lützenkirchen et al. 2008) (**red and brown colored symbols**), estimated from surface complexation modeling. The lower and upper **full lines** show the DDL thickness or Debye length (κ^{-1}), multiplied by respectively 0.17 and 0.5. The data for silica cannot be rationalized similarly, see text. [Figure 22a is reprinted with permission from Langmuir 5:5942–5955 Copyright 1999 American Chemical Society]

[25] Ahualli et al. (2025, next volume) in their chapter consider the distance of 6 nm which they obtain in their calculations as a reasonable estimate.

ionic strength, suggest that the slip-plane is significantly away from the head end of the DDL where water is significantly ordered (Fig. 6). In that case, the type of solid may be subordinate, meaning that materials may behave similarly in this respect. This aligns with part of the data. However, there are also substantial deviations in which the interpretation of measurements point to a 3 times lower value (Fig. 22b). This is found for hydrophobic interfaces of oil, diamond, and Teflon, suggesting either another water structure in the DDL than for oxides or it is related to the data collection (i.e., slip boundary condition in zeta-potentials). For SiO_2 (not shown), no single value of κ^{-1} is found, showing that the picture may be richer than suggested in Figure 22. In general, the quality of the zeta-potentials can be questioned given the many assumptions involved (again see Ahualli et al. 2025, next volume).

One important conclusion is reiterated namely that more systematic studies are required using and combinations of methods should be employed. Thus, the use of force measurements to obtain additional estimates for the diffuse layer potentials (Trefalt et al. 2016) and measurements of counter-ion adsorption (Porus et al. 2011) complimentary with surface and zeta-potentials on the (very) same surface, ideally combined with surface diffraction and non-linear optics as well as other applicable spectroscopic methods could provide very important insight.

MODUS OPERANDI FOR A START

The modus operandi of surface complexation modeling requires insight into the various choices and decisions that need to be made. This section will start by highlighting the challenges of characterizing surface composition, choosing surface reactions, and deciding on EDL models. Two contrasting systems with the same type of material are used for illustration.

Figure 23 shows an example of model options that users or developers can choose from, depending on the complexity of the material. The example involves two extreme cases of possible modeling a system of gibbsite particles (α-$Al(OH)_3$), with contrasting morphologies. In the upper case, hexagonal crystals can be seen, while in the lower case, the morphology is fairly undefined. For the latter, the challenge is to incorporate as many as possible insights that can be gained from studying well-defined systems and attempt to grasp the most crucial factors and identify the deviations from the behavior of well-defined particles that can be modeled based on molecular scale insights.

The application aspect of SCMs concerns natural materials, i.e., even gibbsite particles in the lower part of Figure 23 might be too well defined in that sense. One should not believe that the use of seemingly mechanistic insights will *a priori* improve the predictive aspect of models since the transfer of molecular scale insights to the larger scales usually involves a significant number of assumptions. An example where the transfer is possible and successful is ferrihydrite as illustrated in a separate chapter in this volume.

Reactive sites and densities. Independent of the heterogeneity of the system and the habit of the particles, insight into the composition of the surfaces in terms of the kinds of sites and their respective site densities is essential for the use of any SCM. For well-defined particles, one may go into astonishing details, whereas for ill-defined samples going into such details is probably something for the future and involves significant efforts, with promising results, as shown in the papers by Livi and co-workers for goethite (Livi et al. 2017, 2023).

Site densities cannot be derived reliably from proton ad- or desorption data, but insight can be gained from spectroscopic data as shown in Boily (2025, this volume). The reason is that the electrostatic feedback of protons is enormous, preventing full titration of sites. As discussed above in the section about the charging of gibbsite, at $\Delta pH = 6$, only about 1/3 of the titration maximum is reached. Nevertheless, acid–base titration can be valuable for estimating surface

Complexity/ Heterogeneity	Surface sites: nature & density	Protonation: 1-pK, 2-pK	Electrostatics
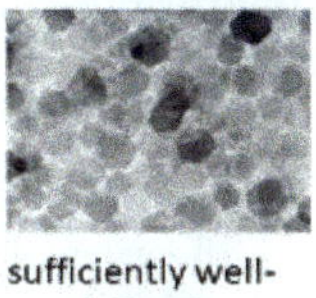 sufficiently well-defined: crystal plane contributions known	mechanistic, based on: morphology, crystallography, surface diffraction	idealized, MUSIC, structure based: proton affinities & partial charges (predicted), „accurate" EDL model ion-pair formation (fitted)	BS, (E)TL, TP, FL water dipoles, geometry, heterogeneity (regulated) charge distribution
 undefined: generic sites	adjusted/fitted/ guessed: currently too complex for mechanistic modelling	simplifying & parsimonious: 1-pK (pzc) & ion pair formation (fitted) simple treatment of electrostatics	non-electrostatic; electrostatic: Nernst, (E)CC, DDL, BS

Figure 23. Model options for two gibbsite systems. **Upper part:** gibbsite particles with a narrow size distribution and well-defined platelets, where the contributions from edges and planes can be experimentally determined. **Lower part:** gibbsite particles with a wide particle distribution and undefined and varying morphologies.

properties by combining it with information collected for well-defined systems. This includes parameters such as ion pair formation constants and capacitance values. With this information, site densities can be assessed with a good set of titration data. This approach can be further constrained by using an additional ion as a probe. For instance, one may measure the adsorption of a high-affinity oxyanion that is also studied for well-defined gibbsite. In the case of gibbsite, the choice of some ions should be avoided as they may form co-precipitates, e.g., PO_4^{3-}, and $Si(OH)_4$ or enhance mineral dissolution, e.g., F^-, AsO_4^{3-} or CrO_4^{2-} might be a good candidate for probing singly-coordinated groups. The presence of patches of $\equiv Al_2OH$ basal faces could perhaps be assessed by measuring the Cm adsorption involving TRLFS spectroscopy which is highly sensitive. In combination, such a tedious effort may lead to an improved guess about the reactive site densities. Again, a combination of methods is the key to better understanding.

Model choice. Starting from the habit of the particle, if the morphology is too complex to quantify or even identify crystal faces, it is possible to opt for a generic model that does not have the ambition of being mechanistic but merely uses it descriptively. However, a preferable option is a mechanistic approach that allows for the identification of similarities and differences with a well-defined reference. Assuming that the chemical composition of the mineral of interest is in principle sound for comparison, it makes sense to combine a surface chemistry model with a physically realistic double layer model. The use of a model without a Stern layer will not be a very wise choice. As discussed by Venema et al. the proton charge development, neither as a function of pH nor as a function of electrolyte level, can be described if *a priori* a Stern layer is excluded (Venema et al. 1996a). Moreover, ions that form IS and OS surface complexes are *by definition* an inherent part of the surface when a Stern layer is absent, leading on forehand to a hypothetical surface speciation.

Proton reaction equations. Another choice is the definition of the protonation reaction, i.e., the use of a 1-pK or 2-pK model. From the knowledge gained from first-principle molecular dynamics such as the study by Liu et al. (2013) and subsequent ones, it has become evident that the ΔpK value of the $\equiv AlOH$ groups pairwise available on the edge faces, hardly differ in proton affinity. This advocates the 1-pK model. If the 2-pK model is used, the ΔpK value should be fixed to a small near-zero value, thereby avoiding too much flexibility. The advantage of choosing a

1-pK model is that monodentate and bidentate complexes can be defined consistently, whereas with the 2-pK, the link with the microscopic reality is unnecessarily lost or at least complicated. Moreover, the use of the 1-pK model allows direct comparison with the information collected in well-defined and well-studied gibbsite systems. This comparison then reveals in which aspects the ill-defined material may differ. This can be the starting point for new research to better understand the material and identify underlying factors for differences.

Ion pair formation. In a mechanistic and realistic approach, quantification of ion-pair formation may be needed. Usually, it is done by fitting acid–base titration data, creating uncertainty because parameter values such as the proton affinity constant, $\log K_H$, ion pair formation constant, $\log K_{ip}$, and the Stern layer capacitance (C_1) value are correlated[26]. This is particularly difficult for SiO_2. If the ion pair formation of cations and anions differs strongly, it will lead to a shift in the IEP (Kosmulski 1998; Kosmulski et al. 2002) that can be tested with modeling (Fig. 22).

Experimental approaches for quantifying ion pair formation are scarce. An early attempt is from Sprycha (1984, 1989b). Later, XPS on fast-frozen samples has been advocated as an option (Shchukarev and Sjöberg 2005), which has been applied to various metal (hydr)oxide particles (silica, gibbsite, manganite, goethite) with results confirming the conventional counter-ion adsorption ideas. In more recent years, optical reflectivity has been applied to planar SiO_2 surfaces (Porus et al. 2011; Wang et al. 2015) with results again in line with the current view of counter-ion adsorption, but with the advantage of being in-situ and quantitative in nature.

Stern layer capacitance. Establishing interfacial capacitance value(s) is rather difficult. In fitting approaches, the value is generally correlated with other parameters in the model, particularly ion pair formation. If ion pair formation is ignored, the capacitance value will vary for each type of ion, as has been well demonstrated for silica (Sonnefeld 1995).

Even for the assumption of a constant value of C (Eqn. 16), evaluation of a fitted value is difficult, particularly, because the relative dielectric constant ε_r is not accessible at present, though attempts exist to evaluate them (Bonthuis et al. 2012; Zarzycki 2023). The general detailed picture involved is a capacitance that is conditional and depends on the structure of the EDL, which may be pH and/or salt-level dependent. The dependence of the capacitance on charge has been discussed in electrochemical contexts (Velikonja et al. 2014) and is highlighted in the chapter by Boily (2025, this volume). Non-linear optical methods suggest that the amount of ordered water varies with pH and salt level (Zhang et al. 2008) following the interfacial electrical field. Ultimately, a formalism will have to be developed that calculates the (mean) capacitance(s) based on the interfacial electrical field, the presence of adsorbed electrolytes ions, and the interfacial dielectric constant, which in itself is seen by some as an ill-defined concept.

Innersphere complexation. The option chosen for interfacial charging preferably starts from evaluating an ill-defined material using a well-defined counterpart as a reference. Next, one may attempt to model the target ion adsorption behavior. The precise structure of the surface complexes involved in mechanistic models for well-defined surfaces depends on the available information from spectroscopy. For ill-defined systems, such information can be obtained and used as a guide to explore with modeling the possible species. Options could be mono- versus multidentate, IS *versus* OS, *etc.* With spectroscopy, the information is averaged, so it may be difficult to separate IS and OS or identify the denticities if various forms are simultaneously present.

From all the information collected and evaluated for ion adsorption phenomena in well-defined systems, it is evident that the implementation of charge distribution is the best and easiest choice if the aim is to conserve the link with observable structures. The required charge distributions as input parameters can be assessed from the geometry of the surface complexes. Even if it is used as a fitting parameter, it will not necessarily lead to a larger number of

[26] See the chapter by Jordan et al. (2025, this volume)

adjustable parameters. If the CD principle is omitted, generally one or more extra surface species are needed to correctly assess the proton co-adsorption as discussed previously. The more simplifications are made, the more the corresponding surface speciation will become virtual and largely lead to degeneration towards a descriptive model.

Denticity. Problems remain in the treatment of multidentate surface complexes (Benjamin 2002; Wang et al. 2013; Lützenkirchen et al. 2015). For undefined systems, available spectroscopic data can be used in defining stoichiometries, but the resulting models will remain speculative due for example to the uncertainty of site distribution. If sites are present and react in pairs ($n=2$), as for the singly coordinated groups on the edge faces of gibbsite, the reaction equation is properly formulated with $(1-\theta_A)^{n=2}$ in the denominator of Equation (2), from the perspective of lattice statistics (Hiemstra and Van Riemsdijk 1990b). However, when the singly coordinated groups lay in a row, the expression is in principle invalid and will affect the site competition, particularly near site saturation. Fortunately, the competition is usually predominantly determined by the electrostatic potentials, particularly that of the 1-plane. This holds equally for well-defined and ill-defined systems. Moreover, in environmental systems, the major interest will be in trace concentrations.

Surface area and site density. Traditionally, surface area is measured with N_2- or Kr-BET. The values do not necessarily agree with surface areas obtained with TEM, as noticed for instance for gibbsite (Parfitt et al. 1977b; Hiemstra et al. 1999) and goethite (Livi et al. 2023). This may be due to, e.g., face-to-face contact, surface roughness, defects, and porosity of the solid. Moreover, all particles in the particle distribution should be well represented in TEM work, a correct morphology model is to be applied, and the translation of the particle volume to a mass requires the mass density, all creating uncertainty.

Surfaces may differ strongly in the reactive site density. For goethite, conventional TEM has been used to assess the reactive surface area (A_{TEM}). In addition, the crystal face distribution has been measured, revealing a large variation in the overall site density of singly and triply coordinated groups due to steps and defects at the surfaces. The site density information collected by high-resolution scanning transmission spectroscopy (Livi et al. 2017, 2023) has been used to consistently model the proton adsorption of the corresponding goethites (Martinez et al. 2023). However, no attempts were made to also scale to A_{TEM}. Instead, A_{BET} was used. The capacitance values obtained with modeling decrease with A_{BET} from ~1.0 and ~0.8 $F{\cdot}m^{-2}$. If A_{TEM} had been used, one may expect lower values, particularly for well-crystallized goethite samples with $A_{TEM} \gg A_{BET}$. According to the present authors, the introduction of detailed morphology analysis at the nanoscale level may contribute to closing the gap between well-defined and ill-defined materials presented as two extremes in Figure 24, and at the same time, it puts into question the ideal substrate character of some thought to be well-defined systems.

Case study: Silica[27]

Although silica and quartz have a relatively simple structure and principally expose only two types of groups ($\equiv Si_2O$ and $\equiv SiOH$) and even though the available materials have been widely studied, it is rather difficult to establish the surface speciation and charging behavior, both experimentally and by surface complexation modeling. The various choices that are possible will be illustrated and the difficulties in pinpointing parameter values will be addressed. In comparison to well-defined gibbsite, surface complexation modeling of silica and quartz becomes more complicated because of the 0/-1 transition, which implies that the $\log K_H$ does not follow *a priori* from the PZC as in the case of gibbsite, but requires fitting. This increases the uncertainty as the $\log K_H$ is correlated with ion pair formation and capacitance value(s).

[27] This case study exemplifies our route into the unknown at (re)analyzing for this paper data about the charging behavior of silica, sheading new light on long-steading issues.

First of all, we like to differentiate between crystalline SiO_2 (like quartz) and more amorphous silica nanoparticles (NPs). The former frequently has a bimodal charging behavior, while the latter in most cases does not. The occurrence or not of the bimodal character is a puzzle for the future. In the following, we will discuss the modeling of amorphous silica nanoparticles that do not show bimodal behavior.

Many silica NPs have a very high surface area in the range from 180 $m^2 \cdot g^{-1}$ (Bolt 1957) to 360 $m^2 \cdot g^{-1}$ (Goel and Lützenkirchen 2022). At the high end, it may set limits to the quantification with the BET method, but may alternatively be assessed from the TEM particle size distribution, assuming non-porosity and a (un)known mass density of the amorphous structure, all making precise scaling to charge density uncertain, which ultimately affects e.g., the fitted capacitance. Very well-defined Stoeber-type silica particles in terms of size and shape may be used but these contain ammonium contamination and it may be very tedious to remove that (Dunstan 1994).

Primary charge. Measuring surface charge for silica NPs can be challenging. Several authors have used commercial charge-stabilized silica NPs. Bolt (1957) dialyzed the material to pH 4.5 before use to remove the excess electrolyte ions, because the native suspensions contain more Na^+ ions than are needed for surface charge compensation. This extra amount in the stock suspensions ranges from ~150–250 (!) mM. Such excess affects the ionic strength and ion-specific interactions, thereby influencing the proton titrations and their interpretation (Goel and Lützenkirchen 2022), if not removed in advance.

In addition, it is noted that often a 0.1 M acid solution is used for the proton titrations. This will lead to an increase in the ionic strength if the studied electrolyte concentration levels are less than 0.1 M. The proton of the added HCl will neutralize the surface groups, releasing an equivalent amount of the Na^+ counter ions that form a 0.1 NaCl solution in the volume of the acid added. Instead, titrations for silica should be done with acid solutions in which the anion concentrations are equal to the electrolyte concentration, but this becomes difficult if the charging at a low ionic strength is studied. Alternatively, one may add OH^- with coulomb titrations (Rosenqvist et al. 2002). At a typical solid concentration of 5% silica with 9 nm particles (Brown et al. 2016a,b), the total surface area is ~13000 $m^2 \cdot L^{-1}$ and ~ 0.2 L acid of 0.1 M per liter suspension is needed for neutralization. It implies that at the end point, the ionic strength is augmented by 20 mM. The interpretation of such data will be unreliable if not corrected for.

Another point of concern is the use of charge-stabilized silica NP for measuring the surface charge in the presence of various alkali ions such as Cs^+, Rb^+, K^+, and Li^+. For the same conditions as above, the solution contains 0.02 M Na^+ as counter ions (apart from excess salt). Brown et al. (2016b) added a 0.05 M solution with alkali ions either Cs^+, K^+, or Li^+. It is obvious that the resulting systems will be a mixture of ions which complicates a straightforward interpretation.

Sonnefeld et al. used another type of silica NP (Sonnefeld 1995; Sonnefeld et al. 1995), not subject to the above problems. Charge-free silica was used to which alkali hydroxide was added to reach pH ~ 8. Additionally, alkali chloride was added to cover the electrolyte concentration range of 0.003 to 0.3 $mol \cdot L^{-1}$. The experiments were limited to the pH 4–8 to avoid variation in ionic strength. Dissolution at high pH also will affect proton balances, once dissolved silicic acid starts to release protons. Sonnefeld and co-workers have studied the charging of silica NP in the presence of 5 different types of electrolyte ions (Li^+, Na^+, K^+, Rb^+, Cs^+). The observed trend in the surface charge development follows the same order as found by others (Abendroth 1970; Dove and Craven 2005). The data of Sonnenfeld et al. will be used collectively for our surface complexation modeling as it is one the most extensive and reliable data presently available in the literature.

Ion pair formation. As discussed above, quantifying ion pair formation experimentally is difficult. It generally follows from model interpretations, where different choices can be made.

Sonnefeld explains his data without ion pair formation but instead, he allows the Stern layer capacitance to vary with the type of electrolyte involved (Sonnefeld 1995). In addition, the proton affinity constant was fitted, leading to $\log K_H = 7.00$ for a realistic site density of 5 nm^{-2}, as presently derived (Fig. 8b). A spherical double layer approximation was used in combination with the Basic Stern model. The fitted Stern layer thicknesses ranged from $\delta = 0.64$ to 0.17 nm in the electrolyte series Li^+, Na^+, K^+, Rb^+, and Cs^+, assuming a relative dielectric constant of water ($\varepsilon_r = 80$) for the Stern layer. If lower values for ε_r are used, δ will be lower (Eqn. 16). The calculated variation in size is larger than the physical size with a single layer of coordinating water molecules (0.32–0.34 nm), derived from available data (Marcus 1988). If the large variation found by the modeling is realistic, it would imply that the Li^+ ions have a full second layer of hydration water $d = 0.32 + 0.28 = 0.60$ nm when located at the minimum distance of approach at the head end of the DDL, while in contrast Cs^+ ions would partially dehydrate and form innersphere complexes, because the Pauling radius of Cs^+ is $r = 0.17$ nm. This appears unlikely.

As an alternative to the above approach, the data can be modeled considering outersphere complexation. Using the Basic Stern model for a single-site approach with a site density of $N_s = 5.0$ nm^{-2}, the fitted Stern layer capacitance is 1.07 ± 0.03 F·m^{-2}, and the proton affinity is $\log K_H = 6.75 \pm 0.03$ (Table 4).

In Figure 24a, the lines represent the modeling results for the single-site option. With parameters derived for the Aerosil 300 silica (Sonnefeld et al. 1995), the data of the Ludox NPs (Bolt 1957) can be described, but only for the data for $I \leq 0.1$ M. At higher ionic strength, higher proton-related surface charge density was measured than is predicted. Introducing Na^+-IS complexation, the entire set can be described. Remarkably, the obtained CD coefficient is very high ($\Delta z_0 \sim 1$), suggesting the incorporation of the Na^+ ion into the surface, which is in line with MD simulations (Döpke et al. 2019). Reevaluation of the whole set, fixing the values of $\log K_{os}$ and the capacitance C_1, leads to $\log K_H = 6.92$ (Table 4). It is not evident that the high-ionic strength behavior of this Ludox is universal or that it is, for instance, induced by the extensive dialysis process performed. Moreover, there is significant uncertainty about the activity corrections for the high ionic strength conditions used in the modeling and, there might even be uncertainty in the experimental data for the 4 M data set.

The charge-stabilized silica nanoparticle suspensions generally have large quantities of excess salt, increasing the ionic strength I relative to the electrolyte level intended by the addition of electrolyte (Goel and Lützenkirchen 2022). In the present modeling, we have accounted for this using the excess concentration specified by the manufacturers. In addition,

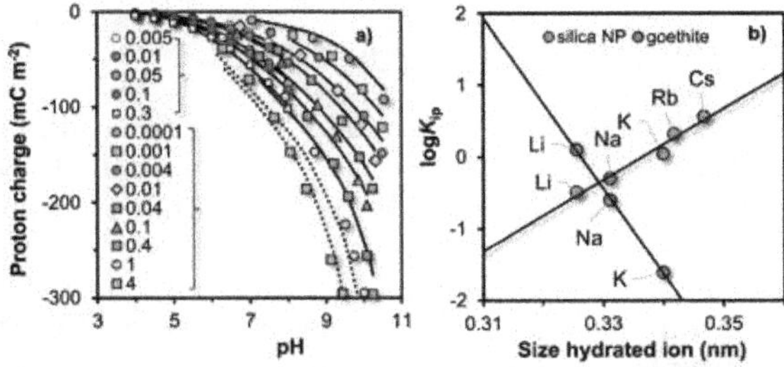

Figure 24. a) Charging of SiO_2 nanoparticles (**symbols**) in a series of NaCl solutions covering the pH range of 4–8 (Sonnefeld et al. 1995) and the pH range of ~7–11 (Bolt 1957). The model lines have been calculated using a total site density of $N_s = 5$ nm^{-2} and fitted ion pair formation constants. The model lines at high ionic strength are **dotted**, as the activity corrections are uncertain. **b)** Ion pair formation constants for silica, given with **blue symbols** using low $\log K_H$ (6.75). For comparison, the $\log K_{ip}$ values for goethite (**orange**) are shown (Hiemstra and Van Riemsdijk 2006), illustrating the very different trend with the diameter of the hydrated cations.

we have corrected for the use of 0.1 M HCl in the titrations, which will affect the ionic strength for $I \neq 0.1$ M. The fitted parameters are given in Table 4. The silica NPs have on average a proton affinity of $\log K_H = 6.8 \pm 0.2$ and a Stern layer capacitance of $C_1 = 1.0 \pm 0.1$ F·m^{-2}, as for the data of Sonnefeld et al. (1995). The mean $\log K_H$ is surprisingly close to the very first estimate of Schindler and Kamber (1968).

For the silica nanoparticles of Table 4, the surface charge data can be well-described assuming ion pair formation. Ideally, ion pair formation should be quantified independently. For Ludox-TM50 silica, Si/Na and Si/Cl ion ratios on fast-frozen samples have been measured with XPS (Cardenas 2005) in an attempt to quantify ion pair formation. In the samples, chloride ions were observed by XPS even for the highest pH values on silica. Likely, these anions reside further away in the double layer (Pfeiffer-Laplaud et al. 2016). The excess of Na^+ over Cl^- was quantified, and expressed as a charge per unit surface area, excellent agreement was found with the surface charge data, measured potentiometrically. This suggests that the XPS results involve all counter- and co-ions including the diffuse layer contribution. It implies that in principle no new information in terms of ion-pair formation is generated.

Table 4. Parameters for describing the charging of nano-silica in NaCl with a single-site Basic Stern model ($N_s = 5$ nm^{-2}, see Fig. 8b), assuming OS ($\log K_{os}$) and IS ($\log K_{is}$) surface complexation. The latter only occurs at very high ionic strength (≥ 0.4 M Na^+). The Stern layer capacitance (C_1) and ion pair formation ($\log K_{os}$) derived for the extensive data set of Aerosil were used in the modeling as reference (≡) and only adapted if needed. The quality of the description is in all cases very good ($R^2 \geq 0.993$).

	Aerosil 300[a]	**Ludox B[b]**	**Ludox SM[c]**	**Bindzil[d]**	**Bindzil[d]**	**Klebesol[d]**	**Ludox SM[d]**	**Mean**
Size (nm)	8	15	10	10	10	9	10	10±3
A (m^2·g^{-1})	266	180	282	360	360	300	345	290±60
SiO_2 g/100g	–	–	5%	5%	1%	5%	5%	–
Excess S (mM)*	–	–	26.9	28.7	5.7	18.9	26.9	–
$\log K_H$	6.75	6.92	6.55	6.74	6.97	6.95	7.05	6.8±0.2
$\log K_{Na\text{-}os}$ **	–0.30#	≡–0.30	≡–0.30	≡ –0.30	≡ –0.30	≡ –0.30	≡ –0.30	–0.3
$\log K_{Na\text{-}is}$	–	–3.56	–	–	–	–	–	–3.5
C_1 (F·m^{-2})	1.07	≡1.07	1.17	0.82	1.1	0.95	0.94	1.0±0.1

Notes:
a–d a) Sonnefeld et al. (1995), b) Bolt (1957), c) Brown et al. (2016), d) Goel and Lützenkirchen (2022)
b–d b) Dialyzed sample Ludox B, precise sample not specified. c-d) Samples are non-dialyzed.
* Excess salt in stock in mM, from (Goel and Lützenkirchen 2022)
** For the other alkali ions: $\log K(Li^+) = -0.49$, $\log K(K^+) = +0.05$, $\log K(Rb^+) = +0.32$, $\log K(Cs^+) = +0.5$
# Fitted value in agreement with data of internal reflection ellipsometry (Wang et al. 2015)

Wang et al. (2015), using internal reflection ellipsometry for a silica coating, quantified the Na^+ ion pair formation for the concentration range from 0.01–1.0 M at pH 10. They described their data with a traditional TL model, involving (in our view) an unrealistically low capacitance value for the second Stern layer ($C_2 = 0.2$ F·m^{-2}). With $\log K_H = 7.0$ and $N_s = 8$ nm^{-2}, the Na^+ OS complexation constant was $\log K_{os} = -1.65 \pm 1$. Translated to a Basic Stern model with $N_s = 5$ nm^{-2}, the value becomes $\log K_{os} = -0.35 \pm 1.0$. This value is within all uncertainties equal to our $\log K_{os}$ value of -0.30 ± 0.12, illustrating that our modeled $\log K_{os}$ value agrees with the experimental data using internal reflection ellipsometry.

Figure 24b gives the relation between the ion pair formation constants and the size of the primary hydrated ions, calculated from published hydrated ionic radii (Marcus 1988). The ion pair formation on silica follows an affinity order that is opposite to those derived for goethite (Hiemstra and Van Riemsdijk 2006), ferrihydrite (Mendez and Hiemstra 2020c), anatase and rutile (Bérubé and De Bruyn 1968; Sprycha 1984; Bourikas et al. 2001), some plane of α-Al_2O_3 (Lützenkirchen 2013), and hematite (Shimizu et al. 2012). Silica in this sense seems to be an exception, which previously has been attributed to the difference in surface hydration energy (Dumont et al. 1990). Surfaces that have a very low heat of immersion in water (< 0.2 J·m^{-2}), such as AgI, graphite, and silica exhibit charging in the series $Cs^+ > Li^+$, while the surfaces of TiO_2, Fe_2O_3, and Al_2O_3, with high heats of immersion ($\sim$ 0.5–0.8 J·m^{-2}), exhibit the opposite trend. For oxides, the heat of immersion in water is simultaneously correlated with the PZC.

Surface potentials. Surfaces of single crystals are suitable for measuring surface potentials. For Fe-oxide colloids, surface potentials can be measured by first coating these on a Pt electrode, followed by heating and etching with concentrated acid (Penners et al. 1986). The measurement revealed a Nernstian behavior for the Fe-oxide coating. Unfortunately, the invasive pretreatments limit a full comparison with the behavior of the original particles.

No pretreatment is needed for the *in-situ* XPS measurement of Si2p band shifts of SiO_2-suspended nanoparticles (Brown et al. 2016b). In the approach, the observed band shifts were interpreted as solely due to variation in the surface potential. An extremely large variation with the type of cation (Li^+, Na^+, K^+, Cs^+) was obtained (~150 mV at pH 10), which is not observed for SiO_2 ISFETS (Fig. 13). If the obtained surface potentials (ψ_o) are combined with the corresponding experimental surface charge expressed in a group density [$\equiv SiO^-$], one can calculate the intrinsic $\log K_H$, according to $\log K_H = \log([\equiv SiOH^o]/[\equiv SiO^-]) + pH + 2.3F\psi_o/RT$. With this for the data collected in 0.05 M NaCl for the pH range 3-8, the mean $\log K_H$ is calculated to be ~6.2 ± 0.3. Above pH 8, the same kind of calculation results in the $\log K_H$ values that strongly drop reaching a value of $\log K_H = 4.0$ at pH 10 (Fig. 25a). This is remarkable, as it suggests that sites with a low proton affinity ($\log K_H$) become reactive at high pH (10). This puts into doubt the validity of the XPS approach, particularly the calibration in an ultra-high vacuum that uses a field with a linearly decreasing potential, which is quite different from a diffuse double layer (Fig. 2).

With the increase in cation affinity from Li^+ to Cs^+, the linear drop of the XPS-derived potential over the Stern layer increases and the contribution of the diffuse double layer decreases. This implies that one increasingly approaches the condition of (ex-situ) calibration, explaining the decrease of the apparent surface potential. Calculations of the apparent $\log K_H$ using the revised approach reveal a linear increase of the electrolyte ion affinity ($\log K_{ip}$) from $\log K_H = 3.5$ for Li^+ to $\log K_H = 6.2$ for Cs^+ (Fig. 25b). From this, we conclude that the surface potentials will be in line with potentials obtained with ISFETs and COS devices (Fig. 13) if the effect of the field on the Si2P energy is correctly calibrated. This may open the possibility of deriving the $\log K_H$ independently. Currently, it can be stated, that there is only one site required and that the expected $\log K_H$ is ≥ 6.5 for these silica nanoparticles.

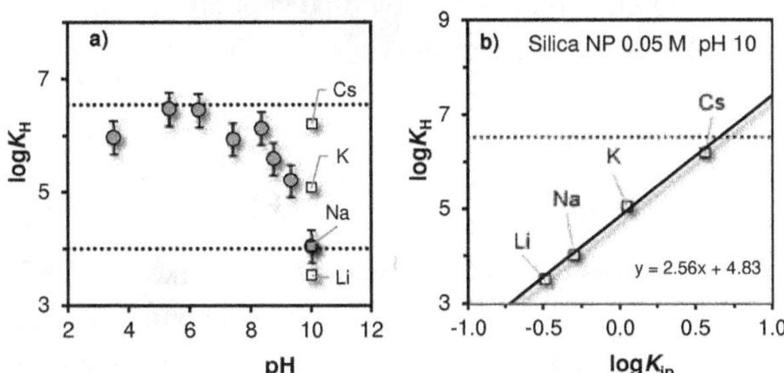

Figure 25. Proton affinity constant ($\log K_H$) calculated using surface potential data of silica NP derived with XPS (Brown et al. 2016b). **a)** A strong decrease in $\log K_H$ at high pH, implying that at high pH low affinity sites would determine the charging behavior, which is odd. **b)** A strong increase of $\log K_H$ for silica NP at pH with the affinity of the electrolyte ions forming ion pairs. Likely, this artifact can be removed by adapting the (ex-situ) calibration applied in the XPS method (see text).

Concluding remarks

Although many methods are available, they are hardly applied collectively and consistently for studying the interfacial and surface chemical behavior of a selected material. Probably, the most extensive comprehensive study on a single well-defined material still is the rutile sample (Předota et al. 2004, 2016; Ridley et al. 2004, 2009, 2012; Zhang et al. 2004, 2006, 2007; Fitts et al. 2005; Machesky et al. 2008; Rosenqvist et al. 2009). Progress in the work of understanding interfaces must involve comprehensive interdisciplinary studies, ideally combining information collected at similar conditions with a large variety of techniques for selected, identical systems in close collaboration rather than fragmentarily producing data that make comparison and integration complicated or impossible. The seemingly trivial example of the silica that should be dialyzed before studying ion-specific effects shows what kinds of experimental pitfalls can hamper the comparison of published data. For single crystals, the pre-treatment like heating can be crucial as already a few references for one single crystal show (Chandrasekharan et al. 2008; Yang et al. 2010; Hütner et al. 2024).

For analyzing wide ranges of experimental observations in combination with (AI)MD simulations and explaining them self-consistently, advanced surface complexation modeling can contribute to integrating microscopic and macroscopic data. This should also be done for natural systems, which is particularly challenging as there are many unanswered questions for these systems that have their own complexity. In this respect, the reactive surface area and the competitive presence of adsorbed organic matter are two very critical parameters[28]. The application of the available tools to gather as much self-consistent experimental information about a sample prepared in a defined way may not only help to improve the understanding of that very system (and thus interfacial chemistry and physics in general) but also improve the interpretation of results from the methods applied. SCMs can also integrate many of the available data and outliers in every sense can be identified. The most advanced SCMs are built from molecular scale insight and can be applied as powerful tools.

The application of SCMs to many if not most environmentally relevant systems is difficult to achieve at the same level and simplifications and often unverifiable assumptions are required. Analyzing big data and designing robust SCMs beyond purely diffuse layer models should be one path that may be open through artificial intelligence and machine learning. However, the careful use of experimental methods, starting from solution chemistry, titration, batch adsorption experiments and electrokinetics should be kept alive, because careful preparation (and this involves control of solution chemistry) is key to the experiments with state-of-the-art spectroscopic methods.

ACKNOWLEDGEMENTS

The authors wish to express their sincere gratitude to Mike Machesky for his inspiring work, which showcased the power of interdisciplinary collaboration, integrating a plethora of knowledge collected by scientists from various fields into surface complexation modeling. We deeply regret his passing and that he was unable to witness the latest developments and interpretations to which his contributions have been so pivotal.

[28] See the chapter of Xu et al. (2025, this volume) on surface complexation modeling of the adsorption of NOM and that of Hiemstra et al. (2025, this volume) on surface complexation modeling of ferrihydrite and its application to natural systems.

REFERENCES

Abbas M, Rao BP, Naga SM, Takahashi M, Kim C (2013) Synthesis of high magnetization hydrophilic magnetite (Fe_3O_4) nanoparticles in single reaction—Surfactantless polyol process. Ceram Int 39:7605–7611, https://doi.org/10.1016/j.ceramint.2013.03.015

Abdala DB, Northrup PA, Arai Y, Sparks DL (2015) Surface loading effects on orthophosphate surface complexation at the goethite/water interface as examined by extended X-ray Absorption Fine Structure (EXAFS) spectroscopy. J Colloid Interface Sci 437:297–303, https://doi.org/10.1016/j.jcis.2014.09.057

Abendroth RP (1970) Behavior of a pyrogenic silica in simple electrolytes. J Colloid Interface Sci 34:591–596, https://doi.org/10.1016/0021-9797(70)90223-7

Adekola F, Fédoroff M, Geckeis H, Kupcik T, Lefèvre G, Lützenkirchen J, Plaschke M, Preocanin T, Rabung T, Schild D (2011) Characterization of acid–base properties of two gibbsite samples in the context of literature results. J Colloid Interf Sci 354:306–317, doi:https://doi.org/10.1016/j.jcis.2010.10.014

Ahrland S, Grenthe I, Norén B, Levitin NE, Westin G (1960a) The ion exchange properties of silica gel. I. The sorption of Na^+, Ca^{2+}, Ba^{2+}, UO_2^{2+}, Gd^{3+}, Zr(IV), Nb, U(IV) and PU(IV). Acta Chem Scand 14:1059–1076

Ahrland S, Grenthe I, Norén B, Levitin NE, Westin G (1960b) The ion exchange properties of silica gel. II. Separation of plutonium and fission products from irradiated uranium. Acta Chem Scand 14:1077–1090

Ahualli S, et al. (2025) Electrokinetics in relation to Surface Complexation Modelling. Rev Mineral Geochem 91B (in press)

Aquino AJA, Tunega D, Haberhauer G, Gerzabek MH, Lischka H (2008) Acid–base properties of a goethite surface model: A theoretical view. Geochim Cosmochim Acta 72:3587–3602

Atkinson RJ, Posner AM, Quirk JP (1967) Adsorption of potential-determining ions at the ferric oxide–aqueous interface. J Phys Chem 71:550–558

Audry MC, Piednoir A, Joseph P, Charlaix E (2010) Amplification of electro-osmotic flows by wall slippage: Direct measurements on OTS-surfaces. Faraday Discuss 146:113–124, https://doi.org/10.1039/b927158a

Axe L, Bunker GB, Anderson PR, Tyson TA (1998) An XAFS analysis of strontium at the hydrous ferric oxide surface. J Colloid Interface Sci 199:44–52

Azam MS, Weeraman CN, Gibbs-Davis JM (2012) Specific cation effects on the bimodal acid–base behavior of the silica/water interface. J Phys Chem Letters 3:1269–1274, https://doi.org/10.1021/jz300255x

Balajka J, Hines MA, DeBenedetti WJI, Komora M, Pavelec J, Schmid M, Diebold U (2018) High-affinity adsorption leads to molecularly ordered interfaces on TiO_2 in air and solution. Science 361:786–788, https://doi.org/10.1126/science.aat6752

Banik NL, Marsac R, Lützenkirchen J, Marquardt CM, Dardenne K, Rothe J, Bender K, Geckeis H (2017) Neptunium sorption and redox speciation at the illite surface under highly saline conditions. Geochim Cosmochim Acta 215:421–431, https://doi.org/10.1016/j.gca.2017.08.008

Bañuelos JL, Borguet E, Brown Jr GE, Cygan RT, DeYoreo JJ, Dove PM, Gaigeot MP, Geiger FM, Gibbs JM, Grassian VH, Ilgen AG (2023) Oxide– and silicate–water interfaces and their roles in technology and the environment. Chem Rev 123:6413–6544, https://doi.org/10.1021/acs.chemrev.2c00130

Bargar JR, Towle SN, Brown GE, Jr., Parks GA (1996) Outer-sphere lead(II) adsorbed at specific surface sites on single crystal α–alumina. Geochim Cosmochim Acta 60:3541–3547

Bargar JR, Brown GE, Jr., Parks GA (1997a) Surface complexation of Pb(II) at oxide–water interfaces: I XAFS and bond-valence determination of mono- and polynuclear Pb(II) sorption products on aluminum oxides. Geochim Cosmochim Acta 61:2617–2637

Bargar JR, Brown GE, Jr., Parks GA (1997b) Surface complexation of Pb(II) at oxide–water interfaces: II XAFS and bond-valence determination of mononuclear Pb(II) sorption products and surface functional groups on iron oxides. Geochim Cosmochim Acta 61: 2639–2652

Bargar JR, Brown GE, Jr., Parks GA (1998) Surface complexation of Pb(II) at oxide–water interfaces: III XAFS determination of Pb(II) and Pb(II)-chloro adsorption complexes on goethite and alumina. Geochim Cosmochim Acta 62:193–207

Barrón V, Torrent J (1996) Surface hydroxyl configuration of various crystal faces of hematite and goethite. J Colloid Interface Sci 177:407–410

Barrow NJ (1987) Reactions with Variable-Charge Soils. Martinus Nijhof Publishers, Dordrecht

Barrow NJ, Bowden JW (1987) A comparison of models for describing the adsorption of anions on a variable charge mineral surface. J Colloid Interface Sci 119: 236–250

Barrow NJ, Cox VC (1992) The effects of pH and chloride concentration on mercury sorption. I By goethite. J Soil Sci 43:437–450

Barrow NJ, Bowden JW, Posner AM, Quirk JP (1980) An objective method for fitting models of ion adsorption on variable charge surfaces. Aust J Soil Res 18:395–404

Beattie JK, Djerdjev AM (2004) The pristine oil/water interface: Surfactant-free hydroxide-charged emulsions. Angew Chem-Int Edit 43:3568–3571, https://doi.org/10.1002/anie.200453916

Benjamin MM (2002) Modeling the mass-action expression for bidentate adsorption. Environ Sci Technol 36:307–313

Bérubé YG, De Bruyn PL (1968) Adsorption at the rutile–solution interface. II Model of the electrochemical double layer. J Colloid Interface Sci 28:92–105

Bickmore BR, Tadanier CJ, Rosso KM, Monn WD, Eggett DL (2004) Bond-valence methods for pK(a) prediction: Critical reanalysis and a new approach. Geochim Cosmochim Acta 68:2025–2042
Bijsterbosch BH, Lyklema J (1978) Interfacial electrochemistry of silver-iodide. Adv Colloid Interface Sci 9:147–251
Bleam WF (1993) On Modeling Proton Affinity at the Oxide/Water Interface. J Colloid Interface Sci 159:312–318
Blok L, De Bruyn PL (1970) The ionic double layer at the ZnO/ solution interface. II Composition model of the surface. J Colloid Interface Sci 32:527–532
Blute I, Pugh RJ, van de Pas J, Callaghan I (2007) Silica nanoparticle sols 1. Surface chemical characterization and evaluation of the foam generation (foamability). J Colloid Interface Sci 313:645–655, https://doi.org/10.1016/j.jcis.2007.05.013
Bockris JOM, Reddy AKN (1976a) Modern Electrochemistry 2. Plenum, New York
Bockris JOM, Reddy AKN (1976b) Modern Electrochemistry 1. Plenum, New York
Boily J-F (1999) The Surface Complexation of Ions at the Goethite (α-FeOOH)/ Water Interface: a Multisite Complexation Approach. Umeå University, Sweden
Boily J-F (2025) Molecular controls on complexation reactions and electrostatic potential development at mineral surfaces. Rev Mineral Geochem 91A:105–148
Boily JF, Felmy AR (2008) On the protonation of oxo- and hydroxo-groups of the goethite (alpha-FeOOH) surface: A FTIR spectroscopic investigation of surface O–H stretching vibrations. Geochim Cosmochim Acta 72:3338–3357
Boily J-F, Chatman SM, Rosso KM (2011) Inner-Helmholtz potential development at the hematite (α-Fe_2O_3) (0 0 1) surface. Geochim Cosmochim Acta 75:4113–4124
Bolt GH (1957) Determination of the charge density of silica sols. J Phys Chem 61:1166–1169. ibid. 1162: 1608
Bolt GH, Van Riemsdijk WH (1982) Ion adsorption on inorganic variable charge constituents. *In*: Soil Chemistry. Vol. B. Bolt GH (ed) Elsevier, Amsterdam, p 459–504
Bonthuis DJ, Gekle S, Netz RR (2012) Profile of the static permittivity tensor of water at interfaces: consequences for capacitance, hydration interaction and ion adsorption. Langmuir 28:7679–7694, https://doi.org/10.1021/la2051564
Borges PD, Scolfaro LMR, Alves HWL, da Silva EF (2010) DFT study of the electronic, vibrational, and optical properties of SnO_2. Theor Chem Acc 126:39–44, https://doi.org/10.1007/s00214-009-0672-3
Borkovec M (1997) Origin of 1-pK and 2-pK models for ionizable water–solid interfaces. Langmuir 13:2608–2613
Borkovec M, Rusch U, Westall JC (1998) Modeling of competitive ion binding to heterogeneous materials with affinity distributions. *In*: Adsorption of Metals by Geomedia Variables, Mechanisms and Model Applications. Jenne EA, (ed) Academic Press, p 467–482
Bourikas K, Hiemstra T, Van Riemsdijk WH (2001) Ion pair formation and primary charging behaviour of titanium oxide (anatase and rutile). Langmuir 17:749–756
Bousse L, De Rooij NF, Bergveld P (1983) The influence of counter-ion adsorption on the psi-0 /pH characteristics of insulator surfaces. Surf Sci 135:479–496
Bousse LJ, Mostarshed S, Hafeman D (1992) Combined measurement of surface potential and zeta potential at insulator/electrolyte interfaces. Sensors Actuators B 10:67–71, https://doi.org/10.1016/0925-4005(92)80013-N
Bowden JW (1973) Models for Ion Adsorption on Mineral Surfaces. PhD dissertation, University of West Australia
Bowden JW, Posner AM, Quirk PJ (1977) Ionic adsorption on variable charge mineral surfaces: theoretical-charge development and titration curves. Aust J Soil Res 15:121–136
Bowden JW, Nagarajah S, Barrow NJ, Posner AM, Quirk PJ (1980) Describing the adsorption of phosphate, citrate and selenite on a variable-charge mineral surface. Aust J Soil Res 18:49–60
Bradbury MH, Baeyens B (1997) A mechanistic description of Ni and Zn sorption on Na-montmorillonite Part II: modelling. J Contam Hydrol 27:223–248, https://doi.org/10.1016/S0169-7722(97)00007-7
Bradbury MH, Baeyens B (2005) Modelling the sorption of Mn(II), Co(II), Ni(II), Zn(II), Cd(II), Eu(III), Am(III), Sn(IV), Th(IV), Np(V) and U(VI) on montmorillonite: linerar free energy relationships and estimates of surface binding constants for some selected heavy metal ions and actinides. Geochim Cosmochim Acta 69:875–892
Bragg WL (1912) The specular reflection of X-rays. Nature 90:410–410, https://doi.org/10.1038/090410b0
Breeuwsma A, Lyklema J (1973) Physical and chemical adsorption of ions in the electrical double layer. J Colloid Interface Sci 43:437–448
Brkljača Z, Namjesnik D, Lützenkirchen J, Předota M, Preočanin T (2018) Quartz/aqueous electrolyte solution interface: molecular dynamic simulation and interfacial potential measurements. J Phys Chem C 122:24025–24036
Brown ID (1978) Bond valences—A simple structural model for inorganic chemistry. Chem Soc Rev 7:359–376
Brown ID (1992) Chemical and steric constraints in inorganic solids. Acta Cryst B48:553–572
Brown ID (2009) Recent developments in the methods and applications of the bond valence model. Chem Rev 109:6858–6919
Brown ID, Altermatt D (1985) Bond-valence parameters obtained from a systematic analysis of the inorganic crystal structure database. Acta Cryst B41:244–247
Brown MA, Goel A, Abbas Z (2016a) Effect of electrolyte concentration on the stern layer thickness at a charged interface. Angew Chem-Int Edit 55:3790–3794, https://doi.org/10.1002/anie.201512025

Brown MA, Abbas Z, Kleibert A, Green RG, Goel A, May S, Squires TM (2016b) Determination of surface potential and electrical double-layer structure at the aqueous electrolyte–nanoparticle interface. Phys Rev X 6:011007, https://doi.org/10.1103/PhysRevX.6.011007

Bullard JW, Cima MJ (2006) Orientation dependence of the isoelectric point of TiO_2 (rutile) surfaces. Langmuir 22:10264–10271, https://doi.org/10.1021/la061900h

Cardenas JF (2005) Surface charge of silica determined using X-ray photoelectron spectroscopy. Colloids Surf A 252:213–219, https://doi.org/10.1016/j.colsurfa.2004.10.085

Catalano JG (2011) Weak interfacial water ordering on isostructural hematite and corundum (001) surfaces. Geochim Cosmochim Acta 75:2062–2071

Catalano JG, Trainor TP, Eng PJ, Waychunas GA, Brown GE (2005) CTR diffraction and grazing-incidence EXAFS study of U(VI) adsorption onto alpha-Al_2O_3 and alpha-Fe_2O_3 (1–102) surfaces. Geochim Cosmochim Acta 69:3555–3572

Catalano JG, Park C, Zhang Z, Fenter P (2006) Termination and water adsorption at the alpha-Al_2O_3 (012)—Aqueous solution interface. Langmuir 22:4668–4673

Catalano JG, Park C, Fenter P, Zhang Z (2008) Simultaneous inner- and outer-sphere arsenate adsorption on corundum and hematite. Geochim Cosmochim Acta 72:1986–2004, https://doi.org/10.1016/j.gca.2008.02.013

Cerník M, Borkovec M (1996) Affinity distribution of competitive ion binding to heterogeneous materials. Langmuir 12:6127–6137

Chandrasekharan R, Zhang L, Ostroverkhov V, Prakash S, Wu Y, Shen YR, Shannon MA (2008) High-temperature hydroxylation of alumina crystalline surfaces. Surf Sci 602:1466–1474, https://doi.org/10.1016/j.susc.2008.02.009

Chapman DL (1913) A contribution to the theory of electrocapillarity. Philos Mag 6:475–481

Charlet L, Manceau AA (1992) X-ray absorption spectroscopic study of the sorption of Cr(III) at the oxide–water interface. J Colloid Interface Sci 148:443–458

Chatman S, Zarzycki P, Rosso KM (2013) Surface potentials of (001), (012), (113) hematite (α-Fe_2O_3) crystal faces in aqueous solution. Phys Chem Chem Phys 15:13911–13921, https://doi.org/10.1039/C3CP52592A

Cheng J, Sprik M (2010) Acidity of the aqueous rutile TiO_2(110) Surface from density functional theory based molecular dynamics. J Chem Theory Comput 6:880–889, https://doi.org/10.1021/ct100013q

Cheng J, Sprik M (2014) The electric double layer at a rutile TiO_2 water interface modelled using density functional theory based molecular dynamics simulation. J Phys Condens Matter 26:11, https://doi.org/10.1088/0953-8984/26/24/244108

Cheng L, Fenter P, Nagy KL, Schlegel ML, Sturchio NC (2001) Molecular-scale density oscillations in water adjacent to a mica surface. Phys Rev Lett 87:156103, https://doi.org/10.1103/PhysRevLett.87.156103

Cheng J, Liu XD, VandeVondele J, Sprik M (2015) Reductive hydrogenation of the aqueous rutile TiO_2(110) surface. Electrochim Acta 179:658–667, https://doi.org/10.1016/j.electacta.2015.03.212

Christl I, Kretzschmar R (1999) Competitive sorption of copper and lead at the oxide–water interface: Implications for surface site density. Geochim Cosmochim Acta 63:2929–2938

Collins CR, Sherman DM, Ragnarsdottir KV (1999a) Surface complexation of Hg^{2+} on goethite: Mechanism from EXAFS spectroscopy and density functional calculations. J Colloid Interface Sci 219:345–350

Collins CR, Ragnarsdottir KV, Sherman DM (1999b) Effect of inorganic and organic ligands on the mechanism of cadmium sorption to goethite. Geochim Cosmochim Acta 63:2989–3002

Combes JM, Chrisholm-Brause CJ, Brown GE, Jr., Parks GA, Conradson SD, Eller PG, Triay IR, Hobart DE, Meijer A (1992) EXAFS spectroscopic study of neptunium(V) sorption at the α-FeOOH / water interface. Environ Sci Technol 26:376–382

Connor PA, Dobson KD, McQuillan AJ (1999) Infrared spectroscopy of the TiO_2/aqueous solution interface. Langmuir 15:2402–2408

Davis JA, Leckie JO (1978) Surface ionization and complexation at the oxide/water interface. II Surface properties of amorphous iron oxyhydroxide and adsorption of metal ions. J Colloid Interface Sci 67:90–107

Davis JA, Leckie JO (1980) Surface ionization and complexation at the oxide/water interface. III Adsorption of anions. J Colloid Interface Sci 74:32

Davis JA, James R, Leckie JO (1978) Surface ionization and complexation at the oxide/water interface. I Computation of electrical double layer properties in simple electrolytes. J Colloid Interface Sci 63:480–499

Diot JL, Joseph J, Martin JR, Clechet P (1985) pH-dependence of the Si/SiO_2 interface state density for EOS systems: Quasi-static and AC conductance methods. J Electroanal Chem 193:75–88, https://doi.org/10.1016/0022-0728(85)85053-1

Döpke MF, Lützenkirchen J, Moultos OA, Siboulet B, Dufrêche J-F, Padding JT, Hartkamp R (2019) Preferential adsorption in mixed electrolytes confined by charged amorphous silica. J Phys Chem C 123:16711–16720, https://doi.org/10.1021/acs.jpcc.9b02975

Dove PM, Craven CM (2005) Surface charge density on silica in alkali and alkaline earth chloride electrolyte solutions. Geochim Cosmochim Acta 69:4963–4970, https://doi.org/10.1016/j.gca.2005.05.006

Du XL, Shao WZ, Bao CL, Zhang LF, Cheng J, Tang F J (2024) Revealing the molecular structures of α-Al_2O_3(0001)–water interface by machine learning based computational vibrational spectroscopy. J Chem Phys 161:124702

Dugger DL, Stanton JH, Irby BN, McConnell BL, Cummings WW, Maatman RW (1964) The exchange of twenty metal ions with the weakly acidic silanol group of silica gel1,2. J Phys Chem 68:757–760, https://doi.org/10.1021/j100786a007

Dumont F, Watillon A (1971) Stability of ferric oxide hydrosols. Discuss Faraday Soc 52:352–360

Dumont F, Warlus J, Watillon A (1990) Influence of the point of zero charge of titanium-dioxide hydrosols on the ionic adsorption sequences. J Colloid Interface Sci 138:543–554, https://doi.org/10.1016/0021-9797(90)90236-h

Dunstan DE (1994) Comparison of the electrokinetic properties of the silica surface. J Chem Soc Faraday Trans 90:1261–1263, https://doi.org/10.1039/ft9949001261

Duval Y, Mielczarski JA, Pokrovsky OS, Mielczarski E, Ehrhardt JJ (2002) Evidence of the existence of three types of species at the quartz–aqueous solution interface at pH 0–10: XPS surface group quantification and surface complexation modeling. J Phys Chem B 106:2937–2945

Dzombak DA, Morel FMM (1990) Surface Complexation Modeling: Hydrous Ferric Oxide. John Wiley & Sons, New York

Eng PJ, Trainor TP, Brown Jr. GE, Waychunas GA, Newville M, Sutton SR, Rivers ML (2000) Structure of the hydrated Al_2O_3 (0001) surface. Science 288:1029–1033, https://doi.org/10.1126/science.288.5468.1029

Farley KJ, Dzombak DA, Morel FMM (1985) A surface precipitation model for the sorption of cations on metal oxides. J Colloid Interface Sci 106:226–242, https://doi.org/10.1016/0021-9797(85)90400-X

Felmy AR, Rustad JR (1998) Molecular statics calculations of proton binding to goethite surfaces: thermodynamic modeling of surface charging and protonation of goethite in aqueous solution. Geochim Cosmochim Acta 62:25–31

Fendorf S, Eick MJ, Grossl P, Sparks DL (1997) Arsenate and Chromate retention mechanism on goethite. 1 Surface structure. Environ Sci Technol 31:315–320

Fenter P, Sturchio NC (2004) Mineral–water interfacial structures revealed by synchrotron X-ray scattering. Prog Surf Sci 77:171–258

Fenter P, Cheng L, Rihs S, Machesky M, Bedzyk MJ, Sturchio NC (2000) Electrical double-layer structure at the rutile–water interface as observed in situ with small-period X-ray standing waves. J Colloid Interface Sci 225:154–165

Filius JD, Hiemstra T, Van Riemsdijk WH (1997) Adsorption of small weak organic acids on goethite: Modeling of mechanisms. J Colloid Interface Sci 195:368–380

Filius JD, Meeussen JCL, van Riemsdijk WH (1999) Transport of malonate in a goethite–silica sand system. Colloid Surf A 151:245–253

Filius JD, Lumsdon DG, Meeussen JCL, Hiemstra T, Van Riemsdijk WH (2000) Adsorption of fulvic acid on goethite. Geochim Cosmochim Acta 64:51–60

Fitts JP, Machesky ML, Wesolowski DJ, Shang XM, Kubicki JD, Flynn GW, Heinz TF, Eisenthal KB (2005) Second-harmonic generation and theoretical studies of protonation at the water/alpha-TiO_2 (110) interface. Chem Phys Lett 411:399–403

Fokkink LGJ, De Keizer A, Lyklema J (1987) Specific ion adsorption on oxides: Surface charge adjustment and proton stoichiometry. J Colloid Interface Sci 118:454–462

Franks GV, Gan Y (2007) Charging behavior at the alumina–water interface and implications for ceramic processing. J Am Ceram Soc 90:3373–3388, https://doi.org/10.1111/j.1551-2916.2007.02013.x

Gaboriaud F, Ehrhardt J (2003) Effects of different crystal faces on the surface charge of colloidal goethite (alpha-FeOOH) particles: An experimental and modeling study. Geochim Cosmochim Acta 67:967–983

Gan Y, Franks GV (2006) Charging behavior of the gibbsite basal (001) surface in NaCl solution investigated by AFM colloidal probe technique. Langmuir 22:6087–6092

Gao Y, Mucci A (2001) Acid base reactions, phosphate and arsenate complexation, and their competitive adsorption at the surface of goethite in 0.7 M NaCl solution. Geochim Cosmochim Acta 65:2361–2378

García D, Lützenkirchen J, Petrov V, Siebentritt M, Schild D, Lefèvre G, Rabung T, Altmaier M, Kalmykov S, Duro L, Geckeis H (2019) Sorption of Eu(III) on quartz at high salt concentrations. Colloid Surf A 578:123610, https://doi.org/10.1016/j.colsurfa.2019.123610

García D, Lützenkirchen J, Huguenel M, Calmels L, Petrov V, Finck N, Schild D (2021) Adsorption of strontium onto synthetic iron(iii) oxide up to high ionic strength systems. Minerals 11:1093, https://doi.org/10.3390/min11101093

Geysels B, Hiemstra T, Groenenberg JE, Comans RNJ (2024) Glyphosate binding and speciation at the water–goethite interface: a surface complexation model consistent with IR spectroscopy and MO/DFT. Water Res 273:123031, https://doi.org/10.1016/j.watres.2024.123031

Gilbert GA, Rideal EK (1944) The combination of fibrous proteins with acids. Proc R Soc London Ser A 182:335–346

Girvin DC, Ames LL, Schwab AP, McGarrah JE (1991) Neptunium adsorption on synthetic amorphous iron oxyhydroxide. J Colloid Interface Sci 141:67–78

Gittus OR, von Rudorff GF, Rosso KM, Blumberger J (2018) Acidity constants of the hematite–liquid water interface from ab initio molecular dynamics. J Phys Chem Letters 9:5574–5582, https://doi.org/10.1021/acs.jpclett.8b01870

Goel A, Lützenkirchen J (2022) Relevance of colloid inherent salt estimated by surface complexation modeling of surface charge densities for different silica colloids. Colloids Interfaces 6:23, https://doi.org/10.3390/colloids6020023

Goldberg S, Sposito G (1984) A chemical model of phosphate adsorption by soils I. Reference oxide materials. Soil Sci Soc Am J 48:772–778

Gouy G (1910) Sur la constitution de la charge électrique à la surface d'un électrolyte. J Phys 9:457–468

Grahame DC (1947) The electrical double layer and the theory of electrocapillarity. Chem Rev 41:441–501

Gunneriusson L (1994) Composition and stability of Cd(II)–chloro and Cd(II)–hydroxo complexes at the goethite (α-FeOOH)/water interface. J Colloid Interface Sci 163:484–492

Gunneriusson L, Sjöberg S (1993) Surface complexation in the H^+–goethite (α-FeOOH)–Hg(II)—chloride system. J Colloid Interface Sci 156:121–128

Gunneriusson L, Lövgren L, Sjöberg S (1994) Complexation of Pb(II) at the goethite (α-FeOOH) / water interface: The influence of chloride. Geochim Cosmochim Acta 58: 4973–4983

Hair ML, Hertl W (1970) Acidity of surface hydroxyl groups. J Phys Chem 74:91–94

Hamadi F, Latrache H, Zekraoui M, Ellouali M, Bengourram J (2009) Effect of pH on surface energy of glass and Teflon and theoretical prediction of *Staphylococcus aureus* adhesion. Mater Sci Eng C 29:1302–1305, https://doi.org/10.1016/j.msec.2008.10.023

Han J, Katz LE (2019) Capturing the variable reactivity of goethites in surface complexation modeling by correlating model parameters with specific surface area. Geochim Cosmochim Acta 244:248–263, https://doi.org/10.1016/j.gca.2018.09.008

Hartley GS, Roe JW (1940) Ionic concentrations at interfaces. Trans Faraday Soc 35:101–109, https://doi.org/10.1039/TF9403500101

Hass KC, Schneider WF, Curioni A, Andreoni W (2000) First-principles molecular dynamics simulations of H_2O on α-Al_2O_3 (0001). J Phys Chem B 104:5527–5540, https://doi.org/10.1021/jp000040p

Hayes KF, Leckie JO (1987) Modeling ionic strength effects on cation adsorption at hydrous oxide/solution interfaces. J Colloid Interface Sci 115: 564–572

Hayes KF, Redden G, Ela W, Leckie JO (1991) Surface complexation models: An evaluation of model parameter estimation using FITEQL and oxide mineral titration data. J Colloid Interface Sci 142:448–469, https://doi.org/10.1016/0021-9797(91)90075-J

Hayes KF, Roe AL, Brown GE, Jr., Hodgson K, Leckie JO, Parks GA (1987) In-situ X-ray absorption study of surface complexes: Selenium oxyanions on α-FeOOH. Science 238:783–785

Healy TW, Fuerstenau DW (2007) The isoelectric point/point-of zero-charge of interfaces formed by aqueous solutions and nonpolar solids, liquids, and gases. J Colloid Interface Sci 309:183–188, https://doi.org/10.1016/j.jcis.2007.01.048

Heberling F (2025) Machine learning applications for Surface Complexation Models. Rev Mineral Geochem 91B (in press)

Hiemstra T (2012) Variable charge and electrical double layer of mineral–water interfaces: silver halides versus metal (hydr)oxides. Langmuir 28:15614–15623, https://doi.org/10.1021/la303584a

Hiemstra T (2013) Surface and mineral structure of ferrihydrite. Geochim Cosmochim Acta 10:316–325

Hiemstra T (2018) Ferrihydrite interaction with oxyanions: Silicate polymerization and competition with phosphate, arsenate, and arsenite. Geochim Cosmochim Acta 238: 453–476

Hiemstra T, Van Riemsdijk WH (1990a) The kinetics of dissolution reactions of Al(hydr)oxides: role of surface structure and multisite surface speciation. Trans 14th Int Congr Soil Sci Kyoto Vol. 2:28–33

Hiemstra T, Van Riemsdijk WH (1990b) Multiple activated complex dissolution of metal (hydr)oxides: A thermodynamic approach applied to quartz. J Colloid Interface Sci 136:132–150

Hiemstra T, Van Riemsdijk WH (1991) Physical chemical interpretation of primary charging behaviour of metal (hydr)oxides. Colloids Surfaces 59:7–25

Hiemstra T, Van Riemsdijk WH (1996) A surface structural approach to ion adsorption: The Charge Distribution (CD) model. J Colloid Interface Sci 179:488–508

Hiemstra T, Van Riemsdijk WH (1999) Surface structural ion adsorption modeling of competitive binding of oxyanions by metal (hydr)oxides. J Colloid Interface Sci 210:182–193

Hiemstra T, Van Riemsdijk WH (2000) Fluoride adsorption on goethite in relation to different types of surface sites. J Colloid Interface Sci 225:94–104

Hiemstra T, Van Riemsdijk WH (2006) On the relationship between charge distribution, surface hydration and the structure of the interface of metal hydroxides. J Colloid Interface Sci 301:1–18

Hiemstra T, Van Riemsdijk WH (2007) Adsorption and surface oxidation of Fe(II) on metal (hydr)oxides. Geochim Cosmochim Acta 71:5913–5933

Hiemstra T, Van Riemsdijk WH (2009) A surface structural model for ferrihydrite I: Sites related to primary charge, molar mass, and mass density. Geochim Cosmochim Acta 73:4423–4436

Hiemstra T, Van Riemsdijk WH, Bruggenwert MGM (1987) Proton adsorption mechanism at the gibbsite and aluminum oxide solid/solution interface. Neth J Agric Sci 35:281–293

Hiemstra T, Van Riemsdijk WH, Bolt GH (1989a) Multisite proton adsorption modeling at the solid/solution interface of (hydr)oxides: A new approach. I. Model description and evaluation of intrinsic reaction constants. J Colloid Interface Sci 133:91–104

Hiemstra T, De Wit JCM, Van Riemsdijk WH (1989b) Multisite proton adsorption modeling at the solid/solution interface of (hydr)oxides: A new approach. II. Application to various important (hydr)oxides. J Colloid Interface Sci 133:105–117

Hiemstra T, Venema P, Van Riemsdijk WH (1996) Intrinsic proton affinity of reactive surface groups of metal (hydr) oxides: The bond valence principle. J Colloid Interface Sci 184:680–692

Hiemstra T, Han Yong, Van Riemsdijk WH (1999) Interfacial charging phenomena of aluminum (hydr)oxides. Langmuir 15:5942–5955

Hiemstra T, Rahnemaie R, Van Riemsdijk WH (2004) Surface complexation of carbonate on goethite: IR spectroscopy, structure and charge distribution. J Colloid Interface Sci 278:282–290

Hiemstra T, Van Riemsdijk WH, Rossberg A, Ulrich KU (2009) A surface structural model for ferrihydrite II: adsorption of uranyl and carbonate. Geochim Cosmochim Acta 73:4437–4451

Hiemstra T, Hofmann A, Mendez JC, Bai Y (2025) Surface complexation and reactivity of ferrihydrite in relation to its surface and mineral structure, with applications to natural systems. Rev Mineral Geochem 91A:175–228

Higgins SSA, Eggleston C, Dos Santos Afonso M (1998) Proton and ligand adsorption at silica- and alumina–water interfaces studied by optical Second Harmonic Generation (SHG). Mineral Mag 62A:616–617, https://doi.org/10.1180/minmag.1998.62A

Hingston FJ, Posner AM, Quirk JP (1968) Adsorption of selenite by goethite. *In*: Symposium on adsorption from aqueous solution. Advances in Chemical Series. Vol 70. p 82–90

Hingston FJ, Posner AM, Quirk JP (1972) Anion adsorption by goethite and gibbsite. I. The role of the proton in determining adsorption envelopes. J Soil Sci 23:177–192

Hingston FJ, Atkinson RJ, Posner AM, Quirk JP (1967) Specific adsorption of anions. Nature 215:1459–1461

Horn RG, Clarke DR, Clarkson MT (1988) Direct measurement of surface forces between sapphire crystals in aqueous solutions. J Mater Res 3:413–416, https://doi.org/10.1557/JMR.1988.0413

Horrocks WD, Sudnick DR (1979) Time-resolved europium(III) excitation spectroscopy: A luminescence probe of metal ion binding sites. Science 206:1194–1196, https://doi.org/10.1126/science.505007

Huang C-P, Stumm W (1973) Specific adsorption of cation on hydrous γ-Al_2O_3. J Colloid Interface Sci 43:409–420

Hug SJ (1997) In situ Fourier transform infrared measurements of sulfate adsorption on hematite in aqueous solutions. J Colloid Interface Sci 188:415–422

Huittinen N, Rabung T, Lützenkirchen J, Mitchell SC, Bickmore BR, Lehto J, Geckeis H (2009) Sorption of Cm(III) and Gd(III) onto gibbsite, α-$Al(OH)_3$: A batch and TRLFS study. J Colloid Interface Sci 332:158–164, https://doi.org/10.1016/j.jcis.2008.12.017

Hunter RJ, Ottewill RH, Rowell RL (2013) Zeta Potential in Colloid Science: Principles and Applications. Academic Press

Hütner JI, Conti A, Kugler D, Mittendorfer F, Kresse G, Schmid M, Diebold U, Balajka J (2024) Stoichiometric reconstruction of the Al_2O_3 (0001) surface. Science 385:1241–1244, https://doi.org/10.1126/science.adq4744

Israelachvili JN, Adams GE (1976) Direct measurement of long range forces between two mica surfaces in aqueous KNO_3 solutions. Nature 262:774–776, https://doi.org/10.1038/262774a0

Israelachvili JN, Wennerstrom H (1996) Role of hydration and water structure in biological and colloidal interactions. Nature 379:219–225

Jodin MC, Gaboriaud F, Humbert B (2005) Limitations of potentiometric studies to determine the surface charge of gibbsite gamma-Al(OH)(3) particles. J Colloid Interface Sci 287:581–591

Joly L, Ybert C, Trizac E, Bocquet L (2004) Hydrodynamics within the electric double layer on slipping surfaces. Phys Rev Lett 93:257805, https://doi.org/10.1103/PhysRevLett.93.257805

Jones P, Hockey JA (1971) Infra-red studies of rutile surfaces II. Hydroxylation, hydration and structure of rutile surfaces. Trans Faraday Soc 67:2679–2685

Jordan N, Heberling F, Kelling J, Lützenkirchen J (2025) History, algorithms, model uncertainty, and common pitfalls of traditional SCM fitting procedures. Rev Mineral Geochem 91A:383-412

Kallay N, Colic M, Fuerstenau DW, Jang HM, Matijevic E (1994) Lyotropic effect in surface charge, electrokinetics and coagulation of a rutile suspension. Colloid Polym Sci 272:554–561

Kallay N, Preočanin T, Supljika H (2008) Measurement of surface potential at silver chloride aqueous interface with single-crystal AgCl electrode. J Colloid Interface Sci 327:384–387

Kanematsu M, Waychunas GA, Boily J-F (2018) Silicate binding and precipitation on iron oxyhydroxides. Environ Sci Technol, https://doi.org/10.1021/acs.est.7b04098

Karamalidis AKD, Dzombak DA (2011) Surface Complexation Modeling: Gibbsite. John Wiley & Sons

Kataoka S, Gurau MC, Albertorio F, Holden MA, Lim SM, Yang RD, Cremer PS (2004) Investigation of water structure at the TiO_2/aqueous interface. Langmuir 20:1662–1666

Kavanagh BV, Posner AM, Quirk JP (1975) Effect of polymer adsorption on properties of electrical double-layer. Faraday Discuss 59:242–249, https://doi.org/10.1039/dc9755900242

Kershner RJ, Bullard JW, Cima MJ (2004) Zeta potential orientation dependence of sapphire substrates. Langmuir 20:4101–4108, https://doi.org/10.1021/la036268w

Kimura T, Choppin GR (1994) Luminescence study on determination of the hydration number of Cm(III). J Alloys Compounds 213–214:313–317, https://doi.org/10.1016/0925-8388(94)90921-0

Klaassen A, Liu F, van den Ende D, Mugele F, Siretanu I (2017) Impact of surface defects on the surface charge of gibbsite nanoparticles. Nanoscale 9:4721–4729, https://doi.org/10.1039/C6NR09491K

Klačić T, Katić J, Kovačević D, Namjesnik D, Abdelmonem A, Begović T (2025) Measurements of the electrostatic potential at the mineral/electrolyte interface. Rev Mineral Geochem 91A:295-336

Koca Fındık B, Jafari M, Song LF, Li Z, Aviyente V, Merz KM, Jr. (2024) Binding of phosphate species to Ca^{2+} and Mg^{2+} in aqueous solution. J Chem Theory Comput 20:4298–4307, https://doi.org/10.1021/acs.jctc.4c00218

Kolthoff IM (1935) Adsorption on Ionic Lattices. J Phys Chem 40:1027–1040

Kolthoff IM, Stenger VA (1932) The adsorption of cations from ammoniacal solution by silica gel. J Phys Chem 36:2113–2126, https://doi.org/10.1021/j150338a001

Kolthoff IM, Stenger VA (1933) II. The adsorption of calcium and copper from ammoniacal medium by silica gel. J Phys Chem 38:475–486

Kolthoff IM, Stenger VA (1934) I. The adsorption of alkali hydroxides by silica gel in the presence of ammonia and ammonium salts. J Phys Chem 38:249–258, https://doi.org/10.1021/j150354a001

Kooner ZS (1993) Comparative study of adsorption behavior of copper, lead, and zinc onto goethite in aqueous systems. Environ Geol 21:242–250

Kosmulski M (1998) Positive electrokinetic charge of silica in the presence of chlorides. J Colloid Interface Sci 208:543–545, https://doi.org/10.1006/jcis.1998.5859

Kosmulski M (2001) Chemical properties of material surfaces. Marcel Dekker, New York [etc.]

Kosmulski M (2003) Comment on "Point of zero charge of a corundum–water interface probed with optical second harmonic generation (SHG) and atomic force microscopy (AFM): new approaches to oxide surface charge" by A. G. Stack, S. R. Higgins, and C. M. Eggleston. Geochim Cosmochim Acta 67:319–320, https://doi.org/10.1016/S0016-7037(02)01034-7

Kosmulski M (2004) pH-dependent surface charging and points of zero charge II. Update. J Colloid Interface Sci 275:214–224, https://doi.org/10.1016/j.jcis.2004.02.029

Kosmulski M (2006) pH-dependent surface charging and points of zero charge III. Update. J Colloid Interface Sci 298:730–741, https://doi.org/10.1016/j.jcis.2006.01.003

Kosmulski M (2009) pH-dependent surface charging and points of zero charge. IV. Update and new approach. J Colloid Interface Sci 337:439–448, https://doi.org/10.1016/j.jcis.2009.04.072

Kosmulski M (2011) The pH-dependent surface charging and points of zero charge V. Update. J Colloid Interface Sci 353:1–15, https://doi.org/10.1016/j.jcis.2010.08.023

Kosmulski M (2014) The pH dependent surface charging and points of zero charge. VI. Update. J Colloid Interface Sci 426:209–212, https://doi.org/10.1016/j.jcis.2014.02.036

Kosmulski M (2016) Isoelectric points and points of zero charge of metal (hydr)oxides: 50 years after Parks' review. Adv Colloid Interface Sci 238:1–61, https://doi.org/10.1016/j.cis.2016.10.005

Kosmulski M (2018) The pH dependent surface charging and points of zero charge. VII. Update. Adv Colloid Interface Sci 251:115–138, https://doi.org/10.1016/j.cis.2017.10.005

Kosmulski M (2020) The pH dependent surface charging and points of zero charge. VIII. Update. Adv Colloid Interface Sci 275:102064, https://doi.org/10.1016/j.cis.2019.102064

Kosmulski M (2021) The pH dependent surface charging and points of zero charge. IX. Update. Adv Colloid Interface Sci 296:102519, https://doi.org/10.1016/j.cis.2021.102519

Kosmulski M (2023) The pH dependent surface charging and points of zero charge. X. Update. Adv Colloid Interface Sci 319:102973, https://doi.org/10.1016/j.cis.2023.102973

Kosmulski M, Maczka E, Rosenholm JB (2002) Isoelectric points of metal oxides at high ionic strengths. J Phys Chem B 106:2918–2921, https://doi.org/10.1021/jp013942e

Kummert R, Stumm W (1980) The surface complexation of organic acids on hydrous γ-Al_2O_3. J Colloid Interface Sci 75:373–385

Kupcik T (2011) Wechselwirkung von dreiwertigen Lanthaniden und Actiniden mit Aluminiumoxiden und -hydroxiden. PhD dissertation, Ruprecht-Karls-Universität Heidelberg

Kupcik T, Rabung T, Lützenkirchen J, Finck N, Geckeis H, Fanghänel T (2016) Macroscopic and spectroscopic investigations on Eu(III) and Cm(III) sorption onto bayerite (β-$Al(OH)_3$) and corundum (α-Al_2O_3). J Colloid Interface Sci 461:215–224, https://doi.org/10.1016/j.jcis.2015.09.020

Kurbatov IW (1931) Adsorption of thorium X by ferric hydroxide at different pH. J Phys Chem 36:1241–1247, https://doi.org/10.1021/j150334a014

Kurbatov MH (1949) Rate of adsorption of barium ions in extreme dilution, by hydrous ferric oxide. J Am Chem Soc 71:858–863

Kurbatov JD, Kulp JL, Mack E (1945) Adsorption of strontium and barium ions and their exchange on hydrous ferric oxide. J Am Chem Soc 67:1923–1929, https://doi.org/10.1021/ja01227a014

Kurbatov MH, Kurbatov JD (1947) Absorption isotherm for determination of barium in quantities as low as 10^{-10} gram atom. J Am Chem Soc 69:438–441, https://doi.org/10.1021/ja01194a071

Kurbatov MH, Wood GB, Kurbatov JD (1951a) Application of the mass law to adsorption of divalent ions on hydrous ferric oxide. J Chem Phys 19:258–259, https://doi.org/10.1063/1.1748186

Kurbatov MH, Wood GB, Kurbatov JD (1951b) Isothermal adsorption of cobalt from dilute solutions. J Phys Chem 55:1170–1182

Leung K, Criscenti LJ (2012) Predicting the acidity constant of a goethite hydroxyl group from first principles. J Phys Condens Matter 24:124105

Leung K, Nielsen IMB, Criscenti LJ (2009) Elucidating the bimodal acid–base behavior of the water–silica interface from first principles. J Am Chem Soc 131:18358–18365, https://doi.org/10.1021/ja906190t

Levine PL, Bell GM (1980) Forces between plates with surface dipole layers. J Colloid Interface Sci 74:530–548, https://doi.org/10.1016/0021-9797(80)90222-2

Li JQ, Sun Y, Cheng J (2023) Theoretical investigation on water adsorption conformations at aqueous anatase TiO_2/ water interfaces. J Mater Chem A 11:943–952, https://doi.org/10.1039/d2ta07994a

Liu XD, Cheng J, Sprik M, Lu XC, Wang RC (2013) Understanding surface acidity of gibbsite with first principles molecular dynamics simulations. Geochim Cosmochim Acta 120:487–495, https://doi.org/10.1016/j.gca.2013.06.043

Livi KJT, Schaffer B, Azzolini D, Seabourne CR, Hardcastle TP, Scott AJ, Hazen RM, Erlebacher JD, Brydson R, Sverjensky DA (2013) Atomic-scale surface roughness of rutile and implications for organic molecule adsorption. Langmuir 29:6876–6883, https://doi.org/10.1021/la4005328

Livi KJT, Villalobos M, Leary R, Varela M, Barnard J, Villacis-Garcia M, Zanella R, Goodridge A, Midgley P (2017) Crystal face distributions and surface site densities of two synthetic goethites: implications for adsorption capacities as a function of particle size. Langmuir 33:8924–8932, https://doi.org/10.1021/acs.langmuir.7b01814

Livi KJT, Villalobos M, Ramasse Q, Brydson R, Salazar-Rivera HS (2023) Surface site density of synthetic goethites and its relationship to atomic surface roughness and crystal size. Langmuir 39:556–562, https://doi.org/10.1021/acs.langmuir.2c02818

Lövgren L, Sjöberg S, Schindler PW (1990) Acid/base reactions and aluminium (III) complexation at the surface of goethite. Geochim Cosmochim Acta 54:1301–1306

Lützenkirchen J, Boily JF, Lovgren L, Sjoberg S (2002) Limitations of the potentiometric titration technique in determining the proton active site density of goethite surfaces. Geochim Cosmochim Acta 66:3389–3396

Lützenkirchen J (1998) Comparison of 1-pK and 2-pK versions of surface complexation theory by the goodness of fit in describing surface charge data of (hydr)oxides. Environ Sci Technol 32:3149–3154

Lützenkirchen J (2013) Specific ion effects at two single-crystal planes of sapphire. Langmuir 29:7726–7734, https://doi.org/10.1021/la401509y

Lützenkirchen J, Preočanin T, Kallay N (2008) A macroscopic water structure based model for describing charging phenomena at inert hydrophobic surfaces in aqueous electrolyte solutions. Phys Chem Chem Phys 10:4946–4955, https://doi.org/10.1039/B807395C

Lützenkirchen J, Zimmermann R, Preočanin T, Filby A, Kupcik T, Küttner D, Abdelmonem A, Schild D, Rabung T, Plaschke M, Brandenstein F (2010) An attempt to explain bimodal behaviour of the sapphire *c*-plane electrolyte interface. Adv Colloid Interface Sci 157:61–74, https://doi.org/10.1016/j.cis.2010.03.003

Lützenkirchen J, Marsac R, Kulik DA, Payne TE, Xue Z, Orsetti S, Haderlein SB (2015) Treatment of multi-dentate surface complexes and diffuse layer implementation in various speciation codes. Appl Geochem 55:128–137, https://doi.org/10.1016/j.apgeochem.2014.07.006

Lyklema J, Overbeek JTG (1961) Electrochemistry of silver iodide. The capacity of the double layer at the silver iodide–water interface. J Colloid Interface Sci 16:595–608

Machesky ML, Wesolowski DJ, Palmer DA, Ichiro-Hayashi K (1998) Potentiometric titrations of rutile suspensions to 250 °C. J Colloid Interface Sci 200:298–309

Machesky ML, Předota M, Wesolowski DJ, Vlcek L, Cummings PT, Rosenqvist J, Ridley MK, Kubicki JD, Bandura AV, Kumar N, Sofo JO (2008) Surface Protonation at the Rutile (110) Interface: Explicit incorporation of solvation structure within the refined MUSIC model framework. Langmuir 24:12331–12339

Malloum A, Fifen JJ, Conradie J (2021) Determination of the absolute solvation free energy and enthalpy of the proton in solutions. Journal of Molecular Liquids 322, https://doi.org/10.1016/j.molliq.2020.114919

Manceau A, Charlet L, Boisset MC, Didier B, Spadini L (1992) Sorption and speciation of heavy metals on hydrous Fe and Mn oxides. From microscopic to macroscopic. Appl Clay Sci 7:201–223

Manning BA, Fendorf SE, Goldberg S (1998) Surface structure and stability of arsenic (III) on goethite: spectroscopic evidence for innersphere complexes. Environ Sci Technol 32:2383–2388

Marcus Y (1988) Ionic radii in aqueous solutions. Chem Rev 88:1475–1498, https://doi.org/10.1021/cr00090a003

Marsac R, Banika NL, Lützenkirchen J, Diascorn A, Bender K, Marquardt CM, Geckeis H (2017) Sorption and redox speciation of plutonium at the illite surface under highly saline conditions. J Colloid Interface Sci 485:59–64, https://doi.org/10.1016/j.jcis.2016.09.013

Marshall K, Ridgewell GL, Rochester CH, Simpson J (1974) The acidity of surface silanol group on silica. Chem Ind 19:775–776

Martinez RJ, Villalobos M, Loredo-Jasso AU, Cruz-Valladares AX, Mendoza-Flores A, Salazar-Rivera H, Cruz-Romero D (2023) Towards building a unified adsorption model for goethite based on direct measurements of crystal face compositions: I. Acidity behavior and As (V) adsorption. Geochim Cosmochim Acta 354:252–262, https://doi.org/10.1016/j.gca.2023.06.021

Mathur SSD, Dzombak DA (2006) Surface complexation modeling: Goethite. *In*: Interface Science and Technology. Lützenkirchen J (ed) Elsevier, p 443–468

Mcdonald RS (1958) Surface functionality of amorphous silica by infrared spectroscopy. J Phys Chem 62:1168–1178

McLaren ADB, Babcock KL (1959) Some characteristics of enzyme reactions at surfaces. *In*: Subcellular particles. T. H, (ed) New York: Ronald Press;, p 23–36

Mendez JC, Hiemstra T (2020a) Ternary complex formation of phosphate with ca and mg ions binding to ferrihydrite: experiments and mechanisms. ACS Earth Space Chem 4:545–557, https://doi.org/10.1021/acsearthspacechem.9b00320

Mendez JC, Hiemstra T (2020b) High and low affinity sites of ferrihydrite for metal ion adsorption: Data and modeling of the alkaline-earth ions Be, Mg, Ca, Sr, Ba, and Ra. Geochim Cosmochim Acta 286:289–305, https://doi.org/10.1016/j.gca.2020.07.032

Mendez JC, Hiemstra T (2020c) Surface area of ferrihydrite consistently related to primary surface charge, ion pair formation, and specific ion adsorption. Chem Geol 532:119304, https://doi.org/10.1016/j.chemgeo.2019.119304

Michael HL, Williams DJA (1984) Electrochemical properties of quartz. J Electroanal Chem Interfacial Electrochem 179:131–139, https://doi.org/10.1016/S0022-0728(84)80282-X

Michel FM, Ehm L, Antao SM, Lee PL, Chupas PJ, Liu G, Strongin DR, Schoonen MAA, Phillips BL, Parise JB (2007) The structure of ferrihydrite, a nanocrystalline material. Science 316:1726–1729

Michel FM, Barron V, Torrent J, Morales MP, Serna CJ, Boily JF, Liu QS, Ambrosini A, Cismasu AC, Brown GE (2010) Ordered ferrimagnetic form of ferrihydrite reveals links among structure, composition, and magnetism. PNAS 107:2787–2792

Molleman B, Hiemstra T (2015) Surface structure of silver nanoparticles as a model for understanding the oxidative dissolution of silver ions. Langmuir 31:13361–13372, https://doi.org/10.1021/acs.langmuir.5b03686

Molleman B, Hiemstra T (2017) Time, pH, and size dependency of silver nanoparticle dissolution: the road to equilibrium. Environ Sci Nano 4:1314–1327, https://doi.org/10.1039/c6en00564k

Morag J, Dishon M, Sivan U (2013) The governing role of surface hydration in ion specific adsorption to silica: An AFM-based account of the Hofmeister Universality and its reversal. Langmuir 29:6317–6322, https://doi.org/10.1021/la400507n

Morel F, Morgan J (1972) Numerical method for computing equilibriums in aqueous chemical systems. Environ Sci Technol 6:58–67, https://doi.org/10.1021/es60060a006

Morelová N, Finck N, Lützenkirchen J, Schild D, Dardenne K, Geckeis H (2020) Sorption of americium/europium onto magnetite under saline conditions: Batch experiments, surface complexation modelling and X-ray absorption spectroscopy study. J Colloid Interface Sci 561:708–718, https://doi.org/10.1016/j.jcis.2019.11.047

Murashov VV, Demchuk E (2005) Surface sites and unrelaxed surface energies of tetrahedral silica polymorphs and silicate. Surf Sci 595:6–19, https://doi.org/10.1016/j.susc.2005.07.030

Neumann J, Brinkmann H, Britz S, Lützenkirchen J, Bok F, Stockmann M, Brendler V, Stumpf T, Schmidt M (2021) A comprehensive study of the sorption mechanism and thermodynamics of *f*-element sorption onto K-feldspar. J Colloid Interface Sci 591:490–499, https://doi.org/10.1016/j.jcis.2020.11.041

Nilsson N, Persson P, Lovgren L, Sjoberg S (1996) Competitive surface complexation of *o*-phthalate and phosphate on goethite (alpha-FeOOH) particles. Geochim Cosmochim Acta 60:4385–4395, https://doi.org/10.1016/s0016-7037(96)00258-x

Ninham BW, Parsegian VA (1971) Electrostatic potential between surfaces bearing ionizable groups in ionic equilibrium with physiologic saline solution. J Theor Biol 31:405–428, https://doi.org/10.1016/0022-5193(71)90019-1

Ohshima H (2006) Chapter 3 - Diffuse double layer equations for use in surface complexation models: Approximations and limits. *In*: Interface Science and Technology. Vol 11. Lützenkirchen J, (ed) Elsevier, p 67–87

Ohshima H, Healy TW, White LR (1982) Accurate analytic expressions for the surface charge density/surface potential relationship and double-layer potential distribution for a spherical colloidal particle. J Colloid Interface Sci 90:17–26, https://doi.org/10.1016/0021-9797(82)90393-9

Ong S, Zhao X, Eisenthal KB (1992) Polarization of water molecules at a charged interface: second harmonic studies of the silica/water interface. Chem Phys Lett 191:327–335, https://doi.org/10.1016/0009-2614(92)85309-X

Ostergren JD, Bargar JR, Brown GE, Jr., Parks GA (1999b) Combined EXAFS and FTIR investigation of sulfate and carbonate effects on Pb(II)sorption to goethite (α-FeOOH). J Synchotron Rad 6:645–647

Ostroverkhov V, Waychunas GA, Shen YR (2005) New information on water interfacial structure revealed by phase-sensitive surface spectroscopy. Phys Rev Lett 94:046102

Parfitt RL, Russell JD (1977) Adsorption on hydrous oxides. IV. Mechanisms of adsorption of various ions on goethite. J Soil Sci 28:297–305

Parfitt RL, Atkinson RJ, Smart RSC (1975) The mechanism of phosphate fixation by iron oxides. Soil Sci Soc Am Proc 39:837–841

Parfitt RL, Russell JD, Farmer VC (1976) Confirmation of the structures of goethite (α-FeOOH) and phosphated goethite by infrared spectroscopy. J Chem Soc Faraday Trans I 72:1082–1087

Parfitt RL, Fraser AR, Russell JD, Farmer VC (1977a) Adsorption on hydrous oxides. I. Oxalate, benzoate on goethite. J Soil Sci 28:29–39

Parfitt RL, Fraser AR, Russell JD, Farmer VC (1977b) Adsorption on hydrous oxides. II. Oxalate, benzoate and phosphate on gibbsite. J Soil Sci 28:40–47

Parfitt RL, Fraser AR, Farmer VC (1977c) Adsorption on hydrous oxides. III Fulvic acid and humic acid on goethite, gibbsite and imogolite. J Soil Sci 28:289–296

Parkman RH, Charnock JM, Bryan ND, Vaughan DJ (1999) Reactions of copper and cadmium ions in aqueous solution with goethite, lepidocrocite, mackinawite, and pyrite. Am Mineral 84:407–419

Parks GA (1965) The isoelectric points of solid oxides, solid hydroxides, and aqueous hydroxo complex systems. Chem Rev 65:177–198

Parks GA (1967) Aqueous surface chemistry of oxides and complex oxide minerals. Isoelectric point and zero point of charge. *In*: Equilibrium Concepts in Natural Water Systems. Vol 67. Gould RF, (ed) American Chemical Society,

Parks GA, De Bruyn PL (1962) The zero point of charge of oxides. J Phys Chem 66:967–973

Partana CF, Suwardi, Salim A (2019) Structure and dynamics of Hg^{2+} in aqueous solution: an Ab Initio QM/MM molecular dynamics study. J Phys Conf Ser 1156:012012, https://doi.org/10.1088/1742-6596/1156/1/012012

Pashley RM, Israelachvili JN (1984) Molecular layering of water in thin films between mica surfaces and its relation to hydration forces. J Colloid Interface Sci 101:511–523

Pauling L (1929) The principles determining the structure of complex ionic crystals. J Am Chem Soc 51:1010–1026

Peak D, Ford RG, Sparks DL (1999) An in-Situ ATR-FTIR investigation of sulfate bonding mechanisms on goethite. J Colloid Interface Sci 218:289–299

Penners NHG, Koopal LK, Lyklema J (1986) Interfacial electrochemistry of haematite (α-Fe_2O_3): homodisperse and heterodisperse sols. Colloids Surfaces 21:457–468, https://doi.org/10.1016/0166-6622(86)80109-3

Peri JB, Hensley SL, Jr. (1968) The surface structure of silica gel. J Phys Chem 72:2926–2933

Perona MJ, Leckie JO (1985) Proton stoichiometry for the adsorption of cations on oxide surfaces. J Colloid Interface Sci 106:65–69

Peskleway CD, Henderson GS, Wicks FJ (2003) Dissolution of gibbsite: Direct observations using fluid cell atomic force microscopy. Amer Min 88:18–26

Pfeiffer-Laplaud M, Costa D, Tielens FJ, Gaigeot M-P, Sulpizi M (2015) Bimodal acidity at the amorphous silica/water interface. J Phys Chem C 119:27354–27362

Pfeiffer-Laplaud M, Gaigeot M-P, Sulpizi M (2016) pK_a at quartz/electrolyte interfaces. J Phys Chem Lett 7:3229–3234, https://doi.org/10.1021/acs.jpclett.6b01422

Porus M, Labbez C, Maroni P, Borkovec M (2011) Adsorption of monovalent and divalent cations on planar water–silica interfaces studied by optical reflectivity and Monte Carlo simulations. J Chem Phys 135:064701, https://doi.org/10.1063/1.3622858

Předota M, Bandura AV, Cummings PT, Kubicki JD, Wesolowski DJ, Chialvo AA, Machesky ML (2004) Electric double layer at the rutile (110) surface. 1. Structure of surfaces and interfacial water from molecular dynamics by use of ab initio potentials. J Phys Chem B 108:12049–12060

Předota M, Cummings PT, Wesolowski DJ (2007) Electric double layer at the rutile (110) surface. 3. Inhomogeneous viscosity and diffusivity measurement by computer simulations. J Phys Chem C 111:3071–3079

Předota M, Machesky ML, Wesolowski DJ (2016) Molecular origins of the Zeta potential. Langmuir 32:10189–10198, https://doi.org/10.1021/acs.langmuir.6b02493

Prélot B, Villiéras F, Pelletier M, Gérard G, Gaboriaud F, Ehrhardt JJ, Perrone J, Fedoroff M, Jeanjean J, Lefèvre G, Mazerolles L (2003) Morphology and surface heterogeneities in synthetic goethites. J Colloid Interface Sci 261:244–254

Preočanin T, Namjesnik D, Brown MA, Lützenkirchen J (2017) The relationship between inner surface potential and electrokinetic potential from an experimental and theoretical point of view. Environ Chem 14:295–309, https://doi.org/10.1071/en16216

Rabung T, Stumpf T, Geckeis H, Klenze R, Kim JI (2000) Sorption of Am(III) and Eu(III) onto γ-alumina:: experiment and modelling. Radiochim Acta 88:711–716, https://doi.org/10.1524/ract.2000.88.9-11.711

Rabung T, Schild D, Geckeis H, Klenze R, Fanghänel T (2004) Cm(III) sorption onto sapphire (α-Al_2O_3) single crystals. J Phys Chem B 108:17160–17165, https://doi.org/10.1021/jp040342h

Rahnemaie R, Hiemstra T, Van Riemsdijk WH (2006) A new structural approach for outersphere complexation, tracing the location of electrolyte ions. J Colloid Interface Sci 293:312–321

Rahnemaie R, Hiemstra T, Van Riemsdijk WH (2007) Geometry, charge distribution and surface speciation of phosphate on goethite. Langmuir 23:3680–3689

Randall SR, Sherman DM, Ragnarsdottir KV (1999) The mechanism of cadmium surface complexation on iron oxyhydroxide minerals. Geochim Cosmochim Acta 63:2971–2987

Ray KC, Sen PK (1974) Correlation of zero points of charge of solid oxides and hydroxides with partial charge of surface oxygen atom. Indian J Chem 12:170–173

Ridley MK, Machesky ML, Wesolowski DJ, Palmer DA (2004) Modeling the surface complexation of calcium at the rutile–water interface to 250 degrees C. Geochim Cosmochim Acta 68:239–251

Ridley MK, Hiemstra T, van Riemsdijk WH, Machesky ML (2009) Inner-sphere complexation of cations at the rutile–water interface: A concise surface structural interpretation with the CD and MUSIC model. Geochim Cosmochim Acta 73:1841–1856, https://doi.org/10.1016/j.gca.2009.01.004

Ridley MK, Lischka H, Tunega D, Aquino AJA (2019) Solvent effect on Al(III) hydrolysis constants from density functional theory. Molecular Physics 117:1507–1518, https://doi.org/10.1080/00268976.2019.1567846

Ridley MK, Hiemstra T, Machesky ML, Wesolowski DJ, van Riemsdijk WH (2012) Surface speciation of yttrium and neodymium sorbed on rutile: Interpretations using the charge distribution model. Geochim Cosmochim Acta 95:227–240, https://doi.org/10.1016/j.gca.2012.07.033

Rietra RPJJ, Hiemstra T, Van Riemsdijk WH (1999) The relationship between molecular structure and ion adsorption on variable charge minerals. Geochim Cosmochim Acta 63:3009–3015

Rietra RPJJ, Hiemstra T, Van Riemsdijk WH (2000) Electrolyte anion affinity and its effect on oxyanion adsorption on goethite. J Colloid Interface Sci 229:199–206

Rietra RPJJ, Hiemstra T, Van Riemsdijk WH (2001a) Interaction of calcium and phosphate adsorption on goethite. Environ Sci Technol 35:3369–3374

Rietra RPJJ, Hiemstra T, Van Riemsdijk WH (2001b) Comparison of selenate and sulfate adsorption on goethite. J Colloid Interface Sci 240:384–390

Rochester CH, Topham SA (1979) Infrared study of surface hydroxyl on goethite. J Chem Soc Faraday Trans I 75:591–602

Röntgen WC (1898) Ueber eine neue Art von Strahlen. Ann Phys 300:12–17, https://doi.org/10.1002/andp.18983000103

Rose J, Flank A, Masion A, Bottero J, Elmerich P (1997) Nucleation and growth mechanisms of Fe oxyhydroxide in the presence of PO_4 ions. 2. P *K*-edge EXAFS study. Langmuir 13:1827–1834

Rosenqvist J, Persson P, Sjoberg S (2002) Protonation and charging of nanosized gibbsite (alpha-Al(OH)(3)) particles in aqueous suspension. Langmuir 18:4598–4604

Rosenqvist J, Machesky ML, Vlcek L, Cummings PT, Wesolowski DJ (2009) Charging properties of cassiterite (α-SnO_2) surfaces in NaCl and RbCl ionic media. Langmuir 25:10852–10861, https://doi.org/10.1021/la901396w

Rowlands WN, O'Brien WR, Hunter RJ, Patrick V (1997) Surface properties of aluminum hydroxide at high salt concentration. J Colloid Interface Sci 188:325–335

Rustad JR, Felmy AR, Hay BP (1996a) Molecular statics calculations for iron oxide and oxyhydroxide minerals: Toward a flexible model of the reactive mineral–water interface. Geochim Cosmochim Acta 60:1553–1562

Rustad JR, Wasserman E, Felmy AR, Wilke C (1998) Molecular dynamics study of proton binding to silica surfaces. J Colloid Interface Sci 198:119–129

Sahai N (2000) Estimating adsorption enthalpies and affinity sequences of monovalent electrolyte ions on oxide surfaces in aqueous solution. Geochim Cosmochim Acta 64:3629–3641, https://doi.org/10.1016/s0016-7037(00)00431-2

Sahai N, Sverjensky DA (1997) Evaluation of internally consistent parameters for the triple-layer model by the systematic analysis of oxide surface titration data. Geochim Cosmochim Acta 61:2801–2826

Scheinost AC, Abend S, Pandya KI, Sparks DL (2001) Kinetic controls on Cu and Pb sorption by ferrihydrite. Environ Sci Technol 35:1090–1096, https://doi.org/10.1021/es000107m

Schindler PW, Gamsjäger H (1972) Acid–base reactions of the TiO_2 (anatase)–water interface and the point of zero charge of TiO_2 suspensions. Kolloid Polymere 250:759–763

Schindler PW, Kamber HR (1968) Die Acidität von Silanolgruppen. Helv Chim Acta 15:1781–1786

Schlegel ML, Nagy KL, Fenter P, Sturchio NC (2002) Structures of quartz (10(1)-0) and (10(1)-1)–water interfaces determined by X-ray reflectivity and atomic force microscopy of natural growth surfaces. Geochim Cosmochim Acta 66:3037–3054, https://doi.org/10.1016/s0016-7037(02)00912-2

Schnurr A, Marsac R, Rabung T, Lützenkirchen J, Geckeis H (2015) Sorption of Cm(III) and Eu(III) onto clay minerals under saline conditions: Batch adsorption, laser-fluorescence spectroscopy and modeling. Geochim Cosmochim Acta 151:192–202, https://doi.org/10.1016/j.gca.2014.11.011

Schudel M, Behrens SH, Holthoff H, Kretzschmar R, Borkovec M (1997) Absolute aggregation rate constants of hematite particles in aqueous suspensions: a comparison of two different surface morphologies. J Colloid Interface Sci 196:241–253

Seigel MD, Ward DB, Bryan C (1995) Batch and column studies of adsorption of Li, Ni and Br by a reference sand for contaminant transport experiments. Yuca Mountain Characterization Report SAND 95-0591, Sandia National Labs, https://doi.org/10.2172/114552

Shchukarev A, Sjöberg S (2005) XPS with fast-frozen samples: A renewed approach to study the real mineral/solution interface. Surf Sci 584:106–112, https://doi.org/10.1016/j.susc.2005.01.060

Shen YR, Ostroverkhov V (2006) Sum-frequency vibrational spectroscopy on water interfaces: Polar orientation of water molecules at interfaces. Chem Rev 106:1140–1154

Shiao S-Y, Meyer RE (1981) Adsorption of inorganic ions on alumina from salts solutions: Correlations of distribution coefficients with uptake of salt. J Inorg Nucl Chem 43:3301–3307

Shimizu K, Shchukarev A, Kozin PA, Boily JF (2012) X-ray photoelectron spectroscopy of fast-frozen hematite colloids in aqueous solutions. 4. Coexistence of alkali metal (Na^+, K^+, Rb^+, Cs^+) and chloride ions. Surf Sci 606:1005–1009, https://doi.org/10.1016/j.susc.2012.02.018

Sigg L, Stumm W (1981) The interaction of anions and weak acids with the hydrous goethite (α-FeOOH) surface. Colloids Surfaces 2:101–117

Siretanu I, Ebeling D, Andersson MP, Stipp SLS, Philipse A, Stuart MC, van den Ende D, Mugele F (2014) Direct observation of ionic structure at solid–liquid interfaces: a deep look into the Stern Layer. Sci Rep 4:4956, https://doi.org/10.1038/srep04956

Siu WM, Cobbold RSC (1979) Basic properties of the electrolyte—SiO_2–Si system: Physical and theoretical aspects. IEEE Trans Electron Devices 26:1805–1815, https://doi.org/10.1109/T-ED.1979.19690

Smit W, Holten CLM (1980) Zeta potential and radiotracer adsorption measurements on EFG α-Al_2O_3 single crystal in NaBr solution. J Colloid Interface Sci 78:1–14

Smith W, Pouvreau M, Rosso K, Clark AE (2022) pH dependent reactivity of boehmite surfaces from first principles molecular dynamics. Phys Chem Chem Phys 24:14177–14186, https://doi.org/10.1039/d2cp00534d

Sonnefeld J (1995) Surface charge density on spherical silica particles in aqueous alkali chloride solutions. Part 2 Evaluation of the surface charge density constants. Colloid Polym Science 275:932–938

Sonnefeld J, Göbel A, Vogelsberger W (1995) Surface charge density on spherical silica particles in aqueous alkali chloride solutions. Part 1 Experimental results. Colloid Polym Science 275:926–931

Spadini L, Manceau A, Schindler PW, Charlet L (1994) Structure and stability of Cd^{2+} surface complexes on ferric oxides. 1 Results from EXAFS spectroscopy. J Colloid Interface Sci 168:73–86

Sposito G (1984) The Surface Chemistry of Soils. Oxford University Press, New York

Sposito G (1998) On points of zero charge. Environ Sci Technol 32:2815–2819, https://doi.org/10.1021/es9802347

Sprycha R (1984) Surface charge and adsorption of background electrolyte ions at anatase/electrolyte interface. J Colloid Interface Sci 102:173–185

Sprycha R (1989b) Electrical double layer at alumina/electrolyte interface II. Adsorption of supporting electrolyte ions. J Colloid Interface Sci 127:12–25

Stanton J, Maatman RW (1963) The reaction between aqueous uranyl ion and the surface of silica gel. J Colloid Sci 18:132–146, https://doi.org/10.1016/0095-8522(63)90003-5

Steinhardt J, Fugitt CH, Harris M (1940) Combination of wool protein with acid and base: The effect of temperature on the titration curve. Textile Res 11:72–94, https://doi.org/10.1177/004051754001100203

Stern O (1924) Zur Theory der electrolytischen Doppelschicht. Z Electrochem 30:508–516

Stolze L, Wagner JB, Damsgaard CD, Rolle M (2020) Impact of surface complexation and electrostatic interactions on pH front propagation in silica porous media. Geochim Cosmochim Acta 277:132–149, https://doi.org/10.1016/j.gca.2020.03.016

Stumm W, Huang CP, Jenkins SR (1970) Specific chemical interaction affecting the stability of dispersed systems. Croat Chem Acta 42:223–245

Stumpf T, Hennig C, Bauer A, Denecke MA, Fanghänel T (2004) An EXAFS and TRLFS study of the sorption of trivalent actinides onto smectite and kaolinite. Radiochim Acta 92:133–138, https://doi.org/10.1524/ract.92.3.133.30487

Su S, Siretanu I, van den Ende D, Mei B, Mul G, Mugele F (2021) Facet-dependent surface charge and hydration of semiconducting nanoparticles at variable pH. Adv Mater 33:2106229, https://doi.org/10.1002/adma.202106229

Su S, Siretanu I, van den Ende D, Mei B, Mul G, Mugele F (2024) Nanometer-resolved operando photo-response of faceted $BiVO_4$ semiconductor nanoparticles. J Am Chem Soc 146:2248–2256, https://doi.org/10.1021/jacs.3c12666

Sulpizi M, Gaigeot MP, Sprik M (2012) The silica–water interface: how the silanols determine the surface acidity and modulate the water properties. J Chem Theory Comput 8:1037–1047, https://doi.org/10.1021/ct2007154

Sverjensky DA (1994) Zero-point-of-charge prediction from crystal chemistry and solvation theory. Geochim Cosmochim Acta 58:3123–3129

Sverjensky DA (2005) Prediction of surface charge on oxides in salt solutions: Revisions for 1:1 (M+L-) electrolytes. Geochim Cosmochim Acta 69:225–257

Tejedor-Tejedor MI, Anderson MA (1990) Protonation of phosphate on the surface of goethite as studied by CIR-FTIR and electrophoretic mobility. Langmuir 6:602–611

Thomas JM (2012) The birth of X-ray crystallography. Nature 491:186–187, https://doi.org/10.1038/491186a

Tian CS, Shen YR (2009) Structure and charging of hydrophobic material/water interfaces studied by phase-sensitive sum-frequency vibrational spectroscopy. PNAS 106:15148–15153, https://doi.org/10.1073/pnas.0901480106

Toney MF, Howard JN, Richer J, Borges GL, Gordon JG, Melroy OR, Wiesler DG, Yee D, Sorensen LB (1995) Distribution of water molecules at Ag(111)/electrolyte interface as studied with surface X-ray scattering. Surf Sci 335 326–332

Tournassat C, Tinnacher RM, Grangeon S, Davis JA (2018) Modeling uranium(VI) adsorption onto montmorillonite under varying carbonate concentrations: A surface complexation model accounting for the spillover effect on surface potential. Geochim Cosmochim Acta 220:291–308, https://doi.org/10.1016/j.gca.2017.09.049

Trainor TP, Templeton AS, Brown GE, Parks GA (2002) Application of the long-period X-ray standing wave technique to the analysis of surface reactivity:: Pb(II) sorption at α-Al_2O_3/aqueous solution interfaces in the presence and absence of Se(VI). Langmuir 18:5782–5791, https://doi.org/10.1021/la015740f

Trefalt G, Behrens SH, Borkovec M (2016) Charge regulation in the electrical double layer: ion adsorption and surface interactions. Langmuir 32:380–400, https://doi.org/10.1021/acs.langmuir.5b03611

Van Hal REG, Eijkel JCT, Bergveld P (1996) A general model to describe the electrostatic potential at the electrolyte oxide interfaces. Adv Colloid Interface Sci 69:31–62

Velikonja A, Gongadze E, Kralj-Iglič V, Iglič A (2014) Charge dependent capacitance of stern layer and capacitance of electrode/electrolyte interface. Int J Electrochem Sci 9:5885–5894, https://doi.org/10.1016/S1452-3981(23)10856-X

Venema P, Hiemstra T, Van Riemsdijk WH (1996a) Comparison of different site binding models for cation sorption: Description of pH dependency, salt dependency, and cation–proton exchange. J Colloid Interface Sci 181:45–59

Venema P, Hiemstra T, Van Riemsdijk WH (1996b) Multisite adsorption of cadmium on goethite. J Colloid Interface Sci 183:515–527

Venema P, Hiemstra T, Van Riemsdijk WH (1998) Intrinsic proton affinity of reactive surface groups of metal (Hydr) oxides: Application to iron (hydr) oxides. J Colloid Interface Sci 198:282–295

Villalobos M, Leckie JO (2000) Carbonate adsorption on goethite under closed and open CO_2 systems. Geochim Cosmochim Acta 64:3787–3802

Villalobos M, Cheney MA, Alcaraz-Cienfuegos J (2009) Goethite surface reactivity: II. A microscopic site-density model that describes its surface area-normalized variability. J Colloid Interface Sci 336:412–422, https://doi.org/10.1016/j.jcis.2009.04.052

Vlcek L, Zhang Z, Machesky ML, Fenter P, Rosenqvist J, Wesolowski DJ, Anovitz LM, Předota M, Cummings PT (2007) Electric double layer at metal oxide surfaces: Static properties of the cassiterite–water interface. Langmuir 23:4925–4937

Vogel P, Palberg T (2024) Electrokinetic effects of ambient and excess carbonization of dielectric surfaces in aqueous environments. J Colloid Interface Sci 656:280–288, https://doi.org/10.1016/j.jcis.2023.10.056

Vogel P, Möller N, Qaisrani MN, Bista P, Weber SAL, Butt H-J, Liebchen B, Sulpizi M, Palberg T (2022) Charging of dielectric surfaces in contact with aqueous electrolytes–The influence of CO_2. J Am Chem Soc 144:21080–21087, https://doi.org/10.1021/jacs.2c06793

von Helmholtz H (1879) Studien über electrische Grenzschichten. Ann Physik Chem 7:337–382

Vuceta J, Morgan JJ (1978) Chemical modeling of trace-metals in fresh waters–Role of complexation and adsorption. Environ Sci Technol 12:1302–1309, https://doi.org/10.1021/es60147a007

Wagner M, Meyer B, Setvin M, Schmid M, Diebold U (2021) Direct assessment of the acidity of individual surface hydroxyls. Nature 592:722–725, https://doi.org/10.1038/s41586-021-03432-3

Waite TD, Davis JA, Payne TE, Waychunas GA, Xu N (1994) Uranium(VI) adsorption to ferrihydrite: Application of a surface complexation model. Geochim Cosmochim Acta 58:5465–5478

Wang X, Liu F, Tan W, Li W, Feng X, Sparks DL (2013) Characteristics of phosphate adsorption-desorption onto ferrihydrite: Comparison with well-crystalline Fe (hydr)oxides. Soil Sci 178:1–11, https://doi.org/10.1097/SS.0b013e31828683f8

Wang L, Zhao C, Duits MHG, Mugele F, Siretanu I (2015) Detection of ion adsorption at solid–liquid interfaces using internal reflection ellipsometry. Sensors Actuators B 210:649–655, https://doi.org/10.1016/j.snb.2014.12.127

Wang Y, Persson POÅ, Michel FM, Brown G (2016) Comparison of isoelectric points of single-crystal and polycrystalline α-Al_2O_3 and α-Fe_2O_3 surfaces. Am Mineral 101:2248 - 2259

Wang X, Kubicki JD, Boily J-F, Waychunas GA, Hu Y, Feng X, Zhu M (2018) Binding geometries of silicate species on ferrihydrite surfaces. ACS Earth Space Chem 2:125–134, https://doi.org/10.1021/acsearthspacechem.7b00109

Waychunas GA, Rea BA, Fuller CC, Davids JA (1993) Surface chemistry of ferrihydrite. 1. EXAFS studies of the geometry of coprecipitated and adsorbed arsenate. Geochim Cosmochim Acta 57:2251–2269

Weidler PG, Schwinn T, Gaub HE (1996) Vicinal faces on synthetic goethite observed by atomic force microscopy. Clays Clay Mineral 44:437–442

Weidler PG, Hug SJ, Wetche TP, Hiemstra T (1999) Determination of growth rates of 100 and 110 faces of synthetic goethite by scanning force microscopy. Geochim Cosmochim Acta 62:3407–3412

Weiss L (1963) The pH value at the surface of *Bacillus subtilis*. Microbiology 32:331–340, https://doi.org/10.1099/00221287-32-3-331

Wen B, Andrade MFC, Liu LM, Selloni A (2023) Water dissociation at the water–rutile TiO_2(110) interface from ab initio-based deep neural network simulations. PNAS 120, https://doi.org/10.1073/pnas.2212250120

Wendelbow R (1987) Sulfate ion interaction with clays. PhD dissertation, University of Oslo

Westall JC (1979) MICROQL. I. A chemical equilibrium program in BASIC. II. Computation of adsorption equilibria in BASIC. EAWAG

Westall JC (1982) FITEQL: A Computer Program for Determination of Chemical Equilibrium Constants from Experimental Data. Department of Chemistry, Oregon State University

Westall J, Hohl H (1980) A comparison of electrostatic models for the oxide/solution interface. Adv Colloid Interface Sci 12:265–294

Westall JC, Zachary JL, Morel F (1976) MINEQL: A Computer Program for the Calculation of Chemical Equilibrium Composition of Aqueous Systems. I. Vol report 18. Water Quality Laboratory, Ralph M. Parsons Laboratory for Water Resources and Environmental Engineering, Dept. of Civil Engineering, Massachusetts Institute of Technology

Wijnja H, Schulthess CP (2000) Vibrational spectroscopy study of selenate and sulfate adsorption mechanisms on the Fe and Al (hydr)oxide surfaces. J Colloid Interface Sci 229:286–297

Xu TY, Stubbs JE, Eng PJ, Catalano JG (2019) Comparative response of interfacial water structure to pH variations and arsenate adsorption on corundum (012) and (001) surfaces. Geochim Cosmochim Acta 246:406–418, https://doi.org/10.1016/j.gca.2018.12.006

Xu Y, Bai Y, Hiemstra T, Weng, L (2025) Ligand and charge distribution modeling of natural organic matter adsorption on metal (hydr)oxides: State-of-the-art. Rev Mineral Geochem 91A:229-250

Yang D, Krasowska M, Sedev R, Ralston J (2010) The unusual surface chemistry of α-Al_2O_3 (0001). Phys Chem Chem Phys 12:13724–13729, https://doi.org/10.1039/c001222j

Yates DE (1975) The structure of the oxide/aqueous electrolyte interface. PhD dissertation, University of Melbourne

Yates DE, Levine S, Healy TW (1974) Site-binding model of the electrical double layer at the oxide/water interface. J Chem Soc Faraday Trans I 70:1807–1818

Yeganeh MS, Dougal SM, Pink HS (1999) Vibrational spectroscopy of water at liquid / solid interfaces: Crossing the isoelectric point of a solid surface. Phys Rev Lett 83:1179–1182

Yoon RH, Salman T, Donnay G (1979) Predicting points of zero charge of oxides and hydroxides. J Colloid Interface Sci 70:483–493

Yuan K, Rampal N, Irle S, Criscenti LJ, Lee SS, Adapa S, Stack AG (2024) Variations in proton transfer pathways and energetics on pristine and defect-rich quartz surfaces in water: Insights into the bimodal acidities of quartz. J Colloid Interface Sci 666:232–243, https://doi.org/10.1016/j.jcis.2024.03.144

Zarzycki P (2023) Distance-dependent dielectric constant at the calcite/electrolyte interface: Implication for surface complexation modeling. J Colloid Interface Sci 645:752–764, https://doi.org/10.1016/j.jcis.2023.04.169

Zarzycki P, Rosso KM (2010) Molecular dynamics simulation of the AgCl/electrolyte interfacial capacity. J Phys Chem C 114:10019–10026

Zarzycki P (2025) Molecular modeling support for SCMs. Rev Mineral Geochem 91B (in press)

Zhang Z, Fenter P, Cheng L, Sturchio NC, Bedzyk MJ, Předota M, Bandura A, Kubicki JD, Lvov SN, Cummings PT, Chialvo AA (2004) Ion adsorption at the rutile–water interface: Linking molecular and macroscopic properties. Langmuir 20:4954–4969

Zhang Z, Fenter P, Kelly SD, Catalano JG, Bandura AV, Kubicki JD, Sofo JO, Wesolowski DJ, Machesky ML, Sturchio NC, Bedzyk MJ (2006) Structure of hydrated Zn^{2+} at the rutile TiO_2 (110)–aqueous solution interface: Comparison of X-ray standing wave, X-ray absorption spectroscopy, and density functional theory results. Geochim Cosmochim Acta 70:4039–4056

Zhang Z, Fenter P, Sturchio NC, Bedzyk MJ, Machesky ML, Wesolowski DJ (2007) Structure of rutile TiO_2 (110) in water and 1 molal Rb^+ at pH 12: Inter-relationship among surface charge, interfacial hydration structure, and substrate structural displacements. Surf Sci 601:1129–1143

Zhang L, Tian C, Waychunas GA, Shen YR (2008) Structures and charging of α-alumina (0001)/water interfaces studied by sum-frequency vibrational spectroscopy. J Am Chem Soc 130:7686–7694, https://doi.org/10.1021/ja8011116

Zhang YC, Liu XD, Cheng J, Lu XC (2021) Interfacial structures and acidity constants of goethite from first-principles Molecular Dynamics simulations. Am Mineral 106:1736–1743, https://doi.org/10.2138/am-2021-7835

Zhang YC, Zhuang YB, Liu XD, Cheng J, Lützenkirchen J, Lu XC (2024) Physical adsorption of OH- causes anomalous charging at oxide–water interfaces. Chem Commun 60:9113–9116

Zhmud BV, House WA (1997) Electrical double layer and charge regulation for a homotattic surface. J Colloid Interface Sci 187:509–514, https://doi.org/10.1006/jcis.1996.4726

Zhmud BV, Meurk A, Bergström L (2000) Application of charge regulation model for evaluation of surface ionization parameters from atomic force microscopy (AFM) data. Colloids Surf A: 164:3–7

Zhuravlev LT (2000) The surface chemistry of amorphous silica. Zhuravlev model. Colloid Surf A 173:1–38, https://doi.org/10.1016/s0927-7757(00)00556-2

Reviews in Mineralogy & Geochemistry
Vol. 91A pp. 85–104, 2025

3

Impedance Spectroscopy of the Mineral–Electrolyte Interfaces

Youzheng Qi, Yuxin Wu

Lawrence Berkeley National Laboratory, Berkeley, CA, USA

YouzhengQi@lbl.gov, YWu3@lbl.gov

INTRODUCTION

This chapter centers on the interactions between the low-frequency electrical and electromagnetic (E&EM) field (1 mHz–10 kHz) and geomaterials, viz the low-frequency impedance spectroscopy (IS), for its merit of interrogating the mineral–electrolyte interfaces (Dukhin 1993). Although the term IS is used here, it may appear as induced polarization (Sumner 1976), low-frequency dielectric dispersion (Schwarz 1962), and complex conductivity/resistivity methods (Börner 2009) in the literature of other disciplines. The IS method investigates the mineral–electrolyte system across micro-, meso- and macro-scales. The micro-scale electrical double layer (EDL), whose physicochemical properties are depicted by surface complexation models (Leroy and Revil 2009), is the root cause of IS. However, those micro-scale EDLs cannot be directly measured because IS itself is performed at the macro-scale of laboratory and field (Binley and Slater 2020). Accordingly, the meso-scale of effective medium models is needed to bridge between micro- and macro-scales (Qi and Wu 2024). Therefore, this chapter is organized following these three different scales to elaborate their respective fundamentals and applications.

MICRO-SCALE: ELECTRICAL DOUBLE LAYER (EDL) POLARIZATION

IS method was pioneered by the Schlumberger brothers in the 1910s who clearly measured decaying potential curves in the earth after the passage of an electrical current during their mineral exploration, which leads to the first-ever IS publication of Schlumberger (1920). Although they could not fully explain those IS phenomena during that time, it was suggested that the IS mechanism should be caused by some micro-scale actions in the subsurface. Nowadays the community is able to explain most prominent phenomena and has solidly built the IS method on its theoretical foundations of electrochemistry and Maxwell's equations, out of which the micro-scale EDL is the key to understanding the polarization mechanism.

Structure of the EDL

Although charges at the mineral–electrolyte interface are ubiquitous, their respective origin can be quite distinct between different systems, which includes (Delgado and Arroyo 2002):

(i). Charge-defective lattice of isomorphous substitution;

This is typically the dominant mechanism for the basal surface charging of clay minerals where a certain number of Al^{3+} are substituted by Mg^{2+} or Fe^{2+} ions that have lower charge with almost the same size. As a consequence, the crystal will be negatively charged. This structural charge is compensated for by cations but they are still exchangeable.

1529-6466/25/0091A-0003$05.00 (print)
1943-2666/25/0091A-0003$05.00 (online)
http://dx.doi.org/10.2138/rmg.2025.91A.03

(ii). Direct dissociation and ionization of surface groups;

This is the mechanism through which carbonate and silica minerals obtain their charges where hydroxyl groups are responsible for the charges. When the pH is above the pK_a of dissociation of these groups, they will ionize, yielding negative surfaces, which means that their surface charges are much more pH-dependent than the above clay minerals.

(iii). Adsorption and desorption of lattice ions.

The most prominent example for this mechanism is silver iodide in Ag^+ or I^- solutions, where the crystal lattice ions, being called potential-determining ions (PDI), can easily find ways into the crystal sites and become part of the surface.

(iv). Specific adsorption of ions in solution.

This is, for example, the case of ionic surfactant adsorption. The dissolved charged entities must have a high affinity for the surface in order to overcome electrostatic repulsion by already adsorbed ions. This chemical adsorption is distinct from physical adsorption.

Despite the fact that different mechanisms lead to the emergence of overall net surface charges, they are compensated by the counterions from the electrolyte to maintain electroneutrality, establishing the structure of EDL at mineral–electrolyte interfaces. It is difficult to illustrate a generic EDL for all the aforementioned four solid-solution systems because the EDL structure depends heavily on the physicochemical properties of both the solid and the solution (Bockris et al. 2000). Throughout the chapter, the system of clay and NaCl solution belonging to the above-mentioned category (I) will be chosen as the representative mineral–electrolyte interface, for it embodies the most common features of the EDL.

Clay minerals are the most abundant natural reactive solids on the Earth's surface and are major components of many types of rocks and soils, such as shale, mudstone, and vertisol (Schroeder 2018). The strong reactivity of clay minerals stems from two characteristics: high surface area and net negative surface charges, the former being from its extremely tiny size (<2 μm) and the latter from its chemical structure. Clay minerals are hydrous aluminum phyllosilicates made of two fundamental building blocks: the tetrahedral sheet between O^{2-} and Si^{4+} and the octahedral sheet between OH^- and Al^{3+} (or Mg^{2+}, Fe^{2+} with isomorphous substitution). This specific chemical structure induces permanent negative charges on the basal surfaces. Though permanently charged, clay minerals in the dry state are not conductive as the charges are fixed, which is completely different from the electronic conduction of metals (Winsauer and McCardell 1953).

Nonetheless, when the dry clay minerals transit into a wet state with ionic solutions, they become conductive, owing to the establishment of the EDL at the clay particle surface with mobile counterions once exposed to the electrolyte solution as shown in Figure 1 (Bockris et al. 2000). Nearest to the clay surface, there is always a hydration sheath because of the strong polarity of water molecules and these water molecules arrange down their orientations for the electrical force of the charged surface. The first row of hydration sheath, sometimes accompanied by specifically adsorbed ions, is named the inner Helmholtz plane (IHP). The second row is largely reserved for solvated cations and this position in the EDL is called the outer Helmholtz plane (OHP). IHP and OHP are together named as the Stern layer or compact layer. Outside the Stern layer, the remaining electric force and thermal jostling both take effect, which leads to the diffuse layer that consists of cations and anions. Beyond the diffuse layer, it is the free electrolyte where cation and anions distribute equally. Note that since the Stern layer cannot completely compensate for the surface potential effect, cations still exceed anions in the diffuse layer compared to the free electrolyte, albeit not in a dominant pattern as the Stern layer (see Fig. 1).

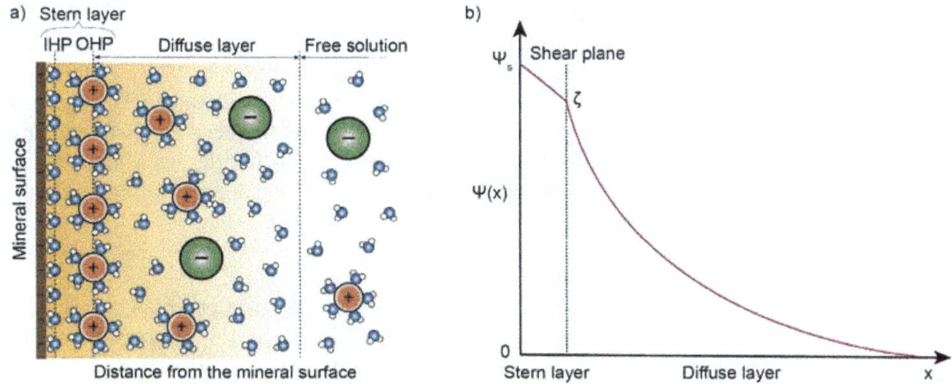

Figure 1. EDL structure and its potential drops from the surface. The shear plane (i.e., the slipping plane) between the compact and mobile layer is theoretically within the diffuse layer and thus not exactly the plane between Stern and diffuse layers, but in most cases the difference between these two planes is negligible so that they are always treated as the same plane (Lyklema 1995, 2002). Note that different minerals have different Zeta potentials that can be depicted via surface complexation models (Leroy and Revil 2009).

The discrimination between Stern and diffuse layers is based on their distinct ionic distribution patterns. Another differentiation is from their electrokinetic properties; that is, their electrical responses under fluid motion in the vicinity of the EDL and vice versa, a vivid example for the former being the streaming potential method and the latter being electroosmosis method. Taking the streaming potential as an example (Fig. 2), when the bulk fluid flows in a pore space under an external pressure, both the cations and anions move and electroneutrality is still respected because of their homogenous distribution. However, this does not apply to the mineral–electrolyte interface where counterions outcompete coions within the EDL. The viscous stress between water molecular drives more counterions than coions in the diffuse layer and consequently creates a streaming current of the micro-scale which leads to a measurable streaming potential of the macroscale. Both theories and experiments show that the plane between the immobile and mobile ions (i.e., the slipping plane) lies in the diffuse layer, hence the EDL can also be divided into a stagnant layer and a mobile layer, between which the potential is named the electrokinetic potential, more commonly known as the Zeta potential ζ. Different minerals intrinsically have different Zeta potentials whose dependence

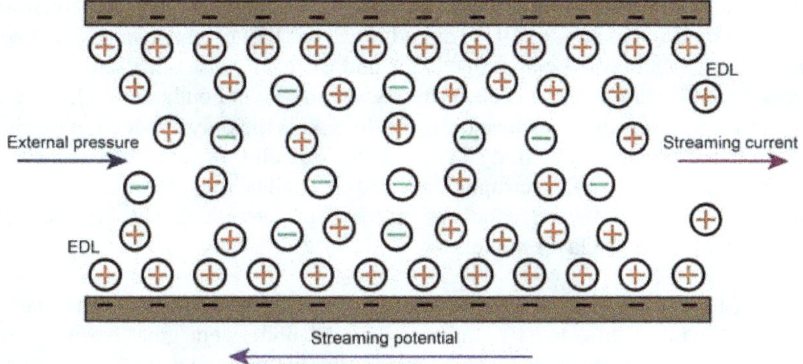

Figure 2. Streaming current and streaming potential within a capillary. Different to the situation of the homogenous bulk fluid that the external pressure drives the cations and anion equally, in the diffuse layer the pressure drives more cations than anions for a negatively charged mineral surface, which results in a streaming current that is measured as the streaming potential for the electrodes at two end-points of the capillary.

on the physiochemical properties of mineral–electrolyte interfaces can be depicted via surface complexation models (Leroy and Revil 2009). Note that although the slipping plane is in the diffuse layer, its distance from the OHP is expected to be negligible. Mostly we can equate the slipping plane to the OHP and treat the diffuse and the mobile layers as identical, which is especially true for low and mediate electrolyte concentrations and has been adopted by many IS practitioners (Lyklema 2002; Leroy et al. 2008). However, for high salinities where the diffuse layer thickness is much reduced, this equivalence begins to fall short of expectations and cautions must be taken (Delgado and Arroyo 2002).

Diffuse layer polarization model

Given that the diffuse layer is subjected to much weaker surface potential restrictions than the Stern layer because of its distance from the mineral surface, it is straightforward to first ascribe this surface conduction to the diffuse layer (i.e., the dynamic diffuse layer model). Here we discuss the diffuse layer conduction and diffuse layer polarization models separately because of their different developmental trajectory.

Surface conduction as a phenomenon was first realized 120 years ago by von Smoluchowski (1905) who explicitly pointed out in addition to the interstitial electrolyte, the excess surface conduction within the diffuse layer contribute to the measured electrical conductivity. Three decades later, Bikerman (1933) developed the first mechanistic model to depict such diffuse layer conductivity Σ_d by integrating the ionic concentration c with the ionic mobility of free solution setting up from the OHP to the bulk solution (e.g., 3~4 Debye length or infinite) as

$$\Sigma_d = \frac{4\mathrm{F}^2 c z^2 D_d}{\mathrm{R}T\kappa}\left(1+\frac{3m}{z^2}\right)\left[\cosh\left(\frac{z\mathrm{F}\zeta}{2\mathrm{R}T}\right)-1\right] \tag{1}$$

with

$$m = \left(\frac{RT}{\mathrm{F}}\right)^2 \frac{2\varepsilon_0\varepsilon_r}{3\eta D_d} \tag{2}$$

where D_d is the ionic diffusion coefficient of diffuse layer, ε_0 and ε_r are vacuum permittivity and the relative permittivity of the electrolytic solution, respectively, η is the viscosity of the solution, z is the ionic valency of the electrolyte, F is the Faraday constant[1], R is the gas constant, and T is absolute temperature. When the low-frequency electrical field moves the ions in the diffuse layer, the viscous stress from the ionic solvation also moves the fluid which creates a fluid flow petering out into the bulk solution, which is known as the electro-osmosis phenomenon (Fig. 3a). Bikerman (1933) already realized that electro-osmosis can contribute to diffuse layer conduction in positive feedback and the term m of Equation (2) accounts for this process. From Equation (1), it is clear that the diffuse layer conduction relates to the Zeta potential ζ, which means that if diffuse layer conduction is the only conduction mechanism of the EDL, the specific EDL conductivity Σ_d measured from all these electrokinetic phenomena, such as the streaming potential, electrophoresis, and IS, shall be the same or at least comparable with each other because of their dependence on the Zeta potential ζ. This will be revisited in the later section of the Stern layer model.

In contrast to the diffuse layer conduction model developed in the early 1930s (Bikerman 1933), the diffuse layer polarization model underwent great hardship. Although IS practitioners during that time realized that a complex-valued specific EDL conductivity Σ_s^* (i.e., the specific surface admittance, in S) must be introduced to explain the low-frequency

[1] Note that Roman print indicates a constant and italics a scalar variable, so, F is the Faraday constant and F the formation factor (see later in this chapter)

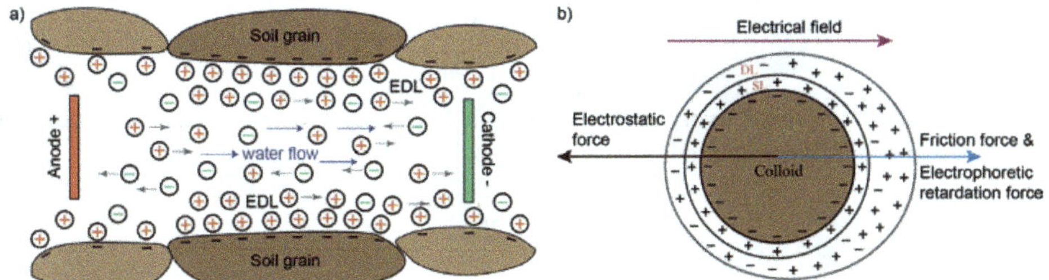

Figure 3. Electro-osmosis and electrophoresis. **(a)** depicts the mechanism of electro-osmosis where the external electric field moves the counterions in the diffuse layer, these mobile ions also drag the surrounding fluid molecules due to viscous forces, which results in the bulk movement of the liquid in the capillary. **(b)** illustrates the mechanism of electrophoresis where an external electric field drags a charged particle; besides the electrostatic and the friction forces directly applied to the particle by the Stokes law, there is an additional retardation force from the migration of the counterions in the diffuse layer that opposes the particle movement through viscous stress.

dielectric permittivity (Fricke and Curtis 1937), this was just a mathematical treatment and the underlying physical mechanism was debated with the possibility of electrophoresis and electro-osmosis. The former was easily excluded because the retardation force is opposite to the electrical field such that the phase of the polarization is positive rather than negative (Fig. 3b), while the latter is too small to explain such a huge polarization (Schwan et al. 1962). Indeed, from Equation (2), in many cases the electro-osmotic effect is much smaller in comparison to the ionic migration in the diffuse layer (Dukhin and Shilov 2002)

In the 1960s, Schwarz (1962) analytically calculated the response of a sphere with a counterion shell under electrical field to demonstrate that the origin of a low-frequency polarization is from the displacement of the counterions in the EDL that introduces an induced dipole moment of the particle, rendering the EDL as a leaky-capacitor under external field (Fig. 4). In the ideal situation where the EDL is infinitely thin in comparison with the spherical grain's radius Λ and the counterions only move tangentially, it was theoretically proven by Schwarz (1962) that the sphere with EDL would perform as the Debye model

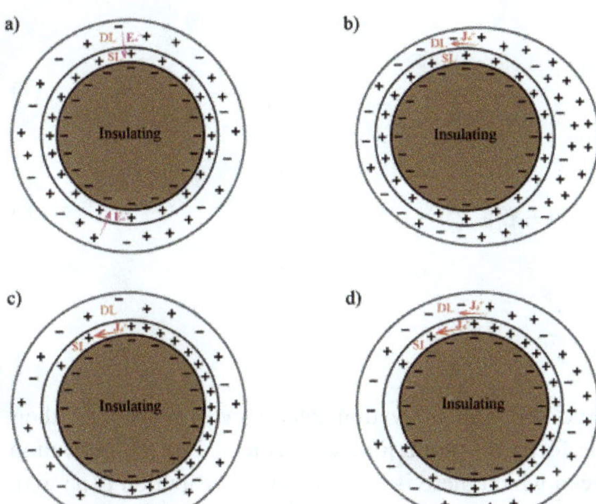

Figure 4. Equilibrium EDL and EDL polarizations. **(a)** Equilibrium EDL, **(b)** Diffuse layer polarization, **(c)** Stern layer polarization, and **(d)** Stern and diffuse layer polarization. SL and DL are the abbreviations for the Stern and diffuse layers, respectively. $\mathbf{E}_n^+$ in the equilibrium EDL is the force from the surface potential that is electrostatic in nature. $\mathbf{J}_d^+$ is the back-diffusion current caused by the surface potential counteracting the ionic migration in the EDL to bring it back to the equilibrium state.

$$\sigma_c^* = \sigma_c^\infty + \frac{\sigma_c^0 - \sigma_c^\infty}{1 + i\omega\tau_c} \tag{3}$$

where $\sigma_c{}^0$ and $\sigma_c{}^\infty$ are respectively the low- and high-frequency asymptotic conductivities that links to the specific EDL conductivity $\Sigma_s{}^*$ via

$$\sigma_c^* = \frac{2\Sigma_s^*}{\Lambda} \tag{4}$$

$\sigma_c{}^*$ corresponds to what the non-thickness EDL conductivity $\Sigma_s{}^*$ (in S) represents in 3D (in S/m). τ_c the characteristic relaxation time is expressed as

$$\tau_c = \frac{\Lambda^2}{2D_c} \tag{5}$$

where D_c denotes the diffusion coefficient within the EDL as

$$D_c = u_c \mathrm{k_B} T \tag{6}$$

where u_c is the mechanical mobility of the ions (i.e., velocity per unit force), $\mathrm{k_B}$ is the Boltzmann constant and T is absolute temperature. Because of the surface potential, the ionic mobility u_c within the EDL is smaller than that of free solution u_0 so that D_c is smaller than the bulk solution diffusion coefficient D_0 (Schwarz 1962).

Despite the fact that an infinitely thin EDL behaves as the ideal Debye model, this behavior is hardly found in the real world. Grain size distribution is a factor that can broaden the ideal Debye spectrum, but even for grains with exactly the same size, this ideally narrow spectrum cannot be measured as well because, as pointed out by Schwarz (1962), counterions shall distribute in a diffuse manner due to the balance between electrostatic attraction and thermal motion. In lieu of researching the diffuse behavior with Poisson–Boltzmann models, Schwarz (1962) solves this issue from the perspective of the activation energy for its merit of simplicity with fewer parameters. To move along the surface, counterions are subject to an additional surface activation energy α so that the above-mentioned ionic mobility u_c links with the mobility u_0 in free solution as

$$u_c = u_0 \, \mathrm{e}^{-\alpha/\mathrm{k_B}T} \tag{7}$$

Thus, the aforementioned relaxation time τ becomes

$$\tau = \tau_0 \, \mathrm{e}^{-\alpha/\mathrm{k_B}T} \tag{8}$$

with

$$\tau_0 = \frac{\Lambda^2}{2u_0 \mathrm{k_B} T} \tag{9}$$

Since different counterions undergo different activation energy with respect to their distances from the mineral surface, the furthest counterions are subjected to the activation energy $\overline{\alpha}$ and the nearest $\overline{\alpha} + \Delta\alpha$, between which there is an activation distribution $\rho(\alpha)$ with the relation

$$\int_0^{\infty} \rho(\alpha) d\alpha = 1 \tag{10}$$

Although it is obvious that the diffusive character of the EDL broadens the Debye model in physics, it is immensely difficult to develop a generic model to describe such an effect mathematically, as different mineral surface intrinsically bears distinct activation energy distributions. For practical purposes, it is always useful to employ a micro-scale Cole–Cole model (Cole and Cole 1941)

$$\sigma_c^* = \sigma_c^{\infty} + \frac{\sigma_c^0 - \sigma_c^{\infty}}{1 + (i\omega\tau_c)^{c_c}} \tag{11}$$

to describe EDL polarization, the advantage being that the Cole–Cole model accounts for the broadness with just one effective parameter c_c. Example curves of the ideal Debye model and Cole–Cole model are shown in Figure 5, illustrating how the effective parameter c_c flattens the ideally narrow Debye model of $c_c = 1$.

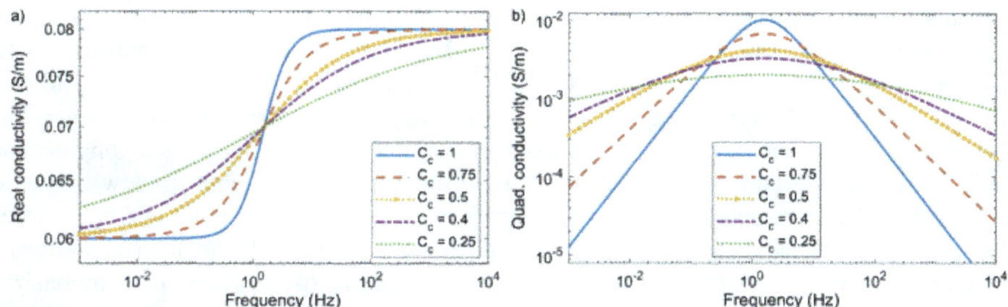

Figure 5. Cole–Cole model curves under different exponent c. Note that $c = 1$ and $c = 0.5$ denotes the Debye and Warburg models, respectively.

The EDL polarization mechanism of Schwarz (1962) predicts the characteristic peak frequency of quadrature conductivity reasonably and effectively explains the main results of experiments, which is considered to be the cornerstone of the IS edifice that brings the IS community from the qualitative to the quantitative epoch (cf. Schwan 2000). Although many other EDL polarization models currently exist to consider some second-order IS effects, as it turns out in the following sections, they mostly improve the seminal model of Schwarz (1962) to account for those effects.

Stern layer polarization model

With the success in explaining the mechanisms of streaming potential, electrophoresis, and electro-osmosis, IS practitioners also attempted to employ the electrokinetic potential (i.e., the Zeta potential ζ) to explain surface conduction; that is, using the Bikerman model (Eqn. 1) that treats the diffuse layer as the only conductive layer to interpret the surface conductivity Σ_S. However, substantial disagreement was found for such a diffuse layer conduction mechanism. For example, Kruyt et al. (1952) studied the specific surface conductivity Σ_s of glass beads with well-defined shapes and sizes and found that their magnitude, depending on the electrolytic salinity, ranges between 10^{-9} and 10^{-6} S that is almost 10 to 1000 times as high as the prediction of the Bikerman model. Even with some experimental uncertainty within those data, it strongly suggests that the Bikerman model underestimates surface conduction significantly, because of which, this measured high surface conduction used to be called 'the anomalous surface conduction'.

Due to the solid foundation of the Bikerman model that has been validated by the aforementioned electrokinetic phenomena of streaming potential, electrophoresis, and electro-osmosis, IS practitioners began to revise the underlying mechanism of surface conduction, such as to consider the possibility of electrical conduction in the Stern layer. Such a dynamic Stern layer model was first presented by Zukoski and Saville (1986) demonstrating the introduction of the Stern layer conduction can significantly decrease the apparent Zeta potential ζ so that the true ζ is comparable with that produced by electrophoresis or electro-osmosis. Subsequently, Stern layer conduction began to be recognized by the IS community not only to happen but also always to exceed and even dominate the surface conduction of the whole EDL in many scenarios (Lyklema 1995). Hence, the historical 'anomalous surface conduction' is not anomalous at all. The term was abandoned after the discovery of the Stern layer conduction (Lyklema 2002)

While the discovery of Stern layer conduction is from the mismatch between experiments and the Bikerman model, here we provide another intuitive explanation for the Stern layer conduction from the perspective of different physical forces. As is known, four fundamental forces govern the interactions of materials at different scales. For the electrokinetic phenomena such as the streaming potential, electrophoresis, and electro-osmosis, they originate from the fluid viscosity where the underlying physical forces are intermolecular forces, including Van der Waals forces, hydrogen bonds, and cohesive forces, which are electromagnetic forces at the molecular level that is too short-range to affect the Stern layer. By contrast, for the electrical conduction of the low-frequency IS, they are caused by the external E&EM field between 1 mHz and 10 kHz that are sufficiently long-range to force an ionic movement within the Stern layer. Thus, the neglect of Stern layer conduction in the IS history is mainly because the IS practitioners attempted to equate the fluidic mechanical force with the low-frequency E&EM fields. Since these two forces are distinct in physics and are governed by different PDEs (Partial Differential Equations) in mathematics, as expected, the hydrodynamically stagnant layer does not coincide with the Stern layer. This means, as previously emphasized, the slipping plane is not exactly the OHP and there is a difference between the diffuse layer potential Φ_d and the Zeta potential ζ. That being said, in most scenarios these two layers can be treated as the same with negligible errors (Delgado and Arroyo 2002). Hence we will not distinguish them in the following sections.

As it is solidly proven that Stern layer can conduct, it is reasonable to deduce that the Stern layer also polarizes because of the counteraction in response to the surface potential. Since Schwarz (1962) theoretically proved that a spherical grain with infinitely thin EDL behaves as the Debye model, this perfectly applies to Stern layer polarization, for the size of the Stern layer is much smaller than the salinity-dependent diffuse layer (i.e., the Debye length) and can be readily seen as infinitely thin with respect to the pore/grain size. The dynamic Stern layer polarization model was first developed by Leroy et al. (2008) and is accepted as one of the most popular polarization models in the IS community (Kemna et al. 2012; Binley and Slater 2020). Because the Stern layer is the only polarizable layer within the EDL for the Stern polarization model, it postulates that the low-frequency asymptotic (i.e., DC) surface conductivity is exclusively implemented by the diffuse layer without interactions between Stern and diffuse layers (Fig. 4c). On the grounds that the non-thickness Stern layer perform as the ideal Debye model (Schwarz 1962), the popularly used Debye decomposition models, introduced into the IS community by Tarasov and Titov (2007) and Nordsiek and Weller (2008) in the respective time and frequency domains, are implicitly Stern polarization models as well. Although the Stern polarization model adopts some approximations, such as the neglect of diffuse layer polarization, it explains many experimental phenomena such as the causation and correlation between the main pore/grain size and the characteristic relaxation time (Leroy et al. 2008). However, due to those approximations, there remains some phenomena that cannot be fully explained. For example, the dynamic Stern layer polarization model cannot predict the salinity dependency of the characteristic relaxation time that may decrease (Kemna et al. 2005) or increase (Mendieta et al. 2021) with increasing electrolytic concentration, which necessarily leads to the polarization model of both Stern and diffuse layers that we will proceed with.

Stern and diffuse layer polarization model

Pros and cons exist in either the Stern layer model or the diffuse layer model. In other words, both the Stern and the diffuse layer models can explain some phenomena in the IS experiments but not all of them and, more importantly, there exists no theory to exclude either of these two models. Though admittedly in some situations, such as at low ionic concentrations, the Stern layer model dominates because the diffuse layer is of rather low charge density and similar to the bulk solution, this does not imply that Stern layer always dominates, which is more evident in the high electrolytic concentration where the diffuse layer is rather thin. Therefore, mostly the conduction and polarization of both the Stern and the diffuse layers have to be taken into consideration to explain the phenomena in a comprehensive way.

Developing such a mechanistic model for the EDL is non-trivial and indeed complicated. Stern and diffuse layer modeling is not just an arithmetic summation of the two models because neither Stern nor diffuse layer stands on its own feet and there exist ionic and field interactions between these two layers (Fig. 4d). The first theoretical work to develop the Stern and diffuse layer model appears in Lyklema et al. (1983) that explains the measured first- and second-order phenomena within the IS community. For example, as to the above-mentioned salinity-dependence of the characteristic relaxation time that cannot be described by the Stern layer polarization model, Lyklema et al. (1983) derived it as

$$\tau_c = \frac{\Lambda^2}{2MD_c} \tag{12}$$

with

$$M = \frac{2}{\cosh(\mathrm{F}\Phi_d / 2\mathrm{R}T)}\left[\frac{Q_f \kappa}{4\mathrm{F}c} - \sinh\left(\frac{\mathrm{F}\Phi_d}{2\mathrm{R}T}\right)\right] \tag{13}$$

where Q_f is the excess surface charge density and Φ_d is the diffuse layer potential that approximates the Zeta potential ζ, as previously discussed. As a matter of fact, the factor M in Equation (13) corrects the Stern layer model for the influence of the diffuse layer within the EDL (Lyklema et al. 1983).

As expected, when diffuse layer polarization comes into play, even the EDL of a perfect spherical grain does not behave as the ideal Debye model anymore (Schwarz 1962; Lyklema et al. 1983), which makes the Debye decomposition model arguable. Instead of the Debye decomposition model, it may be suggested that the Warburg decomposition model shall be employed as a substitution (Revil et al. 2014), the Warburg model being the Cole–Cole model with the exponent of 0.5 (see Fig. 5). Although the Warburg decomposition model supports the idea that the EDL does not behave as the ideal Debye model, the Warburg decomposition model also gets into the same argument as the Debye decomposition model, for there is also no theory showing that EDL performs exactly as the Warburg model. The limited basis underpinning the Warburg decomposition model is that when applied to IS data, the Warburg decomposition model can output a narrower relaxation time distribution (RTD) than the Debye decomposition model (Revil et al. 2014). However, this is actually just a mathematical consequence because using a kernel function with more broadness inescapably leads to a narrower RTD. In other words, employment a Cole–Cole model with the exponent of 0.4 can produce an even narrower RTD than the Warburg decomposition model, but there is no physical corroboration that the Cole–Cole model with the exponent of 0.4 shall be the kernel function. In this respect, the Cole–Cole model seems to provide an alternative way to circumvent such sophisticated circumstances with just one effective exponent, as suggested by Schwan (1962) and Lyklema et al. (1986). That being said, no matter what model is employed, the RTD should not be over-interpreted, for it is demonstrated that there are many factors that can widen the ideal Debye model, such as the diffuse layer (Schwarz 1962; Lyklema et al. 1983); the grain size distribution (Schwan 1962; Schwarz 1962); the surface roughness (Leroy et al. 2008; Zibulski and Klitzsch 2023), the overlapping of membrane polarization (Marshall and Madden 1959; Bücker and Hördt 2013), and different ionic types (Lyklema 2002; Tournassat et al. 2013). Therefore, the RTD can only be qualitatively interpreted unless one broadening factor controls the dispersion.

MESO-SCALE: EFFECTIVE MEDIUM MODEL

Geophysical methods, such as the E&EM and IS methods, are performed at the macro-scale that have sensitivity but limited resolution at the micro-scale.Thus, effective medium models are generally applied to bridge the gap between micro- and macro-scales with the provision of the parameters for the composite rocks and soils. Although effective medium theory (EMT) inevitably simplifies the geometry of real-world geomaterials to make the development of analytic effective medium models possible, the developed effective medium models are based on their governing PDEs, such as Maxwell's equations (Sihvola 1999). Effective medium models may be labeled as petrophysical models in the literature, but they are actually different terminologies. Petrophysical models literally denote the models of rocks that can be empirical or mechanistic, but, in contrast, effective medium models are underpinned by EMT that specifically derives the effective properties (conductivity, permittivity, permeability, etc.) of a composite material out of the respective properties of its components, which must be mechanistic rather than empirical. That is to say, effective medium model is a mandate to develop a mechanistic model between the physico-chemical parameters of the components and the bulk parameters that geophysical methods can measure. Figure 6 illustrates the developing scheme of a mechanistic model and the central position of effective medium models. The procedure includes the assembly of the micro-scale physico-chemical parameters of the water and EDL (leftmost of Fig. 6) into an effective medium model (middle of Fig. 6) that finally produces the non-linear IS curves of the macro-scale geomaterials (rightmost of Fig. 6).

Effective medium models play a pivotal role in geophysical data interpretation, as they directly relate to how the physico-chemical parameters are assembled into mechanistic models (Fig. 6). Different effective medium models may supply distinct parameters for the geophysicists both in the laboratory and field (Cosenza et al. 2009; Glover 2015). The most commonly used

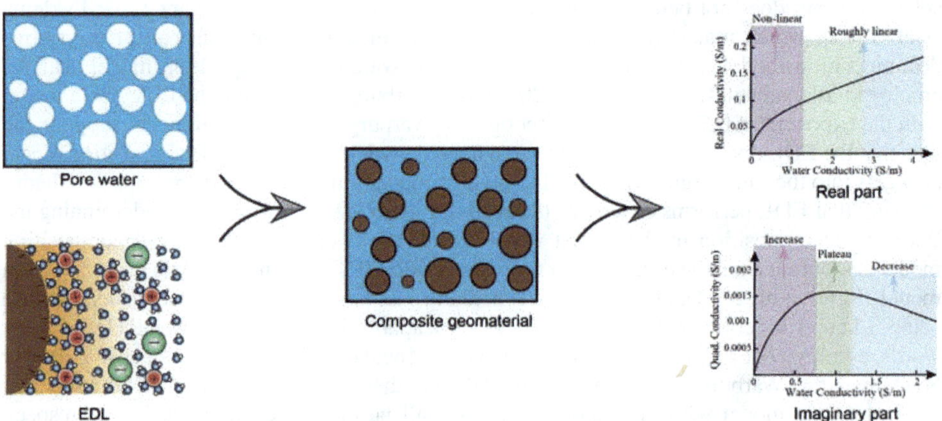

Figure 6. Diagram of the mechanistic IS model development.

effective medium model is Archie's law (Archie 1942) that has been successfully used for more than 80 years for "clean" rocks and soils. Regarding "dirty" ones, e.g., clay, carbonate, and humus, which is omnipresent in the critical zone (Schroeder 2018), things become more complicated, for their high specific surface areas and excess surface charges generate forceful EDLs at the mineral–solution interface that nullifies Archie's law (Winsauer and McCardell 1953). Glover (2015) provides a review list of effective medium models in this aspect with an abundance of 12 models and in a recent analysis by Qi and Wu (2024) 4 models are found to be applicable: the three-resistor model, the Weller–Slater model, the Bruggeman–Hanai–Sen (BHS) model, and the model of Qi and Wu (2024). However, before delving into these four models, the empirical Waxman-type models will be first revisited because of their wide application in the IS community.

Waxman–Smits and Vinegar–Waxman models

Through a massive experiment of various shaly samples, Waxman and Smits (1968) empirically developed an electrical conductivity model and Vinegar and Waxman (1984) extended it into IS as

$$\sigma^* = \frac{\sigma_w}{F} + \frac{BQ_v}{F} + i\frac{\lambda Q_v}{F\phi} \tag{14}$$

where F is the true formation factor and ϕ is the porosity, $F\phi$ being the so-called electrical tortuosity. To make it non-linear, they postulate that the ionic mobility B and λ increase exponentially with salinity, which conflicts with electrochemistry (Johnson and Sen 1988). Although some (Johnson et al. 1986; Revil 2013) attempted to theoretically re-derive Waxman-type models, they also treat EDL and water as two insulating parallel conductors (Fig. 7). This involves the same assumption as Waxman and Smits (1968) and thus cannot be seen as independent proof.

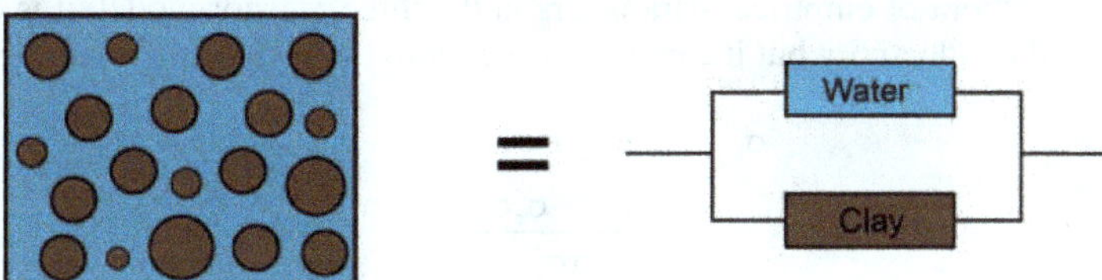

Figure 7. Sketch of the Waxman-type model.

The true formation factor F is a parameter characterizing the pore water topology and, for dirty geomaterials, F denotes the value if those dirty materials were replaced by a geometrically identical but insulating matrix (Worthington 1985), which makes the inclusion of F or $F\phi$ in σ_s arguable. As a matter of fact, there are a myriad of facts demonstrating that such a simple summation within the Waxman-type models is inappropriate, which includes but is not limited to:

(i). Theory: Maxwell's equations dictate that the EDL presence disturbs the E&EM fields in water such that they cannot be treated as two parallel conductors (Dukhin and Derjaguin 1974);

(ii). Experiment: The attribution of the ionic mobility increase with increasing salinity from Waxman and Smits (1968) not only conflicts with the electrochemical theory (Johnson and Sen 1988) but also cannot explain the measured IS data (Lévy et al. 2018);

(iii). Formula: An analytic equation for spherical colloid suspensions evidences that the bulk conductivity σ^* does not follow such a simple summation strategy (cf. Equation 3.4.6 in Dukhin 1993);

(iv). Numerical: Simulations demonstrate that the effective bulk conductivity does not equal to the summation of its two components no matter whether F or $F\phi$ is added or not (Qi and Wu 2024);

(v). Applicability: Waxman-type models are merely used in the geophysics community but seem to be avoided by other communities, such as the electrochemistry and colloid science communities; for example, in the pioneering paper of Schwarz (1962) that explains the IS mechanism, an effective medium model (Maxwell Garnett equation) rather than the Waxman-type model is employed to calculate the bulk conductivity therein.

Surprisingly, despite the above-mentioned obvious drawbacks, the Waxman-type model still dominates the IS community as the most popular model. In fact, as early as 50 years ago, Dukhin and Derjaguin (1974) explicitly pointed out that the empirical Waxman-type model contradicts the E&EM theories and neglects the direct coupling between EDL and water, after which the employment of Waxman-type models in the colloid science community has been hardly seen.

Three-resistor and Weller–Slater models

With the recognition of the impossibility that all the electrical currents conduct separately in the clay and water, Wyllie and Southwick (1954) first developed the three-resistor model, viz the water resistor σ_1, the clay-and-water resistor σ_2, and the clay resistor σ_3 (Fig. 8), to explain the measured nonlinearity of the electrical conductivity curves of dirty material. Soon this model was verified by Sauer et al. (1955) and many others. After 35 years, Schwartz et al. (1989) used the Padé approximant to develop a similar model (i.e., the PA model), and recently Lévy et al. (2018) employed the classic lumped circuit theory to construct their own model independently. However, it has been shown that these two models are the same as the TR model (Revil et al. 2019). Thus, these two models are categorized as the three-resistor model here. Due to a plethora of empirical parameters in the three-resistor model, it is only used in the real electrical conductivity but it can be extended into the complex domain for IS as

$$\begin{aligned}\sigma^* &= \sigma_1 + \sigma_2^* + \sigma_3^* \\ &= \frac{\sigma_w}{F} + \frac{\sigma_c^*\sigma_w}{x\sigma_c^* + y\sigma_w} + \frac{\sigma_c^*}{z} \\ &= \frac{\sigma_w}{F} + \frac{(y+z)\sigma_c^*\sigma_w + x\sigma_c^{*2}}{xz\sigma_c^* + yz\sigma_w}\end{aligned} \tag{15}$$

where x, y, and z are empirical geometric factors. Rigorously speaking, the three-resistor model is not an effective medium model as it is derived from the lumped circuit theory rather than Maxwell's equation. However, this model includes the EDL-and-water coupling in their development through the clay-and-water series σ_2, which makes it applicable.

Weller and Slater (2012) generalized the Stern layer polarization model of Skold et al. (2011) to develop a quadrature conductivity model and found it to explain the measured

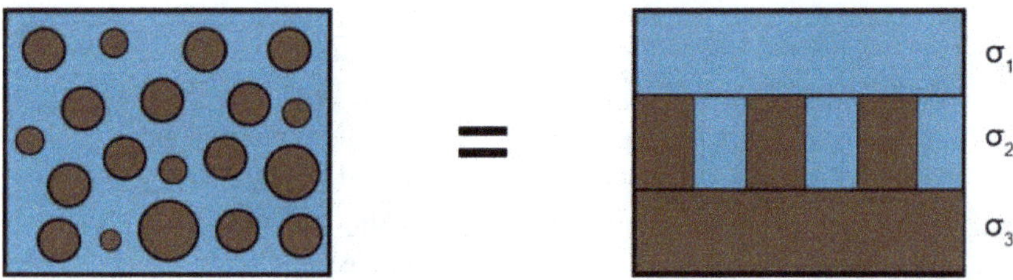

Figure 8. Conceptual schema of the three-resistor model.

quadrature conductivities excellently (Weller and Slater 2012). Although this model previously ascribes the non-linearity to the EDL variation like Skold et al. (2021), Weller et al. (2015) discovered that this model can produce parametric correlations that contradict the original model of Skold et al. (2011). This paradox was later explained by Qi and Wu (2024) that the Weller–Slater model *de facto* performs like an effective medium model rather than a dynamic Stern layer model. Similar to the extension of the three-resistor model, it can also be extended into the complex domain for IS via

$$\sigma^* = \frac{\sigma_w}{F} + c_s^* + (a_s^* \frac{\sigma_w}{b_s^* + \sigma_w}) \tag{16}$$

where a_s, b_s, and c_s are fitting parameters (Weller and Slater 2012). It is worth noting that after some algebraic operations the three-resistor and Weller–Slater models are exactly the same (Qi and Wu 2024).

Bruggeman–Hanai–Sen (BHS) model

The BHS model defined by

$$\sigma^* = \sigma_w \phi^m \left(\frac{1 - \sigma_c^* / \sigma_w}{1 - \sigma_c^* / \sigma^*} \right)^m \tag{17}$$

was developed by Bruggeman (1935), Hanai (1960) and Sen et al. (1981) and can explain real and quadrature nonlinearities concurrently and it can downscale to Archie's law for the one-phase situation. BHS is a well-established effective medium model that has its roots in the differential effective medium (DEM) theory where Sen et al. (1981) use the Maxwell Garnett equation as the basic equation to describe a grain coated with a water shell (water being the host and solid being the guest) and then infinitely add such a grain-with-shell in a differential form (i.e., DEM) to finally develop the BHS model (Fig. 9).

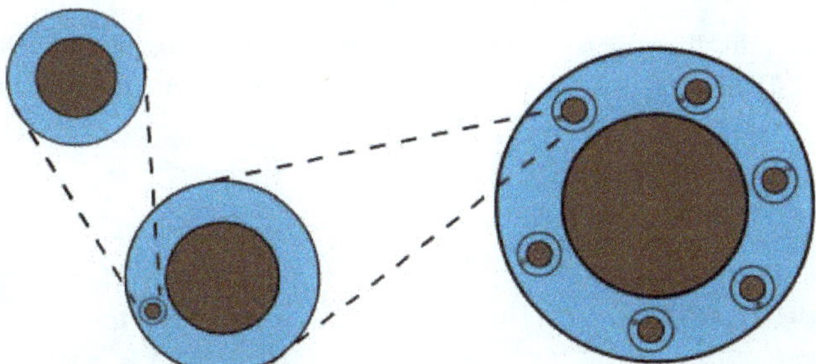

Figure 9. Diagram of the DEM theory from which the BHS model is developed. Note that this diagram just shows three sequences of embedding the grain with water shell and DEM theory indeed employs such a continuous embedment in an infinite means.

Different to the three-resistor and Weller–Slater models, the BHS model is developed from Maxwell's equations and thus has the advantages of EMT, such as using parameters with clear physical meaning. However, it is an implicit function in mathematics that can only be solved numerically (Lesmes and Morgan 2001; Glover et al. 2010), which causes the complex surface conductivity σ_s^* not to appear directly as the three-resistor and Weller–Slater models. It is also found that its cementation factor m differs from that of Archie's law (Friedman 2005; Niwas et al. 2006). Despite those mathematical and physical disadvantages, BHS still represents itself as one of the most successful effective medium models in the past decades and has been employed by many IS practitioners that continuously advances the IS community (Lesmes and Morgan 2001; Leroy et al. 2008).

Qi and Wu (2024) model

Inspired by the fact that three-resistor and BHS models can respectively account for the bulk conductivity σ^* explicitly and implicitly, Qi and Wu (2022) realized that surface conductivity σ_s must be a hybrid parameter and there should be an effective medium model that not only has an explicit from like the three-resistor model but also uses the same number of parameters as the implicit BHS model. With this they developed an effective medium model for electrical conductivity (Qi and Wu 2022), and recently Qi and Wu (2024) extended it into the complex domain for IS via

$$\sigma^* = \frac{\sigma_w}{F} + \frac{(2\xi_w + 1)\sigma_c^*\sigma_w + 2(1-\xi_w)\sigma_c^{*2}}{(2+\xi_w)\sigma_c^* + (1-\xi_w)\sigma_w} \tag{18}$$

where ξ_w is the apparent surface fraction of water in such an EDL-and-water pattern. It is found that this model produces much the same effective EDL conductivity σ_c^* as the BHS model but also generates the same formation factor as Archie's law like three-resistor and Weller–Slater models (Qi and Wu 2024).

It testifies that the aforementioned models, except Waxman-type models, all embrace the EDL-and-water coupling pattern and therefore, these four models can be employed to overarch the meso-scale to bridge between the micro-scale EDL conduction/polarization and the macro-scale IS measurements in the laboratory and field. Taking the model of Qi and Wu (2024) as an example, as shown in Figure 10, if the micro-scale Cole–Cole model is embedded within the effective medium model of Qi and Wu (2024), it becomes

$$\sigma^* = \frac{\sigma_w}{F} + \frac{(2\xi_w + 1)\sigma_c^*\sigma_w + 2(1-\xi_w)\sigma_c^{*2}}{(2+\xi_w)\sigma_c^* + (1-\xi_w)\sigma_w}, \quad \sigma_c^* = \sigma_c^\infty + \frac{\sigma_c^0 - \sigma_c^\infty}{1+(i\omega\tau_c)^{c_c}} \tag{19}$$

In much the same vein, the macro-scale Cole–Cole model can be included in three-resistor, Weller–Slater, and BHS models as well, which all work for IS data interpretation. The choice then just depends on user preferences.

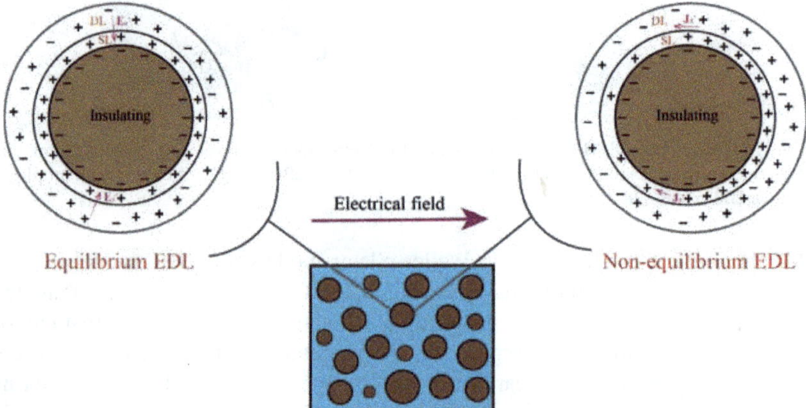

Figure 10. Diagram for the concept of assembling the micro-scale EDL model into an effective medium model. This conceptual diagram evidences that the micro-scale EDL polarization is different from the measured macro-scale polarization of the geomaterial, albeit the former is the original cause of the later.

MACRO-SCALE: FROM LABORATORY TO FIELD

The macro-scale is the scale where the geophysical instrumentation is applied. Like the aforementioned two scales, the macro-scale is also a relative terminology that depends on many factors such as the instrumentation type, the configuration, the deployment density, and the platform, which extends from centimeters or decimeters in the laboratory to kilometers and beyond for airborne or satellite-borne measurements in the field. That being said, the smallest unit of the macro-scale cannot be smaller than the meso-scale of representative element volume (REV), the smallest volume over which a geophysical measurement can be made to yield a representative value of the whole. Otherwise it spans the meso-scale that effective medium models take charge of. Mostly, for the laboratory IS measurement, the samples are considered homogenous as the REV and the measured responses are model-fitted, while for the field IS measurement, it is heterogenous with a multitude of REVs and thus the measured IS responses have to be inverted for interpretation, where, as is underscored, REV will be the smallest scale that field data inversion can reach.

Laboratory experiments

Laboratory IS experiments are not simple connecting the three objects of the IS instrument, the IS setup, and the sample together. Instead, there are much more professional skills behind this. Indeed, there has been a long tortuous journey for IS practitioners to obtain high-quality IS data, among which, two issues immensely decrease the quality of measured data: (I) IS instrumentation and (II) electrode polarization. Regarding the former, because of the poorer electrode-fluid contact in comparison with the metallic wire connection, the instruments designed for electrical engineering cannot work for IS, as the electrical engineering ones mostly hold a large input capacitance (e.g., > 40 pF). This large input capacitance is not a problem with the metallic connection in electrical engineering but when it couples with the electrode-electrolyte contact resistance in the IS experiment, it generates unacceptably huge phase errors for the IS data (cf. Zimmermann et al. 2008; Huisman et al. 2016). Normally used IS instruments hold an input capacitance less than 10 pF; for example, the input capacitance of the commercial PSIP instrument developed by Ontash and Ermac, Inc. is around 8.9 pF (Wang and Slater 2019). Because of the importance of the input capacitance of the IS instrumentation, more advanced techniques are introduced into the hardware development by IS engineers. One outstanding case is worth noting is that Breede et al. (2011) made an extremely high-quality IS instrument that has an input capacitance below 0.1 pF.

Besides the input capacitance of the IS instrument, the other significant issue is the influence of electrode polarization, which would dominate the signal of weak polarizable materials so that the data become useless. One straightforward way of excluding the main electrode polarization is to use the so-called four-electrode mode rather than the normal two-electrode mode (Vinegar and Waxman 1984), where two current electrodes (A and B) and two potential electrodes (M and N) are utilized for current injection and voltage acquisition, respectively (Fig. 11). Although the four-electrode setup improves data quality for the decrease of polarization magnitude in the potential electrodes, it still suffers from electrode polarization, as the M and N potential electrodes undergo redox reactions to pick up the signals from the electrolyte (Zimmerman and Huisman 2024). To eliminate this remaining electrode polarization, Lesmes and Frye (2001) retracted the M and N potential electrodes in small fluid-filled chambers to connect with the samples electrolytically, as adopted by Slater and Lesmes (2002) and Ulrich and Slater (2004), and now becomes one popular laboratory technique to exclude electrode polarization. However, the electrode retraction inevitably increases the contact resistance that couples with the aforementioned input capacitance of the IS instrument (Zimmermann et al. 2008), which makes the relatively high-frequency (>100 Hz) data even worse (Huisman et al. 2016). To make those above 100 Hz data usable, Wang and Slater (2019) developed a procedure based on the different retraction distances from the sample to exclude high-frequency contamination. Similarly but

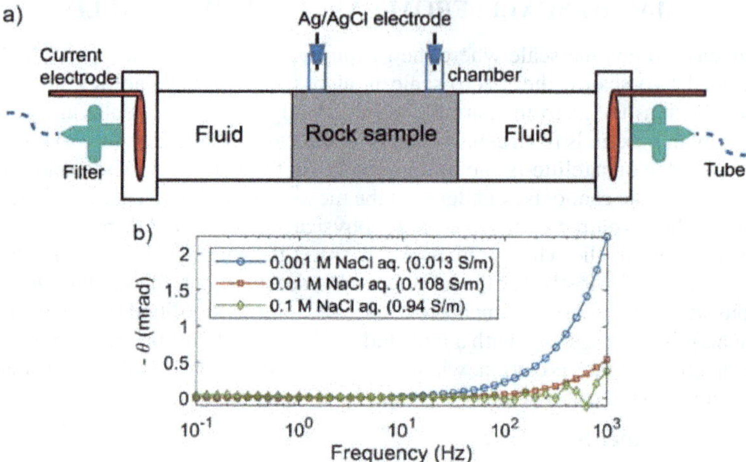

Figure 11. Scheme of an IS setup in the authors' lab for consolidated rocks and its measured phases when the column is filled only by NaCl solutions.

in an algorithm means, Qi and Wu (2024) introduced a wedge term based on the electrode polarization model from the electrochemistry community to exclude this high-frequency noise. That being said, all these hardware and algorithm techniques are only applicable to specific setups and the IS community seems to still have the "above 100 Hz headache", which means that more cautions must be exercised to interpret the IS data above 100 Hz unless they are proven to be free of countable electrode polarization.

One main aspiration for conducting laboratory IS experiments is to obtain sample properties, such as the porosity and permeability, the former assessed through the formation factor F and the latter estimated via S_{por} (the surface area per pore volume) or Λ (the characteristic pore/grain size) within σ_c*. However, as mathematically evidenced in Equation (18), F and σ_c^* cannot be accurately model-fitted without at least three pore water conductivities including the non-linear portion, which implies that multi-salinity measurements are mandatory to get the formation factor F and ξ_w. Otherwise they have to be estimated empirically. In comparison to the multi-salinity routines for clean geomaterials, taking the nonlinearity into consideration in the IS measurements does not change the routine multi-salinity measurement except that the salinities include some relatively lower ones such that the non-linear portion can be acquired. One caveat is that those salinities should be chosen neither too low nor too high, as at extremely low salinity (e.g., <0.05 S/m) EDL varies violently and at exceedingly high salinity (e.g., >1 S/m) a strong decrease of σ_s* occurs (see Fig. 6).

Field measurements

In contrast to the wide application of the frequency-domain IS (FDIS, also known as spectral IS) in the laboratory, time-domain IS (TDIS) is more commonly used in the field. Although in view of the Fourier transform, FDIS and TDIS are exactly the same in the least-squares sense based on the Parseval's theorem, they are quite different in engineering. This results from their different transmitting and receiving patterns. For FDIS, each frequency is emitted into the Earth and the response is measured, which is straightforward and has a high signal to noise ratio (SNR). However, this pattern does not exploit the full capacity of IS instrumentation because of the energy redundancy. In other words, we do not need such high SNR that is far above the sensitivity of modern IS instrumentation. By contrast, TDIS immediately turns off a square waveform and then logs its decay curves (Fig. 12). Since a turn-off signal ideally contains all the spectra ranging from DC to infinity, measuring the decay

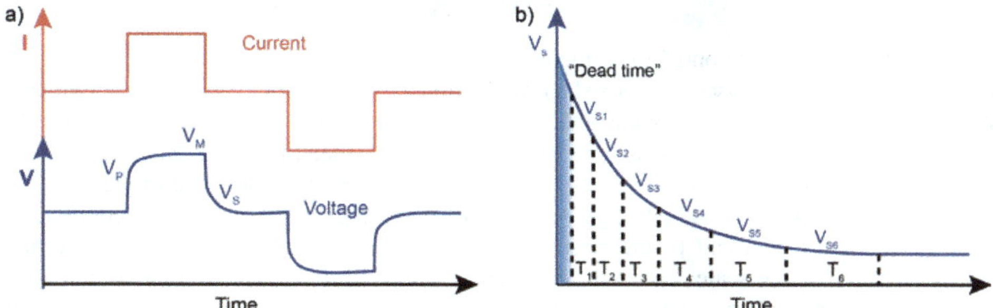

Figure 12. Applied transmitted current waveform and the recorded decay voltages of TDIS. **(a)** In one cycle TDIS measurement a bipolar waveform is used to subtract symmetric ambient noises. V_P, V_S, and V_M denote the primary, secondary, and maximum voltages, respectively. **(b)** Decay curve during the turn-off period. T_1–T_6 indicate the sequential measuring time windows and V_{S1}–V_{S6} are the integrated voltage in each window. Note that a "dead-time" such as 1 ms is always pre-set before the first window to exclude the EM coupling.

curve is much the same as acquiring the responses from one frequency to another, the cost being that the SNR of the TDIS data is always lower than the FDIS, especially for the very late-time decays in noisy circumstances (Martin et al. 2020). With the advances in IS instrumentation and for the pursuit of efficiency, nowadays TDIS measurements are much more commonly used in the field than the FDIS (Maurya et al. 2018), which is different from laboratory experiments.

Similar to the laboratory experiments, the field measurements also have the "above 100 Hz headache" (cf. Zimmerman et al. 2019). Different from the coupling between contact resistance and input capacitance of the instrument in the laboratory (Huisman et al. 2016; Wang and Slater 2019), in the field this E&EM coupling results mainly from capacitive and inductive coupling between transmitting wires and ground (Zimmerman et al. 2019). Despite the concerted effort from IS practitioners to decrease such E&EM couplings in field application (Weigand et al. 2022; Wang et al. 2024), the developed strategies mostly apply to specific IS configurations. Therefore, in many situations the most generic means to acquire accurate IS response is still limited to below 100 Hz where E&EM couplings are negligibly small (Kemna et al. 2012). With the equivalence between early time and high frequency (Qi et al. 2019), it implies that the early time windows in the TDIS measurements are prone to be contaminated by such E&EM coupling as well. Consequently, appropriate caution and carefulness are required during field data processing.

Unlike the standard multi-salinity experiments in the laboratory, routine to acquire the true formation factor F is impractical in the field as the large field cannot be flushed with brines as the small samples, which makes the hydraulic property assessment from the inverted tomograms challenging. In contrast to the conventional E&EM methods, the IS method provides a promising way to solve this problem thanks to its measurement of quadrature conductivity σ'' and normalized chargeability M_n. Based on the ratios between those two parameters with surface conductivity (i.e., $l = \sigma''/\sigma_s$, $l_{mn} = M_n/\sigma_s$), the surface conductivity σ_s can be gauged and excluded such that the true formation factor F can be assessed through the inverted conductivity tomogram. This approach can be backtracked to the work of Börner (1992) and Weller et al. (2013). From the aforementioned effective medium models, it transpires that the two ratios are compounds of both EDL and water properties, hence these two ratios must be calibrated with the representative samples in the laboratory rather than treating these two ratios as universal constants, otherwise the estimated hydraulic properties from the inverted tomograms would be unreliable or even erroneous.

SUMMARY

This chapter focused on IS fundamentals and applications at three scales, viz the micro-, the meso-, and the macro-scales, where landmark science and techniques and their evolutionary path are elaborated. The micro-scale is the origin of the IS signal with EDL polarization at the center; meso-scale is the REV scale that bridges the gap between the micro- and macro-scales where effective medium models are key; macro-scale is where laboratory and field work are conducted and thus a myriad of approaches is involved to exclude the E&EM coupling to acquire quality IS data and to interpret the inverted tomograms with hydraulic properties. Given that these three scales cover the most areas where IS practitioners work, our chapter provides an integrated perspective to understand the IS theories that may help to improve its quantitative applications at these three scales in the future.

REFERENCES

Archie GE (1942) The electrical resistivity log as an aid in determining some reservoir characteristics. Trans AIME 146:54–62, https://doi.org/10.2118/942054-G

Bikerman JJ (1933) Ionentheorie der Elektrosmose, der Strömungsströme und der Oberflächenleitfähigkeit. Z Phys Chem 163:378–394, https://doi.org/10.1515/zpch-1933-16333 (In German)

Binley A, Slater L (2020) Resistivity and induced polarization: Theory and applications to the near-surface Earth. Cambridge University Press

Bockris JO'M, Reddy AKN, Gamboa-Aldeco M (2000) Modern Electrochemistry: Fundamentals of Electrodics (2nd ed., Vol. 2A) Kluwer Academic Publishers

Börner FD (1992) Complex conductivity measurements of reservoir properties. Advances in Core Evaluation III (Reservoir Management) Gordon and Breach Science Publishers, London

Börner F (2009) Complex conductivity measurements. In: R. Kirsch (Ed.) Groundwater Geophysics: A Tool for Hydrogeology. Springer, p 119–153

Breede K, Kemna A, Esser O, Zimmermann E, Vereecken H, Huisman JA (2011) Joint measurement setup for determining spectral induced polarization and soil hydraulic properties. Vadose Zone J 10:716–726, https://doi.org/10.2136/vzj2010.0110

Bruggeman DAG (1935) Berechnung verschiedener physikalischer Konstanten von heterogenen Substanzen. I. Dielektrizitätskonstanten und Leitfähigkeiten der Mischkörper aus isotropen Substanzen. Ann Phys 416:636–664, https://doi.org/10.1002/andp.19354160705 (In German)

Bücker M, Hördt A (2013) Analytical modelling of membrane polarization with explicit parametrization of pore radii and the electrical double layer. Geophys J Int 194:804–813, https://doi.org/10.1093/gji/ggt136

Cole KS, Cole RH (1941) Dispersion and absorption in dielectrics I. Alternating current characteristics. J Chem Phys 9:341–351, https://doi.org/10.1063/1.1750906

Cosenza P, Ghorbani A, Camerlynck C, Rejiba F, Guérin R, Tabbagh A (2009) Effective medium theories for modelling the relationships between electromagnetic properties and hydrological variables in geomaterials: a review. Near Surf Geophys 7:563–578, https://doi.org/10.3997/1873-0604.2009009

Delgado ÁV, Arroyo FJ (2002) Electrokinetic phenomena and their experimental determination: an overview. *In:* ÁV Delgado (Ed.) Interfacial Electrokinetics and Electrophoresis, CRC Press, p 1–54

Dukhin SS (1993) Non-equilibrium electric surface phenomena. Adv Colloid Interface Sci 44:1–134, https://doi.org/10.1016/0001-8686(93)80021-3

Dukhin SS, Derjaguin BV (1974) Electrokinetic Phenomena. *In:* E Matijević (Editor) Surface and Colloid Science (Vol. 7) Wiley

Dukhin SS, Shilov VN (2002) Nonequilibrium electric surface phenomena and extended electrokinetic characterization of particles. *In:* ÁV Delgado (Ed.) Interfacial Electrokinetics and Electrophoresis. CRC Press, p 55–85

Friedman SP (2005) Soil properties influencing apparent electrical conductivity: A review. Comput Electron Agric 46:45–70, https://doi.org/10.1016/j.compag.2004.11.001

Fricke H, Curtis HJ (1937) The dielectric properties of water–dielectric interphases. J Phys Chem, 41:729–745, https://doi.org/10.1021/j150383a011

Glover PWJ (2015) Geophysical properties of the near surface Earth: Electrical properties. *In:* G Schubert (Editor-in-Chief) Treatise on Geophysics: Resources in the Near-Surface Earth, 2nd ed., Vol. 11. Elsevier, p 87–137

Glover PWJ, Ransford TJ, Auger G (2010) A simple method for solving the Bussian equation for electrical conduction in rocks. Solid Earth:85–91, https://doi.org/10.5194/se-1-85-2010

Hanai T (1960) Theory of the dielectric dispersion due to the interfacial polarization and its application to emulsions. Kolloid Z 171:23–31, https://doi.org/10.1007/BF01520320

Huisman JA, Zimmermann E, Esser O, Haegel FH, Treichel A, Vereecken H (2016) Evaluation of a novel correction procedure to remove electrode impedance effects from broadband SIP measurements. J Appl Geophys 135:466–73, https://doi.org/10.1016/j.jappgeo.2015.11.008

Johnson DL, Sen PN (1988) Dependence of the conductivity of a porous medium on electrolyte conductivity. Phys Rev B 37:3502, https://doi.org/10.1103/PhysRevB.37.3502

Johnson DL, Koplik J, Schwartz LM (1986) New pore-size parameter characterizing transport in porous media. Phys Rev Lett 57:2564, https://doi.org/10.1103/PhysRevLett.57.2564

Kemna A, Münch HM, Titov K, Zimmermann E, Vereecken H (2005) Relation of SIP relaxation time of sands to salinity, grain size and hydraulic conductivity. Near Surface 2005-11th European Meeting of Environmental and Engineering Geophysics

Kemna A, Binley A, Cassiani G, Niederleithinger E, Revil A, Slater L, Williams KH, Orozco AF, Haegel F, Hördt A, Kruschwitz S, Leroux V, Kitov K, Zimmermann E (2012) An overview of the spectral induced polarization method for near-surface applications. Near Surf Geophys 10:453–468, https://doi.org/10.3997/1873-0604.2012027

Kruyt HR, Jonker GH, Overbeek J (1952) Colloid Science. Vol. 1. Irreversible Systems. Elsevier

Leroy P, Revil A (2009) A mechanistic model for the spectral induced polarization of clay materials. J Geophys Res: Solid Earth 114:B10202, https://doi.org/10.1029/2008JB006114

Leroy P, Revil A, Kemna A, Cosenza P, Ghorbani A (2008) Complex conductivity of water-saturated packs of glass beads. J Colloid Interface Sci 321:103–117, https://doi.org/10.1016/j.jcis.2007.12.031

Lesmes DP, Frye KM (2001) Influence of pore fluid chemistry on the complex conductivity and induced polarization responses of Berea sandstone. J Geophys Res: Solid Earth 106(B3):4079–4090, https://doi.org/10.1029/2000JB900392

Lesmes DP, Morgan FD (2001) Dielectric spectroscopy of sedimentary rocks. J Geophys Res: Solid Earth 106(B7):13329–13346, https://doi.org/10.1029/2000JB900402

Lévy L, Gibert B, Sigmundsson F, Flóvenz ÓG, Hersir GP, Briole P, Pezard PA (2018) The role of smectites in the electrical conductivity of active hydrothermal systems: electrical properties of core samples from Krafla volcano, Iceland. Geophys J Int 215:1558–1582, https://doi.org/10.1093/gji/ggy342

Lyklema J (1995) Fundamentals of Interface and Colloid Science, Vol II: Solid Liquid Interfaces. Academic Press

Lyklema J (2002) The role of surface conduction in the development of electrokinetics. *In:* ÁV Delgado (Ed.) Interfacial Electrokinetics and Electrophoresis. CRC Press, p 87–97

Lyklema J, Dukhin SS, Shilov VN (1983) The relaxation of the double layer around colloidal particles and the low-frequency dielectric dispersion: Part I. Theoretical considerations. J Electroanal Chem Interfacial Electrochem 143:1–21, https://doi.org/10.1016/S0022-0728(83)80251-4

Lyklema J, Springer MM, Shilov VN, Dukhin S S (1986) The relaxation of the double layer around colloidal particles and the low-frequency dielectric dispersion: Part III. Application of theory to experiments. J Electroanal Chem Interfacial Electrochem 198:19–26, https://doi.org/10.1016/0022-0728(86)90022-7

Marshall DJ, Madden TR (1959) Induced polarization, a study of its causes. Geophysics 24:790–816, https://doi.org/10.1190/1.1438659

Martin T, Günther T, Orozco AF, Dahlin T (2020) Evaluation of spectral induced polarization field measurements in time and frequency domain. J Appl Geophys 180:104141, https://doi.org/10.1016/j.jappgeo.2020.104141

Maurya PK, Fiandaca G, Christiansen AV, Auken E (2018) Field-scale comparison of frequency-and time-domain spectral induced polarization. Geophys J Int 214:1441–1466, https://doi.org/10.1093/gji/ggy218

Mendieta A, Jougnot D, Leroy P, Maineult A (2021) Spectral induced polarization characterization of non-consolidated clays for varying salinities-An experimental study. J Geophys Res: Solid Earth 126:e2020JB021125, https://doi.org/10.1029/2020JB021125

Nordsiek S, Weller A (2008) A new approach to fitting induced-polarization spectra. Geophysics 73:F235–F245, https://doi.org/10.1190/1.2987412

Niwas S, Gupta PK, de Lima OA (2006) Nonlinear electrical response of saturated shaley sand reservoir and its asymptotic approximations. Geophysics 71:G129–G133, https://doi.org/10.1190/1.2196031

Qi Y, Wu Y (2022) Electrical conductivity of clayey rocks and soils: A non-linear model. Geophys Res Lett 49:e2021GL097408, https://doi.org/10.1029/2021GL097408

Qi Y, Wu Y (2024) Induced polarization of clayey rocks and soils: Non-linear complex conductivity models. J Geophys Res: Solid Earth 129:e2023JB028405, https://doi.org/10.1029/2023JB028405

Qi Y, El-Kaliouby H, Revil A, Ahmed AS, Ghorbani A, Li J (2019) Three-dimensional modeling of frequency- and time-domain electromagnetic methods with induced polarization effects. Comput Geosci 124:85–92, https://doi.org/10.1016/j.cageo.2018.12.011

Revil A (2013) Effective conductivity and permittivity of unsaturated porous materials in the frequency range 1 mHz–1GHz. Water Resour Res 49:306–327, https://doi.org/10.1029/2012WR012700

Revil A, Florsch N, Camerlynck C (2014) Spectral induced polarization porosimetry. Geophys J Int 198:1016–1033, https://doi.org/10.1093/gji/ggu180

Revil A, Qi Y, Ghorbani A, Coperey A, Ahmed AS, Finizola A, Ricci T (2019) Induced polarization of volcanic rocks. 3. Imaging clay cap properties in geothermal fields. Geophys J Int 218:1398–1427, https://doi.org/10.1093/gji/ggz207

Sauer MC, Southwick PF, Spiegler KS, Wyllie MRJ (1955) Electrical conductance of porous plugs-ion exchange resin-solution systems. Ind Eng Chem 47:2187–2193, https://doi.org/10.1021/ie50550a044

Sen PN, Scala C, Cohen MH (1981) A self-similar model for sedimentary rocks with application to the dielectric constant of fused glass beads. Geophysics 46:781–795, https://doi.org/10.1190/1.1441215

Schlumberger C (1920) Study of Underground Electrical Prospecting. Gauthier-Villars

Schwan HP (2000) Dielectric spectroscopy of biological materials and field interactions: the connection with Gerhard Schwarz. Biophys Chem 85:273–278, https://doi.org/10.1016/S0301-4622(00)00148-4

Schwan HP, Schwarz G, Maczuk J, Pauly H (1962) On the low-frequency dielectric dispersion of colloidal particles in electrolyte solution. J Phys Chem 66:2626–2635, https://doi.org/10.1021/j100818a066

Schwarz G (1962) A theory of the low-frequency dielectric dispersion of colloidal particles in electrolyte solution. J Phys Chem 66:2636–2642, https://doi.org/10.1021/j100818a067

Schroeder PA (2018) Clays in the Critical Zone. Cambridge University Press

Sihvola A (1999) Electromagnetic mixing formulas and applications. IET

Skold M, Revil A, Vaudelet P (2011) The pH dependence of spectral induced polarization of silica sands: Experiment and modeling. Geophys Res Lett 38:L12304, https://doi.org/10.1029/2011GL047748

Slater LD, Lesmes D (2002) IP interpretation in environmental investigations. Geophysics 67:77–88, https://doi.org/10.1190/1.1451353

Sumner JS (1976) Principles of induced polarization for geophysical exploration. Elsevier

Tarasov A, Titov K (2007) Relaxation time distribution from time domain induced polarization measurements. Geophys J Int 170:31–43, https://doi.org/10.1111/j.1365-246X.2007.03376.x

Tournassat C, Grangeon S, Leroy P, Giffaut E (2013) Modeling specific pH dependent sorption of divalent metals on montmorillonite surfaces. A review of pitfalls, recent achievements and current challenges. Am J Sci 313:395-451, https://doi.org/10.2475/05.2013.01

Ulrich C, Slater LD (2004) Induced polarization measurements on unsaturated, unconsolidated sands. Geophysics 69:762–771, https://doi.org/10.1190/1.1759462

von Smoluchowski M (1905) Zur Theorie der elektrischen Kataphoress und der Oberflachenleitung. Phys Z 6:529–531 (In German)

Vinegar HJ, Waxman MH (1984) Induced polarization of shaly sands. Geophysics 48:1267–1287, https://doi.org/10.1190/1.1441755

Wang C, Slater LD (2019) Extending accurate spectral induced polarization measurements into the kHz range: Modelling and removal of errors from interactions between the parasitic capacitive coupling and the sample holder. Geophys J Int 218:895–912, https://doi.org/10.1093/gji/ggz199

Wang H, Huisman JA, Zimmermann E, Vereecken H (2024) Tackling capacitive coupling in broad-band spectral electrical impedance tomography (sEIT) measurements by selecting electrode configurations. Geophys J Int 238:187–198, https://doi.org/10.1093/gji/ggae154

Waxman MH, Smits LJM (1968) Electrical conductivities in oil-bearing shaly sands. Soc Pet Eng J 8:107–122, https://doi.org/10.2118/1863-A

Weigand M, Zimmermann E, Michels V, Huisman JA, Kemna A (2022) Design and operation of a long-term monitoring system for spectral electrical impedance tomography (sEIT) Geosci Instrum Methods Data Syst 11:413–433, https://doi.org/10.5194/gi-11-413-2022

Weller A, Slater L (2012) Salinity dependence of complex conductivity of unconsolidated and consolidated materials: Comparisons with electrical double layer models. Geophysics 77:D185–D198, https://doi.org/10.1190/geo2012-0030.1

Weller A, Slater L, Nordsiek S (2013) On the relationship between induced polarization and surface conductivity: Implications for petrophysical interpretation of electrical measurements. Geophysics 78:D315–D325, https://doi.org/10.1190/geo2013-0076.1

Weller A. Zhang Z, Slater L (2015) High-salinity polarization of sandstones. Geophysics 80:D309–D318, https://doi.org/10.1190/geo2014-0483.1

Winsauer WO, McCardell WM (1953) Ionic double-layer conductivity in reservoir rock. J Pet Technol 5:129–134, https://doi.org/10.2118/953129-G

Worthington PF (1985) The evolution of shaly-sand concepts in reservoir evaluation. Log Anal 26:23–40

Wyllie MRJ, Southwick PF (1954) An experimental investigation of the SP and resistivity phenomena in dirty sands. J Pet Technol 6:44–57, https://doi.org/10.2118/302-G

Zibulski E, Klitzsch N (2023) Influence of inner surface roughness on the spectral induced polarization response—A numerical study. J Geophys Res: Solid Earth 128:e2022JB025548, https://doi.org/10.1029/2022JB025548

Zimmermann E, Huisman JA (2024) The effect of heterogeneous contact impedances on complex resistivity measurements. Geophys J Int 236:1234–1245, https://doi.org/10.1093/gji/ggad477

Zimmermann E, Kemna A, Berwix J, Glaas W, Münch HM, Huisman JA (2008) A high-accuracy impedance spectrometer for measuring sediments with low polarizability. Meas Sci Technol 19:105603, https://doi.org/10.1088/0957-0233/19/10/105603

Zimmermann E, Huisman JA, Mester A, Van Waasen S (2019) Correction of phase errors due to leakage currents in wideband EIT field measurements on soil and sediments. Meas Sci Technol 30:084002, ://doi.org/10.1088/1361-6501/ab1b09

Zukoski IV CF, Saville DA (1986) The interpretation of electrokinetic measurements using a dynamic model of the stern layer: I. The dynamic model. J Colloid Interface Sci 114:32–44, https://doi.org/10.1016/0021-9797(86)90238-9

Reviews in Mineralogy & Geochemistry
Vol. 91A pp. 105–148, 2025

4

Molecular Controls on Complexation Reactions and Electrostatic Potential Development at Mineral Surfaces

Jean-François Boily

Department of Chemistry, Umeå University, SE 901 87 Umeå, Sweden

jean-francois.boily@umu.se

INTRODUCTION

Adsorption reactions on minerals play crucial roles in regulating concentrations of dissolved metal ions and (in)organic ligands in natural waters. Models that accurately predict reaction products and yields over a wide range of geochemical conditions are strongly needed to understand the fate, transport and cycling of elements, contaminants and nutrients in nature (Sposito 1984, 2005; Stumm 1992; Brown et al. 1999; Al-Abadleh and Grassian 2003; Bañuelos et al. 2023). They are also needed to advance fundamental ideas on geochemically important mechanisms, including mineral dissolution, crystal growth, particle aggregation/flocculation, and interfacial electron transfer (Bañuelos et al. 2023).

Surface Complexation Models (SCMs) predict concentrations of chemical species at mineral surfaces in equilibrium with those in aqueous solutions (Davis and Kent 1990; Dzombak and Morel 1991; Stumm 1992; Lützenkirchen 2006b). The approach (Fig. 1) is a natural extension of aqueous speciation modeling methods (Morel and Hering 1993), with the exception that adsorption free energies are offset by electrostatic potentials generated by charged surface species. These potentials are obtained using an Electrical Double Layer (EDL) model, which accounts for the distribution of charges across the mineral/water interface. SCMs offer a clear advantage over traditional adsorption isotherm models (Sing et al. 1985; Do 1998) by predicting the surface speciation of minerals over a wide range of pH, and when applicable, ionic strength and even temperature (Machesky et al. 2015). Predictions can explain electrokinetic properties of colloidal suspensions (Hunter 1981), and can even be coupled to hydrological transport models (Parkhurst 1995).

Following the site-binding model (Schindler and Kamber 1968; Yates et al. 1974; Davis et al. 1978; Stumm 1992), SCMs predict adsorption using complexation reactions involving surface functional groups. Those exposed at surfaces of metal (Me) (oxy)(hydr)oxides include bare ≡Me sites alongside a variety of ≡Me-bound oxygen sites ($\equiv Me_nO$; Figs. 1a,b). As liquid water favors O-terminated surfaces (Henderson 2002):

$$\equiv Me + \equiv MeO + H_2O \rightleftharpoons 2 \equiv MeOH \tag{1}$$

SCMs have more conveniently been formulated with the generic ≡MeOH site (Fig. 1a), which can be used to predict the pH- and ionic strength-dependent surface charge ($\equiv MeO^{1-} + 2H^+ \rightleftharpoons \equiv MeOH^0 + H^+ \rightleftharpoons \equiv MeOH_2^+$), as well as metal(loid) and ligand adsorption. For example, the formation of a five-membered chelate between oxalate ($C_2O_4^{2-}$) and ≡Me can be modeled with:

$$\equiv MeOH^0 + H^+ + C_2O_4^{2-} \rightleftharpoons \equiv MeO_2C_2O_2^{1-} + H_2O \tag{2}$$

where water is the leaving unit (Fig. 1a).

1529-6466/25/0091A-0004$05.00 (print)
1943-2666/25/0091A-0004$05.00 (online)
http://dx.doi.org/10.2138/rmg.2025.91A.04

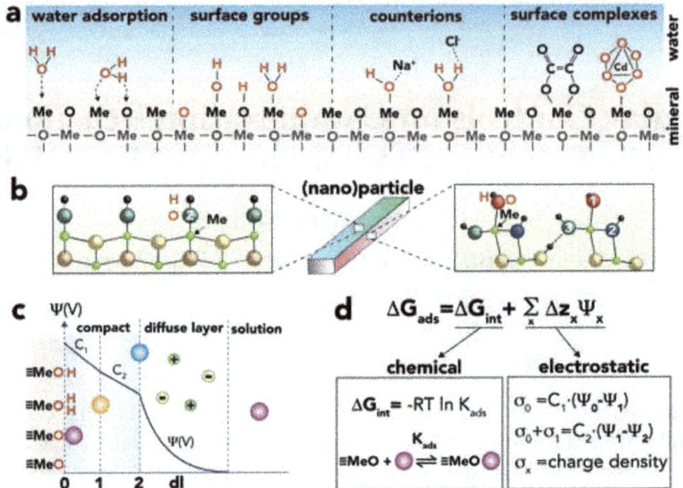

Figure 1. Key features in surface complexation modeling. **(a)** Generic metal oxide surface (MeO) exposed to liquid water produce amphoteric O-bearing functional groups (e.g., $\equiv MeO^{1-}$, $\equiv MeOH^0$, $\equiv MeOH_2^{1+}$) with charges balanced by counterions (e.g., Na^+ and Cl^-). These groups can form surface complexes with (in)organic species (e.g., oxalate and Cd^{2+}). **(b)** In reality, single (nano)particles can expose contrasting populations of surface O(H) groups (O coordinated with '1', '2', or '3' underlying Me sites) on different faces. The case shown here is a simplified representation of a rod-shaped lepidocrocite (γ-FeOOH) nanoparticle. **(c)** An Electrical Double Layer (EDL) model, here showing the electrostatic potential (Ψ) drop across compact planes (with capacitances C_1 and C_2) and the diffuse layer. **(d)** Adsorption data are modeled through a linear combination of an intrinsic chemical (ΔG_{int}) and an electrostatic ($\sum_x \Delta z_x \Psi_x$) term.

Just as in aqueous speciation modeling, SCMs require a solid set of modeling parameters, which can be extracted by modeling experimental adsorption data. The data used to extract SCM parameters are typically adsorption edges over pH, and at different ionic strengths. Proton exchange data can help tremendously in the search for species combinations (Öhman and Sjöberg 1996; Öhman 1998). This search can also be assisted using species identified by spectroscopy or by molecular modeling (Bañuelos et al. 2023).

To appreciate the challenges faced by surface complexation modelers in extracting modeling parameters from these types of data, let us begin by considering challenges already present in modeling relatively simpler homogeneous aqueous solutions. Here, equilibrium constants for ligand protonation, metal hydrolysis or metal–ligand complexation are often extracted by modeling potentiometric titration data, which provide information on the partitioning between free and complexed protons. Despite precise information on solution composition and reaction mechanisms, stringent experimental protocols are needed to obtain reliable equilibrium constants (e.g., $pK \pm 0.01$ (3σ)). For example, drifts in solution pH on the order of < 0.002 pH·h^{-1} (i.e., 0.1 mV·h^{-1}) were the hallmark of high-precision potentiometric titration studies, following practices of the Sillén school of thought (Öhman and Sjöberg 1996; Öhman 1998). This level of precision was often needed to study (e.g., metal–organic) complexation in multicomponent systems, else fictious species would have been needed to account for departures in the proton budget. This vulnerability can be likened to constructing a house of cards, where the entire structure can be compromised by just one weak card (i.e., p*K* value). Still, the pay-off for these stringent experimental conditions can be high, especially where models align with molecular-level information.

This vulnerability is not only carried over to SCMs, which relies on a reliable aqueous speciation model, but it is also compounded by uncertainties regarding the identity and spatial dispositions of reactive ≡MeOH species (Yates 1975; Hiemstra et al. 1989a,b, 1996; Villalobos et al. 2009; Song and Boily 2011a). This especially challenges surface complexation modelers wishing to incorporate molecular-scale information on the coordination environments and geometries of surface species (O'Day 1999; Brown and Sturchio 2002; Fenter and Sturchio 2004; Waychunas et al. 2005; Trainor et al. 2006; Catalano et al. 2008; Lefèvre et al. 2012; Henderson et al. 2014; Mudunkotuwa et al. 2014; Avena 2022). It can, however, be partially alleviated by *a priori* knowledge—or more realistically *assumptions*—of the identity, disposition, and surface densities of O(H) groups. This can be best achieved by working with minerals of low solubility and well-defined crystal habits (Fig. 1b).

An additional source of uncertainty in the extraction of SCM parameters relates to the implementation of an appropriate EDL model (Fig. 1c), which is crucial for modeling electrostatic potentials needed to predict adsorption free energies (Fig. 1d). Challenges include the search for molecularly-sound distributions of charges and compact layer capacitance values (e.g., C_1 and C_2 in Fig. 1c). These challenges underscore the difficulty for SCMs in achieving the same level of precision as in aqueous speciation models, risking a high level of intercorrelation between modeling parameters (Lützenkirchen 1999, 2006; Boily et al. 2000, 2001; Hwang and Lenhart 2008; Bompoti et al. 2018). As a result, researchers face a more intricate task when dealing with SCMs, necessitating careful consideration of these complexities in their modeling approaches.

This Chapter details key elements crucial for the advancement of tomorrow's SCMs needed to predict surface complexation reactions and EDL potential development by consideration of molecular-scale structure of mineral surfaces. Emphasizing essential aspects, the study cases featured in this Chapter predominantly focus on the geochemically important (nano)minerals of the iron oxide family (Schwertmann and Cornell 2003). This choice is motivated by the low solubilities of the minerals of this family, and possibilities in working with nanoparticles, microparticles and even larger specimens of well-defined crystal habits.

The Chapter begins with a section dedicated to the concept of surface site identity in multifaceted iron (oxy)(hydr)oxide minerals (see the *Surface Site Identity and Reactivity* section). It is followed by a section on mineral surface site hydration in electrolytes, and the adsorption of potential-determining ions (see *The Hydrated Interface* section), which are information also needed in the extraction of EDL parameters. The subsequent section addresses the development of electrostatic potentials and outlines a SCM framework that decouples electrostatic potentials between different crystallographic faces (see *The Charged Interface* section). Finally, the last section explores connections between EDLs used in SCMs and those in electrochemical studies of semi-conducting minerals (see *The Electrochemical Interface* section). Collectively, these four sections cover a range of concepts that must be interconnected to formulate modern-day SCMs needed to explain molecular-level processes at the mineral/water interface.

SURFACE SITE IDENTITY AND REACTIVITY

Mineral surfaces exposed to oxygen- and water-rich environments of Earth's upper crust and atmosphere are predominantly terminated by O-bearing functional groups. These groups saturate the coordination environment of metal ions at mineral surfaces into various O species (Fig. 2). These are the cornerstone species that drive the great majority of mineral-driven catalytic reactions of importance to environmental sciences.

Early work on mineral surface O species forged a foundation of understanding that is widely used today in SCM studies (Boehm 1966, 1971; Peri and Hensley 1968; Armistead et al. 1969; Cornell et al. 1974). While early SCMs made use of the generic ≡MeOH site presented

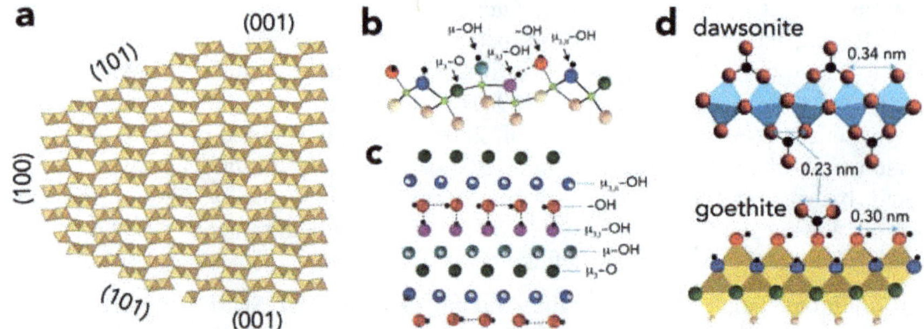

Figure 2. Schematic representation of goethite (α-FeOOH; *Pnma* space group) nanoparticle surfaces and of plausible carbonate surface complexes. **(a)** Dominant crystallographic faces. Modified from (Boily 2012). **(b)** Slice along the (101) face exhibiting singly (–OH), doubly, and three types of triply- (μ_3–OH) coordinated sites, according to Venema et al. (1998), and commonly used in the SCM literature. **(c)** Top view of (b). **(d)** Comparison of a portion of the structure of dawsonite ($NaAlCO_3(OH)_2$) along chains of carbonate-bound edge-sharing Al^{3+} octahedra with that of goethite with one mononuclear monodentate carbonate complex.

in the Introduction of this Chapter, these species are today more commonly distinguished by the number of underlying coordinated metal ions (Me^{z+}). These include singly- (–O), doubly- (μ–O) or triply- (μ_3–O) coordinated (Connelly et al. 2005) surface oxygens. In the context of SCMs, these sites are generally expressed with $\equiv Me_n{}^{nz/CN-2}$ (Hiemstra et al. 1989b), where n is the coordination number of O to Me^{z+} of charge z, and CN is the coordination number (e.g., predominantly 6 but in some cases 4 for iron-(oxy)(hydr)oxides). For this reason, modern SCMs for most minerals of the Fe^{3+}-oxide family consider the species $-O^{1.5-}$ ($\equiv FeO^{1.5-}$) for singly-, $\mu-O^{1.0-}$ ($\equiv Fe_2O^{1-}$) for doubly-, and $\mu_3-O^{0.5-}$ ($\equiv Fe_3O^{0.5-}$) for triply-coordinated oxygens.

Spatial distributions of sites on ideal vs. real particles

When exposed to water these groups speciate into hydroxylated (e.g., –OH) or aquo surface (e.g., $-OH_2$) species, the relative proportions of which are determined by their respective proton binding strengths:

$$-O^{1.5-} + H^+ \rightleftharpoons -OH^{0.5-} \tag{3}$$

$$-OH^{0.5-} + H^+ \rightleftharpoons -OH_2{}^{0.5+} \tag{4}$$

$$\mu-O^{1.0-} + H^+ \rightleftharpoons \mu-OH^{0.0} \tag{5}$$

$$\mu-OH^{0.0} + H^+ \rightleftharpoons \mu-OH_2{}^{1.0+} \tag{6}$$

$$\mu_3-O^{0.5-} + H^+ \rightleftharpoons \mu_3-OH^{0.5+} \tag{7}$$

The $\mu_3-OH^{0.5+}$ cannot typically protonate further because the oxygen is coordinatively saturated. Note that we can generally disregard quadruply-coordinated O sites, whenever present, in surface reactions. It also follows that, by virtue of their contrasting charges and coordination environments, these species have contrasting complexation strengths and ligand exchange capabilities.

Still, traditional SCMs typically only consider singly-coordinated sites because these are the predominantly active sites in the typical range of pH achieved in fresh and saline waters. While this tends to be an oversimplification, given evidence for the proton activity and hydrogen bond forming capabilities of other sites under these conditions, it is a sound approximation for strong metal-binding and ligand exchange reactions. This explains why metal and ligand sorption maxima are only a fraction (e.g., ⅓–¼) of the available O density on Fe^{3+}-oxide surfaces, which is about 10–15 O/nm^2, depending on mineral O packing patterns. On the other hand, complexation under more extreme environments, or strongly-binding ligands, should include binding with other groups. An example of a departure to this statement includes ligand exchange reactions between fluoride and μ–O groups on Al-hydroxide surfaces (Nordin et al. 1999).

Surface complexation modelers can greatly benefit from knowledge of the dominant crystallographic faces of (nano)mineral particles, that is when working with well-defined particles. This idea was considered in the early days of SCM research (cf. D.E. Yates' Ph.D. thesis; Yates 1975 on a variety of minerals; e.g., Si-, Ti-, Fe-oxides), and later in the first set of the MUltiSite Complexation (MUSIC) model papers (Hiemstra et al. 1989a,b).

A sound strategy is to build an O-terminated surface from a slice along a given plane (Figs. 2a–c), a task that completes the coordination environment of surface cations (e.g., typically 6 bonds per Al^{3+} and Fe^{3+} in their respective (oxy)(hydr)oxide phases). Next, a neutrally-charged surface, upon which surface complexation reactions can be described, can be achieved by protonating O sites according to their relative proton affinities. This practice is often guided by pK_a estimations using a correlation with bond valences (Hiemstra et al. 1996; Venema et al. 1998), but can also involve molecular modeling methods that allow O–H bond formation/breakage (e.g., *ab initio* Molecular Dynamics), or energy calculations (Watson et al. 1996) on preselected configurations. The constants must also align with known values of point of zero charge (*pzc*) and isoelectric point (*iep*) (Kosmulski 2016).

Using the neutrally-charged (101) face of goethite (α-FeOOH) as a flagship example for this Chapter (Figs. 2 a–c), a bond valence approach (Hiemstra et al. 1996; Venema et al. 1998) predicts coexisting $-OH^{0.5-}$ (~3 sites/nm^2), $\mu-OH^{0.0}$ (~3 sites/nm^2), $\mu_3-O^{0.5-}$ (~3 sites/nm^2), and $\mu_3-OH^{0.5+}$ (6 sites/nm^2) sites. Other, yet relatively similar, species combinations were obtained by *ab initio* Molecular Dynamics (Aquino et al. 2008; Leung and Criscenti 2012; Zhang et al. 2021). Still, regardless of the protonation state, the crystallographic termination model implies that complexation reactions should predominantly proceed along rows of evenly-spaced $-O^{1.5-}$ sites. Plausible binding geometries must therefore align with the relatively fixed intersite O–O (and underlying Fe–Fe) distances of ~0.30 nm (Fig. 2d). For instance, relative loadings of carbonate ion forming mononuclear monodentate

$$-OH^{0.5-} + H^+ + CO_3^{2-} \rightleftharpoons -OCO_2^{1.5-} + H_2O \tag{8}$$

vs. binuclear bidentate complexes

$$2{-}OH^{0.5-} + 2H^+ + CO_3^{2-} \rightleftharpoons -O_2CO^{1-} + 2H_2O \tag{9}$$

via ligand exchange could be directly controlled by this intersite spacing. Here, the limited range of O–O distances (0.23 nm) achieved by the carbonate ion can be used as an argument against the formation of binuclear bidentate complexes. Such an argument can also be supported by reference to the crystalline structure of chemical and structurally related materials. For instance, binuclear bidentate carbonate complexes of the dawsonite ($KAl(CO_3)(OH)_2$) bulk are only possible along rows of –OH groups because pairs of Al^{3+} octahedra are tilted towards one another (Fig. 2d). A similar finding was also obtained by Density Functional Theory calculations of carbonate-bearing Fe^{3+} dimers used to emulate mineral surfaces (Bargar et al. 2005; Kubicki et al. 2007). Because such tilting is an unlikely scenario for the (101) face of goethite, as it would require extreme changes along edges of double Fe octahedra, one could argue for the predominance of mononuclear monodentate complexes.

Crystallographic termination models can nonetheless break down when considering less ideally shaped or crystalline mineral particles. For instance, goethite nanoparticles made by the fast addition of reagents are considerably rougher than those made by slow additions under regular and sustained stirring (Hiemstra and van Riemsdijk 1991; Boily et al. 2001; Lützenkirchen et al. 2008). (cf. Sugimoto (2001) for an in depth account of physicochemical factors affecting metal oxide nanoparticle size, morphology and surface roughness.) Still, it is possible that nanoparticles produced by two contrasting synthetic methods have the same overall crystal habit, yet these expose different populations of O(H) groups.

Variations in reactive site density was notably inferred by differences in maximal chromate (Villalobos et al. 2009) and arsenate (Martínez et al. 2023) surface loading values on different types of goethite. This led these authors to model variations in reactivity between different particles by normalizing for specific surface area, while at the same time accounting for variations in crystalline face distributions evidenced by microscopy (Livi et al. 2017, 2023). This work suggested that tangential faces exposing high densities of sites are important contributors to surface complexation reactions. These faces can, as such, offer different types of coordination environments than those on dominant crystallographic faces. This concept could even help explain SCMs invoking low and high affinity sites, both in terms of the different steric constraints they offer for coordinating ions and in terms of surface loading-dependent electrostatics (Benjamin and Leckie 1981; Dzombak and Morel 1991; Mendez and Hiemstra 2020).

This idea of mixed sites disposition on minerals is a highly relevant one to consider for tackling the immensely complex question of element transport in Earth's subsurface, especially that it applies to the great majority of (non-ideally shaped) natural particles. Such particles are unlikely to have the well-expressed faces considered in the most ideal nanocrystals used in the research laboratories. In this sense, perhaps one of the most quintessential — and geochemically important for low temperature environments—of these minerals is ferrihydrite (Fig. 3). This nanosized mineral is highly reactive due to its small particle sizes and large concentrations of reactive –OH groups. Because its particle surface structure cannot be measured or yet clearly imaged, one cannot take any *a priori* assumptions on plausible adsorption geometries, such as in our idealized case of the (101) face of goethite. For this reason, SCMs predicting, for example, oxyanion binding, must heavily rely on additional spectroscopic information (e.g., Extended X-ray Fine Structure Absorption Spectroscopy) to formulate surface complexation reactions that best reflect coordination geometries (e.g., mononuclear monodentate *vs.* binuclear bidentate). In doing so, the model can be used to justify an *a posteriori* notion of the spatial dispositions of the exchangeable –OH groups.

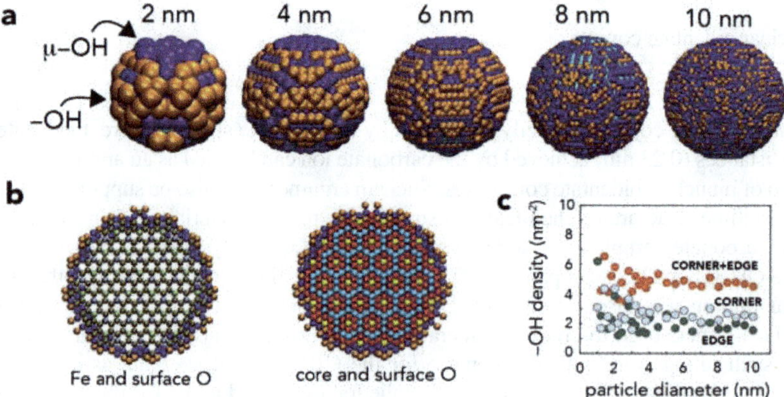

Figure 3. **(a)** Model ferrihydrite (2–10 nm wide) nanoparticles produced by spherical cuts of the crystallographic structure of Michel et al. (2007, 2010). **(b)** A cross section of one nanoparticle showing **(left)** Fe and surface O and **(right)** core and surface O. **(c)** Size-dependent surface density of –OH sites at edges and corners of Fe octahedra. Redrawn from Boily and Song (2020).

While resolving the identity of reactive sites at rough surfaces remains difficult, plausible configurations can still be explored with non-ideal cuts of crystallographic structures. For example, we explored OH populations on ferrihydrite nanoparticles spanning a wide range of sizes, Fe vacancies, and atomic displacements (Figs. 3a,b). This was facilitated by developing a code that generated nanoparticles by spherical cuts of the Michel et al. (2007, 2010) structure. The code then automatically healed the resulting surface for undercoordinated Fe and O sites and protonated O sites according to theoretical pK_a predictions, producing charge-neutral surfaces. The code also implemented the *Surface Depletion* model of Hiemstra (2013), in which two ferrihydrite Fe sites—one in tetrahedral, another in octahedral coordination—are depleted at particles surfaces. Also exploring the impact *Surface Depletion* depth into nanoparticles, the code generated virtual nanosized ferrihydrite particles with relatively high –OH densities (Fig. 3c), while yielding expected size-dependent mass densities (Michel et al. 2010). It also predicted the predominance of two types of –OH (corner and edges) and μ–OH sites, which were also detected by vibrational spectroscopy (Boily and Song 2020). Whole particle Molecular Dynamics simulations of these virtual particles showed that one set of –OH was isolated while another was strongly hydrogen bonded to adjacent μ–OH sites. Simulations also revealed that the high surface curvature of particles smaller than ~4 nm disfavored intersite hydrogen bonds, a finding that could possibly explain previous observations for the size-dependent uptake of phosphate/proton exchange (Wang et al. 2013) or even possibly pK_a values. Of note, it may be beneficial to consider a competing model for ferrihydrite (Drits et al. 1993) in the same fashion, especially considering recent accounts suggesting that contrasting formation conditions may affect structure (Sassi et al. 2021).

Crystallographic, tangential and spherical cuts of minerals of known bulk structure can thus be sound strategies to begin identifying plausible sites outcropping mineral particle surfaces. These can, additionally, provide a basis of interpretation for understanding the relationship between experimental values of maximal cation/ligand loadings and spectroscopically-derived coordination geometries. In the eventuality that these cannot be supported by theoretical cuts, other surface models can be proposed. Alternatively, severely roughened surfaces may be better represented using continuous distributions of affinity constants (Cernik et al. 1995; 1996; Rudzinski and Panczyk 2000).

Spectroscopic markers for site identity

A complementary, and independent, approach for identifying mineral surface OH groups is by tracking their O–H bond strengths using vibrational spectroscopy. This approach has been highly successful in the study of catalytic alumina (Knozinger and Ratnasamy 1978; Chandrasekharan et al. 2008), zinc oxides (Schiek et al. 2006; Raymand et al. 2011) and zeolites (Berlier et al. 2010). It has additionally benefited the study of soil-building Fe- and Al- (oxy)hydroxide nanoparticles through a string of papers by Farmer, Parfitt Russell and co-workers (Russell et al. 1974; Parfitt and Atkinson 1976; Parfitt and Russell 1977; Parfitt et al. 1977a–c; Lewis and Farmer 1986). We later explored these possibilities for a range of Fe^{3+} (oxy)(hydr)oxide nanominerals, and assisted these experimental efforts by molecular simulations (Boily and Felmy 2008; Rustad and Boily 2010; Song and Boily 2011a,b; Ding et al. 2012; Boily and Song 2020).

This approach works on the principle that O-H bond strengths of surface OH groups are affected by the number and strength of (i) Me–O and (ii) hydrogen bonds. The response is so sensitive that the O–H stretching frequency scales in the order of 150–175 cm^{-1} per pm (Ohno et al. 2005; Rustad and Boily 2010). (cf. work by K. Hermansson (e.g., Kebede et al. 2018) for a more detailed account on this response from a quantum mechanical viewpoint). Because metal coordination exerts strong effects on bond strength, O-H stretching frequencies (ν) typically follow the order ν(–OH) > ν(μ–OH) > ν(μ_3–OH). Also, because sites tend to be isolated (not hydrogen-bonded) or less strongly hydrogen-bonded than in the bulk of metal (oxy)hydroxides, their stretching frequencies are typically on the higher end of those of OH groups (e.g., ~3400–3900 cm^{-1}). Their spectral signatures can therefore be well-separated from bulk OH groups (Fig. 4).

Three conditions must be met to detect surface OH species using conventional, bench-top, vibrational spectroscopic techniques. The samples must (i) permit sufficient signal throughput, (ii) be free of water, and (iii) be of free impurities. When these conditions are met, (*nearly*) degenerate surface sites generate sharp O–H stretching bands of low full width at half maximum (Fig. 4c):

i) The *first condition*—sufficient throughput—relates to the number of OH groups needed to generate sufficiently intense bands of high signal-to-noise ratio. This can be appreciated by the size of surface bands in relation to bulk OH group in (oxy)(hydr) oxide minerals (Fig. 4). The commonly used Fourier Transform Infrared (FTIR) technique requires, as such, nanosized particles to generate bands of sufficiently high resolution. Typically, iron oxyhydroxide particles must have Brunauer–Emmett–Teller (Brunauer et al. 1938) specific surface area exceeding ~40–50 m^2/g. Work with larger particles require, in contrast with other methods of measurements, such as Infrared Reflection-Absorption Spectroscopy (IRRAS) or Sum Frequency Generation (SFG) spectroscopy (Geiger 2009; Covert and Hore 2016; Boily et al. 2019; Backus et al. 2021; Gonella et al. 2021; Piontek and Borguet 2022). Of these two methods, SFG has gained more interest for geochemical research because it is highly surface sensitive and because IRRAS is more generally limited to conductive materials. SFG studies appear, however, to have been mostly focused on transparent/ transluscent/white (e.g., CaF_2, SiO_2, α-Al_2O_3, MgO, muscovite) materials. This may possibly be because SFG measurements are often taken in total internal reflection mode, and in part to minimize beam damage in darker materials. SFG is nonetheless offering great insight into the populations and reactivity of mineral surface hydroxo groups. Still, caution should be exerted in comparing the chemistry of macro-sized sample surfaces used by SFG with those of nanoparticles. Concerns include sample preparation (e.g., mechanical and chemical polishing) which can lead to contaminations, variation in (molecular- to meso-scale) step densities, as well as variations in hydration structure (Lützenkirchen et al. 2018).

ii) The *second condition*—water-free systems—is crucial for producing a sufficiently large number of OH sites of (nearly) degenerate O–H bond strength, which are needed to generate narrow (small full width at half maximum). While this second condition may appear to be a major setback in the context of SCM development, as it intrinsically applies to all forms of vibrational spectroscopy, the study of mineral–gas systems still offers much needed insight. This experimental strategy for detecting OH populations provides the *primary knowledge upon which SCMs are developed*, as it can help validate populations inferred by structural models.

iii) Finally, the *third condition*—that surfaces are free of impurities—relates to the second one because it requires that OH populations reflect those of the most ideal crystallographic terminations. Achieving this facilitates a more direct comparison of spectral signatures to those that could be anticipated by theoretical predictions of crystallographically pristine mineral surfaces. For example, exchange of OH groups by traces of strongly nucleophilic ligands (e.g., phosphate or trace atmospheric organic compounds), can readily alter OH populations. Additionally, caution must be taken to remove residual salts which, under dry conditions, can ligand-exchange and produce metal-bonded anion complexes (e.g., –Cl instead of –OH) (Song and Boily 2011a, 2012a, 2013a).

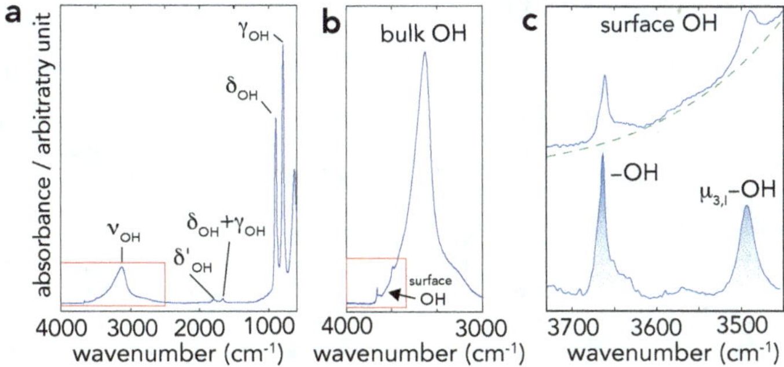

Figure 4. Fourier transform infrared spectrum of N_2(g)-dried synthetic nanosized goethite (α-FeOOH). **(a)** The entire 1000–4000 cm^{-1} region showing bulk OH groups (ν_{OH} = O–H stretching, δ_{OH} and γ_{OH} = O–H bending vibrations), with the red box highlighting the ν_{OH} region. **(b)** Close-up of the ν_{OH} region (**red box of (a)**) showing bulk and, in the **red box**, surface OH groups. **(c)** Close-up of the ν_{OH} region for surface OH group (**red box of (b)**) showing **(top)** the raw spectrum and a polynomial function (**green, dashed**) used to obtain **(bottom)** a spectrum of OH functional groups on goethite. Data were taken and redrawn from Song and Boily (2011a).

Lastly, it should be noted that Resonant Inelastic X-ray Scattering (RIXS) is another vibrational spectroscopic technique that could possibly probe surface OH groups in minerals (Harada et al. 2013). Additionally, although not a vibrational spectroscopic technique, X-ray photoelectron spectroscopy (XPS) can detect surface OH groups. It cannot, however, detect OH groups in different metal-coordination and hydrogen bond environments.

Case example: OH groups on iron (oxy)(hydr)oxide nanoparticles

Building upon earlier work by Farmer, Parfitt, Russell and co-workers (Russell et al. 1974; Parfitt and Atkinson 1976; Parfitt and Russell 1977; Parfitt et al. 1977a–c; Lewis and Farmer 1986), we resolved the spectral signatures of a variety of OH groups on minerals of the iron oxide family (Song and Boily 2011a, 2012b, 2016; Ding et al. 2012; Boily et al. 2015; Boily and Song 2020). These efforts identified common and distinctive attributes that could be explained by the spatial distributions of sites on the dominant crystallographic faces of synthetic nanoparticles. These were, in turn, resolved by Transmission Electron Microscopy and Selected Area Electron Diffraction.

The case of two types lepidocrocite (γ-FeOOH) nanoparticles of contrasting habits (Fig. 5) is a good starting point to explain how OH functional groups can be identified by vibrational spectroscopy. Here, we can compare the spectral signatures of rod-shaped lepidocrocite (RL), which is terminated by equal proportions of the (010) and (001) faces, with lath-shaped lepidocrocite (LL), which preferentially exposes the (010) face, as well as the minor (001) at particle edges.

From the crystallographic model of lepidocrocite, the (010) face is ideally terminated by only a two-dimensional array of μ–OH groups, while the (001) face is terminated by equal loadings of –OH, μ–OH and μ_3–OH each disposed along rows (Fig. 5). These differences were directly manifested in the relative intensities/area of the O–H stretching bands between these two sets of particles. The spectrum of LL particles contains a major band from μ–OH sites (3625 cm^{-1}) of the (010) face, alongside smaller bands from –OH (3667 cm^{-1}) and μ_3–OH (3552 cm^{-1}, 3534 cm^{-1}) sites of the (001) face. Those bands were, on the other hand, of comparable intensities/area in the RL particles as they exposed equivalent areas of the (010) and (001) faces. Note that our band assignment followed the aforementioned expectations for O–H frequencies in the order –OH > μ–OH > μ_3–OH, and was validated by proton binding, ligand exchange and temperature programmed desorption, as detailed in previous work

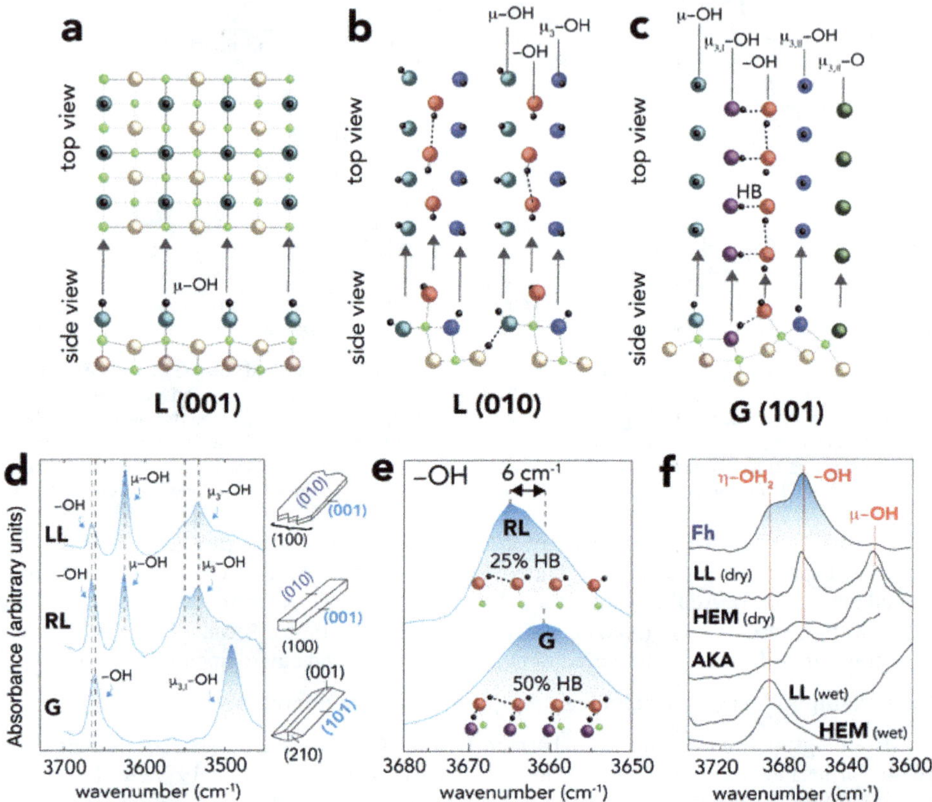

Figure 5. Surface O(H) groups on dominant crystallographic faces of **(a, b)** lepidocrocite (L) and **(c)** goethite (G) nanoparticles and **(d–e)** Fourier transform infrared spectra of corresponding dry synthetic materials. Intersite hydrogen bonds (HB) are shown as **dashed lines** in **(b)** and **(c)**. The right-hand side of (d) shows idealized nanoparticle habits with dominant crystallographic faces for Lath Lepidocrocite (LL), Rod Lepidocrocite (RL) and Goethite (G). **(e)** Close up of the O–H stretching bands of –OH sites on RL and G to highlight the 6 cm^{-1} difference in frequency caused by differences in the hydrogen bonding environments. Data from Song and Boily (2012b, 2013b). **(f)** Comparison of O–H stretching bands of dry and hydrated ('wet', i.e., exposed to water vapor) mineral surfaces with those of ferrihydrite (Fh) (HEM=hematite, AKA=akaganéite), as redrawn from Boily and Song (2020).

(Song and Boily 2012b, 2016). These assignments were validated further in an additional study (Kozin et al. 2014) showing that RL grew into LL by means of oriented attachment through the (010) face. This attachment was driven by the establishment of hydrogen bonds between μ–OH groups of approaching faces, which are the same bond types holding FeOOH sheets in the lepidocrocite bulk. Growth consequently decreased the proportions of surface μ–OH groups in relation to –OH.

Work with other minerals, including akaganéite (β-FeOOH) and hematite (α-Fe_2O_3) (Song and Boily 2011b; Boily et al. 2015), uncovered common spectral signatures for –OH, μ–OH and μ_3–OH sites for minerals of the iron oxide family (Fig. 5f). Minerals exposed to water vapor or reacted with proton loadings (Song and Boily 2011a, 2013c) also revealed chemisorbed water molecules onto bare Fe surface sites (η–OH_2). These Lewis acid centers can be at the apex (≡Fe–OH_2) or at equatorial edges (≡Fe–$(OH_2)_2$) of an undercoordinated Fe octahedron. Most often, these sites are exposed at the tips of acicular FeOOH nanoparticles.

The consistency in band assignments between iron (oxy)(hydr)oxide phases (Fig. 5f) served as a strong impetus for identifying OH groups on the less well-defined surfaces of ferrihydrite nanoparticles (Boily and Song 2020). This comparison supported the concept that ferrihydrite exposed singly-coordinated groups (–OH), doubly-coordinated groups (μ–OH), and chemisorbed water (η–OH_2).

Spectroscopic markers for protonation

Further validation in these band assignments was given by study of the sensitivity of band intensities to proton loadings (Fig. 6). This was achieved by measurements of dry particles that were previously equilibrated in aqueous suspensions over a range of pH values. Because drying concentrates surface loadings of H^+, OH^- and counterions, the approach amplifies the response to proton binding ($-OH^{0.5-} + H^+ \rightleftharpoons -OH_2^{0.5+}$) or release ($-OH^{0.5-} \rightleftharpoons -O^{1.5-} + H^+$), compared to aqueous suspensions.

Work on lepidocrocite (Song and Boily 2012b) revealed that –OH groups (3667 cm^{-1}) responded more strongly to proton loadings than the μ–OH group on the (010) face, and that μ_3–OH group were insensitive to protonation. In the case of goethite, protonation of –OH (3661 cm^{-1}) to –OH_2 ruptured the donating hydrogen bond by underlying μ_3–OH groups (3491 cm^{-1}; Fig. 5c). This rupture was revealed by a blue-shift in the μ_3–OH band (cf. 'no HB' in Fig. 6a), signaling a strengthening of the O-H bond strength.

These spectral markers thus provided a direct and *independent* experimental verification (Fig. 6b) for the relative affinity of surface O groups for protons. These were, additionally, extended further to the cases akaganéite (Song and Boily 2011a), and ferrihydrite (Boily and Song 2020).

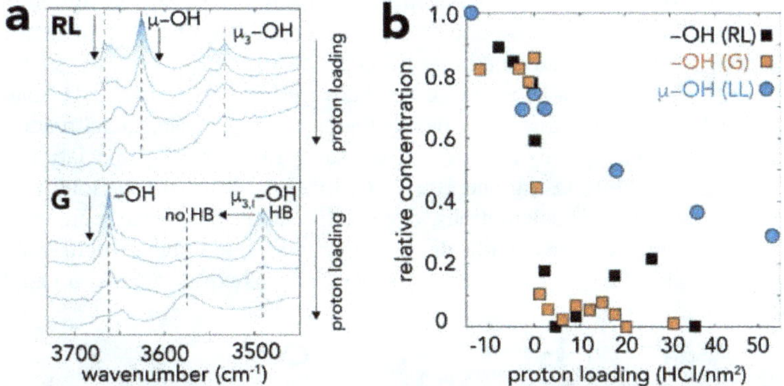

Figure 6. Vibrational spectral evidence for the protonation of surface OH groups on Rod Lepidocrocite (RL) and goethite (G). Data from (Song and Boily 2012b). **(a)** Spectra of N_2(g)-dried RL and G after reaction in HCl-bearing aqueous suspensions. **(b)** Relative concentrations of selected spectral components of RL and G in relation to total HCl. These results revealed a stronger response of –OH groups to protonation than μ–OH.

Spectroscopic markers for site-specific binding

Spectral markers can also be used to identify sites involved in surface complexation reactions. This can be achieved in samples (i) that were previously reacted in aqueous media, or (ii) directly exposed to a stream of gas(es) (e.g., CO_2, H_2O, CO_2+H_2O).

A primary example of site-specific binding was observed for fluoride binding (Ding et al. 2012), where band intensities of –OH were readily attenuated in samples that were previously reacted in fluoride-bearing solutions. This was explained in terms of the ligand exchange reaction:

$$-\mathrm{OH}^{0.5-} + \mathrm{F}^- \rightleftharpoons -\mathrm{F}^{0.5-} + \mathrm{OH}^- \quad (10)$$

leaving μ–OH and μ_3–OH sites unaltered. High fluoride loadings nonetheless exchange with μ–OH groups of the (010) face of lepidocrocite:

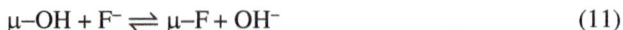

$$\mu\text{–OH} + \mathrm{F}^- \rightleftharpoons \mu\text{–F} + \mathrm{OH}^- \quad (11)$$

as can be understood by the strong ligand exchange capability of the fluoride anion. A similar reaction was resolved for the case of fluoride-binding on Al-(oxy)(hydroxides), and was identified as a pathway towards ligand-promoted dissolution (Nordin et al. 1999). For reference, even phosphate cannot ligand exchange μ-OH groups (Ding et al. 2012). This consequently supports the concept that –OH is the primary ligand-exchangeable site that should be used to develop SCMs for Al(III)- and Fe(III) (oxy)(hydr)oxide surfaces. Still, this does not exclude the necessity for SCMs to account for neighboring groups involved stabilizing (e.g., via hydrogen bonding) large molecules (e.g., multifunctional organics).

An additional case revealing site-specific reactions was studied by monitoring CO_2(g) binding reactions on lepidocrocite (Fig. 7). These reactions were manifested through the preferential attenuation of the O-H stretching band of –OH, while the bands of μ–OH and μ_3–OH sites were unaltered (Fig. 7 a,b). This supported the concept that μ–OH and μ_3–OH sites cannot react with CO_2. It can even imply that that (bi)carbonate species are chiefly located at particle edges and terminations, and not on the (010) face where μ–OH is the predominant species (Fig. 7c). In reality, defect –OH sites on this plane (~0.9 site/nm^2) are also likely CO_2-binding sites (Boily and Kozin 2014).

Goethite –OH sites also responded strongly to CO_2 binding (Fig. 7), although it should be noted that important shifts in the band of the neighboring μ_3–OH sites were not the result of complexation. Shifts rather indicated that μ_3–OH was a stronger donor to –O–CO_2 species than to –OH. This can be taken as evidence for the sensitive response of O–H bond strengths to hydrogen bonding, and one that deepens our understanding of interfacial reactions. Bands of (bi)carbonate species (Fig. 7b) revealed, at the same time, striking speciation differences in lepidocrocite and goethite (Song and Boily 2013b). Based on comparisons with (Density Functional Theory) frequency calculations, dry goethite appears to have a greater propensity for stabilizing carbonate species while dry lepidocrocite can stabilize both carbonate and bicarbonate species (Fig. 7d). This difference in speciation was explained by contrasting intersite

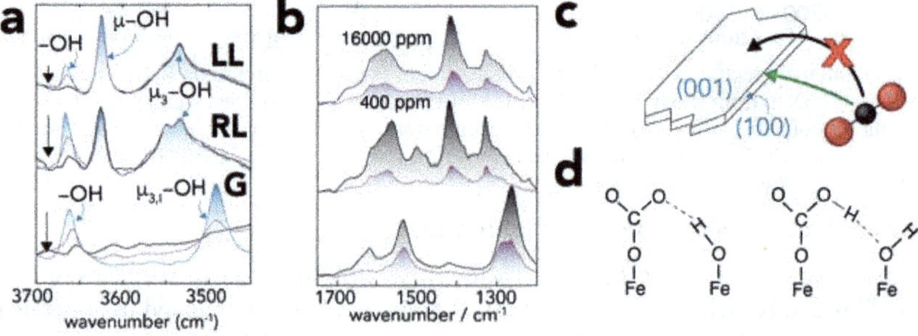

Figure 7. Vibrational spectral signature of **(a)** surface OH groups and **(b)** (bi)carbonate surface complexes on lath lepidocrocite (LL), rod lepidocrocite (RL) and goethite (G) formed by exposure to 0 (**blue**), 400 ppm (**pink**) and 16000 ppm (**black**) CO_2(g) (pH_2O=0 kPa). Data taken and redrawn from (Song and Boily 2013b). Reactions consumed only –OH groups, possibly implying that **(c)** the pristine (001) face of lepidocrocite is impervious to carbonate species formation. In reality, defect –OH sites on this plane (~0.9 site/nm^2) are likely CO_2-binding sites (Boily and Kozin 2014). **(d)** Carbonate (left) and bicarbonate (species) formed along rows of –OH groups.

hydrogen bond populations along rows of –OH sites on both minerals. The predominance of carbonate species on lepidocrocite can be explained by the greater number of isolated –OH sites (~¾) available for stabilizing the charge of the $–O–CO_2$ species via hydrogen bonding. Bicarbonate species are, in contrast, more predominant on goethite because fewer free –OH sites are available for hydrogen bonding. These contrasting results for lepidocrocite and goethite thereby signal the considerable role hydrogen bonding plays in stabilizing species, one that warrants similar lines of thinking when developing SCMs for fully hydrated systems. These results also open a path towards describing a mechanistic pathway for carbonate formation via attachment of CO_2 onto Fe-bound OH groups. Most notably, hydrogen bonding is likely the primary mechanism inducing O–C–O axis bending needed for nucleophilic attack of CO_2 by –OH sites. This happens to be one of many strongly overlapping areas of interest in mineral surface geochemistry and catalysis (Taifan et al. 2016).

Spectroscopic markers for site-specific hydration

The vibrational spectral response of surface OH groups can also be used to track mineral-water interactions, which can be beneficial in advancing SCM concepts. This can be effectively achieved by tracking the response of OH groups as water vapor condenses on minerals in the form of thin, nanometric, water films. These films are primarily held by surface OH groups via hydrogen bonding (Figs. 8–10).

Measurements on rod lepidocrocite (RL; Fig. 5d) revealed preferential interactions with –OH groups of the (001) face rather than μ–OH groups of the basal (010) face (Figs. 8 a,b). Still, most band intensities were completely lost when nanoparticles were covered by ~1 monolayer water (i.e., ~10–12 H_2O/nm^2). Hydration also generated the aforementioned (η–OH) species, which are likely to occur at terminations (e.g., (100) face) of RL nanoparticles. Of note, comparable bands for η–OH were detected in hydrated akaganéite (Song and Boily 2012a; Kozin and Boily 2013a), hematite (Boily et al. 2015), and ferrihydrite (Boily and Song 2020). Possible first layer water structures simulated by Molecular Dynamics (Fig. 8 c,d) revealed a considerably more complex network on the (100) face. Simulations also showed that first layer water structure and hydrogen bonding were preserved in RL exposed to liquid water (Boily 2014).

Water films formed on goethite induced important changes in the hydrogen bond environment of –OH groups (Fig. 9a). Molecular simulations suggested that these groups accepted about twice as many hydrogen bonds than they donated to first layer water

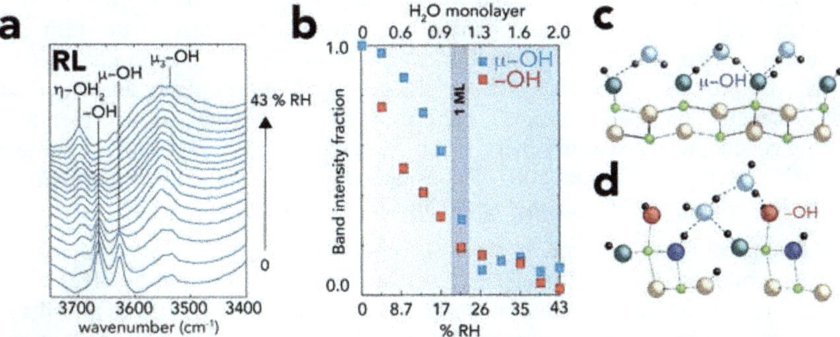

Figure 8. Water films on rod lepidocrocite (RL). **(a)** Hydrogen bonds formed between OH groups and water vapor (0–43 % relative humidity, RH) alter the spectral signatures of O–H stretching bands. **(b)** Response of normalized band intensities of –OH and μ–OH sites with relative humidity (%RH) and water loadings, showing that free O–H bonds disappear at ~1 water monolayer (ML). **(c–d)** Molecular Dynamics revealed hydrogen bonding environments of first layer water molecules on the **(c)** (010) and **(d)** (001) faces of lepidocrocite. Data taken and redrawn from Song and Boily (2013c).

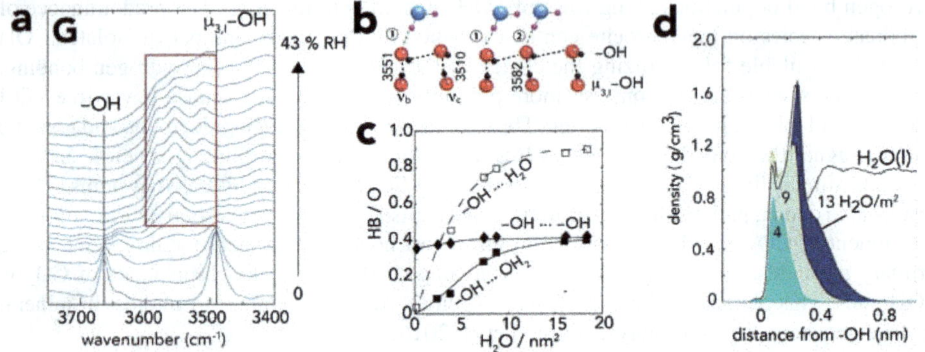

Figure 9. **(a)** Hydrogen bonds formed between surface hydroxo groups and water vapor condensed on goethite (G) alter spectral signatures of surface OH groups. **(b)** Response of bands intensities of –OH and $\mu_{3,I}$–OH sites with relative humidity (%RH). These show that free O–H bonds –OH are attenuated by adsorbed water. The band of $\mu_{3,I}$–OH, which is a hydrogen bond donor to –OH, unravelled into a blue-shifted triplet as the hydrogen bond was weakened by first water layer bound to –OH. (b) A schematic representation of the formation of the triplet. **(c)** Hydrogen bonding analysis obtained by Molecular Dynamics simulations showing that water adsorption does not perturb the intersite –OH hydrogen bonds. Hydrogen bond populations reach a maximum at ~15 H_2O/nm^2, which is roughly the equivalent of a single water monolayer on goethite. **(d)** Water density profiles obtained by Molecular Dynamics for 3 water densities and for liquid water contacted with the (101) face of goethite. All data are taken and redrawn from Song and Boily (2013a).

molecules (Figs. 9b,c). This can be understood by noting that the aforementioned network of intersite –OH···–OH bonds remained unperturbed with hydration (Fig. 9c), a configuration that preferentially exposed oxygen valence electron pairs to first layer water molecules. Interestingly, these hydrogen bonds weaken, or even rupture, hydrogen bonds from the adjacent μ_3–OH sites, producing three new blue-shifted bands (cf. red box in Fig. 9a) indicative of three specific hydration states along rows of –OH groups. At the same time, molecular simulations revealed an intricate arrangement of water molecules, and an interfacial water structure that directly relates to the one of liquid water (Boily 2012). This structure consists of 3 water monolayers within ~0.35 nm from the topmost –OH sites (Fig. 9d).

It is worth noting that these multilayered water films can mediate surface complexation reactions akin to those taking place in aqueous suspensions of minerals. Examples can be found in the petroleum science literature where mechanistic SCMs were linked to wettability models (Korrani and Jerauld 2019; Khurshid and Al-Shalabi 2022), or in vadose zone studies where pore size, mineral type and solution chemistry are thought to control water film thickness (Nishiyama and Yokoyama 2021). Note that the establishment of EDLs in nanometric water films was even previously demonstrated for the case of forsterite using an electrochemical method (Placencia-Gómez et al. 2020). These are certainly an avenue worth developing further considering geochemical and atmospheric chemical research on mineral-supported water films (Ewing 2006; Petrik and Kimmel 2007; Rubasinghege and Grassian 2013; Tang et al. 2016; Placencia-Gómez et al. 2020; Yalcin et al. 2020; Luong et al. 2022).

One example of such systems of interest to geochemistry and to atmospheric sciences includes the formation of (bi)carbonate surface complexes when CO_2(g) dissolves in mineral-supported water films, formed by the condensation of air moisture (Fig. 10). CO_2(g) dissolution in water films held by goethite nanoparticles produced (bi)carbonate species that could be monitored through C–O stretching bands but not through O–H stretching bands, as these were attenuated by hydrogen bonding with first layer water molecules (Yesilbas et al. 2020). These C–O spectroscopic markers indicated that the equivalent of only ~5 H_2O/nm^2 (i.e., ~0.4 monolayer) were needed to fully hydrate (bi)carbonate species. This level of hydration even increased carbonate loadings by facilitating complexation reactions with unreacted

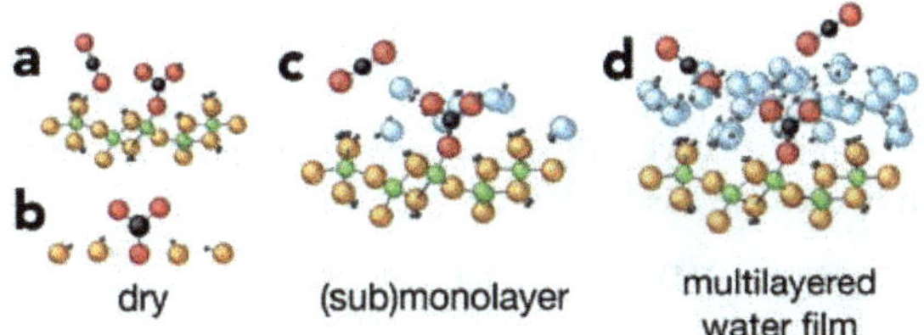

Figure 10. Schematic representation of CO_2 conversion to carbonate species on **(a–b)** dry, and **(c–d)** hydrated goethite surfaces. **(b)** Depiction of a monodentate mononuclear carbonate species bound along rows of –OH sites. Modified from (Yesilbas et al. 2020). (**yellow** = goethite oxygens; **red** = carbonate and carbon dioxide oxygens; **green** = iron; **turquoise** = water oxygens; **black** = carbon; **grey** = protons).

–OH sites. In contrast, carbonate loadings became lower in multilayered water films as free water facilitated the production of carbonic acid, lowering film pH. It could have also been possible that these acidic conditions promoted surface charge ($-OH^{0.5-} + H^+ \rightleftharpoons -OH_2^{0.5+}$) and consequently an EDL with (bi)carbonate species as counterions.

Although only one example was cited here, surface complexation modeling of multilayered water films on minerals remains a largely unexplored avenue of great importance to environmental sciences. It also represents a transitional regime between dry minerals, where surface OH groups can be tracked by vibrational spectroscopy, and hydrated minerals, where spectral signatures are completely attenuated. Fully hydrated environments are then best studied using other techniques (e.g., X-ray reflectivity, Extended X-ray Fine Structure Absorption Spectroscopy, Crystal Truncation Rod).

THE HYDRATED INTERFACE

Expanding on the subject of site-specific reactions on nanominerals exposed to (moist) air, this section delves into the fundamental aspects of hydration and reactivity of fully hydrated mineral surfaces in aqueous media. It explores the connection between these processes and the development of modern-day SCMs, particularly emphasizing how the interfacial hydration of minerals influences the generation of surface charge—a key phenomenon altering the free energies of interfacial reactions predicted by SCMs.

In essence, all SCMs for oxide minerals are built upon reactions that involve the adsorption of potential-determining ions. In aqueous solutions, these reactions involve H^+ adsorption on surface (hydr)oxo groups (e.g., $\equiv FeO^{1.5-} + H^+ \rightleftharpoons \equiv FeOH^{0.5-}$) and OH^- adsorption on Lewis acid sites (e.g., $\equiv Fe^{0.5+} + OH^- \rightleftharpoons \equiv FeOH^{0.5-}$). It is important to note that the notation '≡' is preferred here to emphasize the presence of underlying metal sites, such as Fe^{3+}, as is usually employed in SCMs. In this fashion the –OH notation used in the previous section for singly-coordinated sites, becomes $\equiv FeOH^{0.5-}$ in the context of SCMs. Likewise, μ–OH is $\equiv Fe_2OH^{0.0}$ and μ_3–OH is $\equiv Fe_3OH^{0.5+}$.

The connection between these reactions, surface charge, metal hydrolysis and even mineral solubility in aqueous solutions was made in two landmark papers in the 1960's (Parks and Bruyn 1962; Parks 1965). These papers highlighted the link between the hydrolysis of metal ions (Me^{z+}):

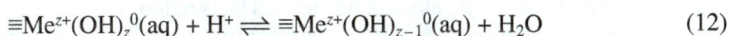

$$\equiv Me^{z+}(OH)_z^0(aq) + H^+ \rightleftharpoons \equiv Me^{z+}(OH)_{z-1}^0(aq) + H_2O \tag{12}$$

and proton adsorption:

$$\equiv MeOH^0 + H^+ \rightleftharpoons \equiv Me(OH)_2^+ \qquad K_1 \qquad (13)$$

which is related to the *pzc* and *iep* of mineral surfaces. Kosmulski later published a historical perspective on Park's seminal work, 50 years after its publication (Kosmulski 2016). The aqueous analogy with hydroxide adsorption is described through:

$$\equiv Me^{z+}(OH)_z^0(aq) + OH^- \rightleftharpoons \equiv Me^{z+}(OH)_{z+1}^-(aq) \qquad (14)$$

and the corresponding surface reaction is:

$$\equiv MeOH^0 + OH^- \rightleftharpoons \equiv MeO^- + H_2O \qquad (15a)$$

namely:

$$\equiv MeOH^0 \rightleftharpoons \equiv MeO^- + H^+ \qquad K_2 \qquad (15b)$$

Using $\equiv MeOH^0$ as the quintessential mineral surface site, Equations (13) and (15b) collectively represent the 2-pK model (Yates et al. 1974; Chan et al. 1975; Davis et al. 1978; Davis and Leckie 1978b), which dominated the SCM literature until at least the early 2000's, if not later. Because the model invoked two successive protonation steps on a single oxygen site within unreasonably small steps in pK_a units, it was suggested to account for the two-step protonation of two oxo sites at metal ion octahedral edges of the type $\mathrm{Me}^{\mathrm{OH}}_{\mathrm{OH}}$ (van Riemsdijk et al. 1987), such that the first protonation step produces $\mathrm{Me}^{\mathrm{OH_2}}_{\mathrm{OH}}$ and the second $\mathrm{Me}^{\mathrm{OH_2}}_{\mathrm{OH_2}}$. In truth, these more likely represent equatorial O on Me sites, which tend to be exposed at tips of acicular/tabular Fe^{3+}-(oxy)(hydr)oxide particles rather than on major crystallographic faces. Considering the previous section on site identity, SCMs should rather adopt speciation models that align with realistic surface sites, and these predominantly involve dangling O groups of the type $\equiv Me_nO^{nz/CN-2}$ and $\mathrm{Me}^{\mathrm{OH}}_{\mathrm{OH}}$, where applicable.

On the other hand, SCMs for mineral nanoparticles of unknown crystal habit or for other reasons (e.g., when coupling with reactive transport models, as in Luo et al. 2021) require a simplified system of equations. One approach is the 1 pK model (Bolt and van Riemsdijk 1982), whereby a generic singly-coordinated groups of the type $\equiv MeOH^{0.5-}$ is the sole site invoked in the model. Here, the pK_a of the site is the pH of the point of zero charge (pK_a = pH$_{pzc}$) or of the isoelectric point (pK_a = pH$_{iep}$). Alternatively, a continuous distribution of pK_a values could express site heterogeneity (Prélot et al. 2002).

For a complete description of all possible reactions, the MUSIC model offers tremendous advantages by accounting for site-specific pK_a values, which collectively explain crystallographic face dependent point of zero charge (Hiemstra et al. 1989a,b; Hiemstra and van Riemsdijk 1996). The MUSIC model also helps a SCM modeler interpret adsorption, plausible coordination geometries, using *a priori* information on the spatial disposition of sites (e.g., parallel rows of $-OH^{0.5-}$, μ–OH^0 and μ_3–$OH^{0.5-}$ sites on the (101) face of goethite, Fig. 2). It offers ideas on plausible coordination geometries of adsorbed metal and ligands, and especially along rows of the reactive $-OH^{0.5-}$ sites. Using more representative pK_a values can also improve the ability of a model to account for pH- and ionic strength-dependent speciation, such as that of carboxylic acids bound as metal- and hydrogen-bonded complexes on minerals. Metal-bonded complexes are favored chiefly under conditions of low pH and high ionic strength as they promote chemisorbed water ($-OH_2^{0.5+}$), which are more readily ligand–exchanged than metal–bonded hydroxide ions ($-OH^{0.5-}$). As a result, properly accounting for the pH–dependent speciation of $-OH^{0.5-}$ is a key goal for models predicting molecular–scale processes. This should require detailed information on how the stability of interfacial species is coupled to local hydration environments and to the interfacial distribution of charge-neutralizing (e.g., electrolyte) ions. This section will highlight selected molecular-scale mechanisms that explain the proton affinity of surface O groups involved in various metal coordination and hydrogen bond environments at mineral surfaces.

Hydration controls on protonation constants

The *intrinsic* proton affinity of surface O(H) groups is a site-specific property, which is formally defined as the negative of the enthalpy change of the reaction. The Gibbs free energy of the reaction needed to obtain protonation constants ($\Delta G_{int} = -RT \ln K_a$) is, however, also affected by the entropy of formation and mixing. It is therefore sensitive to the hydration environment of surface species. In this sense, the great advances made in understanding interfacial hydration of mineral surfaces (Fenter and Sturchio 2004; Backus et al. 2021; Bañuelos et al. 2023) can guide ideas on how first layer water molecules affect the protonation constants of surface O(H) groups.

Of the various approaches that can be used to predict pK_a values (Ho 2014) for O sites on minerals, *ab initio* Molecular Dynamics (Cheng and Sprik 2010; Leung and Criscenti 2012; Tournassat et al. 2016; Zhang et al. 2021; Gao et al. 2023) and the underbonding approach (Venema et al. 1998) have received the most attention in geochemical research. The latter stands out in the context of the MUSIC model for its ability to simply capture proton binding through the concept of bond valences. Briefly, the model achieves this through a linear combination of terms accounting for bond valences to an O^{2-} site from (i) coordinating metal ions ($s_{Me\text{-}O}$), and protons that (ii) donate (s_D) and/or (iii) accept (s_A) hydrogen bonds (Venema et al. 1998). To this effect, the pK_a of an O^{2-} site can be expressed as:

$$\mathrm{p}K_a = 19.8\,(-2 + \sum_{i}^{n_{Me\text{-}O}} s_{Me\text{-}O,i} + \Sigma n_D \cdot s_{D,i} + \Sigma n_A \cdot s_{A,i}) \tag{16}$$

with the summations taken over $n_{Me\text{-}O}$ interactions (e.g., $n_{Me\text{-}O}=3$ for a μ_3–O site), and the number of donating (n_D) and accepting (n_A) hydrogen bonds per O site. The residual charge on O(H) groups thus relates to pK_a, such that a site with a small residual charge is more acidic than one of a larger charge. It also implies that sites accepting hydrogen bonds from neighboring sites (e.g., –OH···–OH) or first layer hydration water molecules would become more acidic.

Because this underbonding approach directly relates to metal–oxygen, oxygen–hydrogen and hydrogen bond strength, it can also benefit from knowledge of atomic distances and interfacial structures. Notably, Machesky et al. (2008) used structures derived from Molecular Dynamics simulations to evaluate the acidity constants of OH groups on rutile. This includes actual bond valences for relaxed metal–O bond lengths (in Å):

$$s_{Me\text{-}O} = e^{\frac{r_0 - r}{0.37}} \tag{17}$$

where r is the average bond length and r_0 a reference value for the relevant metal ion from the bond valence work of (Brown and Altermatt 1985). While one widely used version of the MUSIC model considers that each accepted hydrogen bond from water contributes to 0.2 valence units to underbonding, (Machesky et al. 2008) used a refined expression for expressing hydrogen bond valences:

$$s_A = 1.55 - 1.06 r_{HA} + 0.186 r_{HA}^2 \tag{18}$$

Here, s_A is the bond valence of a hydrogen bond acceptor for a given hydrogen bond length (r_{HA}). The bond valence the donating OH species is typically taken as $s_D = 0.8$ valence unit but can also possibly be:

$$s_D = 1 - s_A \tag{19}$$

Note that s_D and s_A terms are for two different hydrogen bonds involving a given site.

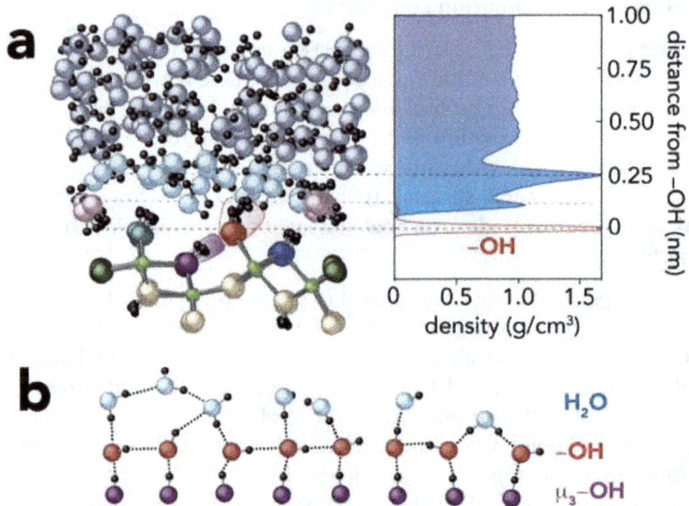

Figure 11. Hydrated surface of (101) face of goethite. Molecular Dynamics simulation results redrawn from Boily (2012). **(a)** Snapshot of the interface (left) and –OH and water density profiles (right), and **(b)** snapshot along a single row of –OH groups showing showing hydrogen bonds.

The approach of Machesky et al. (2008) was later applied to the case of goethite (Boily 2012), and guided using experimentally-determined water structures by Ghose et al. (2010). Here, Molecular Dynamics simulations of the (101) face (Fig. 11) suggested that each –OH site along the aforementioned rows was involved in four hydrogen bonds: (i) intersite –OH···–OH (0.54 HB/site), (ii) donating to 1st layer water molecule (0.28 HB/site), (iii) accepting from 1st layer water molecule (0.64 HB/site), and (iv) accepting from underlying μ_3–OH site (0.97 HB/site). Bond valences of the corresponding interactions can be obtained by Equation (18) through r_{HA}. Together with a $s_{Fe\text{–}O}$ bond valence of 0.51 (r = 0.200 nm), these values predict a pK_a of 6.4. In contrast, analogous rows of –OH groups on the (100) face of goethite have a pK_a of 7.3. This more basic site can chiefly be explained by a smaller number of intersite HB (0.47 HB/site *vs.* 0.54 HB/site). Provided that all other bonds (i.e., Fe-O, –OH···μ_3–OH and donating –OH···OH_2) are relatively unchanged, the net effect translates to a decrease in site acidity. The same approach was applied to all other O(H) groups of various goethite faces.

Because protonation constants are highly sensitive to hydration environments, it is also possible that a single type of surface O(H)—for instance, with a given metal-oxygen bond and intersite hydrogen bond strengths—may have different protonation constants. This was notably illustrated through the *Edge Effect* (Fig. 12a; Rustad and Felmy 2005). Molecular Dynamics simulations using a self-dissociable water model (Rustad and Felmy 2005) suggested that –OH sites at intersections of crystallographic faces were more proton active than those in the middle of faces. This can be attributed to the greater hydration of –OH groups (cf. dielectric effect), as the wettable surface area at the intersections of faces is larger. Differences in metal–oxygen and intersite hydrogen bond strengths can contribute to this variation. Additionally, greater densities of the proton-active –OH groups are exposed at edges. These greater densities were later explored in the context of water condensation in multifaceted hematite nanoparticles (Boily et al. 2015). More recently, Zarzycki et al. (2019) suggested that adjacent faces on nanoparticles can have interdependent structures. Extending these findings to the Edge Effect can thus imply the possibility that the preferential uptake of protons at edges could alter the long-range lateral network of hydrogen bonds at neighboring surfaces.

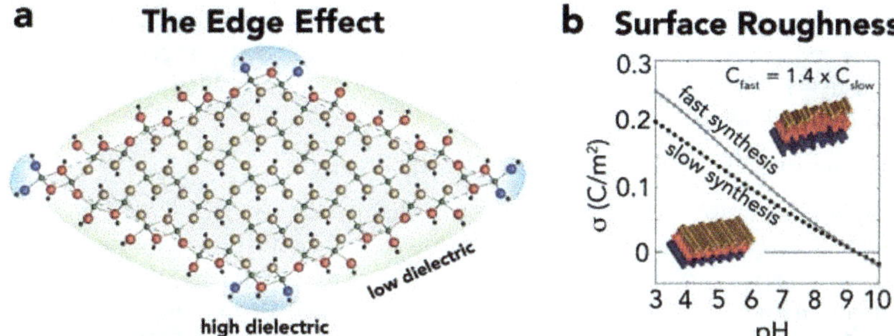

Figure 12. Inhomogeneous distributions of reactive surface OH groups here explained by **(a)** the *Edge Effect* (Rustad and Felmy 2005) and **(b)** surface roughness (Hiemstra and van Riemsdijk 1991; Boily et al. 2001; Lützenkirchen et al. 2008). **(a)** The Edge Effect illustrated here was redrawn from (Rustad and Felmy 2005) and shows a cross-section of a goethite nanoparticle terminated by four adjacent (101) faces, each exposing surface O(H) groups (**blue and red**). Molecular Dynamics simulations using a self-dissociable water model pointed to the preferential proton exchange involving OH groups at particle edges, here shown in **blue regions**. This was explained by the preferential hydration (high dielectric) of these regions, where the finite size of the nanoparticles exposed –OH groups. **(b)** Goethite synthesis initiated by the fast neutralization of Fe^{3+}-bearing solutions produce rougher goethite particles than those by slow neutralization (Hiemstra and van Riemsdijk 1991; Boily et al. 2001; Lützenkirchen et al. 2008). Rough surfaces acquire more surface charge, and this was modeled using a larger Stern layer capacitance value. Predictions shown in (b) were taken and redrawn from Boily et al. 2001.

Acknowledging this sensitivity to hydration can lead be extended to the case of –OH sites at non-crystallographic and/or roughened faces. In particular, molecular-scale consideration of roughened surfaces has yet to be investigated in the context of (hydr)oxo group protonation, extending beyond studies focused on wettability and hydration structure (Döpke et al. 2019; Sun and Bourg 2020; Hubao et al. 2023). This is especially relevant to amorphous or roughened surfaces. These surfaces tend to expose a greater diversity and density of O(H) groups, likely resulting in a broader distribution of pK_a values and reactivity for metal and ligand adsorption (Villalobos et al. 2009). Rough and smooth surfaces of mineral particles produced by nearly identical synthetic methods (Fig. 12b) can even acquire contrasting surface charge (Hiemstra and van Riemsdijk 1991; Boily et al. 2001; Lützenkirchen et al. 2008). Because variations in site density cannot solely account for these changes, SCMs can more effectively model these differences by using larger compact plane capacitances (cf. *The Charged Interface* section) while keeping all other modeling parameters (site densities, pK_a's, ion adsorption constants) unchanged (Fig. 12b). This modeling approach could thus be taken as a suggestion that compact planes are thinner on rough surfaces (cf. Eqn. 33 in *The Charged Interface* section), and therefore more effective at neutralizing electric potentials.

The issue on non-ideal/poorly-characterizable surfaces also applies to curved particles. Here, a surface can be represented as continuous series of tangential cuts of a bulk mineral structure. For instance, spherical cuts of ferrihydrite revealed contrasting hydroxo populations than those at crystallographic faces (Hiemstra 2013; Boily and Song 2020). Additionally, Molecular Dynamics simulations of such nanoparticles unveiled size-dependent populations of hydrogen bonds, strongly suggesting that size-dependent pK_a values (Boily and Song 2020). However, testing this idea awaits whole-particle simulations in liquid water, potentially with reactive models (Aryanpour et al. 2010). At the same time, *ab initio* Molecular Dynamics methods (Leung and Criscenti 2012; Sulpizi et al. 2012; Liu et al. 2013; Gittus et al. 2018), which effectively allow for O–H bond dissociation, have provided insight on local-scale controls on proton affinity. These methods nonetheless continue to be computationally prohibitive for single particle simulations for the foreseeable future, posing a limitation in addressing the impact of surface roughness and curvature on surface complexation reactions.

As the field advances, SCMs developed to account for molecular-scale processes may increasingly require a diversity of –OH sites of unequal reactivity. The same holds true for μ–OH and μ_3–OH sites, which are neighboring sites of considerable importance for charge balance and coordination of larg(er) molecules.

Loadings and interfacial distributions of electrolyte ions

Strongly-related to surface hydration lies the important question of the interfacial distribution of electrolyte ions. Without electrolyte counterions, mineral surfaces have no possibility to generate surface charge (Fig. 13a), a phenomenon that has long been recognized using potentiometric titrations of mineral suspensions. However, to even begin discussing the question of surface charge—the object of the following section—knowledge is needed on the ability of counterions to bind on nanomineral surfaces. This poses a key challenge as the inherently weak nature of the interactions has prevented reliable analytical detection of surface loadings in mineral suspensions, save more rarely-used methods including negative adsorption (Quirk 1960), isotope tracing (Sprycha 1989), optical reflectivity (Porus et al. 2011), or internal reflection ellipsometry (Wang et al. 2015).

The situation is more advantageous for large planar specimens where various methods (X-ray reflectivity, X-ray standing wave-fluorescence spectroscopy, standing wave ambient pressure photoelectron spectroscopy) have provided insight into loadings and vertical spatial distributions of electrolyte ions across the mineral/water interface (Bedzyk and Cheng 2002; Trainor et al. 2006; Lee et al. 2013; Nemšák et al. 2014). Nonlinear optics (Second Harmonic Generation, Sum Frequency Generation) methods have also provided insight into the impact of electrolyte ion identity and on water ordering (Azam et al. 2012; DeWalt-Kerian et al. 2017; Ober et al. 2023; Zelenka and Backus 2024). Cryogenic X-ray photoelectron spectroscopy, in contrast to these techniques, can provide electrolyte ion loadings on nanoparticles (Shchukarev and Ramstedt 2017). Measurements on wet centrifuged pastes of iron oxyhydroxide nanoparticles (Kozin et al. 2013) have provided relationships on electrolyte loadings with surface charge (Fig. 13b). Such relationships are, however, only semi-quantitative because loadings must be normalized to Fe or O atomic ratios, a fraction of which belong to the bulk. Still, they provide a window into charge-dependent loadings that can be achieved by electrolyte ions. Also, when combined with molecular simulations of plausible electrolyte surface complexes (Figs. 13c–d) and (X-ray reflectivity, X-ray photoelectron spectroscopy, non-linear optics) spectroscopy, molecular-scale SCMs can be developed to account for the

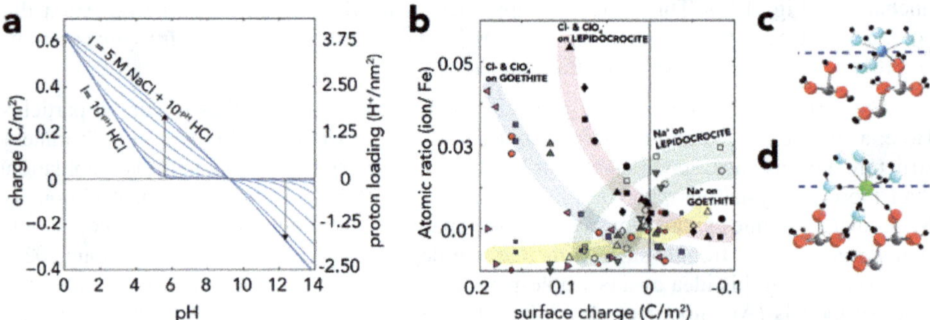

Figure 13. Electrolyte adsorption on metal (oxy)(hydr)oxides. **(a)** Typical ionic strength (*I*) dependent surface charge predicted by a SCM (1-p*K* model with p*K* = 9.4). For simplicity, activities were here generated using the Davies equation, although alternative models (e.g., Pitzer) should be used to account for the higher ionic strengths. **(b)** XPS-derived Na^+, Cl^- and ClO_4^- loadings normalized for Fe with respect to surface charge. Data were replotted from Kozin et al. (2013). **(c–d)** Bound **(c)** Na^+ and **(d)** Cl^- on the (100) face of lepidocrocite obtained by molecular modeling, all taken and redrawn from (Boily 2014).

interfacial distribution of charges by electrolyte ions. This knowledge is important given the impact that electrolyte ions have on the attenuation of electric potentials across the mineral/water interface, and therefore on the extraction of SCM parameters. The following section will link these molecular-scale concepts of site identity, acidity constants and electrolyte ion loadings to generate SCMs that predict surface charge.

THE CHARGED INTERFACE

Knowledge of the identity and spatial disposition of O(H) groups on minerals of pK_a values and of electrolyte binding allows for the formulation of surface complexation reactions that can help bridge macroscopic adsorption with molecular-scale phenomena. A generic SCM for charge development at metal (Me) (oxy)(hydr)oxide surfaces can be expressed as:

$$\equiv \mathrm{Me}_n\mathrm{O}^{0.5n-2} + p\,\mathrm{H}^+ + q\,\mathrm{X_A}^- + r\,\mathrm{X_C}^+ \rightleftharpoons \equiv \mathrm{Me}_n\mathrm{OH}_p{}^{0.5n-2+p}\cdots(\mathrm{X_A}^-)_q\,(\mathrm{X_C}^+)_r \tag{20}$$

Here, n is the number of coordinating metal cations, X_{A^-} is a background electrolyte anion (A) and X_{C^+} the cation (C), and p, q, r are stoichiometric coefficients. The equilibrium constant of the reaction is $K_{n,p,q,r}$. Combined with an EDL model, reactions of this type can predict mineral surface charge (C/m^2) over a wide range of pH and ionic strengths. An example of such predictions is shown in Figure 13a.

Models that aim to account for multisite binding of, say, ligand L^{z-}, may instead require the following expression:

$$\begin{gathered} a \equiv \mathrm{MeO}^{1.5-} + b \equiv \mathrm{Me_2O}^{1.0-} + c \equiv \mathrm{Me_3O}^{0.5-} + p\,\mathrm{H}^+ + q\,\mathrm{L}^{z-} \rightleftharpoons \\ [(\equiv\mathrm{MeO})_a(\equiv\mathrm{Me_2O})_b(\equiv\mathrm{Me_3O})_c\mathrm{H}_p\mathrm{L}_q]^{-1.5a-b-0.5c+p-qz} \end{gathered} \tag{21}$$

with equilibrium constant $K_{a,b,c,p,q}$. A mass action equation of this type can be readily adapted to express any kind of multisite/multidentate coordination environment when independently resolved by spectroscopy or inferred by simulations. It is also most convenient to calculate the corresponding equilibrium constant for this reaction with an expression of the type:

$$\frac{\left[(\equiv\mathrm{MeO})_a(\equiv\mathrm{Me_2O})_b(\equiv\mathrm{Me_3O})_c\mathrm{H}_p\mathrm{L}_q\right]^{(-1.5a-b-0.5c+p-qz)}}{\left[\equiv\mathrm{Me}_{a,b,c}\mathrm{O}\right]\cdot\left[\mathrm{H}^+\right]^p\cdot\left[\mathrm{L}^{z-}\right]^q} = K_{\mathrm{a,c,b,p,q}} \tag{22}$$

Here, the term [$\equiv Me_{a,b,c}O$] implies that ligand binding to a group of adjacent sites counts as a 1:1 interaction. For example, a ligand simultaneous binding to two adjacent $\equiv MeOH_2^{0.5+}$ and $\equiv Me_2OH^0$ sites (e.g., bidendate binuclear complex) could be modeled with Equation (21) using $K_{1,1,0,3,1}$, however with [$\equiv Me_{a,b,0}O$] = [$\equiv MeO^{1.5-}$] = [$\equiv Me_2O^{1.0-}$]. In this fashion, the surface sites concentration is always to the power of 1 in the denominator of Equation (22), and therefore independent of the mineral suspension density (Davis and Leckie 1978a). Work on the thermodynamic implications of this formulation showed this to hold below site saturation (Wang and Giammar 2013).

Multi-site complexation modeling has gained popularity in the quest of linking SCM predictions to the molecular scale. However, a much less developed area in this field aims to tie nanoparticle face-specific reactions which, as covered throughout the previous section, expose contrasting 2D arrays of reactive O(H) groups with distinct coordination numbers with underlying metal sites and distinct hydrogen bonding networks. To this end, the following sections detail the framework, needs and modeling predictions of such an approach.

Multi-face complexation modeling

Typically, SCMs distribute charge over the entire surface of the particles. In this fashion, the electrostatic field generated by potential-determining ions on one plane contribute, albeit nominally, to the effective free energy of adsorption of an ion approaching on another plane. This approach ideally holds for particles with spatially homogeneous distributions of charges. Examples include roughened and/or spheroidal nanoparticles or multifaceted nanoparticles dominated by faces terminated by reactive O(H) groups. It certainly simplifies the treatment of electrostatics for complex and/or poorly-characterizable systems.

This traditional SCM approach may, however, face shortcoming for describing molecular-scale phenomena. In essence modern-day SCMs can require a framework where charges of a crystallographic face do not affect adsorption energies on another face of the same particle. For instance, a composite SCM-Donnan diffusion model was developed to predict charge on contrasting faces of akaganéite (β-FeOOH) nanoparticles, while at the same time accounting for bulk H^+ and Cl^- diffusion into the bulk through ~0.4 nm-wide nanochannels exposed at the tips of the nanoparticles. (Kozin and Boily 2013; Fig. 14).

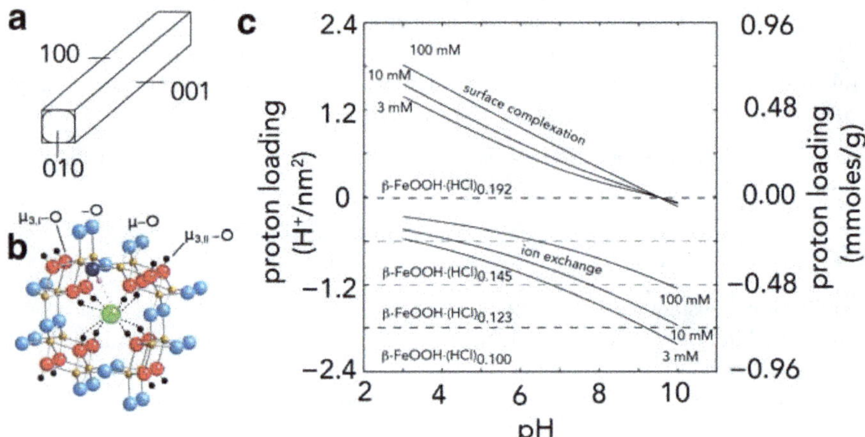

Figure 14. Proton binding on **(a)** crystallographic faces of acicular akaganéite (β-FeOOH) nanoparticles through modeled by concomitant (a,b) face-specific adsorption of potential-determining ions (H^+, OH^-) on dominant crystallographic faces ('surface complexation') and **(b)** H^+ and Cl^- diffusion ('ion exchange') through tips of acicular particles, here taken as the (010) of face. **(c)** A composite SCM-Donnan diffusion model explaining the pH and ionic strength-dependent surface change and bulk ion exchange. Redrawn from Kozin and Boily (2013b).

Although models that isolate adjacent faces remove any notion of cooperativity, such as in charge transfer or propagation of hydrogen bonding network (Zarzycki et al. 2019), they can have numerous advantages. One example is modeling ligand exchange reaction at nanoplatelet edges while excluding reactions on basal faces terminated by non-reactive μ–OH sites. Mineral nanoplatelets where such models could be needed include lepidocrocite, hematite, green rusts, gibbsite and clays (Eggleston et al. 2004; Trainor et al. 2004; Siretanu et al. 2014). Developing such models can also have important repercussions on the ability to explain reactions affected by bulk electron transport between different faces (Eggleston et al. 2004; Yanina and Rosso 2008). At the same time, needs for predicting face-specific electrostatic potentials can be beneficial for understanding particle-particle interactions, for example, in predicting coagulation in relation to colloidal stability, or even mineral growth through orientated aggregation (De Yoreo et al. 2015; Salzmann et al. 2021). Similar models could even be readily developed to account for the Edge Effect (Fig. 12; Rustad and Felmy 2005), simply by defining the fraction of surface area of sites exposed at edges. Such a framework that enables face-specific SCM can be useful to describe a wide range of phenomena, as well as to bridge molecular-scale (molecular modeling, spectroscopy) with macroscopic adsorption data (e.g., potentiometry, batch adsorption, electrophoresis). The following section details an approach for such a face-specific SCM.

The framework for a face-specific SCM

A SCM framework that accounts for crystal face-specific electrostatic potentials (Ψ) relates Gibbs free energy of adsorption of species *j* on a given *hkl* face ($\Delta G_{\text{ads},j,hkl}$) to its intrinsic value at $\Psi = 0$ V ($\Delta G_{\text{int},j,hkl}$) with:

$$\Delta G_{\text{ads},j,hkl} = \Delta G_{\text{int},j,hkl} + \sum_i \Delta z_i \, \text{F} \, \psi_{i,hkl} \tag{23}$$

Here, $\Delta G = -RT \ln K$ (e.g., from reaction of Eqn. 20), F is Faraday's constant, Δz_i is the change in charge and $\Psi_{i,hkl}$ the surface potential at the adsorption plane *i*. The latter is applied to any selected EDL model (e.g., Fig. 15). An example of a Matlab (The Mathworks, Inc.) code implementing this multi-face SCM framework can be found in (Boily 2014).

Using the case of the o- and β-planes of the Basic Stern model (BSM) as an example, surface charges (σ; in $\text{C}\cdot\text{m}^{-2}$) on any given *hkl* faces are expressed through:

$$\sigma_{\text{o},hkl} + \sigma_{\beta,hkl} + \sigma_{\text{dl},hkl} = 0 \tag{24}$$

Here, charge neutrality on face *hkl* is achieved by the distribution of potential-determining ions in the o-plane, and electrolyte counterions in the β-plane as well as diffuse layer (dl). In this framework, the sum of charges for all surface species is normalized with respect to the surface area (s_{hkl}; $\text{m}^2\cdot\text{L}^{-1}$) of the face on which they occur. For instance, charge densities of potential-determining ion adsorbed at the o-plane in the context of the mass action equation of Equation (20) are determined with:

$$\sigma_{\text{o},hkl} = \frac{F}{s_{hkl}} \left(\sum_j (0.5n - 2 + p)[\equiv\text{Me}_n\text{OH}_p]^{0.5n-2+p} \ldots (\text{X}_{\text{A}^-})_q (\text{X}_{\text{C}^+})_r \right) \tag{25}$$

A portion of this charge is then balanced by those of electrolyte counterions, here placed in the β-plane of the BSM:

$$\sigma_{\beta,hkl} = \frac{F}{s_{hkl}} \left(\sum_j (r - q)[\equiv\text{Me}_n\text{OH}_p]^{0.5n-2+p} \ldots (\text{X}_{\text{A}^-})_q (\text{X}_{\text{C}^+})_r \right) \tag{26}$$

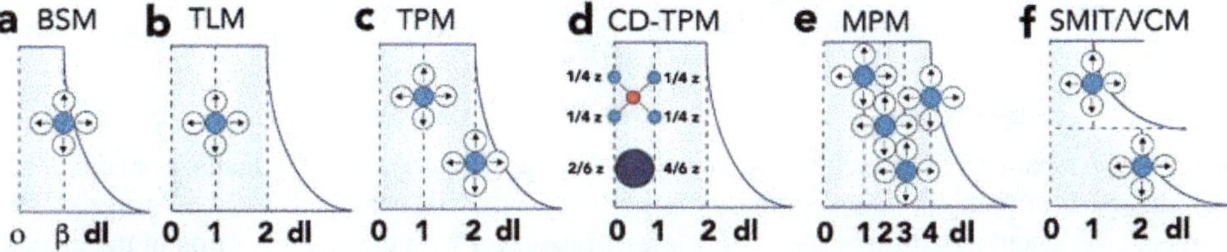

Figure 15. Six EDL structures that can be readily adapted to SCMs. **(a)** Basic Stern Model (BSM) (Stern 1924), **(b)** Triple-Layer Model (TLM) (Yates et al. 1974; Davis et al. 1978), **(c)** Three Plane Model (TPM), **(d)** Charge-Distribution-TPM (Hiemstra and van Riemsdijk 1996), **(e)** Multi-Plane Model (MPM), e.g., with four planes as in Bowden et al. (1980), **(f)** SMIT Model (Smit 1986a,b) and related Variable Capacitance Model (VCM) (Boily 2014). See Westall and Hohl (1980), Hunter (1981) or Bockris and Reddy (1970) for accounts of earlier models.

or in the outer-Helmholtz planes of Triple and Three Layer Models. It then follows that the remaining electrolyte charge in the diffuse layer ($\sigma_{dl,hkl}$) obtained with Equation (24), just as in conventional SCMs. The total charge on the *entire* nanoparticle on adsorption on plane *i* (o, β, dl) can then be obtained through the sum of charge densities taken over *m hkl* faces:

$$\sigma_{i,\,tot} = \sum_{m} f_{m,hkl} \cdot \sigma_{i,hkl} \tag{27}$$

Here, $f_{m,hkl} = s_{hkl}/s_{tot}$ is the fraction of the area of face *hkl* (s_{hkl}) in relation to the total area (s_{tot}):

$$s_{tot} = \sum s_{hkl} \tag{28}$$

such that $1 = \sum_{m} f_{m,hkl}$. Because all charges in adsorption plane *i* over the *m* different *hkl* faces are additive, the model has the same functionality as models that rather uniformly spread charges over the entire surface. In the same token, the model offers the possibility of distinguishing charges allocated to contrasting, coexisting, EDL structures either on separate or even the same faces, as will be detailed in the next sections.

The resulting charges generate face-specific electrostatic potentials (Ψ). Potentials can be generated using face-specific capacitance (C_{hkl}) values:

$$\sigma_{o,\,hkl} = C_{hkl} \cdot (\Psi_{o,\,hkl} - \Psi_{\beta,\,hkl}) \tag{29}$$

In this fashion, the effective capacitance (C_{eff}) of the entire nanoparticle relates to C_{hkl} with:

$$C_{eff} = \sum f_{m,hkl} \cdot C_{hkl} \tag{30}$$

This approach thus implies that charges of an ion adsorbing on one crystallographic face will not affect adsorption energies on another face of the same particle, and vice-versa.

As in the majority of SCMs, the face-specific diffuse layer potential ($\Psi_{dl,\,hkl}$) is given by the potential at the head of the most outer compact plane (e.g., $\Psi_{\beta,\,hkl} = \Psi_{dl,\,hkl}$ for the BSM). The diffuse layer potential ($\Psi_{dl,\,hkl}$) of a given *hkl* face is then related to diffuse layer charge ($\sigma_{dl,\,hkl}$) through Gouy–Chapman equation (Bolt 1955):

$$\sigma_{dl,hkl} = -\sqrt{8000\,I\,\varepsilon\varepsilon_0 RT}\;\sinh\left(\frac{F \cdot \psi_{dl,hkl}}{2RT}\right) \tag{31}$$

Here, *I* is the molar ionic strength, ε the dielectric constant of water, ε_0 the permittivity of vacuum (8.85×10^{-12} $C^2 \cdot J^{-1} \cdot m^{-1}$), R the molar gas constant (8.31 $J \cdot K^{-1} \cdot mol^{-1}$) and *T* temperature. It then follows that electrostatic potentials can be known for any given distance (Δ) from the head of the diffuse layer through of a *hkl* face through (Hunter 1981):

$$\Delta_{hkl} = -\kappa^{-1} \ln\left(\tanh^{-1}\left(\frac{F \cdot \psi_{dl,hkl}}{4RT}\right) \cdot \tanh\left(\frac{F \cdot \psi_{\Delta,hkl}}{4RT}\right)\right) \tag{32}$$

where κ^{-1} is Debye length $\left(\kappa^{-1} = \sqrt{2000F^2 / \varepsilon\varepsilon_0 RT} \cdot \sqrt{I}\right)$. $\Psi_{\Delta,hkl}$ values can be related to ζ potentials, to infer on the location of the shear plane in electrophoretic mobility measurements. For reference, the shear (or slipping) plane separates the hydrodynamically stagnant layer of water of the compact plane and a portion of the head of the diffuse layer with the active portion of the diffuse layer. This can be of particular interest in the context of streaming potential measurements, which can readily be achieved on single oriented crystallographic surfaces (Preočanin et al. 2017; Brkljača et al. 2018; Lützenkirchen et al. 2018), and can be related to Derjaguin–Landau–Verwey–Overbeek (DLVO) theory (Derjaguin 1941; Verwey et al. 1948) to calculate interaction free energies between two oriented surfaces. Other examples of applications include Atomic

Force Microscopy measurements on approach/separation of two oriented surfaces, as well as the study of colloidal stability involving multifaceted mineral suspensions. The latter can help resolving lamination (e.g., clay/clay) and aggregation (e.g., kaolinite/clay, lepidocrocite/lepidocrocite) through basal faces of nanoparticles. It can also help explain mineral growth by oriented attachment (De Yoreo et al. 2015; Salzmann et al. 2021).

Before moving to a deeper discussion on face-specific capacitances, it should be noted that predictions of diffuse layer potentials might need other theoretical frameworks, rather than the widely used Gouy–Chapman model. Notably, accounting for the finite size of ions might become of increasing importance for SCMs to keep in pace with progress made by molecular studies. Examples of alternative models include a modification of the Poisson-Boltzmann model (Kornyshev 2007) and Hypernetted Chain Theory (Larson and Attard 2000; Attard 2006). The latter accounts for both ion size and correlation, and can be especially needed at high surface charge densities (Lützenkirchen 2002). These approaches could provide an alternative basis for bridging SCM- with electrochemically-derived capacitances (Shimizu et al. 2012; Shimizu and Boily 2014, 2015), a topic to be covered in the last section of this Chapter, or bridging experimental results with molecular simulations of (ζ-like) potentials (Předota et al. 2016).

The dynamic compact plane

As a central parameter to SCMs, the compact plane capacitance plays a key role on predictions of electrostatic potentials. It was originally implemented by von Helmholtz (von Helmholtz 1879) who envisioned charged surfaces covered by a stationary layer of solvated ions with opposite charges. Stern later integrated elements of this model with Gouy and Chapman's concept of the diffuse layer (Gouy 1910; Chapman 1913), in which ions are mobile as a result of the balance between electrostatic attractive and diffusion forces. This model, and a later adaptation incorporating two adjacent (inner- and outer-Helmholtz) compact planes (Grahame 1947), thus predicts electrostatic potentials using compact plane capacitance.

While capacitance has primarily been utilized as an adjustable value to predict surface charge data, efforts have also been made to attribute physical meanings to its values. At its core, capacitance is no more than a scalar proxy describing the ability of a system to store electric charge per unit voltage, and is expressed in Farad (F). In SCMs, however, the mineral/water interface is treated as a molecular capacitor because it contains molecular adsorption planes functioning as a parallel-plate capacitor. The intervening layers of water between these planes form the dielectric, which attenuates the electrostatic potential across the mineral/water interface. Capacitance values could, as such, be strongly affected by mineral surface structure and composition.

Capacitance is intricately related to both the thickness (δ) and permittivity ($\varepsilon\varepsilon_o$) of the dielectric:

$$C_{\mathrm{eff},hkl} = \frac{\varepsilon\varepsilon_o}{\delta} \tag{33}$$

Classically, the thickness of the dielectric (e.g., Stern Layer) is said to be about the radius of the solvated ion. However, it can now be better estimated with experimental and/or theoretical evaluations of interfacial water density profiles. In this equation, charge separation inversely scales with capacitance, meaning that systems with counterions at small distances of approach contribute to a larger interfacial capacitance. As a simplified case scenario, a compact plane with SCM-derived capacitance value of 1 F/m^2 could be ~0.7 nm-thick (~2–3 monolayers, depending on packing patterns), if the dielectric constant were to remain as in liquid water (i.e., ε = 78.5 at 25 °C). In reality, the dielectric constant of water is generally expected to be lowered near solid surfaces (Conway et al. 1951; Hiemstra 2012; Fumagalli et al. 2018; Zarzycki 2023). While calculations of plausible values from molecular simulations may be challenged by limitations in defining dielectric constant in molecularly-thick water layers, consideration of these phenomena is needed in the search for interfacial capacitance values.

On (hydr)oxo-terminated mineral surfaces, a drop in dielectric constant can be understood by the alignment of the molecular dipoles of water and thus the decrease in the rotational freedom of water, which is caused by hydrogen bonding to mineral surface sites. This can also be understood in terms of the alignment of water (i) in the direction of the electric field in the Stern/inner-Helmholtz layer, and (ii) toward ions in the outer-Helmholtz layers. These considerations, initially raised by Bockris et al. (1965), are in favour of models that would account for electric field- and ion-loading dependent dielectric constants, and therefore compact plane capacitance. While the great majority SCMs have simplified this view by means of a constant capacitance (e.g., Constant Capacitance Model at high salt content; Schindler and Gamsjäger 1972), consideration of a dynamic capacitance is needed as molecular-level studies continue to influence SCM development. The following two sub-sections highlight in greater detail the impact that (i) electric field strengths, and (ii) interfacial hydration and counterion structure play on capacitance.

Electric field strength controls on capacitance. The coupling of the electric field to the dipole of water molecules can be a source of important dielectric variations. This coupling can be understood in terms of the alignment of compact plane water molecules to counteract the electric field generated by potential-determining ions (Bockris and Reddy 1970). From classical bulk water work (Booth 1951), ordering is manifested in fields exceeding $\sim 10^8$ V/m through a drop in the dielectric constant from ~80 down to ~6 at high field strengths ($>10^{10}$ V/m). Values can be predicted using the Kirkwood–Booth equation (Kirkwood 1939; Fröhlich 1948; Booth 1951).

We can also evaluate field strength based on estimates of surface potential and compact plane thickness. To illustrate this point, consider the typical case of goethite nanoparticles ($pH_{pzc} = 9.4$) which acquired ~1.3 H^+/nm^2 ($\sigma_o = 0.2$ C/m^2) and a ζ-potential of ~60 mV when equilibrated at pH 3 in 0.1 M NaCl (Boily et al. 2001). Assuming a typical capacitance of ~1 F/m^2 for SCMs, Equation (29) predicts a potential difference of ~0.2 V between the o- and β-planes. Adding 60 mV from the ζ-potential to this potential difference, the inner-Helmholtz potential should be of at least Ψ_o ~260 mV. Now, assuming that the compact plane dielectric medium contains ~2 water monolayers ($\delta = 0.56$ nm), we would expect a field strength of $\sim 10^{8.6}$ V/m. From classical work on bulk water dielectric constants (Booth 1951), we can expect that this field strength induced a dielectric constant of ~30, namely a ~2.6-fold decrease in capacitance compared to interfaces with a weak field strength. For reference, a compact plane thickness of at least ~8 water monolayers would be necessary to correspond a dielectric constant of ~80. This, in itself, can serve as a basis to justify the use of electric-field dependent dielectric constants under conditions of high charge density, which are favored under conditions of extreme pH and/or salinity (Hatlo et al. 2012; Nakayama and Andelman 2015). For context, examples of such conditions in geochemically-relevant environments include acid mine drainage systems, acidic soils, alkaline lakes and nuclear waste disposal.

Another consequence of electric field (**E**) can be manifested by an increase in the viscosity (η) of interfacial water, such as seen in the case of water on rutile (Předota et al. 2007). A shift in viscosity of a solvent (η_o) can be expressed through (Andrade and Dodd 1939, 1946, 1951):

$$\eta(E) = \eta_o\left(1 + f\left|\mathbf{E}\right|^2\right) \tag{34}$$

using viscoelectric coefficient $f = 10^{-15}$ m^2·V^{-1} for water (Lyklema and Overbeek 1961; Webb et al. 1974; Hunter and Leyendekkers 1978; Jin et al. 2022). This viscoelectric effect alters chemical speciation at the interface, which can be understood in terms of variations in properties including hydration environment, equilibrium/rate constants and solubility. It also plays an important role on the position of the shear plane in the diffuse layer, and therefore ζ potentials. To this end, expressions for field strength-dependent ζ potentials were devised through the viscoelectric coefficient parameter f of Equation (34) (Lyklema 1994). Functions of this type essentially predict a lowering of electrostatic potentials as the hydrodynamically stagnant layer becomes thicker, thus shifting the shear plane towards the tail of the diffuse layer, where potentials are smaller.

Accounting for such field-driven effects can be important in the development of SCM when considering ζ potentials as (semi)quantitative measured to test the validity of a species combination. Additional caution must be taken in cases where surface conductivity, namely the lateral mobility of ions behind the shear plane (Bikerman 1973), lowers electrostatic potentials. This can be addressed in modified SCMs using the Dynamic Stern Layer model (Zukoski and Saville 1986). In short, consideration of these effects in the calibration of SCMs can tighten the link between adjustable modeling parameters, and the physicochemical properties (e.g., capacitance) they are chosen to emulate.

Interfacial hydration and counterion structure controls on capacitance. These controls can be understood in terms of the concerted actions of surface O(H) groups and ion hydration on (i) water ordering and (ii) compact plane thickness. These are two key parameters affecting compact plane capacitance (Eqn. 33). Mineral surface controls are directly linked to the 2D to 3D ordering of water, as imposed by the spatial distribution of surface O(H) groups, again a mineral- and face-specific property. Counterion controls pertain to the binding sites of (e.g., β-plane) counterions. For example, molecular simulations revealed that counterions can sit directly on these molecular layers, incorporating 1-2 water molecules into their first hydration sheath (e.g., $–OH\cdots(OH_2)\cdots Cl^-\cdots(H_2O)_5$) (Fig. 16). However, others may destabilize water structure, depending on whether they are structure-making (kosmotropic) or structure-breaking (chaotropic). In this sense, one school of thought in the SCM community adheres to the idea that ions adsorb more strongly onto mineral surfaces that induce comparable effects on water structure (cf. review by (Bañuelos et al. 2023)). Related considerations have even motivated predictions relating bulk dielectric constants of minerals to proton and electrolyte binding constants and to EDL capacitance (Sverjensky 2001). It remains, however, unclear how bulk mineral dielectric constants physically relate (face-specific) mineral hydrophilicity and to interfacial water structure.

Additionally, we note that the basic premise that electrolyte ions bind to a single β- or outer-Helmholtz plane does not necessarily hold for all electrolyte ions. For example, some ions (e.g., Na^+) can transect the dielectric and incorporate surface (hydro)oxo groups (e.g., $–OH\cdots Na^+(H_2O)_5$) into their hydration sheath (Fig. 16). Polyatomic counterions (e.g., nitrate or perchlorate) and more strongly binding adsorbates (e.g., oxyanions) have, in turn, the ability to lodge one oxygen group closer to the o-plane, yet the remaining charges lie towards the bulk solution. Various other configurations are also possible, such as on phyllosilicates where cation binding into or near ditrigonal cavities is size-dependent (Bourg et al. 2017). To this end, the Charge Distribution (CD) model can be of great utility to convey these ideas by splitting charge of molecules across adjacent EDL compact planes, while still using a manageable number of adjustable parameters (Hiemstra and van Riemsdijk 1996).

Alternatively, SCMs can envision surfaces in terms of coexisting patches of contrasting composition, charge density, and interfacial water structure (Fig. 15f) (Smit 1986a,b; Boily 2014). In essence, such models can be proxies for mean-field effects that adsorbates exert on the charge-storage capability of mineral surfaces. This idea is supported by related work (Boily 2007) emphasizing the need for representing charge in adsorbates in terms of a continuous intramolecular distribution of electrons, rather than discrete values split between distinct adsorption planes. Models of this type can consequently be more readily connected

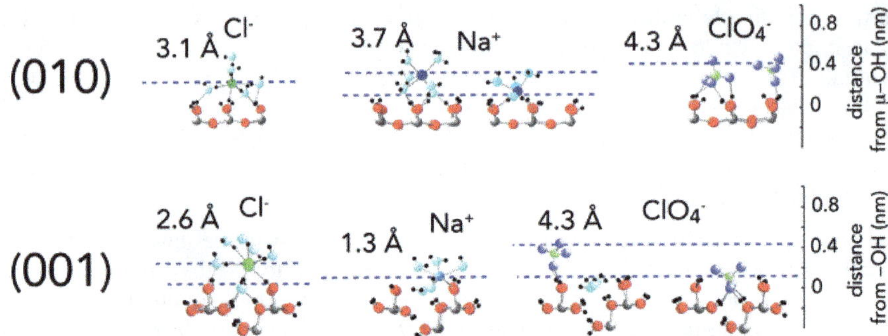

Figure 16. Interfacial structure of Cl^-, Na^+, and ClO_4^- on the (**top**) (010) and (**bottom**) (001) faces of lepidocrocite surfaces exposed to liquid water (not shown, except for first hydration shell of selected ions). Geometries were taken from a snapshot of MD simulations from Boily (2014).

to molecular simulations/measurements unravelling, for instance, (re)organized (e.g., 1st and 2nd shell) water and counterions in the vicinity of adsorbed molecules. The following section details additional approaches that allow modeling of co-existing EDL structures with ion- and face-specific capacitance values.

The Variable Capacitance Model

The Variable Capacitance Model (VCM) was developed to account for coexisting EDLs of contrasting structures on a single mineral surface (Boily 2014). This model was inspired by work by Smit (1986a,b), who expressed single surfaces in terms of co-existing patches of EDLs (Fig. 15f). The model can represent the overall effects on capacitance caused by alterations in water structure caused by variations in pH, salt composition and surface charge.

In the VCM, the overall effective compact plane capacitance is expressed in terms of a linear combination of capacitances of electrolyte- or ion-specific EDLs. The capacitances are electrolyte-loading dependent, and can be made face-specific (Eqn. 30) or can hold for the entire nanoparticle surface. The general equation for face- and ion- or electrolyte-specific capacitances is:

$$C_{\text{eff}} = \sum_{m}\sum_{j} f_{m,hkl} \cdot f_{j,hkl} \cdot C_{j,hkl} \tag{35}$$

with the summation taken over m crystallographic faces and j co-existing ions or electrolytes. In this equation, $f_{m,hkl}$ is the fraction of the area of face m on the nanoparticle as for Equation (30), and $f_{j,hkl}$ is the mole fraction of adsorbed ion or electrolyte j. This equation can thus be directly incorporated into a model for face-specific capacitances (m >1), or used in cases where charge is evenly distributed over a single nanoparticle (m = 1).

In its simplest form, the VCM can be applied to the entire surface of a nanoparticle (m = 1) and for a binary electrolyte (j = 2). This works under the assumption that all crystallographic faces are equivalent in terms of charge-storage capability. The model reduces to:

$$C_{\text{eff}} = f_{X_{A^-}} \cdot C_{X_{A^-}} + f_{X_{C^+}} \cdot C_{X_{C^+}} \tag{36}$$

where X_{A^-} denotes the anion and X_{C^+} the cation, just as in the generic mass action equation of Equation (20). Values of $f_{X_{A^-}}$ and $f_{X_{C^+}}$ are obtained directly by the SCM from the fraction of the concentrations of adsorbed electrolyte ions.

$$f_{A^-} = \frac{\left[\equiv Me_nOH_p{}^{0.5n-2+p} \cdots X_{A^-}\right]}{\left[\equiv Me_nOH_p{}^{0.5n-2+p} \cdots X_{A^-}\right] + \left[\equiv Me_nOH_p{}^{0.5n-2+p} \cdots X_{C^+}\right]} \tag{37}$$

As an example, Equation (36) was used to model charge development on hematite nanoparticles in NaCl (Fig. 17) by also taking into account the pH-dependent loadings of Na^+ and Cl^- ions measured by cryogenic X-ray photoelectron spectroscopy (Boily and Shchukarev 2010). The model made use of ion-specific capacitance values (C_{Na^+}= 3.8 F/m^2, C_{Cl^-}= 1.3 F/m^2) which, by virtue of Equation (33), emulated the expected thinner compact plane thickness in patches of Na^+ complexes than those with Cl^-.

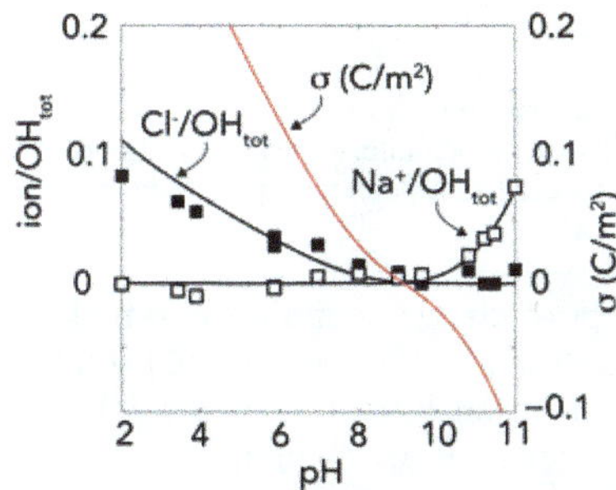

Figure 17. SCM prediction surface charge (σ) and ion loadings compared and X-ray photoelectron spectroscopy-derived Cl and Na atomic % on hematite in 10 mM NaCl. Experimental on loadings were normalized for OH content determined by X-ray photoelectron spectroscopy. Data were taken and redrawn from (Boily and Shchukarev 2010).

This methodology can also be developed to account for contrasting charge-storage capabilities of co-existing faces (m > 1) on a nanoparticle. For example, charge development on lath-shaped lepidocrocite in separate $NaClO_4$ and NaCl solutions was modeled by proton adsorption to the (100) and (001) edges (Boily and Kozin 2014). The dominant (001) basal face is ideally proton inactive because it is populated by μ–OH groups (Fig. 16) but, for reasons detailed in (Hiemstra and van Riemsdijk 2007; Boily and Kozin 2014), the model also included contributions from a lower density of defect –OH sites. VCM modeling of charge uptake considered the contrasting EDL structures developed by the molecular structures resulting from mineral–water, mineral–electrolyte ion, and water–electrolyte interactions (Fig. 16). Through Equation (36), capacitance values expressing the pH and salt-dependent surface charge were related to the ion-dependent compact plane thickness, inferred through water density profiles. In this fashion, sodium- and perchlorate-specific capacitances were expressed through the capacitance for chloride (C_{Cl^-}) via the ratios of their respective compact plane thickness (e.g., $C_{Na^+} = \delta_{Cl^-} \cdot \left(\delta_{Na^+}\right)^{-1} \cdot C_{Cl^-}$), obtained by molecular simulations (e.g., Fig. 16). The resulting model predicted zeta potential values through face-specific electrostatic potentials, using ion loading- and pH-dependent effective capacitance values (Fig. 18).

Lastly, in another adaptation of Equation (35), the VCM was used to predict the surface charge of goethite in mixed solutions of NaCl and $NaClO_4$ (Kozin and Boily 2014). Due to the high point of zero charge of goethite, charge development under circumneutral to acidic conditions is largely controlled by differences in the charge-neutralization capabilities of the Cl^- and ClO_4^- ions. Because Cl^- has a greater size-to-charge ratio, and achieves greater surface densities than ClO_4^-, charge densities of potential-determining ions are greater in solutions of NaCl. For instance, the charge density achieved on goethite at pH 3 and I = 0.1 M is ~0.5 H^+/nm^2 (~0.08 C/m^2) greater in NaCl than in $NaClO_4$. This difference amounts to ~1/6 of the density of the reactive –OH sites, which are chiefly responsible for charge uptake.

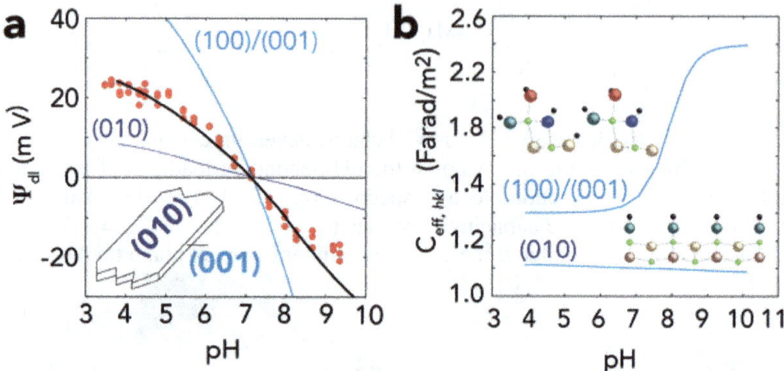

Figure 18. VCM prediction of face-specific and electrolyte loading-dependent **(a)** diffuse layer potential (ψ_{dl}) **(b)** capacitance values on lath lepidocrocite. Potentials in (a) predicted particle-averaged values and corresponding experimentally-measured zeta potentials (data points). Data and models were taken and redrawn from Boily and Kozin (2014).

Closely-tied to the concept of charge-neutralization, molecular simulations revealed two distinct planes of adsorption for Cl^- and Na^+ ions in NaCl solutions. In contrast, the interfacial distribution of charge was wider in $NaClO_4$ solutions not only because ClO_4^- is larger but also because it induced a broader interfacial distribution of Na^+ ions. This can be linked back to Equation (33), where a wider compact plane thickness (δ) relates to lower interfacial capacitance, and therefore lower charge. The $C_{j,hkl}$ term of Equation (35) can thus relate to the electrolyte-specific charge-neutralization capability of the interface. The term is a proxy for synergetic effects of size-to-charge ratio effects, interfacial anion and cation distribution, water ordering, and compact plane thickness. As a result, the approach is an alternative strategy to the Charge Distribution model (Hiemstra and van Riemsdijk 1996) where the charges of the ClO_4^- ion would have instead been separated between adsorption planes (e.g., ¼ in the o- and ¾ in the β-plane).

Model-wise, Equation (35) applied to a mixed electrolyte system can be simplified to a single surface ($m = 1$) when assuming a uniform distribution of charge, which is applicable to the case of nanogoethite (Fig. 19). The equation for a one face ($m=1$), two electrolyte ($j=2$) system then reduces to:

$$C_{\mathrm{NaCl/NaClO_4,eff}} = f_{\mathrm{NaCl}} \cdot C_{\mathrm{NaCl}} + f_{\mathrm{NaClO_4}} \cdot C_{\mathrm{NaClO_4}} \tag{38}$$

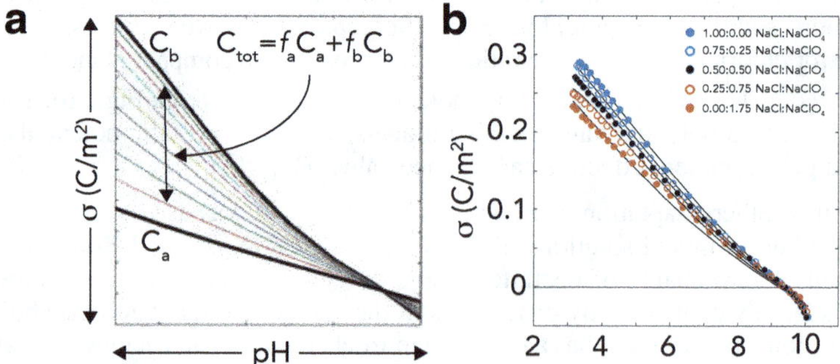

Figure 19. Mineral surface charge (σ) development in mixed electrolytes, predicted using a linear combination of capacitance values for single electrolytes scaled for electrolyte ratio (Eqn. 38). **(a)** Generalized concept Equation (38). **(b)** Experimental charge on goethite for electrolytes of varied NaCl:$NaClO_4$ composition. Data taken and redrawn from Kozin and Boily (2014).

In this equation, C_{NaCl} and C_{NaClO_4} values were derived by surface complexation modeling of the separate electrolyte systems over a range of ionic strengths. Values of f_{NaCl} and f_{NaClO_4} were taken at those of the total mole fraction of the electrolytes, and not those of the SCM-derived individual ion loadings. This simplification was justified by noting the smaller role that Na^+ played under the experimental window considered for that work. This approach thus represents another example of how the VCM can be adapted for predicting the pH-dependent surface charge for any electrolyte mixture ratio using end-member modeling parameters for the separate electrolytes.

THE ELECTROCHEMICAL INTERFACE

As emphasized in the previous section, capacitance is a key parameter denoting the charge-storage capability of a solid surface exposed to an electrolyte solution. In the SCM literature, capacitance is chiefly inferred by modeling ionic strength dependent surface charge from H^+/OH^- adsorption data. This is typically achieved by co-optimizing additional SCM parameters, including protonation constants, electrolyte adsorption constants, and even site density. Capacitance values can thereby still be highly model-dependent, and strongly correlated to other parameters (Westall and Hohl 1980; Lützenkirchen 1999, 2005, 2006; Boily et al. 2001; Payne et al. 2013). This co-dependency can, however, be attenuated by locking selected parameters to known values, such as site identity and density, protonation constant, and/or even (semi-quantitative) electrolyte ion loadings.

Direct electrochemical measurements of capacitance are also possible, however mostly for conductive materials (e.g., Hg, AgI) where *both* potential and charge can be simultaneously determined. These measurements are achieved by applying a voltage (ψ) and measuring the resulting charge (σ), or vice versa (Bockris and Reddy 1970). A *differential capacitance* can thus be readily obtained by taking the gradients in charge ($d\sigma$) and potential $d\psi$ ($C_{diff} = d\sigma \cdot (d\Psi)^{-1}$). Because both charge and potential are linearly related in the compact (e.g., Stern) layer, the differential capacitance is identical to the *integral capacitance* ($C_{int} = \sigma \cdot \Psi^{-1}$). This is, however, not the case for diffuse layer capacitances because gradients in charge and potential are not linearly related (Bockris and Reddy 1970).

Compact layer capacitance values for metal (iodide) electrodes (e.g., Hg, AgI) are typically in the order of a few hundred $\mu F \cdot cm^{-2}$. In contrast, SCM-derived values for metal (oxy)(hydr) oxides are readily 10–100 times greater, in the order of 1 $F \cdot m^{-2}$. Fokkink et al. (1989) explained this disparity by noting that metal (oxy)(hydr)oxides are more hydrophilic than metal (iodide) electrodes. Metal (oxy)(hydr)oxides are thus more readily covered by a stable layer of water that acts as the charge-free layer of the compact (e.g., Stern) layer. Linking back to the previous section, the structure, viscosity, dielectric constant of the interface and its capacitance can be strongly affected by crystallographic orientation. In contrast, water layers on the more hydrophobic metal (iodide) surfaces are more strongly oriented, and thereby form a charge-free layer of lower dielectric constant, and therefore of smaller charge storage capabilities.

Because surface potential and capacitance are intricately linked to the assumed EDL structure, a crucial objective in advancing SCM research is to independently measure EDL properties of metal (oxy)(hydr)oxide surfaces. To this and related ends, electrochemical methods have also been used to explore the electrostatic potential development on semi-conductive minerals, including hematite, galena and pyrite (Kallay and Preočanin 2008; Yanina and Rosso 2008; Zarzycki et al. 2010, 2011; Boily et al. 2011; Shimizu et al. 2012, 2015; Chatman et al. 2013a,b; Lützenkirchen et al. 2013; Renock et al. 2013; Shimizu and Boily 2014, 2015; Yuan et al. 2015, 2022; Preočanin et al. 2017; Cook et al. 2019). These electrodes typically consist of sub-centimeter-sized specimens (Fig. 20 a), often with (mechanically- and chemically-) polished surfaces. Work on hematite (α-Fe_2O_3) has particularly been beneficial to the field because it is sufficiently conductive (~2.0–2.2 eV band gap) to be used as electrodes, and its properties can be tied to those of other geochemically important minerals of the iron oxide family. It also parallels similar work on doped hematite-bearing materials in the context of photoelectrochemical water splitting (Sivula et al. 2011).

Studies on the pH- and ionic-strength dependent open circuit potentials (OCPs) of hematite electrodes revealed proxy values for inner-Helmholtz potentials, which were generated by the adsorption of potential-determining ions (Fig. 20b). While most results from different research laboratories (cf. aforementioned references) showed non-Nernstian charging and small response to ionic strength, different electrodes/set-ups generated contrasting OCP values, often not crossing to zero potential value at the expected *pzc* (Zarzycki and Preočanin 2012; Chatman et al. 2013a,b; Shimizu and Boily 2014, 2015). This was ascribed to variations in electrochemical cell composition (Shimizu and Boily 2014), variations in preparation/annealing/ageing methods which altered surface chemistry (Lützenkirchen et al. 2013), variations in accepted equilibrations times (Shimizu and Boily 2015), kinetic effects that could potentially relate to electrostatic screening (Chatman et al. 2013b), and to differential electrochemical charging (Chatman et al. 2013a). Despite these uncertainties, OCP measurements have still been used to identify the *pzc* of hematite, for instance through the inflection point (Lützenkirchen et al. 2013), common intersection points (Zarzycki and Preočanin 2012), and from Maxwell construction of pH-dependent data (Zarzycki et al. 2011) (Fig. 20c).

Measurements of hematite electrodes with different crystallographic orientations (Fig. 20b) also revealed unexpected trends in potential development (Chatman et al. 2013a; Shimizu and Boily 2015). Thus, while the low pH charging of the basal (001) face of hematite was explained by the protonation of its dominant μ-OH sites, even lower potential values were obtained for the high index (012). This was explained by shifts in the acidity constant of functional groups caused by electrostatic repulsion with other groups. Still, SCMs predicting OCPs on contrasting crystallographic faces offered possibilities for testing MUSIC protonation constant values (Boily et al. 2011; Chatman et al. 2013a; Lützenkirchen et al. 2013). However, as surface charge cannot be determined on these electrodes, capacitance remained an adjustable parameter in SCMs for pH- and ionic strength-dependent OCP values.

In an attempt to independently measure EDL capacitance values, hematite electrodes were studied by Electrochemical Impedance Spectroscopy (EIS; Fig. 21; Shimizu et al. 2012, 2015; Shimizu and Boily 2014, 2015; Lucas and Boily 2015, 2017). The approach involves applying a small amplitude alternating current (AC) signal to the hematite electrode over a range of frequencies, while measuring the resulting voltage response. The impedance (complex

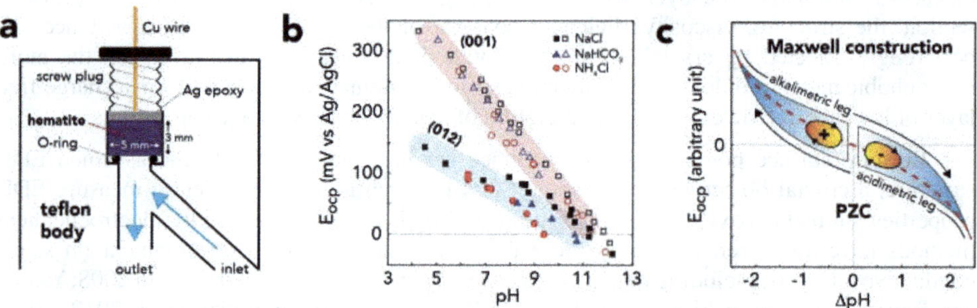

Figure 20. Open circuit potential (OCP) measurement on hematite single crystals. **(a)** Schematic representation of electrode embedded in a flow-through cell after (Shimizu and Boily 2014) (n.b. reference and auxiliary electrodes not shown). **(b)** OCP developed on the (001) and (012) faces of hematite in solutions of NaCl, $NaHCO_3$ and NH_4Cl (redrawn from Shimizu and Boily 2015). **(c)** Maxwell construction of hysteresis in OCPs (redrawn from Zarzycki et al. 2011).

resistance) of the system can then be used to distinguish slow chemical reactions (ion transfer across compact and diffuse layers) from fast electrostatic (space charge and bulk electron conduction) reactions. These phenomena can be captured by equivalent circuit modeling of impedance data (Fig. 21).

Equivalent circuit model parameters that relate to the slower response of charge carriers pertain to ions of the EDL. These include (i) capacitance (C_{ad}) and internal resistance (R_{ad}) terms relating to potential-determining ion. (H^+, OH^-) storage in the compact plane, and (ii) a capacitance (C_{dl}) from electrolyte ions in the diffuse layer. Notable findings from these efforts include (i) identification of minima in pH-resolved C_{ad} values near the pH of *pzc* (Shimizu et al. 2012), and (ii) larger C_{ad} and C_{dl} and lower R_{ad} values on the (012) than on the basal (001) faces (Shimizu et al. 2012; Shimizu and Boily 2014, 2015). These findings thus added support to the concept for a dynamic compact capacitance with responses to changes in electric fields and electrolyte ion loadings. They also aligned with contrasting charge-storage capabilities of contrasting crystallographic orientations. This links back to the first section (*Surface Identity and Reactivity*) where important differences in the populations and reactivity of proton active and electrolyte-binding functional groups and water structure on these surfaces is expected on hematite (Trainor et al. 2004; Waychunas et al. 2005; Catalano et al. 2007a,b, 2008, 2009; Tanwar et al. 2007a,b; Catalano 2010, 2011; Kerisit 2011).

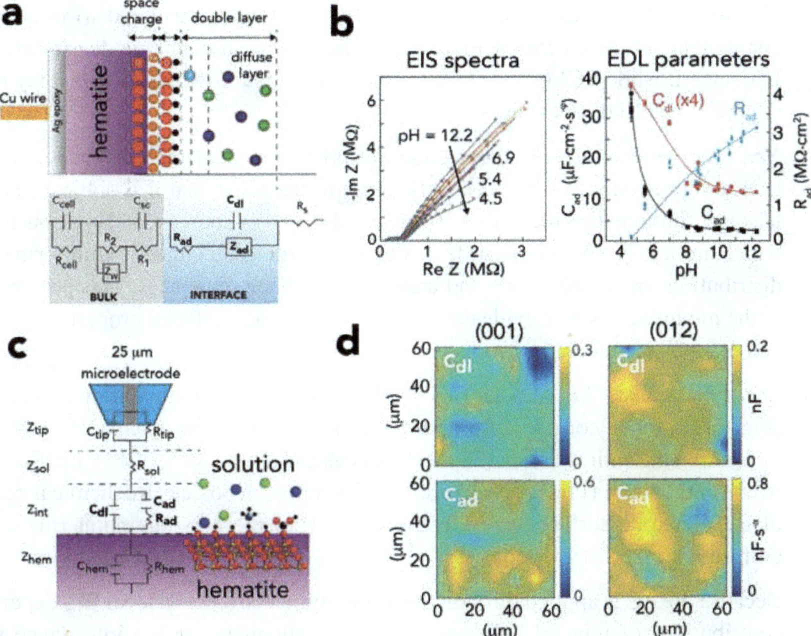

Figure 21. Electrochemical Impedance Spectroscopy (EIS) on oriented hematite electrodes. **(a)** Set-up for Global EIS measurement on mm-sized electrode, showing interfacial distribution of charge carriers, alongside equivalent electric circuit model used for the extraction of capacitance (C_{dl}, CPE) terms. Redrawn from Shimizu and Boily (2015). **(b)** Experimental global EIS spectra of imaginary (Im) and real (Re) impedance (*Z*) for the (012) face of hematite **(left)**. EDL parameters **(right)** for double layer capacitance (C_{dl}), adsorption capacitance (C_{ad}) and resistance (R_{ad}) were obtained by equivalent electric circuit modeling. These revealed a systematic response to pH. All data in (b) were taken and redrawn from Shimizu and Boily (2015). **(c)** Set-up for local EIS using a 25 μm-wide Pt ultramicroelectrode, and representative equivalent circuit. **(d)** Microscale spatial distribution of adsorption (C_{ad}) and double layer (C_{dl}) capacitance extracted by local EIS on hematite electrodes on the (001) and (012) faces. (c) and (d) were redrawn from Lucas and Boily (2015).

The magnitudes of the capacitance values for macro-sized hematite electrode were, however, on the same order of magnitude as those of metal (iodide) electrodes (Shimizu et al. 2012; Shimizu and Boily 2014, 2015). This raises an important problem as SCM-derived capacitances for hematite nanoparticles are considerably higher. It should also be noted that this situation is not unique to hematite. Similar EIS efforts on TiO_2 electrodes report similar ranges of capacitance values (Baram and Ein-Eli 2010; Prasannalakshmi et al. 2016), while SCMs for TiO_2 nanoparticles imply considerably larger capacitance values. Although the source of this disparity is currently unclear, the following considerations may guide the path to future research in this area:

i) *Electrode wettability*. Polished metal oxide electrodes may possibly be less hydrophilic than nanoparticles counterparts. These polished surfaces could thereby have low interfacial capacitances comparable to those of metal (iodide) electrodes. This may align with previous accounts for the effects of hydroxylation by ageing on OCP development (Lützenkirchen et al. 2013), and on the wetting characteristics of hematite (Shrimali et al. 2016).

ii) *EIS* vs. *SCM capacitance*. EIS-derived capacitance terms do not necessarily directly relate to those implied by SCM. While EIS-derived terms are detached from EDL models used in SCMs, they are, in turn, strongly coupled to the chosen equivalent circuit model. In EIS, several capacitance values are considered to treat various components of the electrochemical set-up but are blind to charge density and OCP. In contrast, while SCM capacitances consider the latter two quantities, they rely on the presumed EDL structure.

iii) *Non-ideal capacitance*. Without going into the specific electrochemical details, the EDL on electrodes is not necessarily ideal, in the sense that it is subject to current leakage through the dielectric (Brug et al. 1984). Sources of this non-ideality, which are less likely to be manifested in nanoparticles, include non-uniform current distributions on the electrode and meso-scale surface roughness. This possibly adds to the mounting body of evidence for the contrasting interfacial properties of macro- and nano-sized minerals.

iv) *Local vs. global EIS*. Local EIS work using 25 μm-wide ultramicroelectrodes (Fig. 21c) probed considerably smaller portions of hematite electrodes (~490 μm^2), rather than the entire surface (25 mm^2) (Lucas and Boily 2015, 2017). LEIS mapping over a larger area (Fig. 21d) also revealed heterogeneous electrochemical response of the electrode surface, a result linking back to the previous point (iii) on non-uniform current distributions

v) Because the AC is applied through the ultramicroelectrode in a local EIS experiment, contributions of mineral bulk conduction are attenuated in the impedance signal. As an interesting outcome, probe area-normalized capacitances become closer in magnitude, yet still not in the same range, to those derived by SCM.

Resolution of these issues may thus help reconcile ongoing disparities in capacitance values extracted by electrochemical methods and SCMs models. They may also open new possibilities in the development of novel surface complexation modeling frameworks, which will likely be needed as tomorrow's experimental and computational methods shed more light on molecular- to meso-scale processes at the mineral/water interface.

SUMMARY

This Chapter highlighted four key areas necessary for the development of SCMs, with the aim of replicating the wealth of information provided by molecular-scale studies. The *Surface Site Identity and Reactivity* section underscored the significance of understanding populations and spatial distribution of (hydr)oxo functional groups across diverse crystallographic faces of multifaceted particles. This understanding is indispensable for formulating surface complexation reactions that accurately depict coordination environments and geometries, aligning with experimental and/or simulation studies. The section elucidated both numerical and experimental tools applicable to identifying sites on both ideal and less ideal mineral surfaces.

Building upon this foundation, *The Hydrated Interface* section presented essential aspects necessary for understanding mineral surface hydration, counterion adsorption, and links to protonation of surface O(H) groups. Emphasizing the distinct hydration and coordination environments of various crystallographic terminations, it also highlighted the roles of local hydration at intersections of adjacent crystallographic faces.

The implications of interfacial hydration on charge development were then presented in *The Charged Interface* section. A SCM framework was introduced to predict charge development and surface complexation reactions on complex mineral particles. This framework was extended to dynamic capacitance approaches, incorporating the effects of electric fields and interfacial ion solvation effects on the charge-storage capability of the mineral/water interface.

Finally, in *The Electrochemical Interface* section, efforts to independently verify EDL properties using electrochemical methods were presented. Addressing the disparity between properties inferred by SCMs and those measured electrochemically, this section raised concerns about extrapolating knowledge form of macro-scale mineral specimens to nanoparticles.

Interconnecting these four aspects—*site identity*, *hydration*, *surface charge*, *surface potential*—is crucial for the evolving field of SCM research. As the demand for SCMs reflecting molecular-scale observations will intensify in the years to come, a comprehensive understanding of these domains will be pivotal in shaping the future of surface complexation modeling.

ACKNOWLEDGMENTS

This work was supported by the Swedish Research Council (2020-04853), FORMAS (2022-012146), the Kempe Foundation and the Carl Tryggers Foundation. J. Lützenkirchen and an anonymous reviewer are thanked for constructive comments leading to the final version of this Chapter.

REFERENCES

Al-Abadleh HA, Grassian VH (2003) Oxide surfaces as environmental interfaces. Surf Sci Rep 52:63–161

Andrade ENDC, Dodd C (1939) Effect of an electric field on the viscosity of liquids. Nature 144:117–118

Andrade ENDC, Dodd C (1946) The effect of an electric field on the viscosity of liquids. Proc R Soc London Ser A 187:296–337

Andrade ENDC, Dodd C (1951) The effect of an electric field on the viscosity of liquids. II. Proc R Soc London Ser A 204:449–464

Aquino AJA, Tunega D, Haberhauer G, Gerzabek MH, Lischka H (2008) Acid–base properties of a goethite surface model: A theoretical view. Geochim Cosmochim Acta 72:3587–3602

Armistead CG, Tyler AJ, Hambleton FH, Mitchell SA, Hockey JA (1969) Surface hydroxylation of silica. J Phys Chem 73:3947

Aryanpour M, van Duin ACT, Kubicki JD (2010) Development of a reactive force field for iron-oxyhydroxide systems. J Phys Chem A 114:6298–6307

Attard P (2006) Chapter 4–Fits to hypernetted chain calculations for electrostatic potential and ion concentrations for use in surface complexation. *In*: Interface Science and Technology. Vol 11. Lützenkirchen J, (ed) Elsevier, p 88–111

Avena M (2022) The reactivity of the metal oxide–water and mineral–water interfaces—An inorganic/coordination viewpoint. Eur J Inorg Chem 2022

Azam MS, Weeraman CN, Gibbs-Davis JM (2012) Specific cation effects on the bimodal acid–base behavior of the silica/water interface. J Phys Chem Lett 3:1269–1274

Backus EHG, Schaefer J, Bonn M (2021a) Probing the mineral–water interface with nonlinear optical spectroscopy. Angew Chem Int Ed 60:10482–10501

Bañuelos JL, Borguet E, Brown Jr GE, Cygan RT, DeYoreo JJ, Dove PM, Gaigeot MP, Geiger FM, Gibbs JM, Grassian VH, Ilgen AG (2023) Oxide– and silicate–water interfaces and their roles in technology and the environment. Chem Rev 123:6413–6544

Baram N, Ein-Eli Y (2010) Electrochemical impedance spectroscopy of porous TiO_2 for photocatalytic applications. J Phys Chem C 114:9781–9790

Bargar JR, Kubicki JD, Reitmeyer R, Davis JA (2005) ATR–FTIR spectroscopic characterization of coexisting carbonate surface complexes on hematite. Geochim Cosmochim Acta 69:1527–1542

Bedzyk MJ, Cheng L (2002) X-ray standing wave studies of minerals and mineral surfaces: principles and applications. Rev Mineral Geochem 49:221–266

Benjamin MM, Leckie JO (1981) Multiple-site adsorption of Cd, Cu, Zn, and Pb on amorphous iron oxyhydroxide. J Colloid Interface Sci 79:209–221

Berlier G, Lamberti C, Rivallan M, Mul G (2010) Characterization of Fe sites in Fe-zeolites by FTIR spectroscopy of adsorbed NO: are the spectra obtained in static vacuum and dynamic flow set-ups comparable? Phys Chem Chem Phys 12:358–364

Bikerman JJ (1973) Electric properties of foams. *In*: Foams. Bikerman JJ, (ed) Springer Berlin Heidelberg, Berlin, Heidelberg, p 223–230

Bockris JOM, Reddy AKN (1970) Modern electrochemistry: an introduction to an interdisciplinary area. Plenum Press, New York

Bockris JOM, Devanathan MAV, Müller K (1965) On the structre of charged interfaces ††Reprinted with minor changes from Proc R Soc, A, 274, 55–79, 541 (1963) *In*: Electrochemistry. Friend JA, Gutmann F, (eds) Pergamon, p 832–863

Boehm HP (1966) Functional groups on surfaces of solids. Angew Chem 5:533

Boehm HP (1971) Acidic and basic properties of hydroxylated metal-oxide surfaces. Disc Farad Trans 52:264–275

Boily JF (2007) Charge localization in cation–sulfate complexes: Implications for thermodynamic surface complexation models of the mineral/water interface. J Phys Chem C 111:1299–1306

Boily JF (2012) Water structure and hydrogen bonding at goethite/water interfaces: Implications for proton affinities. J Phys Chem C 116:4714–4724

Boily JF (2014) The variable capacitance model: A strategy for treating contrasting charge-neutralizing capabilities of counterions at the mineral water interface. Langmuir 30:2009–2018

Boily JF, Felmy AR (2008) On the protonation of oxo- and hydroxo-groups of the goethite (alpha-FeOOH) surface: A FTIR spectroscopic investigation of surface O–H stretching vibrations. Geochim Cosmochim Acta 72:3338–3357

Boily JF, Kozin PA (2014) Particle morphological and roughness controls on mineral surface charge development. Geochim Cosmochim Acta 141:567–578

Boily JF, Shchukarev A (2010) X-ray photoelectron spectroscopy of fast-frozen hematite colloids in aqueous solutions. 2. Tracing the relationship between surface charge and electrolyte adsorption. J Phys Chem C 114:2613–2616

Boily JF, Song X (2020) Direct identification of reaction sites on ferrihydrite. Nat Comm Chem 3:79

Boily JF, Persson P, Sjoberg S (2000) Benzenecarboxylate surface complexation at the goethite (alpha-FeOOH)/ water interface–III. The influence of particle surface area and the significance of modeling parameters. J Colloid Interface Sci 227:132–140

Boily JF, Lützenkirchen J, Balmès O, Beattie J, Sjöberg S (2001) Modeling proton binding at the goethite (alpha-FeOOH)–water interface. Coll Surf A 179:11–27

Boily JF, Chatman S, Rosso KM (2011) Inner-Helmholtz potential development at the hematite (alpha-Fe_2O_3) (001) surface. Geochim Cosmochim Acta 75:4113–4124

Boily JF, Yeşilbaş M, Md Musleh Uddin M, Baiqing L, Trushkina Y, Salazar-Alvarez G (2015) Thin water films at multifaceted hematite particle surfaces. Langmuir 31:13127–13137

Boily JF, Fu L, Tuladhar A, Lu Z, Legg BA, Wang ZMM, Wang HF (2019) Hydrogen bonding and molecular orientations across thin water films on sapphire. J Colloid Interface Sci 555:810–817

Bolt GH (1955) Analysis of the validity of the Gouy–Chapman theory of the electric double layer. J Colloid Sci 10:206–218

Bolt GH, van Riemsdijk WH (1982) Developments in soil science 5B *In*: Soil Chemistry. Bolt GH, (ed) Elsevier, Amsterdam

Bompoti N, Chrysochoou M, Machesky M (2018) Assessment of modeling uncertainties using a Multistart optimization tool for surface complexation equilibrium parameters (MUSE). ACS Earth Space Chem 3:473–483

Booth F (1951) The dielectric constant of water and the saturation effect. J Chem Phys 19:391–394

Bourg IC, Lee SS, Fenter P, Tournassat C (2017) Stern layer structure and energetics at mica–water interfaces. J Phys Chem C 121:9402–9412

Bowden J, Nagarajah S, Barrow N, Posner A, Quirk J (1980) Describing the adsorption of phosphate, citrate and selenite on a variable-charge mineral surface. Soil Research 18:49–60

Brkljača Z, Namjesnik D, Lützenkirchen J, Předota M, Preočanin T (2018) Quartz/aqueous electrolyte solution interface: molecular dynamic simulation and interfacial potential measurements. J Phys Chem C 122:24025–24036

Brown ID, Altermatt D (1985) Bond-valence parameters obtained from a systematic analysis of the inorganic crystal-structure database. Acta Cryst Sect B 41:244–247

Brown GE, Sturchio NC (2002) An overview of synchrotron radiation applications to low temperature geochemistry and environmental science. *In*: Applications of Synchrotron Radiation in Low-Temperature Geochemistry and Environmental Sciences. Vol 49. Fenter PA, Rivers ML, Sturchio NC, Sutton SR (eds) p 1–115

Brown Jr GE, Henrich V, Casey W, Clark D, Eggleston C, Andrew Felmy AF, Goodman DW, Gratzel M, Maciel G, McCarthy MI, Nealson KH (1999) Metal oxide surfaces and their interactions with aqueous solutions and microbial organisms. Chem Rev 99:77–174

Brug GJ, Vandeneeden ALG, Sluytersrehbach M, Sluyters JH (1984) The analysis of electrode impedances complicated by the presence of a constant phase element. J Electroanal Chem 176:275–295

Brunauer S, Emmet PH, Teller A (1938) Adsorption of gases in multimolecular layers. J Am Chem Soc 60:309–319

Catalano JG (2010) Relaxations and interfacial water ordering at the corundum (110) surface. J Phys Chem C 114:6624–6630

Catalano JG (2011) Weak interfacial water ordering on isostructural hematite and corundum (001) surfaces. Geochim Cosmochim Acta 75:2062–2071

Catalano JG, Fenter P, Park C (2007a) Interfacial water structure on the (012) surface of hematite: Ordering and reactivity in comparison with corundum. Geochim Cosmochim Acta 71:5313–5324

Catalano JG, Zhang Z, Park CY, Fenter P, Bedzyk MJ (2007b) Bridging arsenate surface complexes on the hematite (012) surface. Geochim Cosmochim Acta 71:1883–1897

Catalano JG, Park C, Fenter P, Zhang Z (2008) Simultaneous inner- and outer-sphere arsenate adsorption on corundum and hematite. Geochim Cosmochim Acta 72:1986–2004

Catalano JG, Fenter P, Park C (2009) Water ordering and surface relaxations at the hematite (110)–water interface. Geochim Cosmochim Acta 73:2242–2251

Cernik M, Borkovec M, Westall JC (1995) Regularized least-squares methods for the calculation of discrete and continuous affinity distributions for heterogeneous sorbents. Environ Sci Technol 29:413–425

Cernik M, Borkovec M, Westall JC (1996) Affinity distribution description of competitive ion binding to heterogeneous materials. Langmuir 12:6127–6137

Chan D, Perram JW, White LR, Healy TW (1975) Regulation of surface potential at amphoteric surfaces during particle–particle interaction. J Chem Soc Farad Trans 1 71:1046–1057

Chandrasekharan R, Zhang L, Ostroverkhov V, Prakash S, Wu Y, Shen YR, Shannon MA (2008) High-temperature hydroxylation of alumina crystalline surfaces. Surf Sci 602:1466–1474

Chapman (1913) A contribution to the theory of electrocapillarity. Phil Mag 25:475

Chatman S, Zarzycki P, Rosso KM (2013a) Surface potentials of (001), (012), (113) hematite (alpha-Fe_2O_3) crystal faces in aqueous solution. Phys Chem Chem Phys 15:13911–13921

Chatman S, Zarzycki P, Preoanin T, Rosso KM (2013b) Effect of surface site interactions on potentiometric titration of hematite (alpha-Fe_2O_3) crystal faces. J Colloid Interface Sci 391:125–134

Cheng J, Sprik M (2010) Acidity of the aqueous rutile TiO_2 (110) surface from density functional theory based molecular dynamics. J Chem Theory Comput 6:880–889

Connelly NG, Damhus T, Hartshorn RM, Hutton AT (2005) Nomenclature of Inorganic Chemistry—IUPAC Recommendations 2005. RSC Publishing

Conway BE, Bockris JO, Ammar IA (1951) The dielectric constant of the solution in the diffuse and hemlholtz double layers at a charged interface in aqueous solution. Trans Farad Soc 47:756–766

Cook P, Kim Y, Yuan K, Marcano MC, Becker U (2019) Electrochemical, spectroscopic, and computational investigations on redox reactions of selenium species on galena surfaces. Minerals 9:437

Cornell RM, Posner AM, Quirk JP (1974) Crystal morphology and dissolution of goethite. J Inorg Nucl Chem 36:1937–1946

Covert PA, Hore DK (2016) Geochemical insight from nonlinear optical studies of mineral–water interfaces. Annu Rev Phys Chem 67:233–257

Davis JA, Leckie JO (1978a) Surface ionization and complexation at oxide–water interface. 2. Surface Properties of amorphous iron oxyhydroxide and adsorption of metal ions. J Colloid Interface Sci 67:90–107

Davis JA, Leckie JO (1978b) Surface ionixation and complexation at oxide–water interface. 2. Surface Properties of amorphous iron oxyhydroxide and adsorption of metal ions. J Coll Interface Sci 67:90–107

Davis JA, James RO, Leckie JO (1978) Surface ionization and complexation at oxide–water interface. 1. Computation of electric double-layer properties in simple electrolytes. J Colloid Interface Sci 63:480–499

De Yoreo JJ, Gilbert PU, Sommerdijk NA, Penn RL, Whitelam S, Joester D, Zhang H, Rimer JD, Navrotsky A, Banfield JF, Wallace AF (2015) Crystallization by particle attachment in synthetic, biogenic, and geologic environments. Science 349:aaa6760

Derjaguin BV (1941) Theory of the stability of strongly charged lyophobic sol and of the adhesion of strongly charged particles in solutions of electrolytes. Acta Phys Chim URSS 14:633

DeWalt-Kerian EL, Kim S, Azam MS, Zeng H, Liu Q, Gibbs JM (2017) pH-dependent inversion of Hofmeister trends in the water structure of the electrical double layer. J Phys Chem Lett 8:2855–2861

Ding X, Song X, Boily J-F (2012) Identification of fluoride and phosphate binding sites at FeOOH surfaces. J Phys Chem C 116:21939–21947

Do DD (1998) Adsorption Analysis: Equilibria and Kinetics. Imperial College Press, London

Döpke MF, Lützenkirchen J, Moultos OA, Siboulet B, Dufrêche J-F, Padding JT, Hartkamp R (2019) Preferential adsorption in mixed electrolytes confined by charged amorphous silica. J Phys Chem C 123:16711–16720

Drits VA, Sakharov BA, Salyn AL, Manceau A (1993) Structural model for ferrihydrite. Clay Minerals 28:185–207

Dzombak DA, Morel FMM (1991) Surface Complexation Modeling: Hydrous Ferric Oxide. John Wiley & Sons, Inc., New York

Eggleston CM, Stack AG, Rosso KM, Bice AM (2004) Adatom Fe(III) on the hematite surface: Observation of a key reactive surface species. Geochem Trans 5:33–40

Ewing GE (2006) Ambient thin film water on insulator surfaces. Chem Rev 106:1511–1526

Fenter P, Sturchio N (2004) Mineral–water interfacial structures revealed by synchrotron X-ray scattering. Prog Surf Sci 77:171–258

Fokkink LGJ, Dekeizer A, Lyklema J (1989) Teperature-dependence of the electrical double-layer on oxides–Rutile and hematite. J Coll Interface Sci 127:116–131

Fröhlich H (1948) General theory of the static dielectric constant. Trans Faraday Soc 44:238–243

Fumagalli L, Esfandiar A, Fabregas R, Hu S, Ares P, Janardanan A, Yang Q, Radha B, Taniguchi T, Watanabe K, Gomila G (2018) Anomalously low dielectric constant of confined water. Science 360:1339–1342

Gao PY, Liu XD, Guo ZJ, Tournassat C (2023) Acid–base properties of cis-vacant montmorillonite edge surfaces: A combined first-principles molecular dynamics and surface complexation modeling approach. Environ Sci Technol 57:1342–1352

Geiger FM (2009) Second harmonic generation, sum frequency generation, and χ^3: Dissecting environmental interfaces with a nonlinear optical Swiss army knife. Annu Rev Phys Chem 60:61–83

Ghose SK, Waychunas GA, Trainor TP, Eng PJ (2010) Hydrated goethite (alpha-FeOOH) (100) interface structure: Ordered water and surface functional groups. Geochim Cosmochim Acta 74:1943–1953

Gittus OR, von Rudorff GF, Rosso KM, Blumberger J (2018) Acidity constants of the hematite–liquid water interface from ab initio molecular dynamics. J Phys Chem Lett 9:5574–5582

Gonella G, Backus EH, Nagata Y, Bonthuis DJ, Loche P, Schlaich A, Netz RR, Kühnle A, McCrum IT, Koper MT, Wolf M (2021) Water at charged interfaces. Nat Rev Chem 5:466–485

Gouy M (1910) Sur la constitution de la charge électrique à la surface d'un électrolyte. J Phys Appl 9: 457–468

Grahame DC (1947) The electrical double layer and the theory of electrocapillarity. Chem Rev 41:441–501

Harada Y, Tokushima T, Horikawa Y, Takahashi O, Niwa H, Kobayashi M, Oshima M, Senba Y, Ohashi H, Wikfeldt KT, Nilsson A (2013) Selective probing of the OH or OD stretch vibration in liquid water using resonant inelastic soft-X-ray scattering. Phys Rev Lett 111:193001

Hatlo MM, van Roij R, Lue L (2012) The electric double layer at high surface potentials: The influence of excess ion polarizability. Europhys Lett 97:28010

Henderson MA (2002) The interaction of water with solid surfaces: fundamental aspects revisited. Surf Sci Rep 46:1–308.

Henderson GS, Neuville DR, Downs RT (eds) (2014) Spectroscopic Methods in Mineralogy and Material Sciences. Reviews in Mineralogy and Geochemistry, Vol 78, Mineralogical Society of America

Hiemstra T (2012) Variable charge and electrical double layer of mineral–water interfaces: Silver halides versus metal (hydr)oxides. Langmuir 28:15614–15623

Hiemstra T (2013) Surface and mineral structure of ferrihydrite. Geochim Cosmochim Acta 105:316–325

Hiemstra T, van Riemsdijk WH (1991) Physical chemical interpretation of primary charging behaviour of metal (hydr) oxides. Colloids Surf 59:7–25

Hiemstra T, van Riemsdijk WH (1996) A surface structural approach to ion adsorption: The charge distribution (CD) model. J Colloid Interface Sci 179:488–508

Hiemstra T, van Riemsdijk WH (2007) Adsorption and surface oxidation of Fe(II) on metal (hydr)oxides. Geochim Cosmochim Acta 71:5913–5933

Hiemstra T, van Riemsdijk WH, Bolt GH (1989a) Multisite proton adsorption modeling at the solid/solution interface of (hydr)oxides: A new approach: I. Model description and evaluation of intrinsic reaction constants. J Colloid Interface Sci 133:91–104

Hiemstra T, De Wit JCM, van Riemsdijk WH (1989b) Multisite proton adsorption modeling at the solid/solution interface of (hydr)oxides: A new approach: II. Application to various important (hydr)oxides. J Colloid Interface Sci 133:105–117

Hiemstra T, Venema P, van Riemsdijk WH (1996) Intrinsic proton affinity of reactive surface groups of metal (hydr) oxides: The bond valence principle. J Colloid Interface Sci 184:680–692

Ho JM (2014) Predicting pK_a in implicit solvents: Current status and future directions. Austr J Chem 67:1441–1460

Hubao A, Yang Z, Hu R, Chen Y-F (2023) Molecular origin of wetting characteristics on mineral surfaces. Langmuir 39:2932–2942

Hunter RJ (1981) Zeta Potential in Colloid Science: Principles and Applictions. Academic Press

Hunter RJ, Leyendekkers JV (1978) Viscoelectric coefficient for water. J Chem Soc Farad Trans 1 74:450–455

Hwang YS, Lenhart J (2008) The dependence of hematite site-occupancy standard state triple-layer model parameters on inner-layer capacitance. J Colloid Interface Sci 319:206–213

Jin D, Hwang Y, Chai L, Kampf N, Klein J (2022) Direct measurement of the viscoelectric effect in water. PNAS 119:e2113690119

Kallay N, Preočanin T (2008) Measurement of the surface potential of individual crystal planes of hematite. J Colloid Interface Sci 318:290–295

Kebede GG, Mitev PD, Briels WJ, Hermansson K (2018) Red-shifting and blue-shifting OH groups on metal oxide surfaces—Towards a unified picture. Phys Chem Chem Phys 20:12678–12687

Kerisit S (2011) Water structure at hematite–water interfaces. Geochim Cosmochim Acta 75:2043–2061

Khurshid I, Al-Shalabi EW (2022) New insights into modeling disjoining pressure and wettability alteration by engineered water: Surface complexation based rock composition study. J Petrol Sci Eng 208:109584

Kirkwood JG (1939) The dielectric polarization of polar liquids. J Chem Phys 7:911–919

Knozinger H, Ratnasamy P (1978) Catalytic aluminas–surface models and characterization of surface sites. Cat Rev Sci Eng 17:31–70

Kornyshev AA (2007) Double-layer in ionic liquids: Paradigm change? J Phys Chem B 111:5545–5557

Korrani AKN, Jerauld GR (2019) Modeling wettability change in sandstones and carbonates using a surface-complexation-based method. J Petrol Sci Eng 174:1093–1112

Kosmulski M (2016) Isoelectric points and points of zero charge of metal (hydr)oxides: 50 years after Parks' review. Adv Colloid Interface Sci 238:1–61

Kozin PA, Boily J-F (2013) Proton binding and ion exchange at the akaganéite/water interface. J Phys Chem C 117:6409–6419

Kozin PA, Boily JF (2014) Mineral surface charge development in mixed electrolyte solutions. J Coll Interface Sci 418:245–253

Kozin PA, Shchukarev A, Boily JF (2013) Electrolyte ion binding at iron oxyhydroxide mineral surfaces. Langmuir 29:12129–12137

Kozin PA, Salazar-Alvarez G, Boily JF (2014) Oriented aggregation of lepidocrocite and impact on surface charge development. Langmuir 30:9017–9021

Kubicki JD, Kwon KD, Paul KW, Sparks DL (2007) Surface complex structures modelled with quantum chemical calculations: carbonate, phosphate, sulphate, arsenate and arsenite. Eur J Soil Sci 58:932–944

Larson I, Attard P (2000) Surface charge of silver iodide and several metal oxides. Are all surfaces nernstian? J Colloid Interface Sci 227:152–163

Lee SS, Fenter P, Nagy KL, Sturchio NC (2013) Changes in adsorption free energy and speciation during competitive adsorption between monovalent cations at the muscovite (001)–water interface. Geochim Cosmochim Acta 123:416–426

Lefèvre G, Preočanin T, Lützenkirchen J (2012) Attenuated total reflection–infrared spectroscopy applied to the study of mineral–aqueous electrolyte solution interfaces: A general overview and a case study. *In*: Infrared Spectroscopy. Theophile T, (ed) IntechOpen, Rijeka, p Ch. 5

Leung K, Criscenti LJ (2012) Predicting the acidity constant of a goethite hydroxyl group from first principles. J Phys: Condens Matter 24:124105

Lewis DG, Farmer VC (1986) Infrared-absorption of surface hydroxyl-groups and lattice-vibrations on lepidocrocite (gamma-FeOOH) and boehmite (gamma-AlOOH) Clay Minerals 21:93–100

Liu X, Cheng J, Sprik M, Lu X, Wang R (2013) Understanding surface acidity of gibbsite with first principles molecular dynamics simulations. Geochim Cosmochim Acta 120:487–495

Livi KJT, Villalobos M, Leary R, Varela M, Barnard J, Villacís-García M, Zanella R, Goodridge A, Midgley P (2017) Crystal face distributions and surface site densities of two synthetic goethites: implications for adsorption capacities as a function of particle size. Langmuir 33:8924–8932

Livi KJT, Villalobos M, Ramasse Q, Brydson R, Salazar-Rivera HS (2023) Surface site density of synthetic goethites and its relationship to atomic surface roughness and crystal size. Langmuir 39:556–562

Lucas M, Boily JF (2015) Mapping electrochemical heterogeneity at iron oxide surfaces: A local electrochemical impedance study. Langmuir 31:13618–13624

Lucas M, Boily JF (2017) Electrochemical response of bound electrolyte ions at oriented hematite surfaces: A local electrochemical impedance spectroscopy study. J Phys Chem C 121:27976–27982

Luo T, Xu J, Cheng W, Zhou L, Marsac R, Wu F, Boily J-F, Hanna K (2021) Interactions of anti-inflammatory and antibiotic drugs at mineral surfaces can control environmental fate and transport. Environ Sci Technol 56:2378–2385

Luong TN, Ilton ES, Shchukarev A, Boily J-F (2022) Water film-driven Mn (oxy)(hydr)oxide nanocoating growth on rhodochrosite. Geochim Cosmochim Acta 329:87–105

Lützenkirchen J (1999) Parameter estimation for the constant capacitance surface complexation model: analysis of parameter interdependencies. J Colloid Interface Sci 210:384–390

Lützenkirchen J (2002) Surface complexation models of adsorption: a critical surface in the context of experimental data. *In*: Adsorption: Theory, Modeling, and Analysis. Vol 107. J. T, (ed) Marcel Dekker, p 631–710

Lützenkirchen J (2005) On derivatives of surface charge curves of minerals. J Colloid Interface Sci 290:489–497

Lützenkirchen J (2006) Parsons–Zobel plots: An independent way to determine surface complexation parameters? J Colloid Interface Sci 303:214–223

Lützenkirchen J, Boily JF, Gunneriusson L, Lovgren L, Sjoberg S (2008) Protonation of different goethite surfaces–Unified models for $NaNO_3$ and NaCl media. J Colloid Interface Sci 317:155–165

Lützenkirchen J, Preočanin T, Stipic F, Heberling F, Rosenqvist J, Kallay N (2013) Surface potential at the hematite (001) crystal plane in aqueous environments and the effects of prolonged aging in water. Geochim Cosmochim Acta 120:479–486

Lützenkirchen J, Franks GV, Plaschke M, Zimmermann R, Heberling F, Abdelmonem A, Darbha GK, Schild D, Filby A, Eng P, Catalano JG (2018) The surface chemistry of sapphire-c: A literature review and a study on various factors influencing its IEP. Adv Colloid Interface Sci 251:1–25

Lyklema J (1994) On the slip process in electrokinetics. Colloids Surf A 92:41–49

Lyklema J, Overbeek JTG (1961) On the interpretation of electrokinetic potentials. J Colloid Sci 16:501–512

Machesky ML, Předota M, Ridley MK, Wesolowski DJ (2015) Constrained surface complexation modeling: rutile in RbCl, NaCl, and $NaCF_3SO_3$ Media to 250 °C. J Phys Chem C 119:15204–15215

Machesky ML, Predota M, Wesolowski DJ, Vlcek L, Cummings PT, Rosenqvist J, Ridley MK, Kubicki JD, Bandura AV, Kumar N, Sofo JO (2008) Surface protonation at the rutile (110) interface: explicit incorporation of solvation structure within the refined MUSIC model framework. Langmuir 24:12331–12339

Martínez RJ, Villalobos M, Loredo-Jasso AU, Cruz-Valladares AX, Mendoza-Flores A, Salazar-Rivera H, Cruz-Romero D (2023) Towards building a unified adsorption model for goethite based on direct measurements of crystal face compositions: I. Acidity behavior and As (V) adsorption. Geochim Cosmochim Acta 354:252–262

Mendez JC, Hiemstra T (2020) High and low affinity sites of ferrihydrite for metal ion adsorption: Data and modeling of the alkaline-earth ions Be, Mg, Ca, Sr, Ba, and Ra. Geochim Cosmochim Acta 286:289–305

Michel FM, Ehm L, Antao SM, Lee PL, Chupas PJ, Liu G, Strongin DR, Schoonen MAA, Phillips BL, Parise JB (2007) The structure of ferrihydrite, a nanocrystalline material. Science 316:1726–1729

Michel FM, Barron V, Torrent J, Morales MP, Serna CJ, Boily JF, Liu Q, Ambrosini A, Cismasu AC, Brown GE, Jr. (2010) Ordered ferrimagnetic form of ferrihydrite reveals links among structure, composition, and magnetism. PNAS 107:2787–2792

Morel FMM, Hering JG (1993) Principles and Applications of Aquatic Chemistry. Wiley

Mudunkotuwa IA, Al Minshid A, Grassian VH (2014) ATR–FTIR spectroscopy as a tool to probe surface adsorption on nanoparticles at the liquid–solid interface in environmentally and biologically relevant media. Analyst 139:870–881

Nakayama Y, Andelman D (2015) Differential capacitance of the electric double layer: The interplay between ion finite size and dielectric decrement. J Chem Phys 142:044706

Nemšák S, Shavorskiy A, Karslioglu O, Zegkinoglou I, Rattanachata A, Conlon CS, Keqi A, Greene PK, Burks EC, Salmassi F, Gullikson EM (2014) Concentration and chemical-state profiles at heterogeneous interfaces with sub-nm accuracy from standing-wave ambient-pressure photoemission. Nat Commun 5:5441

Nishiyama N, Yokoyama T (2021) Water film thickness in unsaturated porous media: effect of pore size, pore solution chemistry, and mineral type. Water Resour Res 57:e2020WR029257

Nordin JP, Sullivan DJ, Phillips BL, Casey WH (1999) Mechanisms for fluoride-promoted dissolution of bayerite beta-$Al(OH)_3$(s) and boehmite gamma-AlOOH:F-19-NMR spectroscopy and aqueous surface chemistry. Geochim Cosmochim Acta 63:3513–3524

O'Day PA (1999) Molecular environmental geochemistry. Rev Geophys 37:249–274

Ober P, Kolbinger SH, Backus EHG, Bonn M (2023) Ion-specific interactions at a mineral–water interface revealed by surface-sensitive spectroscopy under flow conditions. J Phys Chem C 127:13005–13010

Öhman LO (1998) Experimental determination of stability constants of aqueous complexes. Chem Geol 151:41–50

Öhman LO, Sjöberg S (1996) The experimental determination of thermodynamic properties for aqueous aluminium complexes. Coord Chem Rev 149:33–57

Ohno K, Okimura M, Akai N, Katsumoto Y (2005) The effect of cooperative hydrogen bonding on the OH stretching-band shift for water clusters studied by matrix-isolation infrared spectroscopy and density functional theory. Phys Chem Chem Phys 7:3005–3014

Parfitt RL, Atkinson RJ (1976) Phosphate adsorption on goethite (alpha-FeOOH) Nature 264:740–742

Parfitt RL, Russell JD (1977) Adsorption on hydrous oxides. 4. Mechanisms of adsorption of various ions on goethite. J Soil Sci 28:297–305

Parfitt RL, Fraser AR, Farmer VC (1977a) Adsorption on hydrous oxides. 3. Fulvic-acid and humic-acid on goethite, gibbsite and imogolite. J Soil Sci 28:289–296

Parfitt RL, Farmer VC, Russell JD (1977b) Adosrption on hydrous oxides. 1. Oxalate and benzoate on goethite. J Soil Sci 28:29–39

Parfitt RL, Fraser AR, Russell JD, Farmer VC (1977c) Adsorption on hydrous oxides. 2. Oxalate, benzoate and phosphate on gibbsite. J Soil Sci 28:40–47

Parkhurst DL (1995) User's guide to PHREEQC : A computer program for speciation, reaction-path, advective-transport, and inverse geochemical calculations. Lakewood, Colo. : U.S. Dept. of the Interior, U.S. Geological Survey, Denver, CO: Earth Sci Info Center, Open-File Reports Section [distributor], 1995

Parks GA (1965) The isoelectric points of solid oxides, solid hydroxides, and aqueous hydroxo complex systems. Chem Rev 65:177–198

Parks GA, Bruyn PLD (1962) The zero point of charge of oxides. J Phys Chem 66:967–973

Payne TE, Brendler V, Ochs M, Baeyens B, Brown PL, Davis JA, Ekberg C, Kulik DA, Lützenkirchen J, Missana T, Tachi (2013) Guidelines for thermodynamic sorption modelling in the context of radioactive waste disposal. Environ Model Soft 42:143–156

Peri JB, Hensley AL (1968) Surface structure of silica gel. J Phys Chem 72:2926

Petrik NG, Kimmel GA (2007) Hydrogen bonding, H–D exchange, and molecular mobility in thin water films on TiO2(110) Phys Rev Lett 99

Piontek SM, Borguet E (2022) Vibrational dynamics at aqueous–mineral interfaces. J Phys Chem C 126:2307–2324

Placencia-Gómez E, Kerisit SN, Mehta HS, Qafoku O, Thompson CJ, Graham TR, Ilton E, Loring JS (2020) Critical water coverage during forsterite carbonation in thin water films: activating dissolution and mass transport. Environ Sci Technol 54:6888–6899

Porus M, Labbez C, Maroni P, Borkovec M (2011) Adsorption of monovalent and divalent cations on planar water-silica interfaces studied by optical reflectivity and Monte Carlo simulations. J Chem Phys 135:064701

Prasannalakshmi P, Shanmugam N, kumar AS, Kannadasan N (2016) Phase-dependent electrochemistry of TiO_2 nanocrystals for supercapacitor applications. J Electroanal Chem 775:356–363

Předota M, Cummings PT, Wesolowski DJ (2007) Electric double layer at the rutile (110) Surface. 3. Inhomogeneous viscosity and diffusivity measurement by computer simulations. J Phys Chem C 111:3071–3079

Předota M, Machesky ML, Wesolowski DJ (2016) Molecular origins of the zeta potential. Langmuir 32:10189–10198

Prélot B, Charmas R, Zarzycki P, Thomas F, Villiéras F, Piasecki W, Rudziński W (2002) Application of the theoretical 1-p*K* approach to analyzing proton adsorption isotherm derivatives on heterogeneous oxide surfaces. J Phys Chem B 106:13280–13286

Preočanin T, Namjesnik D, Brown MA, Lützenkirchen J (2017) The relationship between inner surface potential and electrokinetic potential from an experimental and theoretical point of view. Environ Chem 14:295–309

Quirk JP (1960) Negative and positive adsorption of chloride by kaolinite. Nature 188:253–254

Raymand D, van Duin ACT, Goddard WA, Hermansson K, Spangberg D (2011) Hydroxylation structure and proton transfer reactivity at the zinc oxide–water interface. J Phys Chem C 115:8573–8579

Renock D, Mueller M, Yuan K, Ewing RC, Becker U (2013) The energetics and kinetics of uranyl reduction on pyrite, hematite, and magnetite surfaces: A powder microelectrode study. Geochim Cosmochim Acta 118:56–71

Rubasinghege G, Grassian VH (2013) Role(s) of adsorbed water in the surface chemistry of environmental interfaces. Chem Commun 49:3071–3094

Rudzinski W, Panczyk T (2000) Kinetics of isothermal adsorption on energetically heterogeneous solid surfaces: A new theoretical description based on the statistical rate theory of interfacial transport. J Phys Chem B 104:9149–9162

Russell JD, Parfitt RL, Fraser AR, Farmer VC (1974) Surface structures of gibbsite goethite and phosphated goethite. Nature 248:220–221

Rustad JR, Boily JF (2010) Density functional calculation of the infrared spectrum of surface hydroxyl groups on goethite (alpha-FeOOH). Am Mineral 95:414–417

Rustad JR, Felmy AR (2005) The influence of edge sites on the development of surface charge on goethite nanoparticles: A molecular dynamics investigation. Geochim Cosmochim Acta 69:1405–1411

Salzmann BBV, van der Sluijs MM, Soligno G, Vanmaekelbergh D (2021) Oriented attachment: From natural crystal growth to a materials engineering tool. Acc Chem Res 54:787–797

Sassi M, Chaka AM, Rosso KM (2021) Ab initio thermodynamics reveals the nanocomposite structure of ferrihydrite. Commun Chem 4:134

Schiek M, Al-Shamery K, Kunat M, Traeger F, Woll C (2006) Water adsorption on the hydroxylated H-(1×1) O–ZnO(000$\bar{1}$) surface. Phys Chem Chem Phys 8:1505–1512

Schindler PW, Gamsjäger H (1972) Acid-base reactions of TiO_2 (anatase)–water interface and point of zero charge of TiO_2 suspensions. Kolloid Z Z Polym 250:759

Schindler P, Kamber HR (1968) Die Acidität von Silanolgruppen. Vorläufige Mitteillung. Helv Chim Acta 51:1781–1786

Schwertmann U, Cornell RM (2003) The Iron Oxides: Structure, Properties, Reactions, Occurrences and Uses. Wiley-VCH, Weinheim

Shchukarev A, Ramstedt M (2017) Cryo-XPS: probing intact interfaces in nature and life. Surf Interface Anal 49:349–356

Shimizu K, Boily JF (2014) Electrochemical properties and relaxation times of the hematite/water interface. Langmuir 30:9591–9598

Shimizu K, Boily JF (2015) Electrochemical signatures of crystallographic orientation and counterion binding at the hematite/water interface. J Phys Chem C 119:5988–5994

Shimizu K, Lasia A, Boily JF (2012) Electrochemical impedance study of the hematite/water interface. Langmuir 28:7914–7920

Shimizu K, Nyström J, Geladi P, Lindholm-Sethson B, Boily JF (2015) Electrolyte ion adsorption and charge blocking effect at the hematite/aqueous solution interface: an electrochemical impedance study using multivariate data analysis. Phys Chem Chem Phys 17:11560–11568

Shrimali K, Jin J, Hassas BV, Wang X, Miller JD (2016) The surface state of hematite and its wetting characteristics. J Colloid Interface Sci 477:16–24

Sing KSW, Everett DH, Haul RAW, Moscou L, Pierotti RA, Rouquerol J, Siemieniewska T (1985) Reporting physisorption data for gas solid systems with special reference to the determination of surface area and porosity (Recommendations 1984) Pure Appl Chem 57:603–619

Siretanu I, Ebeling D, Andersson MP, Stipp S, Philipse A, Sutart M, van den Ende DFM (2014) Direct observation of ionic structure at solid-liquid interface: A deep look into the Stern layer. Nature 4:4956–4963

Sivula K, Le Formal F, Gratzel M (2011) Solar water splitting: Progress using hematite (alpha-Fe_2O_3) Photoelectrodes. Chemsuschem 4:432–449

Smit W (1986a) A site-binding model requiring low (outer) compact layer capacitances. J Coll Interface Sci 109:295–298

Smit W (1986b) Surface complexation constance of the site binding model. J Colloid Interface Sci 113:288–291

Song X, Boily JF (2011a) Surface hydroxyl identity and reactivity in akaganeite. J Phys Chem C 115:17036–17045

Song XW, Boily JF (2011b) Surface hydroxyl identity and reactivity in akaganeite. J Phys Chem C 115:17036–17045

Song X, Boily JF (2012a) Competitive ligand exchange on akaganeite surfaces enriches bulk chloride loadings. J Colloid Interface Sci 376:331–333

Song X, Boily JF (2012b) Structural controls on OH site availability and reactivity at iron oxyhydroxide particle surfaces. Phys Chem Chem Phys 14:2579–2586

Song X, Boily JF (2013a) Water vapor adsorption on goethite. Environ Sci Technol 47:7171–7177

Song X, Boily JF (2013b) Carbon dioxide binding at dry FeOOH mineral surfaces: Evidence for structure-controlled speciation. Environ Sci Technol 47:9241–9248

Song X, Boily JF (2013c) Water vapor interactions with FeOOH particle surfaces. Chem Phys Lett 560:1–9

Song X, Boily J-F (2016) Surface and bulk thermal dehydroxylation of FeOOH polymorphs. J Phys Chem A 120:6249–6257

Sposito G (1984) The Chemistry of Soils. Oxford University Press

Sposito G (2005) The Surface Chemistry of Natural Particles. Oxford University Press

Sprycha R (1989) Electrical double-layer at alumina electrolyte interface. 2. Adsorption of supporting electrolyte ions. J Coll Interface Sci 127:12–25

Stern O (1924) Zur Theorie der elektrolytischen Doppelschicht. Z Electrochem 30:508

Stumm W (1992) Chemistry of the Solid–Water Interface. Processes at the Mineral–Water and Particle–Water Interface in Natural Systems. Wiley

Sugimoto T (2001) Monodispersed Particles. Elsevier, Amsterdam

Sulpizi M, Gaigeot M-P, Sprik M (2012) The silica–water interface: how the silanols determine the surface acidity and modulate the water properties. J Chem Theory Comput 8:1037–1047

Sun EW-H, Bourg IC (2020) Molecular dynamics simulations of mineral surface wettability by water versus CO_2: thin films, contact angles, and capillary pressure in a silica nanopore. J Phys Chem C 124:25382–25395

Sverjensky DA (2001) Interpretation and prediction of triple-layer model capacitances and the structure of the oxide–electrolyte–water interface. Geochim Cosmochim Acta 65:3643–3655

Taifan W, Boily J-F, Baltrusaitis J (2016) Surface chemistry of carbon dioxide revisited. Surf Sci Rep 71:595–671

Tang M, Cziczo DJ, Grassian VH (2016) Interactions of water with mineral dust aerosol: water adsorption, hygroscopicity, cloud condensation, and ice nucleation. Chem Rev 116:4205–4259

Tanwar KS, Catalano JG, Petitto SC, Ghose SK, Eng PJ, Trainor TP (2007a) Hydrated alpha-Fe_2O_3($1\bar{1}02$) surface structure: Role of surface preparation. Surf Sci 601:L59–L64

Tanwar KS, Lo CS, Eng PJ, Catalano JG, Walko DA, Brown GE, Waychunas GA, Chaka AM, Trainor TP (2007b) Surface diffraction study of the hydrated hematite ($1\bar{1}02$) surface. Surf Sci 601:460–474

Tournassat C, Davis JA, Chiaberge C, Grangeon S, Bourg IC (2016) Modeling the acid–base properties of montmorillonite edge surfaces. Environ Sci Technol 50:13436–13445

Trainor TP, Chaka AM, Eng PJ, Newville M, Waychunas GA, Catalano JG, Brown GE (2004) Structure and reactivity of the hydrated hematite (0001) surface. Surf Sci 573:204–224

Trainor TP, Templeton AS, Eng PJ (2006) Structure and reactivity of environmental interfaces: Application of grazing angle X-ray spectroscopy and long-period X-ray standing waves. J Electron Spectrosc Relat Phenom 150:66–85

van Riemsdijk WH, de Wit JCM, Koopal LK, Bolt GH (1987) Metal ion adsorption on heterogeneous surfaces: Adsorption models. J Colloid Interface Sci 116:511–522

Venema P, Hiemstra T, Weidler PG, van Riemsdijk WH (1998) Intrinsic proton affinity of reactive surface groups of metal (hydr)oxides: application to iron (hydr)oxides. J Colloid Interface Sci 198:282–295

Verwey EJW, Overbeek JThG, van Nes K (1948) Theory of the Stability of Lyophobic Colloids: The Interaction of Sol Particles Having an Electric Double Layer. Elsevier, Amsterdam

Villalobos M, Cheney MA, Alcaraz-Cienfuegos J (2009) Goethite surface reactivity: II. A microscopic site-density model that describes its surface area-normalized variability. J Colloid Interface Sci 336:412–422

von Helmholtz HLF (1879) Studien über electrische Grenzschichten. Ann Physik 7:337

Wang ZM, Giammar DE (2013) Mass action expressions for bidentate adsorption in surface complexation modeling: Theory and practice. Environ Sci Technol 47:3982–3996

Wang X, Li W, Harrington R, Liu F, Parise JB, Feng X, Sparks DL (2013) Effect of ferrihydrite crystallite size on phosphate adsorption reactivity. Environ Sci Technol 47:10322–10331

Wang L, Zhao C, Duits MHG, Mugele F, Siretanu I (2015) Detection of ion adsorption at solid–liquid interfaces using internal reflection ellipsometry. Sensors Actuators B 210:649–655

Watson GW, Kelsey ET, deLeeuw NH, Harris DJ, Parker SC (1996) Atomistic simulation of dislocations, surfaces and interfaces in MgO. J Chem Soc-Faraday Trans 92:433–438

Waychunas G, Trainor T, Eng P, Catalano J, Brown G, Davis J, Rogers J, Bargar J (2005) Surface complexation studied via combined grazing-incidence EXAFS and surface diffraction: arsenate an hematite (0001) and ($10\bar{1}2$) Anal Bioanal Chem 383:12–27

Webb JT, Bhatnagar PD, Williams DG (1974) The interpretation of electrokinetic potentials and the inaccuracy of the DLVO theory for anatase sols. J Colloid Interface Sci 49:346–361

Westall J, Hohl H (1980) A comparison of electrostatic models for the oxide/solution interface. Adv Colloid Interface Sci 12:265–294

Yalcin SE, Legg BA, Yeşilbaş M, Malvankar NS, Boily J-F (2020) Direct observation of anisotropic growth of water films on minerals driven by defects and surface tension. Sci Adv 6:eaaz9708

Yanina S, Rosso K (2008) Linked reactivity at mineral-water interfaces through bulk crystal conduction. Science 320:218–222

Yates DE (1975) The structure of the oxide/aqueous electrolyte interface. Ph.D. University of Melbourne, Melbourne

Yates DE, Levine S, Healy TW (1974) Site-binding model of electrical double-layer at oxide–water interface. J Chem Soc Faraday Trans I 70:1807–1818

Yesilbas M, Song XW, Boily JF (2020) Carbon dioxide binding in supercooled water nanofilms on nanominerals. Environ Sci Nano 7:437–442

Yuan K, Ilton ES, Antonio MR, Li ZR, Cook PJ, Becker U (2015) Electrochemical and spectroscopic evidence on the one-electron reduction of U(VI) to U(V) on magnetite. Environ Sci Technol 49:6206–6213

Yuan K, Antonio MR, Ilton ES, Li ZR, Becker U (2022) Pentavalent uranium enriched mineral surface underelectrochemically controlled reducing environments. ACS Earth Space Chem 6:1204–1212

Zarzycki P (2023) Distance-dependent dielectric constant at the calcite/electrolyte interface: Implication for surface complexation modeling. J Colloid Interface Sci 645:752–764

Zarzycki P, Preočanin T (2012) Point of zero potential of single-crystal electrode/inert electrolyte interface. J Colloid Interface Sci 370:139–143

Zarzycki P, Rosso KM, Chatman S, Preočanin T, Kallay N, Piasecki W (2010) Theory, experiment and computer simulation of the electrostatic potential at crystal/electrolyte interfaces. Croat Chem Acta 83:457–474

Zarzycki P, Chatman S, Preočanin T, Rosso KM (2011) Electrostatic potential of specific mineral faces. Langmuir 27:7986–7990

Zarzycki P, Colla CA, Gilbert B, Rosso KM (2019) Lateral water structure connects metal oxide nanoparticle faces. J Mater Res 34:456–464

Zelenka M, Backus EHG (2024) Investigating aqueous mineral interfaces using sum frequency generation spectroscopy. *In*: Encyclopedia of Solid–Liquid Interfaces (First Edition) Wandelt K, Bussetti G, (eds) Elsevier, Oxford, p 148–157

Zhang YC, Liu XD, Cheng J, Lu XC (2021) Interfacial structures and acidity constants of goethite from first-principles molecular dynamics simulations. Am Mineral 106:1736–1743

Zukoski CF, Saville DA (1986) The interpretation of electrokinetic measurements using a dynamic model of the stern layer: I. The dynamic model. J Colloid Interface Sci 114:32–44

Reviews in Mineralogy & Geochemistry
Vol. 91A pp. 149–174, 2025

5

Surface Complexation at Charged Organic Surfaces

Maryam Salehi

Department of Civil and Environmental Engineering, University of Missouri, Columbia, Missouri 65201, U.S.A.

and

Missouri Water Center, Columbia, MO, USA

mshfp@missouri.edu

INTRODUCTION

Organic matter (OM), originating from sources such as plant residues, organisms, microbial exudates, decomposed biomass, and humic substances, or synthesized by humans, provides ligands that coordinate heavy metal (HM) ions through surface complexation reactions. The interactions between sites on organic surfaces and metal ions in aqueous systems influence numerous processes including contaminant transport and transformation, nutrient availability, and chemical reactions occurring at the interfaces between solids and water. Understanding surface complexation phenomena is essential for comprehending and predicting the behavior of HMs within aqueous and terrestrial environments. Moreover, organic surfaces can bind metal ions through complexation, stabilizing organic carbon compounds and preventing their rapid degradation (Clarholm et al. 2015).

Surface Complexation Models (SCMs) offer a robust quantitative methodology for modeling HM adsorption from an aqueous solution onto various solid surfaces. Initially developed for metal oxide surfaces, these models have subsequently, been adopted to study many other systems, including organic materials and microbial surfaces. The main principle of SCMs is that ion adsorption onto the surface sites is analogous to forming soluble metal complexes. Moreover, SCMs are found as effective approaches to predicting the acid/base behavior of many environmental surface reactions (Daughney and Fein 1998; Yee and Fein 2001; Daughney et al. 2002). The overall success of SCMs is achieved by consideration of aqueous HM speciation, pH variation, and ratios of adsorbent to adsorbate while relying on balanced chemical reactions. Assuming thermodynamic equilibrium, SCMs account for the chemical speciation of solute, the nature and concentration of the distinct reactive sites on the adsorbents that are present in the system, and the formation of surface complexes characterized by well-defined stoichiometric relationships (Alam et al. 2018a). Through SCM, numerous molecular-scale reactions can be considered, which makes them more data-intensive and complicated compared to the empirical adsorption models. The required modeling parameters include surface site concentrations, acidity constants, stability constants for individual surface complexes, and capacitance values (depending on the model).

Certainly, while SCMs have found widespread applications in understanding the interactions between minerals, HMs, and nutrients, their application to OM is less developed. This may be related to the complexity of organic compounds and the challenges they pose in terms of modeling and characterization. Most of the literature on HMs adsorption by organic adsorbents such as biochar (BC) employ the empirical adsorption isotherms (e.g., Langmuir, Freundlich) to fit the experimental adsorption data (Fang et al. 2014; Gao et al. 2020a; Salehi et al. 2021; Herath et al. 2022; Sharafoddinzadeh et al. 2022), which fail to consider the impacts of HM speciation. Instead, SCMs provide an alternative mechanistic modeling approach by considering (i) HM species present in the aqueous system, (ii) type and concentration of available sites on the surface of adsorbents, (iii) formation of surface

1529-6466/25/0091A-0005$05.00 (print)
1943-2666/25/0091A-0005$05.00 (online)
http://dx.doi.org/10.2138/rmg.2025.91A.05

complexes, and (iv) presence of the electrical double layer (DL) at the biological surface. Since common empirical adsorption isotherms are only useful for the specific conditions of the experiments, factors such as pH, ionic strength, sorbate-to-sorbent ratios, and temperature can significantly influence their reliability and applicability (Bethke and Brady 2000).

Objectives

This chapter focuses on the utilization of SCMs for HM adsorption by charged organic surfaces, including microbiological matter, BC, mineral–organic composites, and cellulose-based and polymeric materials. To begin with, the surface characteristics of these materials are discussed, including their primary surface functional groups involved in complexation reactions and their respective densities. Additionally, relevant surface acidity constants for a variety of materials are presented. Subsequently, the surface complexation reactions that have been suggested to occur among these organic surfaces, and the corresponding modeling are presented. Lastly, the chapter addresses the impacts of key water chemistry factors, including pH, ionic strength, and initial metal concentration on metal complexation reactions.

CHARGED ORGANIC SURFACES AND THEIR CHARACTERISTICS

SCMs aim to predict the extent of metal adsorption by a diverse range of organic surfaces in response to variations in pH, adsorbent-to-metal ratio, and ionic strength. The surface functional groups present in organic materials, such as carboxyl, hydroxyl, phosphoryl, amide, amine, sulfonyl, and sulfhydryl, can all be involved in HM binding in various settings (Fig. 1). These surface functional groups typically contain coordinating atoms such as N, O, P, or S. They also exhibit acid–base behavior (protonation/deprotonation) that resembles mineral surface sites. To develop a reasonable SCM for HM adsorption, it is important to collect information on surface functional groups, their density, and pK_a values, which indicate the surface's characteristics. This information helps build models that predict HM adsorption on organic surfaces. In this section, we discuss the key characteristics of OM studied through SCM, including functional groups and pK_a values, and their role in understanding HM adsorption.

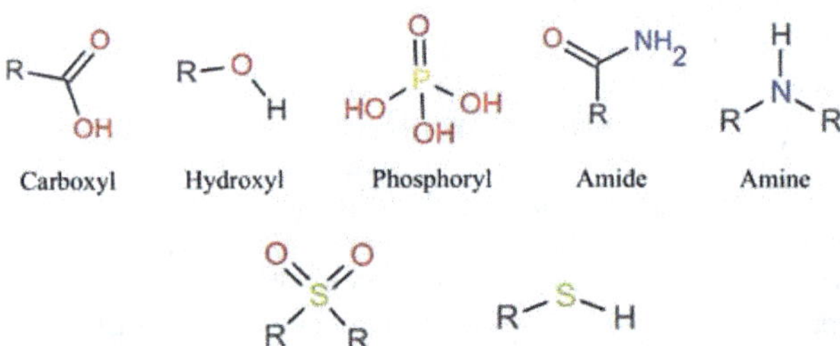

Figure 1. Common functional groups present on organic surfaces that may participate in complexation reactions.

Microbial matter

Microorganisms adhere to surfaces and generate an Extracellular Polymeric Substance (EPS) rich matrix, acting as the structural backbone for biofilm formation. These three-dimensional matrices, created by microorganisms on surfaces, involve bacterial communities associated with EPS (Ghoochani et al. 2023). Bacteria cell walls, EPS, and their metabolites (e.g., organic acids) involve functional groups such as sulfhydryl and carboxyl groups. These functional groups can form complexes with HMs, influencing their solubility and mobility in the environment (Hadiuzzaman et al. 2023a). As eutrophication expands and chlorophyte blooms become more prevalent, there is a growing need to investigate the interactions of HMs with chlorophytes, including macroalgae (Zoll and Schijf 2012). The widespread occurrence of such interactions underscores the significance of the relationship between microbial matter and diverse aqueous metal cations in various environmental settings (Ahamed et al. 2020).

A certain interest in applying the SCMs to biological materials and bacterial cells arises from the fact that the bioavailable fraction of the total dissolved HM is more important than the total amount. Moreover, the adsorption of dissolved metal species onto the bacterial cell impacts their biomineralization and their global cycling (Yee and Fein 2001). The affinity of biological matter to bond metals, their elevated surface area to volume ratios, and their abundance in near-surface geological systems all contribute to their crucial impacts on the distribution of HMs within the environment (Wei et al. 2011). Additionally, the speciation and binding of HMs to the bacterial surfaces may impact their toxicity (Flynn et al. 2014). However, modeling these interactions is challenging due to their binding capability with the many other unspecified organic ligands and inorganic materials. The relevant studies mostly applied SCMs for bacterial species and marine algae are listed in Table 1. Model variants that include three and four adsorption sites were mostly applied to biological matter. The three site models involve carboxyl, phosphate, and hydroxyl/amine as major functional groups. The pK_a values for carboxyl sites varied from 3.7 to 5.4, for phosphate from 6.0 to 7.2, and for hydroxyl/amine from 9.0 to 10.4 (Table 1). Four-site models were also applied to biological matter (Ginn and Fein 2008). *B. subtilis* cells were chosen as model organisms due to their ability to adsorb metals similar to other types of bacterial cells (Daughney and Fein 1998; Yee and Fein 2001; Daughney et al. 2002). Carboxyl, phosphoryl, and amino groups were considered the major surface functional sites present in the *B. subtilis* cells (Daughney et al. 2002).

The adsorption of HMs on biomass has also been studied to develop biosorbents or biofilters for metal pollution in a wide range of aquatic environments and associated SCMs were proposed (Zoll and Schijf 2012). For instance, the study conducted by Zoll and Schijf (2012) applied a non-electrostatic SCM for yttrium and rare earth element uptake onto an *Ulva lactuca* tissue homogenate that has been dried and resembled materials used in chemical engineering investigations to create sustainable biosorbents. Through alkalimetric titration, three monoprotic groups on fresh *U. lactuca* were obtained with approximate pK_a of 4.0, 6.0, and 9.5 tentatively representing carboxyl, phosphate, and hydroxyl/amine groups (Zoll and Schijf 2012). Yee and Fein (2001) have suggested a metal adsorption process that did not consider individual bacterial species and developed a generic and simplified model to study the influence of bacterial adsorption on the transport of dissolved metal within the environment (Yee and Fein 2001). A wide range of site concentrations was reported for different microbiological matter, as shown in Table 1. Most studies reported greater site densities for carboxyl and hydroxyl/amine compared to the phosphate groups.

Biochar

Biochar (BC) is a heterogenous carbonaceous material that is generated by thermal alteration (pyrolysis) of organic materials (biomass) under oxygen-limited conditions (Vithanage et al. 2015). During ashing, the hydrolysis of silicon dioxide produces silanol surface groups with weak acidity (Alam et al. 2018a). BC contains a range of surface functional groups as illustrated in Figure 2 (Vithanage et al. 2015; Xiao et al. 2018). In recent years, several studies showed that HMs are effectively bound to specific active sites on BC, such as carboxyl groups, amine, aromatic >C–O< groups, anomeric >O–C–O< carbons, and O-alkylated groups (Li et al. 2013; Fang et al. 2014b). The negative surface charge of BC limits the removal of oxyanions, while it promotes the removal of HM cations. The choice of raw materials and production processes impact the adsorption capacity of BC via the resulting surface area and functional groups. Under similar pyrolysis conditions, the plant-based BC materials have a greater carbon content, lower cation exchange capacity, and larger surface area compared to those generated from livestock manure (Li et al. 2019). The most abundant

Table 1. An overview of recent studies on surface complexation on microbiological matter NE-SCM: None-electrostatic SCM, CCM: Constant capacitance.

Specie	Target elements (salt, ionic strength, M)	pK_a	SCM	Sorption site	Site density (mmol/g dry mass)	References
Ulva lactuca	Yttrium and rare elements (NaCl, 0.5)	3.8	NE-SCM	Carboxyl	0.5	Zoll and Schijf (2012)
		6.0		Phosphate	0.5	
		9.4		Phenol/Amine	0.5	
E. coli	Cd ($NaNO_3$, 0.1)	4.9	CCM	Carboxyl	2.2	Yee and Fein (2001)
		NA[a]		NA	NA	
		NA		NA	NA	
P. aeruginosa	Cd ($NaNO_3$, 0.1)	5.1	CCM	Carboxyl	3.1	
		NA		NA	NA	
		NA		NA	NA	
B. megaturium	Cd ($NaNO_3$, 0.1)	5.2	CCM	Carboxyl	1.9	
		7.6		Phosphate	0.8	
		19.3		Hydroxyl	2.9	
B. subtilis	C ($NaNO_3$, 0.1)	4.8	CCM	Carboxyl	1.2	
		6.9		Phosphate	0.4	
		9.4		Hydroxyl	1.2	
Untreated *B. subtilis*	Cd (KNO_3, 0.01)	4.5	NE-SCM	Carboxyl	1.2	Wei et al. (2011)
		7.2		Phosphate	0.7	
		10.3		Hydroxyl	1.0	
EPS-free *B. subtilis*	Cd (KNO_3, 0.01)	4.7	NE-SCM	Carboxyl	0.4	
		6.8		Phosphate	0.3	
		9.1		Hydroxyl	0.4	
Untreated *P. putida*	Cd (KNO_3, 0.01)	4.5	NE- SCM	Carboxyl	0.9	
		6.6		Phosphate	0.4	
		9.4		Hydroxyl	0.5	

Table 1 (cont'd). An overview of recent studies on surface complexation on microbiological matter NE-SCM: None-electrostatic SCM, CCM: Constant capacitance.

Specie	Target elements (salt, ionic strength, M)	pK_a	SCM	Sorption site	Site density (mmol/g dry mass)	References
EPS-free *P. putida*	Cd (KNO_3, 0.01)	4.8	NE-SCM	Carboxyl	0.5	
		6.8		Phosphate	0.3	
		10.4		Hydroxyl	0.3	
Ulva lactuca	None (NaCl, 0.1)	4.0	NE-SCM	Carboxyl	0.5	Schijf and Ebling (2010)
		6.1		Phosphate	0.5	
		9.5		Amine	0.5	
Synechococcus sp.	Cd ($NaNO_3$, 0.01)	5.4	NE-SCM	Carboxyl	12.8	Liu et al. (2015)
		7.4		Phosphoryl	6.6	
		9.9		Amino	14.1	
Shewanella oneidensis (WT strains)	Zn, Pb ($NaNO_3$, 0.1)	4.3	CCM	Carboxyl	0.10	Ha et al. (2010)
		6.8		Phosphory	0.05	
		9.7		Amide	0.10	
Shewanella oneidensis ΔEPS strains	Zn, Pb ($NaNO_3$, 0.1)	4.2	CCM	Carboxyl	0.09	
		6.5		Phosphory	0.05	
		9.7		Amide	0.10	
D. radiodurans	Cd, Pb ($NaClO_4$, 0.1)	3.2	NE-SCM	NA	0.12[b]	Ginn and Fein (2008)
		4.7		NA	0.12[b]	
		6.5		NA	0.06[b]	
		9.2		NA	0.07[b]	
T. thermophilus	Cd, Pb ($NaClO_4$, 0.1)	3.4	NE-SCM	NA	0.09[b]	Ginn and Fein (2008)
		4.9		NA	0.05[b]	
		6.7		NA	0.02[b]	
		9.2		NA	0.03[b]	

Table 1 (cont'd). An overview of recent studies on surface complexation on microbiological matter
NE-SCM: None-electrostatic SCM, CCM: Constant capacitance.

Specie	Target elements (salt, ionic strength, M)	pK_a	SCM	Sorption site	Site density (mmol/g dry mass)	References
T. thermophilus	Cd, Pb ($NaClO_4$, 0.1)	3.4	NE-SCM	NA	0.12[b]	Ginn and Fein (2008)
		4.9		NA	0.09[b]	
		6.7		NA	0.06[b]	
		9.2		NA	0.08[b]	
Untreated *P. putida*	Cd (KNO_3, 0.01)	4.5	NE-SCM	Carboxyl	0.9	Fowle and Fein (2000)
		6.6		Phosphate	0.4	
		9.4		Hydroxyl	0.5	

Notes:
[a]NA: not available
[b] mmole per gram wet bacteria mass

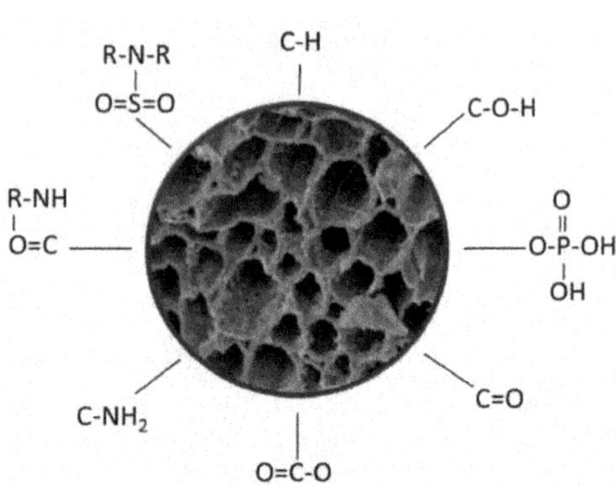

Figure 2. Major functional groups that can be present in BC, [the SEM image of BC used by permission of IOP Science from Wongrod et al. (2020), IOP Conf. Ser. Earth Environ. Sci., 463, Fig. 1. CC-BY 3.0].

acidic functional groups on BC are reported to be carboxyl (pK_H = 1.7–4.7), lactonic (pK_H = 6.37–10.25), and phenolic-OH sites (pK_H=9.5–13) (Francesca et al. 2008). However, the structural diversity of BC results in a wide variation of pK_a and makes it even more challenging to designate specific functional groups for each site solely based on pK_a (Li et al. 2014).

To enhance HMs adsorption characteristics of BC, certain chemical modifications have been applied to generate the active functional groups on its surface. For instance, the addition of carboxyl functional groups has been considered an effective approach. The strong affinity of metal ions to carboxylic groups is via complexation, ion exchange, and electrostatic interaction (Zhou et al. 2018). Two-site and three-site SCMs were applied for HM adsorption by BC. In the study by Xiong et al. (2021), sulfhydryl groups were grafted onto wood and palm BC to enhance their lead adsorption characteristics. SCM modeling suggested that Pb binds to low-proton affinity sites at low pH, or to high-proton affinity sites at high pH. The pK_a values for site 1 (4.96–5.17) and site 2 (8.30–8.56) corresponding to carboxyl and lactonic groups were determined. Sulfhydryl modification did not significantly change these pK_a values (Xiong et al. 2021). The study conducted by Alam et al. (2018a) applied three discrete adsorption sites assigned to carboxyl, lactonic, and phenolic hydroxyl groups (Alam et al. 2018a). An overview of major studies that applied SCM for metal adsorption onto BC is provided in Table 2.

Natural organic matter (NOM)

NOM consists of complex molecules that are heterogeneous and have polyelectrolyte and polyfunctional behavior. They provide an important suit of ligands to form complexes with metals. NOMs can be adsorbed onto various mineral surfaces such as clay, iron, aluminum hydroxide, and/or amorphous silica. Acidic pH creates a positive surface charge for some minerals and promotes NOM adsorption (El-sayed et al. 2019). The polyfunctional nature of NOM results in maintaining its ability to form complexes with metals even after adsorption to the mineral surface. Soil NOM content ranges from 0.1% to 10% by weight, and NOM can be found in concentrations of (0.1–200 mg/L) in many natural waters. Increased concentration of OM in soil raises the risk of the transport of contaminants in the natural environment. Additionally, the presence of dissolved organic matter (DOM) or NOM can promote microbial respiration, which increases the production of greenhouse gases.

The major functional groups of NOM are carboxylic, amino, and phenolic (Kantar 2007). Among NOMs, the humic substances have been described as either a combination of known ligands with similar binding constants and structures, hypothetical ligands exhibiting continuous or discrete distribution of binding constants, or one of the ligands involving an electrostatic energy term (Westall et al. 1995). Humic substances are complex organic compounds generated by the decomposition and transformation of residual biomolecules from plant and animal matter. Based on their solubility in alkalis, humic substances are categorized as humin (insoluble), humic acids (soluble, precipitated at $pH < 2$), and fulvic acids (soluble at all pH values; Zavarzina et al. 2021). Fulvic acids are often thought to be smaller molecules than humic acids.

The primary functional groups associated with NOM including carboxyl and phenolic groups as shown in Figure 3, and exhibit a high affinity for metals depending on pH. In addition, amine functional groups can be present in humic substances but may exhibit lower 'steric efficiency' compared to carboxylates due to their larger size, causing them to bump into other ligands when occupying coordination sites around a metal. This collision often results in decreased stability. Furthermore, humic substances contain aldehydes, ketones, esters, and sulfur functional groups (e.g., thiol, thiophene, sulfoxide) in lower abundance (Adusei-Gyamfi et al. 2019). The affinity of different metals for specific functional groups in humic substances varies. For example, at low pH values, calcium tends to bind with the carboxylic groups, whereas zinc (II) shows a preference for amine groups. In general, most cations bind with multiple functional groups. The reaction of NOM with cations present in drinking water significantly influences the efficiency of water treatment processes such as coagulation, adsorption, ion exchange, and membrane filtration (Adusei-Gyamfi et al. 2019).

Studies that investigated HM uptake and transport in the presence of NOM have considered single and multicomponent systems. For instance, to examine arsenic (III) interaction with thiols (–SH) in NOM, a thiol-functionalized ion exchange resin named Ambersep GT74 was used. A single component system involving thiol functional groups was considered and the pK_a value for thiol function was found as 9.6 and thiol binding site density was found as 0.34 mmol/g of Ambersep (Sanchez 2021). Due to the polyfunctional nature of NOM, even after adsorption to the mineral phase, they can form complexes with metal ions depending on the system characteristics (e.g., type of organic ligands, pH). Organics with multiple functional groups have been reported to enhance HM adsorption by minerals. However, those organics with one or two functional groups have not changed or even decreased metal adsorption (Davis and Leckie 1978). The binding of cobalt (II) to leonardite humic acid has been studied considering four acidic sites with pK_a values of 4.0, 6.0, 8.0, and 10.0 (Westall et al. 1995). Understanding the complexation of metal ions with NOMs, especially on mineral surfaces, is essential for predicting the migration of metal species. The varied affinities of different metals for specific functional groups in NOM further underscore the complexity of these interactions.

Table 2. A summary of major recent studies on surface complexation of heavy metals with bichar

BC Type	Target elements (salt, ionic strength, M)	pK_a	Surface site	Site density (mmol/g dry mass)	Reference
AD-char (from anaerobically digested garden waste)	Cu(II) ($NaNO_3$, 0.01)	5.75–10.20	Carboxyl	0.95	Zhang and Luo (2014)
			Lactonic	0.18	
			Phenolic	0.78	
E-char (from Eucalyptus leaves)		2.15–10.70	Carboxyl	0.40	
			Lactonic	0.93	
			Phenolic	0.02	
Wood BC	Pb(II) (KCl, 0.01)	5.17 8.56	Carboxyl	0.015	Xiong et al. (2021)
			Lactone	0.075	
			Phenolic	0.059	
Palm BC		4.82 8.51	Carboxyl	0.019	
			Lactone	0.167	
			Phenolic	0.184	
Sulfhydryl-modified wood BC		4.96 8.30	Carboxyl	0.019	
			Lactone	0.167	
			Phenolic	0.184	
Sulfhydryl-modified palm BC		4.96 8.47	Carboxyl	0.252	
			Lactonic	0.013	
			Phenolic	0.265	
Wheat straw BC	Ni(II), Zn(II) ($NaNO_3$, 0.01)	4.24	Carboxyl	0.75	Alam et al. (2018a)
		5.82	Lactonic	0.12	
		7.87	Phenolic Hydroxyl/silanol	0.97	
Wood pin chips BC		4.23	Carboxyl	0.98	
		6.18	Lactonic	0.11	
		7.94	Phenolic Hydroxyl/silanol	0.14	
Sewage sludge BC		4.50	Carboxyl	1.1	
		7.00	Lactonic	0.33	
		8.42	Phenolic Hydroxyl/silanol	0.41	

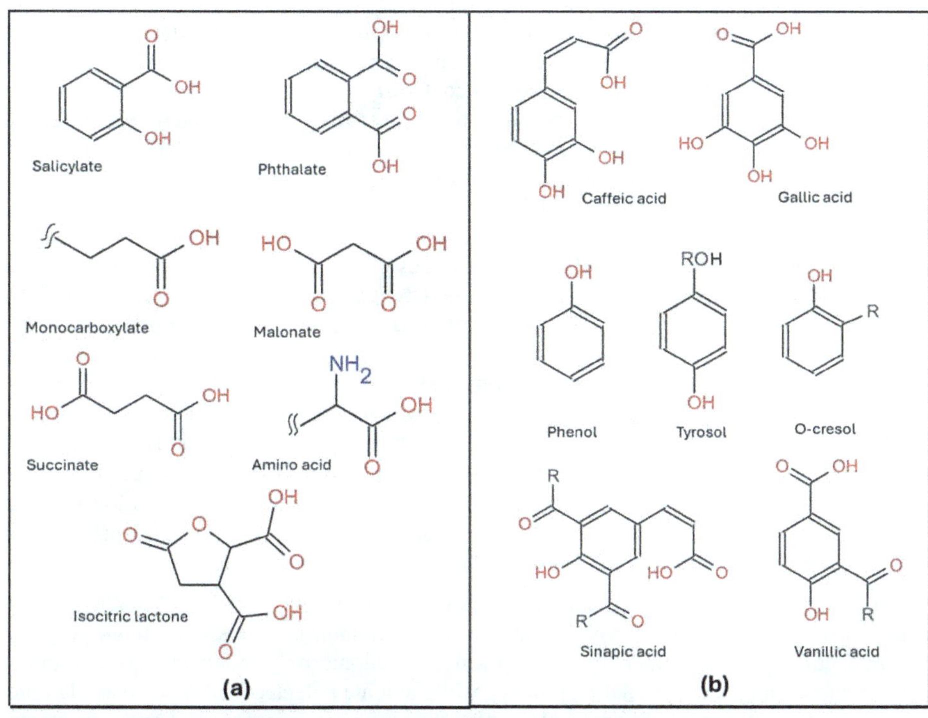

Figure 3. **(a)** Carboxyl [redrawn with permission from Elsevier from Deshmukh et al. (2007), Geochimica et Cosmochimica Acta 71:3533–3544, Fig.1] and **(b)** phenolic group structures suggested for humic substances [redrawn with permission from John Wiley and Sons for Rappoport (2003), The Chemistry of Phenols, The Chemistry of Functional Groups].

Other types of organic surfaces

Many other organic materials, including carbonaceous materials, polymers, and cellulose can provide high affinity surface functional groups for forming complexes with HM cations. These diverse organic matrices, each possessing a unique chemical composition, again provide various types of ligands to bind metals. Carbonaceous matter, consisting primarily of carbon, involves a range of materials such as activated carbon and carbon nanofibers (CNFs). Few studies have applied SCMs for carbonaceous materials considering surface hydroxyls and carboxyls as major functional groups participating in the complexation reactions. An early study conducted by Corapcioglu and Huang (1987) investigated the adsorption of Cu(II), Pb(II), Ni(II), and Zn(II) onto the surface of diverse brands of hydrous activated carbon. In that study, in addition to the free metal ions (M^{2+}), inorganic species such as MOH^+, $M(OH)_2$, and $M(OH)_3^{2-}$ were considered as the main species involved in the adsorption process. This investigation suggested varying pK_{a1} values for activated carbon of Nuchar, Filtrasorb, and Darco series, ranging from 2.3–7.7, 6.8–8.6, and 4.2–9.7, respectively. Similarly, the pK_{a2} values for activated carbon from these series exhibited disparities, ranging from 5.4–10.4, 9.6–12.2, and 6.1–11.8, respectively. The distinctive adsorption characteristics observed in these activated carbons were attributed to the presence of phosphoryl groups in the materials. In the presence of these groups, HMs demonstrated a greater propensity to form surface complexes compared to hydroxyl groups, highlighting the critical interplay between the activated carbon composition and the adsorption behavior of HM ions (Corapcioglu and Huang 1987). In the study conducted by Sun et al. (2014), the adsorption of Cu(II) onto carbon nanotubes (CNTs) functionalized with COOH groups was compared to those functionalized with OH groups. The investigation revealed a significantly greater Cu(II) adsorption capacity for CNTs–COOH in comparison to CNTs–OH. At the experimental pH of 6.0, where Cu(II) is the dominant species in the system, the deprotonated carboxyl groups on CNT effectively facilitated Cu(II) adsorption by forming complexes. In contrast, CNTs–OH might not undergo deprotonation at this pH but instead remain polarized. Consequently, the primary mechanism driving Cu(II) adsorption for CNTs-OH is electrostatic attraction. This distinction in the adsorption behavior sheds light on the pH-dependent interplay between the surface functionalization of CNTs and the adsorption mechanisms governing HM removal (Sun et al. 2014). This consideration agrees with the SCMs for solute adsorption to oxides. The SCM was also applied to study the adsorption mechanism of uranium (VI) to carbonaceous nanofibers (CNFs). This study found CNFs as promising candidates to adsorb the radionuclides from water (Hu et al. 2017). Chen et al. (2008) studied the adsorption of strontium (II) and europium (III) onto oxidized multiwall carbon nanotubes (MWCNTs) using SCM, and revealed the surface site densities for carboxyl and hydroxyl groups on oxidized MWCNTs as 2.94×10^{-4} mol/g and 1.30×10^{-4} mol/g, respectively. This study further determined intrinsic acidity constants for carboxyl protonation (K_{S-a1}^{int}) and deprotonation (K_{S-a2}^{int}) as $10^{0.16}$ and $10^{-7.60}$, respectively. Additionally, the intrinsic acidity constants for hydroxyl groups, K_{w-a1}^{int} and K_{w-a2}^{int} were determined as $10^{4.12}$ and$10^{-5.4}$, respectively. This comprehensive investigation provides crucial insights into the surface chemistry of oxidized MWCNTs and their interactions with Sr(II) and Eu(III) ions (Chen et al. 2008).

Polymeric substances are yet another class of OM that may form complexes with HMs. In addition to the metal hydroxide precipitates (iron or aluminum), cationic and anionic polymers are often utilized during water treatment practices. Cationic polyelectrolytes are commonly employed to enhance the coagulation process, while anionic polyelectrolytes serve as filter aids and contribute to sludge conditioning. Regarding the land application of solid residuals created by water treatment utilities, a study conducted by Butkus et al. (1998) investigated the binding of phosphate to a matrix composed of a quaternary polyamine (Fig. 4) and ferric hydroxide. In this study, it was assumed that both components interacted with phosphate. In addition to modeling the complexation of phosphate with iron, its complexation with the functional groups of polyamine was studied (Butkus et al. 1998). Moreover, a study conducted by Saha and Streat (2005) applied SCM to multiple polymeric resins to examine their Cu(II), Zn(II), and Ni(II) adsorption characteristics. The SCM model considers resins to be plain surfaces with a uniform distribution of functional groups. This surface was assumed to contain a series of capacitors with no electrostatic interaction among adjacent surface groups (Saha and Streat 2005).

As synthetic polymers, natural polymers can also form complexes with metals. These organic compounds, derived from natural sources, possess surface functionalities that make them prone to interactions with metal ions. For instance, lignin is a complex natural polymer found in plant cell walls that provides structural support and rigidity to plants. Its chemical structure is shown in Figure 4b. It contains a variety of functional groups including hydroxyl, carboxyl, carbonyl, aliphatic phenolic, and methoxyl groups (El Mansouri and Salvadó 2007). In the study conducted by Lv et al. (2011) a nonelectrostatic SCM was applied for Hg(II) adsorption by lignin, and included three reactive sites on the lignin surface, phenolic, aliphatic carboxylic, and aromatic carboxylic. The authors reported the pK_{a1} (5.16–5.27), pK_{a2} (6.78–6.91), and pK_{a3} (9.22–9.26) corresponding to aliphatic carboxylic-, aromatic carboxylic-, and phenolic-type sites, respectively (Lv et al. 2012). In a more recent study conducted by Xiao et al. (2019) surface complexation was identified as the primary mechanism for removing Cr(VI) and Cu(II) from an aqueous solution on chitosan combined with magnetic loofah BC as an adsorbent. The phenolic hydroxyl and carboxyl groups present in this adsorbent were found to contribute to the formation of complexes with these metals (Xiao et al. 2019). Kongdee and Bechtold (2008) conducted a study on the complexation of Co(II), Cu(II), and Zn(II) by various

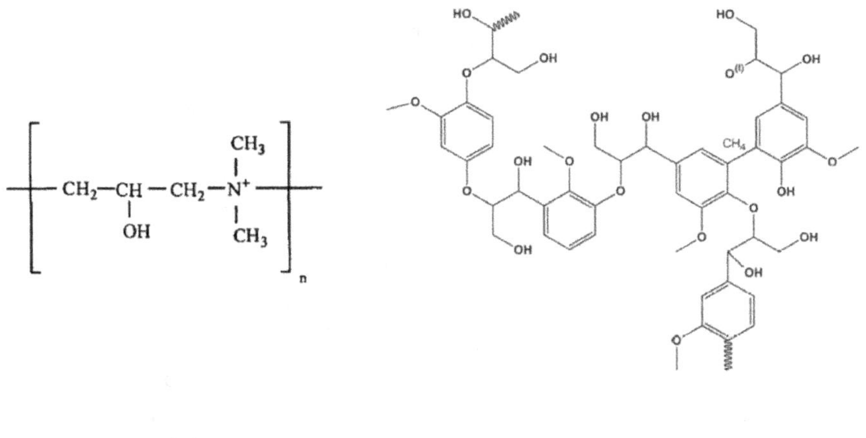

(a) (b)

Figure 4. Chemical structure of **(a)** a quaternary polyamine [reprinted with permission from John Wiley and Sons, Butkus et al. (1998), Journal of Environmental Quality 27:1055–1063, Fig.1] and **(b)** lignin [reprinted with permission from MPDI for Sarika et al. (2020), Polymers 12, Fig.2. CC BY 4.0]

cellulosic fibers including cotton, lyocell, and viscose. Hence, D-gluconate (DGL) and glycine (GLY) were selected as complexing agents, and metal adsorption onto the cellulosic materials was investigated in the presence of these complexing agents in the aqueous solution. Ligand exchange between DGL/GLY and cellulose occurs due to the similarity in their structures, and results in the adsorption of HMs onto the cellulosic fibers (Kongdee and Bechtold 2009).

In addition to the solely organic materials, SCMs were applied to mineral–organic composite materials such as layered double hydroxides (LDH)/BC and montmorillonite/BC composites. The LDH is constructed from positively charged metal hydroxide layers consisting of cation, e.g., Ni(II), Co(II), Cu(II), Zn(II), and counterbalancing anions, e.g., CO_3^{2-}, SO_4^{2-}, Cl^-. They may be utilized for the adsorption of anions, e.g., PO_4^{3-}, As, through electrostatic interaction, anion exchange, and complexation (Li et al. 2020). LDHs, including Mg–Al–LDH, Zn–AL–LDH, and Cu–Al–LDH, were used for the functionalization of corn stalk BC to enhance its anion exchange capacity, surface hydroxyl groups, and consequently promote As(V) removal from contaminated soil. The surface functional groups were identified as >COO and >OH (Gao et al. 2020b).

Hudcová et al. (2022) applied a non-electrostatic SCM for Mg–Fe LDH-coated BC for the removal of Zn(II) and As(V) from aqueous solution. These LDH–BC composites have shown promising metal adsorption characteristics in a broad range of pHs and loadings. At low Zn and As concentrations inner-sphere surface complexation controlled the adsorption process, but with increasing the concentrations, Zn precipitation and generation of ferric arsenate contributed to the metal removal process. The pK_a values were varied as 3.25–5.60 for the first site, 5.98–6.87 for the second site, and 8.90–9.57 for the third site. The site concentrations were highly variable among the amended BCs with different origins as 0.03–0.23 mM for the first site, 0.09–0.16 mM for the second site, and 0.13-0.15 mM for the third site. That study showed that complexation and precipitation contributed to HMs removal by Mg-Fe LDH-coated BC, and its results were validated by a three-site non-electrostatic model (Hudcová et al. 2022). A study conducted by Dano et al. (2020) applied SCMs for two types of BCs generated from bamboo biomass and its composite with montmorillonite K10 in the removal of technetium (VII) from liquid waste. The presence of montmorillonite resulted in faster kinetics and greater uptake of Tc(VII). The model considered edge and layer sites. In that study, three SCMs including two electrostatic and one nonelectrostatic model were applied (Daňo et al. 2020).

A limited number of studies on SCM-type approaches applied to phenomena such as metal ion uptake on plastics as an omnipresent scavenger of pollutants. A recent paper on Pb uptake on nanoplastics (Blancho et al. 2024) is an example, where a geochemical (PHREEQC) and an extrapolation program (Phreeplot) were applied with models originally developed for oxidic surfaces. Another class of interfaces is "inert" surfaces such as Teflon, diamond, gold, or gases. They have in common that they do not expose specific surface functional groups to aqueous solutions but nevertheless exhibit pH-dependent charges (Healy and Fuerstenau 2007). Some modeling attempts for describing the observed behaviors are available (Lützenkirchen et al. 2008). Unaged plastics can also be placed in this group. Both plastics and surfaces like gold, however, can in principle change with time, and then surface functional groups may be present on the surfaces of plastics (Johansson 2017) or gold (Tabor et al. 2011), with the concomitant changes in their surface chemistries. Prior investigations revealed the formation of carbonyl and carboxyl functional groups on polyethylene due to the photodegradation and chemical degradation processes (Herath and Salehi 2022; Hadiuzzaman et al. 2023b; Herath et al. 2023).

From the above survey, it can be concluded that the application of SCMs to organic material in many communities does not use the full extent of the possibilities offered by the state-of-the-art approaches. This would require indeed more experimental efforts as well. All the techniques and approaches are available, but the trend seems to be fast publication.

TYPES OF SURFACE COMPLEXES AT CHARGED ORGANIC SURFACES AND SCMS

In this section, we discuss the surface complexation process, focusing on the interaction between diverse organic substances (e.g., microbial cells, BC, and NOM) and HM ions. The focus is on the distinctive features of each class of organic substances and their interactions with a range of HMs. The aim is to provide a comprehensive understanding of the complexities involved in the surface complexation processes between these organic entities and HM ions.

Metal complexation with microbial matter

The microbial cells, whether Gram-negative or Gram-positive, are characterized by proton-active surface functional groups. Consequently, their exposure to adsorbing cations or aqueous ligands leads to the dissociation of protons from their surfaces and the subsequent possible creation of a surface charge. The chemical reaction (1) can be used to model their acidity, where R is the bacterial cell with an attached functional group (L_n), where n depends on the type of surface functional group ($n = 1, 2, 3$). Through the deprotonation process, the species $R\text{–}L_n^-$ forms. An equilibrium constant (K) is defined by Equation (2).

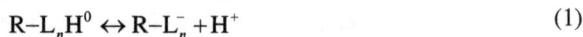

$$R\text{–}L_nH^0 \leftrightarrow R\text{–}L_n^- + H^+ \tag{1}$$

$$K = \frac{\left[R\text{–}L_n^-\right]\left[H^+\right]}{\left[R\text{–}L_nH^0\right]} \tag{2}$$

The deprotonation of biological matter is highly pH-dependent. As the pH of the surrounding water increases, surface deprotonation and the formation of anionic metal binding sites are favored. For most biological matter, carboxyl groups are primarily responsible for deprotonation at lower pH levels (4.0–6.0). However, as the pH increases within the range of 5.0 to 8.0, the phosphoryl group starts to play a significant role in the deprotonation process. Furthermore, as the pH continues to rise, reaching values between 8.0 and 10.0, hydroxyl and amino groups become increasingly crucial in the deprotonation process. Reactions (3–6) show the relevant proton dissociation reactions (Liu et al. 2015).

$$\text{R–COOH} + \text{OH}^- \leftrightarrow \text{R–COO}^- + \text{H}_2\text{O} \tag{3}$$

$$\text{R–PO}_4\text{H}_2 + \text{OH}^- \leftrightarrow \text{R–PO}_4\text{H}^- + \text{H}_2\text{O} \tag{4}$$

$$\text{R–NH}_3^+ + \text{OH}^- \leftrightarrow \text{R–NH}_2 + \text{H}_2\text{O} \tag{5}$$

$$\text{R–OH} + \text{OH}^- \leftrightarrow \text{R–O}^- + \text{H}_2\text{O} \tag{6}$$

The interaction between aqueous metal species and deprotonated solid surfaces can be described by reaction (7). Then partitioning of metal between aqueous and solid phases can be calculated using Equation (8) (Yee and Fein 2001).

$$\text{M}^{m+} + \text{R–L}_n^- \leftrightarrow \text{R–L}_n\left(\text{M}\right)^{(m-1)+} \tag{7}$$

$$K_{\text{ads}} = \frac{\left[\text{R–L}_n\left(\text{M}\right)^{(m-1)+}\right]}{\left[\text{M}^{m+}\right]\left[\text{R–L}_n^-\right]} \tag{8}$$

Square brackets indicate molar concentrations, implying that the reaction quotient is conditional. This involves the absence of both corrections for aqueous activity coefficients and smeared-out (mean-field) interfacial electrostatics. At infinite dilution and zero surface charge/potential, the reaction quotient equals the intrinsic equilibrium constant. Among other potential additional components of non-ideality, the activity coefficients of the surface species are neglected because these coefficients cannot be measured experimentally. As a result, the activity coefficients for surface species are assumed to have a ratio of unity, which simplifies the calculations by canceling out this effect in most cases. However, this simplification may not be valid for multi-dentate surface species.

Adsorption of cadmium onto microbial cells has been frequently simulated through SCMs. The study conducted by Liu et al. (2015) applied a non-electrostatic SCM to examine Cd(II) adsorption onto a common marine phytoplankton. A three-site model well described the experimental data. The adsorption reaction was written as (9) and the Cd-ligand stability constant was defined as (10), where $a_{\text{Cd}^{2+}}$ is the activity of Cd(II) in buffer solution. This study only considered the free metal cation and not its other forms such as hydrolyzed species (Liu et al. 2015). The highly pH-dependent Cd(II) adsorption is impacted by the speciation of the bacterial cell surface functional groups. The full protonation of surface functional groups at low pH hinders Cd(II) adsorption; however, by deprotonation of surface functional groups at higher pH, and formation of anionic metal binding sites, Cd(II) adsorption increased (Yee and Fein 2001).

$$\text{R–L}_n^- + \text{Cd}^{2+} \leftrightarrow \text{R–L}_n\text{Cd}^+ \tag{9}$$

$$K_{\text{L}_n\text{–Cd}^+} = \frac{\left[\text{R–L}_n\text{–Cd}^+\right]}{\left[\text{R–L}_n^-\right].a_{\text{Cd}^{2+}}} \tag{10}$$

To simulate the adsorption of yttrium (Y) and rare earth metals (YREEs) onto a dehydrated tissue of a marine macroalga (*Ulva lactuca*), three pK_a values corresponding to the surface functional groups of carboxyl, phosphate, and hydroxyl/amine were considered. At elevated pH, surface functional groups are being deprotonated, and the hydrolysis process controls YREE speciation. It was suggested that YREE-hydroxide complexes only bind to one of the surface functional groups present on *Ulva lactuca.* Reactions (11) and (12) show the YREE hydrolysis and chloride complexation. At elevated pH, the first YREE-hydrolysis species

(MOH^{2+}) was associated with the least acidic group. The YREE-chloride species may also form some complexes with the surface functional groups present on *Ulva lactuca*.

$$M^{3+} + H_2O \leftrightarrow MOH^{2+} + H^+ : \beta_1^* = \frac{[MOH^{2+}][H^+]}{[M^{3+}]} \tag{11}$$

$$M^{3+} + Cl^- \leftrightarrow MCl^{2+} : \beta_1 = \frac{[MCl^{2+}]}{[M^{3+}][Cl^-]} \tag{12}$$

In the SCM, each deprotonated functional group attached to the algae surface can form complexes with YREE cation and its hydroxide complexes as shown in reactions (13) and (14).

$$R\text{–}L_n^- + M^{3+} \leftrightarrow R\text{–}L_nM^{2+} : L_n\beta_1 = \frac{[R\text{–}L_nM^{2+}]}{[R\text{–}L_n^-][M^{3+}]} \tag{13}$$

$$R\text{–}L_3^- + MOH^{2+} \leftrightarrow S\text{–}L_3MOH^+ : L_3\beta_1^* = \frac{[R\text{–}L_3MOH^+]}{[R\text{–}L_3^-][MOH^{2+}]} = \frac{[R\text{–}L_3MOH^+][H^+]}{\beta_1^* \times [R\text{–}L_3^-][M^{3+}]} \tag{14}$$

It was assumed that the concentration of each functional group is equal to the summation of its protonated and deprotonated forms and log ${}_iK_s$ can be calculated according to Equation (15) (Zoll and Schijf 2012). However, it should be noted that this assumption is not required when using speciation codes, as the problem can also be solved with electrostatic models.

$$\log {}_iK_s = \log \frac{1}{3} \left\{ \frac{10^{(\log L_1\beta_1 - pK_{a1} + pH)}}{1 + 10^{(pH - pK_{a1})}} + \frac{10^{(\log L_2\beta_1 - pK_{a2} + pH)}}{1 + 10^{(pH - pK_{a2})}} + \frac{10^{(\log L_3\beta_1 - pK_{a3} + pH)} + 10^{(\log L_3\beta_1^* + \log\beta_1^* - pK_{a3} + 2 \times pH)}}{1 + 10^{(pH - pK_{a3})}} \right\} \tag{15}$$

The study conducted by Daughney et al. (2002) applied SCM to simulate the adsorption of mercury chloride and hydroxide complexes by *Bacillus subtilis* cells (Daughney et al. 2002). In the model, it was assumed that carboxyl, phosphoryl, and amino are the functions present on the bacterial surface, and different species of Hg(II) react with one or more bacterial surface sites to form the surface complexes. The significant impacts of solution characteristics on Hg(II) adsorption can be handled by the SCM model. For instance, the pH dependence of Hg(II) adsorption indicated that its adsorption is controlled by speciation of the bacterial surface. In the range of pH 6.0 to 8.0, Hg(II) adsorption decreased as phosphoryl surface sites (pK_a=7.0) were deprotonated. Thus, in the absence of chloride, the major dissolved species, $Hg(OH)_2$, is more strongly bound to protonated than to deprotonated surface sites. The SCM modeling that was based on a single adsorption reaction did not show a good correspondence to the experimental data. However, those three SCMs that considered two reactions matched in a reasonable way to the experimental data. They included one of the side reactions of (16) and (17) in addition to $Hg(OH)_2$ adsorption by protonated phosphoryl sites.

$$Hg(OH)_2^0 + R\text{–}OPO_3H_2 \leftrightarrow R\text{–}OPO_3H_2 - Hg(OH)_2^0 \tag{16}$$

$$Hg(OH)_2^0 + R\text{–}COO^- \leftrightarrow R\text{–}COO - Hg(OH)_2^- \tag{17}$$

Moreover, the SCM has been extended to account for chloride complexes. Below pH 8.0, the dominant forms of Hg(II) were chloride complexes. Under the studied conditions, two additional reactions (18) and (19) were included in the model explaining the adsorption of HgOHCl and $HgCl_2$ by protonated phosphoryl sites. Thus, four adsorption reactions and their stability constants accurately predicted the Hg(II) adsorption results acquired through the experimental work (Daughney et al. 2002).

$$HgCl_2^0 + R{-}OPO_3H_2^0 \leftrightarrow ROPO_3H_2 - HgCl_2^0 \tag{18}$$

$$HgOHCl^0 + R{-}OPO_3H_2^0 \leftrightarrow ROPO_3H_2 - HgOHCl^0 \tag{19}$$

The successful fitting of the experimental data has been attributed to not considering the reduced sulfur as the complexation agent. Moreover, the presence of trace metals in the field, competing with Hg(II) species for adsorption surface sites, could have led to an over-prediction by the model (Daughney et al. 2002).

Metal complexation with biochar

The presence of functional groups on the BC surface results in the adsorption of metal ions via surface complexation reactions. For modeling the surface complexation process, it is important to consider the speciation of metal ions in the aqueous system. For instance, the presence of carbonate (CO_3^{2-}), calcium, and hydroxyl (OH^-) ions in natural waters affect the speciation of metals present in these systems. A study conducted by Alam et al. (2018b) applied a non-electrostatic SCM for studying the U(VI) species onto BC using extended X-ray absorption fine structure (EXAFS) structural investigation. For this purpose, Equation (20) was used to determine the U(VI)–ligand stability constants, where a denotes the activity of the subscribed species. The aqueous U species that were considered in this study included its complexes with OH^-, CO_3^{2-}, and Ca^{2+}. The results suggested UO_2^{2+} as the dominant U species at pH<5, adsorbing onto deprotonated carboxyl (>COOH) surface sites (sites 1) as shown in reaction (21).

$$K_{U^{6+}} = \frac{\left[R - L_i\left(U^{6+}\right)^{n-1}\right] a_{H^+}}{\left[R - L_i^-\right] a_{U^{+6^n}}} \tag{20}$$

$$R{-}L_1H + UO_2^{2+} \leftrightarrow R{-}L_1\left(UO_2^+\right) + H^+ : \quad K_{UO_2^{2+}} = \frac{\left[R{-}L_1UO_2^+\right] a_{H^+}}{\left[R{-}L_1^-\right] a_{UO_2^{2+}}} \tag{21}$$

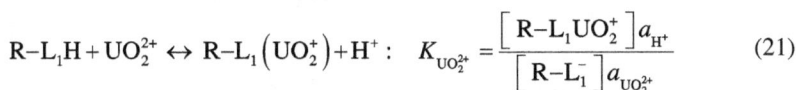

The performance of this model reduced as pH increased from 4.0 to 8.0. This is attributed to the occurrence of aqueous uranyl carbonate and hydroxide complexes that are dominant in this pH range, leading to their adsorption onto site 2, as shown in the reaction (22). However, for pH>8, bidentate and tridentate surface complexes were accounted for additionally as shown in Equations (23) and (24). While the success of the bidentate model was comparable to the monodentate model, the tridentate model did not converge. The EXAFS investigation was consistent with monodentate bonding.

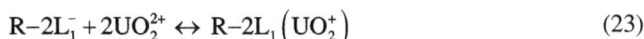

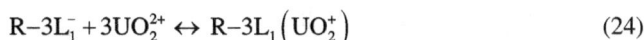

$$R{-}L_2^- + UO_2CO_3^0 \leftrightarrow R{-}L_2UO_2CO_3^- \tag{22}$$

$$R{-}2L_1^- + 2UO_2^{2+} \leftrightarrow R{-}2L_1\left(UO_2^+\right) \tag{23}$$

$$R{-}3L_1^- + 3UO_2^{2+} \leftrightarrow R{-}3L_1\left(UO_2^+\right) \tag{24}$$

In non-electrostatic models typically the required adsorption reactions mirror much more closely the aqueous solution speciation than in electrostatic models, which entails more reactions and adjustable parameters. There is a study conducted by Alam et al. (2018a) applied a non-electrostatic SCM to study the Ni(II) and Zn(II) adsorption onto BC. It was assumed that Zn(II), Ni(II), and proton interact with negatively charged organic acid surface functional groups of BC. The model considering Ni(II) and Zn(II) adsorption onto the two sites of a three-site model best fits the experimental data. The hydroxyl surface sites on BC are associated with inorganic silanol groups and organic phenols. It was indicated that carboxyl and hydroxyl are predominant surface sites on the BC, that could coordinate with Ni(II) and Zn(II) (Alam et al. 2018a).

The study conducted by Vithanage et al. (2015) applied the diffuse double layer model (DDLM) and the triple layer model (TLM) to examine antimony (III and V) adsorption onto soybean stover-derived BC. Only the phenolic and carboxylic sites were considered in the DDLM, and $Sb(OH)_6^-$, $Sb(OH)_3$, and $Sb(OH)_4^-$ were included as aqueous Sb species for modeling purposes as shown in reactions (25–27).

$$Sb(OH)_5 + H_2O \leftrightarrow Sb(OH)_6^- + H^+ \tag{25}$$

$$Sb(OH)_3 + H_2O \leftrightarrow Sb(OH)_4^- + H^+ \tag{26}$$

$$Sb(OH)_2^+ + H_2O \leftrightarrow Sb(OH)_3 + H^+ \tag{27}$$

The model finally suggested the formation of monodentate complexes and successfully applied a TLM for Sb adsorption by soybean stover-derived BC in the pH range of 4.0–9.0. Considering the amphoteric sites of hydrated BC in that study, the surface reactions are shown as (28) and (29), where L_i is the reactive active site that includes graphitic carbon. The fitting of experimental data was possible through reactions (30) and (31). The study considered carboxyl, lactonic, and phenolic groups on the BC as active surface sites. The adsorption of Sb onto the BC at pH levels below its precipitation point occurs through the formation of either monodentate or bidentate complexes. The TLM model well fitted the Sb(V) and Sb(III) data, while DDLM only fit the experimental data at low Sb concentrations (4.1–41 μM) (Vithanage et al. 2015).

$$-L_iOH + H^+ \leftrightarrow -L_iOH_2^+ \tag{28}$$

$$-L_iOH \leftrightarrow -L_iO^- + H^+ \tag{29}$$

$$(-L_i)_2OH + Sb(OH)_3 \leftrightarrow (-L_i)O\text{–}Sb(OH)_2^0 + H_2O \tag{30}$$

$$-L_iOH + Sb(OH)_6^- \leftrightarrow L_iOSb(OH)_5^- + H_2O \tag{31}$$

A non-electrostatic SCM with two sites has been applied by Xiong et al. (2021) to simulate Pb(II) adsorption onto raw and modified wood and palm BC. Two types of surface sites, low and high-affinity types of acidic surface functional groups were considered in this study. Site in their modeling refers to a group of functional groups with similar acidity constants rather than distinct functional groups. Thus, the carboxyl, carbonyl, phenolic hydroxyl, and ester groups present on BC were divided into phenolic groups ($pK_a = 9.5$–13), lactonic groups ($pK_a = 6.37$–10.25), and carboxylic groups ($pK_a = 1.7$–4.7) according to their dissociation constants. The results showed that pK_a values for sites 1 and 2 were close to pK_a values for carboxylic and lactonic groups. Moreover, the sulfhydryl modification of BC did not alter K_a and indicated that proton affinity remained unchanged following this modification (Xiong et al. 2021).

Metal complexation with natural organic matter (NOM)

Carboxylic and phenolic functional groups are the most important cationic binding sites present in NOMs, with carboxylic groups being more prevalent than phenolic groups. In particular, fulvic acids are reported to gain 78–90% of their total acidity from carboxylic groups, while humic acids derive 69–82% of their acidity from this functional group (Adusei-Gyamfi et al. 2019).

A study conducted by Zhou et al. (2004) examined the competitive complexation of Ni(II), Ca(II), and Al(III) with humic substances. It was assumed that both phenolic and carboxylic functional groups are present at low and high pH conditions. Moreover, this study assumed that only monodentate binding occurs. Humic substances were generically considered as a collection of a series of ligands without any interaction among the binding sites. Thus, the protonation and deprotonation reactions can be presented according to Equations (32–35). WOH, WO^- and WOH_2^+ represent the neutral, negatively, and positively charged species of weak acidic functional groups. SOH, SO^-, and SOH_2^+ represent the neutral, negatively, and positively charged species of strong acidic functional groups. $H^+(S)$ is the proton concentration at the surface of humic substances.

$$\mathrm{WOH} + \mathrm{H}^+(\mathrm{S}) \leftrightarrow \mathrm{WOH}_2^+ \tag{32}$$

$$\mathrm{WOH} \leftrightarrow \mathrm{WO}^- + \mathrm{H}^+(\mathrm{S}) \tag{33}$$

$$\mathrm{SOH} + \mathrm{H}^+(\mathrm{S}) \leftrightarrow \mathrm{SOH}_2^+ \tag{34}$$

$$\mathrm{SOH} \leftrightarrow \mathrm{SO}^- + \mathrm{H}^+(\mathrm{S}) \tag{35}$$

In electrostatic models, the interfacial profile refers to the diffuse part of the EDL, which represents a continuous distribution of charged species. Species that have the same sign of the surface in terms of charge are repelled and have lower concentration at the interface than in bulk solution. Species with a different sign compared to the mean-field surface potential are attracted and their concentrations are enhanced relative to the solution. The reaction quotient can then be written with the surface concentration of the adsorbing species as below. This treatment does not consider local effects such as discrete charge, ion size, hydrogen bonding, or water restructuring but represents a mean-field electrostatic, point charge view similar to most EDLs. Accordingly, the acidity constants were defined using Equations (36–39). The metal and proton surface activities cannot be measured directly, so they were assumed to be a function of their activities in the bulk solution according to Equations (40) and (41). In which Ψ_0 is mean-field potential at the surface of the adsorbent, R and F are the ideal gas constant and Faraday constant, and T is the absolute temperature. Using these values, the intrinsic surface complexation constants were determined as Equations (42) and (43) (Zhou et al. 2005).

$$K_{\mathrm{S-a1}}^{\mathrm{int}} = \frac{[\mathrm{SOH}][\mathrm{H}^+]_s}{[\mathrm{SOH}_2^+]} \tag{36}$$

$$K_{\mathrm{S-a2}}^{\mathrm{int}} = \frac{[\mathrm{SO}^-][\mathrm{H}^+]_s}{[\mathrm{SOH}]} \tag{37}$$

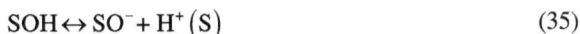

$$K_{\mathrm{W-a1}}^{\mathrm{int}} = \frac{[\mathrm{WOH}][\mathrm{H}^+]_s}{[\mathrm{WOH}_2^+]} \tag{38}$$

$$K^{\text{int}}_{\text{W-a2}} = \frac{\left[\text{WO}^-\right]\left[\text{H}^+\right]_s}{\left[\text{WOH}\right]} \tag{39}$$

$$\left[\text{M}^{n+1}\right]_s = \left[\text{M}^{n+}\right]\left[\exp\left(-\frac{\psi_0 \text{F}}{\text{R}T}\right)\right]^n \tag{40}$$

$$\left[\text{H}^+\right]_s = \left[\text{H}^+\right]\exp\left(-\frac{\psi_0 \text{F}}{\text{R}T}\right) \tag{41}$$

$$K^{\text{int}}_{\text{S-Me}^{n+}} = \frac{\left[\text{SOM}^+\right]\left[\text{H}^+\right]}{\left[\text{SOH}\right]\left[\text{M}^{n+}\right]\exp\left(-\psi_0 \text{F}/\text{R}T\right)^{n-1}} \tag{42}$$

$$K^{\text{int}}_{\text{W-Me}^{n+}} = \frac{\left[\text{WOM}^+\right]\left[\text{H}^+\right]}{\left[\text{WOH}\right]\left[\text{M}^{n+}\right]\exp\left(-\psi_0 F/\text{R}T\right)^{n-1}} \tag{43}$$

The study conducted by Liu et al. (2000) reported the simulation of Pb(II), Cu(II), and Cd(II) adsorption onto humic acid surfaces using a similar approach that applied the DL theory combined with Poisson–Blotzman equation. The results showed the order of Pb > Cu > Cd for the strength of binding. Similar to the Ping et al. (2004) study two types of surface sites with weak and strong acidities were considered to avoid the full complexityof the HA chemical structure (Liu and Gonzalez 2000).

Metal complexation with other organic matter

The complexation of metals with available surface sites on other OM such as carbonaceous substances has also been studied. The study conducted by Chen et al. (2008) applied the DDLM to experimental data for Sr(II) and Eu(III) adsorption onto oxidized MWCNTs. It was assumed that only monodentate binding sites are available on MWCNTs, and carboxyl and hydroxyl groups were considered as available adsorption sites on MWCNTs. The DDLM provided a good fit to acid–base characteristics of oxidized MWCNTs. For pH < 7.0, the surface complex of $-ROHSr^{2+}$ is dominant while at pH > 8.0, $-ROSr^{+}$ predominates. The surface species $-ROHEu^{3+}$ and $-ROEu^{2+}$ are correspondingly dominant as a function of pH (Chen et al. 2008). The adsorption of Cu(II), Ni(II) and Pb(II) on activated carbon cloths (ACCs) has also been studied using DDLM approach (Faur-Brasquet et al. 2002). The intrinsic surface acidity constants considering the electrostatic correction terms were calculated using Equations (44) and (45), where $\{H^+\}$, $\{-ROH\}$, and $\{-RO^-\}$ are the activities of H^+, protonated, and deprotonated ACCs, respectively.

$$K_{\text{a1}} = \frac{\left\{-\text{ROH}\right\}\left\{\text{H}^+\right\}}{\left\{-\text{ROH}_2^+\right\}}\exp\left(-\frac{\text{F}\psi_0}{\text{R}T}\right) \tag{44}$$

$$K_{\text{a2}} = \frac{\left\{-\text{RO}^-\right\}\left\{\text{H}^+\right\}}{\left\{-\text{ROH}\right\}}\exp\left(-\frac{\text{F}\psi_0}{\text{R}T}\right) \tag{45}$$

In addition to the free metal ion M^{2+}, other species, e.g., $M(OH)^+$, $M(OH)_2$, were included in the equilibrium constants calculation. In the applied model, in the pH below precipitation of solid $M(OH)_2$ (s), it was assumed that all metal ions that will be adsorbed are forming monodentate complexes with available surface sites. A worse fit for Ni(II) adsorption was found, compared to the other two metals. As hydrolysis of Ni(II) has been shifted to a higher pH (>7.0) than Pb(II) (pH > 5.0) and Cu(II) (pH > 6.0), the affinity between Ni(II) and ACCs is lower in this pH range because of the pH_{PZC} value of ACC.

In addition to OM, SCMs have been specifically applied for materials composed of mineral and organic substances. Two approaches have been suggested for the application of SCM for multi-sorbent systems including (1) conducting the general composite modeling which accounts for the generic functional groups that may be present on all surfaces and (2) component additivity (CA) modeling that considers a known set of site densities and constants for each surface. The advantage of CA is that it enables the identification of the contribution of each individual site in surface complex formation (Hudcová et al. 2022). Following the strategy, the study conducted by Hudcová et al. (2022) applied the SCM to Zn(II) and As(V) adsorption onto Mg–Fe LDH/BC composites. For this purpose, initially, the single sorbent SCM was applied for LDH and BC using the metal aqueous species concentrations, site concentrations, and protonation/deprotonation constants. Then the calculated acidity constants for each sorbent were used to develop a model through surface complexation and component additivity approach. By conducting the initial step for the target LDH, a better fit was found for monodentate complexes such as $-ROZn^+$ and $-ROAsO_3H^-$, compared to bidentate complexes. However, for BC the results showed the best fit by invoking a bidentate complex (e.g., $(-RO)_2Zn$) on site 1, and a monodentate complex ($-ROZn^+$) on site 2. In the second step, the surface speciation and relevant stability constants were utilized from single sorbent models. Subsequently, surface site concentration was determined considering the contents of BC and LDH in this composite. The prediction of this model was in agreement with the experimental data generated for BC/LDH composite as a function of ionic strength and pH. Minor overestimation was found for Zn(II) and As(V) adsorption for all BC/LDH composites potentially due to the blockage of surface binding sites due to the aggregation of BC and LDH components. Moreover, it was found that Zn(II) adsorption onto BC/LDH is mostly controlled by the LDH component, and only 1–5% of adsorption is caused by BC. This finding aligns with the single sorbent analysis, which demonstrated a higher Zn(II) affinity for LDH compared to BC (Hudcová et al. 2022).

Only a few studies applied SCM for multisorbent systems including NOM. Alam et al. (2018c) applied a non-electrostatic SCM to model Cd(II) and Se(VI) adsorption onto agricultural soil that was amended by BC. Stability constants and site concentrations were initially estimated from single metal and single sorbent experiments. The resulting SCM parameters were utilized successfully to model more complex systems containing multiple metals and sorbents. Between one to four distinct sites were tested to fit the experimental titration data. The model invoking two types of surface complexes best described selenate adsorption via reactions (46) and (47), and corresponding equilibrium constants also defined by Equations (48) and (49). Cd(II) adsorption was modeled considering Equations (50) and (51). The model ultimately predicted metal distribution in a multisorbent mixture, successfully (Alam et al. 2018c).

$$-\text{R–O–H}^0 + \text{HSeO}_4^- + \text{H}^+ \leftrightarrow -\text{R–HSeO}_4^0 + \text{H}_2\text{O} \tag{46}$$

$$-\text{R–O–H}^0 + \text{HSeO}_4^- \leftrightarrow -\text{R–SeO}_4^- + \text{H}_2\text{O} \tag{47}$$

$$K^{\text{int}}_{\text{W-Me}^{n+}} = \frac{[\text{WOM}^+][\text{H}^+]}{[\text{WOH}][\text{M}^{n+}]\exp(-\psi_0 F / RT)^{n-1}} \tag{48}$$

$$K^{\text{int}}_{\text{W-Me}^{n+}} = \frac{[\text{WOM}^+][\text{H}^+]}{[\text{WOH}][\text{M}^{n+}]\exp(-\psi_0 F / RT)^{n-1}} \tag{49}$$

$$-\text{R}-\text{L}_i^- + \text{Cd}^{2+} \leftrightarrow -\text{R}-\text{L}_i-\text{Cd}^+ \tag{50}$$

$$K_{i-\text{Cd}} = \frac{\left[-\text{R}-\text{L}_i-\text{Cd}^+ \right]}{\left[-\text{R}-\text{A}_i^- \right] a_{\text{Cd}^{2+}}} \tag{51}$$

As a conclusion, the formalism that has been developed for oxidic metal surfaces works well when it comes to simulating inorganic solute adsorption to organic surfaces. The current level of modeling is not as advanced as in other chapters in this volume. Diversity of the approaches on the one hand and the scarcity of directly comparable studies make an integrative survey difficult at present. While experimental tools could improve the modeling and incorporate more molecular details, challenges remain due to the complexity of functional groups and interactions in multi-adsorbent systems. The modeling of NOM has advanced beyond the approaches outlined here, as discussed in the chapter by Xu et al. (2025, this volume). In contrast to simpler models, NOM modeling addresses multi-site/heterogeneity through alternative models for charge distribution on humic and fulvic acids. Future models will need to include more detailed experimental characterizations.

MAJOR FACTORS INFLUENCE METAL COMPLEXATION WITH ORGANICS

Aqueous solution chemistry factors including pH, ionic strength, and metal concentrations critically influence the complexation of metals with organic substances. The solution pH controls the protonation status of binding sites or ligands present on the surface of organic materials, concurrently influencing their capability to create stable complexes with metal ions. Under low pH conditions, ligands typically undergo protonation, decreasing their availability for metal binding. Conversely, at elevated pH levels, deprotonation tends to promote the complexation of ligands with metal ions. The pH-dependent characteristics of ligand protonation play a critical role in determining the extent of metal and OM interactions. The pH conditions play an important role in determining adsorbed metal speciation, concurrently impacting their affinity to form complexes with the binding sites on organic surfaces. Understanding this pH-dependent speciation is fundamental to understanding the interaction between metal ions and organic substances. The study conducted by Daughney et al. (2002) reported pH dependent adsorption of Hg(II) by bacterial matter. Increasing the pH from 6.0 to 8.0 resulted in a reduction in metal adsorption. This decrease can be attributed to the deprotonation of phosphoryl surface sites, which, in turn, reduced the affinity of Hg(II) to bind with these specific surface sites (Daughney et al. 2002). Wei et al. (2011) also reported increasing Cd(II) adsorption by untreated (4–18 fold) and EPS-free (1–13 fold) bacteria with increasing pH from 3.0 to 5.0 (Wei et al. 2011). At pH>7.0, enhanced formation of YREE hydroxide complexes resulted in increased YREE sorption by *U. lactuca* (Zoll and Schijf, 2012). The surface charge of wheat straw and wood pin chip BC becomes more negative by increasing pH from 2.0 to 6.0, and U(VI) positive species (e.g., UO_2^{2+}, UO_2OH^+] are dominant below pH 6.0. Accordingly, by increasing pH from 2.0 to 6.0, U(VI) adsorption increases due to the enhanced electrostatic attraction (Alam et al. 2018). Studies on Cr(VI) removal by BC showed that its surface charge becomes negative above the isoelectric point (pH 3.5). As the pH increases above this point, electrostatic repulsion occurs between the BC surface functional groups and dominant aqueous species (e.g., CrO_4^{2-},$HCrO_4^-$), in the system (Alam et al. 2020b).

The ionic strength of an aqueous solution may significantly influence the complexation of metals with organic materials. With the increase in ionic strength and the higher concentration of ions in the solution, competition increases between metal ions and other ions present in the solution for binding sites on organic materials. This increased competition can impact the formation and stability of metal–organic complexes. Elevated ionic strength results in a stronger shielding effect, in which more charged sites on organic molecules shield the charged sites from the electric field in the surrounding solution. This shielding effect can reduce the electrostatic interactions between metals and OM, subsequently influencing the formation of complexes. The study conducted by Schijf and Ebling (2010) on *U. lactuca* acid revealed highly ionic strength dependent pK_a values that followed the extended Debye–Huckel relation Liu et al. (2015) reported a reduction in total site density with increased ionic strength from 0.01 M to 0.56 M, with amino groups being the most affected surface sites on cyanobacterium. Notably, this study found that 50% of Cd(II) uptake by cyanobacterium occurred at pH 5.0 at 0.01 M ionic strength, shifting to pH 6.0 when ionic strength was 0.56 M (Liu et al. 2015). The ionic strength does not always influence metal sorption onto the OM. For example, a study of Zn(II) adsorption onto LDH-coated BC revealed no significant influences by varying ionic strength, confirming the inner sphere complexation. Yet, a slight ionic strength dependency was found for As(V) adsorption by LDH-coated BC at $pH \geq 8.5$, indicating outer-sphere complexation (Hudcová et al. 2022). A similar finding by Vithanage et al. (2015) highlighted the ionic strength-independent adsorption of antimony (III and V) to soybean stover-derived BC, confirming inner sphere complexation (Vithanage et al. 2015). This also concurs with the known phenomena on oxidic surfaces (Hayes and Leckie 1987) and again confirms that the phenomena on the various surfaces have comparable underlying origins. In the same line of reasoning, the modeling via SCMs should be feasible involving all the possible levels of detail, including the very advanced models for oxides.

The initial concentration of metal in a solution may also play a critical role in influencing the extent of metal sorption onto OM, by overcoming the resistance to mass transfer across the interface of the solid adsorbent and aqueous solution. It is the solute to surface site ratio that is the important parameter, so at constant solid concentration, the influence of solute concentration can be discussed. With increasing the initial ion concentration, the monolayer capacity towards full coverage of the organic surface will be reached. Following this, a condensation phenomenon in the form of multi-layer complexation, corresponding to precipitation or the formation of a three-dimensional phase, may occur. For oxides, this has been described by the surface precipitation model (Farley et al. 1985). In this respect, maybe organic surfaces can be a convenient adsorbent to study surface precipitation since the more favorable contrast between the adsorbent and the surface precipitate can be used in optical measurements. Consequently, the increase of the initial ion concentration results in an increased quantity of adsorbed ions. An example of this phenomenon is found in the study of Cd(II) adsorption by BC, where surface loading significantly increases with the increased Cd(II) initial concentration, ultimately reaching the maximum adsorption capacity (Zhou et al. 2018). Characterization of such behaviors often involved adsorption isotherms, including the commonly applied Langmuir and Freundlich isotherms. These models are frequently applied to describe variations in metal adsorption by solid surfaces as the initial ion concentration increases (Wei et al. 2011), but SCMs are more advanced tools in this respect since they are not conditional.

SUMMARY

This chapter attempts to provide an overview of the composition and surface characteristics of common organic materials such as microbiological matter, BC, and NOM, found in aquatic environments and focuses on surface complexation reactions with HMs. The application of SCMs is extensively discussed by including active surface functional groups and their acidities, which may ultimately lead to a mechanistic understanding of HM interactions with them. This chapter has discussed the challenges of modeling metal adsorption onto biological materials and emphasized the importance of SCMs in understanding metal adsorption phenomena and predicting adsorption at the solid–liquid interface. Moreover, the critical role of NOM in metal ion complexation was addressed, insinuating their heterogeneity, polyfunctionality, and polyelectrolyte behavior. Additionally, the interaction of metals with other organic materials, such as carbonaceous matter and polymers was discussed, shedding light on their potential for forming complexes with metal cations. Studies applying SCMs to organic materials may provide insight into their adsorption characteristics, emphasizing the importance of better understanding and describing these complex interactions for water treatment applications and contaminant fate and transport studies. Furthermore, the chapter explores the significant influences of pH, ionic strength, and initial metal concentrations on surface complexation and metal adsorption onto OM. In this review, it has become clear that the state-of-the-art approaches of surface complexation models that are described in other chapters of this volume have not reached many communities dealing with adsorption from aqueous solutions. As an example, the modeling of NOM has reached significant sophistication and can help to explain real-world phenomena (Xu et al. 2025, this volume). It is expected that in the future, these and other approaches that combine molecular level aspects and electrical DL theories will also be applied to the surfaces discussed in this chapter.

ACKNOWLEDGMENT

This work was supported by US National Science Foundation (NSF) grants CBET-2305189 and CBET-2309475, and US Environmental Protection Agency (EPA) EM840651.

REFERENCES

Adusei-Gyamfi J, Ouddane B, Rietveld L, Cornard JP, Criquet J (2019) Natural organic matter-cations complexation and its impact on water treatment: A critical review. Water Res 160:130–147, https://doi.org/10.1016/j.watres.2019.05.064

Ahamed T, Brown SP, Salehi M (2020) Investigate the role of biofilm and water chemistry on lead deposition onto and release from polyethylene: an implication for potable water pipes. J Hazard Mater 400:123253, https://doi.org/10.1016/j.jhazmat.2020.123253

Alam MS, Gorman-Lewis D, Chen N, Flynn SL, Ok YS, Konhauser KO, Alessi DS (2018a) Thermodynamic analysis of Nickel(II) and Zinc(II) adsorption to biochar. Environ Sci Technol 52:6246–6255, https://doi.org/10.1021/acs.est.7b06261

Alam MS, Gorman-Lewis D, Chen N, Safari S, Baek K, Konhauser KO, Alessi DS (2018b) Mechanisms of the removal of U(VI) from aqueous solution using biochar: A combined spectroscopic and modeling approach. Environ Sci Technol 52:13057–13067, https://doi.org/10.1021/acs.est.8b01715

Alam MS, Swaren L, von Gunten K, Cossio M, Bishop B, Robbins LJ, Hou D, Flynn SL, Ok YS, Konhauser KO, Alessi DS (2018c) Application of surface complexation modeling to trace metals uptake by biochar-amended agricultural soils. Appl Geochem 88:103–112, https://doi.org/10.1016/j.apgeochem.2017.08.003

Alam MS, Bishop B, Chen N, Safari S, Warter V, Byrne JM, Warchola T, Kappler A, Konhauser KO, Alessi DS (2020) Reusable magnetite nanoparticles–biochar composites for the efficient removal of chromate from water. Sci Rep 10:19007, https://doi.org/10.1038/s41598-020-75924-7

Alessi DS, Fein JB (2010) Cadmium adsorption to mixtures of soil components: Testing the component additivity approach. Chem Geol 270:186–195, https://doi.org/10.1016/j.chemgeo.2009.11.016

Bethke C, Brady P (2000) How the K_d approach undermines ground water cleanup. Groundwater 38:324–474, https://doi.org/10.1111/j.1745-6584.2000.tb00230.x

Blancho F, Davranche M, Leon A, Marsac R, Reynaud S, Grassl B, and Gigault J. (2024) Mechanistic description of lead sorption onto nanoplastics. Environ Sci: Nano, https://doi.org/10.1039/D3EN00677H

Butkus MA, Grasso D, Schulthess CP, Wijnja H (1998) Surface complexation modeling of phosphate adsorption by water treatment residual. J Environ Qual 27:1055–1063, https://doi.org/10.2134/jeq1998.00472425002700050010x

Chen C, Hu J, Xu D, Tan X, Meng Y, Wang X (2008) Surface complexation modeling of Sr(II) and Eu(III) adsorption onto oxidized multiwall carbon nanotubes. J Colloid Interface Sci 323:33–41, https://doi.org/10.1016/j.jcis.2008.04.046

Clarholm M, Skyllberg U, Rosling A (2015) Organic acid induced release of nutrients from metal-stabilized soil organic matter—The unbutton model. Soil Biol Biochem 84:168–176, https://doi.org/10.1016/j.soilbio.2015.02.019

Corapcioglu MO, Huang CP (1987) The adsorption of heavy metals onto hydrous activated carbon. Water Res 21: 1031–1044, https://doi.org/10.1016/0043-1354(87)90024-8

Daňo M, Viglašová E, Galamboš M, Štamberg K, Kujan J (2020) Surface complexation models of pertechnetate on biochar/montmorillonite composite-batch and dynamic sorption study. Materials 13:3108, https://doi.org/10.3390/ma13143108

Daughney CJ, Fein JB (1998) The effect of ionic strength on the adsorption of H^+, Cd^{2+}, Pb^{2+}, and Cu^{2+} by *Bacillus subtilis* and *Bacillus licheniformis*: A surface complexation model. J Colloid Interface Sci 198:53–77, https://doi.org/10.1006/jcis.1997.5266

Daughney CJ, Siciliano SD, Rencz AN, Lean D, Fortin D (2002) Hg(II) adsorption by bacteria: A surface complexation model and its application to shallow acidic lakes and wetlands in Kejimkujik National Park, Nova Scotia, Canada. Environ Sci Technol 36:1546–1553, https://doi.org/10.1021/es010713x

Davis J, Leckie J (1978) Effect of adsorbed complexing ligands on trace metal uptake by hydrous oxides. Environ Sci Technol 12:1309–1315, https://doi.org/10.1021/es60158a002

Deshmukh, A.P., Pacheco, C., Hay, M.B., Myneni, S.C.B., 2007. Structural environments of carboxyl groups in natural organic molecules from terrestrial systems. Part 2: 2D NMR spectroscopy. Geochim Cosmochim Acta, 71:3533–3544. https://doi.org/10.1016/j.gca.2007.03.039

El Mansouri, Salvadó J (2007) Analytical methods for determining functional groups in various technical lignins. Ind Crops Prod 26:116–124, https://doi.org/10.1016/j.indcrop.2007.02.006

El-sayed MEA, Khalaf MMR, Gibson D, Rice JA (2019) Assessment of clay mineral selectivity for adsorption of aliphatic/aromatic humic acid fraction. Chem Geol 511:21–27, https://doi.org/10.1016/j.chemgeo.2019.02.034

Fang Q, Chen B, Lin Y, Guan Y (2014) Aromatic and hydrophobic surfaces of wood-derived biochar enhance perchlorate adsorption via hydrogen bonding to oxygen-containing organic groups. Environ Sci Technol 48:279–288, https://doi.org/10.1021/es403711y

Farley KJ, Dzombak, DA, Morel FM (1985) A surface precipitation model for the sorption of cations on metal oxides. J Colloid Interface Sci 106, 226–242, https://doi.org/10.1016/0021-9797(85)90400-X

Faur-Brasquet C, Reddad Z, Kadirvelu K, Le Cloirec P(2002) Modeling the adsorption of metal ions (Cu^{2+}, Ni^{2+}, Pb^{2+}) onto ACCs using surface complexation models. Appl Surf Sci 196:356–365, https://doi.org/10.1016/S0169-4332(02)00073-9

Flynn SL, Szymanowski JES, Fein JB (2014) Modeling bacterial metal toxicity using a surface complexation approach. Chem Geol 374–375:110–116, https://doi.org/10.1016/j.chemgeo.2014.03.010

Fowle DA, Fein JB (2000) Experimental measurements of the reversibility of metal–bacteria adsorption reactions. Chem Geol 168:27–36, https://doi.org/10.1016/S0009-2541(00)00188-1

Francesca P, Sara M, Luigi T (2008) New biosorbent materials for heavy metal removal: Product development guided by active site characterization. Water Res 42:2953–2962, https://doi.org/10.1016/j.watres.2008.03.012

Gao, X, Peng, Y, Guo L, Wang Q, Guan CY, Yang F, Chen Q (2020) Arsenic adsorption on layered double hydroxides biochars and their amended red and calcareous soils. J Environ Manage 271:111045, https://doi.org/10.1016/j.jenvman.2020.111045

Ghoochani S, Hadiuzzaman M, Mirza N, Brown SP, Salehi M (2023) Effects of water chemistry and flow on lead release from plastic pipes versus copper pipes, implications for plumbing decontamination. Environ Pollut 337:122520, https://doi.org/10.1016/j.envpol.2023.122520

Ginn BR, Fein JB (2008) The effect of species diversity on metal adsorption onto bacteria. Geochim Cosmochim Acta 72(16): 3939–3948, https://doi.org/10.1016/j.gca.2008.05.063

Ha J, Gélabert A, Spormann A M, Brown GE (2010) Role of extracellular polymeric substances in metal ion complexation on *Shewanella oneidensis*: Batch uptake, thermodynamic modeling, ATR-FTIR, and EXAFS study. Geochim Cosmochim Acta 74:1–15, https://doi.org/10.1016/j.gca.2009.06.031

Hadiuzzaman M, Mirza N, Brown SP, Ladner DA, Salehi M (2023a) Lead (Pb) deposition onto new and biofilm-laden potable water pipes. Chemosphere 342:14035, https://doi.org/10.1016/j.chemosphere.2023.140135

Hadiuzzaman M, Ladner DA, Salehi M (2023b) Impact of the surface aging of potable water plastic pipes on their lead deposition characteristics. Environ Sci: Water Res Technol 9:2501–2514, https://doi.org/10.1039/D3EW00043E

Hayes KF, Leckie JO (1987) Modeling ionic strength effects on cation adsorption at hydrous oxide/solution interfaces. J Colloid Interface Sci 115:564–572, https://doi.org/10.1016/0021-9797(88)90039-2

Healy TW, Fuerstenau DW (2007) The isoelectric point/point-of zero-charge of interfaces formed by aqueous solutions and nonpolar solids, liquids, and gases. J Colloid Interfac Sci 309:183–188, https://doi.org/10.1016/j.jcis.2007.01.048

Herath A, Salehi M (2022) Studying the combined influence of microplastics' intrinsic and extrinsic characteristics on their weathering behavior and heavy metal transport in storm runoff. Environ Pollut 308:119628, https://doi.org/10.1016/j.envpol.2022.119628

Herath A, Salehi M, Jansone-Popova S (2022) Production of polyacrylonitrile/ionic covalent organic framework hybrid nanofibers for effective removal of chromium(VI) from water. J Hazard Mater 427:128167, https://doi.org/10.1016/j.jhazmat.2021.128167

Herath A, Datta DK, Bonyadinejad G, Salehi M (2023) Partitioning of heavy metals in sediments and microplastics from stormwater runoff. Chemosphere 332:138844, https://doi.org/10.1016/j.chemosphere.2023.138844

Hu B, Hu Q, Xu D, Chen C (2017) The adsorption of U(VI) on carbonaceous nanofibers: A combined batch, EXAFS and modeling techniques. Sep Purif Technol 175:140–146, https://doi.org/10.1016/j.seppur.2016.11.025

Hudcová B, Fein JB, Tsang DCW, Komárek M (2022) Mg–Fe LDH-coated biochars for metal(loid) removal: Surface complexation modeling and structural change investigations. Chem Eng J 432:134360, https://doi.org/10.1016/j.cej.2021.134360

Johansson KS (2017) Surface modification of plastics. *In:* Applied Plastics Engineering Handbook. Elsevier, p 443–487

Kantar C (2007) Heterogeneous processes affecting metal ion transport in the presence of organic ligands: Reactive transport modeling. Earth Sci Rev 81:175–198, https://doi.org/10.1016/j.earscirev.2006.11.001

Kongdee A, Bechtold T (2009) Influence of ligand type and solution pH on heavy metal ion complexation in cellulosic fibre: Model calculations and experimental results. Cellulose 16:53–63, https://doi.org/10.1007/s10570-008-9248-y

Li X, Shen Q, Zhang D, Mei X, Ran W, Xu Y, Yu G (2013) Functional groups determine biochar properties (pH and EC) as studied by two-dimensional ^{13}C NMR correlation spectroscopy. PLoS ONE 8:e65949, https://doi.org/10.1371/journal.pone.0065949

Li Liu, Q, Lou Z, Wang Y, Zhang Yaping, Qian G (2014) Method to characterize acid–base behavior of biochar: site modeling and theoretical simulation. ACS Sustainable Chem Eng 2:2501–2509, https://doi.org/10.1021/sc500432d

Li F, Jin J, Shen Z, Ji H, Yang M, Yin Y (2020) Removal and recovery of phosphate and fluoride from water with reusable mesoporous Fe_3O_4@mSiO_2@mLDH composites as sorbents. J Hazard Mater 388:121734, https://doi.org/10.1016/j.jhazmat.2019.121734

Li S, Harris S, Anandhi A, Chen G (2019) Predicting biochar properties and functions based on feedstock and pyrolysis temperature: A review and data syntheses. J Cleaner Product 215:890–902, https://doi.org/10.1016/j.jclepro.2019.01.106

Liu A, Gonzalez RD (2000) Modeling adsorption of copper(II), cadmium(II) and lead(II) on purified humic acid. Langmuir 16:3902–3909, https://doi.org/10.1021/la990607x

Liu Y, Alessi DS, Owttrim GW, Petrash DA, Mloszewska AM, Lalonde SV, Martinez RE, Zhou Q, Konhauser KO (2015) Cell surface reactivity of Synechococcus sp. PCC 7002: Implications for metal sorption from seawater. Geochim Cosmochim Acta 169:30–44, https://doi.org/10.1016/j.gca.2015.07.033

Lützenkirchen J, Preočanin T, Kallay N. (2008) A macroscopic water structure based model for describing charging phenomena at inert hydrophobic surfaces in aqueous electrolyte solutions. Phys Chem Chem Phys 10:4946-4955, https://doi.org/10.1039/B807395C

Lv J, Luo L, Zhang J, Christie P, Zhang S (2012) Adsorption of mercury on lignin: Combined surface complexation modeling and X-ray absorption spectroscopy studies. Environ Pollut 162:255–261, https://doi.org/10.1016/j.envpol.2011.11.012

Rappoport Z (Ed.) (2003) The Chemistry of Phenols, The Chemistry of Functional Groups. Wiley, Hoboken, NJ

Saha B, Streat M (2005) Adsorption of trace heavy metals: Application of surface complexation theory to a macroporous polymer and a weakly acidic ion-exchange resin. Ind Eng Chem Res 44:8671–8681, https://doi.org/10.1021/ie048848

Salehi M, Sharafoddinzadeh D, Mokhtari F, Esfandarani MS, Karami S (2021) Electrospun nanofibers for efficient adsorption of heavy metals from water and wastewater. Clean Technol Recycling 1:1–33, https://doi.org/10.3934/ctr.2021001

Sanchez AA (2021) Characterization of arsenite interactions with the thiol-functionalized resin AmbersepGT74 as a model for Arsenite interactions with thiols in natural organic matter. Master's Thesis, Cornell University, Ithaca, NY

Sarika PR, Nancarrow P, Khansaheb A, Ibrahim T (2020) Bio-based alternatives to phenol and formaldehyde for the production of resins. Polymers 12:1–24, https://doi.org/10.3390/polym12102237

Schijf J, Ebling AM (2010) Investigation of the ionic strength dependence of *Ulva lactuca* acid functional group p*K* as by manual alkalimetric titrations. Environ Sci Technol 44:1644–164, https://doi.org/10.1021/es9029667

Sharafoddinzadeh D, Salehi M, Jansone-Popova S, Herath A, Bhattacharjee L (2022) Advance modification of polyacrylonitrile nanofibers for enhanced removal of hexavalent chromium from water. J Appl Polymer Sci 139:52169, https://doi.org/10.1002/app.52169

Sun WL, Xia J, Shan YC (2014) Comparison kinetics studies of Cu(II) adsorption by multi-walled carbon nanotubes in homo and heterogeneous systems: Effect of nano-SiO_2. Chem Eng J 250:119–127, https://doi.org/10.1016/j.cej.2014.03.094

Tabor RF, Morfa AJ, Grieser F, Chan DYC, Dagastine RR (2011) Effect of gold oxide in measurements of colloidal force. Langmuir 27:6026–6030, https://doi.org/10.1021/la200166r

Vithanage M, Rajapaksha AU, Ahmad M, Uchimiya M, Dou X, Alessi DS, Ok YS (2015) Mechanisms of antimony adsorption onto soybean stover-derived biochar in aqueous solutions. J Environ Management 151:443–449, https://doi.org/10.1016/j.jenvman.2014.11.005

Wei X, Fang L, Cai P, Huang Q, Chen H, Liang W, Rong X (2011) Influence of extracellular polymeric substances (EPS) on Cd adsorption by bacteria. Environ Pollut 159:1369–1374, https://doi.org/10.1016/j.envpol.2011.01.006

Westall JC, Jones JD, Turner GD, Zachara JM (1995) Models for association of metal ions with heterogeneous complexation of Co(II) by Leonardite humic acid as a function of pH and NaClO* concentration. Environ Sci Technol 29(4):951–959, https://doi.org/10.1021/es00004a015

Wongrod S, Watcharawittaya A, Vinitnantharat S (2020) Recycling of nutrient-loaded biochars produced from agricultural residues as soil promoters for Gomphrena growth. IOP Conf Ser: Earth Environ Sci 463:012099, https://doi.org/10.1088/1755-1315/463/1/012099

Xiao X, Chen B, Chen Z, Zhu L, Schnoor JL (2018) Insight into multiple and multilevel structures of biochars and their potential environmental applications: a critical review. Environ Sci Technol 52:5027–5047, https://doi.org/10.1021/acs.est.7b06487

Xiao F, Cheng J, Cao W, Yang C, Chen J, Luo Z (2019) Removal of heavy metals from aqueous solution using chitosan-combined magnetic biochars. J Colloid Interface Sci 540:579–584, https://doi.org/10.1016/j.jcis.2019.01.068

Xiong J, Zhou M, Qu C, Yu D, Chen C, Wang M, Tan W (2021) Quantitative analysis of Pb adsorption on sulfhydryl-modified biochar. Biochar 3:37–49, https://doi.org/10.1007/s42773-020-00077-9

Xu Y, Bai Y, Hiemstra T, Weng, L (2025) Ligand and charge distribution modeling of natural organic matter adsorption on metal (hydr)oxides: State-of-the-art. Rev Mineral Geochem 91A:229–250

Yee N, Fein J (2001) Cd adsorption onto bacterial surfaces: A universal adsorption edge? Geochim Cosmochim Acta 65:2037–2042, https://doi.org/10.1016/S0016-7037(01)00587-7

Zavarzina AG, Danchenko NN, Demin VV (2021) Humic substances: hypotheses and reality (a review) Eurasian Soil Sci 54:1826–1854, https://doi.org/10.1134/S1064229321120164

Zhang Y, Luo W (2014) Adsorptive removal of heavy metal from acidic wastewater with biochar produced from anaerobically digested residues: kinetics and surface complexation modeling. Bioresources 9:2484–2499

Zhou P, Yan H, Gu B (2005) Competitive complexation of metal ions with humic substances. Chemosphere 58:1327–1337, https://doi.org/10.1016/j.chemosphere.2004.10.017

Zhou X, Zhou J, Liu Y, Guo J, Ren J, Zhou F (2018) Preparation of iminodiacetic acid-modified magnetic biochar by carbonization, magnetization and functional modification for Cd(II) removal in water. Fuel 233:469–479, https://doi.org/10.1016/j.fuel.2018.06.075

Zoll AM, Schijf J (2012) A surface complexation model of YREE sorption on *Ulva lactuca* in 0.05–5.0M NaCl solutions. Geochim Cosmochim Acta 97:183–199, https://doi.org/10.1016/j.gca.2012.08.022

Reviews in Mineralogy & Geochemistry
Vol. 91A pp. 175–228, 2025

6

Surface Complexation and Reactivity of Ferrihydrite in Relation to its Surface and Mineral Structure, with Applications to Natural Systems

Tjisse Hiemstra

Wageningen University, Department of Soil Chemistry, Droevendaalsesteeg 3a, 6708 PB Wageningen, The Netherlands

tjisse.hiemstra@wur.nl

Annette Hofmann

University Lille, CNRS, University Littoral Côte d'Opale, UMR 8187, LOG, Laboratoire d'Océanologie et de Géosciences, F 59000 Lille, France

annette.hofmann@univ-lille.fr

Juan C. Mendez

Agronomic Research Center and Faculty of Agronomy, University of Costa Rica, San Pedro de Montes de Oca, San José, Costa Rica

juancarlos.mendez@ucr.ac.cr

Yilina Bai

Wageningen University, Department of Soil Chemistry, Droevendaalsesteeg 3a, 6708 PB Wageningen, The Netherlands

yilina.bai @wur.nl

INTRODUCTION AND HISTORICAL NOTE

Ferrihydrite (Fh) is the most important iron (hydr)oxide from the perspective of regulating the bioavailability and mobility of ions in the natural environment. Its existence was already known in the nineteenth century. Van Bemmelen and Klobbie (1892) studied the composition of what they called in French "Oxyde Ferrique Humide Amorphe" (Van Bemmelen and Klobbie 1896), being different from "Hydroxyde Ferrique Cristallin". At about the same time (1895), X-rays were discovered by Wilhelm Conrad Röntgen, but their use for unraveling the structures of crystalline Fe (hydr)oxides and ferrihydrite was yet to come. In 1912, Max von Laue reported that X-rays can be diffracted by a crystal and behave as waves. Later that year, Laurence Bragg and his father William interpreted this diffraction as a reflection of X-rays. Their works were the birth of crystallography (Thomas 2012). Atoms were found to be regularly ordered in crystalline materials. The underlying rules for a stable atomic arrangement were formulated by Linus Pauling in 1929 (Pauling 1929). For hematite, the structure was resolved in 1925 (Pauling and Hendricks 1925), and for goethite in 1935 (Goldsztaub 1935). In contrast to "Hydroxyde Ferrique Cristallin", it took about another 80 years before the structure of "Oxyde Ferrique Humide Amorphe" was elucidated (Michel et al. 2007, 2010); this time by X-ray scattering.

Van Bemmelen and Klobbie (1896) used the word "Cristallin" for a material where one could detect crystals with a light microscope. If not, a prepared material was classified as amorphous. In their work, the focus was on the water content of Fe (hydr)oxides, putting forward the hypothesis that crystalline Fe oxide hydrates should remain unchanged at an increase in temperature and a lowering of the partial pressure up to certain values in contrast to amorphous materials.

1529-6466/25/0091A-0006$10.00 (print)
1943-2666/25/0091A-0006$10.00 (online)
http://dx.doi.org/10.2138/rmg.2025.91A.06

Van Bemmelen and Klobbie (1892) prepared a colloidal Fe oxide material using diluted $FeCl_3$ solutions. The material could not be kept in a stable equilibrium state. Over time, the composition, expressed as $Fe_2O_3{\cdot}nH_2O$, continuously changed. A value of $n \sim 7$ was reported for the freshly prepared moist product. After exposure to air of 15 °C and a relative humidity of 70 % for 2 months, it changed to $n \sim 4$–5, and in air with zero humidity to $n \sim 2$. Drying the initial product at 100 °C for a few hours led to $n \sim 1$. Interpreting the latter ($n = 1$) as predominantly due to chemisorbed water, the initial product formed by precipitation from a diluted $FeCl_3$ solution must have had a specific surface area of about 600 $m^2{\cdot}g^{-1}$ or so, meaning that the particles were very small having a typical size in the order of 2 nm.

Van Bemmelen and Klobbie (1892) reported that freezing and thawing did not alter the composition of the gel, at most, its appearance under a microscope, as with freezing–thawing one can create intentionally dense Fh aggregates (Hofmann et al. 2004; van Beinum et al. 2005). In contrast to amorphous Fe oxide hydrate, crystalline Fe oxide monohydrate (goethite) remained unchanged according to van Bemmelen and Klobbie up to a temperature of about 300 °C, where it lost nearly all its hydration water at a constant rate. The composition of goethite was presented by van Bemmelen and Klobbie (1892) as "$Fe_2O_3{\cdot}H_2O$", but there is no H_2O group in the structure, as established by Goldsztaub (1935), who suggested writing the composition as "FeO·OH", based on the crystal structure of goethite that he resolved.

For a long time, Fe (hydr)oxide materials, precipitated by neutralizing a Fe(III) solution, were known as hydrous ferric oxide (HFO). If naturally formed, it was named ferrihydrite (Chukhrov 1973; Schwertmann and Fischer 1973). In 1837, Bunsen and Berthold (1837) already noticed for arsenite the chemical reactivity of freshly precipitated "iron oxide hydrate", as it was called in translation. The remarkable ability of hydrous ferric oxide to adsorb ions was highlighted about a century later by Boswell and Dickson (1918), Hahn (1924), Yoe (1930), and Kurbatov (1931). Over the next five decades, numerous studies on ion adsorption continued to explore these intriguing phenomena, critically reviewed by Dzombak and Morel (1990). In the seventies of the twentieth century, Davis and Leckie (1978) strongly contributed to our understanding of the ion adsorption to HFO by interpreting the proton adsorption characteristics. Their work pointed to a remarkably high specific surface area for HFO and the existence of very tiny particles, as small as ~2 nm. Consequently, a relatively large fraction of the Fe ions is exposed to the surface, where the local environment differs from the bulk (Hiemstra 2013). This notion is key to understanding the properties and structure of ferrihydrite.

Many researchers have contributed to unraveling the structure of ferrihydrite, deploying a variety of techniques and tools, such as X-ray diffraction, EXAFS, Mossbauer spectroscopy, FTIR, Thermal Gravimetric Analysis (TGA), or MO/DFT computations, which led to a diversity of ideas and models on the structure and composition of ferrihydrite, as reviewed by Harrington et al. (2011) and Manceau (2011). In 2007, based on the measurement of the total scattering of high-energy X-rays (HEXS), Michel et al. (2007) proposed a novel structure for Fh. In this newly proposed structure, the ideal mineral core of ferrihydrite ("ferri-ferrihydrite") has a composition close to that of a Fe-oxide, i.e., $FeO_{1.4}(OH)_{0.2}$. However, ferrihydrite material as a whole has an apparent composition that is notably water-rich, as was already reported by Van Bemmelen and Klobbie (1892) in their early work.

The overall composition of air-dry Fh can be written as $FeO_{1.4}(OH)_{0.2}{\cdot}nH_2O{\cdot}mH_2O$, in which n is the amount of water attributable to the formation of surface groups, and m signifies the amount of physio-sorbed water. Both values (n, m) depend on the particle size (Hiemstra 2013). The value of m will also depend on the drying conditions and approaches a zero value at a temperature near 125 °C (Xu et al. 2011). The reason for the difference between the composition of the particle ($Fe_{1.4}OH_{0.2}{\cdot}nH_2O$) and that of the core ($FeO_{1.4}OH_{0.2}$) of dried Fh is the huge contribution of the surface that acts as an "inter-phase". This implies that the mineral core of Fh is H-poor, while the particle as a whole is H_2O-rich (Hiemstra 2013). In fact, the historical name for ferrihydrite (Fh), i.e., Hydrous Ferric Oxide (HFO), fits beautifully with the contemporary understanding of the structure of Fh nanoparticles.

Elucidating the crystal structure of Fh has been crucial in comprehending its surface structure (Hiemstra 2013). This paved an avenue for mechanistic surface complexation modeling, providing a firm foundation for applying this knowledge to understand the reactivity of the natural iron (hydr)oxide fractions in soils and elsewhere. The present contribution intends to synthesize new insights, articulate and exemplify the approaches, and shed light on their implications for understanding soil chemical processes involving iron (hydr)oxides. Part of the current work includes original, yet unpublished experimental results and calculations.

MINERAL STRUCTURE OF FERRIHYDRITE

Ferrihydrite family

As already noticed by Van Bemmelen and Klobbie (1892), the overall composition of Fh is dynamic and changes with conditions. Using the structure of Michel et al. (2010) as the basis, all Fh particles can be built according to the same "aufbau" principle. Still, the particles differ strongly in size, composition, and many properties, together forming a diverse family of nanoparticles (Hiemstra 2018a). Figure 1 illustrates this nanoparticle family, showing the polyhedral structure for Fh particles of different sizes.

The nanoparticles depicted in Figure 1 have sheets with regular Fe octahedra (yellow) that are connected by Fe in tetrahedral (light blue) and distorted octahedral (dark blue) configurations. The regular (yellow) octahedra are relatively stable, protruding from the core. The contribution of these surface-oriented octahedra to the overall composition is size-dependent. With the increase of the particle size, the overall composition gradually converges to that of the mineral core/bulk, as the fraction of surface-oriented Fe octahedra decreases with increasing particle size. In the Fh family, the number of diffraction lines increases with size, leading to products, referred to as two-line Fh (2LFh) or 6-line Fh (6LFh) etc. This size-dependent interplay between surface and core components adds complexity to understanding the ferrihydrite composition and structure within this nanoparticle family.

Relative presence and sharing of Fe polyhedra. Interpretation of HEXS data suggests that the relative abundance of Fe at lattice position numbers two (Fe2) and three (Fe3) in the Fh structure is lower than expected for the ideal Fh core (Michel et al. 2010; Wang et al. 2016). The relative number is size-dependent as shown in Figure 2a (green symbols). The observed lower values for Fe2 and Fe3 compared to Fe1 were attributed to the prevalence of numerous defects in the core leading to a water-rich mineral core (Michel et al. 2010). Two-line ferrihydrite (2LFh) can be largely dehydrated without significant structural changes (Xu et al. 2011). This sets an important constraint on structural models with a significant number of structural OH groups (Drits et al. 1993; Hiemstra and Van Riemsdijk 2009; Manceau et al. 2014) or vacancies and defects that create -OH or $-OH_2$ (Drits et al. 1993; Michel et al. 2010; Gilbert et al. 2013). The work of Xu et al. (2011) is an ingenious example of the validation of the early hypothesis of van Bemmelen and Klobbie (1892).

The molar OH/Fe ratio at which restructuring of 2LFh becomes significant at dehydration is about 0.2 (Xu et al. 2011), aligning well with the theoretical value for a defect-free $FeO_{1.4}(OH)_{0.2}$ core. It strongly suggests that the observed deviations are not related to the bulk when we follow the above-mentioned hypothesis of Van Bemmelen and Klobbie (1892). The release of water is predominantly linked to the presence of a surface, as emphasized by Hiemstra (2013).

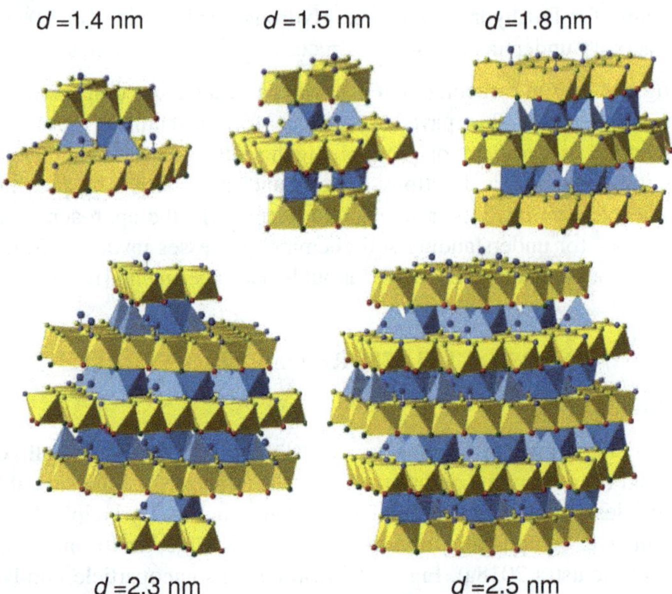

Figure 1. Polyhedral structure of the family of small ferrihydrite particles (Hiemstra 2018a) built from parallel sheets of regular Fe octahedra (**yellow**) interconnected by Fe, organized in tetrahedra (**light blue**) and distorted octahedra (**dark blue**). The octahedra in the parallel sheets carry singly coordinated ≡FeOH(H) groups when present at the surface (protons not shown). These surface groups handle most of the reactivity of Fh (Hiemstra and Zhao 2016; Boily and Song 2020). The other polyhedra (**blue**) do not form singly coordinated groups at the surface for stability reasons. Consequently, Fh nanoparticles have by definition less of these (Fe2 and Fe3) than the regular ones (Fe1), which is called surface depletion. Forced adsorption of Fe1 to the stoichiometric Fh core is the other side of the same coin. For the above Fh series with 23–193 Fe per particle, the equivalent particle size ranges from 1.4 to 2.5 nm. The smallest particles may form at stepwise loading of ferritin with Fe (Hiemstra and Zhao 2016). Co-precipitation of Fe/Si may lead to slightly larger particles (Hiemstra 2018a), followed by particles formed at ultra-fast neutralization of Fe (Hiemstra et al. 2019). Freshly-prepared Fh has a typical size of nearly 2.3 nm (Michel et al. 2010), and aging for 4 hours at room temperature leads to a diameter d of ~2.5 nm (Hiemstra et al. 2019). [Reproduced from Hiemstra (2018a) with permission from the Royal Society of Chemistry.]

The size dependency of the occupancy of the lattice positions Fe2 and Fe3 (green symbols in Fig. 2a) in the structure of Michel et al. (2010) can be explained by a fundamentally different polyhedral composition at the surface compared to the mineral core. The observed particle size-dependency of the pair distribution function (Michel et al. 2010) can be rationalized by the surface depletion (SD) model, which states that Fe2 and Fe3 polyhedra cannot form singly-coordinated FeOH(H) groups at the surface, as this destabilizes their coordination environments (Hiemstra 2013). Just applying this singular rule in constructing Fh particles with software results in a good prediction of the relative number of Fe2 and Fe3 for the overall structure (open symbols in Fig. 2a) of Michel et al. (2010).

Deriving the Fh structure, the material has been considered as a single-phase material. As the surface and core of Fh differ structurally, the material has been treated as a nanocomposite (Funnell et al. 2020) with crystalline domains embedded in an amorphous matrix of polyhedra without including surface water. However, it is more realistic to consider Fh as an aggregated collection of distinct crystalline nanoparticles having a significant surface contribution that is responsible for the presence of chemisorbed and physisorbed water, the development of an electrical double layer with a high concentration of counter- and co-ions, and a very high capacity for binding cations, anions, and organic acids.

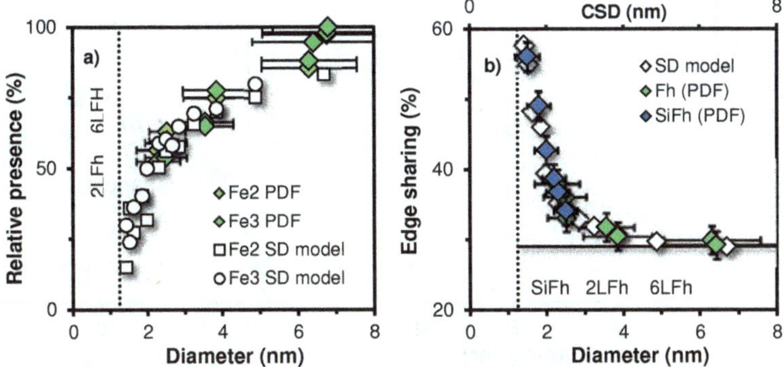

Figure 2. a) Relative abundance of Fe in Fe2- and Fe3-lattice positions (**green symbols**) measured with high-energy total X-ray scattering as a function of the mean particle size measured with HR-TEM (Michel et al. 2010), showing that in these Fh particles, up to ~30% of Fe in the lattice position numbers Fe2 and Fe3 can be absent in 2 nm particles. **Open spheres and squares** refer to the polyhedral composition calculated with the surface depletion (SD) model using a whole-particle approach (Hiemstra 2013). **b)** Relative contribution of edge-sharing of Fe polyhedra as a function of particle size. The **green symbols** correspond to the data from the time series of Michel et al. (2010), produced and aged in the presence of some citrate. The **blue points** are for Fh produced in the presence of silicic acid (Cismasu et al. 2014). The values for siliceous ferrihydrite (SiFh) are given as a function of the size of the coherent scattering domain (CSD) measured with HEXS. The **open symbols** show the percentage of edge-sharing in Fh particles constructed with the surface depletion model. The **horizontal line** gives the percentage of edge-sharing in the mineral bulk of Fh. The **dotted vertical lines** in a) and b) represent the minimum size ($d \sim 1.25$ nm) of an Fh particle that can be produced during nucleation (Michel et al. 2010; Hiemstra and Zhao 2016; Weatherill et al. 2016). These nuclei with as few as ~ 13 Fe ions may bind some additional ones before being stabilized and fixed by adsorbed Si. In the absence of interfering oxyanions, Fh readily forms ~1.7 nm or larger particles (Hiemstra et al. 2019). Organic matter may also reduce the size of the Fh particles when present during the formation (see Fig. 18). [Reproduced from Hiemstra (2018a) with permission from the Royal Society of Chemistry.]

The pair distribution function also allows an assessment of the edge and corner-sharing ratio of the polyhedra in the Fh structure. Figure 2b illustrates this for the Fh series (green symbols) from published work (Michel et al. 2010) and for a series of siliceous Fh samples (blue symbols) (Cismasu et al. 2014). The open symbols are predictions of the SD model (Hiemstra 2018a). Details are provided in the legend of the figure.

Size-dependent Fh properties. The change of the mean size of the Fh particles during aging, measured by Michel et al. (2010) for their preparation shows that at a specific moment of aging at 175 °C, rather suddenly, the mean particle size almost doubles (Fig. 3a). Remarkably, this does not occur for the small particles in the particle distribution, leading to a deviation of the initial ratio between the minimum and mean size of 2/3, as illustrated in the inset of Figure 3a. These observations can be explained by a dual-step mechanism for particle growth (Burleson and Penn 2006) in which particles of sufficient size form doublets by oriented particle attachment (Li et al. 2012; De Yoreo et al. 2015), whereas the small particles still change in size through atom-by-atom growth. The process of oriented-attachment may introduce mismatches, and create defects and vacancies (Banfield et al. 2000), that will affect the magnetic properties of Fh and align with the results of structural modeling of 6LFh (Gilbert et al. 2013).

Ferrihydrite can behave ferrimagnetically at low temperatures (Michel et al. 2010; Hiemstra 2018a) where the spins of the unpaired electrons in the magnetic domains can be aligned in a specific way by an external magnetic field up to a saturation value. Figure 3b shows the magnetic saturation of Fh from the series of Michel et al. (2010) as a function of the mean particle size, because Fe at the surfaces can not sufficiently contribute to the alignment. The gradual increase of the magnetic saturation with larger particles (Michel et al. 2010) is due to

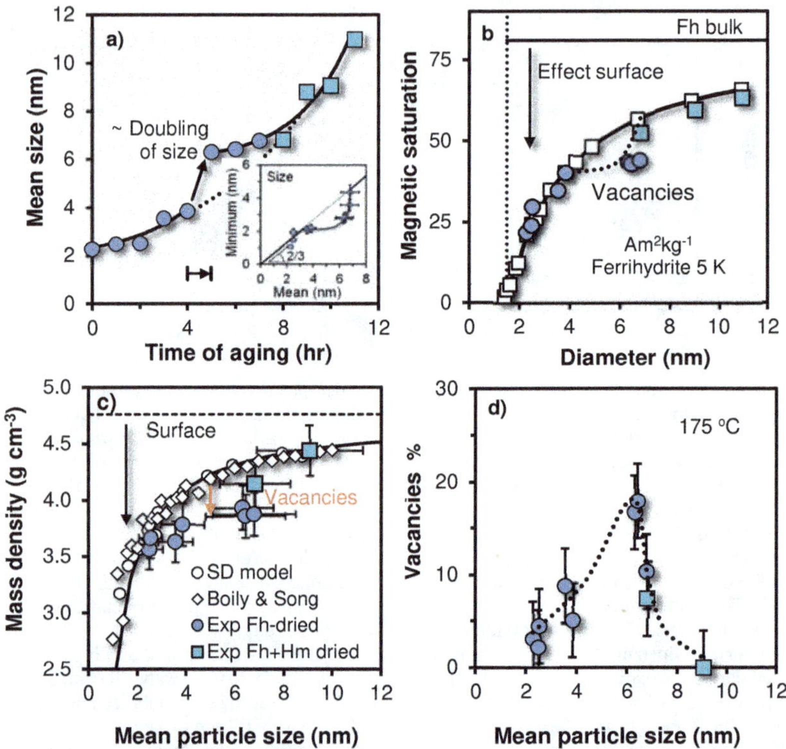

Figure 3. a) Change of the mean particle size of ferrihydrite upon aging at 175 °C (Michel et al. 2010) in systems with only Fh (**colored spheres**) or with additional hematite (**squares**), showing an abrupt increase of the mean size once the particle reached a mean size of ~ 4 nm. **Inset:** Minimum and mean sizes in the particle size distribution of Fh (Michel et al. 2010), with an intrinsic ratio of 2/3 (Hiemstra 2015) that is interrupted by a sudden increase in the mean size, supporting a dual-step mechanism for particle growth (Burleson and Penn 2006) by oriented particle attachment (Li et al. 2012; De Yoreo et al. 2015). **b)** Magnetic saturation (Michel et al. 2010) versus the mean particle size (Hiemstra 2018a), illustrating the strong influence of the surface (**arrow**) on this property and a further deviation due to defects (**dotted curved line**). The **open squares** are model values excluding Fe1 in excess, and additionally, Fe in a surface layer with a thickness $L = 0.2$ nm (Hiemstra 2018a). **c)** Calculated mass densities (**open symbols**) using whole-particle constructions, applying the SD rule (Hiemstra 2013), or using a layer approach (**open diamonds**) (Boily and Song 2020) and the mass density (**colored symbols**) of dried-Fh, recalculated from the experimental value (Michel et al. 2010) that includes physisorbed water. The mismatch (**open versus colored symbols**) can be attributed to the presence of vacancies. The line has been calculated using the set of equations (Eqns.1–3) (Hiemstra et al. 2019; Mendez and Hiemstra 2020c). **d)** Calculated relative number of vacancies (2–20%) explaining the difference between the theoretical and experimental mass density as a function of the mean particle size. [Reproduced (modified) from Hiemstra (2018a) with permission from the Royal Society of Chemistry.]

an enlargement of the magnetic core. A core–shell model (Hiemstra 2018a) allows quantification of this increase (open symbols). Initially, a fair minority of Fe (~20 %) forms the magnetic core, but the magnetic saturation at 5 K steadily approaches the theoretical maximum (horizontal full line) that can be calculated for Fh as bulk material (Hiemstra 2018a). When the mean size doubles ($t >$ ~4 h), the measured magnetic saturation is lower than expected, indicating the presence of vacancies or defects that disturb the alignment of the spins of the unpaired electrons. Interestingly, for Fh, ferrimagnetic as well as anti-ferromagnetic behaviors have been reported. The energy difference between both states is tiny, about 1 kJ·mol^{-1} Fe (Pinney et al. 2009), implying that a small difference in the condition of preparation, including doping with citrate or silicate, may change the surface energy (Hiemstra 2018a) or even change the core.

With the change in particle size, the macroscopic properties of Fh, such as the mass density, change (Fig. 3c). The open spheres in Figure 3c are predictions of the SD model calculated by constructing a family of particles of different sizes with software as exemplified in Figure 1 (Hiemstra 2013). The open diamonds have been derived with a very similar approach by Boily and Song (2020). The line in Figure 3c has been calculated using a consistent set of equations (Hiemstra 2018b; Hiemstra et al. 2019; Mendez and Hiemstra 2020c). The mean mass density ρ_{nano} ($g{\cdot}m^{-3}$) can be calculated as a function of the mean particle size d (m), using the number of O per Fe in the mineral core (n_O =1.6), the lattice volume expressed per mol oxygen $V_O = 10.7 \times 10^{-6}$ $m^3{\cdot}mol^{-1}$ O, and the site density of surface groups, expressed in a water content of $N_{H_2O} = 12.6\ 10^{-6}$ $mol{\cdot}m^{-2}$, according to:

$$\rho_{nano} = \left(\frac{M_{core}}{n_O V_O}\right) - \left(\frac{M_{core}}{n_O} - M_{H_2O}\right)\frac{6}{d} N_{H_2O} \tag{1}$$

where M_{core} is the molar mass of the bulk mineral (81.65 $g{\cdot}mol^{-1}$Fe), and M_{H_2O} is the molar mass of water (18 $g{\cdot}mol^{-1}$). The first term ($M_{core}/(n_O V_O)$) represents the mass density of the core (ρ_{core}) derived with crystallographic methods. Using the above parameters, $\rho_{core} = 4.77 \times 10^6$ $g{\cdot}m^{-3}$.

With the calculated ρ_{nano}, the corresponding mean molar mass of the particles M_{nano} ($g{\cdot}mol^{-1}$ Fe) becomes:

$$M_{nano} = \frac{M_{core}}{\left(1 - \left(\frac{6}{\rho_{nano}\, d}\right) N_{H_2O} M_{H_2O}\right)} \tag{2}$$

and the corresponding specific surface area A ($m^2{\cdot}g^{-1}$), according to:

$$A = \frac{6}{\rho_{nano}\, d} \tag{3}$$

The above equations can be applied, for instance, to scale and interpret ion adsorption data with surface complexation modeling in an internally consistent manner (Hiemstra et al. 2019). This is essential for consistently comparing data from various Fh preparations with different particle sizes (Hiemstra et al. 2019). The equations can also be used for other nano-materials, such as silica nanoparticles (Hiemstra and Lützenkirchen 2025) or nano-$Al(OH)_3$, which can be relevant for scaling ion adsorption data of Al-rich soils (Mendez et al. 2020). An equivalent set of equations, using A rather than d as input, is given in Mendez and Hiemstra (2020c).

Equation (1) provides a robust description of the size dependency on the mass density of Fh particles obtained with particle construction using software, applying the SD model (Fig. 3c, solid line). However, the mass densities, measured by Michel et al. (2010) for their series (colored symbols), are lower than the theoretical values, particularly at the moment of aging when the mean particle size doubles. Lower mass densities than predicted by theoretical calculations may indicate that the actual Fh structure contains relatively fewer heavy elements (Fe) and more light elements (H) at a given oxygen density (Hiemstra and Van Riemsdijk 2009), which may point to vacancies. To address this mismatch, we performed new calculations to assess the number of Fe vacancies in the structure of Michel et al. (2010) and explain the experimental mass density (Fig. 3d). Initially, the number of vacancies is minimal (5%) in the Fh series of Michel et al. (2010). Still, it significantly increases when the Fh material undergoes the formation of doublets through oriented attachment at a size of about 4 nm (Fig. 3d). A positive trend with an increase in the mean particle size is also found when fitting X-ray total scattering data of another series of Fh preparations (Wang et al. 2016). However, their fitting yields more vacancies. Possible explanations could be a difference in the method of Fh preparation or the fitting method.

SURFACE AREA OF FERRIHYDRITE AND ITS DYNAMICS

Specific surface area

Regardless of modeling, the specific surface area (SSA) of Fh is pivotal for consistently scaling ion adsorption data, as it determines the total number of sites available in a system (Bompoti et al. 2019) and the strength of the electrostatic field. The SSA of Fh has often been measured with gas adsorption, relying on the multiple layer theory of Brunauer, Emmett, and Teller (BET). The BET values obtained for Fh are lower than the physical surface area of Fh, as illustrated in Figure 4b for the Fh series of Michel et al. (2010) as an example.

Michel et al. (2010) employed high-resolution transmission electron microscopy (HR-TEM) to measure the particle size distribution of freshly prepared Fh and samples aged at 175 °C. The particle size distributions were summarized by reporting the mean particle diameters, the standard deviations of the distributions as well as the minimum and maximum sizes of the particles (Fig. 4a). These HR-TEM data are used here for a novel assessment of the specific surface area by calculating the total surface area numerically (Fig. 4b), distinguishing in the size distribution a discrete set of classes of particles, each with a specific size. In these calculations, the molar mass and mass density of each particle size fraction were adapted to the corresponding values using the above set of equations (Eqns. 1–3). The resulting physical surface area offers a more comprehensive understanding of the available surface sites in the system than the BET surface area.

According to the analyzed HR-TEM data from Michel et al. (Fig. 4a), the prepared starting material of 2LFh has a remarkable specific surface area of at least 640 $m^2 \cdot g^{-1}$ (Fig. 4b). The calculations assume spherical geometry. If the particles exhibit a prolate shape, the SSA will be even higher, e.g., 690 $m^2 \cdot g^{-1}$ at an aspect ratio of $c/a = 2$. However, as time progresses, the SSA decreases due to particle growth. After approximately 5 or more hours of aging at 175 °C, the physical surface area approaches the value calculated with gas adsorption (Fig. 4b). During this period, the particles have grown sufficiently large, exceeding 6–7 nm, and may change shape due to doublet formation, resulting in less contact area and a gas-inaccessible dead volume that is sufficiently small, leading to a physical surface area close to the values calculated from the gas adsorption.

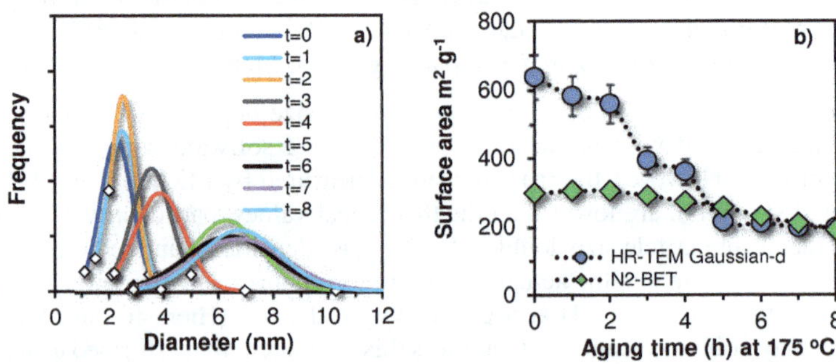

Figure 4. a) Evolution of the particle size distribution of Fh with time *t* (h) (Hiemstra 2015), measured with HR-TEM by Michel et al. (2010). The minimum and maximum values in the distribution are indicated with diamonds. **b)** Specific surface area of Fh measured with BET-N_2 gas adsorption (Michel et al. 2010), showing relatively low numbers and little change over time (**green diamonds**). The surface areas can also be calculated using the particle size distributions (**blue spheres**). In the calculations, full consistency between size, molar mass, and mass density was used applying the appropriate set of equations (Eqns. 1–3). If present as spherical particles, the initial preparation of Michel et al. (2010) has an SSA of 640 $m^2 \cdot g^{-1}$, which decreases gradually through particle growth, showing consistency with the N_2-BET data after 5 or more hours of aging at high temperatures. [Figure 4a (adapted) reproduced from Hiemstra (2015) with permission of Elsevier.]

Surface area and particle growth. In the context of employing ferrihydrite as an adsorbent in ion adsorption studies, comprehending the process of aging of ferrihydrite is crucial due to the rapid changes in properties, particularly in the initial stages of growth. Figure 5a illustrates the change of the reactive surface area for suspensions (colored symbols) kept at various pH values during formation and growth (Hiemstra et al. 2019) at ambient laboratory conditions (20 °C).

The change in surface area (Fig. 5) has been monitored using phosphate as a probe ion (Hiemstra et al. 2019). In the experiments depicted in Figure 5a. the colored symbols refer to measurements of the SSA using PO_4 as a probe ion (Hiemstra et al. 2019) applying the CD model with the parameters of Hiemstra and Zhao (2016). In the experiment of Figure 5a, different organic pH-buffer solutions were used, which suppressed the growth rate, however. Therefore, additional data were collected for Fh in the absence of such a pH buffer (Fig. 5b).

It is also possible to use other ions to track the changes in SSA. The open symbols in Figure 5b are for the SSA probed with proton titrations, applying the parameters derived by Mendez and Hiemstra (2020c). Both methods demonstrate internal consistency, disclosing the specific surface area during Fh particle growth.

The experimental results of Figure 5a reveal that the Fh material, formed immediately after (pre)nucleation, contains particles with *on average* about 45 Fe. The corresponding reactive surface area is ~1100 $m^2 \cdot g^{-1}$, equivalent to a mean spherical size of d = 1.65 nm. The above number of Fe ions per ferrihydrite particle (n_{Fe}) can be calculated from the mean particle size using consistently the molar mass (M_{nano}) and mass density (ρ_{nano}) according to:

$$n_{Fe} = \frac{\rho_{nano}}{M_{nano}} \frac{\pi d^2}{6} N_{Av} \tag{4}$$

where N_{av} is Avogadro's number.

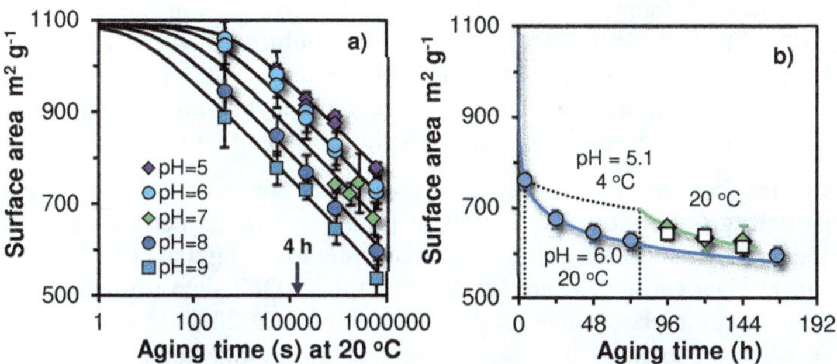

Figure 5. a) Evolution of the specific surface area (SSA) of Fh measured with phosphate as a probe ion in systems with a fixed pH, maintained by different organic pH buffers. The lines were calculated with a model for particle growth ($R = k\ (Q_{so}/K_{so})^2$) in which the size-dependent supersaturation (Q_{so}/K_{so}) is the driving force. The initial particles, formed upon neutralization of a Fe(III) solution, have a size of ~1.7 nm, contain ~45 Fe atoms, and have a SSA of almost 1100 $m^2 \cdot g^{-1}$ (Hiemstra et al. 2019). **b)** Change of the surface area of Fh at 20 °C in a system with a 0.01 M $NaNO_3$ solution of pH 6 in the absence of an organic pH buffer increasing the rate of aging. The SSA was measured by probing the surface with either phosphate ions (**colored symbols**) or protons (**open squares**) in 0.01 M $NaNO_3$ (Mendez and Hiemstra 2020c). The latter data were collected by storing a part of the prepared suspension at a low temperature (4 °C) and low pH (5.1) to suppress the aging process temporarily. For this material, the SSA was also probed with PO_4 (**green symbols**). Both types of probes (H, PO_4) give the same result, showing internal consistency. The lines are calculated with the same model for particle growth, using a rate constant of $\log k = 9.1$ at 20 °C (pH = 6) and an activation energy of $E_{act} = 68 \pm 4$ kJ mol^{-1}. [Reproduced from Hiemstra et al. (2019) with permission from the Royal Society of Chemistry.]

Data analysis of the aging kinetics reveals that the particle diameter scales with $t^{1/4}$ (Hiemstra et al. 2019). Consequently, the surface area changes dramatically over time. For pH = 6, the surface area changes during the first minutes to hours from initially ~1100 $m^2 \cdot g^{-1}$ to ~675 $m^2 \cdot g^{-1}$ (Fig. 5b). In the subsequent days, the change in reactive surface area is much lower, only ~ 10 %.

Model for the rate of particle growth. The lines in Figure 5a have been calculated based on a growth model in which the rate R ($mol \cdot m^{-2}$) is proportional to the square of the oversaturation of the solution relative to the bulk at infinite size, according to (Hiemstra et al. 2019):

$$R = k\left(\frac{Q_{so}}{K_{so}}\right)^2 \tag{5}$$

in which k is a temperature and pH-dependent constant ($E_{act} = 68 \pm 4$ kJ mol^{-1}), and Q_{so}/K_{so} is the ratio of solubility product of Fh at finite and infinite size. Importantly, this relative solubility product is not constant, but changes rapidly over time with particle size since the equilibrium depends on the surface curvature of the particles. The relative solubility can be related to the surface area of Fh using the Ostwald–Freundlich equation (Hiemstra 2015):

$$RT\,\ln\frac{Q_{so}}{K_{so}} = \frac{2}{3}\gamma A^*_{critical} = \frac{2}{3}\gamma\phi A^*_{mean} = \frac{2}{3}\gamma\phi M_{nano}\,A_{mean} \tag{5b}$$

in which R is the gas constant ($J \cdot mol^{-1} \cdot K^{-1}$) and T is the temperature (K). In the equation, γ is the surface Gibbs free energy ($\gamma = 0.186 \pm 0.01$ $J \cdot m^{-2}$), and $A^*_{critical}$ is the surface area ($m^2 \cdot mol^{-1}$ Fe) of the smallest particles in the size distribution. The smallest particles have a critical energy state, allowing spontaneous dissolution (Hiemstra 2015). In the equation, ϕ is the ratio between the surface area of the smallest and mean size of the particles in the distribution ($A^*_{critical} / A_{mean}$). For small 2LFh particles, the ratio ϕ has a value of about 3/2 (Hiemstra 2015). In the above equation, the mean surface area A^*_{mean} in $m^2 \cdot mol^{-1}$ Fe can be rewritten to A_{mean} the unit $m^2 \cdot g^{-1}$ Fh by introducing the mean molar mass of Fh, M_{nano} expressed in g $Fh \cdot mol^{-1}$ Fe. This translation is done by applying the consistent set of mathematical relationships presented in Equations (1–3). Introducing the ratio $\phi = 3/2$, the above equation is coincidentally equal to the Ostwald equation (Hiemstra 2015). The intrinsic solubility of Fh as virtual bulk material, isostructural to akdalaite, is $\log K_{so} = -40.6 \pm 0.2$ (Pinney et al. 2009; Hiemstra 2015). With this information, the lines in Figures 5a,b have been calculated, showing close agreement with the data. With the model, the observed time dependency of $t^{1/4}$ can be rationalized.

Surface Gibbs free energy of Fh. The surface Gibbs free energy (γ) of Fh plays a crucial role in the particle growth model presented in Equation (5). Its determination involved the integration of experimental data for the dissolution enthalpy (Majzlan et al. 2004), temperature dependency of the heat capacity (Snow et al. 2013), MO-DFT data (Pinney et al. 2009), and experimental information about interfacial entropy (Hiemstra 2015). Notably, surface entropy significantly contributes to the total surface Gibbs free energy and cannot be disregarded for Fh.

Figure 6a illustrates that Fh exhibits an exceptionally low surface Gibbs free energy compared to all other Fe (hydr)oxides. The energy associated with surface formation is directly linked to the stability of the lattice, expressed in the intrinsic solubility ($\log K_{so}$). The relationships between the lattice and surface stability are depicted in Figure 6a using lines. Fe-oxides with an oxygen coordination number (CN) of four in the bulk (i.e., 4 Fe–O bonds) have a slightly higher surface energy than Fe oxyhydroxides with a mean CN value of 3 (i.e., 3 Fe–O bonds). This is because, on average, fewer Fe–O bonds participate in surface formation. According to this depiction, a relatively high surface Gibbs free energy would be expected for the $FeO_{1.4}(OH)_{0.2}$ structure of Fh (~0.55 J m^{-2}). However, experimental results reveal that the surface energy of Fh is much smaller (0.186 J m^{-2}) than that, and all other Fe oxides and oxyhydroxides (Fig. 6a). This discrepancy can be attributed to an excess presence of relatively stable Fe1 octahedra at the surface, which lowers the surface Gibbs free energy and stabilizes the core.

An alternative to the $FeO_{1.4}(OH)_{0.2}$ structure proposed by Michel et al. (2010) is the FeOOH structure suggested by Manceau et al. (2014). In this structure, only octahedral Fe is present. First-principle calculations of Sassi et al. (2021) indicate that the FeOOH structure has a lower enthalpy ($\Delta = -4.5$ kJ·mol^{-1}$FeO_{3/2}$) compared to the $FeO_{1.4}(OH)_{0.2}$ structure. As this structure is a Fe oxyhydroxide (FeOOH), it also has a different standard entropy of formation ΔS_f^o. Using the average ΔS_f^o values of α-FeOOH and γ-FeOOH from Hiemstra (2015), the assessed value is ΔS_f^o ($FeO_{3/2}$) = 27.4±2.7 J·mol^{-1}. In combination with the standard enthalpy of formation ΔH_f^o ($FeO_{3/2}$) = −406.1 − 4.5 = −410.6 kJ·mol^{-1}, the intrinsic solubility constant can be calculated, resulting in $\log K_{so} = -40.6 \pm 0.3$ at 298 K. This value coincides with the previously derived (Hiemstra 2015) for the Fh with the $FeO_{1.4}(OH)_{0.2}$ structure according to Michel et al. (2010), indicating that both cores have the same intrinsic stability.

The pertinent question arises whether both materials also share the same surface Gibbs free energy. In the study of Sassi et al. (2021), a potential difference between the surface Gibbs free energy of both Fh structures has not been considered, while it is pivotal for a reliable interpretation of the formation and relative stability of both types of nanoparticles. As indicated in Figure 6a with a horizontal arrow, Fe oxyhydroxides (FeOOH) with the same solubility as Fh may exhibit a substantially higher surface Gibbs free energy (0.44 J·m^{-2}). Sassi et al. (2021) calculated a surface enthalpy of > 0.40 J·m^{-2} for the FeOOH structure of Manceau et al. This aligns well with the value suggested with the arrow in Figure 6a. For the $FeO_{1.4}(OH)_{0.2}$ structure of Michel et al., the calculations of Sassi et al. (2021) led to instabilities pointing to surface reconstruction. This supports the surface depletion model for the Fh structure of Michel et al. (2010). However, definite conclusions are pending. Another approach to discriminate between Fh structures is by studying the dehydration using TGA combined with HEXS, X-ray diffraction, and Fourier-transformed infrared spectroscopy. Such an approach, akin to the work of Xu et al. (2011), is particularly useful when studying Fh preparations for which the structure of Manceau et al. is claimed. $FeO_{1.4}(OH)_{0.2}$ and FeOOH are expected to react distinctly differently.

The relative stability of the core and surface of Fe (hydr)oxides can be translated into the size dependency of the solubility, as depicted in Figure 6b. In the calculations, we considered

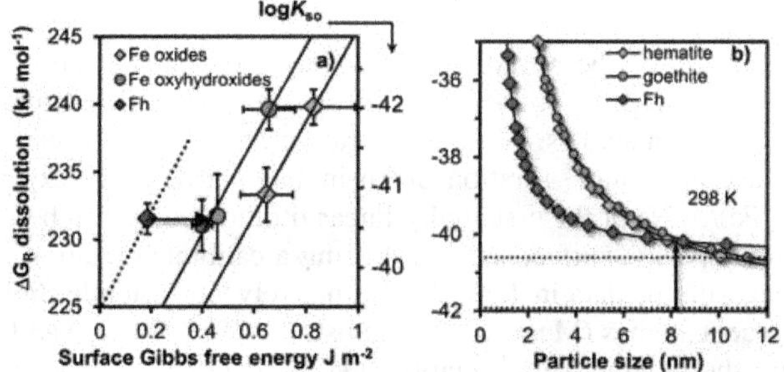

Figure 6. a) Relation between the surface Gibbs free energy of Fe (hydr) oxide minerals and the stability of the bulk minerals at 298 K expressed in the Gibbs free energy change at the dissolution of bulk $FeO_{3/2}$ (**left *y*-axis**) and the corresponding intrinsic solubility $\log K_{so}$, defined as $\log K_{so} = \log(Fe^{3+}) + 3\log(OH^-)$ (**right *y*-axis**). Oxide and oxyhydroxide data (**blue symbols**) are from Navrotsky et al. (2008). The **full lines** give the fitted relationship between the lattice stability and the surface Gibbs free energy. Fh has an exceptionally low surface Gibbs free energy (Hiemstra 2015), considering that the mineral core is nearly an oxide ($FeO_{1.4}(OH)_{0.2}$). The low value implies that the surface can be easily formed, making the Fh particle overall very stable at small sizes compared to all other Fe (hydr)oxides (Hiemstra 2013). The **horizontal arrow** indicates that a FeOOH mineral with the same solubility as Fh may have a higher surface Gibbs free energy (0.44 J·m^{-2}) if not stabilized by surface depletion. **b)** Size dependency of the solubility ($\log Q_{so}$) of Fh, goethite, and hematite, calculated with the Ostwald–Freundlich equation (Eqn. 5b) combined with Equations (1–3) for consistency reasons. Fh has the highest intrinsic solubility ($\log K_{so}$) but the lowest solubility ($\log Q_{so}$) when the particles are small. Goethite and/or hematite can spontaneously form for Fh≥8 nm, given with the **vertical arrow**. The **dotted horizontal lines** refer to the intrinsic solubilities of the bulk materials. [Used by permission of Elsevier from Hiemstra (2015).]

the particle size dependency of the mass density (ρ_{nano}) for translation of the particle diameter (d) to the corresponding surface area (A), using Equations (1–3). In the very nanometer range, Fh has by far the lowest solubility (Fig. 6b). This characteristic can be attributed to the relatively low Gibbs free surface energy ($\gamma = 0.186\ J \cdot m^{-2}$ at 298 K) needed for the formation of the surface of $FeO_{1.4}(OH)_{0.2}$. Consequently, this material is the initial product formed during Fe(III) precipitation as it has relatively the lowest solubility when particles are very small (Fig. 6b).

Over time, the particles will grow and eventually transform into more stable Fe (hydr) oxides. Fh may spontaneously transform into hematite or goethite when its size reaches approximately 8–9 nm (Hiemstra 2015), as indicated with an arrow in Figure 6b. Given that Fh has a size distribution (Fig. 4a), nano goethite or hematite may be detected at smaller mean sizes. This aligns with the observations of Michel et al., who reported the presence of some hematite in suspensions with a mean particle size of about 6 nm and larger. However, we note that their aging is at 175 °C in the presence of some citrate and both may influence the surface Gibbs free energy. The results of Figure 6b illustrate the classical rule of Ostwald expressing his experience that a material with the highest solubility will precipitate first, followed by transformation into more stable products (Ostwald 1897; Cardew 2023).

Evaluating the specific surface area of Fh. The above discussion highlights the inherent instability of Fh under standard conditions, introducing uncertainties in the scaling of ion adsorption data. It prevents a consistent comparison among different investigations. By measuring the specific surface area (SSA) using for instance the proton as a probe ion, this may be overcome. When reported proton adsorption data of freshly prepared Fh are interpreted in a consistent manner, the reactive surface area varies largely among authors showing a typical range from ~550–700 $m^2 \cdot g^{-1}$ (Table 1). The underlying primary experimental titration data are basically expressed per mole Fe and need translation to the unit g^{-1} using the molar mass (M_{Fh}). Traditionally, a value of $M_{Fh} = 89\ g \cdot mol^{-1}$ Fe has been chosen (Davis and Leckie 1978). However, the molar mass varies with size from, e.g., 86 $g \cdot mol^{-1}$ for 6 nm particles to 101 $g \cdot mol^{-1}$ for 2 nm particles, as can be calculated iteratively (Mendez and Hiemstra 2020c) using Equations (1–3), in which molar mass (M_{Fh}), mass density (ρ), specific surface area (A), and mean particle size (d) are linked.

Table 1 illustrates that the variation in surface area can be large, emphasizing the importance of characterizing the specific surface area of prepared ferrihydrite (Fh) batches intended for ion adsorption studies. To facilitate future work, Mendez and Hiemstra (2020c) have proposed to use of a single titration of Fh in 1 M $NaNO_3$. The experimental slope S (Δ(mmol H mol^{-1} Fe)/ΔpH) of the essentially linear titration curve can be directly translated into a corresponding specific surface area (A_H) using a calibrated relation ($A_H = 94.0–15.61\ S$) that is consistent with the data in Table 1. Alternatively, consistently probing Fh with PO_4 has been used in recent studies (Mendez and Hiemstra 2019, 2020a,c; Van Eynde et al. 2020, 2022). Interpreting thermogravimetric analysis (TGA) data creates uncertainty in the SSA derived, as it is difficult to precisely discriminate between physisorbed and chemisorbed water. Additionally, there may be some overestimation of the reactive surface area (Mendez and Hiemstra 2020c), as Fh particles can potentially have vacancies (Fig. 3d).

Antelo et al. (2010, 2015) have used freeze-dried Fh in their investigations of the phosphate/arsenate adsorption hoping that the surface area of this material would be more stable than the surface area of freshly prepared ferrihydrite. The freeze-dried material had a surface area of 230 $m^2 \cdot g^{-1}$ when measured with gas adsorption, but 350 $m^2 \cdot g^{-1}$ when measured with proton adsorption using the approach of Davis and Leckie (1978). The latter value for the surface area is much lower than the typical values found for freshly prepared Fh (Table 1), suggesting some irreversible aggregation of these freeze-dried aggregates. In a more recent study, it was demonstrated that the reactive surface area of resuspended freeze-dried Fh was not constant, but varied with the pH of the suspension (Gustafsson and Antelo 2022). It highlights the limited value of using freeze-dried material in ion adsorption studies.

Table 1. Overview of the specific surface area *A*, particle size *d*, and point of zero salt effect of freshly prepared Fh kept in the wet state before performing adsorption experiments (Hiemstra 2018b; Mendez and Hiemstra 2020c). The specific surface areas have been derived with H^+ as the probe ion (Davis and Leckie 1978) applying consistently a set of size-dependent values for the molar mass *M* (Eqn. 2), mass density ρ (Eqn. 1), and capacitances values C_r (Eqn. 6) for the Extended Stern layer model (Hiemstra and Zhao 2016) in an iterative procedure (Hiemstra 2018b; Mendez and Hiemstra 2020c) and using ion pair formation (Fig. 8).

Study	Method	Aging	Molar mass $g·mol^{-1}$	Mass density $g·cm^{-3}$	Electrolyte	PZSE*	*A* $m^2·g^{-1}$	*d* nm
Pivovarov (2009)	b	2 days pH = 5	97.3	3.68	$NaNO_3$/NaCl	8.1	710	~2.3
Hsi and Langmuir (1985)	c	4 h pH = 7	96.5	3.72	$NaNO_3$	8.0	680	~2.4
Jain et al. (1999)	a	<10 d at 2 °C	96.0	3.74	NaCl	$8.^5$	660	~2.4
Fukushi et al. (2013)	c**	4 h	95.8	3.75	NaCl	8.2	650	~2.5
Davis (1977)	c	4 h	94.7	3.81	$NaNO_3$	7.9	610	~2.6
Hiemstra (2018b)	d	4 h	94.7	3.81	$NaNO_3$	8.1	610	~2.6
Nagata et al. (2009)	c	4 h	94.7	3.81	$NaNO_3$	7.9	610	~2.6
Kinniburgh et al. (1975)$	e	< few days at 5°C	94.2	3.83	$NaNO_3$	8.1	585	~2.7
Moon and Peacock (2013)	a	5 days	93.4	3.88	$NaNO_3$	8.0	550	~2.8
Dyer et al. (2003)	a	24 h	93.4	3.88	$NaNO_3$	7.9	550	~2.8
Girvin et al. (1991)	f	24 h pH=7	92.8	3.91	$NaNO_3$	8.0	530	~2.9

Notes:
* The PZSE is close to the PZC when using electrolyte ions with near equal affinities.
$ From the H^+ consumption between pH = 8 and pH = 5.5 in 1 M $NaNO_3$, reported by Kinniburgh and Jackson (1982).
a Schwertmann and Cornell (1991) with 0.2 M $Fe(NO_3)_3$.
b Neutralization of 0.1 M $Fe(NO_3)_3$ to only pH = 5.5–6.0.
c Davis and Leckie (1978) with 0.1 M $Fe(NO_3)_3$ or (c**) with 0.1 M $FeCl_3$.
d Neutralization of 0.01 M $Fe(NO_3)_3$ in 0.01 M HNO_3 with 0.02 M NaOH to pH = 8.5.
e Neutralization of probably 1 M $Fe(NO_3)_3$ with 1.5 M NaOH to pH = 7, maintaining this pH for 1 h. The resulting suspension contained 0.33 M Fe in 1 M $NaNO_3$ and was stored at 5 °C before use.
f Benjamin (1983) with 0.1 M $Fe(NO_3)_3$ and aging at pH = 7

VARIABLE PROTON CHARGE OF FERRIHYDRITE

Reactive sites of ferrihydrite

Spectroscopic measurements of the structure of surface complexes reveal that ions can react with surface groups of various kinds, forming a variety of surface complexes. For a surface complexation model (SCM), the challenge is to link such microscopic information to macroscopic adsorption data and *vice versa*, bridging the gap between detailed structural insights and the broader picture of ion adsorption behavior, ensuring a comprehensive understanding of the surface processes involved.

In early models for surface complexation of Fh, goethite was used as a structural proxy. This approach was initiated by Manceau and co-workers (Spadini et al. 1994; Manceau and Gates 1997) and subsequently adapted by Hiemstra and Van Riemsdijk (2009). In the latter model, two types of singly coordinated surface groups were introduced because modeling of the adsorption of uranyl to Fh in the absence and presence of carbonate ions was only successful with a 2-site approach for singly coordinated surface groups (Hiemstra et al. 2009). The uranyl ions reacted with one specific type of singly coordinated groups, forming mononuclear bidentate edge-sharing complexes (Waite et al. 1994; Rossberg et al. 2009; Foerstendorf et al. 2012; Dublet et al. 2017) while the carbonate ions were allowed to react with another type of singly coordinated surface groups allowing the formation of double corner (DC) binuclear bidentate complexes (Hiemstra et al. 2004). Attempts to simplify the model by not differentiating singly coordinated groups led to an incoherent description of the experimental adsorption data. It shows that *the interaction of ions is not only regulated by electrostatics but can also have site limitations.* The latter occurs for ions with a relatively high affinity that can only bind to a limited number of sites that are a subset of the total.

In the model of Gustafsson and co-workers (Gustafsson 2003; Tiberg et al. 2013; Tiberg and Gustafsson 2016; Larsson et al. 2017; Gustafsson and Antelo 2022), no differentiation is made between types of singly coordinated sites. Therefore, it is not generally applicable to the adsorption of all ions as it omits structural information collected with spectroscopy. This sub-model has been used for describing the surface complexation of Cu(II) by edge-sharing (Tiberg et al. 2013) as well as Cd(II) double corner-sharing (Tiberg and Gustafsson 2016), while both ions use a different subset of sites. Another example of ion binding to a different set of singly coordinated sites of Fh is edge-sharing by As(III) (Ona-Nguema et al. 2005; Gao et al. 2013; Hiemstra 2018b) and double corner-sharing by As(V).

After resolving the mineral structure of Fh (Michel et al. 2007) and the introduction of surface depletion (Hiemstra 2013), the surface structure of Fh has been analyzed to identify the types of sites and their surface densities (Hiemstra and Zhao 2016). This analysis involved the construction of spherical Fh particles with Fe2 and Fe3 polyhedra that do not form singly coordinated surface groups, as stated by the SD model. The result of this evaluation is presented in Fig. 7 for the singly-coordinated surface groups (Hiemstra and Zhao 2016). The total surface density of ≡FeOH(H) sites is on average about $N_s \sim 5.8\ nm^{-2}$ (blue squares in Fig. 7) of which only a part can form binuclear bidentate or DC corner complexes (blue spheres in Fig. 7).

The singly-coordinated surface groups are either present as $\equiv FeOH^{-1/2}$ or $\equiv FeOH_2^{+1/2}$, which is directly observable with spectroscopy (Boily and Song 2020). Doubly coordinated surface groups ($\equiv Fe_2OH^0$) are considered non-reactive in the ordinary pH range (Hiemstra et al. 1996; Boily and Song 2020). The surfaces of Fh also expose triply coordinated groups ($\equiv Fe_3O(H)$) that differ in proton affinity. These sites are particularly relevant for creating additional surface charge. The effective site density is low ($N_s \sim 1.4 \pm 0.5\ nm^{-2}$) when fitted from PO_4 adsorption data (Hiemstra and Zhao 2016). The value is consistent with the value suggested by Boily and Song (2020).

PRIMARY CHARGE AND ION PAIR FORMATION OF FERRIHYDRITE

Proton charge and ion pairs. The proton adsorption and ion pair formation of Fh have been extensively investigated (Mendez and Hiemstra 2020c) using an internally consistent approach for acid-base titrations, developed earlier by Rahnemaie et al. (2006). The measurements conducted for Fh in various types of background electrolytes (NaCl,

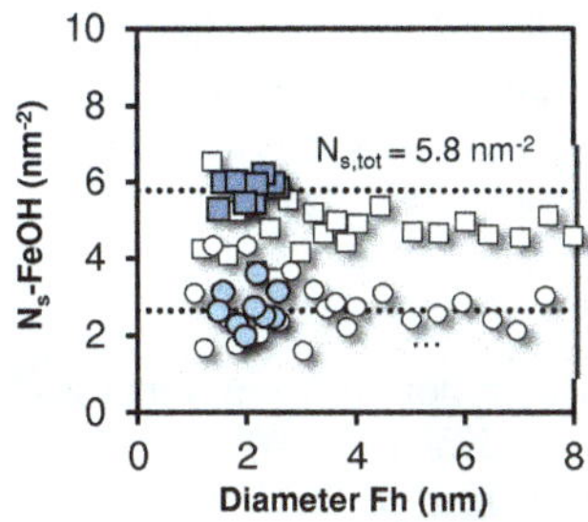

Figure 7. Site density of the singly-coordinated surface groups ≡FeOH(H) of ferrihydrite (**colored symbols**), derived with particle construction, applying the SD rule stating that Fe2 and Fe3 polyhedra cannot form singly coordinated surface groups (Hiemstra and Zhao 2016). The singly coordinated $\equiv FeOH_b$ groups that can form double corner (DC) complexes with, e.g., PO_4^{3-}, SeO_3^{2-} or CO_3^{2-} have a density of 2.8±0.6 nm^{-2}. The total number of singly coordinated FeOH(H) found is on average about 5.8±0.3 nm^{-2} or 9.6 $\mu mol \cdot m^{-2}$. The difference is the density of the $\equiv FeOH_a$ groups (3.0±0.3 nm^{-2}) that can form bidentate surface complexes by edge-sharing, e.g., with UO_2^{2+}, Cu(II), and $As(OH)_3$. The **open symbols** have been reported by Boily and Song (2020) using a slightly different SD approach. For $\equiv FeOH_b$ (**open spheres**), a very good agreement is found. For $\equiv FeOH_a$ (**open squares**), the density is slightly lower. In the approach of Boily and Song, the numbers depend on the chosen thickness (δ) of the layer of depletion. At a slight increase of the chosen value of δ, their data will fully match with the data derived by applying the SD rule in the particle construction. The Fh surface has additionally triply coordinated surface groups ($\equiv Fe_3O(H)$) with an effective surface density of 1.4±0.5 nm^{-2}. As akdalaite ($AlO_{1.4}(OH)_{0.2}{\cdot}nH_20$) is isostructural (Parise et al. 2019) to Fh ($FeO_{1.4}(OH)_{0.2}{\cdot}nH_20$), the sites of densities of both minerals will be nearly identical.

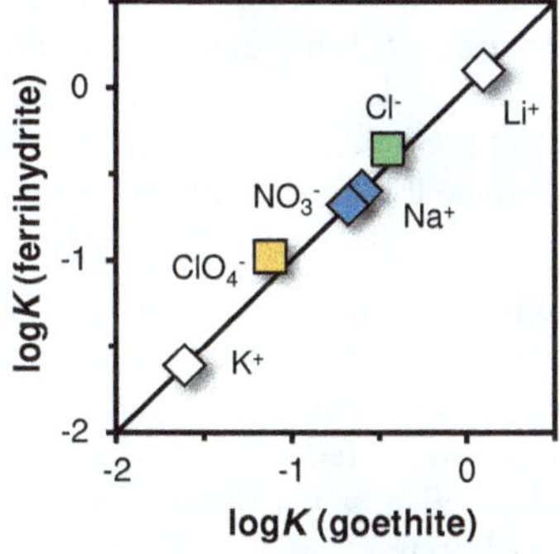

Figure 8. Relationship between the ion pair formation constants (**colored symbols**) of Fh and goethite, using in the analysis of the acid–base titration data common values for Na^+ and NO_3^- (**blue diamonds**). By adding the log*K* values of K^+ and Li^+ for goethite (**open diamonds**) the wide range of values of affinities of monovalent electrolyte ions is illustrated. [Reproduced from Mendez and Hiemstra (2020c) by permission of Elsevier.]

$NaNO_3$, $NaClO_4$), show that the resulting variation in charging can be effectively explained using the ion pair formation constants (Fig. 8), previously determined for goethite (Hiemstra and Van Riemsdijk 2006).

Point of zero charge. The point of zero charge (PZC) for ferrihydrite is generally reported to be ~ 8.1±0.2 for ferrihydrite (Table 1), which is about one unit lower than the value for goethite. Hiemstra and Van Riemsdijk (2009) have suggested that the PZC values of both materials might be related through the ratio of the singly and triply coordinated surface groups. Ferrihydrite is known to have predominantly singly-coordinated groups (Hiemstra and Zhao 2016; Boily and Song 2020) resulting in a lower PZC (Hiemstra and Van Riemsdijk 2009), while for well-crystallized goethite, the relative contribution of ≡FeOH(H) is much lower, leading to a higher PZC. The difference in $\log K_H$ values of ≡FeOH ($\log K_H$ ~7.7) and $\equiv Fe_3O$ ($\log K_H$ ~ 11.7) according to the MUSIC model (Hiemstra et al. 1996) has been used to rationalize the striking difference in PZC of goethite and ferrihydrite (Hiemstra and Van Riemsdijk 2009). This insight indirectly supports a non-zero value for $\Delta \log K_H$ between both types of groups as predicted by the MUSIC model (Hiemstra et al. 1996). Recently, this notion has been used to unify ion adsorption data for goethite preparations with different crystal face contributions and corresponding differences in the ratio of the mentioned reactive surface groups (Han and Katz 2019; Martinez et al. 2023). A recent development is a hydride use of Machine learning (ML) and CD-MUSIC modeling to derive a unifying parameter set (Chen et al. 2024) integrating a large number of data sets.

A higher PZC or PZSE has been reported in certain cases for Fh. This discrepancy can be attributed to the use of electrolytes such as KCl (Wang et al. 2013), KNO_3 (Antelo et al. 2010), or NaCl (Jain and Loeppert 2000) instead of $NaNO_3$ (Mendez and Hiemstra 2020c). There is no reason to attribute the higher PZC of PZSE to aging (Bompoti et al. 2017). The use of KCl or KNO_3 introduces strongly asymmetrical ion pair formation (Fig. 8). For titration curves, this leads to a shift of the common intersection point (CIP) or point of zero salt effect (PZSE) towards higher values than found for the pristine point of zero charge (Mendez and Hiemstra 2020c). This phenomenon results in a predicted CIP value of 8.6 for KNO_3, a value of 8.7 for KCl, and a value of 8.4 for NaCl rather than 8.1 for $NaNO_3$, all in excellent agreement with the various observations (Mendez and Hiemstra 2020c). Note that the PZSE or CIP is often interpreted as the PZC value. However, these are not the same (Sposito 1984) as the PZSE differs by definition from the PZC if the electrolyte ion pair affinity is asymmetrical.

Double layer structure

Spherical double layer. Ferrihydrite consists of ultra-small nanoparticles, resulting in an extreme surface curvature. When these particles are charged, they radiate an electrostatic field. The corresponding field strength will not only diminish because of the neutralization by electrolyte ions in the double layer, but also because the field is flaring-out in three dimensions. A spherical field is less strong than a linear field at the same charge density. As an electrostatic field gives negative feedback on the surface proton adsorption at pH < PZC and proton desorption at pH > PZC, a spherical field will less suppress the development of the proton surface charge. The role of surface curvature will be evaluated here with a novel set of calculations and new experiments.

Theory. Figure 9a illustrates the charging of a single spherical particle as a function of the particle diameter in electrolyte solutions of 4 different ionic strengths (symbols and full lines in Fig. 9a), calculated with spherical double layer theory. For the model calculations, the Extended Stern model was used, and ion pair formation was included. The adsorbed electrolyte ions neutralize part of the surface charge, which makes the influence of the spherical DDL less than without these ions. This is most significant at high ionic strength.

In Figure 9a, the horizontal dotted lines are the levels of charging at $r \to \infty$ (flat surface). As the surface curvature increases (i.e., smaller particles), the surface charge becomes more positive, because the field is less repulsive for protons at pH < PZC (full lines with symbols). The most significant differences occur at a low ionic strength because the spherical field extends most strongly out into the solution under these conditions.

In Figure 9b, the development of the pH-dependent charge is depicted for four electrolyte levels. The lines were calculated assuming for the DDL either a spherical (dotted lines) or a planar field (full lines). In the latter case, the charging is significantly suppressed, particularly at low ionic strength. Figure 9b illustrates that the electrical double-layer model may have a substantial impact on the shape of the charging behavior of nanoparticles. The calculations indicate that the difference in charging behavior between the use of a linear and spherical field is most pronounced at low ionic strength, while at high ionic strength, the influence of the type of DDL diminishes.

Experimental information. The selection of the most adequate double-layer model for Fh can be evaluated experimentally. This has been done with pH-stat measurements (unpublished

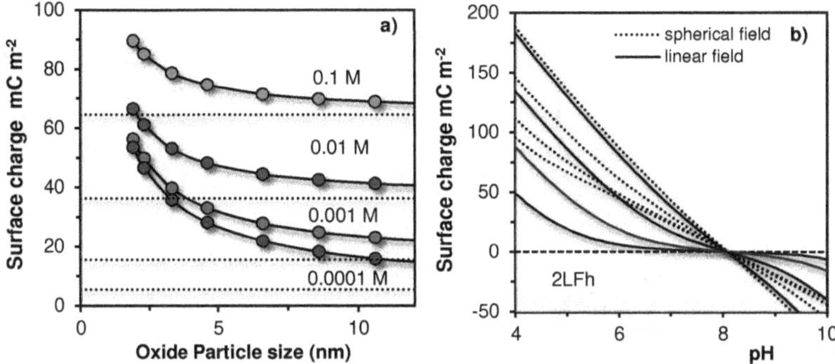

Figure 9. a) Calculated surface charge ($mC \cdot m^{-2}$) of a single Fh particle as a function of the particle size for four different values of the ionic strength ($NaNO_3$) at $\Delta pH \equiv PZC - pH = 2$, calculated applying spherical double layer theory in the Extended Stern model. The total thickness of the Stern layers is $\Delta r = 0.7$ nm. The Stern layer capacitances vary with the particle size (see text). Electrolyte ion pair formation with the surface groups is assumed ($\log K_{Na} = -0.6$ and $\log K_{NO_3} = -0.68$). The site density is set to 7.2 nm^{-2}. **b)** Comparison of the pH-dependent charging behavior of a 2.6 nm Fh particle for two different DDL models, i.e., the spherical DDL model (**dotted lines**) and the planar DDL model (**full lines**). The Stern layer capacitance values for Fh of this size are $C_1 = 1.14\ F \cdot m^{-2}$ and $C_2 = 0.90\ F \cdot m^{-2}$. For the spherical DDL calculations, a dedicated version of ECOSAT was used as a subroutine, in which the Poisson–Boltzmann equation for a radial geometry was solved numerically. In further calculations, the electroneutrality was calculated by scaling the spherical double layer charge to the surface of the nanoparticle, according to $\sigma_{DDL} \times A_{DDL}/A_{ox}$, because the oxide surface (m^2 oxide) of spherical particles differs from the surface area at the head-end of the DDL (m^2 DDL). In addition, the charge of both Stern planes was scaled to A_{ox}, and we applied the electrostatic equations $\sigma_o = C_1 (\psi_0 - \psi_1)$ and $(\sigma_o A_{ox}/A_1 + \sigma_1) = C_2(\psi_1 - \psi_2)$. The principles of these adaptations were pointed out to us by Johannes Lützenkirchen[1], referring to the work of Ohshima (2006).

data) of the surface charge as a function of the ionic strength (Fig. 10). For these experiments, a Fh batch was prepared according to the method of Schwertmann and Cornell (1991). The excess electrolyte was removed from the stock suspension through repeated washing and resuspension with CO_2-free ultra-pure water to a final conductivity of 6–9 $\mu S \cdot cm^{-1}$, using high-speed centrifugation. The final suspension was stored at 4 °C to minimize aging. Subsamples were used for a set of pH-stat titrations in which $NaNO_3$ was added in steps.

As illustrated in Figure 10a, an increase in the ionic strength at a given pH led to the development of surface charge for pH ≠ PZC. This development cannot be effectively described with a model assuming the presence of a spherical DDL (dotted lines in Fig. 10a). Particularly, at extremely low ionic strength, the deviations between the experimental data and the model become evident.

The pH-stat charging behavior of Fh can be well-described with the Extended Stern layer model assuming a linear electrostatic field for the DDL, rather than a spherical field. The ferrihydrite suspension does not behave as a collection of individual Fh nanoparticles, each surrounded by a spherical DDL. The observed deviation from the spherical double layer theory can be attributed to the pronounced aggregation of the Fh nanoparticles, commonly observed with TEM. It is also evident from the settling in the Fh suspension at low pH (5) and very low ionic strength (10^{-4} M), shown in Figure 10b. When Fh is present as an aggregated collection of particles, an extensive spherical DDL cannot develop, and double-layer overlap occurs. This creates a much stronger electrostatic field, best described by the classical diffuse double layer theory for planar surfaces (full lines in Fig. 10a).[1]

[1] In another contribution, more details are given by Hiemstra and Lützenkirchen (2025, this volume).

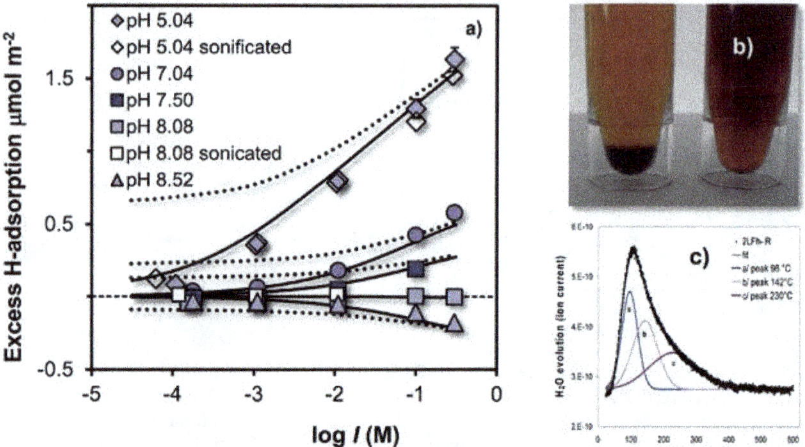

Figure 10. a) Excess proton surface charge of Fh, prepared according to Schwertmann and Cornell (1991), as a function of the ionic strength at pH-stat conditions, for A = 600 m^2·g^{-1}. The lines have been calculated with the extended Stern model with C_1 = 1.14 F·m^{-2} and C_2 = 0.90 F·m^{-2} using a spherical DDL (**dotted lines**) or a planar DDL (**full lines**) with $\log K_H = 8.1$, $\log K_{Na} = -0.6$, and $\log K_{NO_3} = -0.68$. **b)** picture of test tubes containing Fh suspensions. At low ionic strength (10^{-4} M) and low pH, particles of the opaque and reddish-colored Fh suspension tend to settle in 24 h (**left tube**) unless the colloidal suspension is sonicated in a water bath (30+10 min. at <25 °C), yielding a bright-red transparent solution (**right tube**), showing the breakdown of the aggregated/flocculated structure. However, this treatment did not influence the proton adsorption properties as follows from the **open symbols**, collected for sonicated Fh at pH 5.04, given a). In the sonicated suspension, Fh remains a collection of primary nanoparticles in an aggregated state. **c)** Rate of water release in a thermogravimetric analysis coupled with mass spectrometry (TGA-MS), resolving three populations of water, i.e., physisorbed, chemisorbed, and structural water (Hofmann et al. 2013), in agreement with the work of Xu et al. (2011). The molar mass of the Fh material, after removal of physisorbed water, was $M = 96 \pm 0.7$ g·mol^{-1}. This value is consistent with the calculated molar mass of 2.6 nm Fh particles (Eqs. 1–3) having a surface area of A = 600 m^2·g^{-1} and M = 94.7 g·mol^{-1}, which were used in the modeling of pH-stat data of Figure 10a.

With sonication of Fh at low ionic strength for 30 minutes at <25 °C, the aggregation state can be adapted. Such a suspension changes color and becomes visually transparent (Fig. 10b) and over a long time (months), no significant particle settling was observed. Although this sonication treatment leads to deflocculation, the Fh nanoparticles remain primarily aggregated, because sonication of the suspension did not change the charging behavior measured for pH 5.04 (open symbols in Fig. 10a). The model lines of Figure 10a were calculated with the Extended Stern layer model using a specific surface area of 600 m^2·g^{-1} for 2.6 nm Fh particles. The experimental data were scaled using the corresponding molar mass of M = 94.7 g·mol^{-1} according to Equations (1–3), consistent with the value that can be derived from the water content measured with thermogravimetric analysis coupled with mass spectrometry (Hofmann et al. 2013).

Size-dependency of the Stern layer capacitance. In the above SCM modeling, the Stern layer capacitances were adjusted based on the mean size of the nanoparticles. The capacitance C_r for a single spherical capacitor can be related to the corresponding capacitance C of a flat-plate capacitor according to:

$$C_r = \frac{1}{4\pi r^2}\frac{Q}{\Delta\Psi} = \frac{1}{4\pi r^2}\frac{4\pi\varepsilon_r\varepsilon_o}{\left[\dfrac{1}{r+\Delta r}-\dfrac{1}{r}\right]} = \frac{r+\Delta r}{r}\frac{\varepsilon_r\varepsilon_o}{\Delta r} = \frac{r+\Delta r}{r}C \qquad (6)$$

in which r is the particle radius and Δr the thickness of the surrounding layer (Hiemstra and Van Riemsdijk 2009; Hiemstra and Zhao 2016). The capacitor has a charge density of $Q/4\pi r^2$ and a potential difference $\Delta\psi$ between the plates. At $r \rightarrow \infty$, $C_r = C$.

For well-crystallized goethite, the use of a model with an inner and outer Stern layer is appropriate (Hiemstra and Van Riemsdijk 2006). These Stern layers typically have a total thickness equivalent to the arrangement of water in about 2 or 3 layers (Hiemstra and Van Riemsdijk 2006). For well-crystallized goethite, the capacitance value for the inner Stern layer is $C_1 = 0.9$ F·m^{-2}, and for the outer Stern layer, $C_2 = 0.75$ F·m^{-2}. These Stern layer capacitance values can be translated to values for spherical nanoparticles (C_r). For a Fh particle with a diameter of $d = 2.6$ nm, translation with Equation (6) leads to respectively $C_1 \sim 1.14$ F·m^{-2} and $C_2 \sim 0.9$ F·m^{-2}. When applied in surface complexation modeling of Fh, one basically assumes that these nanoparticles stay sufficiently hydrated for developing a Stern layer around the particles, even in an aggregated state. This water can only be removed thermally. Thermogravimetric analysis (TGA) of our Fh shows three populations of bound water (Fig. 10c), in agreement with observations and structural interpretation (Xu et al. 2011).

It is possible to calculate the amount of water present in a Stern layer and compare it to the amount of chemisorbed and structural water of Fh particles. For non-aged Fh nanoparticles ($d \sim 2.2$ nm), surrounded by an equivalent water layer with a thickness of 0.9 nm, the Fh composition expressed as $FeO_{1.4}(OH)_{0.2} \cdot nH_2O \cdot mH_2O$ will then be $n = 0.9$ and $m = 6.4$. Expressed as an oxide with the composition $Fe_2O_3 \cdot nH_2O$, one gets $n = 1$ and $n = 7.4$ for Fh, without and with physisorbed water, respectively. Both values are largely in agreement with early observations (Van Bemmelen and Klobbie 1892) for freshly prepared moist Fh. These authors also reported a value of $n \sim 4.5$ for "$Fe_2O_3 \cdot nH_2O$" for an air-dried Fh sample, aged for 2 months at 15 °C in air. The latter data suggest a mean particle size of ~3 nm when the Fh particles are surrounded by an equivalent water layer of 0.9 nm.

SURFACE COMPLEXATION OF IONS BY FERRIHYDRITE

Experimental approaches

The ion complexation of ferrihydrite has been studied extensively for almost a century because of its phenomenal ability to bind ions. For this reason, it was also employed in the first pioneering EXAFS study (Hayes et al. 1987) to resolve the structure of the surface complexes of specifically bound ions. Numerous subsequent studies followed using, apart from EXAFS, various other in-situ techniques, including attenuated total reflectance for Fourier transform infrared spectroscopy (ATR-FTIR). Moreover, the increase in computational power and the development of advanced first-principles methods are of great value, tremendously enhancing insight into the ion surface speciation on Fh. However, the identification of surface species as snapshots in terms of conditions alone is insufficient. Ideally, the collected data should offer quantitative information across a range of conditions in particular when surface speciation changes. Such molecular scale information then becomes instrumental in developing and validating SCM. This will be illustrated below in several sections with selected examples, chosen for a cohesive discussion on various crucial aspects of cation and anion adsorption to ferrihydrite. We will start with pseudo-monocomponent adsorption, using Si-adsorption as an illustrative example. Competitive and cooperative ion–ion interactions will be another important theme as well as the heterogeneity in ion affinity of ferrihydrite. Finally, the coprecipitation phenomena will be briefly touched.

Charge distribution model

The electrostatic interaction between adsorbed ions and the surface charge is intricately influenced by the location of the ion charge within the electrical double layer (EDL) profile.

For innersphere complexes, a fraction of the ligands of the central ion is shared with the surface while the other part is outside the surface, in the compact part of the double layer. Consequently, the charge of the adsorbed ion is distributed in the interface between two corresponding electrostatic model positions. This charge distribution also depends on the relative bond lengths within the coordination sphere of the adsorbed ion (see footnote [1]).

In the CD model (Hiemstra and Van Riemsdijk 1996), the electrostatic interaction is calculated from a minimum of two electrostatic energy terms. One term is related to the charge attributed to the surface or 0-plane (Δz_o) and the other term is related to the charge (Δz_1) attributed to the electrostatic 1-plane of the electrostatic module applied to the SCM model. The symbol Δ is given as it is about a change of charge.

The charge, attributed in the CD model to the surface plane (Δz_o), is the most important factor for the electrostatic interaction of an ion with the protons adsorbed on the surface. This interaction has a pronounced influence on the co-adsorption or co-desorption of protons and the corresponding pH dependency of the ion adsorption. The charge attributed to the 1-plane (Δz_1) is most relevant for the concentration dependency of the adsorption, as this charge significantly changes the electrostatic potential of the corresponding 1-plane with an increase in the ion loading. In addition, Δz_1 and the potential ψ_1 are key to competitive and cooperative adsorption of ions. For a more detailed understanding of the CD-MUSIC model, readers are referred to the respective contribution in this volume (Hiemstra and Lützenkirchen 2025, this volume). Its application to Fh is elaborated in much detail elsewhere (Hiemstra and Zhao 2016).

Monocomponent adsorption

In this section, the adsorption of silicic acid in pseudo monocomponent systems is chosen to illustrate the principles of the pH dependency of ion adsorption. Silicic acid is omnipresent in nature and may interact with Fe, leading to the formation of silicious ferrihydrites. These materials carry significant amounts of adsorbed Si and are formed under conditions with minimal competitive adsorption, e.g., of phosphate. Another motivating factor for this choice is its capability to polymerize.

pH dependency. The adsorption edges of H_4SiO_4 exhibit a typical shape, featuring a peak in the adsorption, therefore the pattern has also been termed an adsorption envelope. It is found for the adsorption of siliceous acid by (hydr)oxides in general. For ferrihydrite, this typical adsorption envelope is shown in Figure 11, for different total ratios of Si/Fe. Starting at low pH, the adsorption increases towards a maximum at high pH. As a general rule, the increase of ion adsorption with pH signifies a net release of protons in the overall adsorption reaction and follows from a thermodynamic consistency principle (Perona and Leckie 1985). In its generalized form, the thermodynamic consistency principle can be given as (Rietra et al. 2000):

$$\left(\frac{\partial \Gamma_H}{\partial \Gamma_i} - n_H\right)_{pH} \equiv \left(\chi_H - n_H\right)_{pH} = \left(\frac{\partial \log C_t}{\partial pH}\right)_{\Gamma_i} \tag{7}$$

The expression says that the change in the H adsorption ($\partial\Gamma_H$), attributed to the adsorption of species i ($\partial\Gamma_i$), adjusted for the mean protonation state of the species in solution at a given pH (n_H), equals the change in the logarithm of the total concentration of the species in solution ($\partial \log C_t$) with respect to the change of pH (∂pH) at a given level of the adsorption (Γ_i).

The value of the proton co-adsorption ($\chi_H \equiv \Gamma_H / \Gamma_i$) as well as the mean protonation state of the species in solution (n_H) will depend on the chosen reference state. Often, an unprotonated species is selected as the reference species. However, in the case of the adsorption of Si, it is convenient to use $H_4SiO_4^o$ as the reference species for calculating/measuring the proton co-adsorption (χ_H) as well as the mean excess protonation state of this species in solution ($n_H = 0$).

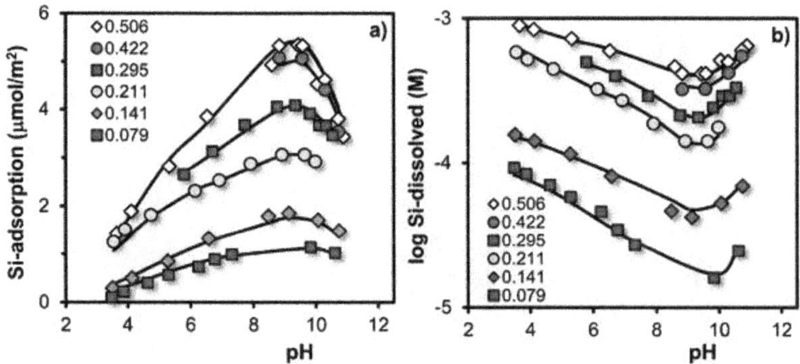

Figure 11. a) pH-dependent Si adsorption in Fh systems with increasing molar Si/Fe ratios and **b)** its corresponding equilibrium concentration in 0.1 or 0.15 M $NaNO_3$ systems (**symbols**). Data from Hiemstra (2018b). The **lines** have been calculated using the CD model. [Reproduced from Hiemstra (2018b) by permission of Elsevier.]

Below pH ~8.5, the excess protonation state of H_4SiO_4 species in solution tends to be near zero ($n_H \sim 0$). The measured/calculated proton co-adsorption at this pH is a negative number ($\chi_H < 0$), i.e., protons are released. According to Equation (7), a negative value for $(\chi_H - n_H)_{pH}$ implies $\partial \log C_{H_4SiO_4,aq}$ with respect to the pH (∂pH) is negative. It means that the slopes of the Si-solubility lines in Figure 11b are negative at this pH condition.

At high pH, aqueous $H_4SiO_4^o$ releases a proton, forming $H_3SiO_4^-$ (aq). Due to this conversion of the dominant solution species, $n_H = -1$ at about pH 11, the value of $\chi_H - n_H$ changes from negative to positive, causing the slopes of the lines in Figure 11b to change from negative to positive values. It illustrates that the reversal of the pH-dependency of the Si adsorption is primarily driven by the change of the speciation in the solution. As illustrated in Figure 11b, the concentration of Si in solution has a minimum, corresponding to a maximum in the Si adsorption (Fig. 11a). At this point, the net number of protons released or consumed in the overall adsorption reaction is zero.

The thermodynamic consistency relation of Equation (7) shows that the pH-dependency of ion adsorption, expressed in $\partial \log C_t / \partial pH$ depends on a surface property (χ_H) as well as the solution speciation (n_H). As n_H may change with pH, so will it change the pH dependency of the adsorption as already demonstrated in 1967 (Hingston et al. 1967).

Si-monomer versus oligomer formation. At a low Si loading, Si is bound as a monomer. At high concentrations, oligomers are formed (Swedlund et al. 2010a,b; Do Hamid et al. 2011; Kanematsu et al. 2018; Wang et al. 2018; Hiemstra 2018b). For the Si monomers, Wang et al. (2018) suggest the formation of a monodentate complex, considering the Fe-Si distance of 331 pm according to differential pair distribution function analysis (d-PDF) and MO/DFT modeling. However, one of the free OH ligands of the Si surface complex forms a robust H-bond with an adjacent surface group (Fig. 12a).

In Figure 12a, the MO/DFT optimized structure (Hiemstra 2018b) with a Fe-Si distance of 333 pm is depicted. Figures 12b,c show the structures of two adsorbed oligomers that have been used in modeling the Si adsorption with the CD-MUSIC model for Fh (Hiemstra 2018b). The MO/DFT-calculated Si–Fe distances in these structures (324 pm) equal the measured distances (Wang et al. 2018).

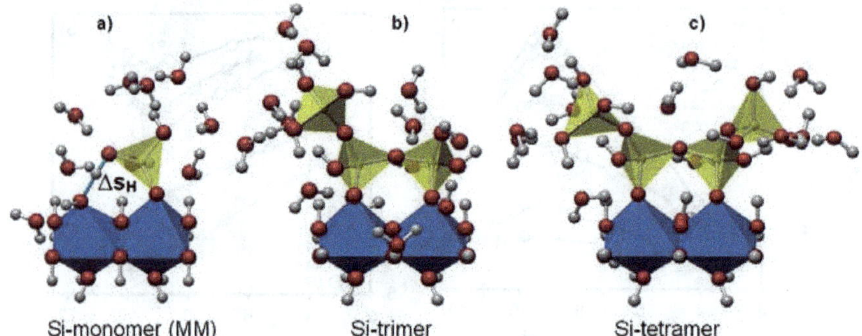

Figure 12. MO/DFT/B3LYP optimized geometries of three Si complexes that have been revealed with CD modeling of the Si adsorption of ferrihydrite (Hiemstra 2018b). **a)** At low loading, monomeric Si is bound as a single-corner mononuclear monodentate complex. In this structure, one of the outer OH ligands strongly interacts with the adjacent ≡FeOH group (Wang et al. 2018) via the formation of a strong H-bond, which transfers charge (Δs_H ~ 0.2 v.u.), changing the CD coefficient and thereby, the pH dependency. **b–c)** At high loading, Si-oligomers are formed (Swedlund et al. 2010b) where two Si tetrahedra link to two Fe octahedra (Fe_2Si_2), additionally binding one or two Si tetrahedra. The MO/DFT calculated Fe–Si distances (d_{FeSi}) are in close agreement with the experimental data. The calculated Si–Si distances (301 ± 1 pm) and the Si–O–Si angles (128° ± 3) of the oligomers closely resemble the values found in the structure of orthopyroxenes. [Reproduced from Hiemstra (2018b) by permission of Elsevier.]

The formation reactions can be formulated as:

$$\begin{gathered}\equiv FeOH_2^{+\frac{1}{2}} + \equiv FeOH^{-\frac{1}{2}} + H_4SiO_4\,(aq) \rightleftharpoons \\ \equiv(FeOHFeO)^p\, Si(OH)_3^q + 1\,H^+(aq) + 1\,H_2O\,(l)\end{gathered} \tag{8}$$

and

$$\begin{gathered}\equiv FeOH^{-\frac{1}{2}} + \equiv FeOH^{-\frac{1}{2}} + n H_4SiO_4\,(aq) \rightleftharpoons \\ \equiv(FeO)_2\, Si_n O_{n-1} (OH)_{2n-m}\, O_m^{-1-m} + m H^+(aq) + (1+n) H_2O(l)\end{gathered} \tag{9}$$

with n = 3 or 4 and m = 1 in Equation (9).

Validation of the Si surface speciation. Surface speciation of Si has been explored both experimentally and through modeling. The charge distribution in the structures presented in Figure 12 has been calculated using a bond valence analysis of the Si–O distances obtained from the MO/DFT optimized structures. As the monomer also has a very strong H-bond, the corresponding bond valence charge transfer (Δs_H = 0.20±0.02 v.u.), derived using various functionals (B3LYP, EDF1, EDF2, ωB97X-D), has been included in the calculated charge distribution of the adsorbed H_4SiO_4 species ($\Delta z_o = 0.48 \pm 0.02$, $\Delta z_1 = -0.48 \pm 0.02$). The change in bond valence in the H-bridge can be assessed with $s_H = \exp(-(d_{O...H} - 0.73) / 0.73)$ in which $d_{O....H}$ is the bond length (Å) of longest part of the H-bridge (Hiemstra 2018b).

A free fit of the adsorption data with the CD model yields $\Delta z_o = 0.55 \pm 0.03$ and $\Delta z_1 = -0.55 \pm 0.03$ for the monomer, confirming the monodentate structure with additional transfer of charge due to strong H-bond formation. Regarding both oligomers bound as bidentates, the MO/DFT-derived Δz_o values are 0.31 and 0.33, illustrating the asymmetry in the coordination sphere of the Si tetrahedra attached to Fe. This asymmetry contributes to a better neutralization of the O-ligands in the Fe–O–Si bonds with the surface.

The Si surface speciation can be calculated using the CD model, guided by spectroscopic information about the types of species. Ideally, the predicted surface speciation is validated by comparing the modeling results with quantitative information independently obtained with spectroscopic methods that have collected data over a wide range of solution conditions by varying parameters such as pH, ionic strength, and ion loading.

For H_4SiO_4 adsorption to Fh, the validation has been done using data (Fig. 13) measured with both ex-situ and in-situ methods under well-specified conditions data (Swedlund et al. 2011; Wang et al. 2018), allowing validation of the SCM. As the Si loading increases, the relative contribution of the adsorbed oligomeric surface species increases too. For Fh in 0.01 M NaCl at pH 8.5, this has been ex-situ measured with IR and XPS (Fig. 13a). The observed change in surface speciation is well-predicted by the CD model (lines in Fig. 13a). Furthermore, the change in surface speciation, measured as a function of the Si loading for pH values ranging from pH 4 to pH 10 using in-situ ATR-FTIR (Fig. 13b), is well reproduced by the model. This also demonstrates the robust quantitative agreement with the model predictions. It implies that the CD model is proficient in making exact predictions under changing solution conditions while providing a realistic picture of the molecular mechanisms of ion adsorption. This model capability is important for generalizing such information as well as for practical applications.

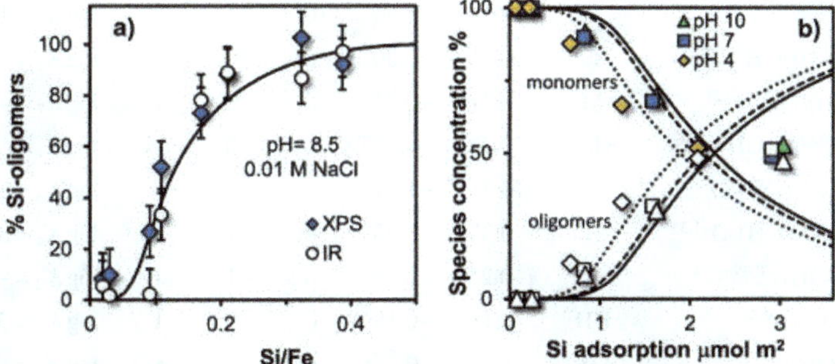

Figure 13. a) Polymerization of adsorbed Si as a function of the Si-loading in ferrihydrite systems for pH = 8.5 and 0.01 M NaCl, according to two different *ex-situ* spectroscopic techniques (Swedlund et al. 2011) (**symbols**) and predicted by the CD model (Hiemstra 2018b) (**line**). **b)** Relative surface speciation of Si (%) derived with (*in-situ*) ATR-IR spectroscopy (Wang et al. 2018) as a function of the Si adsorption at three pH values in 0.01 M $NaNO_3$. **Colored symbols** are for monomeric Si and **open symbols** are for oligomeric Si. The **lines** are corresponding predictions by the CD model (Hiemstra 2018b). [Reproduced from Hiemstra (2018b) with permission of Elsevier.]

Competitive ion–ion interactions

Surface complexation models can be powerful numerical instruments to elucidate ion adsorption behavior in multicomponent systems. This capability to predict changes in ion adsorption is pivotal to practical applications and fundamental research in complex natural systems. In multicomponent systems, the various types of adsorbed ions often predominantly interact via electrostatic interactions. These can be competitive as well as cooperative. Both are discussed below with meaningful examples.

Monocomponent adsorption of oxyanions. The CD-MUSIC model for Fh has been used to describe the competitive adsorption of PO_4, AsO_4 (Hiemstra and Zhao 2016; Hiemstra 2018b), and CO_3^{2-} (Mendez and Hiemstra 2019), as well as the competitive adsorption of $B(OH)_3$ (Van Eynde et al. 2020), $As(OH)_3$ and $H_4Si(OH)_4$ (Hiemstra 2018b). In all systems, the adsorption data were first scaled to the appropriate values for the reactive surface area (RSA) of Fh, using PO_4 as a probe ion. The consistent set of parameters is given in Table 2.

Table 2. A consistent set of parameters for describing the adsorption of selected oxyanions with the CD model, using $\equiv FeOH_a$ (3.0 nm^{-2}), $\equiv FeOH_b$ (2.8 nm^{-2}), and $\equiv Fe_3O$ (1.4 nm^{-2}) sites (Fig. 7) with a protonation constant of $\log K_H = 8.1$ The applied Stern layer model has size-dependent capacitance values (Eqn. 6) with goethite as reference ($C_1 = 0.9$ F·m^{-2} and $C_2 = 0.74$ F·m^{-2}). The ion pair formation constants of Fh as given in Figure 8.

Species	Code #	Δz_0	Δz_1	Δz_2	H⁺	Ion*	log*K*
$\equiv(FeO)_2^{-0.54}PO_2^{-1.46}$	B	0.46	−1.46	0	2	1	$\log K_1 = 28.31 \pm 0.04$
$\equiv(FeO)_2^{-0.35}POOH^{-0.65}$	BH	0.65	−0.65	0	3	1	$\log K_2 = 33.52 \pm 0.13$
$\equiv FeO^{-0.22}PO_2OH^{-1.28}$	MH	0.28	−1.28	0	2	1	$\log K_3 = 26.36 \pm 0.20$
$\equiv FeO^{-0.17}PO_2OH^{-0.33}$	MH_2	0.33	−0.33	0	3	1	$\log K_4 = 29.84 \pm 0.23$
$\equiv(FeO)_2^{-0.53}As(V)O_2^{-1.47}$	B	0.47	−1.47	0	2	1	$\log K_1 = 27.70 \pm 0.03$
$\equiv(FeO)_2^{-0.42}As(V)OOH^{-0.58}$	BH	0.58	−0.58	0	3	1	$\log K_2 = 34.00 \pm 0.09$
$\equiv FeO^{-0.20}As(V)O_2OH^{-1.30}$	MH	0.30	−1.30	0	2	1	$\log K_3 = 23.88 \pm 0.93$
$\equiv FeO^{-0.17}As(V)O_2OH^{-0.33}$	MH_2	0.33	−0.33	0	3	1	$\log K_4 = 29.94 \pm 0.06$
$\equiv(FeO)_2^{-0.33}CO^{-0.67}$	B	0.67	−0.67	0	2	1	$\log K_1 = 21.67 \pm 0.05$
$\equiv(FeO)_2^{-0.33}CONa^{+0.33}$	BNa	0.67	+0.33	0	2	1 + 1	$\log K_2 = 22.49 \pm 0.04$
$\equiv FeO^{-0.26}CO_2^{-1.34}$	M	0.34	−1.34	0	1	1	$\log K_3 = 11.63 \pm 0.02$
$\equiv(FeO)_2^{-0.70}As(III)OH^{-0.30}$	E	0.30	−0.30	0	0	1	$\log K_1 = 5.60 \pm 0.04$
$\equiv FeO^{-0.29}As(III)(OH)_2^{-0.21}$	M	0.21	−0.21	0	0	1	$\log K_2 = 4.81 \pm 0.07$
$\equiv(FeOFeOH)^{-0.53}\ Si(OH)_3^{-0.47}$	M_{mH}	0.47	−0.47	0	0	1	$\log K_1 = 5.59 \pm 0.05$
$\equiv(FeO)_2^{-0.69}\ Si_3O_2(OH)_5O^{-1.31}$	B	0.31	−1.31	0	−1	3	$\log K_3 = 5.37 \pm 0.15$
$\equiv(FeO)_2^{-0.67}\ Si_4O_3(OH)_7O^{-1.33}$	B	0.33	−1.33	0	−1	4	$\log K_4 = 8.69 \pm 0.10$
$\equiv(FeO)_2^{-1.25}B(OH)_2^{-0.75}$	B	−0.25	−0.75	0	−1	1	$\log K_2 = -4.68 \pm 0.20$
$\equiv(FeO)_2^{-0.82}B(OH)^{-0.18}$	BH	0.18	−0.18	0	0	1	$\log K_1 = 3.39 \pm 0.18$
$\equiv FeOH_2^{0.5}$-$B(OH)_3^0$	OS	1	0	0	1	1	$\log K_3 = 9.32 \pm 0.07$
$\equiv Fe_3OH^{0.5}$-$B(OH)_3^0$	OS	1	0	0	1	1	$\log K_3 = 9.32 \pm 0.07$

Notes:

\# B= Binuclear bidentate complex BH = Protonated B-complex, BNa = B-complex with attached Na^+ ion

M = Monodentate complex, MH= protonated M-complex, MH_2 double protonated M complex,

E = edge-sharing complex, M_{mH} Monomeric monodentate complex with a H-bridge to the adjacent ≡FeOH surface group, OS = outersphere complex.

* Reference ion in the reactions: PO_4^{3-}, AsO_4^{3-}, CO_3^{2-}, $As(OH)_3^0$, $H_4SiO_4^0$, $B(OH)_3^0$, and Na^+.

The double corner bidentate complexes (B) in Table 2 are defined by a reaction with two $\equiv FeOH_b$ groups. The single-corner monodentate complexes (M) react with all FeOH surface groups ($\equiv FeOH_a$ and $\equiv FeOH_b$) and the same affinity is assumed for both types of ≡FeOH groups. For $As(OH)_3$, the bidentate complex is formed by edge-sharing (Ona-Nguema et al. 2005; Gao et al. 2013), as discussed in Hiemstra (2018b). The $\equiv FeOH_a$ sites are involved. For monomeric H_4SiO_4, there is ligand exchange with only one $\equiv FeOH_b$ group in combination with the formation of a strong H-bond with an adjacent $\equiv FeOH_b$ group (Fig. 12a). In addition, the formation of Si-polymers involves two $\equiv FeOH_b$ groups (Hiemstra 2018b). $B(OH)_3$ binds to Fh by outer and inner sphere complexation (Van Eynde et al. 2020). The latter is extremely weak. Consequently, it only reacts with so-called high-affinity sites (0.25 ± 0.08 nm^{-2}), as will be discussed later. In the consistent model approach, the chosen surface species are in line with spectroscopic data. The charge distributions (Δz_0, Δz_1) have been calculated independently by optimizing the structure of small, hydrated Fe clusters with the complexing ion using MO/DFT, followed by a bond valence analysis of the calculated bond lengths.

Competitive adsorption with PO_4–CO_3 as an example. In a series of investigations, the competitive interaction of the ions of Table 2 has been validated (Hiemstra and Zhao 2016; Hiemstra 2018b; Mendez and Hiemstra 2019; Van Eynde et al. 2020). Below the competitive adsorption will be exemplified for PO_4–CO_3, because it can be used to assess the reactive surface area of the natural metal(hydr)oxide fraction of soils.

The competitive adsorption of phosphate and carbonate has been studied extensively (Mendez and Hiemstra 2019). In Figure 14a, the interplay of PO_4 and CO_3 in ferrihydrite systems is illustrated with a set of PO_4 adsorption isotherms at various levels of dissolved carbonate for pH 8.7. For comparison, competitive adsorption isotherms are also given for goethite (Fig. 14b), illustrating that carbonate ions bind stronger to ferrihydrite than to goethite leading to more competition with PO_4 ions. In fact, the affinity of CO_3 is notably low compared to PO_4 but nevertheless, a significant desorption of PO_4 is possible if the concentration of CO_3 is sufficiently higher (Figs. 14a,b). This concentration-compensating phenomenon is employed in soil extractions as discussed later.

For Fh systems, the interaction of PO_4 and CO_3 has been systematically examined by varying the total amounts of PO_4 and CO_3. The resulting data have been effectively modeled with the CD model for Fh. The affinity constants for CO_3 (Table 2) were derived exclusively based on the experimental concentrations of PO_4 during modeling. Once derived, the $\log K$ values for CO_3 have been employed to *predict* and verify the adsorption of CO_3 in mono-component systems, demonstrating excellent agreement with the experimental data (Fig. 14c). This suggests a comprehensive understanding of the competitive interaction between PO_4 and CO_3.

Figure 14c illustrates that the adsorption of CO_3 reaches a maximum at pH ~ 6. The shift of the pH dependency of the CO_3 adsorption can be attributed to changes in the solution chemistry ($HCO_3^- + H^+ \rightleftharpoons H_2CO_3$) according to the thermodynamic consistency principle (Eqn. 7). At a circumneutral pH, HCO_3^- is the dominant species. The change of the protonation state of the aqueous carbonate species in solution (H_2CO_3), changes the pH dependency of the CO_3 adsorption (Hingston et al. 1967).

The dominance of dissolved HCO_3^- or H_2CO_3 does not imply that inorganic carbon is bound to Fh as HCO_3^-. It is the CO_3^{2-} ion that interacts with two $\equiv FeOH_2$ groups, releasing two water molecules, and forming the surface species $\equiv Fe_2O_2CO$. In this process, a fair amount of negative charge is introduced on the surface. This leads to a significant

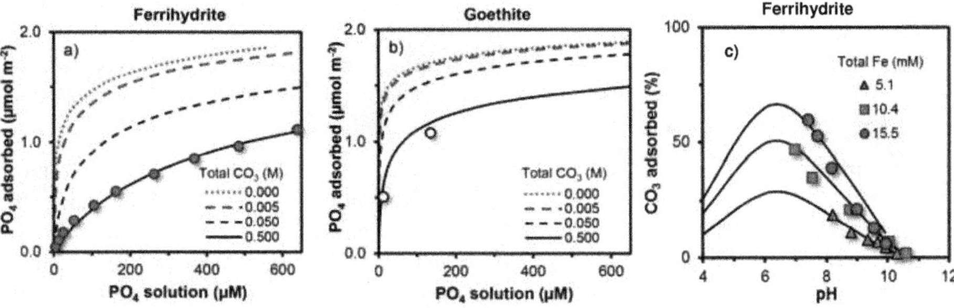

Figure 14. Adsorption isotherms of PO_4 in **a)** ferrihydrite systems and **b)** goethite systems with $CO_{3[T]}$ = 0–0.5 M at pH 8.70±0.0. The **symbols** are experimental data Mendez and Hiemstra (2019) and the lines depict model predictions based on a large collection of competitive adsorption data for Fh (Mendez and Hiemstra 2019) and goethite (Rahnemaie et al. 2007), both not shown. **c)** Adsorption edges of 1 mM (H)CO_3 (Mendez and Hiemstra 2019) at three levels of added ferrihydrite (**symbols**). The **lines** represent pure model predictions using the CD model for Fh, calibrated only with competitive PO_4 adsorption data, underlining the validity of the model. [Reproduced from Mendez and Hiemstra (2019) with permission of the American Chemical Society.]

co-adsorption of protons, causing the overall reaction stoichiometry to approach a ratio of 1 for H/CO_3, as if HCO_3^- ions were adsorbed.

Cooperative ion–ion interactions

In oxide systems having ions of opposite charge, the adsorption of both types of ions can be significantly enhanced compared to their behavior in mono-component systems. This enhancement can arise from mutual electrostatic interactions as well as the formation of ternary surface complexes. Below, this will be exemplified for Fh systems with calcium and phosphate ions as well as for systems with heavy metal ions and phosphate. Ternary complexation can only be revealed with SCM if the undelaying monocomponent adsorption is well understood.

Monocomponent adsorption of Ca^{2+} and Mg^{2+}. In the application of SCM to natural systems, a crucial aspect is understanding the adsorption behavior of Ca^{2+} and Mg^{2+} and their interactions with oxyanions, as both divalent ions typically dominate natural solutions. Mendez and Hiemstra (2006a,b) have studied the adsorption of these ions to Fh.

The relatively large pH dependency of the adsorption of Ca^{2+}, i.e., a relatively high release of protons per adsorbed ion (Eq. 7), suggests that Ca^{2+} predominantly forms bidentate complexes (Fig. 15a, inset). Through MO/DFT optimizations of the structure of a hydrated cluster with two Fe polyhedra and a Ca^{2+} ion forming a double corner complex, the charge distribution (Δz_0, Δz_1) within the coordination spheres of the cation can be derived. The interpretation of the bond length in the coordination sphere is based on the Brown bond valence concept (Brown 2009). Similarly, the CD values for the other ions of the alkaline-earth series can be determined. Comparing the CD values obtained with the Brown and Pauling bond valence concept reveals a systematic deviation for all ions in the alkaline earth ion series (Fig. 15a), pointing to asymmetry in the coordination sphere of the adsorbed cations. In general, more positive charge is allocated to the Fh surface upon adsorption of these cations than predicted by assuming a symmetrical charge distribution (i.e., a Pauling bond valence distribution).

High and low affinity. The variation in the affinity at low loading is a commonly observed phenomenon when investigating the metal ion binding for ferrihydrite (Dzombak and Morel 1990). Describing the adsorption of Ca^{2+} and other ions in the alkaline earth series involves distinguishing between "high and low-affinity sites". The surface of Fh exhibits a notable degree of variation in the affinity to bind ions. This heterogeneity of Fh for binding metal ions is not yet well understood.

The dual-affinity character of ferrihydrite may be due to heterogeneity in the surfaces (Mendez and Hiemstra 2020b). Only a few sites per particle are then involved. Or, the heterogeneity is a continuum, simplified in modeling to a minimum of two types of sites. However, the aggregation of Fh particles, creating internal and external surfaces and sites, is another yet plausible explanation for the variation in ion affinity.

Similarly to the discussion on proton adsorption concerning the double layer model (Figs. 9, 10), the internal surfaces of Fh aggregates have an EDL profile and interfacial water structure that differs from the external surfaces of the aggregates. Internally, double layer overlap amplifies the strength of the electrostatic field and intensifies the repulsion of cations when bound to positively charged surfaces. This may lead internally to sites with a low affinity for cations. The Fh particles on the outer aggregate surface can exhibit a more spherical field, resulting in less repulsion for the cations, leading to relatively high cation adsorption. In that case, the high affinity has an electrostatic rather than an intrinsic chemical origin. This concept explains why Dzombak and Morel (1990) did not distinguish between high and low-affinity sites for anions in contrast to cations. The double-layer overlap in the interior of the aggregates is favorable for the binding of anions. These sites form the majority while the minority with a low affinity will be largely invisible in adsorption phenomena.

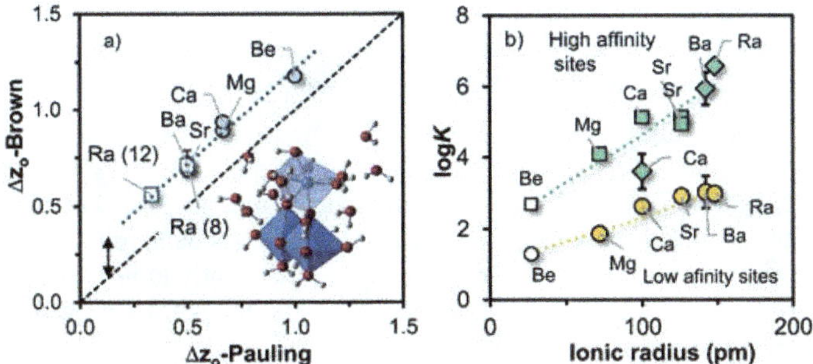

Figure 15. a) CD values for alkaline earth metal ions, according to the Pauling and Brown bond valence approach. Inset displays the MO/DFT-optimized geometry of a hydrated Fe hydroxide cluster forming a double corner complex with Ca^{2+} (Mendez and Hiemstra 2020b). The bond lengths in this structure can be translated to CD values, being systematically different from the CD values based on a Pauling bond valence distribution. With the increase of the coordination number of the metal ions in this series, the attribution of the charge to the common values ligands (Δz_0) decreases. **b)** Increase in the intrinsic affinity of alkaline-earth metal ions with the ionic radius. This trend is found for the M^{2+} binding to both, high-affinity and low-affinity sites. The increase in affinity can be attributed to a water–ligand exchange and restructuring of interfacial water as calculated with an empirical adsorption energy model (**dotted lines**), according to Equation (10). [Reproduced from Mendez and Hiemstra (2020b) with permission of Elsevier.]

Mendez and Hiemstra (2020b) suggested a surface density of 0.3 ± 0.1 nm^{-2} for (*sic*) "high-affinity sites", derived from their modeling of alkali earth ion adsorption, i.e., about 5% of the total amount of sites available. For Fh particles with a SSA of 600 $m^2 \cdot g^{-1}$, this corresponds to an external surface area of the aggregate of 5%, which is approximately $A_a = 30\pm10$ $m^2 \cdot g^{-1}$. From this, one can estimate the aggregate size (d_a), applying the relation $A_a = 6/(\rho_a d_a)$ in which ρ_a is the mass density of the hydrated spherical aggregate. Using a calculated mass density of $\rho_a = 1.82 \times 10^6$ $g \cdot m^{-3}$ for the collection of Fh particles, each surrounded by a water layer equivalent to the thickness of Extended Stern of 0.7 nm with $\rho_{water} \sim 10^6$ $g \cdot m^{-3}$, the derived mean diameter of the aggregates will be $d_a \sim 110\pm30$ nm. For 2LFh, TEM observations (Michel et al. 2010) show loosely aggregated Fh consisting of primary aggregates with a comparable size supporting our novel explanation.

The development of different double layer structures on the outer and inner surfaces of Fh aggregates can influence the structure and density of interfacial water. A variation in water structure may have a large effect on ion affinity, as follows from the evaluation of the affinity trends (Fig. 15b) in the series of alkali earth ions with a model for the adsorption energy (Mendez and Hiemstra 2020b). This aspect is discussed below.

Adsorption energy model. The affinity of the metal ions in the alkali earth series increases with the size of the ions. This is found for both "high-affinity sites" and "low-affinity sites" (Fig. 15b). The increase in affinity with the increase of the ion size can be attributed to energy changes due to 1) partial dehydration of the adsorbing ion (E_{dh}), 2) the formation of bonds between the ion and the surface groups (E_f), and 3) the elimination and restructuring of interfacial water upon attachment of the partial hydrated M^{2+} ion to the surface (E_{inf}). Mathematically, the adsorption energy of alkaline earth ions can be conceptualized as a combination of these three energy terms (Mendez and Hiemstra 2020b), expressed to:

$$\Delta E_{Ads} = E_f + E_{dh} + E_{inf} = \left(f_1 - f_2\right) E_h - k r_M = \Delta f E_h - k r_M \tag{10}$$

in which E_h is the ion hydration energy. In the model, the energy of partial dehydration ($M(H_2O)_m \rightarrow M(H_2O)_{m-2}$) and surface bond formation energy ($M(H_2O)_{m-2} + 2{\equiv}FeOH$) are both assumed to be proportional to E_h, denoted as $(f_1 - f_2)\ E_H = \Delta f E_H$. For $E_f > E_{dh}$, $\Delta f > 0$.

The energy of water removal from the interface upon attachment of the partially hydrated ion is assumed to be proportional to the ionic radius (r_M). The minus sign introduced in Equation (10) indicates that the removal of interfacial water is presumed to result in an energy release.

The dotted lines in Figure 15b have been calculated with the above adsorption energy model. Evaluation of the two-parameter model, using tabulated values for E_h, shows that for a zero value of k, a negative trend with the ion radius size is found for $\Delta f > 0$ or $E_f > E_{dh}$. It results in the ion order of the well-known Hofmeister series which has been reported for other oxide materials. However, the affinity trend reverses, when a sufficiently large value for k is chosen, making the interfacial water energy term relevant. The model can describe the experimental trends with the set parameters of Table 3.

Table 3. Parameters for the adsorption energy model (Eqn.10). For the high and low-affinity sites, the values of Δf and k are promotional and only differ by a single factor 2. The positive value of Δf implies that $E_f > E_{dh}$.

Parameter	Low-affinity sites	High-affinity sites	Ratio
Δf	0.002	0.004	2
k	0.1	0.2	2

For the high- and low-affinity sites, the values of Δf and k are found to be proportional and differ by a single factor of 2. This observed proportionality implies that sites with a high affinity for ions also exhibit a high affinity for adsorbed water and vice versa. This interfacial water is present in the compact part of the electrical double layer. Within the discussed model of hydrated aggregates, our finding suggests that interfacial water associated with the internal surfaces is less strongly bound than water at the external surfaces. This may be due to electrical double layer overlap or differences in the water network structure.

Cooperative adsorption of Ca^{2+} with PO_4. Using the dual-affinity approach for Fh (Mendez and Hiemstra 2020a) an insightful interpretation of the cooperative adsorption involving Ca^{2+}, Mg^{2+}, and PO_4^{3-} is possible. The mutual interaction of divalent cations with PO_4 is influenced, in part, by electrostatic changes. The presence of adsorbed PO_4^{3-} ions introduces a net negative charge that mitigates further binding, while the additional adsorption of Ca^{2+} or Mg^{2+} ions does the opposite. The mutual electrostatic interaction amplifies the adsorption of both ions. The adsorption of the ion with the lowest intrinsic affinity is most affected by the mutual interaction, as illustrated in Figure 16 for the Ca–PO_4 system. A comparison of adsorption in this binary system (full lines) with the adsorption in the single ion systems (dotted lines) reveals an increase in the adsorption of both types of ions, with the most substantial enhancement for Ca^{2+}.

A model evaluation of the adsorption data of PO_4–Ca^{2+} systems shows that the effect of PO_4 on the Ca^{2+} adsorption is larger than can be solely attributed to just electrostatic interactions. The notably higher interaction has been attributed to the formation of ternary complexes. The data are best explained by the formation of a PO_4 surface complex to which a Ca^{2+} ion is bound. Within these complexes, the PO_4 ions form single-corner as well as double-corner complexes with the Fe polyhedra of the surface as illustrated in Figure 17. For the Ca^{2+}, a monodentate interaction with PO_4 has been assumed. However, it can not be excluded that Ca^{2+} interacts by bidentate formation because such species have recently been suggested for aqueous $CaPO_4$ complexes based on molecular modeling (Koca Fındık et al. 2024). If realistic, it implies that the applied CD value for the PO_4 moiety, derived with MO/DFT geometry optimization, will be slightly different, affecting the surface speciation of the ternary complexes.

Cooperative adsorption of $Zn^{2+}$and Cu^{2+} with PO_4. Ternary complex formation may also play a role in other metal ion–phosphate systems with Fh. Extensive investigations of Eynde et al. (2022) have focused on the interaction of phosphate and Zn^{2+} and additionally, on the interaction of phosphate and Cu^{2+}.

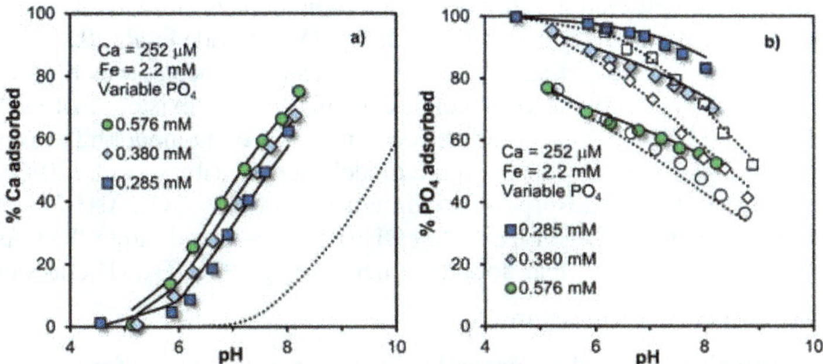

Figure 16. pH-dependent adsorption of Ca **(a)** and PO_4 **(b)** in binary ion systems (**colored symbols**) with ferrihydrite in I = 0.01 M $NaNO_3$ at fixed initial concentrations of Fe and Ca, and three levels of added PO_4, as given. The **full lines** are model predictions for the binary systems. The **dotted lines** depict model predictions for the single ion systems with the same initial concentration as used in the binary systems, illustrating how strong the adsorption changes upon the addition of the competitive ion. These changes are most pronounced for the ion with the lowest affinity (Ca). As the Ca^{2+} adsorption is small at low pH, correspondingly, the effect on the PO_4 adsorption is relatively low. [Reproduced from Mendez and Hiemstra (2020a) by permission of the American Chemical Society.]

The authors found a substantial enhancement of the Zn adsorption in the presence of PO_4. However, the Zn adsorption could be explained adequately from just the electrostatic interactions of adsorbed Zn^{2+} and PO_4 at the surface of Fh, as long as the Zn^{2+} and PO_4 concentrations were moderate. Only at high PO_4 concentrations, ternary complexes formation did become evident, in agreement with differential ATR-FTIR spectroscopy (Liu et al. 2016). Eynde et al. (2022) reported the formation of a monodentate complex (≡FeOH–ZnOH) at a high Zn loading. At low Zn loading, the formation of bidentate double corner complexes was suggested by CD modeling, which may hydrolyze (≡$(FeOH_b)_2$–Zn(OH)).

In the case of Cu(II), bidentate complex formation was found in agreement with EXAFS data (Tiberg et al. 2013). The surface species ≡$(FeOH_a)_2Cu$, formed by edge-sharing, has a planar structure that easily hydrolyses forming ≡$(FeOH_a)_2CuOH$. The hydrolysis is due to the shortening of the Cu–O bonds resulting from the Jahn–Teller distortion. Modeling of the

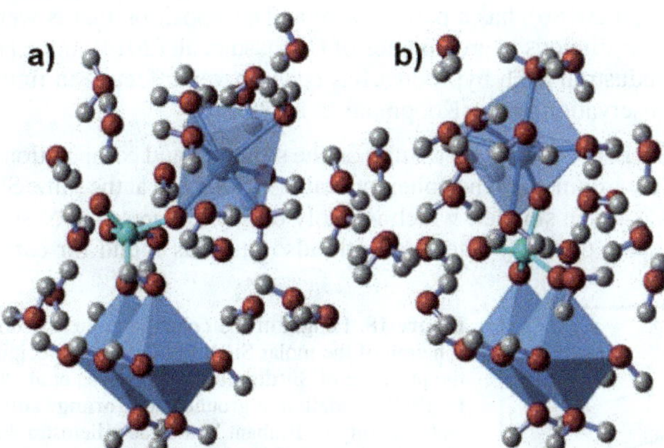

Figure 17. MO/DFT optimized structure of hydrated Fe (hydr)oxide (**both blue octahedra**) cluster representing ternary PO_4–Ca complexes given with respectively a ball-stick representation (PO_4) and an octahedron (Ca^{2+}). The PO_4 forms a single corner **(a)** or double corner **(b)** bond with the Fe polyhedra. For the Ca–PO_4 interaction, a single corner interaction is assumed. The CD values derived for the above complexes were used in the modeling. [Reproduced from Mendez and Hiemstra (2020a) by permission of the American Chemical Society.]

$Cu–PO_4$ interaction also revealed the formation of a monodentate surface complex. No ternary $Cu–PO_4$ complex formation was found with modeling (Van Eynde et al. 2022). This contrasts with the CD modeling of Tiberg et al. (2013), in which such a species was proposed although their EXAFS data does not provide convincing evidence for the existence of such a $Cu–PO_4$ species (Van Eynde et al. 2022). One of the reasons for the different modeling results may be the absence of ≡FeOH sites for edge-sharing in the model used by Tiberg et al. (2013), potentially in line with the finding that the adsorption and interaction of UO_2,CO_3. $As(OH)_3$ etc. at the Fh surface cannot be understood if one type of ≡FeOH sites is assumed rather than distinguishing sites for edge (≡$FeOH_a$) and binuclear double corner-sharing (≡$FeOH_b$) (Hiemstra et al. 2009)

Ion adsorption and co-precipitation

Silicious ferrihydrites. In both nature and industrial settings, Fh may form in the presence of omnipresent ions. Such processes can be studied in synthetic systems (Cismasu et al. 2014; Senn et al. 2015; Ahmad and van Genuchten 2024). An example is the coprecipitation of Fe(III) in the presence of H_4SiO_4 (Cismasu et al. 2014). The co-precipitates formed have been characterized by high-energy X-ray scattering (HEXS). The structural changes of Fh due to the presence of Si can be well-described with the surface depletion model, as illustrated in Figure 2b.

The length of the coherent scattering domain (CSD) decreases with the increase of the applied Si/Fe ratio (Fig. 18). At higher levels of Si, the particles formed become notably smaller. Consequently, the specific surface area of the core increases substantially, which leads to a higher binding capacity for H_4SiO_4 and elevated Si/Fe ratios (Fig. 18). This can be understood from a decrease of the surface Gibbs free energy due to the adsorption of ions (Vayssieres 2005; Hiemstra 2015), lowering the energy barrier for nucleation and suppressing the rate of particle growth due to surface complex formation of the foreign ions. At the highest additions of H_4SiO_4, the Fh particles formed are approximately 1.5 nm in diameter and contain, on average, n_{Fe} ~30 Fe (Eqn. 4), of which the vast majority is directly exposed to the surface. The corresponding specific surface area of these very small particles of the Fh family (Fig. 1) is about ~1300 $m^2 \cdot g^{-1}$. This SSA is similar to the SSA (~1100 $m^2 \cdot g^{-1}$) found for particles immediately after Fh formation by precipitation in the growth experiment of Fig. 5. The increased SSA at co-precipitation stimulates the adsorption of competitive oxyanions (Senn et al. 2015). A decrease in particle size is also found at the co-precipitation of Fe and As (Ahmad and van Genuchten 2024).

A silicious Fh, formed upon Fe(II) oxidation of groundwater for drinking water production (Hiemstra 2018b), has a particle size and composition that is well in line with the data collected for the silicious coprecipitates of Cismasu et al. (2014) in the lab (orange square in Fig. 18). This industrial SiFh byproduct has been proven effective in immobilizing PO_4 in soils in natural conservation areas (Koopmans et al. 2020).

Cismasu et al. (2011) conducted a study on the structure and composition of coprecipitates formed in natural environments. The coherent scattering domain at the same Si/Fe ratio as in the synthetic systems is much smaller, which is likely due to the presence of significant amounts of OM in the samples that may interfere with adsorption as found for carbon-rich biogenic

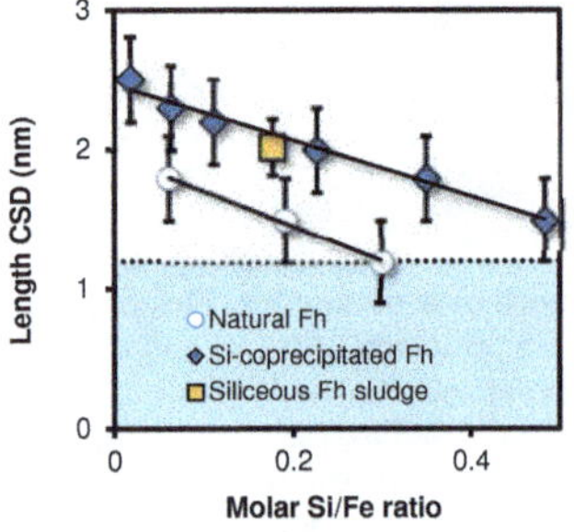

Figure 18. Length of the coherent scattering domain (CSD) as a function of the molar Si/Fe ratio of Fh coprecipitated in the lab in the presence of Si (**diamonds**) (Cismasu et al. 2014), or produced by Fe(II) oxidation in groundwater (**orange square**) in a drinking water facility of Brabant Water Inc (Hiemstra 2018b). For natural Fh (**spheres**), the CSD size is smaller (Cismasu et al. 2011), likely due to the adsorption of organic matter. The **dotted horizontal line** gives the minimum equivalent size of Fh particles corresponding to critical nuclei with around ~13 Fe ions (Hiemstra and Zhao 2016; Weatherill et al. 2016). [Reproduced from Hiemstra (2018b) by permission of Elsevier.]

Fh with coherent scattering domain sizes of 1.2–1.8 nm (Whitaker et al. 2021), similarly as shown in Figure 18. Even in the absence of Si, the natural particles stay smaller in size than in synthetic systems, as follows from the intercepts of the lines of Figure 18 at Si/Fe = 0, showing $d \sim 2 \pm 0.4$ nm. A similar size of primary Fh particles ($d \sim 1.6$ nm) is measured at oxidative coprecipitation of Fe(II) humic acid (Guénet et al. 2017), using small angle X rays scattering. The presence of organic molecules inhibits the growth of small Fh clusters (ThomasArrigo et al. 2019; Vantelon et al. 2019; Whitaker et al. 2021).

For the data of Figure 18, the size seen for the smallest natural Fh particles is close to the minimum size of Fh particles, i.e., the size of the Keggin structure with $n = 13$. Using Equations (1–3), one can derive for this polymer a corresponding equivalent spherical size of $d \sim 1.25$ nm and specific surface area A_{Fh} of ~1750 $m^2 \cdot g^{-1}$ or ~2.3 10^5 $m^2 \cdot mol^{-1}$ Fe. Similar ultra-small moieties have also been identified at coprecipitation of Fe(III) at very acid conditions (Weatherill et al. 2016).

Coprecipitates of Zn and Fe. Additional examples of co-precipitation include the formation of Fh in the presence of U(VI) ions (Ulrich et al. 2006) and Zn(II) ions (Waychunas et al. 2002), both studied with EXAFS. In the case of Zn(II), the authors conducted a comparative study on the Zn adsorption at traditionally formed Fh, revealing similar behavior. In solution, Zn^{2+} ions are hexa-coordinated (CN = 6), while adsorbed Zn(II) ions have a tetrahedral coordination sphere (CN = 4). This change in coordination number stimulates the hydrolysis of the adsorbed ion, forming, e.g., ≡FeOH–ZnOH. It increases the bond valence ($\nu = z/\mathrm{CN}$), stimulating the proton release from one of the coordinating OH_2 ligands.

At low Zn loading, a binuclear bidentate complex is formed (species A in the upper panel of Fig. 19) with the number of metal ions in the second shell of Zn denoted as N = 2 in Figure 19a. The spectroscopic data (Waychunas et al. 2002) given as squares are quantitatively in agreement with the CD-MUSIC modeling (spheres) of Zn adsorption data (Kinniburgh and Jackson 1982; Van Eynde et al. 2022).

As the Zn loading increases, the number of metal ions in the second shell changes towards $N = 1$ according to EXAFS (Waychunas et al. 2002), which can be attributed to the formation of a monodentate complex (species B in the upper panel of Fig. 19). This observation is well-supported by CD modeling. Under these conditions, the CD modeling of the Zn^{2+} adsorption resolved the formation of a hydrolyzed monodentate complex ≡FeOH–ZnOH.

At extremely high Zn loading, the CD modeling suggests that an additional Zn^{2+} ion is bound to the adsorbed monomeric monodentate Zn complex (≡FeOH–ZnOH), forming a dimeric ≡FeOH–$(ZnOH)_2Zn(OH)_2$ (species C in Fig. 19). This process is the onset of the polymerization of Zn at the surface (Fig. 19b) which leads to an increase in the mean metal coordination number (Fig. 19a). Ultimately, this process culminates in the formation of a $Zn(OH)_2$ precipitate in which the Zn ions may have a coordination number of N=4. Figure 19b illustrates the relative surface speciation as a function of the Zn/Fe ratio used in the adsorption experiments at pH 6.5.

APPLICATION OF SURFACE COMPLEXATION MODELS TO SOIL

SCMs can only serve as reliable tools for describing the fate of ions in natural systems if information is available about at least 1) the reactive surface areas of the soil minerals and b) the interaction of these surfaces with ubiquitous organic matter. Both aspects will be comprehensively discussed in this section.

Reactive surface area

For determining the reactive surface area (RSA) of the natural oxide fraction in soils, it is a widespread practice to measure the Fe and Al concentrations in certain soil extracts.

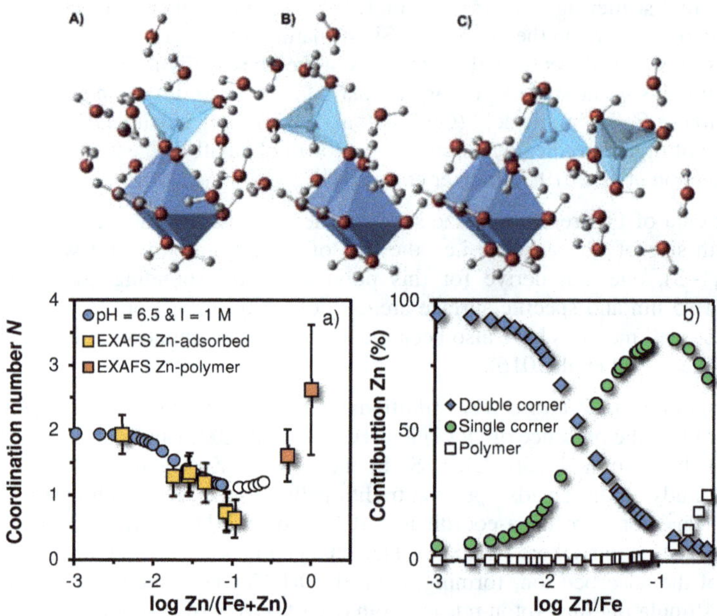

Figure 19. Upper panel: Representation of the MO/DFT-optimized surface species formed in Zn adsorption experiments, involving a binuclear bidentate complex (A), a monodentate complex (B), and a Zn polymer (C). **Lower panel: a)** Number of ions (*N*) in the second shell of Zn (**colored squares**) as determined by EXAFS (Waychunas et al. 2002) and by CD modeling (**spheres**) of the Zn- adsorption experiment of Kinniburgh and Jackson (1982) without (**blue**) and with (**white**) the presence of Zn polymers at pH 6.5 (Van Eynde et al. 2022). **b)** Corresponding surface speciation of Zn at pH 6.5 in 1 M $NaNO_3$. [Reproduced from Van Eynde et al. (2022) by permission of Elsevier.]

The concentrations are then converted to a Fe and Al (hydr)oxide content using an assumed molar mass, usually 89 g·mol^{-1} Fe and 78 g·mol^{-1} Al. Specifically, in the case of an acid ammonium oxalate extraction (AO), which preferentially extracts the fraction of nanocrystalline Fe and Al minerals, a standard assumption involves the specific surface area (SSA) of 600 m^2·g^{-1} for the (hydr)oxides. For crystalline Fe and Al (hydr)oxides, the SSA may be on average about 10 times lower. The above procedure is full of assumptions, which motivates the search for alternatives.

Principle surface probing. As described in an earlier section, the SSA of ferrihydrite can be very well assessed by using a probe ion. In the context of soil, orthophosphate emerges as a suitable candidate because the ion is omnipresent in the natural environment and exhibits a high affinity for Fe and Al (hydr)oxides. Hiemstra et al. proposed the use of a series of equilibrium extraction with 0.5 M $NaHCO_3$ at pH 8.5 to determine a part of the desorption isotherm of PO_4 by varying the soil-to-solution ratio SSR, ρ_{SSR} (Hiemstra et al. 2010a). The measured equilibrium concentrations of PO_4 (c_p) can be translated with the CD model into a phosphate surface loading (Γ_p), expressed in mol·m^{-2}, using in the modeling the final pH of the extract after equilibration.

The choice of using 0.5 M $NaHCO_3$ simplifies the composition of the equilibrium solution as it leads to the precipitation of Ca^{2+} and Mg^{2+} ions under this condition. This simplification allows for the interpretation of the data by considering only CO_3–PO_4 competition. Natural organic matter (NOM) may also compete for PO_4, although the imposed high value for the pH (8.5) will suppress it. To further reduce it, activated carbon, precleaned with ammonium oxalate, is added to absorb significant quantities of NOM. Factors that release NOM are the removal of the bridging Ca and Mg ion, the increase of the charge of the NOM molecules due to a high pH, and the desorption of the functional groups.

Depending on the soil-to-solution ratio (ρ_{SSR}, kg/L) in the extraction series, the total amount of reversibly bound PO_4 (R_{ev} in mol·kg^{-1} soil) is redistributed between the particle surface and the solution term of the mass balance:

$$R_{ev} = A_{ox}\,\Gamma_p + c_p\,\rho_{SSR}^{-1} \tag{11}$$

in which A_{ox} is the reactive oxide surface area (m^2·kg^{-1} soil) and Γ_p is the phosphate surface loading (mol·m^{-2}) calculated with SCM for a measured concentration of phosphate (c_p). R_{ev} is a fixed value for a sampled soil imposed by the field conditions. Applying Equation (11) for a given soil to a set of c_p data collected for a range of SSR values, the amount of reversibly bound PO_4, R_{ev}, and the reactive oxide surface area, A_{ox} (m^2·kg^{-1} soil), can be found by fitting.

At the time of the initial development of the method, goethite was used as the reference oxide, since for this material the CO_3–PO_4 had been measured (Rahnemaie et al. 2007). For a representative series of agricultural topsoils, the interpretation pointed out that the reactive oxide fraction largely consists of oxide nanoparticles, given the notably high specific surface area. This advocates using nanocrystalline ferrihydrite as reference oxide material rather than well-crystalline goethite. Therefore, the CO_3–PO_4 competition was studied and interpreted using the new insights into the surface structure and reactive sites of ferrihydrite (Hiemstra 2013; Hiemstra and Zhao 2016), as outlined elsewhere (Mendez and Hiemstra 2019).

Consistent relationships for soil metal (hydr)oxides. In the approach with Fh as reference nanomaterial, the consistency of the relationships between mean particle size, molar mass, and mass density were adapted to accommodate the differences between Al and Fe nano(hydr) oxides (Mendez et al. 2022). The natural nano-(hydr)oxide fraction was considered as a combination of Al and Fe (hydr) oxides. The molar mass and mass density of the soil (hydr) oxide fraction were found by scaling the molar masses and mass densities of these endmembers to the concentration of Fe_{AO} and Al_{AO} (mol·kg^{-1}) in the ammonium oxalate extracts as follows:

$$M_{soil(hydr)oxide} = \frac{Fe_{AO}}{Fe_{AO} + Al_{AO}}\,M_{nano,Fh} + \frac{Al_{AO}}{Fe_{AO} + Al_{AO}}\,M_{nano,Al(OH)_3} \tag{12}$$

$$\rho_{soil(hydr)oxide} = \frac{Fe_{AO}\,M_{nano,\,Fh} + Fe_{AO}\,M_{nano,\,Al(OH)_3}}{Fe_{AO}M_{nano,\,Fh}\,/\,\rho_{nano,\,Fh} + Al_{AO}\,M_{nano,\,Al(OH)_3}\,/\,\rho_{nano,\,Al(OH)_3}} \tag{13}$$

$M_{nano,Fh}$ and $M_{nano,Al(OH)3}$ are the molar masses of both endmembers (g·mol^{-1} metal ion), and $\rho_{nano,\,Fh}$ and $\rho_{nano,\,Al(OH)3}$ (g·m^{-3}) are the corresponding mass densities that can be calculated using Equations (1–3) and the information in Table 4.

Reference oxide in soil systems. Interpretation of probe ion data demonstrated a significant change in the amount of reversibly bound PO_4 (R_{ev}) if Fh rather than goethite was used as the reference oxide. This can be attributed to the stronger competition of carbonate in Fh systems (Figs. 14a,b). For Fh, this leads to lower PO_4 surface loading (Γ_{PO4}). It decreases the high-affinity character and shape of the PO_4 adsorption isotherm. This affects the fitted amount of reversibly bound PO_4 (R_{ev}). The improvement of the prediction of the amount of reversibly PO_4 is illustrated in Figure 20 for two sets of soils with contrasting properties. The data in Figure 20 show that the natural hydroxide fraction is much better represented by ferrihydrite than by goethite as reference material. The AO-extractable fraction of the natural metal (hydr)oxides behave similarly as the Fh nanoparticles do. The findings suggest that the contribution of the crystalline metal (hydr)oxides to the RSA of soils is relatively small, although these crystalline metal (hydr)oxides dominate on a mass basis, particularly in tropical soils of the sub-Saharan regions that are classified as Acrisols and Ferrosols.

Table 4. Parameters of the mineral core of ferrihydrite (Fh) and Al (hydr)oxide nanoparticles that are used in the mathematical expressions for calculating the size-dependent particle properties (Eqns. 1–3), being the molar mass of the core M_{core}, the molar oxygen metal ions ratio n_o, the volume of the unit cell expressed per mol oxygen (V_O), the excess water content due to the presence of surface groups, N_{H_2O}, and the mass density of the core, ρ_{core} (Mendez et al. 2022).

	Fh ($FeO_{1.4}OH_{0.2}{\cdot}nH_2O$)	**Nano-gibbsite** ($Al(OH)_3{\cdot}nH_2O$)
M_{core} ($g{\cdot}mol^{-1}$ metal ion)	81.65	78.00
n_O (mol $O{\cdot}mol^{-1}$ metal ion)	1.6	3.0
V_O ($m^3{\cdot}mol^{-1}$ O)	$1.07 \times 10^{-5*}$	$1.07 \times 10^{-5*}$
N_{H_2O} (mol $H_2O{\cdot}m^{-2}$)	12.6×10^{-6}	6.3×10^{-6}
ρ_{core} ($g{\cdot}m^{-3}$)	4.77×10^6	2.42×10^6

Notes: * $V_O = M_{core} / (n_O\ \rho_{core})$

Scaling the RSA to the AO-extractable fraction reveals the SSA of the nano metal (hydr) oxide fraction, which can be translated into a corresponding mean particle size. For both soil series, a similar size range of 0.5–5 nm is found. The individual soils vary considerably in SSA ranging from 350–1700 $m^2{\cdot}g^{-1}$. This challenges the traditional assumption of 600 $m^2{\cdot}g^{-1}$ for the AO-extractable metal (hydr)oxides.

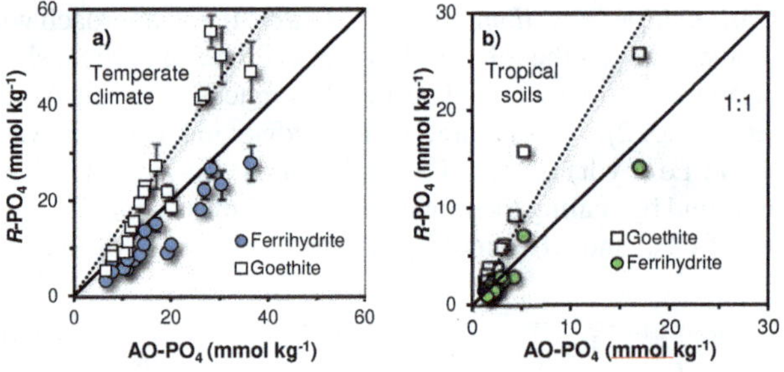

Figure 20. a) Amount of acid ammonium oxalate-extractable PO_4 (AO-PO_4) in relation to the amount reversibly bound PO_4 (R-PO_4) for a series of agricultural top soils of the Copernicus soil series (temperate climate soils), calculated with the CD model using either goethite (**squares**) or ferrihydrite (**circles**) as reference oxide (Mendez et al. 2020). When choosing goethite as reference (hydr)oxide, the fitted value of R_{ev} is too high in comparison to the amount of PO_4 extracted with AO, which can be attributed to the stronger competition of CO_3 with PO_4 adsorbed to ferrihydrite, decreasing the surface loading and changing the shape of the adsorption isotherm (Figs. 14a,b). In the case of the use of ferrihydrite, the data are closer to the 1:1 line. **b)** The same relation, but for weathered tropical soils of the sub-Saharan region with a relatively large fraction of crystalline metal (hydr) oxides (Mendez et al. 2022). In these soils, the metal (hydr)oxide fraction is dominated by Al (hydr)oxides while in the others (a), Fe (hydr)oxides dominate. [Reproduced (adapted) from Mendez et al. (2020, 2022) by permission of Elsevier and the American Chemical Society.]

Reference oxide for specific soil types. The probe ion method is currently implemented to evaluate the RSA of tropical volcanic soils having substantial amounts of variable charge nanomaterials of Al and Fe. Preliminary, Fh is used as reference oxide in CD modeling. With that assumption, the variation of the PO_4 binding capacity in soil sequences, related to altitude, moisture regime, and soil profile depth, can be effectively addressed. Combined with selective dissolution extractions of Fe and Al, using ammonium oxalate and sodium pyrophosphate, it offers insights into the factors determining the variation in the adsorption capacity of PO_4 of soils as well as the solubility of PO_4 in routine soil chemical analysis (e.g., P-Olsen) for assessing the potential availability of P in agricultural systems. These applications show that using Fh as reference oxide can readily answer practical questions related to soil chemical processes in these soil systems.

The RSA values have also been used to scale a set of collected PO_4 adsorption isotherms to the surface area (μmol/m^2) in these volcanic soils. Preliminary results suggest a much higher adsorption density in a subset of soils with predominantly nano Al-silicates and (hydr) oxides. This may be due to the much larger site density of nano-gibbsite ($Al(OH)_3 \cdot nH_2O$) being 10.1 nm^{-2} rather than 5.8 nm^{-2} for Fh ($FeO_{1.4}(OH)_{0.2} \cdot nH_2O$) and nano-akdalaite ($AlO_{1.4}(OH)_{0.2} \cdot nH_20$) (Fig. 7). In soils, it leads to about a 1.5 ± 0.2 times higher PO_4 adsorption at the same ion affinities. Ignoring this may cause a bias in the calculation of the RSA when using only Fh as reference oxide for these soils. This issue is addressed in ongoing research.

Organo-oxide interaction

Once the RSA of a soil sample is established, it becomes important to gain insight into the relation between the presence of natural oxide nanoparticles and natural organic matter (NOM), given that both materials strongly interact and are associated in soils (Mikutta et al. 2006). When relating the RSA to the amount of soil organic carbon (SOC), a notably good correlation is found (Fig. 21a), particularly for the sandy soils (green spheres). For the clay soils (blue squares), a similar relation is found but the SOC content tends to be higher (Fig. 21a), suggesting an association of SOC with clay minerals. However, introducing the SSA of the soil metal (hydr) oxide fraction in addition to the RSA of the soils, alters this view.

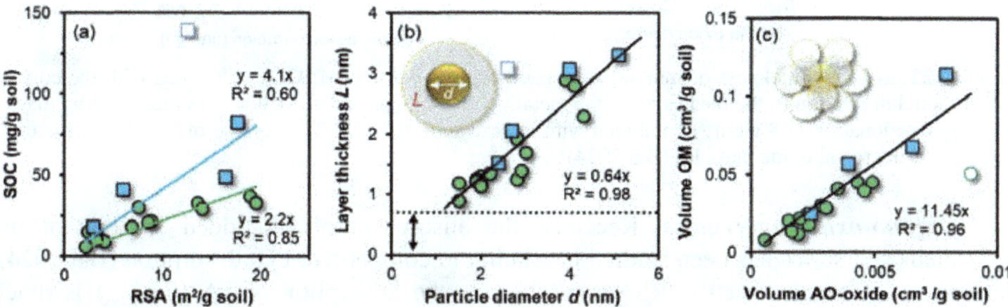

Figure 21. a) Relationship between soil organic carbon (SOC) and the reactive surface area (RSA) for the Copernicus series for soils with a clay content <20% (**circles**) and ≥20% (**squares**). One soil is peaty and therefore given as an **open square**. **b)** Interpretation of SOC content with a core–shell model (**inset**) in which the metal (hydr)oxide core with diameter *d* is surrounded by a SOC layer with a thickness *L*, showing a single relation of all soils. **c)** The association between NOM and oxides also leads to a strong relation between the volume of both materials. The volume of OM is much larger which can be interpreted as the embedding of the oxide nanoparticles in a collection of humic nanoparticles (HNP), forming organo-oxide aggregates (Mendez et al. 2020). [Reproduced from Mendez et al. (2020) by permission of the American Chemical Society.]

In Figure 21b, the amount of OM is represented by a layer around the oxide particles, using a core–shell representation (inset in Fig. 21b). The layer thickness is very well correlated with the oxide particle size, irrespective of the clay content of the soils. It underlines that the mean particle size of the soil oxide fraction is a major difference between the sand and clay soils.

The strong relation between NOM layer thickness and particle size can be interpreted structurally as a collection of oxide and NOM particles of similar mean size. When expressing the amount of NOM and metal (hydr)oxide in a volume, a good correlation is found. The dominance of the volume of OM suggests that oxide nanoparticles are embedded in a collection of humic nanoparticles (HNP) forming organo-oxide aggregates.

Mineral-associated organic carbon. In Figure 21, the total amount of SOC was used in the evaluations although part of SOC is not associated with the mineral fraction but is present as particulate organic matter. The mineral-associated organic carbon (MAOC) fraction can be assessed with mass density fractions combined with intense sonication (Bai 2024). Such MAOC data, expressed as a NOM layer thickness, are depicted in Figure 22, showing similar linear relationships as in Figures 21b,c. Since less OC is involved, the slopes of the lines are lower. Consequently, the calculated volume ratio of organic matter and nano-oxide decreases from approximately 12 (Fig. 21c) to 7 (Fig. 22.b).

The mean layer thickness of MAOC (L) is about half of the mean oxide particle diameter (Fig. 22a). Interpretation of the core–shell results in terms of a collection of nanoparticles, the humic and metal (hydr)oxide particles are approximately equal in size and bound together by particle sharing. Such a structural organization can contribute to the stability of the MAOC fraction under ambient soil conditions.

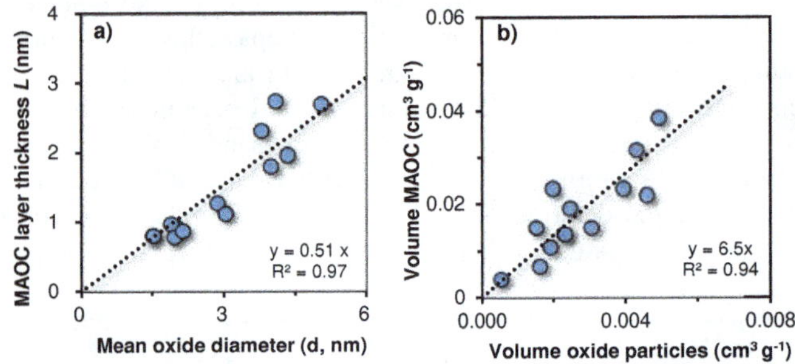

Figure 22. a) Layer thickness of mineral-associated organic carbon (MAOC), calculated with the core–shell model in relation to the mean size of the metal (hydr)oxide particles. MAOC was measured by mass density fractionation (1.8 g·cm^{-3}) combined with intense sonication. **b)** The volume of MAOC versus the volume of the metal oxide particles (Bai 2024).

Organo-oxide aggregates. Recently, the adsorption of FA added to soils of the Copernicus soil series has been studied in relation to competitive PO_4 desorption (Bai 2024). Data analysis reveals that the RSA available for the adsorption of FA (RSA_{FA}) is much smaller than the RSA for the oxyanions. This can be explained by assuming the formation of organo-oxide aggregates. The added FA only interacts with the external surfaces of such aggregates, while PO_4 also may interact with the internal oxide surfaces. The adsorption of added FA and the competitive release of PO_4 can be understood via SCMs as discussed later. The limited reactivity of added FA is due to the relatively large size of the FA molecules, which hampers the penetration of these (added) molecules into the porous structure of organo-oxide aggregates. For the soil series studied, the size of the nano-aggregates is typically in the range of >50–300 nm, as we will discuss later (Fig. 29).

Applications. The probe ion approach for determining the soil RSA proves to be a valuable tool in diverse soil-related studies. One such application involves the addition of Fe-oxide sludge to soils aiming to reduce P-availability (Koopmans et al. 2020). Additionally, this approach has been instrumental in studying the enhancement of P-availability facilitated by earthworms, with observed changes in the interactions between the soil oxide particles and soil organic carbon (Vos et al. 2022). Another example is from the PhD thesis of Bai (2023) where changes in the soil oxide fraction and speciation of organic matter in a long-term field experiment spanning 45 years of P fertilization are described (Fig. 23).

In the results presented in Figure 23 (Bai 2023), the RSAs of soils were tracked with the PO_4 ion probing method to investigate the impact of long-term P fertilization. The results of the long-term P application at different P fertilization rates revealed a threefold change of

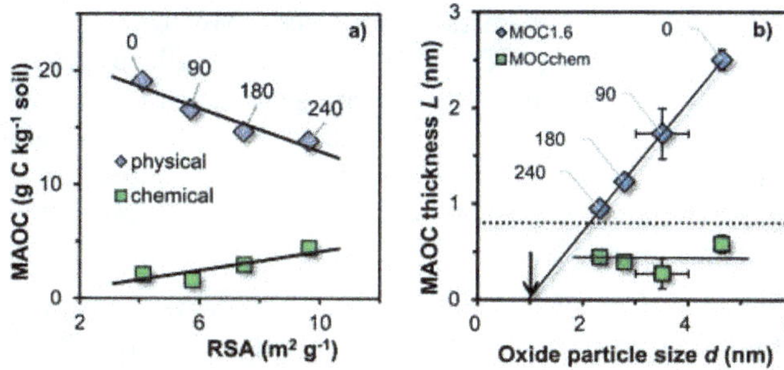

Figure 23. a) Relationship between mineral-associated organic carbon (MAOC) and the reactive surface area (RSA) for soils of a long-term P-fertilization experiment with annual P applications rates of 0–240 kg P_2O_5, yet unpublished work of Bai (2023). MAOC was measured after applying NaClO oxidation or a mass density fractionation using a mass density of 1.6 g·cm^{-3}. The latter mass density was chosen because the overall mass density of the organo-mineral fraction is in these soils lower than usual, due to the dominance of Al in the oxide fraction. **b)** Corresponding thickness of the MAOC layer around the oxide particles of different sizes, according to a core–shell model interpretation (Bai 2023). The **arrow** points to the minimum oxide particle size. The **dotted horizontal line** indicates the thickness of the compact part of the EDL. NaClO-oxidizable OC can be accommodated within the Stern layer region (**green squares**). Depending on the rate of P-fertilization, a large fraction of MAOC is outside this region as follows from the position of the **blue diamonds**. Due to long-term fertilization, the oxide particles become smaller, and the OM is reallocated in thinner layers around the particles. At the highest rate of long-term P fertilization, a significant part is present in the compact part of the EDL. However, the total amount of $MAOC_{phys}$ on a soil mass basis also decreases (Fig. 23a).

the reactive surface area of the soils, resulting in values between 3 and 10 m^2·g^{-1} (Fig. 23a). This increase in the surface area corresponds to a decrease in the mean particle size from approximately 4 nm to about 1.5 nm. Bai's work showed that long-term fertilizer application changes the oxide particles, in line with independently collected data (Mendez and Hiemstra 2020c; Mendez et al. 2022) for two contrasting soil series (Figs. 20a,b) that differ by a factor of ~2 in reversibly bound PO_4 (R_{ev}). The feedback of the PO_4 application on the RSA and SSA has also been reported by Vos et al. (2022) who compared two soil samples of the same origin but with and without long-term P fertilization.

The work of Bai (2023) also revealed an alteration in the status of NOM, specifically mineral-associated organic carbon (MAOC). The characterization of MAOC involved both chemical (NaClO oxidation) and physical (mass density fractionation) methods. The quantity of OM resistant to NaClO oxidation increased linearly with RSA (Fig. 23a), suggesting that this part of OM is protected through adsorption to the oxide surfaces.

The data for MAOC were further interpreted using a core–shell model approach (Fig. 23b). The analysis demonstrated a robust linear relation between the layer thickness with OM present in the heavy fraction (mineral-associated OM) and the mean oxide particle size. In contrast, for the non-oxidizable OM fraction, a near-constant layer thickness was found. The amount of NOM of this chemically stable fraction can be accommodated in the Stern layers around the particles (Fig. 23b). The study emphasizes the dynamic nature of RSA and its intricate association with NOM in response to long-term P fertilization.

NOM-CD model

The above evaluation highlights that in soils the metal (hydr)oxide nanoparticles are embedded in a collection of organic matter nanoparticles, forming nano-aggregates. The close interdependence between both types of materials implies that the reactive metal (hydr)oxide fraction cannot be considered in isolation. The presence of organic matter must also be considered. Understanding the ion adsorption behavior of the organo-oxide fraction needs comprehensive modeling of the competitive interaction of NOM with Fe (hydr)oxides. This encouraged the development of the NOM-CD model (Hiemstra et al. 2010b, 2013).

For humic nanoparticles, the interaction with metal (hydr)oxides can be described with the ligand and charge distribution (LCD) model (Weng et al. 2006b, 2007; Xu et al. 2022). In principle, the model has the capability to relate an aqueous concentration to a surface loading. However, when it comes to soils, it is difficult to relate the adsorbed fraction and type of interacting organic molecules to a solution concentration, because HA becomes insoluble and aggregated in soil systems dominated by divalent ions. Therefore, applying this specific aspect of the LCD model poses difficulties.

To advance in the understanding of the role of NOM in ion adsorption modeling, natural and synthetic systems have been investigated using a virtual component designed to mimic the competitive behavior of functional groups of OM. The first practical attempt by Gustafsson et al. (2006) involved defining a virtual solution component for OM that can bind to the surface through a reaction with an extremely high-affinity constant. Building on this, the method was adapted by defining adsorbed OM as a surface component that redistributes its charge within the interface (Hiemstra et al. 2010b). A subsequent improvement (Hiemstra et al. 2013) involved the introduction of outer- and innersphere complexation to account for spectroscopic information about the binding mechanism (Boily et al. 2000; Johnson et al. 2004; Hanna and Quiles 2011). Additionally, the functional group of the virtual species was allowed to protonate. In that approach, the surface speciation, calculated with the NOM-CD model, was calibrated using data from simulations of the adsorption of FA with the LCD model (Weng et al. 2006b).

Model formulation. In the NOM-CD model (Hiemstra et al. 2013), the defined surface component ($\equiv HNOM^z$) consists of two carboxylic groups of which one reacts with a singly coordinated surface group $\equiv FeOH^{-0.5}$ and redistributes its charge in the interface according to:

$$\equiv FeOH^{-0.5} + \equiv HNOM^z \rightleftharpoons \equiv Fe^{-0.5+z+\Delta z_0}\,NOM^{\Delta z_1+\Delta z_2} + H_2O(l);\ \log K_1 \equiv 0 \qquad (14)$$

In the approach, NOM is represented by a species with two non-protonated $RCOO^-$ groups, implying that the charge of $\equiv HNOM^z$ is equal to $z = -1$. The total surface charge of the reference groups ($\equiv FeOH^{-0.5}$ + $\equiv HNOM^{-1}$) is −1.5 v.u. Upon complex formation (Eqn. 14), this surface charge is neutralized by a shift of positive charge from the 1- and 2-plane to the surface. This leads to $\Delta z_0 = +1.5$ and $\Delta z_1 + \Delta z_2 = -1.5$. In more detail, the distribution of the charge between the 1- and 2-plane is conceptualized based on a molecule picture. In the model, one of the carboxylic groups distributes its charge between the 0-plane and 1-plane while the other carboxylic group distributes its charge between the 1- and 2-plane. As each O-ligand of the $RCOO^-$–$RCOO^-$ groups has a charge of −0.5, one derives $\Delta z_1 = 2 \times -0.5 = -1$ and $\Delta z_2 = -0.5$. In other words, the charge (re)distribution coefficients are $\Delta z_0 = +1.5$, $\Delta z_1 = -1.0$, and $\Delta z_2 = -0.5$ (Table 5). Note that the sum of CD coefficients will be 0 for the formation of the $\equiv FeNOM$ species ($\Sigma \Delta z_i = 0$).

Carboxylic groups of organic molecules may also participate in outersphere complex formation, by switching one of the oxygen ligands of $RCOO^{-1}$ of the above innersphere complex from the 0- to the 1-plane. The outersphere complex formation can be given as:

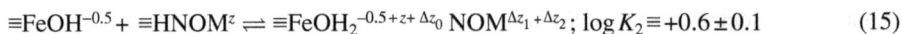

$$\equiv FeOH^{-0.5} + \equiv HNOM^z \rightleftharpoons \equiv FeOH_2{}^{-0.5+z+\Delta z_0}\,NOM^{\Delta z_1+\Delta z_2};\ \log K_2 \equiv +0.6 \pm 0.1 \qquad (15)$$

Compared to innersphere complexation, there is a shift of the common O-ligand of $RCOO^-$ innersphere complex from the surface to the 1-plane. The corresponding shift of charge is −0.5 v.u., leading to $\Delta z_0 = +2.0$, $\Delta z_1 = -1.5$, and $\Delta z_2 = -0.5$ (Table 5). The value of the above formation reaction ($\log K_2$) has been found through calibration to the $RCOO^-$ group speciation calculated with the LCD model (Weng et al. 2006b) as described by Hiemstra et al. (2013). The $\log K$ value is 0.6 ± 0.1 suggesting that intrinsically (no electrostatic contribution), outersphere complexation is preferred over innersphere complexation. However, the overall preference may strongly change due to the electrostatic potentials.

The carboxyl groups of the virtual species may also act at low pH as a proton acceptor. At these conditions, the surface is positively charged which favors the formation of innersphere complexes. The proton is then bound to the RCOO ligand in the second Stern layer. The corresponding reaction can be expressed as follows:

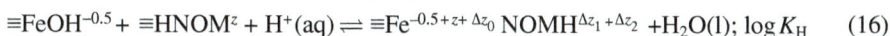

$$\equiv FeOH^{-0.5} + \equiv HNOM^{z} + H^{+}(aq) \rightleftharpoons \equiv Fe^{-0.5+z+\Delta z_0} NOMH^{\Delta z_1 + \Delta z_2} + H_2O(l); \log K_H \quad (16)$$

In the case of an equal distribution of the proton charge over the 1- and 2-plane, one derives the following set of charge distribution coefficients, $\Delta z_0 = +1.5$, $\Delta z_1 = -0.5$, and $\Delta z_2 = 0$ (Table 5).

Table 5. Table defining the three types of NOM species (innersphere IS and outersphere OS) using the surface component ≡HNOM. The surface density of ≡HNOM is $N_{\equiv HNOM}$ = [≡Fe^0NOM] + [≡$FeOH_2^{0.5}$–NOM] + [≡Fe^0NOMH] and is used to define the total amount of organic matter in the compact part of the interface (Extended Stern layer model).

	Surface species	≡$FeOH^{-1/2}$	≡$HNOM^{-1}$	Δz_0	Δz_1	Δz_2	H^+	log *K*
IS	≡$Fe^0NOM^{-1.0,\,-0.5}$	1	1	1.5	−1.0	−0.5	0	≡ 0
OS	≡$FeOH_2^{+0.5}$–$NOM^{-1.5,\,-0.5}$	1	1	2.0	−1.5	−0.5	0	0.6±0.1*
IS+H	≡$Fe^0NOMH^{-0.5,\,-0.0}$	1	1	1.5	−0.5	−0.0	1	3.4±0.5**

Notes:
* Derived by fitting data generated by the LCD model (Hiemstra et al. 2010b)
**Derived by fitting the competitive adsorption of PO_4 in HA-goethite systems (Hiemstra et al. 2013).

Application of the NOM-CD model approach

NOM-CD modeling in oxide systems. The NOM-CD model has been successfully applied to describe the competitive interaction of various natural and pyrogenic humic nanoparticles with oxyanions in goethite systems (Hiemstra et al. 2013; Otero-Fariña et al. 2017; Deng et al. 2019, 2020). In the presence of PO_4, the measured equilibrium concentrations become more than 10–100 times higher (Figs. 24a–c). By comparing the phosphate concentration in solution in the absence and presence of these humic nanoparticles, the total concentration of the NOM surface component ($N_{\equiv HMOM}$) can be fitted. If the amount of the component ($N_{\equiv HNOM}$) is linearly related to the measured fraction (f_{HNP}) of adsorbed HNP, one may derive its maximum value (≡$FeNOM_T$) at 100% adsorption with [$N_{\equiv HNOM}$] = f_{HNP} × [≡$FeNOM_T$]. In that case, [≡$FeNOM_T$] and $\log K_H$ are the two adjustable parameters in the experiments of

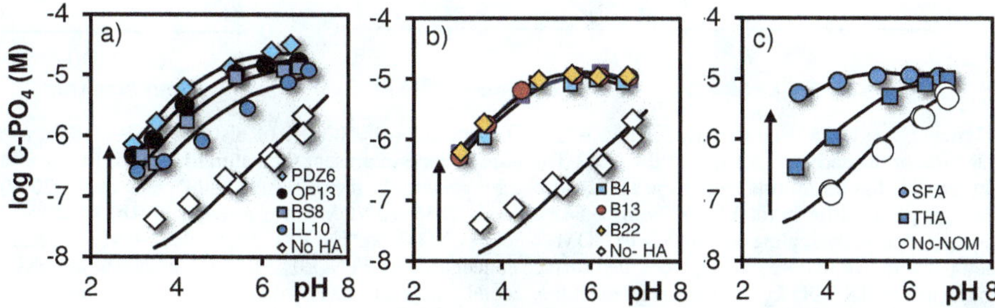

Figure 24. a) Logarithmic concentration of dissolved PO_4 as a function of pH for PO_4–goethite systems with or without various natural HA in 0. M $NaNO_3$ (1.02 g goethite·L^{-1}, A_{BET} = 85 $m^2 \cdot g^{-1}$, PO_4-ini = 0.122 mM, HA-ini = 25 mg C·L^{-1}). Data from Hiemstra et al. (2013). **b)** Same as a), but for pyrogenic humic acids. **c)** Logarithmic concentration of dissolved PO_4 as a function of pH for PO_4–goethite systems with and without FA or HA in 0.01 M $NaNO_3$ (1.0 g goethite·L^{-1}, A_{BET} = 94 $m^2 \cdot g^{-1}$, PO_4-ini = 0.15 mM, HA-ini = 50 mg·L^{-1}). Data of Weng et al. (2008). The **arrows** indicate the increase of the PO_4 concentration due to the competition of adsorbed humic nanoparticles. [Reprinted (adapted) from Hiemstra et al. (2013) with permission. Copyright 2013 American Chemical Society.]

Figures 24a–c. For the latter, an effective $\log K_H$ of 3.4±0.5 was found for the humic acids (Hiemstra et al. 2013). Deng et al. (2019) reported $\log K_H$ values of 2.8±0.3 for their HAs, and Amini et al. (2020) used a value of $\log K_H = 2$.

Natural HAs (Fig. 24a) display a significant variability in their ability to compete with PO_4, a diversity attributed to differences in the site density of carboxylic groups. Pyrogenic humic acids exhibit the strongest competition (Fig. 24b). Fulvic acid (FA) is also a very good competitor, particularly at low pH (Fig. 24c). The latter competitiveness stems from a combination of a high site density of the ROO(H) groups and a low protonation constant ($\log K_H = 2.3$). Consequently, FA particles carry a high negative charge, enabling strong electrostatic interactions with adsorbed PO_4.

The competitive strength of various HNPs is linearly related to the RCOO(H) group density of the HAs used, as illustrated in Figure 25a with the HA competition data of Hiemstra et al. (2013) and Deng et al. (2019) (Fig. 25a). More RCOO(H) groups per kg lead to more competition, resulting in a higher value for $\equiv FeNOM_T$. In both sets of experiments, the HA system loading differed considerably (0.54±0.04 and 0.35 mg·m^{-2} respectively), which is the major explanation for the difference in $\equiv FeNOM_T$ of the data sets presented in Figure 25a. By scaling $\equiv FeNOM_T$ to the total system loading NOM_T (mg·m^{-2}) most of the variation disappears (Fig. 25b). This scaled density ($\equiv FeNOM_T$ / NOM_T) has the unit mol ≡FeNOM·kg^{-1} NOM, but it can also be expressed in the unit RCOO kg^{-1} because the defined ≡FeNOM species have 2 $RCOO^-$ groups, leading to the numbers at the second *y*-axis of Figure 25b. The mean slope of the line in that unit is 1.0±0.1, which implies that the chosen carboxyl density for the ≡FeNOM species (2 $RCOO^-$) very well stands for adsorbed HA. Moreover, it implies that the value fitted for virtual $\equiv FeNOM_T$ can be translated into a physical number.

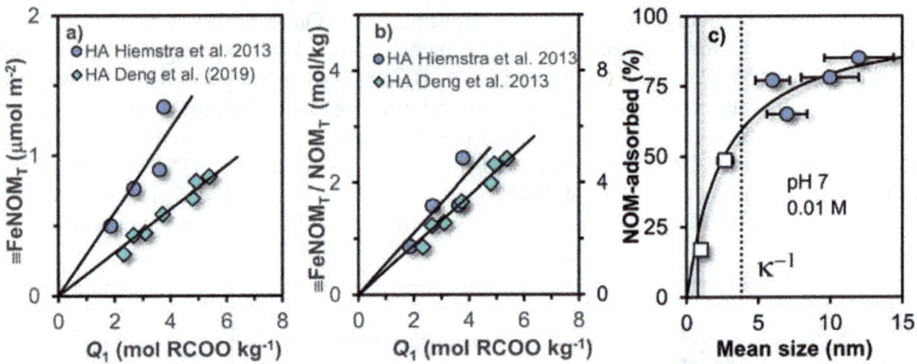

Figure 25. **a)** Calculated surface loading with NOM species ($\equiv FeNOM_T$) of distinct types of HAs as a function of the carboxyl density of the HNP. The **blue spheres** represent the natural HNPs of Figure 24a, the **green-blue diamonds** depict the HA adsorbed by goethite in the experiments of Deng et al. (2019). Both data sets differ in total NOM being 0.54±0.04 and 0.35 mg NOM_T·m^{-2}, respectively. **b)** Most variation disappears by scaling $\equiv FeNOM_T$ to NOM_T (mol $\equiv FeNOM_T$·kg^{-1} NOM). As the ≡FeNOM species are defined with two RCOO groups, the scaled surface concentration $\equiv FeNOM_T$ / NOM_T can also be expressed in the unit mol RCOO·kg^{-1}, which is given at the second *y*-axis. **c)** The HA materials vary in mean particle size, measured with size exclusion chromatography (SEC). The HAs of Hiemstra et al. (**blue spheres**) were not freeze-dried, leading to relatively large mean particle sizes with a low density and open structure. These materials have a higher affinity for binding to goethite, likely because they can more easily change their molecular conformation when bound to the surface. The **dotted line** indicates the Debye length which is a measure for the thickness of the DDL. The **open squares** are for freeze-dried FA and HA acid. The thickness of the Stern layer (~0.8 nm) is shown with the **light-blue rectangle**. This relatively thin Stern layer accommodates most HA when adsorbed, illustrating the substantial change in molecular configuration when HAs are adsorbed, particularly at low pH. [Figure 25c is reprinted (adapted) from Hiemstra et al. (2013) with permission. Copyright 2013 American Chemical Society.]

The HAs of Hiemstra et al. (2013) were prepared without freeze-drying, which may have prevented structural condensation, resulting in a low-dense and open structure of these humic molecules. Indeed, these non-freeze-dried HAs have a large mean particle size (Fig. 25c), as measured with size exclusion chromatography (SEC). The mean size of these HAs (closed spheres) is much larger than the equivalent thickness of the double layer, expressed in the Debye length (κ^{-1}), given in Figure 25c with a dotted line. The open squares in the figure refer to the freeze-dried materials of Weng et al. (2008) evaluated in Figure 24c. The relationship between the mean size and the fraction adsorbed at pH=7 suggests that HAs with an open structure are more effectively bound, which may be due to a greater flexibility of the molecules to change molecular conformation when adsorbed.

HAs data (Fig. 24a) can also be modeled using an LCD approach (Deng et al. 2019). The identified key properties include the site density of carboxylic groups, protonation constants of the carboxylic groups, and particle size or molar mass of NOM, i.e., the same factors as discussed above for the NOM-CD approach. Both models equally well describe the competition with oxyanions (Deng et al. 2019).

NOM-CD modeling for soils. The NOM-CD model has also been applied to soils. In a first approach, a single NOM species was used, combined with goethite as the reference oxide for the natural metal (hydr)oxide fraction (Hiemstra et al. 2010a). In a subsequent approach (Hiemstra et al. 2013), the three-species model, discussed earlier, was implemented in combination with goethite as the reference oxide and was also used by others (Deng et al. 2020). Given that the reactive soil metal (hydr)oxides predominantly consist of small nanoparticles (Figs. 21,22), ferrihydrite was used more recently (Bai 2024) as the reference oxide for calculating the $\equiv FeNOM_T$ loading of soils as a function of pH (Fig. 26). The initial $\equiv FeNOM_T$ loading, presented in Figures 26d–f can be calculated with the NOM-CD model

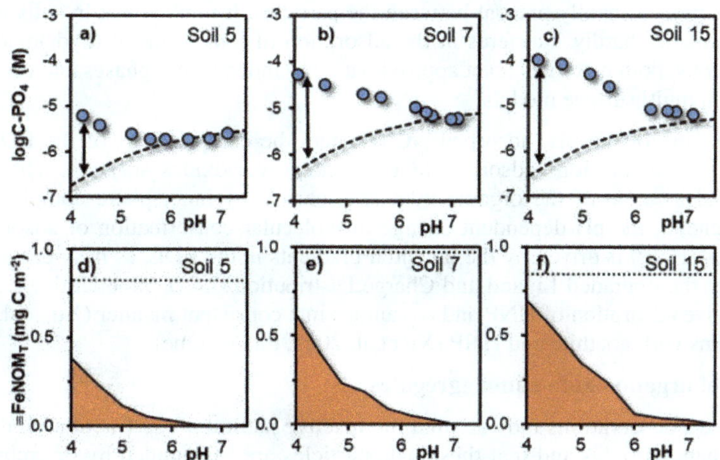

Figure 26. a–c) pH-dependency of the logarithm of the PO_4 concentration in solution (a-c) measured in three soil systems of the Copernicus series with a soil-to-solution ratio of 0.2 kg·L^{-1} and 10 mM $CaCl_2$ (**symbols**) and predicted with the CD model (**dashed lines**) assuming no interaction of NOM (Bai 2024). The model calculations are based on the amount of reversible bound phosphate (R_{ev} mol·g^{-1}) and RSA (m^2·g^{-1}) derived with the $NaHCO_3$ equilibration method. The difference (**arrow**) is due to the competition by adsorbed NOM. **d–f)** Total amount of $\equiv FeNOM_T$ calculated from the difference between the measured and predicted PO_4 concentration, back-calculated to mg C·m^{-2} (Bai 2024). See text. **Dotted lines** give the physical maximum of the NOM surface loading (see text), assuming a Stern layer thickness of 0.8 nm (Hiemstra and Van Riemsdijk 2006). This physical maximum depends on the oxide particle size. [Reproduced from Bai (2024) by permission of Elsevier.]

using the difference between the PO_4 concentration (Figs 26a–c) in the absence (dashed lines) and the presence of competition of adsorbed NOM (blue spheres).

In the NOM-CD model, the ≡FeNOM loading is expressed in $mol \cdot m^{-2}$, but can be translated to a mass in $g \cdot m^{-2}$, see Fig. 25a,b. The total amount of carbon involved in the ≡FeNOM binding can be derived from NOM_{ads} ($mg\,C \cdot m^{-2}$) = 2 × [≡$FeNOM_T$] ($\mu mol \cdot m^{-2}$) / Q_{RCOOH} ($mol \cdot kg^{-1} C$), where Q_{RCOOH} is the generic site density of the carboxylic groups of NOM (Bai 2024). The resulting values are provided in the lower panels of Figures 26d–f.

In addition to the translation of the molar surface concentration of ≡FeNOM to a surface mass density, one can calculate the maximum amount of NOM (≡NOM_{max}) that can be accommodated in the compact part of the EDL with thickness L around an oxide particle (Bai 2024). In a core–shell model (inset Fig. 21b), the volume of NOM in a layer of equivalent thickness L around a particle with a diameter d can be given as $\Delta V = 1/6\pi\ [(d+2L)^3 - d^3]$, which leads to a corresponding mass by multiplying it with a mass density (ρ_{NOM} = 1250 $kg \cdot m^{-3}$). For soils 5, 7, and 15 with a mean oxide particle size of d = 4.3, 2.9, and 4.0 nm respectively, this results in NOM_{max} = 0.82, 0.98, and 0.85 $mg\,C \cdot m^{-2}$, using L=0.8 nm (Hiemstra and Van Riemsdijk 2006). These size-dependent maxima are shown in Figure 26d–f with dotted horizontal lines.

The actual≡$FeNOM_T$ amounts are lower than the calculated adsorption maxima. These amounts vary with the pH. At the highest pH values, the lack of competition with adsorbed PO_4 (Figs. 26a–c), lowers the calculated values of ≡$FeNOM_T$ (Figs. 26d–f). At low pH, about 25 to 75% of the volume of the Stern layer is used by adsorbed OM. Calculation of the underlying surface speciation shows that the reactive RCOO(H) ligands of the adsorbed NOM increasingly form innersphere complexes and that the associated $RCOO^-$ groups become partly protonated (Bai 2024).

The calculated variation in ≡$FeNOM_T$, given in Figures 26d–f, indicates that the RCOO(H) ligands of adsorbed NOM may switch location in the interface, depending on the pH. At high pH, NOM is predominantly present between the particles. It implies that in soils with a high pH, organic matter hardly interferes in the adsorption of ions to metal (hydr)oxides. Under these conditions, both materials react approximately as independent phases and can be treated in that way in multi surface models.

At lowering of the pH, adsorbed NOM enters the compact part of the EDL, leading to interaction with the ions adsorbed to the metal (hydr)oxides surfaces. The change in the interfacial location of the organic ligands is caused by innersphere complexation (Bai 2024). In general, the pH-dependent change in molecular conformation of adsorbed humic nanoparticles (HNP) is driven by the potential gradients in the EDL, as has been shown very recently with the upgraded Ligand and Charge Distribution (LCD_{cc}) model that can describe the competitive adsorption of HNP and oxyanions in a consistent manner (Xu et al. 2024) for model systems with goethite and HNP (Xu et al. 2025, this volume).

Reactivity of organo-oxide nanoaggregates

The above observations indicate that the reactive natural oxide fraction predominantly consists of nanoparticles and that these nanoparticles are surrounded by or embedded in a collection of humic nanoparticles of comparable size. The intimate association of humic and oxide nanoparticles suggests that these nanoparticles are structurally organized into aggregates. The size of these aggregates can be assessed with a novel approach that involves interpreting measurements of the kinetic release of phosphate under zero-sink conditions in combination with equilibrium data about the surface loading of PO_4, obtained with the probe ion method (R_{ev} / RSA).

Dynamic release of PO_4 at zero-sink conditions. The kinetic release of soil phosphate, measured at zero-sink conditions, is commonly described with a two-pool model (Vanderzee et al. 1987; Lookman et al. 1995) in which total PO_4 is divided into a fast and a slow PO_4 pool. The fast pool readily responds to changes in the soil solution and is thought to be bound by the oxide particles at the exterior of the organo-oxide nanoaggregates. The slow pool contributes to P release through the diffusion of PO_4 from sites present in the interior of the nanoaggregates (Lookman et al. 1995). Upon desorption from the internal oxide surfaces, the phosphate ions must pass a zone in which the anions are excluded by the prevailing electrical field between the oxide particles. Due to the presence of organic matter and adsorbed PO_4, this field is repulsive for anions (Fig. 27). Relatively to the free solution, the concentration in the Donnan phase is suppressed by a typical factor of 10 in systems with 0.01 M $CaCl_2$. This will reduce the PO_4 flux from the interior of the organo-oxide aggregates to the free solution. In soil systems with an excess addition of water (low ionic strength), this reduction of the concentration in the Donnan phase can be a factor of up to 100.

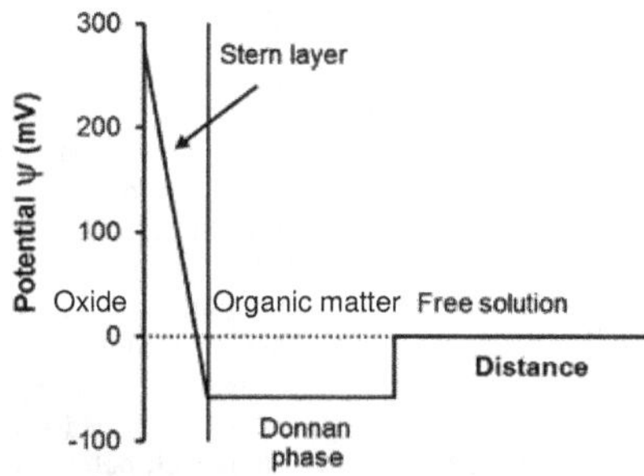

Figure 27. Simplified picture of the electrical double layer structure of an organo-oxide phase. The oxide surface is positively charged. Phosphate ions and $RCOO^-$ ligands are specifically adsorbed, reversing the particle charge and potential. The net negative charge can be attributed to the Donan phase, which has additionally, the charge of the $RCOO^-$ groups of organic matter. The resulting Donnan potential determines the anion exclusion, lowering the concentration in the Donnan phase relative to that of the free solution. This reduced anion concentration (100 times) in the organo-oxide phase will limit its transport rate by diffusion if disequilibrium is created with a zero-sink method.

Fe-oxide-impregnated paper as zero-sink. The size of the fast P pool can be measured with zero-sink methods by perturbing the PO_4 equilibrium upon the addition of a sink that removes the released PO_4. For this, Fe-oxide-impregnated paper has been used (Vanderzee et al. 1987). The fast release can be described with a first-order kinetic model (Vanderzee et al. 1987):

$$Q(t) = Q_{max}(1 - e^{-k_d t}) \tag{17}$$

in which $Q(t)$ is the amount of desorbed PO_4 at a given desorption time t, Q_{max} is the maximum amount of the fast pool, and k_d is the rate constant of desorption.

To determine the maximum desorption capacity of the fast pool, one can measure the amount of PO_4 released at a given time ($Q(t)$) and interpret it with a known rate constant. Such rate constants were measured for nine soils in 0.01 M $CaCl_2$ (Vanderzee et al. 1987). The values vary and the applied experimental method has its limitations due to the adherence of soil particles to the Fe-oxide-impregnated paper (Koopmans et al. 2006; van Rotterdam et al. 2009). To date, our best estimate suggests values in the range of $k_d \sim 0.2$–$0.4\ h^{-1}$.

Excess water as zero-sink. This extraction method (Sissingh 1971) provides a good alternative to the use of Fe-oxide-impregnated paper for studying the kinetic release of PO_4. It is advocated here because it does not have the typical technical complications associated with the use of Fe oxide-impregnated paper (Koopmans et al. 2006; van Rotterdam et al. 2009).

In the extraction method with excess water, a sample is first preequilibrated overnight (22 h) at a high soil-to-water ratio (typically 1:2 kg·L^{-1}). Next, excess water is added leading to a corresponding ratio of 1:72 kg·L^{-1}. This strongly dilutes the overnight-equilibrated solution. After firm shaking for t = 1 h, the solution is appropriately filtered and analyzed for PO_4.

The very strong dilution, decreasing the Ca^{2+} concentration, abruptly shifts the equilibrium condition towards a new equilibrium state with a much higher PO_4 equilibrium concentration that cannot be reached in the short term by PO_4 desorption. As illustrated in Figure 28 with data provided by Gerwin Koopmans, the diluted sample stays well below the equilibrium condition during the first hour of the reaction, implying that practically under these conditions, only a forward desorption reaction occurs that can be interpreted with Equation (17).

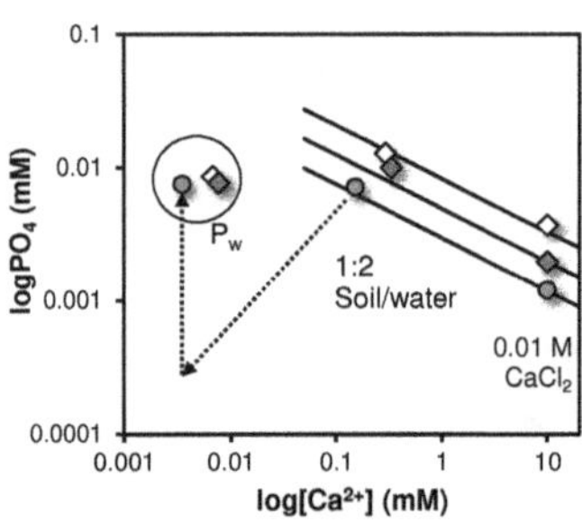

Figure 28. Phosphate and calcium concentrations (**symbols**) in different systems (Unpublished data, courtesy of Gerwin Koopmans). In the 0.01 M $CaCl_2$ and 1:2 soil:water systems, equilibrium is reached, in contrast to the system of the P_w method. In the solutions of the three methods, hugely different Ca concentrations are found. The variation in Ca^{2+} is a key factor in determining the equilibrium state. The **full lines** give the relation between the equilibrium concentration of PO_4 and Ca^{2+} predicted by the NOM-CD model. In the P_w method, a preequilibrated 1:2 soil:water sample was diluted to a corresponding ratio of 1:72 kg·L^{-1}, leading initially to a 36-fold lower concentration of PO_4 and Ca (**dotted line**). Through the desorption during 1 h, the PO_4 concentration increases (**dotted vertical line**), but the concentration remains well below the equilibrium state, illustrating the excellent conditions of this zero-sink method.

A key factor in the PO_4 extraction and equilibration is the concentration of Ca^{2+}, as illustrated in Figure 28, depicting the PO_4 concentration measured in three different extraction solutions as a function of the measured concentration of dissolved Ca^{2+} (symbols). In the 0.01 M $CaCl_2$ systems with a soil-to-solution ratio (SSR) of 0.01 kg·L^{-1}, the lowest PO_4 equilibrium concentrations are found. If these soils are equilibrated with water at a SSR 1:2 kg·L^{-1}, the equilibrium concentrations of PO_4 are much higher and those of Ca^{2+} much lower (< 10^{-3} M). The experimental relation between the decrease of the Ca^{2+} concentration and the increase in PO_4 equilibrium concentration is supported by CD modeling (full lines).

The change in PO_4 and Ca concentrations due to the dilution applied in the dynamic method is illustrated with the dotted arrow for one of the samples (blue sphere). At the onset of the desorption, the PO_4 concentration has decreased to ~0.2 μM and further increases to 80 μM at t = 1 h. This concentration is, by chance, approximately the same as the equilibrium concentration in the 1:2 method before the dilution.

A major difference lies in the much lower Ca level (~ 4 μM) in the P_w solution, implying that the measured PO_4 concentration (80 μM at t = 1 h) is still well below the new equilibrium state. In other words, the PO_4 release in the P_w method is solely due to the forward desorption reaction, making the P_w-method an excellent zero-sink method. Additionally, the excess water addition leads to a strong decrease in the electrolyte concentration of the Donnan phase of the organo-oxide aggregates of the soil. This decrease will increase the negative potential in the Donnan phase, suppressing the corresponding concentration of PO_4 (Fig. 27), and thereby limiting the transport of PO_4 from the interior of the organic-oxide particles to the free solution.

Surface area and aggregate size of organo-oxide aggregates. With the above new insights in the dynamic release of PO_4, opportunities to assess the size of the organo-oxide aggregates by combining this information with the initial PO_4 surface loading, measurable with the $NaHCO_3$ equilibration method.

Rewriting Equation (17), the maximum desorption capacity of phosphate (Q_{max}) can be derived from the amount of desorbed PO_4 measured with the zero-sink (P_w) method. The maximum PO_4 desorption from the fast pool (Q_{max}) is linearly related to the measured PO_4 concentration in the P_w extract (C_{Pw}) at $t=1$ h for a given rate constant k_d and soil-to-solution ratio SSR. Using the mean value of the range of measured rate constants ($k_d = 0.3 \pm 0.1$ h^{-1}), one may write:

$$Q_{max} = \frac{Q(t=1)}{\left(1-e^{-k_d \times 1}\right)} = \frac{C_{Pw}\, SSR^{-1}}{\left(1-e^{-k_d}\right)} = \frac{C_{Pw}\, SSR^{-1}}{0.26 \pm 0.07} \tag{18}$$

The derived value of the fast-pool size Q_{max} (mol·g^{-1}) applying Equation (18) can be used to assess the oxidic surface area available at the exterior of the organo-oxide aggregates (RSA_{ex}). This new approach combines Q_{max} (mol·g^{-1}) with the measured PO_4 surface loading of the soil (hydr)oxides Γ_{PO_4} (mol·m^{-2}) according to:

$$RSA_{ex} = \frac{\Gamma_{PO_4}}{Q_{max}} \tag{19}$$

The application of this approach to samples of the Copernicus soil series unveils an average RSA_{ex} value of 1.8 ± 1.2 m^2·g^{-1}. In contrast to the mean total reactive surface area which is approximately 5 times larger with $RSA_{tot} \sim 9.0 \pm 6.1$ m^2·g^{-1}. RSA_{ex} stands for the surface area associated with the oxide particles situated at the external surfaces of the organo-oxide aggregates.

As a first approximation of the aggregate size using the above value for RSA_{ex}, one may employ a core–shell model. In this model, the external oxide particles with the desorbable PO_4 fraction are in the shell. The relative volume fraction of the external layer, assuming a given shell thickness L, can be expressed as:

$$\frac{V_{ex}}{V_{tot}} = \frac{d_a^3 - (d_a - 2L)^3}{d_a^3} \tag{20}$$

in which d_a is the equivalent diameter of the aggregate. The value L is unknown. In the case of a constant value for thickness L of the PO_4 depletion zone, the aggregate size correlates with the total RSA. This correlation is depicted in Figure 29 for a depletion thickness of $L=3$ nm that corresponds to the average size of the oxide particles in the organo-oxide phase. With the assumption that PO_4 depletion is limited to the first layer of the oxide particles in the aggregates, the equivalent size of the organo-oxide aggregates is in the range of ~50–300 nm (Fig. 29). It is important to note that the actual aggregate size may be larger as nanoaggregates are porous, non-spherical in shape, and exhibit irregular surfaces. TEM observations of organo-oxide aggregates in subsoil samples suggest a typical size range in the order of 100–200 nm (Mikutta et al. 2006), and a similar size range is found for the primary aggregates of synthetic Fe–OC coprecipitates (Guénet et al. 2017).

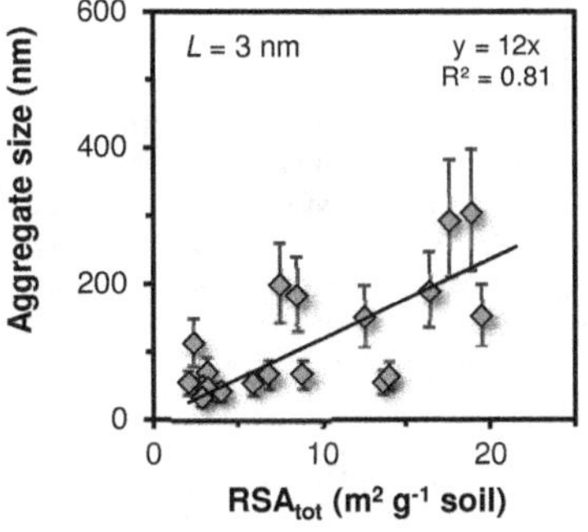

Figure 29. Organo-oxide aggregate size (d_a) as a function of the total reactive surface of the soils of the Copernicus series. The aggregate size was derived from the kinetic release of PO_4 by applying zero-sink extractions using excess water (Eqn. 18). In the applied core–shell model (Eqn. 20), a PO_4-depletion zone L of 3 nm was assumed. The error bars represent the uncertainty related to the applied rate constant k_d for PO_4 desorption.

Reactivity of DOC

Dissolved organic matter can interact with soil and impact the concentration of oxyanions through competitive adsorption and desorption processes. Measured speciation of DOC (van Zomeren and Comans 2007), indicates that fulvic and hydrophilic acids form a significant fraction of DOC. These acid components are known for their interaction with oxides.

In a study involving agricultural top soils, statistical analysis (Hiemstra et al. 2010b), demonstrated that surface loading of oxides with PO_4 (Γ_{PO_4} ~ 1–3 μmol·m^{-2}) is better explained by the logarithmic concentration of DOC (R^2=0.66) than the logarithmic concentration of dissolved PO_4 (R^2=0.31). It was also found that an increase in the DOC concentration, resulting from a slow disintegration of soil organic matter upon long-term storage, coincides with an increase of the PO_4 concentration in the soil solution (Hiemstra et al. 2010a). Koopmans et al. (2006) noticed a simultaneous increase in the PO_4 and DOC concentrations when soils were dried after sampling. These observations collectively underscore the close relationship between the oxyanion (PO_4) and DOC concentration in soil systems.

Interaction of adsorbed PO_4 with FA added to soil. Bai et al. (2024) used fulvic acid (FA) as a first-order proxy for Dissolved Organic Carbon (DOC). The results for the adsorption behavior of added FA for soils of the Copernicus soil series in conjunction with the release of PO_4. The findings indicated that the interaction of added FA is restricted to the external surfaces of organo-oxide particles. The FA adsorption is relatively weak, and isotherms are rather linear (Fig. 30a), both attributable to the competition with natively bound NOM. The extent of FA adsorption (mg C·g^{-1}) appears to be rather independent of soil pH (Fig. 30b), in contrast to the release of PO_4 due to the competitive interaction of FA. The pH independence of the FA adsorption is observed in the Copernicus soil series, where pH has a natural variation (triangles in Fig. 30b), and in a subset of this soil series, where the pH was intentionally altered with the addition of acid or base.

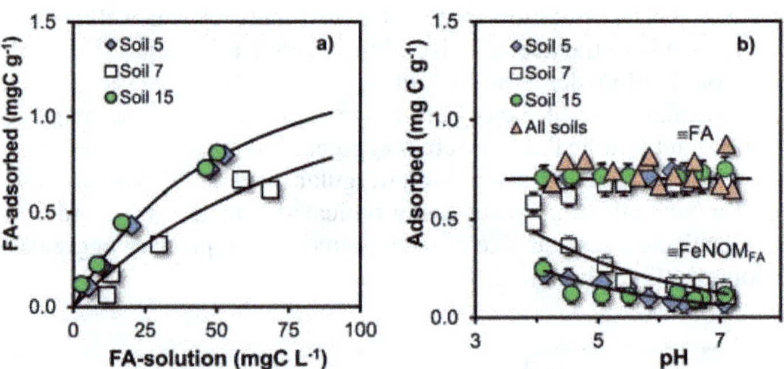

Figure 30. a) Adsorption isotherms of FA added to three soils of the Copernicus series (data of Bai et al. 2024). The isotherms are rather linear, suggesting a relatively low affinity for FA. **b)** The adsorption of added FA as a function of pH shows little pH dependency, in contrast to the amount of added FA present in the compact part of EDL (≡$FeNOM_{FA}$). The latter amount was calculated with the NOM-CD model used in the dual mode, explaining the release of PO_4 upon the addition of PO_4. Similarly to initially adsorbed ≡$FeNOM_T$ of native NOM (Figs. 26d–f), the amount of ≡$FeNOM_{FA}$ increases at lowering the pH, pointing to a shift if the RCOO ligands of adsorbed FA for binding to the metal ions of the oxide. The amount of interfacial FA remains rather small due to the competition with natively adsorbed NOM discussed at Figure 26. Statistical analysis of the FA adsorption data including the RSA and particle size indicates that the amount of added FA that can enter the Stern layers is space-limited. The difference between ≡FA and ≡$FeNOM_{FA}$ is the amount of adsorbed FA present outside the compact part of the EDL. [Reproduced from Bai (2024) by permission of Elsevier.]

Soil factors governing the FA–PO_4 interaction. The soils of the Copernicus series exhibit significant variations in many soil characteristics, encompassing the content of soil organic carbon (SOC), mineral-associated organic carbon (MAOC), AO- and DCB-extractable Fe+Al, soil carbonates, clay content, as well as pH, RSA, mean oxide particle size, and PO_4 surface loading. Remarkably, none of these factors demonstrated a correlation with the quantity of FA adsorbed (Bai 2024). In contrast, the release of PO_4 due to FA adsorption was significantly determined by three factors, i.e., pH ($p = 0.009$), initial DOC concentration ($p = 0.005$), and the maximum amount of NOM ($p = 0.01$) that can be accommodated in the compact part of the double layer around the oxide particles (Max_{NOM}). These findings were derived from data evaluation using a dual-phase approach in the NOM-CD modeling (Fig. 31), where the RSA for the adsorption of FA (RSA_{FA}) was constrained to 2 $m^2 \cdot g^{-1}$ (Bai 2024).

Dual-phase model results. Interpretation of the PO_4-release using the NOM-CD model in a dual-phase mode, as applied by Bai et al. (2024), revealed changes in the surface concentration of ≡ FeNOM, attributable to the adsorption of FA in the compact part of the EDL, denoted as $\equiv FeNOM_{FA}$. The interfacial value of $\equiv FeNOM_{FA}$ in μmol m^{-2} can be converted to the surface concentration in mg C·g^{-1} soil, using the translation method described earlier. The resulting $\equiv FeNOM_{FA}$ is given with symbols in Figure 30b.

The calculated values of $\equiv FeNOM_{FA}$ are found to be lower than the total amount of added FA that is bound. The interpretation of calculated $\equiv FeNOM_{FA}$ data suggests that only a minority of added FA is chemically bound to the oxide surfaces, and that this binding is pH-dependent. This observation indicates that with a decrease in pH, the reactive ligands of adsorbed FA shift in the interface towards the surface for binding to the available ≡FeOH sites of the oxide surface. This shift enhances the competitive desorption of PO_4. In the applied dual-phase approach (Fig. 31), the amount of PO_4 desorbed from the external surfaces of the nano aggregates is redistributed within the system This redistribution occurs through binding to the interior oxide surfaces as well as by a release to the solution.

The above findings suggest that in soils, changes in the DOC concentration may lead to alterations in the solution concentration of PO_4 and other oxyanions. However, for a quantitative understanding, it is important to note that large organic molecules cannot access all the reactive surface areas available for oxyanions, The binding of these large organic molecules is restricted to the external surfaces of soil aggregates.

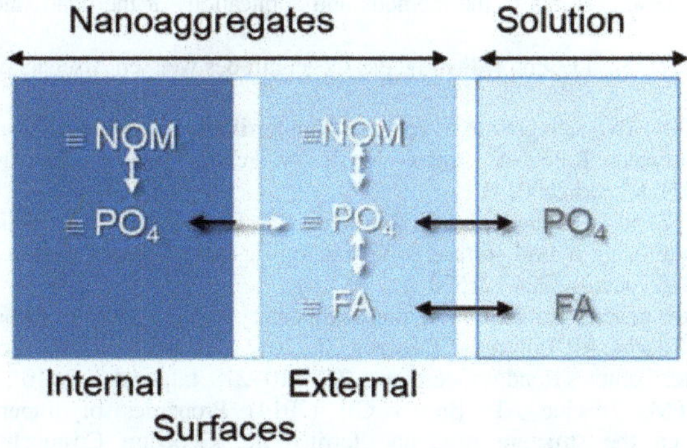

Figure 31. Dual-phase model for equilibration of PO_4 and (added) FA for systems with organo-oxide nanoaggregates (Bai 2024). The reactivity of added FA is restricted to the external surfaces, in contrast to PO_4. At the surfaces in the interior of the nanoaggregates, PO_4 and NOM_{ini} interact competitively, while at the external surfaces, the ligands of adsorbed FA contribute too. [Reproduced from Bai (2024) by permission of Elsevier.]

ACKNOWLEDGMENT

The authors express their gratitude to Johannes Lützenkirchen for his valuable contribution in highlighting the need for a size-dependent scaling of charge in the spherical double layer theory when combined with one or more Stern layers and for pointing to relevant literature. Additionally, we thank Gerwin Koopmans for sharing unpublished data concerning the P_w method and other PO_4 extractions of soil samples.

REFERENCES

Ahmad A, van Genuchten CM (2024) Deep-dive into iron-based co-precipitation of arsenic: A review of mechanisms derived from synchrotron techniques and implications for groundwater treatment. Water Res 249:120970, https://doi.org/10.1016/j.watres.2023.120970

Amini M, Antelo J, Fiol S, Rahnemaie R (2020) Modeling the effects of humic acid and anoxic condition on phosphate adsorption onto goethite. Chemosphere 253:126691, https://doi.org/10.1016/j.chemosphere.2020.126691

Antelo J, Fiol S, Perez C, Marino S, Arce F, Gondar D, Lopez R (2010) Analysis of phosphate adsorption onto ferrihydrite using the CD-MUSIC model. J Colloid Interface Sci 347:112–119, https://doi.org/10.1016/j.jcis.2010.03.020

Antelo J, Arce F, Fiol S (2015) Arsenate and phosphate adsorption on ferrihydrite nanoparticles. Synergetic interaction with calcium ions. Chem Geol 410:53–62, https://doi.org/10.1016/j.chemgeo.2015.06.011

Bai YL (2023) Interactions of phosphate and mineral-associated organic carbon in soils. PhD Wageningen University, Wageningen, The Netherlands

Bai YL, Weng LP, Hiemstra T (2024) Interaction of fulvic acid with soil organo-mineral nano-aggregates and corresponding phosphate release. Geoderma 441:116737

Banfield JF, Welch SA, Zhang HZ, Ebert TT, Penn RL (2000) Aggregation-based crystal growth and microstructure development in natural iron oxyhydroxide biomineralization products. Science 289:751–754, https://doi.org/10.1126/science.289.5480.751

Benjamin MM (1983) Adsorption and surface precipitation of metals on amorphous iron oxyhydroxide. Environ Sci Technol 17:686–692, https://doi.org/10.1021/es00117a012

Boily JF, Song XW (2020) Direct identification of reaction sites on ferrihydrite. Commun Chem 3:79, https://doi.org/10.1038/s42004–020–0325-y

Boily JF, Persson P, Sjöberg S (2000) Benzenecarboxylate surface complexation at the goethite (alpha-FeOOH)/water interface: II Linking IR spectroscopic observations to mechanistic surface complexation models for phthalate, trimellitate, and pyromellitate. Geochim Cosmochim Acta 64:3453–3470

Bompoti N, Chrysochoou M, Machesky M (2017) Surface structure of ferrihydrite: Insights from modeling surface charge. Chemical Geology 464:34–45, https://doi.org/10.1016/j.chemgeo.2016.12.018

Bompoti NM, Chrysochoou M, Machesky ML (2019) A unified surface complexation modeling approach for chromate adsorption on iron oxides. Environ Sci Technol 53:6352–6361, https://doi.org/10.1021/acs.est.9b01183

Boswell MC, Dickson JVE (1918) The adsorption of arsenious acid by ferric hydroxide. J Am Chem Soc 40:1793–1801

Brown ID (2009) Recent developments in the methods and applications of the bond valence model. Chem Rev 109:6858–6919

Bunsen RWB, Berthold AA (1837) Eisenoxydhydrat, das Gegengift des weissen Arseniks oder der arsenigen Säure, Göttingen

Burleson DJ, Penn RL (2006) Two-step growth of goethite from ferrihydrite. Langmuir 22:402–409

Cardew PT (2023) Ostwald Rule of stages—Myth or reality? Cryst Growth Des 23:3958–3969, https://doi.org/10.1021/acs.cgd.2c00141

Chen K, Guo C, Wang C, Zhao S, Xiong B, Lu G, Reinfelder JR, Dang Z (2024) Prediction of Cr(VI) and As(V) adsorption on goethite using hybrid surface complexation-machine learning model. Water Res 256:121580, https://doi.org/10.1016/j.watres.2024.121580

Chukhrov FV (1973) On the genesis problem of thermal sedimentary iron ore deposits. Miner Deposita 8:138–147

Cismasu AC, Michel FM, Tcaciuc AP, Tyliszczak T, Brown GE (2011) Composition and structural aspects of naturally occurring ferrihydrite. Comptes Rendus Geoscience 343:210–218, https://doi.org/10.1016/j.crte.2010.11.001

Cismasu AC, Michel FM, Tcaciuc AP, Brown GE (2014) Properties of impurity-bearing ferrihydrite III Effects of Si on the structure of 2-line ferrihydrite. Geochim Cosmochim Acta 133:168–185, https://doi.org/10.1016/j.gca.2014.02.018

Davis JA (1977) Adsorption of trace metals and complexing ligands at the oxide/water interface. PhD Stanford University, California, USA

Davis JA, Leckie JO (1978) Surface ionization and complexation at the oxide/water interface. II Surface properties of amorphous iron oxyhydroxide and adsorption of metal ions. J Colloid Interface Sci 67:90–107

De Yoreo JJ, Gilbert PU, Sommerdijk NA, Penn RL, Whitelam S, Joester D, Zhang H, Rimer JD, Navrotsky A, Banfield JF, Wallace AF (2015) Crystallization by particle attachment in synthetic, biogenic, and geologic environments. Science 349:6247, https://doi.org/10.1126/science.aaa6760

Deng Y, Weng L, Li Y, Ma J, Chen Y (2019) Understanding major NOM properties controlling its interactions with phosphorus and arsenic at goethite–water interface. Water Res 157:372–380, https://doi.org/10.1016/j.watres.2019.03.077

Deng YX, Weng LP, Li YT, Chen YL, Ma J (2020) Redox-dependent effects of phosphate on arsenic speciation in paddy soils. Environ Pollut 264:114783, https://doi.org/10.1016/j.envpol.2020.114783

Do Hamid R, Swedlund PJ, Song YT, Miskelly GM (2011) Ionic strength effects on silicic acid (H_4SiO_4) sorption and oligomerization on an iron oxide surface: An interesting interplay between electrostatic and chemical forces. Langmuir 27:12930–12937, https://doi.org/10.1021/la201775c

Drits VA, Sakharov BA, Salyn AL, Manceau A (1993) Structural model for ferrihydrite. Clay Min 28:185–207

Dublet G, Pacheco JL, Bargar JR, Fendorf S, Kumar N, Lowry GV, Brown GE (2017) Partitioning of uranyl between ferrihydrite and humic substances at acidic and circum-neutral pH Geochim Cosmochim Acta 215:122–140, https://doi.org/10.1016/j.gca.2017.07.013

Dyer JA, Trivedi P, Scrivner NC, Sparks DL (2003) Lead sorption onto ferrihydrite. 2. Surface complexation modeling. Environ Sci Technol 37:915–922, https://doi.org/10.1021/es025794r

Dzombak DA, Morel FMM (1990) Surface Complexation Modeling: Hydrous Ferric Oxide. John Wiley & Sons: New York:393

Foerstendorf H, Heim K, Rossberg A (2012) The complexation of uranium(VI) and atmospherically derived CO_2 at the ferrihydrite–water interface probed by time-resolved vibrational spectroscopy. J Colloid Interface Sci 377:299–306, https://doi.org/10.1016/j.jcis.2012.03.020

Fukushi K, Aoyama K, Yang C, Kitadai N, Nakashima S (2013) Surface complexation modeling for sulfate adsorption on ferrihydrite consistent with in situ infrared spectroscopic observations. Appl Geochem 36:92–103, https://doi.org/http://dx.doi.org/10.1016/j.apgeochem.2013.06.013

Funnell NP, Fulford MF, Inoué S, Kletetschka K, Michel FM, Goodwin AL (2020) Nanocomposite structure of two-line ferrihydrite powder from total scattering. Commun Chem 3:22, https://doi.org/10.1038/s42004–020–0269–2

Gao XD, Root RA, Farrell J, Ela W, Chorover J (2013) Effect of silicic acid on arsenate and arsenite retention mechanisms on 6-L ferrihydrite: A spectroscopic and batch adsorption approach. Appl Geochem 38:110–120, https://doi.org/10.1016/j.apgeochem.2013.09.005

Gilbert B, Erbs JJ, Penn RL, Petkov V, Spagnoli D, Waychunas GA (2013) A disordered nanoparticle model for 6-line ferrihydrite. Am Mineral 98:1465–1476, https://doi.org/10.2138/am.2013.4421

Girvin DC, Ames LL, Schwab AP, McGarrah JE (1991) Neptunium adsorption on synthetic amorphous iron oxyhydroxide. J Colloid Interface Sci 141:67–78

Goldsztaub MS (1935) Étude de quelques dérivés de l'oxyde ferrique (FeOOH, FeONa, FeOGl) ; Détermination de leurs structures. Bull Soc Fr Mineral 58:6–76

Guénet H, Davranche M, Vantelon D, Gigault J, Prévost S, Taché O, Jaksch S, Pédrot M, Dorcet V, Boutier A, Jestin J (2017) Characterization of iron–organic matter nano-aggregate networks through a combination of SAXS/SANS and XAS analyses: impact on As binding. Environ Sci Nano 4:938–954, https://doi.org/10.1039/c6en00589f

Gustafsson JP (2003) Modelling molybdate and tungstate adsorption to ferrihydrite. Chem Geol 200:105–115

Gustafsson JP (2006) Arsenate adsorption to soils: Modelling the competition from humic substances. Geoderma 136:320–330

Gustafsson JP, Antelo J (2022) Competitive arsenate and phosphate adsorption on ferrihydrite as described by the CD-MUSIC Model. ACS Earth Space Chem 6:1397–1406, https://doi.org/10.1021/acsearthspacechem.2c00081

Hahn O (1924) Untersuchung oberflächenreicher Substanzen nach radioaktiven Methoden und ihre Anwendung auf chemische und radioaktive Probleme. Die Naturwissenschaften 12:1140–1145, https://doi.org/10.1007/BF01504610

Han J, Katz LE (2019) Capturing the variable reactivity of goethites in surface complexation modeling by correlating model parameters with specific surface area. Geochim Cosmochim Acta 244:248–263, https://doi.org/10.1016/j.gca.2018.09.008

Hanna K, Quiles F (2011) Surface complexation of 2,5-dihydroxybenzoic acid (gentisic acid) at the nanosized hematite–water interface: An ATR-FTIR study and modeling approach. Langmuir 27:2492–2500, https://doi.org/10.1021/la104239x

Harrington R, Hausner DB, Xu WQ, Bhandari N, Michel FM, Brown GE, Strongin DR, Parise JB (2011) Neutron pair distribution function study of two-line ferrihydrite. Environ Sci Technol 45:9883–9890, https://doi.org/10.1021/es2020633

Hayes KF, Roe AL, Brown GE, Jr., Hodgson K, Leckie JO, Parks GA (1987) In-situ X-ray absorption study of surface complexes: Selenium oxyanions on α-FeOOH. Science 238:783–785

Hiemstra T (2013) Surface and mineral structure of ferrihydrite. Geochim Cosmochim Acta 10:316–325

Hiemstra T (2015) Formation, stability, and solubility of metal oxide nanoparticles: Surface entropy, enthalpy, and free energy of ferrihydrite. Geochim Cosmochim Acta 158:179–198

Hiemstra T (2018a) Surface structure controlling nanoparticle behavior: Magnetism of ferrihydrite, magnetite, and maghemite. Environ Sci Nano 5:752–764, https://doi.org/ 10.1039/C7EN01060E

Hiemstra T (2018b) Ferrihydrite interaction with oxyanions: Silicate polymerization and competition with phosphate, arsenate, and arsenite. Geochim Cosmochim Acta 238: 453–476

Hiemstra T, Lützenkirchen J (2025) Development and modus operandi relating surface structure and ion complexation modeling for important metal (hydr)oxides. Rev Mineral Geochem 91A:13-84

Hiemstra T, Van Riemsdijk WH (1996) A surface structural approach to ion adsorption: The charge distribution (CD) model. J Colloid Interface Sci 179:488–508

Hiemstra T, Van Riemsdijk WH (2006) On the relationship between charge distribution, surface hydration and the structure of the interface of metal hydroxides. J Colloid Interface Sci 301:1–18

Hiemstra T, Van Riemsdijk WH (2009) A surface structural model for ferrihydrite I: sites related to primary charge, molar mass, and mass density. Geochim Cosmochim Acta 73:4423–4436

Hiemstra T, Zhao W (2016) Reactivity of ferrihydrite and ferritin in relation to surface structure, size, and nanoparticle formation studied for phosphate and arsenate. Environ Sci Nano 3:1265–1279, https://doi.org/10.1039/c6en00061d

Hiemstra T, Venema P, Van Riemsdijk WH (1996) Intrinsic proton affinity of reactive surface groups of metal (hydr) oxides: The bond valence principle. J Colloid Interface Sci 184: 680–692

Hiemstra T, Rahnemaie R, Van Riemsdijk WH (2004) Surface complexation of carbonate on goethite: IR spectroscopy, Structure and Charge Distribution. J Colloid Interface Sci 278:282–290

Hiemstra T, Van Riemsdijk WH, Rossberg A, Ulrich KU (2009) A surface structural model for ferrihydrite II: adsorption of uranyl and carbonate. Geochim Cosmochim Acta 73:4437–4451

Hiemstra T, Antelo J, Rahnemaie R, van Riemsdijk WH (2010a) Nanoparticles in natural systems I: The effective reactive surface area of the natural oxide fraction in field samples. Geochim Cosmochim Acta 74:41–58

Hiemstra T, Antelo J, van Rotterdam AMD, van Riemsdijk WH (2010b) Nanoparticles in natural systems II: The natural oxide fraction at interaction with natural organic matter and phosphate. Geochim Cosmochim Acta 74:59–69

Hiemstra T, Mia S, Duhaut PB, Molleman B (2013) Natural and pyrogenic humic acids interacting with phosphate on goethite: modeling and application to soils. Environ Sci Technol 47:9182–9189

Hiemstra T, Mendez JC, Li J (2019) Evolution of the reactive surface area of ferrihydrite: time, pH, and temperature dependency of growth by Ostwald ripening. Environmental Science: Nano, https://doi.org/10.1039/c8en01198b

Hingston FJ, Atkinson RJ, Posner AM, Quirk JP (1967) Specific adsorption of anions. Nature 215:1459–1461, https://doi.org/10.1038/2151459a0

Hofmann A, Pelletier M, Michot L, Stradner A, Schurtenberger P, Kretzschmar R (2004) Characterization of the pores in hydrous ferric oxide aggregates formed by freezing and thawing. J Colloid Interface Sci 271:163–173, https://doi.org/10.1016/j.jcis.2003.11.053

Hofmann A, Vantelon D, Montargès-Pelletier E, Villain F, Gardoll O, Razafitianamaharavo A, Ghanbaja J (2013) Interaction of Fe(III) and Al(III) during hydroxylation by forced hydrolysis: The nature of Al–Fe oxyhydroxy co-precipitates. J Colloid Interface Sci 407:76–88, https://doi.org/10.1016/j.jcis.2013.06.020

Hsi C-KD, Langmuir D (1985) Adsorption of uranyl onto ferric oxyhydroxides: Application of the surface complexation site-binding model. Geochim Cosmochim Acta 49:1931–1941

Jain A, Loeppert RH (2000) Effect of competing anions on the adsorption of arsenate and arsenite by ferrihydrite. J Environ Qual 29:1422–1430

Jain A, Raven KP, Loeppert RH (1999) Arsenite and arsenate adsorption on ferrihydrite: Surface charge reduction and net OH- release stoichiometry. Environ Sci Technol 33:1179–1184, https://doi.org/10.1021/es980722e

Johnson SB, Yoon TH, Slowey AJ, Brown GE (2004) Adsorption of organic matter at mineral/water interfaces: 3. Implications of surface dissolution for adsorption of oxalate. Langmuir 20:11480–11492

Kanematsu M, Waychunas GA, Boily J-F (2018) Silicate binding and precipitation on iron oxyhydroxides. Environ Sci Technol, https://doi.org/10.1021/acs.est.7b04098

Kinniburgh DG, Jackson ML (1982) Concentration and pH-dependence of calcium and zinc adsorption by iron hydrous oxide gel. Soil Sci Soc Am J 46:56–61

Kinniburgh DG, Syers JK, Jackson ML (1975) Specific adsorption of trace amounts of calcium and strontium by hydrous oxides of iron and aluminum. Soil Sci Soc Am J 39:464–470

Koca Fındık B, Jafari M, Song LF, Li Z, Aviyente V, Merz KM, Jr. (2024) Binding of phosphate species to Ca^{2+} and Mg^{2+} in aqueous solution. J Chem Theory Comput 20:4298–4307, https://doi.org/10.1021/acs.jctc.4c00218

Koopmans GF, Chardon WJ, Dekker PHM, Romkens P, Schoumans OF (2006) Comparing different extraction methods for estimating phosphorus solubility in various soil types. Soil Sci 171:103–116, https://doi.org/10.1097/01.ss.0000187361.00600.d6

Koopmans GF, Hiemstra T, Vaseur C, Chardon WJ, Voegelin A, Groenenberg JE (2020) Use of iron oxide nanoparticles for immobilizing phosphorus in-situ: Increase in soil reactive surface area and effect on soluble phosphorus. Sci Total Environ 711, https://doi.org/10.1016/j.scitotenv.2019.135220

Kurbatov IW (1931) Adsorption of thorium X by ferric hydroxide at different pH: J Phys Chem 36:1241–1247, https://doi.org/10.1021/j150334a014

Larsson MA, Persson I, Sjostedt C, Gustafsson JP (2017) Vanadate complexation to ferrihydrite: X-ray absorption spectroscopy and CD-MUSIC modelling. Environ Chem 14:141–150, https://doi.org/10.1071/en16174

Li D, Nielsen MH, Lee JRI, Frandsen C, Banfield JF, De Yoreo JJ (2012) Direction-specific interactions control crystal growth by oriented attachment. Science 336:1014–1018, https://doi.org/10.1126/science.1219643

Liu J, Zhu RL, Xu TY, Xu Y, Ge F, Xi YF, Zhu JX, He HP (2016) Co-adsorption of phosphate and zinc(II) on the surface of ferrihydrite. Chemosphere 144:1148–1155, https://doi.org/10.1016/j.chemosphere.2015.09.083

Lookman R, Freese D, Merckx R, Vlassak K, van Riemsdijk WH (1995) Long-term kinetics of phosphate release from soil. Environ Sci Technol 29:1569–1575, https://doi.org/10.1021/es00006a020

Manceau A (2011) Critical evaluation of the revised akdalaite model for ferrihydrite. Am Mineral 96:521–533, https://doi.org/10.2138/am.2011.3583

Manceau A, Gates WP (1997) Surface structural model for ferrihydrite. Clays Clay Min 45:448–460

Manceau A, Skanthakumar S, Soderholm L (2014) PDF analysis of ferrihydrite: Critical assessment of the under-constrained akdalaite model. Am Mineral 99:102–108, https://doi.org/10.2138/am.2014.4576

Martinez RJ, Villalobos M, Loredo-Jasso AU, Cruz-Valladares AX, Mendoza-Flores A, Salazar-Rivera H, Cruz-Romero D (2023) Towards building a unified adsorption model for goethite based on direct measurements of crystal face compositions: I Acidity behavior and As (V) adsorption. Geochim Cosmochim Acta 354:252–262, https://doi.org/10.1016/j.gca.2023.06.021

Mendez JC, Hiemstra T (2019) Carbonate adsorption to ferrihydrite: competitive interaction with phosphate for use in soil systems. ACS Earth Space Chem 3:129–141, https://doi.org/10.1021/acsearthspacechem.8b00160

Mendez JC, Hiemstra T (2020a) Ternary complex formation of phosphate with Ca and Mg ions binding to ferrihydrite: experiments and mechanisms. ACS Earth Space Chem 4:545–557, https://doi.org/10.1021/acsearthspacechem.9b00320

Mendez JC, Hiemstra T (2020b) High and low affinity sites of ferrihydrite for metal ion adsorption: Data and modeling of the alkaline-earth ions Be, Mg, Ca, Sr, Ba, and Ra. Geochim Cosmochim Acta 286:289–305, https://doi.org/10.1016/j.gca.2020.07.032

Mendez JC, Hiemstra T (2020c) Surface area of ferrihydrite consistently related to primary surface charge, ion pair formation, and specific ion adsorption. Chem Geol 532, https://doi.org/10.1016/j.chemgeo.2019.119304

Mendez JC, Hiemstra T, Koopmans GF (2020) Assessing the reactive surface area of soils and the association of soil organic carbon with natural oxide nanoparticles using ferrihydrite as proxy. Environ Sci Technol 54:11990–12000, https://doi.org/10.1021/acs.est.0c02163

Mendez JC, Van Eynde E, Hiemstra T, Comans RNJ (2022) Surface reactivity of the natural metal (hydr)oxides in weathered tropical soils. Geoderma 406:115517, https://doi.org/10.1016/j.geoderma.2021.115517

Michel FM, Ehm L, Antao SM, Lee PL, Chupas PJ, Liu G, Strongin DR, Schoonen MAA, Phillips BL, Parise JB (2007) The structure of ferrihydrite, a nanocrystalline material. Science 316:1726–1729

Michel FM, Barron V, Torrent J, Morales MP, Serna CJ, Boily JF, Liu QS, Ambrosini A, Cismasu AC, Brown GE (2010) Ordered ferrimagnetic form of ferrihydrite reveals links among structure, composition, and magnetism. PNAS 107:2787–2792

Mikutta R, Kleber M, Torn MS, Jahn R (2006) Stabilization of soil organic matter: Association with minerals or chemical recalcitrance? Biogeochemistry 77:25–56, https://doi.org/10.1007/s10533–005–0712–6

Moon EM, Peacock CL (2013) Modelling Cu(II) adsorption to ferrihydrite and ferrihydrite-bacteria composites: Deviation from additive adsorption in the composite sorption system. Geochim Cosmochim Acta 104:148–164, https://doi.org/10.1016/j.gca.2012.11.030

Nagata T, Fukushi K, Takahashi Y (2009) Prediction of iodide adsorption on oxides by surface complexation modeling with spectroscopic confirmation. J Colloid Interface Sci 332:309–316, https://doi.org/http://dx.doi.org/10.1016/j.jcis.2008.12.037

Navrotsky A, Mazeina L, Majzlan J (2008) Size-driven structural and thermodynamic complexity in iron oxides. Science 319:1635–1638

Ohshima H (2006) Chapter 3 - Diffuse double layer equations for use in surface complexation models: Approximations and limits. *In*: Lützenkirchen J (ed), Interface Science and Technology, Elsevier, Vol 11, p 67–87

Ona-Nguema G, Morin G, Juillot F, Calas G, Brown GE, Jr. (2005) EXAFS analysis of arsenite adsorption onto two-line ferrihydrite, hematite, goethite, and lepidocrocite. Environ Sci Technol 39:9147–9155

Ostwald W (1897) Studien über die Bildung und Umwandlung fester Körper. 1 Abhandlung: Übersättigung und Überkaltung 22U:289–330, https://doi.org/10.1515/zpch–1897–2233

Otero-Fariña A, Fiol S, Arce F, Antelo J (2017) Effects of natural organic matter on the binding of arsenate and copper onto goethite. Chem Geol 459:119–128, https://doi.org/10.1016/j.chemgeo.2017.04.012

Parise JB, Xia BY, Simonson JW, Woerner WR, Plonka AM, Phillips BL, Ehm L (2019) Structural chemistry of akdalaite, $Al_{10}O_{14}(OH)_2$, the isostructural aluminum analogue of ferrihydrite. Crystals 9:246, https://doi.org/10.3390/cryst9050246

Pauling L, Hendricks SB (1925) The crystal structures of hematite and corundum. J Am Chem Soc 47:781–790, https://doi.org/10.1021/ja01680a027

Perona MJ, Leckie JO (1985) Proton stoichiometry for the adsorption of cations on oxide surfaces. J Colloid Interface Sci 106:65–69

Pinney N, Kubicki JD, Middlemiss DS, Grey CP, Morgan D (2009) Density functional theory study of ferrihydrite and related Fe-oxyhydroxides. Chem Mater 21:5727–5742, https://doi.org/10.1021/cm9023875

Pivovarov S (2009) Diffuse sorption modeling. J Colloid Interface Sci 332:54–59, https://doi.org/10.1016/j.jcis.2008.11.074

Rahnemaie R, Hiemstra T, Van Riemsdijk WH (2006) A new structural approach for outersphere complexation, tracing the location of electrolyte ions. J Colloid Interface Sci 293:312–321

Rahnemaie R, Hiemstra T, Van Riemsdijk WH (2007) Carbonate adsorption on goethite in competition with phosphate. J Colloid Interface Sci 315:415–425

Rietra RPJJ, Hiemstra T, Van Riemsdijk WH (2000) Electrolyte anion affinity and its effect on oxyanion adsorption on goethite. J Colloid Interface Sci 229:199–206

Rossberg A, Ulrich KU, Weiss S, Tsushima S, Hiemstra T, Scheinost AC (2009) Identification of uranyl surface complexes on ferrihydrite: Advanced EXAFS data analysis and CD-MUSIC modelling. Environ Sci Technol 43:1400–1406

Sassi M, Chaka AM, Rosso KM (2021) Ab initio thermodynamics reveals the nanocomposite structure of ferrihydrite. Commun Chem 4:134, https://doi.org/10.1038/s42004–021–00562–7

Schwertmann U, Cornell RM (1991) Iron oxides in the laboratory: Preparation and characterization. Wiley-VCH Verlag GmbH

Schwertmann U, Fischer WR (1973) Natural amorphous ferric hydroxide. Geoderma 10:237–247

Senn AC, Kaegi R, Hug SJ, Hering JG, Mangold S, Voegelin A (2015) Composition and structure of Fe(III)-precipitates formed by Fe(II) oxidation in water at near-neutral pH: Interdependent effects of phosphate, silicate and Ca. Geochim Cosmochim Acta 162:220–246, https://doi.org/10.1016/j.gca.2015.04.032

Sissingh HA (1971) Analytical technique of Pw method, used for assessment of phosphate status of arable soils in Netherlands. Plant Soil 34:483, https://doi.org/10.1007/bf01372800

Snow CL, Lilova KI, Radha AV, Shi Q, Smith S, Navrotsky A, Boerio-Goates J, Woodfield BF (2013) Heat capacity and thermodynamics of a synthetic two-line ferrihydrite, $FeOOH{\cdot}0.027H_2O$. J Chem Thermodyn 58:307–314, http://dx.doi.org/10.1016/j.jct.2012.11.012

Spadini L, Manceau A, Schindler PW, Charlet L (1994) Structure and Stability of Cd^{2+} Surface Complexes on Ferric Oxides. 1 Results from EXAFS spectroscopy. J Colloid Interface Sci 168:73–86

Sposito G (1984) The Surface Chemistry of Soils. Oxford University Press, New York

Swedlund PJ, Hamid RD, Miskelly GM (2010a) Insights into H_4SiO_4 surface chemistry on ferrihydrite suspensions from ATR-IR, Diffuse Layer Modeling and the adsorption enhancing effects of carbonate. J Colloid Interface Sci 352:149–157, https://doi.org/10.1016/j.jcis.2010.08.011

Swedlund PJ, Miskelly GM, McQuillan AJ (2010b) Silicic acid adsorption and oligomerization at the ferrihydrite-water interface: Interpretation of ATR-IR spectra based on a model surface structure. Langmuir 26:3394–3401, https://doi.org/10.1021/la903160q

Swedlund PJ, Sivaloganathan S, Miskelly GM, Waterhouse GIN (2011) Assessing the role of silicate polymerization on metal oxyhydroxide surfaces using X-ray photoelectron spectroscopy. Chem Geol 285:62–69, https://doi.org/http://doi.org/10.1016/j.chemgeo.2011.02.022

Thomas JM (2012) The birth of X-ray crystallography. Nature 491:186–187, https://doi.org/10.1038/491186a

ThomasArrigo LK, Kaegi R, Kretzschmar R (2019) Ferrihydrite growth and transformation in the presence of ferrous iron and model organic ligands. Environ Sci Technol 53:13636–13647, https://doi.org/10.1021/acs.est.9b03952

Tiberg C, Gustafsson JP (2016) Phosphate effects on cadmium(II) sorption to ferrihydrite. J Colloid Interface Sci 471:103–111, https://doi.org/10.1016/j.jcis.2016.03.016

Tiberg C, Sjostedt C, Persson I, Gustafsson JP (2013) Phosphate effects on copper(II) and lead(II) sorption to ferrihydrite. Geochim Cosmochim Acta 120:140–157, https://doi.org/10.1016/j.gca.2013.06.012

Ulrich KU, Rossberg A, Foerstendorf H, Zanker H, Scheinost AC (2006) Molecular characterization of uranium(VI) sorption complexes on iron(III)-rich acid mine water colloids. Geochim Cosmochim Acta 70:5469–5487

Van Beinum W, Hofmann A, Meeussen JCL, Kretzschmar R (2005) Sorption kinetics of strontium in porous hydrous ferric oxide aggregates I The Donnan diffusion model. J Colloid Interface Sci 283:18–28

Van Bemmelen JM, Klobbie EA (1892) Uber das amorphe, wasserhaltige Eisenoxyd, das kristallinische Eisenoxydhydrat, das Kaliumferrit und das Natriumferrit. J Prakt Chem 46

Van Bemmelen JM, Klobbie EA (1896) Sur l'oxyde ferrique humide amorphe, l'hydroxyde ferrique cristallin, les ferrites de potassium et de sodium. Extrait des Archives Néerlandaises XXIX

Van Eynde E, Mendez JC, Hiemstra T, Comans RNJ (2020) Boron adsorption to ferrihydrite with implications for surface speciation in soils: Experiments and modeling. ACS Earth Space Chem 4:1269–1280, https://doi.org/10.1021/acsearthspacechem.0c00078

Van Eynde E, Hiemstra T, Comans RNJ (2022) Interaction of Zn with ferrihydrite and its cooperative binding in the presence of PO_4. Geochim Cosmochim Acta 320:223–237, https://doi.org/10.1016/j.gca.2022.01.010

van Rotterdam AMD, Temminghoff EJM, Schenkeveld WDL, Hiemstra T, van Riemsdijk WH (2009) Phosphorus removal from soil using Fe oxide-impregnated paper: Processes and applications. Geoderma 151:282–289, https://doi.org/10.1016/j.geoderma.2009.04.013

van Zomeren A, Comans RNJ (2007) Measurement of humic and fulvic acid concentrations and dissolution properties by a rapid batch procedure. Environ Sci Technol 41:6755–6761

Vanderzee S, Fokkink LGJ, Vanriemsdijk WH (1987) A new technique for assessment of reversibly adsorbed phosphate. Soil Sci Soc Am J 51:599–604

Vantelon D, Davranche M, Marsac R, La Fontaine C, Guénet H, Jestin J, Campaore G, Beauvois A, Briois V (2019) Iron speciation in iron–organic matter nanoaggregates: A kinetic approach coupling Quick-EXAFS and MCR-ALS chemometrics. Environ Sci Nano 6:2641–2651, https://doi.org/10.1039/C9EN00210C

Vayssieres L (2005) On the thermodynamic stability of metal oxide nanoparticles in aqueous solutions. Int J Nanotechnol 2:411–439, https://doi.org/10.1504/ijnt.2005.008077

Vos HMJ, Hiemstra T, Lopez MP, van Groenigen JW, Voegelin A, Mangold S, Koopmans GF (2022) Earthworms affect reactive surface area and thereby phosphate solubility in iron-(hydr)oxide dominated soils. Geoderma 428, https://doi.org/10.1016/j.geoderma.2022.116212

Waite TD, Davis JA, Payne TE, Waychunas GA, Xu N (1994) Uranium(VI) adsorption to ferrihydrite: Application of a surface complexation model. Geochim Cosmochim Acta 58:5465–5478

Wang X, Liu F, Tan W, Li W, Feng X, Sparks DL (2013) Characteristics of phosphate adsorption–desorption onto ferrihydrite: Comparison with well-crystalline Fe (hydr)oxides. Soil Sci 178:1–11, https://doi.org/10.1097/SS0b013e31828683f8

Wang X, Zhu M, Koopal LK, Li W, Xu W, Liu F, Zhang J, Liu Q, Feng X, Sparks DL (2016) Effects of crystallite size on the structure and magnetism of ferrihydrite. Environ Sci Nano 3:190–202, https://doi.org/10.1039/c5en00191a

Wang X, Kubicki JD, Boily J-F, Waychunas GA, Hu Y, Feng X, Zhu M (2018) Binding geometries of silicate species on ferrihydrite surfaces. ACS Earth Space Chem 2:125–134, https://doi.org/10.1021/acsearthspacechem.7b00109

Waychunas GA, Fuller CC, Davids JA (2002) Surface complexation and precipitate geometry for aqueous Zn(II) sorption on ferrihydrite I: X-ray absorption extended fine structure spectroscopic analysis. Geochim Cosmochim Acta 66:1119–1137

Weatherill JS, Morris K, Bots P, Stawski TM, Janssen A, Abrahamsen L, Blackham R, Shaw S (2016) Ferrihydrite Formation: The Role of Fe–13 Keggin Clusters. Environ Sci Technol 50:9333–9342, https://doi.org/10.1021/acs.est.6b02481

Weng LP, Van Riemsdijk WH, Koopal LK, Hiemstra T (2006a) Adsorption of humic substances on goethite: Comparison between humic acids and fulvic acids. Environ Sci Technol 40:7494–7500

Weng LP, Van Riemsdijk WH, Koopal LK, Hiemstra T (2006b) Ligand and Charge Distribution (LCD) model for the description of fulvic acid adsorption to goethite. J Colloid Interface Sci 302:442–457

Weng LP, Van Riemsdijk WH, Hiemstra T (2007) Adsorption of humic acids onto goethite: Effects of molar mass, pH and ionic strength. J Colloid Interface Sci 314:107–118

Weng LP, Van Riemsdijk WH, Hiemstra T (2008) Humic nano-particles at the oxide–water interface: Interaction with phosphate ion adsorption. Environ Sci Technol 42:8747–8752

Whitaker AH, Austin RE, Holden KL, Jones JL, Michel FM, Peak D, Thompson A, Duckworth OW (2021) The structure of natural biogenic iron (oxyhydr)oxides formed in circumneutral pH environments. Geochim Cosmochim Acta 308:237–255, https://doi.org/10.1016/j.gca.2021.05.059

Xu WQ, Hausner DB, Harrington R, Lee PL, Strongin DR, Parise JB (2011) Structural water in ferrihydrite and constraints this provides on possible structure models. Am Mineral 96:513–520, https://doi.org/10.2138/am.2011.3460

Xu Y, Hiemstra T, Tan WF, Bai YL, Weng LP (2022) Key factors in the adsorption of natural organic matter to metal (hydr) oxides: Fractionation and conformational change. Chemosphere 308:136129, https://doi.org/10.1016/j.chemosphere.2022.136129

Xu Y, Bai YL, Hiemstra T, Weng LP (2024) A new consistent modeling framework for the competitive adsorption of humic nanoparticles and oxyanions to metal (hydr)oxides: Multiple modes of heterogeneity, fractionation, and conformational change. J Colloid Interface Sci 660:522–533

Xu Y, Bai Y, Hiemstra T, Weng, L (2025) Ligand and charge distribution modeling of natural organic matter adsorption on metal (hydr)oxides: State-of-the-art. Rev Mineral Geochem 91A:229-250

Yoe JH (1930) The adsorption of arsenious acid by hydrous ferric oxide. J Am Chem Soc 52:2785–2790

Reviews in Mineralogy & Geochemistry
Vol. 91A pp. 229–250, 2025

7

Ligand and Charge Distribution Modeling of Natural Organic Matter Adsorption on Metal (Hydr)oxides: State-of-the-art

Yun Xu[*,1,2], Yilina Bai[*,1,3], Tjisse Hiemstra[1], Liping Weng[1,4]

[1]Soil Chemistry and Chemical Soil Quality, Wageningen University & Research, 6708 PB, Wageningen, the Netherlands
[2]State Environmental Protection Key Laboratory of Soil Health and Green Remediation, College of Resources and Environment, Huazhong Agricultural University, 430070, Wuhan, China
[3]Chongqing Academy of Agriculture Sciences, Chongqing 40000, China
[4]Agro-Environmental Protection Institute, Ministry of Agriculture, 300191, Tianjin, China

yun.xu@wur.nl/ yunxu@mail.hzau.edu.cn, yilina.bai@wur.nl/ gzyilina@163.com, tjisse.hiemstra@wur.nl, liping.weng@wur.nl

**These authors contributed equally*

INTRODUCTION

Surface complexation modeling (SCM) stands as a valuable tool to elucidate and comprehend interactions between adsorbates and adsorbents. In the realm of natural systems, it has found frequent application in characterizing ion adsorption to both metal (hydr)oxides and natural organic matter (NOM). In soil, metal (hydr)oxides and natural organic matter coexist, are closely associated, and organized in organo-oxide aggregates. The intricacies of the interaction between NOM and metal (hydr)oxide surfaces as well as the corresponding interfacial distribution of NOM remain insufficiently understood. There is a need to develop mechanistic surface complexation models to enhance our comprehension of these complex phenomena, which influence not only the distribution and bioavailability of nutrients and pollutants but also soil structure and carbon sequestration.

A notable characteristic shared by both NOM and metal (hydr) oxides is the development of pH-dependent surface charge. Metal (hydr)oxides exhibit variable charge behavior, a feature effectively described by the Charge Distribution MUlti-SIte Complexation (CD-MUSIC) model (Hiemstra et al. 1989; Hiemstra and van Riemsdijk 1996). This approach typically employs the classical diffuse double layer theory, augmented by one or more Stern layers, to describe the electrostatic interactions. By considering discrete surface sites, derived from the crystallographic surface structure, the CD-MUSIC model is extensively applied to evaluate and understand a wide range of ion adsorption phenomena of metal (hydr)oxides.

In contrast, natural organic matter (NOM) is characterized by a heterogeneous composition of ill-defined organic acids that dissociate protons, leading to variable charge. NOM displays two prominent types of functional groups, namely carboxylic acids (RCOOH) and alcohol or phenolic hydroxyls (ROH). The proton affinity of each type of group varies significantly depending on the position of the groups on the carbonic backbone, resulting in a continuum of affinities. To address the heterogeneity of functional groups, another class of ion adsorption models has been developed with the Langmuir model as the basis but with a distribution of ion affinities.

1529-6466/25/0091A-0007$05.00 (print)
1943-2666/25/0091A-0007$05.00 (online)
http://dx.doi.org/10.2138/rmg.2025.91A.07

A prominent example is the well-known Langmuir–Freundlich (LF) model, which can be derived by introducing a semi-Gaussian distribution of affinities (Sips 1948, 1950; Koopal et al. 2020). Proton binding and dissociation of NOM can be effectively addressed by the Langmuir–Freundlich (LF) approach (van Riemsdijk and Koopal 1992).

When considering proton reactions with NOM, the LF model is expanded to incorporate a description of electrostatic interactions. The electrostatic interactions can be handled with, for instance, a Donnan model as an approximation (Benedetti et al. 1996). For the adsorption of cations by NOM, modifications to the LF model are required, resulting in the NICA (Nonideal consistent Competitive Adsorption) approach (Koopal et al. 1994; Benedetti et al. 1995; Kinniburgh et al. 1999; Milne et al. 2001). This model accommodates variations in the affinity distributions among the various ions involved. The recognition of variable charge phenomena and chemical heterogeneity in the functional groups are pivotal ingredients for understanding the behavior of NOM materials.

Metal (hydr)oxides and natural organic matter are complex materials, each with its specific behavior. In a robust surface complexation model for the adsorption of NOM to metal (hydr) oxides, the major features of both materials must be accommodated. The Langmuir model can be considered the foundation of surface complexation modeling. In the model, unoccupied sites of the absorbent become occupied by an adsorbate, involving a single intrinsic interaction energy as the sites and adsorbate are considered homogeneous. For NOM, at least two levels of heterogeneity must be considered in relation to its adsorption to metal hydroxides. NOM is chemically heterogeneous at the level of the functional groups, but also concerning the NOM particles. This view is supported by model studies (Vermeer and Koopal 1998; Vermeer et al. 1998) in which self-consistent field (SCF) theory was applied to linear polyelectrolytes to gain insight into the factors that may control the adsorption behavior of NOM. These studies showed that the shape of the adsorption isotherms, in particular the absence of a distinct adsorption maximum, is due to polydispersity and a variation in affinity. To address the variation in adsorption affinities among different NOM particles, the LF approach provides a starting point for the development of mechanistic models that can describe the interaction of NOM with metal (hydr)oxide surfaces, as will be shown later.

To comprehend the interaction of NOM with metal (hydr)oxides, the adsorption of well-defined small organic acids to goethite has been investigated as a first step (Filius et al. 1997; Boily et al. 2000). These organic acids were modeled using the charge distribution (CD) model by distinguishing the formation of one or more surface complexes. The adsorption of fulvic acid has been evaluated along similar lines by Filius et al. (2000) by considering the formation of a limited number of discrete species, and more authors have followed this type of approach (Liang et al. 2019; Lodeiro et al. 2019; Chen et al. 2023)

A very different attempt to understand the interaction of NOM with metal (hydr)oxide was suggested by Hiemstra et al. (2010, 2013) by developing the NOM-CD model. These authors focused on the competitive interaction of adsorbed NOM with oxyanions. No relation between the NOM surface loading and a corresponding NOM concentration in solution was considered. In the NOM-CD model developed, adsorbed NOM is represented by a virtual component, denoted as ≡HNOM. This component consists of two carboxylic groups. One of these carboxylic groups reacts with a surface group of metal (hydr)oxides, while the other remains in the compact part of the electrical double layer (EDL). This leads to a competitive interaction with adsorbed oxyanions through site-binding and electrostatic interaction (Hiemstra et al. 2013). In the approach, the surface concentration of the ≡HNOM component is adjusted to fit the oxyanion equilibrium between the surface and solution. The derived surface concentration offers insight into the extent of surface interaction of NOM with the metal (hydr)oxides. It is important to emphasize that this adsorption model is not predictive as there is no thermodynamic relation between adsorbed NOM and NOM in the solution.

Modeling of NOM adsorption to metal (hydr)oxides with a discrete species approach has severe limitations, as the actual number of NOM species that can bind to metal (hydr) oxides is tremendous, frustrating a discrete approach with a manageable number of surface species. To overcome this, Filius et al. (2003) proposed a Ligand and Charge Distribution model approach for the adsorption of FA. This LCD model combines the NICA-Donnan and CD-MUSIC models that have been developed for the adsorption of ions to respective NOM and metal (hydr)oxides. In the LCD model, the speciation and charge of functional groups are the result of a statistical average instead of individual species. The average speciation of adsorbed NOM particles involves innersphere complex formation, (de)protonation of the carboxylic and phenolic groups, and binding of other cations present. It is calculated with the nonideal consistent competitive adsorption model (Kinniburgh et al. 1999) that accommodates the chemical heterogeneities of the reactive groups of NOM, while also offering the ease of using the NICA parameters that are available for a wide range of cations (Milne et al. 2001, 2003)

In the early version of the LCD model for FA adsorption (Filius et al. 2003), a formalism was used in which the intrinsic adsorption energy change was solely based on local interactions of the ligands in the adsorption phase with the surface, whereas thermodynamics considers the difference in the energy states of FA in adsorption phase and the solution phase. To resolve this thermodynamic problem, Weng et al. (2006b, 2007) changed to a formulation in which the adsorption energy is calculated from the change in the ligand and charge distribution of the NOM particle upon adsorption. The model for calculating the adaptation of the ligand speciation and the corresponding free energy change is named ADAPT (ADsorption and AdaPTation) as described in Weng et al. (2006a).

In the LCD approach, natural organic matter (NOM) is represented by a single class of humic nanoparticles (HNPs) with an average molar mass. For these HNPs, the chemical heterogeneity only exists at the level of the functional groups. However, humic and fulvic acids represent a collection of particles of different mass, size, aromaticity, and charge. In addition, ligands are not equally distributed over various particles in terms of types, affinities, and density (Schellekens et al. 2017). This implies that heterogeneity not only exists at the level of the ligands or functional groups but also at the level of the NOM particles. This variation in characteristics leads to preferential adsorption and a corresponding particle fractionation when HNPs are added to the suspension with metal (hydr)oxide. These aspects were not considered in the LCD models of Filius et al. (2003) and Weng et al. (2006b, 2007).

Additionally, in the LCD model of Weng et al. (2006b, 2007), the adsorbed humic nanoparticles are rigid entities with ligands occupying fixed electrostatic positions in the interface at different pH and loading conditions. For relatively large humic nanoparticles (HNP), a significant part of the ligands is assumed to be present outside the compact part of the EDL. However, the work of Hiemstra et al. (2013), using the NOM-CD model, challenged this assumption and demonstrated that at low NOM surface loading of relatively large humic particles, most of the charged ligands reside within the Stern layer region. This observation pointed to a great flexibility of the humic particles, allowing changes in molecular conformation. Recognizing this level of complexity is crucial for a more accurate representation of the interactions between NOM and metal (hydr)oxides.

Addressing the complexities highlighted above has motivated the further development of the LCD model, resulting in a consistent LCD model for the competitive adsorption of HNP, denoted as the LCD_{cc} model (Xu et al. 2024). The new approach incorporates three key features into the LCD framework: i) variation in the mean molar mass (M_w) of adsorbed NOM due to particle fractionation, ii) variation in the spatial distribution of NOM due to changes of the molecular conformation of adsorbed NOM as related to surface and solution conditions, and iii) various modes of the heterogeneity of NOM that contribute to a distribution of the adsorption affinities.

The development of the new LCD model used phosphate as a probe to investigate its competition with adsorbed HNP. This competition provides information about the distribution of ligand charge in the interface. Additionally, with the introduction of PO_4, the electrostatic field can be varied. This may alter the molecular conformation of the adsorbed HNPs as shown in a precursor of the LCD_{cc} model (Xu et al. 2022), revealing that the interfacial distribution of ligands is driven by the gradients of electrostatic potentials in the compact part of the EDL.

The present contribution aims to provide an overview of the current state-of-the-art in modeling the interaction between humic nanoparticles (HNPs) and metal (hydr)oxides, with a critical emphasis on recent advancements in the development of a robust and consistent Ligand and Charge Distribution adsorption model (Xu et al. 2024). The discussion will focus on the interfacial processes underlying the interaction of natural organic matter (NOM) with iron (hydr)oxides. By highlighting the achievements of the new model, this review offers insights into competitive adsorption and provides a perspective on future challenges in advancing the understanding of these intricate interactions at the interface of HNPs and metal (hydr)oxides.

THEORETICAL BACKGROUND OF THE MODELING APPROACH

The Ligand and Charge Distribution (LCD) model is a comprehensive framework that integrates the approaches of the Non-Ideal Competitive Adsorption model (NICA) and Charge Distribution and Multi-Site Ion Complexation model (CD-MUSIC) and that has been further developed by introducing aspects of the heterogeneity of humic nanoparticles that belong to the particle level. The new model (Xu et al. 2024) can describe in a consistent manner the adsorption and competitive behavior of humic nanoparticles. In the subsequent sections, we will dive into the details of the various modeling concepts encompassed within the new LCD_{cc} model framework.

Heterogeneous adsorption model for HNP

NOM encompasses a heterogeneous mixture of organic molecules, varying in chemical composition, types of functional groups and their densities, particle charge and size, as well as molar mass. At the particle level, heterogeneity can be implemented using a semi-Gaussian distribution around a median value for the overall adsorption affinity. Combined with the Langmuir approach, one may derive the well-known Langmuir–Freundlich (LF) equation. In the newly developed LCD_{cc} model, the equilibrium between the solution and the adsorbed fraction of the HNP at the surface is described with this relationship. The approach is fundamentally different from the original LCD model in which only a single intrinsic affinity was considered for an average HNP.

The central adsorption equation of the new LCD_{cc} model can be given as:

$$\frac{\phi_{ads}}{1-\phi_{ads}} = (\tilde{K}_o\, \phi_{sol})^q \tag{1a}$$

In this Langmuir–Freundlich equation, the surface (ϕ_{ads}) and solution (ϕ_{sol}) concentrations of the HNPs are expressed as volume fractions ($m^3\ m^{-3}$). The median value of the affinity is given as $\tilde{K}_o$ and the width of the affinity distribution is represented by the coefficient q. In the case of $q = 1$, the equation reduces to the expression for the Langmuir model. The LF equation may look very simple, but it is a challenge to calculate the median affinity $\tilde{K}_o$. The value of q is found by modeling data.

In Equation (1a), the median affinity reaction coefficient $\tilde{K}_o$ is not constant, but varies with the solution conditions and the surface loading. In the LCD_{cc} model, the value of $\tilde{K}_o$ will be calculated with the ADAPT module in combination with the NICA and CD-MUSIC module for HNPs with an average molar mass (M_w) that may change through selective adsorption.

In these modules, the local electrostatic potentials are key parameters. These potentials are related to solution conditions (pH, ionic strength), surface loading as well as the charge distribution of the ligands in the interface, following from the molecular conformation of the adsorbed HNPs. In the LCD_{cc} model, the molecular conformation is described by two parameters that both depend on the potential gradients in the interface. Both parameters are described with an electrostatic functional that has been developed for the model. These functionals for the molecular conformation are the core elements of the successful assessment of the variable adsorption energy expressed in the value of $\tilde{K}_o$

$$\frac{\phi_{ads}}{1-\phi_{ads}} = = (\tilde{K}_{p,nsp}\ \tilde{K}_{p,sp}\ \phi_{sol})^q \tag{1b}$$

Particle adsorption energy. In the LCD model framework, the adsorption energy of HNP can be calculated based on the change in the free energy state of the particles when transferred from the solution (HNP_{sol}) to the adsorption (HNP_{ads}) phase. As the surface is charge regulated, the adsorbed ligands will experience another electrostatic potential than in solution. This will change the speciation of the functional groups of the HNP and change the free energy state of the adsorbed particles. Below the corresponding expressions for the ADAPT (ADsorption and AdaPTation) model are presented (Weng et al. 2006b) that can be used to calculate the free energy change of the adsorption reaction in combination with the NICA and CD-MUSIC model.

In the adsorption process, charged HNPs are bound to a given surface without the accompanying counter ions, non-specifically bound. Formation of this hypothetical particle HNP^* from the HNP in solution requires the removal of the excess charge of the non-specifically bound counter ions present in the Donnan phase with a volume V_D ($L \cdot g^{-1}$) and potential ψ_D (V). The corresponding free energy of HNP^* formation for the equilibrium reaction $HNP_{sol} \rightleftharpoons HNP^*$ can be given as (Weng et al. 2006a):

$$\Delta F_{p,nsp} = \Delta F_{HNP^*} - \Delta F_{HNP,\,sol} = +RT\sum C_{i,0}(e^{-\frac{z_i\psi_D}{RT}} - 1)V_D\,M_w - 0 \tag{2a}$$

Here, $C_{i,0}$ ($mol \cdot L^{-1}$) represents the concentration of free ion i present in the equilibrium solution, and M_w is the molar mass of the HNP^* ($g \cdot mol^{-1}$). This free energy change can be translated into the corresponding reaction constant $K_{p,nsp}$ for $HNP_{sol} \rightleftharpoons HNP^*$ according to:

$$K_{p,nsp} = e^{-\frac{\Delta F_{p,nsp}}{RT}} = e^{-\sum C_{i,0}(e^{-\frac{z_i\psi_D}{RT}}-1)V_D M_w} \tag{2b}$$

Upon adsorption, the HNP* (without excess counter ions) enters the electrostatic field of the metal (hydr)oxide surface and may change in charge and corresponding chemical potential due to protonation or deprotonation reactions on the functional groups and formation of innersphere (IS) surface complexes. The corresponding specific energy term for the equilibrium reaction $HNP^* \rightleftharpoons HNP_{ads}$ can be calculated with:

$$K_{p,sp} = \prod_{l}\prod_{j=1}^{2}\left\{\left(\frac{1-\sum\theta_{i,j,sol}}{1-\sum\theta_{i,j,ads,l}}\right)^{\frac{N_{max,j,l}}{n_{H,j}\,p_j}}\right\}\left(\frac{B_D}{B_l}\right)^{N_{max,j,l}} \tag{3a}$$

in which $N_{max,j,l}$ is the total or maximum number of functional groups in $mol \cdot mol^{-1}$ for the carboxylic ($j=1$) and phenolic groups ($j=2$), located at a specific electrostatic plane of the interface, i.e., the surface ($l=0$), the first ($l=1$) and second ($l=2$) Stern plane, or at an additional plane of adsorption at a distance d outside the surface ($l=3$). The maximum number of functional groups ($mol \cdot mol^{-1}$) of the type of functional group i per location l can be given as:

$$N_{\max,j,l} = f_l\, Q_{\max,j} M_{\mathrm{w}} \tag{3b}$$

in which f_l is the fraction of the HNPs with a mean molar mass M_w (g·mol^{-1}) attributed to the various electrostatic planes l, and with site densities $Q_{\max,\,j}$ (mol·g^{-1}). In Equation (3a), $n_{\mathrm{H},j}$ and p_j are NICA model parameters that describe respectively the non-ideality and the width of the affinity distribution for ion adsorption; B_{D} is the Boltzmann factor for the Donnan phase of the HNPs in solution, and B_l represents the local Boltzmann factor for each electrostatic plane. In the above equation (Eqn. 3a), the fraction of the functional groups of type j present at location l that are complexed by a component i (protons or sites) is given as $\theta_{i,j,\mathrm{ads},l}$, and the fraction of the functional groups of type j, complexed with ion i (protons), is given as $\theta_{i,j,\mathrm{sol}}$. These fractions are calculated with the NICA approach because the functional groups have a non-ideal ion affinity distribution.

NICA model. In the LCD model framework, the NICA module describes the specific binding of ions or oxide sites to the two types of reactive ligands (carboxylic and phenolic types) of NOM, accounting for the intrinsic chemical heterogeneity of the ligands and ion-specific non-ideality. The ion loadings (θ_i) calculated for the HNPs in the solution and adsorption phase are used in the above ADAPT model (Eqn. 3a).

In the NICA model, the total amount of a specifically bound component i (N_i) is given by:

$$N_i = \sum_{j=1}^{2}\left\{N_{\max,j}\frac{n_{i,j}}{n_{\mathrm{H},j}}\theta_{i,j}\right\} = \sum_{j=1}^{2}\left\{N_{\max,j}\frac{n_{i,j}}{n_{\mathrm{H},j}}\frac{\left(\tilde{K}_{i,j}C_i\right)^{n_{i,j}}}{\Sigma\left(\tilde{K}_{i,j}C_i\right)^{n_{i,j}}}\frac{\left\{\Sigma\left(\tilde{K}_{i,j}C_i\right)^{n_{i,j}}\right\}^{p_j}}{1+\left\{\Sigma\left(\tilde{K}_{i,j}C_i\right)^{n_{i,j}}\right\}^{p_j}}\right\} \tag{4}$$

where $N_{\max,j}$ is the total number of the type j ligands (carboxylic or phenolic type) per mole particles; $\theta_{i,j}$ is the fraction of the type j ligand that is complexed with i; $n_{i,j}$ is the non-ideality parameter for ion i binding to the type j ligand; $n_{\mathrm{H},j}$ is the non-ideality parameter for proton adsorption to the type j ligand; $\tilde{K}_{i,j}$ is the median value of the affinity distribution for ion i; C_i (mol·L^{-1}) is the local concentration of component i; p_j is the width of the affinity distribution for the type j ligand, representing the intrinsic heterogeneity of the ligands. In the above NICA equation for the LCD model, N_i and $N_{\max,j}$ are given in the unit mol·mol^{-1}, whereas generally the unit mol·kg^{-1} is used. This choice is consistent with the unit used in the ADAPT model (Eqn. 3b) to calculate the free energy of HNP* adsorption.

In the application of the NICA model, the parameters for the adsorbed HNPs are assumed equal to those of the HNPs in the solution. For the formation (Eqn. 5) of the innersphere complexes of the carboxylic ligands ($RCOO^-$) of adsorbed HNPs with the metal ion of the singly coordinated surface groups of goethite for example ($\equiv FeOH_2^{+1/2}$), we used an optimized complexation constant of $\log\tilde{K}_{s,1} = -1$ (Weng et al. 2008). The negative charge of the carboxylic ligand ($\equiv RCOO^-$) in the reaction (Eqn. 5) is evenly distributed between the 0- and 1-plane.

$$\equiv FeOH_2^{+0.5} + \equiv RCOO^- \rightleftharpoons \equiv FeOOCR^{-0.5} + H_2O(l);\ \log\tilde{K}_{s,1} = -1 \tag{5}$$

Double layer structure. For large HNPs such as HA, only a part of the reactive ligands can be accommodated in the Stern layers of the electrical double layer (EDL) at high-loading conditions. The remaining part will protrude and enter the diffuse layer (Hiemstra et al. 2013; Deng et al. 2019) depending on the pH- and loading-dependent maximum allowed in the Stern layers (Fig.1), as clarified later (Eqn. 7). In the LCD model, the extended Stern layer model is adapted to allow the accommodation of a part of the ligands beyond the Stern layers. When the ligands enter the diffuse double layer, they contribute to the total charge of the DDL. In principle, the diffuse distribution of ligands and ions can be calculated as demonstrated in Weng et al. (2007). Implementation of these types of calculations showed that similar modeling results can be obtained with the introduction of an additional electrostatic plane at a distance d from the surface or a distance Δx from the second Stern plane. Beyond the d-plane, the potential for the counter- and co-ions adsorption is treated according to Gouy–Chapman's theory (Chapman 1913; Sposito 2018).

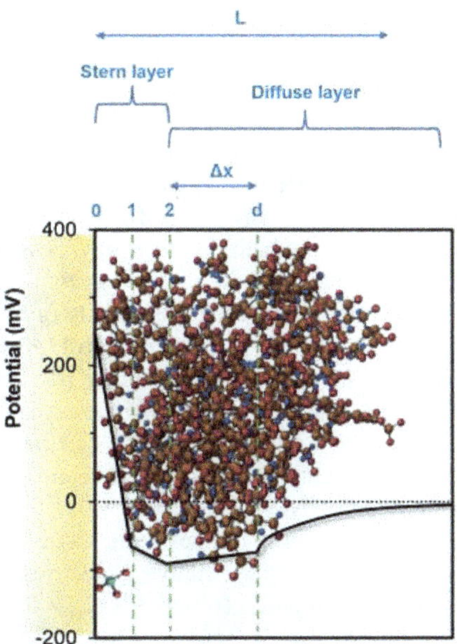

Figure 1. Schematic representation of the interfacial adsorption of a soft HNP and rigid phosphate ion onto a metal (hydr)oxide surface, showing the electrostatic potential profile in the electrical double layer (EDL) for 0.01 M NaCl background solution of pH 4, calculated with the Ligand and Charge Distribution framework for consistent modeling of the competitive adsorption (LCD_{cc}) for a specific loading. The x-axis reflects the distance from the surface. The charge of the HNP ligands is distributed over four electrostatic planes (0, 1, 2, and d-plane). L is the layer thickness of the adsorbed HNP being $L = 2\,\Delta x + d_{ST}$, in which Δx is the distance from the 2-plane to the d-plane and d_{ST} is the thickness of the Stern layer region (~0.8 nm). The typical value of L is about 3 nm. Size exclusion chromatography (SEC), calibrated with polystyrene sulfonate standards of various sizes with an assumed mass density of 800 $kg{\cdot}m^{-3}$, reveals a typical size range of ~1.5–7.5 nm with a mean size ~ 2.5 nm for the HNP used. Translation to a mean molar mass yields 5.7 kDa at an assumed mass density of 1250 $kg{\cdot}m^{-3}$. Most HNP ligands reside within this layer. The hydrodynamic layer thickness may be larger due to some tailing of segments (Vermeer et al. 1998). The amount of electrolyte ions located in the layer between the 2- and d-plane has been calculated using the average potential of both planes. The corresponding charge was attributed to the d-plane.

Figure1 demonstrates the electrostatic potential profile in the EDL for typical mean conditions in our experiments. At low pH, the surface is positively charged. The adsorption of negatively charged ligands of the HNPs and, for instance, abundantly adsorbed oxyanions, leads to a charge reversal and negative potentials at the Stern planes as well as at the additional adsorption plane (d). The potential profile around the center of the adsorbed HNP is rather flat in the given example. This rather constant potential resembles a Donnan phase behavior.

Based on the charge, the potentials at 0-, 1-, and 2-planes (ψ_0, ψ_1, ψ_2) are calculated using the classical Stern layer capacitor formulations. HNPs may also contribute to the diffuse layer (DL) by adding some charge that is attributed to the additionally defined d-plane at a distance Δx from the compact part of the EDL (Fig. 1). The potential at this additional adsorption plane (ψ_d) depends on its location:

$$\left(\frac{\psi_d - \psi_2}{\Delta x}\right) = \frac{\sigma_{DL}}{\varepsilon_r \varepsilon_0} \tag{6a}$$

which can be derived by combining the expressions (Eqn. 6b) for the capacity C of a capacitor related to the plate distance Δx and the potential difference $\Delta\psi$ between the plates in relation to excess charge σ_{DL} in the DL layer between the d-plane and infinity:

$$C_{\Delta x} = \frac{\varepsilon_r \varepsilon_0}{\Delta x} \quad \text{and} \quad C_{\Delta x} = \frac{\sigma_{DL}}{\psi_d - \psi_2} \tag{6b}$$

In our model, the distance Δx has been set at half of the maximum distance of HNPs in the region outside the first two Stern layers (Fig. 1). If L is the maximum layer thickness of adsorbed HNPs, then $\Delta x = ½\,(L - d_{ST})$ in which d_{ST} is the thickness of the Stern layer. The total layer thickness L (m) can be estimated by dividing the maximum loading of adsorbed HNPs ($g{\cdot}m^{-2}$) measured at a given ionic strength by the mass density ($g{\cdot}m^{-3}$) of adsorbed HNPs. The mass density of the adsorbed fraction of HNPs has been set to 1250 $kg{\cdot}m^{-3}$ (Xiong et al. 2018), which is larger than the density of well-hydrated HNPs in the free solution (700 $kg{\cdot}m^{-3}$; de Wit et al. 1993) but smaller than the mass density of HNPs in the dry state (1700 $kg{\cdot}m^{-3}$; Birkett et al. 1997). The in-between value chosen gives credit to the occurrence of some dehydration of HNPs upon adsorption, caused by innersphere complexation and the observed conformational changes, both compressing adsorbed HNPs. The chosen value for a maximum distance L of 3 nm leads to $\Delta x = (L - d_{ST}) / 2 = 1.2$ nm. The resulting value of Δx can be translated to a corresponding capacitance for a value for the relative dielectric constant. Using the relative dielectric constant of water ($\varepsilon_r \sim 80$), Equation (6b) yields $C_{\Delta x} = 0.6$ $F{\cdot}m^{-2}$.

Spatial ligand distribution. In the LCD_{cc} model, the charge of the ligands depends on their location in the electrical double layer. Therefore, the spatial distribution plays a pivotal role in calculating the adsorption affinity of HNPs, with particular significance for large HNPs (Li et al. 2022). The spatial distribution of the ligands of HNPs in the metal (hydr)oxide interface may vary depending on loading and pH conditions. Under conditions of low loading and low pH, HNPs tend to preferentially occupy the compact part of the electric double layer (EDL). Conversely, at higher loading and pH, the HNPs may extend to occupy a portion of the double layer (DL), outside the compact part of the interface (Hiemstra et al. 2013; Deng et al. 2019). Understanding the spatial distribution under different conditions will contribute to an accurate representation of the adsorption behavior of HNPs.

Direct visualization of the spatial distribution of NOM is lacking (Newcomb et al. 2017). Spatial information is typically obtained at the micrometer scale rather than at nanometer scale resolution using in-situ spectroscopic methods (Kleber et al. 2007). This limitation is particularly evident as recent investigations suggest that HNPs are not rigid entities and may alter their molecular conformation upon adsorption (Hiemstra et al. 2013; Xiong et al. 2015). These conformational changes are likely driven by a strong interaction of the functional groups of HNPs with the surface where electrostatic attraction or repulsion plays a prominent role.

The spatial distribution of the HNP is *a priori* unknown but may be assessed through modeling defining a set of distribution equations. The ligands of adsorbed HNPs can be present in three layers and the sum of the fractions in each layer equals one (Eqn. 7a). At low surface loading, the HNPs tend to predominantly occur in the compact part of the EDL (Hiemstra et al. 2013). This is defined in Equation (7b). The HNPs will first occupy the compact part of the EDL ($f_{0+1} + f_{1+2} = 1$) up to a maximum loading allowed in these Stern layers, given as $\Gamma_{MST}/\Gamma_{tot}$. Any excess of ligands is attributed to the additional adsorption d plane beyond the compact part of the EDL, as illustrated in Figure 1.

$$f_{0+1} + f_{1+2} + f_{d} = 1 \tag{7a}$$

$$f_{d} = \frac{\Gamma_{tot} - \Gamma_{MST}}{\Gamma_{tot}} \text{ for } \Gamma_{tot} > \Gamma_{MST};\ f_{d} = 0 \text{ for } \Gamma_{tot} \leq \Gamma_{MST} \tag{7b}$$

$$\theta_{s} = \frac{\Gamma_{MST}}{\Gamma^{o}_{MST}} \tag{7c}$$

$$R = \frac{f_{0+1}}{\left(f_{0+1} + f_{1+2}\right)} \tag{7d}$$

Modeling has indicated that the maximum adsorption in the Stern layers (Γ_{MST}) is conditional and generally lower than the physical adsorption maximum for the Stern layers (Γ^{o}_{MST}). During model development, the ratio θ_s between the maximum loading Γ_{MST} at a given condition and the physical maximum Γ^{o}_{MST} (Eqn. 7c) has been explored. It turned out that this ratio is a function of the potential gradient in the compact part of the EDL ($\psi_0 - \psi_2$), as will be discussed later.

As mentioned, the restriction in filling the Stern layer space (θ_s) is not a constant but varies according to our model evaluation with the solution chemistry that determines the potential profile such as pH, ionic strength, and loading with HNP. At low pH, the molecular conformation is predominantly influenced by surface complexation, leading to a preferential filling of the Stern layer. At these conditions, the relative maximum of the Stern layer occupation (θ_s) is close to 1, aligning with literature data on the adsorption of humic acid (HA), probed with phosphate ions (Hiemstra et al. 2013). Conversely, at high pH, the metal (hydr)oxide surface becomes negatively charged, creating strong repulsive forces on the negatively charged functional groups, resulting in repulsive forces acting on the negatively charged ligands. This repulsion tends to move the ligands towards and into the diffuse layer. Refinement of this approach suggests that at high loading, the molecules exhibit a near-spherical shape, as illustrated later (Fig. 2).

Closer to the surface, fine-tuning of ligand distribution is crucial. Here, the potential gradients are significant, and the ligands will respond to these gradients. The relative number of ligands occupying the inner Stern layer is defined in Equation (7d). This relative amount is conditional and found to be a function of the potential gradient, in this case ($\psi_0 - \psi_1$).

When phosphate ions adsorb to metal (hydr)oxides, they form innersphere complexes that attribute charge to the 0-plane and 1-plane, thereby strongly changing the fraction of functional groups present in the inner Stern layer (R). At high pH or a near-zero gradient ($\psi_0 - \psi_1 \approx 0$), modeling suggests very low values for R, implying almost no innersphere complexes with the metal ions of the (hydr)oxide surface, which also agrees with the quite low total HNP adsorption at these conditions.

Spatial distribution functionals. Through the modeling of individual data points collected over a broad range of conditions, a comprehensive understanding of the interfacial ligand distribution has emerged. This distribution can be effectively described by the two parameters defined in Equations 7(c,d), namely the maximum occupation of the compact part of the EDL (θ_s) and the relative ligand distribution (R), as illustrated in Figure 2 along with an example of the modeling results for the pH-dependent adsorption of HA at three levels of addition to goethite. Both conformational parameters can be linked to potential gradients as illustrated in Figures 2b,c. The lines in these figures represent the calibrated functionals, defined as:

$$\theta_{S} = \theta_{min} + \Delta\theta_{max} \frac{1}{1 + ae^{-b\,\Delta\psi_{02}}} \tag{8a}$$

$$R = R_{\min} + \Delta R_{\max} \frac{1}{1+c\ e^{-d\,\Delta\psi_{01}}} \tag{8b}$$

where θ_{min} and R_{min} are respectively the minimum values of θ_s and R when the electrostatic potential gradients are very small; $\Delta\theta_{max}$ and ΔR_{max} are, respectively, the maximum values of θ_s and R that are reached when the electrostatic field is highly attractive; $\Delta\psi_{02}$ and $\Delta\psi_{01}$ are gradients of electrostatic potential over 0- and 2-plane and 0- and 1-plane, respectively; a, b, c, and d are constants, determining the shape of the sigmoidal curves of θ_s and R.

The relative maximum of occupation of the Stern layer (θ_s) can be interpreted in terms of a mean molecular conformation. At low pH and loading, where the potential gradients over the compact part of the EDL are substantial, the value of θ_s approaches one. This implies that all Stern layer space is available for the ligands as depicted in Figure 2b, allowing stretching of the molecules and maximizing their interaction with the sites of the metal (hydr)oxide surface. Conversely, at high pH and loading, the potential gradients become relatively small, decreasing the relative maximum occupation (θ_s) to values slightly lower than 0.5 (Fig. 2b). It can be demonstrated (Xu et al. 2022) that this aligns with a more spheroidal shape of the HNPs.

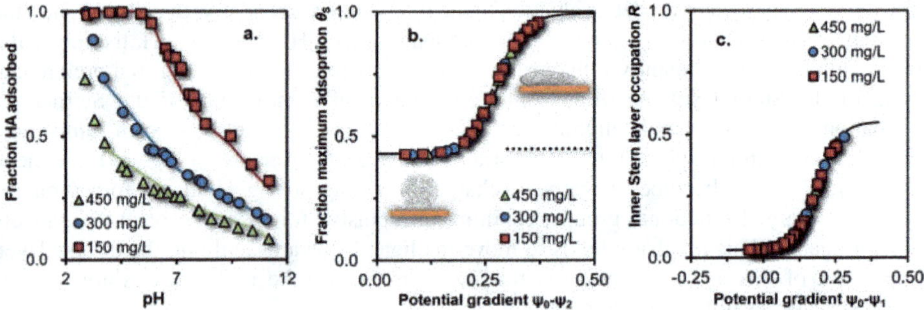

Figure 2. **(a)** pH-dependent HA adsorption to goethite measured by Weng et al. (2007) (symbols) and modeled (lines) with the consistent competitive LCD (LCD_{cc}) model using the heterogeneity coefficient $q = 0.1$. **(b)** Relative maximum occupation of the Stern layer (θ_s, Eqn. 8a) as a function of the potential gradient between the 0- and 2-plane ($\psi_0 - \psi_2$). **(c)** Fraction of reactive ligands in the inner Stern layer over the reactive ligands in the compact part of the EDL (R, Eqn. 8b) as a function of the potential gradient between 0- and 1-plane ($\psi_0 - \psi_1$). The results are for goethite (1 $g \cdot L^{-1}$) in 0.01 M $NaNO_3$ at three initial HA loadings, 150 $mg \cdot L^{-1}$ (**squares**), 300 $mg \cdot L^{-1}$ (**circles**), and 450 $mg \cdot L^{-1}$ (**triangles**). The **dotted line** in (b) refers to the minimum value of θ_s (~ 0.45). The **solid lines** in (b) and (c) have been calculated with the functionals for the ligand distribution of adsorbed HA (Eqns. 8a,b), having minimum and maximum values for θ_s and R, as discussed in the text. Apart from an adjustable conformational change, calculated with the calibrated functionals of Equation (8), the only adjustable parameter in the modeling is the heterogeneity constant ($q = 0.1$) used in the Langmuir–Freundlich approach (Eqn. 1), in which the median affinity constant $\tilde{K}_o$ has a specific and non-specific contribution calculated with ADAPT module of the LCD_{cc} model (Eqn. 2). Both affinity constants are calculated using the output of NICA model calculations for adsorbed and soluble HA.

Fractionation. Due to the polydisperse nature of NOM and the variation in the affinity of the particles for adsorption to mineral surfaces, HNPs such as HA or FA exhibit molecular fractionation when added to a metal (hydr)oxide suspension (Lv et al. 2018). This phenomenon leads to a change in the mean molar mass of the adsorbed humic substances (Ko et al. 2005; Kang and Xing 2008; Janot et al. 2012; Qin et al. 2015; Wang et al. 2019; Xu et al. 2019). Size exclusion chromatography (SEC) data show that for adsorbed HA, the mean molar mass (M_w) decreases due to selective adsorption (Wang et al. 2019; Xu et al. 2022). The decrease of M_w

of adsorbed HA shows a linear relation with the percentage of added HA that was adsorbed, as depicted in Figure 3a for the fractionation of HA by goethite. The data suggest that smaller HA molecules become increasingly important in the adsorption phase as the adsorbed fraction decreases. In contrast, for FA (~1.7 kDa), there is very little adsorptive fractionation in terms of particle size (data not shown).

In an earlier investigation of the adsorption of the same HA to goethite, Xu et al. (2022) expressed the change of M_w (kDa) as a function of its degree of adsorption using a fractionation sensitivity factor *k*.

$$M_w = M_o + k\left(1 - \frac{\rho_{ssr} C_{HA}}{HA_{tot}}\right) \tag{9}$$

where ρ_{ssr} is the solution-to-solid ratio (i.e., 1.0 $L{\cdot}g^{-1}$ goethite), C_{HA} ($mg{\cdot}L^{-1}$) is the HA concentration in solution, and HA_{tot} ($mg{\cdot}g^{-1}$) is the total amount of HA added. In Equation (9), M_o is the molar mass of the most preferred HA particles in the adsorption and *k* is the fractionation factor (kDa). Equation (9) can be used to describe the fractionation as a function of the HA concentration left in solution at different values for the pH and solution-to-solid ratio (SSR), as depicted by the lines in Figure 3b.

Returning to Figure 3a, when the fraction of adsorption is lower, there is a corresponding decrease in the molar mass of adsorbed HA. This suggests a preference of the surface for HA particles of lower molar mass. Extrapolation of the linear relationship to zero values yields a molar mass of only $M_o \sim 3$ kDa. The slope of the line in the plot represents the value *k* of Equation (9). This coefficient expresses the sensitivity of humic material to fractionation. Consistent modeling of adsorption data, collected at three electrolyte levels, indicates that the selectivity coefficient *k* is ionic strength-dependent, as illustrated in Figure 3c.

The observed fractionation can be generalized to systems involving HA and PO_4, thus covering a wider range of fractionations. Integrating all data, the measured M_w of adsorbed HA follows a single linear relationship with the adsorbed fraction of added HA, irrespective of a large variation in factors such as pH and the initial loading with HA, FA, and PO_4 that may all alter the percentage of HA adsorbed.

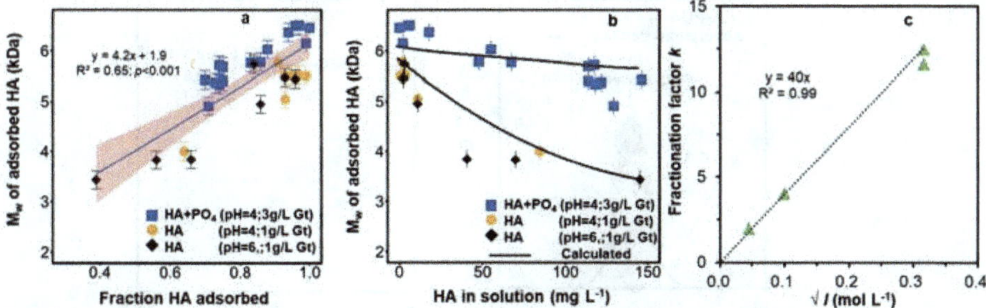

Figure 3. Mean molar mass of adsorbed HA as a function of the fraction adsorbed **(a)** and the HA concentration in the equilibrium solution **(b)** in the system with and without phosphate addition at pH 4 and 6 (background: 1.0 $g{\cdot}L^{-1}$ goethite; 0.01 M NaCl). The mean molar mass of the adsorbed fraction was derived from the difference between the SEC chromatograms (retention time 7.0–15 min, 50 Da–50 kDa) of original HA and HA remaining in solution after adsorption. The **solid line** in Figure 3a is the linear regression for M_w as a function of the fraction of HA adsorbed fitted based on the ordinary least square method (de Souza and Junqueira 2005). The **solid lines** in Figure 1b are the mean molar mass of adsorbed HA calculated with Equation (9). **(c)** Effect of ionic strength on the fractionation factor *k* of HA Equation (9) derive by modeling. The mean molar mass of adsorbed HA at a given fractionation increases with the root of the ionic strength (Weng et al. 2007; Xu et al. 2022). [CC-BY-NC by Xu et al. (2022).]

In our LCD_{cc} model, the above description of the fractionation has been incorporated by using the calculated molar mass of HA in the adsorption phase (Eqn. 9) as input in the calculations. Looking forward, a future challenge lies in the development of a mechanistic model for the process of fractionation, allowing predictions rather than descriptions.

Analysis of the variation in affinity of HNP adsorption. The affinity of NOM for adsorption to metal (hydr)oxides is related to numerous factors. Primary factors are the molar mass and size as well as the density and acidity of the functional groups, as will be shown below, Additionally, hydrophobicity (Ko et al. 2005; Avneri-Katz et al. 2017), aromaticity (Reiller and Moulin 2006; Claret et al. 2008; Janot et al. 2012; Avneri-Katz et al. 2017; Coward et al. 2019), and the presence of aliphatic groups (Rahman et al. 2013; Xu et al. 2019) may also play a role.

When optimizing the LCD model, we found that our data for HA adsorption can be best described using the LF approach (Eqn. 1) with a parameter for the width of the distribution of $q = 0.1$. This rather low q value indicates that the HA particles exhibit a wide range of adsorption affinities, suggesting a very large heterogeneity at the particle level. This heterogeneity can be understood from a variation in molar mass, as measured with SEC, and a variation in the density of functional groups, in particular carboxylic groups among particles, as follows from Figure 4.

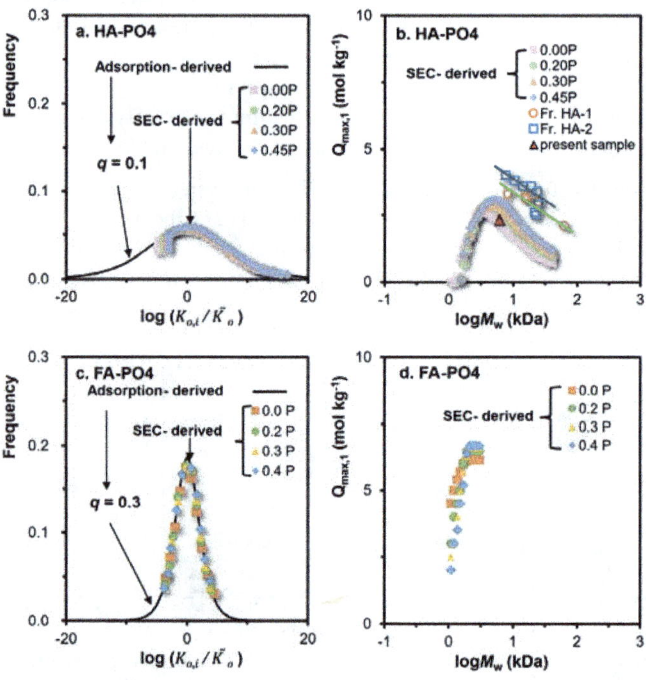

Figure 4. Left panels: Relative affinity distribution according to the adsorption data in 0.01 M $NaNO_3$ interpreted with LF model (Eqn. 1) using $q = 0.1$ (**full line**) compared to the calculated affinity distribution using the molar mass distribution measured with SEC for adsorption series of $x\mathrm{PO_4}$+[450HA] **(a)** and $x\mathrm{PO_4}$+[200FA] **(c)**, in combination with an adaptation for each mass fraction of the carboxylic group density Q_{max} (**symbols**). **Right panels:** Relationship between the optimized carboxylic group density ($Q_{max,1}$) and the SEC-measured molar mass ($\log M_w$) of the HA **(b)** and FA **(d)** in the distribution. The **open symbols** in b) are for fractionated HA by Christl et al. (2000) and Li et al. (2003), showing similar trends. [CC-BY by Xu et al. (2024).]

The line in Figure 4a shows the calculated affinity distribution of HA for a value of $q = 0.1$ found by adsorption modeling. The distribution has been calculated with the expression for the semi-Gaussian distribution of the affinity constant K_i, according to de Wit et al. (1990):

$$f(\log K_i) = \frac{\ln(10)\sin(q\pi)}{\pi\left[\left(\frac{K_i}{\tilde{K}_o}\right)^{-q} + 2\cos(q\pi) + \left(\frac{K_i}{\tilde{K}_o}\right)^{q}\right]} \tag{10}$$

where $\tilde{K}_o$ is the median value of the affinity distribution. This expression in combination with the Langmuir equation has also been used to derive the Langmuir–Freundlich equation (Sips 1948, 1950; de Wit et al. 1990; Koopal et al. 2020; Xu et al. 2024) applied (Eqn. 1).

The distribution of adsorption affinities of HA (Eqn. 10) has been used to evaluate the factors that contribute to the width of the distribution. According to the ADAPT model (Eqn. 2), the adsorption affinity of humic nanoparticles is determined by two major factors, i.e., the molar mass M_w (size) of the particle and the density of the functional groups Q_{max}. Both values are combined in the parameter N_{max} (Eqn. 2c). The variation in molar mass M_w has been measured with SEC. With this factor alone, the width of the distribution cannot be fully explained. To derive a close agreement between the theoretical affinity distribution according to Equation (10) and the distribution expected from the two major factors for the adsorption affinity according to the ADAPT model, the corresponding variation in carboxylic group density (Q_{max}) has been optimized. The results of this optimization are depicted in Figure 4b and indicate that there is a relation with the molar mass or size of the particles.

The calculated variation in carboxylic group density with molar mass can be compared with experimental data for fractioned HA, reported by Christl et al. (2000) and Li et al. (2003). Both papers report a decrease in the carboxylic group density with an increase in the molar mass of separated HA fractions. This negative trend with the molar mass is also found in our evaluation for the major portion of HA in the distribution (solid-colored symbols in Fig. 4b). However, at a low molar mass, the opposite trend is obtained. Then, the values of Q_{max} increase with molar mass, meaning that there is a maximum at intermediate values of the molar mass. For the smallest ones, the carboxylic groups are nearly absent. This suggests the presence of neutrals or low-charged molecules in the HA preparation, which agrees with earlier suggestions (Janot et al. 2012). Maybe these can be removed from a HA stock suspension using XAD resins.

For FA, similar approaches can be done (Figs. 4c,d). The variation in adsorption affinity of FA is less since for the parameter for the width of the distribution a value of $q = 0.3$ is found, indicating that FA is less heterogeneous than HA at the particle level. For FA, a positive correlation between carboxylic group densities and molar mass is found. Smaller-sized FA in the distribution has a lower carboxylic density than larger FA particles.

Vermeer et al. (1998) have applied SCF theory to a linear polyelectrolyte to explore the behavior of a pseudo-humic acid. The linear polyelectrolyte was a combination of neutral, carboxylic ($\log K_H = 4$), and phenolic ($\log K_H = 10$) segments, with a chain length of 200 segments in total. In all cases, high-affinity isotherms with a flat plateau were found to have a maximum that increased with the interaction energy of the carboxylic segments with the surface, the introduction of phenolic segments, and a decrease in solvent interaction, i.e., more hydrophobicity. A steady increase of the adsorption without reaching a clear plateau could not be explained by the mentioned factors used in the SCF approach, while it is typical for natural HNPs. By defining a mixture of two polyelectrolytes with different molar mass and carboxylic density, a steady increase of the adsorption without reaching a plateau could be mimicked (Vermeer and Koopal 1998). This finding aligns with our LCD modeling in which the particle heterogeneity, expressed in q, is due to a combined variation in molar mass and density of carboxylic groups.

APPLICATION AND PRACTICAL IMPLICATIONS

HNPs in natural soil systems

Soil organic carbon (SOC) constitutes a substantial fraction of the organic carbon in the world's terrestrial environment, playing a vital role in the global carbon cycle (Lal 2004). The adsorption of SOC onto soil minerals is considered an effective protective mechanism, regulating the long-term preservation of natural organic carbon and contributing to climate change mitigation. Iron and aluminum (hydr)oxides are recognized as significant soil minerals that bind and stabilize SOC (Kögel-Knabner et al. 2008). Understanding and predicting the adsorption behavior of soil organic carbon is challenging due to the inherent complexity and heterogeneity of both SOC and natural metal (hydr)oxides, as well as the varying influence of environmental conditions and land-use practices.

Organo-oxide aggregates. The dominant binding mechanism of organic matter involves the reaction of the carboxylic/hydroxylic moieties (Vermeer and Koopal 1998; Weng et al. 2008; Hiemstra et al. 2013; Lv et al. 2018; Coward et al. 2019), according to spectroscopic analysis and LCD_{cc} modeling. Outersphere complexation may occur too. These mechanisms are illustrated in Figure 5a, zooming in to the very nanometer scale.

In soils, the reactive surface area is predominantly determined by the fraction of metal (hydr)oxide nanoparticles in the size range of 1.5–5 nm (Mendez et al. 2020b, 2022; Bai 2023). These particles are associated with natural organic matter. Quantification of the fraction of mineral-associated organic carbon (MAOC) can be done with mass density fractionation (Bai 2023; Bai et al. 2024). The connection with the natural metal hydroxide fraction can

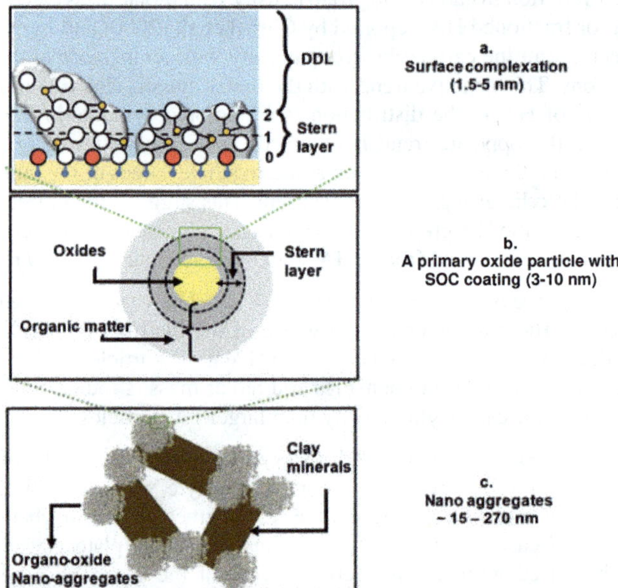

Figure 5. Conceptual diagrams showing the formation of organo-mineral associations at different length scales. In the **upper panel**, surface complexation is portrayed at the scale of the compact part of the interface (1 nm). The **middle panel** illustrates a single primary organo-oxide particle with a metal (hydr)oxide core, enveloped by a coating of organic matter, having in total a typical size in the range of 5–10 nm. The **lower panel** depicts an organo-mineral micro-aggregate consisting of aggregated organo-oxide particles attached to other silicates such as clay minerals, acting as a support. The organo-oxide nano aggregates have a typical size of ~15–270 nm, while micro-aggregates in soil will be larger (Asano et al. 2018; Mendez et al. 2020b; Xu et al. 2022; Bai et al. 2024).

be elucidated by modeling these data with a core–shell approach. In this approach, the primary metal (hydr)oxide particle forms the core and associated organic matter is present in a surrounding shell, as depicted in Figure 5b. Interpretation of the measured amount of associated organic matter indicates a strong correlation between the equivalent thickness of the organic layer L and the mean size of metal (hydr)oxide core (d). Moreover, both are of similar size (Mendez et al. 2020b, 2022; Bai 2023).

The obtained core–shell model results can also be interpreted as the presence of a collection of nano (hydr)oxide and humic nanoparticles of similar size, organized in nano-aggregates. These nano-aggregates represent the primary level in the hierarchy of aggregate formation. The organo-oxide nano-aggregates typically range in size from ~15–270 nm, which can be assessed from phosphate desorption kinetics, as outlined in another contribution to this volume (Hiemstra et al. 2025). Mass density fractionation measurements suggest additionally an association with silicate particles (Bai 2023), creating larger aggregates. Figure 5c depicts an organo-mineral micro-aggregate consisting of aggregated organo-oxide particles attached to clay mineral surfaces that serve as a support.

Organo-oxide interaction. The above structural picture of organo-oxide aggregates, consisting of a collection of metal (hydr)oxide and humic nanoparticles, can be interpreted by comparison with the outcomes of LCD_{cc} modeling. Specifically, we can focus on the attachment of a HNP to a metal (hydr)oxide surface as depicted in Figure 1, representing sub-neutral pH conditions and a relatively high HNP loading. At these conditions, the particle exhibits a rather spheroidal shape. The calculated electrostatic potential profile indicates a relatively constant negative potential coinciding with the center of the adsorbed HNP.

If the attached HNP nanoparticle of Figure 1 is shared between two or more surfaces, the resulting picture resembles the situation of the HNPs in the organic-oxide phase depicted in Figure 5c. The HNPs between the oxide surfaces experience a double-layer overlap, enhancing the potential of the internal section of the HNPs that has a relatively constant electrostatic potential. This internal potential becomes then rather comparable with that of a Donnan phase, see (Hiemstra et al. 2025). The ligands of the HNP at the exterior may interact with the surfaces of the metal (hydr)oxide, causing deviations from this Donnan-like potential.

A notable difference between our model systems and natural organo-oxide systems lies in the presence of Ca^{2+} and Mg^{2+} ions. These ions reduce the electrostatic repulsion between the ligands inside the HNPs, contributing to a stabilization of the spheroidal shape of HNPs. Metal ion bridging may additionally contribute to structural condensation. This view supports the idea that the structure of organic-oxide nanoaggregates comprises a collection of spheroidal HNP particles, chemically interacting with neighboring oxide particles of a similar mean size.

The LCD_{cc} modeling results indicate that the molecular conformation and distribution of ligands of HNP adsorbed to a metal (hydr)oxide surface, are primarily determined by pH, HNP loading, and the presence of adsorbed anions like phosphate. The interaction of ligands with the surface through innersphere complexation plays an important role in the linking of the primary particles within the organo-oxide aggregates. Depending on the pH, the degree of this interaction may vary.

As the pH decreases, the ligands of HNPs undergo a shift of location at the interface, increasingly entering the compact part of the double layer. This shift has implications for the adsorption of oxyanions, to be discussed below, as recently observed in soil studies (Bai et al. 2024).

Understanding competitive interactions of oxyanions with NOM

In complex media such as soil, the fate of ions, nutrients, and pollutants is determined by their interaction with the solid phase. Natural organic matter, metal (hydr)oxides, and clay minerals are key materials for the reaction with organic and inorganic ions. Surface complexation models serve as valuable tools for understanding and predicting their availability and mobility in the natural environment. Natural organic matter and metal (hydr) oxide are important for ion-specific adsorption reactions. Traditionally, the reactivity of both materials is separately treated in surface complexation modeling. However, a comprehensive understanding, description, and prediction of the reactivity is only possible if, in addition, the interaction between both phases is considered. This is particularly important for anions, specifically oxyanions. The LCD_{cc} model contributes to elucidating these interactions.

In soils, oxyanions mainly adsorb to metal (hydr)oxide and (hydr)oxide-like surfaces. This interaction occurs through innersphere complexation. Carboxylic groups of NOM can do the same. Consequently, there is, almost by definition, a competitive interaction in soil. The adsorptive interaction between phosphate and mineral-associated organic carbon (MAOC) can be seen as an example of great agricultural and environmental importance. Insight into the interaction of PO_4 with the organo-oxide phase of soils is required to maintain or enhance soil fertility for plant growth while mitigating eutrophication and climate change.

NOM-oxide model systems. Figure 6a depicts the adsorption isotherm of phosphate to goethite in the presence of variable amounts of either humic (HA) or fulvic acid (FA). The corresponding simultaneous adsorption of either HA or FA is depicted in Figures 6b and 6c, respectively. The solid lines are obtained by applying the LCD_{cc} model using the spatial ligand distribution functionals (θ_s, R), and the affinity distribution of the HA particles (q) as the only two adjustable parameters. These parameters were obtained by simultaneously optimizing the datasets for the competitive adsorption in PO_4–HA and PO_4–FA systems as well as the adsorption of PO_4, HA, and FA in monocomponent systems. Only by introducing heterogeneity at the particle level (q), the model can correctly describe the simultaneous adsorption of either HA or FA with phosphate (Fig. 6).

The competition of FA is more pronounced compared to the competition of HA with phosphate, as illustrated in Figure 6a. According to LCD_{cc} modeling (Xu et al. 2024), the stronger competition of FA with PO_4 can be ascribed to the smaller particle sizes of FA and higher carboxylic group densities. This suggests that the small, highly charged NOM molecules, potentially including simple organic acids, play a relatively significant role in the competition with PO_4. It is important to emphasize that this conclusion is only valid at equal amounts of NOM, present on a mass basis.

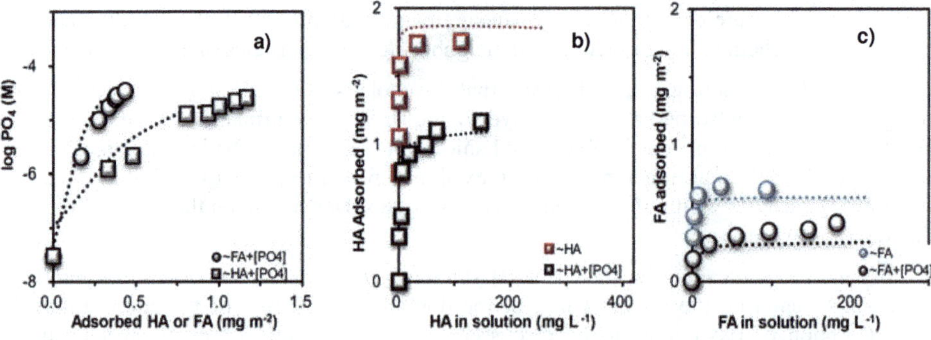

Figure 6. Competitive interaction of phosphate (**a**), HA (**b**), and FA (**c**), measured for goethite (3 g·L^{-1}) in 0.01 M NaCl at pH 4 (**symbols**) and calculated with the LCD_{cc} model (**lines**). Unpublished data of Bai (2023). In the experiments, the HA and FA addition varied respectively from 0 to 500 mg L^{-1} and from 0 to 350 mg·L^{-1}, at a fixed phosphate addition of 0.45 mM. The heterogeneity of HA and FA differs as is evident from the parameter values for the width of the distribution applied in the modeling ($q = 0.1$ for HA, $q = 0.3$ for FA). At the particle level, FA is less heterogeneous than HA.

Comparing the adsorption data of HA in Figure 6b and FA in Figure 6c reveals that, on a mass basis, HA exhibits a much larger adsorption than FA, despite the lower effectiveness of HA in competition with PO_4 (Fig. 6a). This is due to the amount and distribution of the reactive ligands. FA is much smaller in size and is primarily present in the compact part of the EDL. HA is larger occupying more space at high loadings, increasing the amount of adsorbed HA. However, its competition is less effective because a large fraction of the ligands is relatively far from the surface.

Organo-oxides in soils. The interaction of NOM with natural metal (hydr)oxides in the organo-oxide aggregates of soil has not been explored with the LCD_{cc} model, given the complexity of soil. However, this interaction has recently been modeled with a NOM-CD model, which is a simplification of the LCD model framework. Without treating this model in detail, since it is described elsewhere (Hiemstra et al. 2013), the basis of the NOM-CD model is the definition of a NOM component reacting with a singly coordinated ≡FeOH surface group of the metal (hydr)oxide, forming a ≡FeNOM species. This species distributes its charge in the compact part of the interface, where the competition with PO_4 takes place.

The defined surface component comprises two carboxylic groups of which one forms an innersphere complex that distributes its charge over the electrostatic planes of the first Stern layer, while the second one distributes its charge over the planes of the second Stern layer. In the language of the LCD_{cc} model, the fraction of the ligands in the inner Stern layer region is a fixed number in the NOM-CD model, denoted as $R = 0.5$, for the ≡FeNOM species. As follows from the LCD_{cc} modeling (Fig. 2c), such a value of R is applicable in situations with a high potential gradient that is typically observed at high NOM loading, the presence of adsorbed PO_4, and sub-neutral pH conditions. The NOM-CD model is usually applied under such conditions.

Figure 7 illustrates the pH dependency of the PO_4 concentration in solution for a selected soil (Bai et al. 2024), along with the corresponding variation in the surface concentration of the ≡FeNOM species, calculated with the NOM-CD approach. In the soil, the pH was varied by acid-base additions. Depending on the pH, the measured equilibrium concentration of soluble phosphate can differ by nearly two orders of magnitude from the equilibrium concentration of soluble PO_4 in the absence of competition by NOM. At neutral pH, this competitive effect

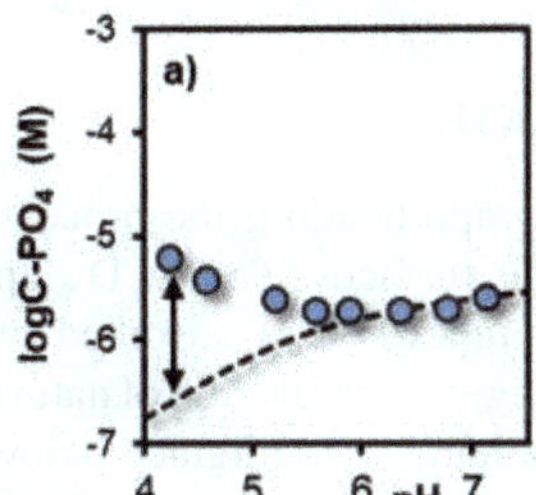

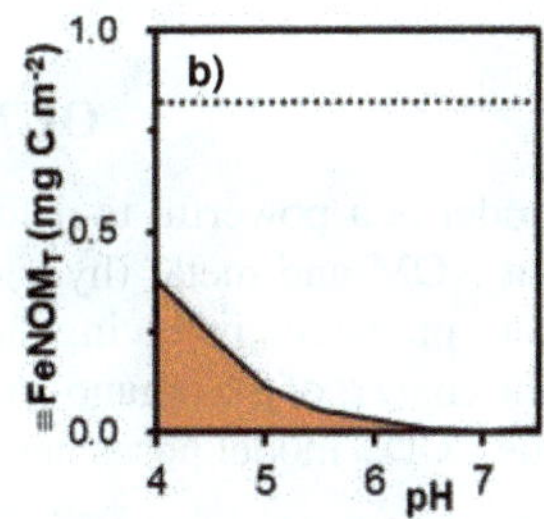

Figure 7. a) Logarithmic of the PO_4 equilibrium concentration as a function of the soil pH, measured in a background of 0.01 M $CaCl_2$ (**symbols**) by Bai et al. (2024). The **dashed line** is the expected PO_4 concentration in the soil solution, calculated with the CD-MUSIC model for ferrihydrite (Mendez and Hiemstra 2020a). At low pH, the mismatch is large due to the competition with adsorbed NOM. **b)** Variation in the total surface concentration of the ≡FeNOM species, calculated with the NOM-CD model. The **dashed line** gives the physical maximum for the occupation of the compact part of the interface by ≡FeNOM species, showing that the adsorption is not space-limited. At high pH, the carboxylic groups prefer to be part of the HNPs of the organo-oxide phase. The physical maximum of the NOM adsorption has been calculated from the volume of the Stern layer ($m^3 \cdot m^{-2}$) multiplied by a scaled mass density ($mg\,C \cdot m^3$). The molar surface concentration of ≡FeNOM ($\mu mol \cdot m^{-2}$) has been converted to a mass concentration ($mg\,C \cdot m^{-2}$) using the generic carboxylic density of NOM ($mol \cdot kg^{-1}$ C) and the number (2) of carboxylic groups per ≡FeNOM, as described by Bai et al. (2024). [CC-BY by Bai et al. (2024).]

disappears (Fig. 7a). The pronounced competition at low pH is due to the adsorption of reactive ligands of the HNPs, forming innersphere complexes which in turn adds negative charge to the compact part of the interface. In this process, there is a pH-dependent shift of the location of the reactive ligands in the interface of the organo-oxide phase, similar to what is found with the LCD_{cc} approach. The corresponding pH-dependent change in the surface concentration of the ≡FeNOM component is shown in Figure 7b.

In Figure 7b, the dashed line represents the maximum physical adsorption of the ≡FeNOM component in the compact part of the interface (Bai et al. 2024). The ≡FeNOM data show that the actual occupation of the Stern layer space, calculated with the NOM-CD model, is rather low for this soil. In the LCD_{cc} model, the relative occupation is described by the functional θ_s (Fig. 2b). The low occupation of the Stern layer derived with the NOM-CD model, implies that carboxylic groups prefer to remain outside the Stern layers, particularly at high pH, rather than being bound by the surface. This modest occupation partly aligns with a relatively low potential gradient according to the functional θ_s in Figure 2b. The actual occupation is lower, possibly influenced by Ca ion-bridging, which may condense the structure of HNPs, making HNPs less flexible than in Na systems. This advocates the organo-oxide phase as a collection of nanoparticles rather than a structure according to a core–shell approach.

In the NOM-CD model, the quantity of ≡FeNOM is expressed in $\mu mol \cdot m^{-2}$ rather than $mg\,C \cdot m^{-2}$. The former is determined through modeling the solid-solution distribution of PO_4, while the latter is experimentally accessible. To facilitate the comparison between the calculated surface loading and the measured amount of adsorbed NOM in $mg\,C \cdot m^{-2}$, the molar surface concentration of ≡FeNOM ($\mu mol \cdot m^{-2}$) was rescaled to a corresponding carbon mass, and the density of the RCOO(H) groups (Bai et al.2024). This scaling is justified as outlined in another contribution to this volume (Hiemstra et al. 2025).

Deng et al. (2019) modeled the competitive adsorption of oxyanions (PO_4, AsO_4, $As(OH)_3$) in goethite systems with added HA with the NOM-CD as well as LCD approach. Both models can equally well describe the competition with oxyanions. The fitted site densities of carboxylic groups of HA in Stern layers were almost equal when using the LCD model and NOM-CD model. Both the NOM-CD and the LCD model show similarity in the surface speciation of HA bound by goethite in the Stern layers and corresponding oxyanion surface species.

OUTLOOK

The LCD_{cc} model is a powerful tool for comprehending the processes involved in the interaction between NOM and metal (hydr)oxide surfaces. The LCD_{cc} approach also sheds light on the intricate processes involving the competition of adsorbed organic matter with oxyanions within the context of the organo-oxide aggregate fraction of natural soils. Improving and generalizing the LCD_{cc} model poses new challenges, highlighted below.

Size fractionation

For a more generalized model with a theoretical basis, a new challenge could be to improve the current *semi-empirical* correlation and predict the molar mass based on information about the initial size distribution of NOM derived from additional spectroscopy data. In addition, our evaluation of the affinity distribution of HA indicates that there is a maximum for the carboxylic group density at an intermediate molar mass. A better molecular characterization of functional groups of NOM in the adsorption phase as a function of the pH, ionic strength, NOM loading, and PO_4 and Ca ions using advanced spectroscopic techniques could also be beneficial for the development of a theoretical model in describing NOM adsorptive fractionation by metal (hydr)oxides.

Variability of NOM materials

The complexity of NOM arises from variations in the origin and its decomposition, resulting in a diversity of properties that influence the interaction of those NOM materials with soil minerals (Xu et al. 2020; Kleber et al. 2021; Bao et al. 2022). The current model development involved just one HA and FA. Other materials may deviate having specific properties not yet grasped. Further model development may benefit from studying the adsorption of NOM of different origins.

Competitive interaction of FA and HA

The current LCD_{cc} model has been developed for the competitive adsorption of either HA or FA with oxyanions. The interaction between HA and FA has not yet been modeled. Data collection became possible through the work of (Xu et al. 2020) who developed a methodology to quantify in a mixture of HA and FA, the concentration of both constituting components. Development of the LCD_{cc} model for describing this interaction, extended with the simultaneous competitive interaction with oxyanions is ongoing. This will contribute to an improved validity of the LCD_{cc} model framework.

NOM-oxide coprecipitates

The current LCD_{cc} model has been developed for competitive adsorption to well-defined crystal surfaces. In natural systems, the interaction of NOM with metal (hydr)oxide is predominantly related to the presence of organo-oxide nano-aggregates. This interaction is presently described with a simplified version of the LCD model, i.e., the NOM-CD model. Validation of that model is lacking. The interaction cannot be studied with the traditional approach of adding HA or FA to a metal (hydr)oxide suspension, because of the very strong aggregation of nano-oxide particles in general, leading to adsorption on the external aggregate surfaces only. A better proxy will be the use of, e.g., NOM-ferrihydrite coprecipitates. A challenge will then be the characterization of the size of the oxide particles formed and their reactive surface area. This will be part of future work.

REFERENCES

Asano M, Wagai R, Yamaguchi N, Takeichi Y, Maeda M, Suga H, Takahashi Y (2018) In search of a binding agent: nano-scale evidence of preferential carbon associations with poorly-crystalline mineral phases in physically-stable, clay-sized aggregates. Soil Systems 2:32, https://doi.org/10.3390/soilsystems2020032

Avneri-Katz S, Young RB, McKenna AM, Chen H, Corilo YE, Polubesova T, Borch T, Chefetz B (2017) Adsorptive fractionation of dissolved organic matter (DOM) by mineral soil: Macroscale approach and molecular insight. Org Geochem 103:113–124, https://doi.org/10.1016/j.orggeochem.2016.11.004

Bai Y (2023) Interactions of phosphate and mineral-associated organic carbon in soils: in relation to long-term phosphorus fertilization. PhD dissertation, WU, Wageningen University, Netherlands

Bai Y, Weng L, Hiemstra T (2024) Interaction of fulvic acid with soil organo-mineral nano-aggregates and corresponding phosphate release. Geoderma 441:116737, https://doi.org/10.1016/j.geoderma.2023.116737

Bao Y, Bolan NS, Lai J, Wang Y, Jin X, Kirkham MB, Wu X, Fang Z, Zhang Y, Wang H (2022) Interactions between organic matter and Fe (hydr)oxides and their influences on immobilization and remobilization of metal(loid)s: A review. Crit Rev Environ Sci Technol 52:4016–4037, https://doi.org/10.1080/10643389.2021.1974766

Benedetti MF, Milne CJ, Kinniburgh DG, van Riemsdijk WH, Koopal LK (1995) Metal ion binding to humic substances: Application of the non-ideal competitive adsorption model. Environ Sci Technol 29:446–457, https://doi.org/10.1021/es00002a022

Benedetti MF, Van Riemsdijk WH, Koopal LK (1996) Humic substances considered as a heterogeneous donnan gel phase. Environ Sci Technol 30:1805–1813, https://doi.org/10.1021/es950012y

Birkett JW, Jones MN, Bryan ND, Livens FR (1997) Computer modelling of partial specific volumes of humic substances. Eur J Soil Sci 48:131–137, https://doi.org/10.1111/j.1365-2389.1997.tb00192.x

Boily J-F, Persson P, Sjöberg S (2000) Benzenecarboxylate surface complexation at the goethite (α-FeOOH)/water interface: II. Linking IR spectroscopic observations to mechanistic surface complexation models for phthalate, trimellitate, and pyromellitate. Geochim Cosmochim Acta 64:3453–3470, https://doi.org/http://dx.doi.org/10.1016/S0016-7037(00)00453-1

Chapman DL (1913) A contribution to the theory of electrocapillarity. Philos Mag 25:475–481, https://doi.org/10.1080/14786440408634187

Chen H, Hou M, He Z, Liang Y, Xu J, Tan W (2023) Adsorption behavior of soil fulvic acid on crystal faces of kaolinite and goethite: Described by CD-MUSIC model. Sci Total Environ 903:165806, https://doi.org/10.1016/j.scitotenv.2023.165806

Christl I, Knicker H, Kögel-Knabner I, Kretzschmar R (2000) Chemical heterogeneity of humic substances: characterization of size fractions obtained by hollow-fibre ultrafiltration. Eur J Soil Sci 51:617–625, https://doi.org/10.1111/j.1365-2389.2000.00352.x

Claret F, Schäfer T, Brevet J, Reiller PE (2008) Fractionation of Suwannee River fulvic acid and Aldrich humic acid on α-Al_2O_3: Spectroscopic evidence. Environ Sci Technol 42:8809–8815, https://doi.org/10.1021/es801257g

Coward EK, Ohno T, Sparks DL (2019) Direct evidence for temporal molecular fractionation of dissolved organic matter at the iron oxyhydroxide interface. Environ Sci Technol 53:642–650, https://doi.org/10.1021/acs.est.8b04687

de Souza SVC, Junqueira RG (2005) A procedure to assess linearity by ordinary least squares method. Anal Chim Acta 552:25–35, https://doi.org/10.1016/j.aca.2005.07.043

de Wit JCM, van Riemsdijk WH, Nederlof MM, Kinniburgh DG, Koopal LK (1990) Analysis of ion binding on humic substances and the determination of intrinsic affinity distributions. Anal Chim Acta 232:189–207, https://doi.org/10.1016/S0003-2670(00)81235-0

de Wit JCM, van Riemsdijk WH, Koopal LK (1993) Proton binding to humic substances. 1. Electrostatic effects. Environ Sci Technol 27:2005–2014, https://doi.org/10.1021/es00047a004

Deng Y, Weng L, Li Y, Ma J, Chen Y (2019) Understanding major NOM properties controlling its interactions with phosphorus and arsenic at goethite–water interface. Water Res 157:372–380, https://doi.org/10.1016/j.watres.2019.03.077

Filius JD, Hiemstra T, van Riemsdijk WH (1997) Adsorption of small weak organic acids on goethite: modeling of mechanisms. J Colloid Interface Sci 195:368–380, https://doi.org/10.1006/jcis.1997.5152

Filius JD, Lumsdon DG, Meeussen JCL, Hiemstra T, van Riemsdijk WH (2000) Adsorption of fulvic acid on goethite. Geochim Cosmochim Acta 64:51–60, https://doi.org/10.1016/S0016-7037(99)00176-3

Filius JD, Meeussen JCL, Hiemstra T, van Riemsdijk WH (2003) Modeling the binding of fulvic acid by goethite: The speciation of adsorbed FA molecules. Geochim Cosmochim Acta 67:1463–1474, https://doi.org/10.1016/S0016-7037(02)01042-6

Hiemstra T, van Riemsdijk WH (1996) A surface structural approach to ion adsorption: the charge distribution (CD) model. J Colloid Interface Sci 179:488–508, https://doi.org/10.1006/jcis.1996.0242

Hiemstra T, van Riemsdijk WH, Bolt GH (1989) Multisite proton adsorption modeling at the solid/solution interface of (hydr)oxides: A new approach. I. Model description and evaluation of intrinsic reaction constants. J Colloid Interface Sci 133:91–104, https://doi.org/10.1016/0021-9797(89)90284-1

Hiemstra T, Antelo J, van Rotterdam AMD, van Riemsdijk WH (2010) Nanoparticles in natural systems II: The natural oxide fraction at interaction with natural organic matter and phosphate. Geochim Cosmochim Acta 74:59–69, https://doi.org/10.1016/j.gca.2009.10.019

Hiemstra T, Mia S, Duhaut P-B, Molleman B (2013) Natural and pyrogenic humic acids at goethite and natural oxide surfaces interacting with phosphate. Environ Sci Technol 47:9182–9189, https://doi.org/10.1021/es400997n

Hiemstra T, Hofmann A, Mendez JC, Bai Y (2025) Surface complexation and reactivity of ferrihydrite in relation to its surface and mineral structure, with applications to natural systems. Rev Mineral Geochem 91A:175-228

Janot N, Reiller PE, Zheng X, Croué JP, Benedetti MF (2012) Characterization of humic acid reactivity modifications due to adsorption onto α-Al_2O_3. Water Res 46:731–740, https://doi.org/10.1016/j.watres.2011.11.042

Kang S, Xing B (2008) Humic acid fractionation upon sequential adsorption onto goethite. Langmuir 24:2525–2531, https://doi.org/10.1021/la702914q

Kinniburgh DG, van Riemsdijk WH, Koopal LK, Borkovec M, Benedetti MF, Avena MJ (1999) Ion binding to natural organic matter: competition, heterogeneity, stoichiometry and thermodynamic consistency. Colloids Surf A 151:147–166, https://doi.org/10.1016/S0927-7757(98)00637-2

Kleber M, Sollins P, Sutton R (2007) A conceptual model of organo-mineral interactions in soils: self-assembly of organic molecular fragments into zonal structures on mineral surfaces. Biogeochemistry 85:9–24, https://doi.org/10.1007/s10533-007-9103-5

Kleber M, Bourg IC, Coward EK, Hansel CM, Myneni SCB, Nunan N (2021) Dynamic interactions at the mineral–organic matter interface. Nat Rev Earth Environ 2:402–421, https://doi.org/10.1038/s43017-021-00162-y

Ko I, Kim J-Y, Kim K-W (2005) Adsorption properties of soil humic and fulvic acids by hematite. Chem Speciation Bioavailability 17:41–48, https://doi.org/10.3184/095422905782774928

Kögel-Knabner I, Guggenberger G, Kleber M, Kandeler E, Kalbitz K, Scheu S, Eusterhues K, Leinweber P (2008) Organo-mineral associations in temperate soils: Integrating biology, mineralogy, and organic matter chemistry. J Plant Nutrition Soil Sci 171:61-82, https://doi.org/10.1002/jpln.200700048

Koopal LK, van Riemsdijk WH, de Wit JCM, Benedetti MF (1994) Analytical isotherm equations for multicomponent adsorption to heterogeneous surfaces. J Colloid Interface Sci 166:51–60, https://doi.org/10.1006/jcis.1994.1270

Koopal LK, Tan W, Avena M (2020) Equilibrium mono- and multicomponent adsorption models: From homogeneous ideal to heterogeneous non-ideal binding. Adv Colloid Interface Sci 280:102138, https://doi.org/10.1016/j.cis.2020.102138

Lal R (2004) Soil carbon sequestration impacts on global climate change and food security. Science 304:1623–1627, https://doi.org/10.1126/science.1097396

Li L, Huang W, Peng Pa, Sheng G, Fu J (2003) Chemical and molecular heterogeneity of humic acids repetitively extracted from a peat. Soil Sci Soc Am J 67:740–746, https://doi.org/10.2136/sssaj2003.7400

Li J, Weng L, Deng Y, Ma J, Chen Y, Li Y (2022) NOM-mineral interaction: Significance for speciation of cations and anions. Sci Total Environ 820:153259, https://doi.org/10.1016/j.scitotenv.2022.153259

Liang Y, Ding Y, Wang P, Lu G, Dang Z, Shi Z (2019) Modeling sorptive fractionation of organic matter at the mineral-water Interface. Soil Sci Soc Am J 83:107–117, https://doi.org/10.2136/sssaj2018.07.0275

Lodeiro P, Martínez-Cabanas M, Herrero R, Barriada JL, Vilariño T, Rodríguez-Barro P, Sastre de Vicente ME (2019) The proton binding properties of biosorbents. Environ Chem Lett 17:1281–1298, https://doi.org/10.1007/s10311-019-00883-z

Lv J, Miao Y, Huang Z, Han R, Zhang S (2018) Facet-mediated adsorption and molecular fractionation of humic substances on hematite surfaces. Environ Sci Technol 52:11660-11669, https://doi.org/10.1021/acs.est.8b03940

Mendez JC, Hiemstra T (2020a) Ternary complex formation of phosphate with Ca and Mg ions binding to ferrihydrite: Experiments and mechanisms. ACS Earth Space Chem 4:545–557, https://doi.org/10.1021/acsearthspacechem.9b00320

Mendez JC, Hiemstra T, Koopmans GF (2020b) Assessing the reactive surface area of soils and the association of soil organic carbon with natural oxide nanoparticles using ferrihydrite as proxy. Environ Sci Technol 54:11990–12000, https://doi.org/10.1021/acs.est.0c02163

Mendez JC, Van Eynde E, Hiemstra T, Comans RN (2022) Surface reactivity of the natural metal (hydr) oxides in weathered tropical soils. Geoderma 406:115517, https://doi.org/10.1016/j.geoderma.2021.115517

Milne CJ, Kinniburgh DG, Tipping E (2001) Generic NICA-Donnan model parameters for proton binding by humic substances. Environ Sci Technol 35:2049–2059, https://doi.org/10.1021/es000123j

Milne CJ, Kinniburgh DG, van Riemsdijk WH, Tipping E (2003) Generic NICA-Donnan model parameters for metal–ion binding by humic substances. Environ Sci Technol 37:958–971, https://doi.org/10.1021/es0258879

Newcomb CJ, Qafoku NP, Grate JW, Bailey VL, De Yoreo JJ (2017) Developing a molecular picture of soil organic matter–mineral interactions by quantifying organo–mineral binding. Nat Commun 8:396, https://doi.org/10.1038/s41467-017-00407-9

Qin X, Liu F, Wang G, Hou H, Li F, Weng L (2015) Fractionation of humic acid upon adsorption to goethite: Batch and column studies. Chem Eng J 269:272–278, https://doi.org/10.1016/j.cej.2015.01.124

Rahman MS, Whalen M, Gagnon GA (2013) Adsorption of dissolved organic matter (DOM) onto the synthetic iron pipe corrosion scales (goethite and magnetite): Effect of pH. Chem Eng J 234:149–157, https://doi.org/10.1016/j.cej.2013.08.077

Reiller P, Moulin C (2006) Sorption of Aldrich humic acid onto hematite: Insights into fractionation phenomena by electrospray ionization with quadrupole time-of-flight mass spectrometry. Environ Sci Technol 40:2235-2241, https://doi.org/10.1021/es0520518

Schellekens J, Buurman P, Kalbitz K, van Zomeren A, Vidal-Torrado P, Cerli C, Comans RNJ (2017) Molecular features of humic acids and fulvic acids from contrasting environments. Environ Sci Technol 51:1330–1339, https://doi.org/10.1021/acs.est.6b03925

Sips R (1948) On the Structure of a Catalyst Surface. J Chem Phys 16:490–495, https://doi.org/10.1063/1.1746922

Sips R (1950) On the Structure of a Catalyst Surface. II. J Chem Phys 18:1024–1026, https://doi.org/10.1063/1.1747848

Sposito G (2018) Gouy–Chapman Theory. *In*: Encyclopedia of Geochemistry. White WM (ed) Springer International Publishing, Cham, p 623–628

van Riemsdijk WH, Koopal LK (1992) Ion binding by natural heterogeneous particles. *In*: Environmental particles. Vol 1. Buffle J, van Leeuwen HP (eds). Lewis Publ., Boca Raton, p 455–495

Vermeer AWP, Koopal LK (1998) Adsorption of humic acids to mineral particles. 2. Polydispersity effects with polyelectrolyte adsorption. Langmuir 14:4210–4216, https://doi.org/10.1021/la970836o

Vermeer AWP, van Riemsdijk WH, Koopal LK (1998) Adsorption of humic acid to mineral particles. 1. Specific and electrostatic interactions. Langmuir 14:2810–2819, https://doi.org/10.1021/la970624r

Wang L, Li Y, Weng L, Sun Y, Ma J, Chen Y (2019) Using chromatographic and spectroscopic parameters to characterize preference and kinetics in the adsorption of humic and fulvic acid to goethite. Sci Total Environ 666:766–777, https://doi.org/10.1016/j.scitotenv.2019.02.235

Weng L, van Riemsdijk WH, Hiemstra T (2006a) Adsorption free energy of variable-charge nanoparticles to a charged surface in relation to the change of the average chemical state of the particles. Langmuir 22:389–397, https://doi.org/10.1021/la051730t

Weng L, van Riemsdijk WH, Koopal LK, Hiemstra T (2006b) Ligand and Charge Distribution (LCD) model for the description of fulvic acid adsorption to goethite. J Colloid Interface Sci 302:442–457, https://doi.org/10.1016/j.jcis.2006.07.005

Weng L, van Riemsdijk WH, Hiemstra T (2007) Adsorption of humic acids onto goethite: Effects of molar mass, pH and ionic strength. J Colloid Interface Sci 314:107–118, https://doi.org/10.1016/j.jcis.2007.05.039

Weng L, van Riemsdijk WH, Hiemstra T (2008) Humic nanoparticles at the oxide–water interface: Interactions with phosphate ion adsorption. Environ Sci Technol 42:8747–8752, https://doi.org/10.1021/es801631d

Xiong J, Koopal LK, Weng L, Wang M, Tan W (2015) Effect of soil fulvic and humic acid on binding of Pb to goethite–water interface: Linear additivity and volume fractions of HS in the Stern layer. J Colloid Interface Sci 457:121–130, https://doi.org/10.1016/j.jcis.2015.07.001

Xiong J, Weng L, Koopal LK, Wang M, Shi Z, Zheng L, Tan W (2018) Effect of soil fulvic and humic acids on pb binding to the goethite/solution interface: Ligand charge distribution modeling and speciation distribution of Pb. Environ Sci Technol 52:1348–1356, https://doi.org/10.1021/acs.est.7b05412

Xu H, Ji L, Kong M, Jiang H, Chen J (2019) Molecular weight-dependent adsorption fractionation of natural organic matter on ferrihydrite colloids in aquatic environment. Chem Eng J 363:356–364, https://doi.org/10.1016/j.cej.2019.01.154

Xu Y, Bai Y, Hiemstra T, Tan W, Weng L (2020) Resolving humic and fulvic acids in binary systems influenced by adsorptive fractionation to Fe-(hydr)oxide with focus on UV–Vis analysis. Chem Eng J 389:124380, https://doi.org/10.1016/j.cej.2020.124380

Xu Y, Hiemstra T, Tan W, Bai Y, Weng L (2022) Key factors in the adsorption of natural organic matter to metal (hydr)oxides: Fractionation and conformational change. Chemosphere 308:136129, https://doi.org/10.1016/j.chemosphere.2022.136129

Xu Y, Bai Y, Hiemstra T, Weng L (2024) A new consistent modeling framework for the competitive adsorption of humic nanoparticles and oxyanions to metal (hydr)oxides: Multiple modes of heterogeneity, fractionation, and conformational change. J Colloid Interface Sci 660:522–533, https://doi.org/10.1016/j.jcis.2024.01.078

Reviews in Mineralogy & Geochemistry
Vol. 91A pp. 251–294, 2025

8

Ion-Dependent Calcium Carbonate Cohesion: Insights from Surface Forces Measured between Calcite Surfaces

Joanna Dziadkowiec

NJORD Centre for Studies of the Physics of the Earth, University of Oslo, Norway

joanna.dziadkowiec@mn.uio.no

Anja Røyne

Department of Physics, University of Oslo, Norway

anja.royne@fys.uio.no

INTRODUCTION

The goals of this review chapter are to summarize the effects of various salt solutions on 1) the mechanical strength of fluid-saturated calcium carbonate rocks and on 2) the nanoscale forces acting between calcite surfaces. Based on these data, we then explore the question of to what extent the salt solution-induced mechanical deformation of calcium carbonate rocks can be explained by the ion-specific changes to surface forces acting between calcite surfaces in aqueous salt solutions. The influence of ion-specific electrostatic or hydration forces on calcium carbonate rocks deformation has been highlighted but the relative importance of these effects remains unclear. Several intrinsic features of calcite convolute our current understanding of the associated phenomena.

Calcite, $CaCO_3$, a mineral with a simple chemical composition, displays a complex and dynamic hydrophilic surface (Stipp 1999), the properties of which largely determine interactions of calcite with the environment. Calcite is a salt mineral with a mixture of ionic and covalent bonds in its structure (Skinner et al. 1994). In comparison with other salts, calcite is relatively insoluble however, as it is built by carbonate anions, its solubility is sensitive to any changes in solution pCO_2, pH, and temperature. This relatively reactive nature of calcite complicates molecular-level description of calcite surfaces. Although the immense progress has been recently made owing to computational and surface-sensitive experimental studies with updated complexation models of calcite surfaces (Heberling et al. 2021; Bonto et al. 2022), it is for example still not systematically known how various ions control the surface charge of calcite (Fu et al. 2021).

Abundance of calcite in the environment generates an immense interest in the properties of this mineral. In natural settings, rocks rich in calcite and other calcium carbonate minerals fix CO_2 (Matter and Kelemen 2009), control the fate of pollutants (Lin et al. 2023), interact with hydrocarbons in limestone-hosted reservoirs (Shepherd 2009), and display a complex seismic response (Prakash et al. 2023). Calcite is also an important mineral resource with estimated 4 billion tonnes of carbonate mined worldwide annually (Tegethoff et al. 2001). It is processed and used for cement production, as a filler and pigment in plastic, paper, and paint production, and as an excipient in pharmaceutical products. Both in nature and industry, there is a high interest to understand the detailed behavior of calcite surfaces, which govern the interactions of calcite with the surroundings.

1529-6466/25/0091A-0008$05.00 (print)
1943-2666/25/0091A-0008$05.00 (online)
http://dx.doi.org/10.2138/rmg.2025.91A.08

In this chapter, we first summarize and discuss cohesion in calcium carbonate rocks. Here, we highlight various mechanisms that affect cohesion in rocks fully or partially saturated with water, including the importance of solution-sensitive surface forces for cohesion. We then briefly discuss the theory of nanoscale surface forces. In the subsequent sections, we review the experimental measurements of mechanical strength of calcium carbonate rocks saturated with various salt solutions and the experimental measurements of surface forces between calcite surfaces immersed in various salts. We then assess the relative importance of ion-specific surface forces in controlling the cohesion and mechanical strength of rocks saturated with salt solutions as sketched in Figure 1.

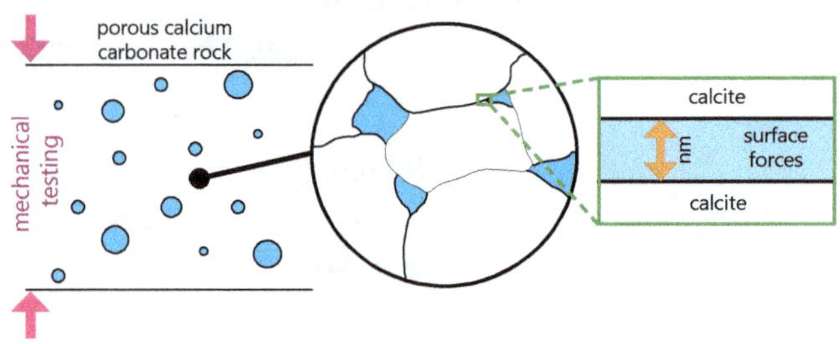

Figure 1. Schematic representation of the chapter contents. Can the repulsive and attractive nanometer range surface forces acting between calcite grains immersed in aqueous salt solutions explain the salt solution-induced changes to mechanical strength of calcium carbonate rocks?

COHESION IN CALCIUM CARBONATE ROCKS

Cohesion in relation to textures of calcium carbonates

Cohesion is a material property of rock, which corresponds to the non-frictional shear resistance of rock that is independent of normal stress. In rock mechanics, this parameter is derived from the Mohr–Coulomb failure criterion (Labuz and Zang 2014). Cohesion originates from bonds that hold together the neighboring mineral grains inside a rock, while frictional strength reflects the resistance of the grains to dislocation and slipping movement (Peng and Cai 2019). Cohesion is thus mainly the effect of contact cementation and short-range attractive surface forces acting between the mineral grains. In addition, other effects such as humidity-dependent capillary forces and changes in pore pressure when an external load is applied to a saturated rock contribute to cohesion.

Calcium carbonate-based rocks are widespread and diverse sediments, which display significantly varying cohesion. The majority of calcium carbonate minerals found in geological environments were precipitated by marine organisms (Wilson 2012). Accumulation and lithification of these minerals in sedimentary basins significantly increased in the Mesozoic (Kuznetsov 2023) and led to the formation of vast limestone deposits. The variety of these deposits, which are estimated to comprise up to 25% of all Earth's sedimentary rocks worldwide (Boggs 2009), is immense and reflects differences in their depositional environments, extent of diagenesis, and deformation histories (Pomar and Hallock 2008). Sedimentary carbonate rocks display mineralogical (mainly calcite, dolomite, and aragonite) and textural differences with varying proportions of the main textural components, which according to a simplified classification by Leighton and Pendexter (1962) include grains, lime mud (micrite), cement, voids, organic framebuilders, and recrystallization calcite. An exhaustive description of the carbonate sedimentary rock constituents is given by Folk (1962). Carbonate rocks can also be grouped based on their depositional textures, with the focus on grain support (Dunham 1962; Embry and Klovan 1971). The depositional textures of carbonates are frequently altered by diagenetic processes, which mainly include cementation, compaction (often by pressure solution), recrystallization, and grain replacement (Wright 1992; Hallsworth and Knox 1999; Lokier and Al Junaibi 2016). In rare cases, carbonate rocks can also form in non-sedimentary settings such as for example magmatic carbonatites (Mitchell 2005) or hydrothermal travertines (Gandin and Capezzuoli 2014). Thus, microstructures of carbonate rocks are highly heterogeneous with strongly varying porosity, degree of cementation, and intergranular cohesion, often with mixed textures within the same formation (Chen et al. 2009; Regnet et al. 2019). Such heterogeneity results in strongly varying cohesive characteristics of calcium carbonate rocks spanning from weak calcium carbonate sands and gravels to very strong non-porous crystalline limestones without any recognizable fragments of biogenic origin, which is in line with the largely varying strain rates in experimental studies on carbonate deformation, depending on their textures (Fruth et al. 1966; Ebhardt 1968; Hellmann et al. 2002; Chuhan et al. 2003; Vajdova et al. 2004; Brantut et al. 2014; Nicolas et al. 2016).

Uniaxial compressive strength (UCS) is a widely reported and thus convenient parameter to compare the mechanical strength and indirectly the cohesion of various carbonate rock types. Following a definition by Bell (2005), UCS represents the maximum stress that a rock sample can withstand before failure when a unidirectional stress is applied in an axial direction to the ends of a cylindrical sample. Figure 2A complies the UCS of dry carbonate rocks with varying textures, compiled from the literature data. According to these data, UCS of different types of carbonates may vary up to 5 times and carbonates may comprise very weak to very strong rock grades. The lowest cohesion and mechanical strength can be attributed to carbonate rocks rich in bioclasts and skeletal fragments, with high porosity and low degree of cementation and diagenesis (wackestones, packstones), which reflects low number of contact points between unaltered bioclasts. In contrast, carbonates mainly composed of recrystallized calcium carbonate grains, with little porosity, and those rich in micrite or sparite types of cement show in general higher mechanical strength. These comprise mainly crystalline calcium carbonates, in which depositional texture is no longer recognizable, some of the mudstone textural types, and metamorphic marble as summarized in Figure 2A. Limestones rich in finer grains will be generally stronger than those abundant in larger, coarser grains (Rutter 1974). Carbonate rocks, which are rich in cement but did not undergo compaction or were cemented in the absence of substantial stress often remain weakly cohesive. This is well illustrated by low mechanical strength of biocemented sands, which show UCS of only several MPa (Liu et al. 2019; Wang and Yin 2021).

As such, cohesion in calcium carbonate rocks is strongly related to the degree of their compaction and diagenesis. During compaction, porosity is progressively reduced as a result of mechanical and chemical compaction. Mechanical compaction involves processes such as grain rearrangement, pore collapse and brittle fracturing of grains or intracrystalline plasticity. Chemical compaction occurs by fluid-assisted grain alteration (Vajdova et al. 2004; Croizé et al. 2013; Brantut et al. 2014) and encompasses several relatively slow deformation processes common to carbonate rocks subject to stress. These mainly include pressure solution with local stress-dependent dissolution and precipitation, mineral phase conversion, and subcritical crack propagation (Maugis 1985; Røyne et al. 2011; Croizé et al. 2013; Brantut et al. 2014). Porosity reduction can additionally occur through precipitation of pore-filling cement around grain contacts (Scholle and Halley 1989). In some cases, however, such cementation may preserve porosity by creating a stress-resistant network of cemented grains (Croizé et al. 2010). The compaction and diagenesis processes, especially pressure solution, allow neighboring calcite particles to increase their contact area and form cohesive bonds originating from attractive short-range surface forces, mainly from attractive van der Waals interactions acting at grain separations of a few nanometers, which is referred to as contact cementation (Fabricius 2003, 2007; Descamps et al. 2017; Nermoen et al. 2018). The degree of contact cementation is often correlated to the strength of carbonate rocks (Alam et al. 2010), while the correlation of strength with porosity is not always evident (Fig. 2B).

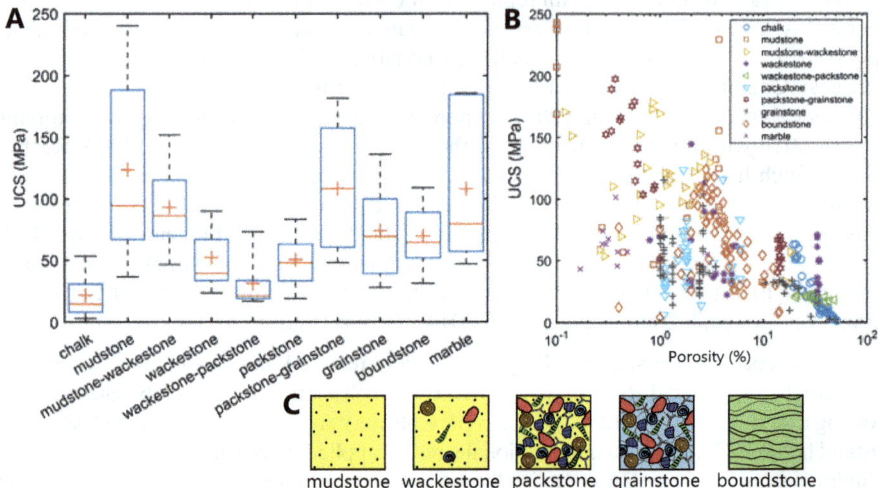

Figure 2. A. Box and whiskers plot comparing uniaxial compressive strength (UCS) of dry carbonate rock samples as a function of textural differences and **B.** UCS of dry carbonate rock samples as a function of their porosity. Data in both panels A and B were grouped according to the Dunham's classification (Dunham 1962). Chalk and marble were additionally specified. Panels compiled from the available literature data (Mortimore et al. 2004; Palchik and Hatzor 2004; Baykasoğlu et al. 2008; Török and Vásárhelyi 2010; You 2011; Mazidi et al. 2012; Rajabzadeh et al. 2012; Cheshomi and Sheshde 2013; Briševac et al. 2014; Sopacı and Akgün 2015; Madhubabu et al. 2016; Ajalloeian et al. 2017; Hebib et al. 2017; Jalali et al. 2017 Heidari et al. 2018; Van Stappen et al. 2019; Abd El-Aal et al. 2021; Mircescu et al. 2022; Sadeghi et al. 2022; Abu-Mahfouz et al. 2023). **C.** Cartoons highlighting differences between main textures of carbonate rocks adapted from Dunham (1962).

Effect of water on carbonate cohesion

As discussed in the previous section, differences in mechanical properties and cohesion of calcium carbonate rocks can be first and foremost explained by the great textural variability related to the biogenic origin of carbonates, characteristics of their depositional environments, and the extent of diagenesis that they have undergone (Harris et al. 1985). Importantly, cohesion is also sensitive to the type of fluid saturating the rock. Calcium carbonate rocks typically show lower cohesion and lower mechanical strength in the presence of water than in a dry state. Porous carbonate rocks such as chalks or calcarenites are especially prone to water-induced weakening (Risnes et al. 2005; Ciantia et al. 2015b; Wong et al. 2016). Figure 3 compares dry and water-saturated UCS for a set of limestone samples, showing a drop of UCS for wet conditions. Although in general Figure 3A shows a relatively small difference between the UCS for dry and saturated limestone samples, the UCS of water-saturated samples is consistently lower for various types of limestone samples. This is illustrated in Figures 3B and C where the UCS for dry and wet limestones is plotted as a function of porosity. The UCS of wet samples is consistently lower than for the dry ones at the same porosity value. There are several mechanisms, which may explain reduced cohesion of calcium carbonate rocks in water. These can be grouped as hydromechanical, chemical, physicochemical, and chemo-mechanical effects.

Hydromechanical effects. Hydromechanical effects affect rocks partially saturated with water and are mainly caused by changing pore pressure during rock compression and the action of the capillary pressure. Pore pressure of fluid within pores is elevated during compaction of undrained and poorly drained porous systems. Pore pressure acts on the rock matrix and the increasing pore pressure decreases the effective stress and the load-bearing capacity of the rocks (Nur and Byerlee 1971; Rutter 1974; Eslami et al. 2010). Capillary effects are especially important at low water saturation levels. With the increasing humidity,

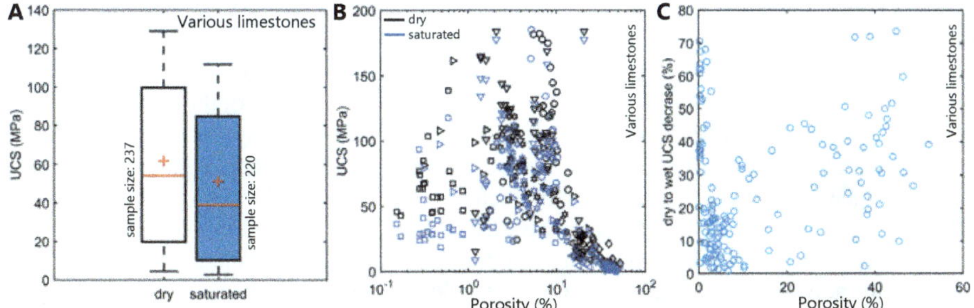

Figure 3. A. Box and whiskers plot comparing UCS of dry and saturated calcium carbonate rocks, using available literature data. In each data set, the same set of samples (or the same type of limestone samples in one case) is compared both in dry and water-saturated conditions. (Kasim and Shakoor 1996; Vásárhelyi 2005; Török and Vásárhelyi 2010; Rajabzadeh et al. 2012; Ercikdi et al. 2016; Daraei and Zare 2018; Davarpanah et al. 2019; Alzabeebee et al. 2022; Davarpanah et al. 2022; Briševac et al. 2023). **B.** USC of dry and saturated calcium carbonate rocks as a function of porosity using arbitrarily chosen literature data for various limestones and marble. Each symbol corresponds to one literature data set. In each data set, the same set of samples (or the same type of limestone samples in one case) is compared both in dry and water-saturated conditions (Vásárhelyi 2005; Török and Vásárhelyi 2010; Rajabzadeh et al. 2012; Ercikdi et al. 2016; Daraei and Zare 2018; Alzabeebee et al. 2022; Briševac et al. 2023). **C.** Percent decrease in UCS from dry to water-saturated state as a function of porosity for various limestones using the data shown in panel **B.** Each data point corresponds to UCS difference for one sample tested both in dry and water-saturated conditions.

the experimentally observed decrease in mechanical strength of partially water-saturated samples can be explained by the destruction of capillary bridges, which may hold mineral grains together by attractive capillary suction (Brignoli et al. 1995; Papamichos et al. 1997; Talesnick et al. 2001). However, the hydromechanical effects of pore pressure and capillary forces have frequently been found insufficient to fully account for the experimentally observed loss of mechanical strength in water-saturated calcium carbonate rocks relative to the dried samples, especially in porous carbonates (Risnes et al. 2003; Heggheim et al. 2005; Risnes et al. 2005; Eslami et al. 2010; Ciantia et al. 2015a; Kibikas et al. 2023).

Chemical effects. In many cases, chemical effects such as mineral dissolution may become important. Calcium carbonate, which is an ionic mineral, has a relatively low solubility in comparison to other ionic salts but it greatly exceeds the solubility of various aluminosilicate minerals that prevail in sedimentary rocks. Dissolution affects the texture of a rock and may decrease the area of contact between constituent mineral grains, which causes degradation of strong diagenetic intergranular bonds (Ciantia et al. 2015b). Such chemically enhanced weakening that originates from changes in the limestone microstructure is especially expected for rocks in contact with reactive brines that are strongly undersaturated with respect to the minerals comprising the rock (Deng et al. 2015; Clark and Vanorio 2016; Li et al. 2018). The timescale of the chemical effects depends on the dissolution rate of calcium carbonate minerals and the timescale of transport of reactive aqueous solutions within the porous matrix of the rock. Chemical dissolution is often enhanced under compression. Increasing pore water pressure has been shown to deteriorate limestone because of increasing chemical dissolution of pore walls and lubrication of mineral grains (Song et al. 2020). However, the dependence of mechanical strength on the type of saturating fluid and the ionic composition of water in the absence of substantial dissolution and precipitation suggests that there are also other physicochemical mechanisms responsible for weakening activated by water (Risnes et al. 2003; Pluymakers et al. 2021). This is also strongly supported by the acoustic measurements of ultrasonic velocity in limestone samples, which show that the decrease of shear modulus in water-saturated samples (after 12 h in distilled water) with respect to the dry ones is reversible (Baechle et al. 2009). This indicates that no major permanent changes to the limestone microstructure are induced by water (Baechle et al. 2009; Rozhko 2020).

Physicochemical effects. Physicochemical effects activated by water can induce carbonate deformation and weakening without the direct action of external pressure and in the absence of mineral dissolution. These effects are mainly attributed to adsorption of water and hydrated counterions on calcium carbonate surfaces and the associated decrease of the surface free energy of mineral surfaces (Orowan 1944; Michalske and Freiman 1982; Røyne et al. 2015; Diao and Espinosa-Marzal 2016) and to repulsive electrostatic forces that act between similarly charged mineral surfaces in aqueous solutions (Israelachvili 2015). Hydrophilic mineral surfaces, such as calcite, adsorb water, experience a decrease in surface energy, and thus experience weaker surface-surface interactions. Such physicochemical effects are instantaneous, which means that they are activated as soon as an aqueous solution gets in contact with mineral grains. The presence of tightly bound water layers and hydrated counterions on hydrophilic surfaces decreases the attractive interactions experienced between similar solid surfaces as it gives rise to repulsive force components (hydration forces) at the smallest surface separations below a few nanometers (Israelachvili 2015; Diao and Espinosa-Marzal 2016; Brekke-Svaland and Bresme 2018). The phenomenon, where the adsorption of various environmental components on solid surfaces decreases their cohesion, affects in general all types of solid surfaces and is sometimes termed the Rehbinder effect or AIRS—adsorption-induced reduction of strength of solids (Rehbinder and Lichtman 1957; Malkin 2012). Later experiments with calcium carbonate surfaces have confirmed that the surface energy of calcite decreases in contact with water and with the increasing water activity, and that the strong repulsive surface forces operate between two calcite surfaces immersed in water (Lomboy et al. 2011; Røyne et al. 2011, 2015). Repulsive electrostatic forces originate from the development of electrical double layers (EDL) on charged solid surfaces immersed in aqueous solutions and are sensitive to ionic solution composition (Israelachvili 2015). These short-range forces become negligible when the surface charge of minerals becomes fully compensated by adsorbed ionic species (point of zero charge). EDL forces have been suggested as a major cause of water weakening observed in chalk in several experimental studies (Megawati et al. 2013; Katika et al. 2015; Nermoen et al. 2018; Geremia et al. 2021; Meireles et al. 2021).

Chemo-mechanical effects. Chemo-mechanical effects encompass all water induced deformation processes activated in rocks subject to externally imposed stress. These mainly include pressure solution with local stress-dependent dissolution and precipitation, subcritical crack propagation, and shear-induced brittle failure in the presence of fluids. Pressure solution is one of the key mechanisms of the chemical compaction in carbonates (Gratier et al. 1999; Budd 2002; Zhang and Spiers 2005a; Zhang et al. 2010) and it is generally a slow (Angheluta et al. 2012), ductile deformation and starts from the dissolution of contacting mineral grains along stressed grain boundaries. This stress-activated dissolution is subsequently followed by diffusion of the dissolved species to pore spaces where the solute concentration is lower, and finally precipitation of the material on the less stressed crystal faces or further transport by diffusion and advection (Weyl 1959; Rutter 1976; Gundersen et al. 2002). Stylolites, clay seams, sutured grain contacts, indented grains, and cemented microcracks are the most typical pressure solution microstructures (Croizé et al. 2013; Rolland et al. 2014; Toussaint et al. 2018). While in general pressure solution acts to increase contact area between dissolving grains, reduce porosity, and thus increase the cohesion of carbonates under compaction, the existence of extended and relatively thick stylolites (Baud et al. 2016) has been shown to weaken the mechanical strength of stylolite-rich carbonates. Water-activated subcritical fracturing leads to brittle failure of calcite grains, in which cracks propagate in a material at stresses below the fracture toughness of a given material (Henry et al. 1977; Røyne et al. 2011; Rostom et al. 2013; Bergsaker et al. 2016; Nara et al. 2017; Ilgen et al. 2018). Physicochemical effects and surface forces acting between grain surfaces immersed in water have been suggested to significantly affect the pressure solution and subcritical grain fracturing (Renard et al. 1997; Alcantar et al. 2003; Rostom et al. 2013).

In rocks, many of these various effects often act together. It has, for example, been shown that the enhanced repulsive surface forces between fine calcite particles in water can mobilize these particles into the percolating pore fluids and weaken the rock in a wet state (Ciantia et al. 2015a; Levenson and Emmanuel 2017). Upon drying, these fine water-suspended particles may accumulate together again in larger pores inside water menisci forming at grain contacts due to the action of capillary forces. As such, they comprise rather weak depositional inter-grain bonds, which can lead to enhanced mechanical strength of granular, porous calcium carbonate rocks in a dry state (Ciantia et al. 2015a,b). In an open system, such fine particles resuspended in water can be irreversibly removed, leading to progressive decrease in mechanical strength during repeated drying–wetting cycles (Andriani and Walsh 2007). In the following sections of this review, we will focus on the more detailed understanding of physicochemical effects and their importance on the mechanical strength of calcium carbonate rocks.

THEORETICAL ASPECTS OF SURFACE FORCES IN BRIEF

Physicochemical effects and disjoining pressure

Changes to intergranular rock cohesion by physicochemical mechanisms require water to act at grain contacts. For that, there must exist a thin liquid film separating two mineral surfaces. This means that the pressure in this film must be higher than the pressure of the surrounding bulk fluid (Weyl 1959; Rutter 1976; Renard et al. 1997). The pressure in the liquid film confined between two solid surfaces is referred to as disjoining pressure and it can be understood as a difference in pressure of a film between two surfaces and the hydrostatic pressure of a bulk fluid. Disjoining pressure (Π) depends on the film thickness (x), which is reflected in the following definition:

$$\Pi = -\frac{1}{A}\frac{\partial G}{\partial x}\bigg|_{A,T,P} \tag{1}$$

where ∂G is the change in Gibbs free energy with distance per constant unit area A, at fixed temperature (T) and external pressure (P) conditions (Butt and Kappl 2018). If the disjoining pressure of a confined water film is positive, it is larger than the hydrostatic fluid pressure and work has to be performed to displace the water from the gap between the surfaces. Negative disjoining pressure means that the fluid film is unstable, and it migrates into the bulk solution. Positive and negative disjoining pressures correspond respectively to repulsive and attractive surface forces. It has been experimentally measured that for smooth, hydrophilic mineral surfaces, the disjoining pressure can be of a significant magnitude at separations below several nanometers (Pashley 1981; Pashley and Israelachvili 1984; Donaldson Jr et al. 2015; Diao and Espinosa-Marzal 2016).

The stability of a liquid film between two surfaces can also be expressed in terms of interfacial energies. Interfacial energy is a change in free energy when the interface separating two immiscible phases is expanded by one unit area (Israelachvili 2015). Thus, the energy change of separating two similar surfaces 1 in liquid medium 3 (W_{131}) can be defined as:

$$W_{131} = \gamma_{13} - \gamma_{11} \tag{2}$$

where γ_{13} is solid–liquid and γ_{11} is solid–solid interfacial energy expressed per unit area, assuming smooth surfaces and no energy dissipation. If $W_{131} < 0$, then two liquid–solid interfaces have smaller energy ($2\gamma_{13}$) than the solid–solid interface γ_{11}, and the liquid medium will spontaneously separate two solid surfaces. This scenario corresponds to the positive disjoining pressure in the liquid film. As such, the properties of the surfaces and a liquid film determine if the disjoining pressure is positive or negative. γ_{11} is also equal to the free surface energy of solid 1 and represents the energy cost to increase the surface of the solid by unit area. The water–solid interfacial energies of hydrophilic minerals are generally much smaller than their surface free energies. This means that once a crystal is cleaved or fractured, a thin layer of water will quickly adsorb on the newly exposed surface, making it difficult to heal the fracture. This is also expressed in the Griffith's formula (Griffith 1921; Malkin 2012) providing a relationship between fracture stress (σ_s) and surface free energy per unit area (γ_s):

$$\sigma_s = \text{const}\sqrt{\frac{E_Y \gamma_s}{L_C}} \tag{3}$$

where E_Y is the Young's modulus and L_C is the initial surface crack length. It has been shown that for ionic solids such as calcite, the activation energy for a fracture in contact with aqueous solutions is up to 8 times lower than the strength of typical bond energies for ionic solids (Malkin 2012).

It has been proposed that the presence of nanometer-thick water films is crucial in rock compaction and weathering processes, largely because such films enable transport of material by keeping the two opposing mineral surfaces separated (Weyl 1959; Rutter 1976; Renard et al. 1997; Alcantar et al. 2003). Importantly, the equilibrium thickness of such confined fluid films depends not only on the applied pressure and surface energies of the confining mineral surfaces but also on the chemical composition of the confined fluid (Pashley and Israelachvili 1984; Israelachvili 2015). The relationship between these parameters can be understood by considering surface forces that act between two surfaces immersed in a solution as described thoroughly in previous chapters. Since the magnitude and range of surface forces strongly depend on type and concentrations of ions adsorbed onto surfaces and squeezed in a confined solution, it has been suggested that a varying balance of attractive and repulsive surface forces must operate at different stages of pressure solution deformation (Alcantar et al. 2003; Diao et al. 2020). Variations in surface forces with time are expected for reactive surfaces such as calcite, because calcite dissolution and precipitation affect the fluid composition near the surfaces. It is thus crucial to understand the main contributions to nanometer-range surface forces acting between mineral surfaces immersed in aqueous solutions.

DLVO theory

The basic DLVO theory describes the total interaction between two surfaces in an aqueous electrolyte solution as a sum of van der Waals (vdW) and electrical double layer (EDL) forces:

$$F_{\text{DLVO}} = F_{\text{vdW}} + F_{\text{EDL}} \tag{4}$$

vdW forces are mainly related to dispersion forces that originate due to instantaneous dipole moments generated by electrons fluctuating in their positions. Instantaneous dipoles can polarize nearby atoms or molecules, leading to attractive forces acting between them (Israelachvili 2015). As such, vdW forces always act also between macroscopic surfaces. vdW forces depend on the geometry of the contact. This is because they can be considered as the total interaction energy of atoms of one surface with the atoms of the second surface, which are in a contact region at nanometer range distances where the interaction force is of a non-negligible magnitude. vdW forces are not additive as interaction energies of each atom pair are affected by the presence of other atoms nearby. The following expression can be used to calculate surface separation (D)-dependent vdW forces (between two planar surfaces per unit area:

$$F_{\text{vdW}} = \frac{-\text{A}}{6\pi D^3} \tag{5}$$

where A is Hamaker constant (Israelachvili 2015). The Hamaker constant can be calculated based on the Lifshitz theory (Lifshitz 1955), using the static dielectric constants and refractive indices of interacting surfaces and of the medium across which the surfaces interact. Thus, vdW forces depend not only on the properties of the surfaces but also on the properties of the solution confined between the surfaces. vdW forces acting between two identical mineral surfaces across water are always attractive. The theoretically determined Hamaker constant of two interacting calcite surfaces in air is $A_{cac} = 10.1 \times 10^{-20}$ J, and it is several times lower in water: $A_{caw} = 1.44 \times 10^{-20}$ J (Bergström, 1997). vdW forces are generally nor largely affected by the salinity of water medium. A typical range of vdW forces for two calcite surfaces interacting in water where the forces are of significant magnitude (> 0.01 mN/m) is approximately 15 nm.

Although vdW forces act between mineral surfaces already at such relatively large separations, the surfaces must be usually places much closer to each other to experience attraction in electrolyte solutions. The reason for that is generally the presence of the electrical double layer (EDL), which is populated by an increased concentration of counterions relative to the bulk solution. Counterions accumulate near surfaces because of the surface charge. Most mineral surfaces become charged when immersed in water or an electrolyte solution. There are three main mechanisms that can cause charging of a mineral surface in the presence of an aqueous solution: 1) dissociation of protruding surface groups; 2) adsorption of ions from the solution; or 3) charge exchange between dissimilar surfaces placed in contact (Israelachvili 2015). The concentration of counterions on a mineral surface (σ) depends on the mineral surface charge (ρ_s) and can be approximated as:

$$\rho_s = \rho_0 + \frac{\sigma^2}{2\varepsilon_0 \varepsilon k T} \tag{6}$$

where ρ_0 is the bulk concentration, ε_0 is the vacuum permittivity, ε is the water dielectric constant, k is the Boltzmann constant, and T is temperature (Israelachvili 2015). It has to be noted that also depends on the solution chemistry. It is a fraction of the maximum possible charge of a given surface (σ_0), and depends on the number of surface sites (α) that are charged in a given solution (Israelachvili 2015):

$$\sigma = \alpha\sigma_0 \tag{7}$$

The density of counterions is the highest on the charged surface and decreases to the bulk solution. In general, the higher concentration of ions near surfaces gives rise to electrostatic repulsion between similarly charged surfaces because when the surfaces approach each other, the regions with a higher concentration of counterions start to overlap. Development of surface charge on calcite is quite complex due to its relatively high reactivity and a complex structure of hydration layer developing on the calcite surface, however, it is thought to be mostly controlled by the concentration of potential determining ions, which mainly include Ca^{2+}, Mg^{2+}, and CO_3^{2-} ions (Stipp and Hochella Jr 1991; Stipp 1999; Wolthers et al. 2008; Heberling et al. 2011; Al Mahrouqi et al. 2017).

If two mineral surfaces are similarly charged in an aqueous solution, the EDL electrostatic forces acting between them are repulsive. In a simple case of a 1:1 electrolyte composed of monovalent ions (z = 1), EDL forces (as a function of separation distance (D) for two flat surfaces per unit area can be estimated with the following expression:

$$F_{EDL} = \frac{\kappa^2}{2\pi} Z e^{-\kappa D} \tag{8}$$

where κ^{-1} is the Debye length and is Z is an interaction constant that depends on the electrolyte valency and surface potential Ψ_0 of isolated surfaces in a given electrolyte solution (Israelachvili 2015)., The relation between surface charge (σ) and surface potential of isolated surfaces for low $\Psi_0 < 25$ mV can be roughly estimated from:

$$\sigma = \varepsilon_0 \varepsilon \kappa \psi_0 \tag{9}$$

where ε_0 and ε are the vacuum permittivity and the water dielectric constant (Israelachvili 2015). In the above simplified equations, it is assumed that surface potential and surface charge do not vary as the separation between surfaces changes. However, this is not the case because the total number of ions in the gap between the interacting surfaces changes with surface separation. Also, the number of the charged surface sites drops as the counterions are forced to adsorb onto the surface sites upon decreasing surface separation. More correct EDL force expressions which account for these effects can be obtained by solving the Poisson–Boltzmann equation (Gray and Stiles 2018). Simplified equations can only estimate the interactions correctly at larger surface separations, typically > 5 nm. The magnitude and range of the EDL forces depend not only strongly on the surface charge and electrolyte concentration but also on the valency of the ionic species in the solution. In general, for many types of mineral surfaces, multivalent ions can bind to charged surface sites more effectively in comparison with the monovalent ions and reduce surface charge to larger extent. Continued adsorption of multivalent ions can sometimes lead to surface charge reversal effect, even at relatively low bulk concentrations (Parsons and Ninham 2010; Israelachvili 2015).

Other forces

In many cases, the DLVO forces cannot fully describe forces between mineral surfaces as surface-adsorbed ions may give rise to other non-DLVO surface forces that act at small surface separations below a few nanometers. These interactions comprise mainly of solvation, hydrophobic, steric, ion correlation, and hydration forces (Donaldson Jr et al. 2015; Israelachvili 2015; Butt and Kappl 2018). The experimental measurements of these interactions show that they can be repulsive, attractive, or oscillatory and are strongly ion-specific and mineral-type specific (Donaldson Jr et al. 2015). Since their magnitude is only dominant at surface separations below a few nanometers, they are highly relevant to interactions between constituent mineral grains in rocks as these grains are often considered to be separated by nanometer-thick water films. Especially hydration forces and ion correlation forces have been inferred or suggested to operate between calcite surfaces (Røyne et al. 2015; Diao and Espinosa-Marzal 2016, 2019; Javadi and Røyne 2018; Dziadkowiec et al. 2021; Fu et al. 2021). Hydration forces are generally repulsive and act at the smallest surface separations below a few nanometers, while ion correlation forces may contribute to enhanced attraction between surfaces.

Hydration forces are short-range structural forces that arise due to adsorption of water and hydrated counterions on hydrophilic mineral surfaces (Donaldson Jr et al. 2015; Israelachvili 2015). These forces act in the very proximity of mineral surfaces. As the water strongly binds to hydrophilic surfaces, it gives rise to repulsion that can significantly exceed the magnitude of the repulsive EDL forces. Computations and experiments indicate the oscillatory nature of hydration forces on calcite (Diao and Espinosa-Marzal 2016; Brekke-Svaland and Bresme 2018). Although the theoretical origin of hydration forces (F_H) is still debated (Parsegian and Zemb 2011), they can be roughly approximated using the exponential force law:

$$F_H = Ce^{-D/D_H} R$$

where C (mN/m) is an experimentally measured force constant and R is a radius of sample curvature (Donaldson Jr et al. 2015). Using this approximation, hydration forces on mineral surfaces can be characterized by exponentially decaying repulsion with decay lengths (D_H) typically lower than 2 nm (Pashley 1981; Pashley and Israelachvili 1984; Grabbe and Horn 1993; Røyne et al. 2011; Diao and Espinosa-Marzal 2016; Heuberger et al. 2017). Based on the different nature of hydration forces measured between two silica or two mica surfaces, a distinction has been made between the primary and secondary hydration forces. The much-shorter ranged primary hydration repulsion ($D_H < 0.5$ nm) typical for silica has been suggested to be due to direct adsorption of water on protruding surface groups, while the longer-ranged secondary hydration typical for mica or calcite was attributed to the hydration of counterions adsorbed onto mineral surfaces as it depended both on ion concentration and ion type (Pashley 1981; Pashley and Israelachvili 1984; Donaldson Jr et al. 2015; Diao and Espinosa-Marzal 2016; Guo and Kovscek 2019).

Ion correlation forces are short range attractive forces that become especially important in the presence of multivalent ions such as Ca^{2+} (Wennerström et al. 1982). Ion correlation attraction has been suggested to cause strong adhesion between clay mineral surfaces in the presence of Ca^{2+} (Kjellander et al. 1988). There has been no simple force law proposed to quantify ion-correlation forces yet. However, this interaction has been suggested to originate due to the fluctuations in the density of mobile counterions in the EDL and can be thought of as a van der Waals-like attraction of double layers (Israelachvili 2015). It is currently not clear to what extent ion correlation forces can act between calcite surfaces (Pourchet et al. 2013; Javadi and Røyne 2018; Liberto et al. 2019; Diao and Espinosa-Marzal 2019; Dziadkowiec et al. 2021).

Roughness and reactivity

Roughness and reactivity of brittle calcite surfaces significantly complicate the understanding of experimentally measured surface forces in carbonate systems. The surface of freshly cleaved calcite is highly dynamic in air and in water (Stipp et al. 1996; Claesson et al. 2024). Surface reactivity of calcite, including precipitation, dissolution, pressure solution, and surface roughening in natural confined settings such as pores and wet grain contacts may modify interactions between surfaces to an unknown extent, depending on the chemistry of water percolating in the pore network (Dziadkowiec et al. 2018, 2019; Diao et al. 2020; Fu et al. 2021). It has also a considerable effect on the forces measured between calcite surfaces.

Surface roughness affects all types of forces. This is because of the variation in surface heights across the contact area and because of the reduced real area of contact with respect to the nominal contact area defined by the geometry of the surfaces at the macroscale. A rough topography affects the distribution of ionic species both adsorbed on the surface and located in the EDL, which smears out the interaction forces. One approach to estimating these roughness effects is to average the total interaction energy (G) or forces measured between two surfaces with respect to the distribution of surface heights in the contact region (Parsons et al. 2014; Eom et al. 2017). The distribution of surface heights for a given surface has to be measured, for example with atomic force microscopy.

There is also another surface effect present in surface force measurements, which is related to the mechanical deformation of surface asperities. This effect is present at the lowest surface separations when the surfaces are being pressed into contact during typical force–distance measurements. Once the asperities of rough surfaces start to contact each other, they deform elastically and/or plastically. Since energy is needed to deform the asperities, this always produces a repulsive force. If the surfaces have a random distribution of surface heights, the force arising due to mechanical deformation of asperities can be approximated as exponentially repulsive. Typically, the effects of the mechanical contact force are present at separations of 3 rms, where rms is the root-mean-square roughness of the surface (Parsons et al. 2014). This means that such roughness effects are typically present at distances of several nanometers for rough mineral surfaces. Their exact magnitude and range depend on the mechanical properties of surfaces (e.g. ,Young's modulus and Poisson's ratio) and the size of the asperities. For rough surfaces, repulsive mechanical force can overlap with other repulsive interactions such as EDL or hydration forces as they give rise to repulsion operating at similar surface separations. Such overlap can make it difficult to isolate the repulsive effects from each other. This is especially important in force measurements with relatively reactive surfaces, such as calcite, which may undergo surface roughening during force measurements.

EXPERIMENTAL MEASUREMENTS OF CALCIUM CARBONATE MECHANICAL STRENGTH

In this section, we will review experimental measurements of fluid-saturated calcium carbonate rocks samples subject to mechanical testing and highlight the main deformation mechanisms in samples saturated with various salt solutions.

Carbonate cohesion in salt solutions

There have been multiple studies in the context of water-induced weakening, in which the loss of cohesion within chalk and limestone rocks in contact with water and fluids of varying salinity has been related to surface forces acting between the contacting calcite grains. It has been shown that the weakening of calcium carbonate rocks is not only sensitive to the type of saturating fluid but also to the ionic composition and concentration of the aqueous solutions saturating the pore space (Risnes and Flaageng 1999; Risnes et al. 2003; Hiorth et al. 2010; Megawati et al. 2013; Nermoen et al. 2018; Geremia et al. 2021; Pluymakers et al. 2021).

There is however no consensus and no systematic picture of to what extent the variations in ion distribution, which lead to the presence of electrostatic and other short-range forces, can contribute to the water-induced weakening of calcium carbonate rocks (Table 1, Figs. 4, 5). It has been observed that various ionic species give rise to varying trends in the experimentally measured mechanical strength of chalk, limestones, and pre-compressed calcite powder (Zhang and Spiers 2005a; Hiorth et al. 2010; Megawati et al. 2013; Pluymakers et al. 2021). In many cases, mineral dissolution and reprecipitation has been suggested to have a more important influence on the deformation processes in chalk rocks saturated with saline solutions than possible electrostatic interactions (Newman 1983; Heggheim et al. 2005; Zhang and Spiers 2005a; Hiorth et al. 2010; Madland et al. 2011). Other works show however the strongest weakening for salts such as $CaCl_2$, in which calcite solubility is low relative to other brines due to common ion effect (Katika et al. 2015; Meireles et al. 2020, 2021). Deformation experiments on porous limestone by Lisabeth and Zhu (2015) have shown that samples compacted in the presence of unsaturated water cause more weakening that in water equilibrated with limestone for 24 h. Interestingly, these experiments indicate that these two different aqueous solutions not only weaken the limestone to a varying extent but also change the main deformation mechanism from plastic deformation by mechanical twinning in distilled water to brittle microcracking dominant in

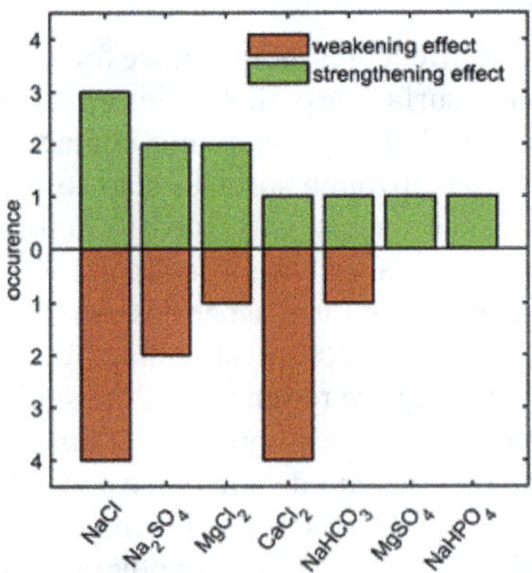

Figure 4. Summary of strengthening or weakening effects of salt solutions on calcium carbonate with respect to formation brine, deionized water or $CaCO_3$-equilibrated water. Data is based on the room temperature studies listed in Table 1. Note that this table is not exhaustive and because of the insufficient statistics is only meant to highlight the ambiguous effect of salts on the mechanical strength of calcium carbonate samples.

Table 1. Effect of salts on the deformation of calcite and carbonate rocks. *T* is temperature, *C* is concentration. Symbol * means that the salt solution was presaturated with respect to the rock or with $CaCO_3$ powder.

Authors	Sample/ Method	T (°C)	Salt	C (M)	Effect	Dominant Mechanism
Newman (1983)	chalk cores/ mechanical testing		NaCl	0.5	weakened with respect to oil and the formation brine	dissolution and pressure solution
Risnes et al. (2003)	chalk cores/ mechanical testing		$CaCl_2$*	0.3–4.8	increased strength with respect to chalk-equilibrated water, $CaCl_2$ strengthens the chalk more than NaCl	decreased water activity in concentrated salt solutions
			NaCl*	1.7–4.3		
Zhang and Spiers (2005a) Zhang and Spiers (2005b)	pre-compacted calcite or crushed limestone aggregates/ mechanical testing		NaCl*	0.1–0.5	weakening with respect to $CaCO_3$-equilibrated water	enhanced dissolution
			$MgCl_2$*	0.01	strengthening with respect to $CaCO_3$-equilibrated water	surface passivation by magnesium or phosphate adsorption that prevents dissolution
			Na_2HPO_4*	10^{-6}–10^{-3}		
Katika et al. (2015)	chalk plugs/ mechanical testing and nuclear magnetic resonance		NaCl*	1.8	weakening in all salts with respect to deionized water; the weakening effect is the strongest for $CaCl_2$ and Na_2SO_4	enhanced electrostatic repulsion between chalk grains in the presence of divalent ions
			$MgCl_2$*	0.6		
		room temp	$CaCl_2$*	0.6		
			Na_2SO_4*	0.6		
Nermoen et al. (2018)	chalk plugs/ mechanical testing		NaCl	0.657	weakening observed in all salts with respect to oil-saturated samples; Na_2SO_4 salt contributes to the strongest weakening and softening	enhanced electrostatic repulsion in the presence of sulphate anions
			Na_2SO_4	0.219		
			$MgCl_2$	0.219		
Diao et al. (2020)	single calcite crystals/ surface forces apparatus		NaCl*	10^{-3}–1	NaCl slows down pressure solution at concentrations below 10^{-2} M, while $CaCl_2$ speeds it up relative to $CaCO_3$-saturated water at all concentrations	ion-specific changes to hydration structure on calcite surfaces
			$CaCl_2$*	10^{-3}–1		
Pluymakers et al. (2021)	limestone cores/ mechanical testing		Na_2SO_4*	0.4	weakening with respect to limestone-equilibrated water	ion-induced alteration of surface charge
			$MgSO_4$	0.4	strengthening with respect to limestone-equilibrated water	

Table 1 (cont´d). Effect of salts on the deformation of calcite and carbonate rocks. *T* is temperature, *C* is concentration. Symbol * means that the salt solution was presaturated with respect to the rock or with $CaCO_3$ powder.

Authors	Sample/ Method	*T* (°C)	Salt	*C* (M)	Effect	Dominant mechanism
Choens et al. (2021)	calcite grain packs/ mechanical testing	room temp	Na_2SO_4	0.5	strengthening with respect to deionized water with the weakest effect for $NaHCO_3$	formation of fracture-tip stabilizing Ca-anion complexes
			NaCl	0.5		
			$NaHCO_3$	0.5		
Meireles et al. (2021)	chalk cores/ mechanical and ultrasonic velocity testing		$CaCO_3$*	10^{-4}	weaker and softer samples in $CaCl_2$ than in calcite-saturated water	stronger electrostatic repulsive forces in the presence of $CaCl_2$
			$CaCl_2$	0.06		
Geremia et al. (2021)	chalk cores/ mechanical testing		NaCl	0.1–1	comparable weakening to that in deionized water for NaCl, $NaHCO_3$, and $CaCl_2$ salts and inhibited water-weakening in the presence of $MgCl_2$ and Na_2SO_4 salts; no clear concentration dependence	ion-specific repulsive forces, which lower the chalk cohesion
			$MgCl_2$	0.05–0.5		
			$NaHCO_3$	0.1–0.6		
			Na_2SO_4	0.07–0.37		
			$CaCl_2$	0.09–0.45		
Korsnes et al. (2008)	chalk cores/ mechanical testing	50–130	seawater	–	decreasing strength with increasing temperature in seawater with the opposite trend observed in deionized water	Mg^{2+} adsorption and chalk dissolution in seawater
Liteanu and Spiers (2009)	pre-compacted calcite aggregates\ mechanical testing	80	NaCl with CO_2	0.6–3	increased strength at 1 M and below, decreased strength at higher concentrations relative to CO_2-saturated fluid with no salt added	reduced intergranular pressure solution at lower salinity and enhanced subcritical cracking in most concentrated salts
			$MgCl_2$ with CO_2	1, 2		
Madland et al. (2011)	chalk cores/ mechanical testing	130	$MgCl_2$	0.1–0.22	strength is low in seawater and $MgCl_2$ relative to distilled water and NaCl	chalk dissolution and precipitation of Mg-bearing minerals
			seawater	–		
			NaCl	0.66		
Megawati et al. (2013)	chalk cores/ mechanical testing	50, 130	Na_2SO_4	0.01–0.2	chalk strength reduced in the presence of sulphate relative to NaCl, this effect is negligible at 50°C	sulphate adsorption increases repulsion between calcite surfaces
			NaCl	0.6		

limestone-equilibrated water. Pluymakers et al. (2021) has in turn shown that $CaCO_3$-saturated water cause less deformation in limestone than deionized water (equilibrated with limestone for 24 h). These results underline that subtle differences in ionic strength of the solutions and calcium carbonate sample type may significantly affect the mechanical strength of limestone rocks. It is therefore vital to systematically review the literature data on the effect of various ions on the mechanical strength of the calcium carbonate rocks as compiled in Table 1.

Table 1 compiles eleven room temperature and four high temperature experimental studies, in which calcium carbonate strength is affected by the ionic composition of inorganic electrolyte solutions. These studies report either weakening or strengthening effects of the used salt solutions relative to oil, formation brine, or $CaCO_3$-equilibrated water, as compiled in Figure 5. It is however not systematic how various salts affect the mechanical strength of calcium carbonate with both strengthening and weakening reported for most salts as shown in Figure 4. The most important weakening mechanisms in calcium carbonate samples proposed in the studies listed in Table 1 (exposed to salt solutions specified in brackets) involve:

- dissolution of constituent calcium carbonate minerals (NaCl, $MgCl_2$),
- precipitation of new minerals ($MgCl_2$),
- enhanced pressure solution (NaCl),
- enhanced subcritical cracking (NaCl, $MgCl_2$),
- ion-specific changes to hydration structure on calcite ($CaCl_2$),
- enhanced electrostatic repulsion associated with the differences in ion adsorption onto calcite and the resultant ion-specific surface charge alteration (NaCl, $MgCl_2$, $CaCl_2$, Na_2SO_4, $NaHCO_3$).

Out of these effects, the most commonly reported were the enhanced electrostatic repulsion in a given electrolyte and calcite dissolution. In turn, the proposed mechanisms responsible for calcium carbonate strengthening included:

- decreased water activity in concentrated salts (NaCl, $CaCl_2$),
- surface passivation by adsorbed ions that prevents calcite dissolution ($MgCl_2$, $NaHPO_4$),
- ion-specific changes to hydration structure on calcite (NaCl),
- weaker electrostatic repulsion due to ion adsorption and the associated surface charge screening ($MgSO_4$, $MgCl_2$, Na_2SO_4),
- formation of Ca-anion complexes that prevent subcritical cracking (Na_2SO_4, NaCl, $NaHCO_3$),
- reduced intergranular pressure solution (NaCl, $MgCl_2$).

Based on these studies, the action of the electrostatic forces has been suggested as relevant in rocks exposed to the following salt solutions: NaCl, $MgCl_2$, $CaCl_2$, Na_2SO_4, $NaHCO_3$ (weakening effect) and $MgSO_4$, $MgCl_2$, Na_2SO_4 (strengthening effect).

Although the number of experimental studies on the influence of salt on calcium carbonate strength is quite limited, with varying types of natural or synthetic calcium carbonate samples, varying methodologies of mechanical testing, and varying concentration of the used salts, below we grouped the weakening and strengthening mechanisms proposed in the studies listed in Table 1 by the type of the used salt solution. In these works, the mechanisms related to the action of surface forces were generally not directly measured but inferred from the observed changes in mechanical strength, while changes in sample microstructure related to pressure solution, subcritical cracking, and mineral growth and dissolution could be detected in relevant cases in post-experimental sample characterization.

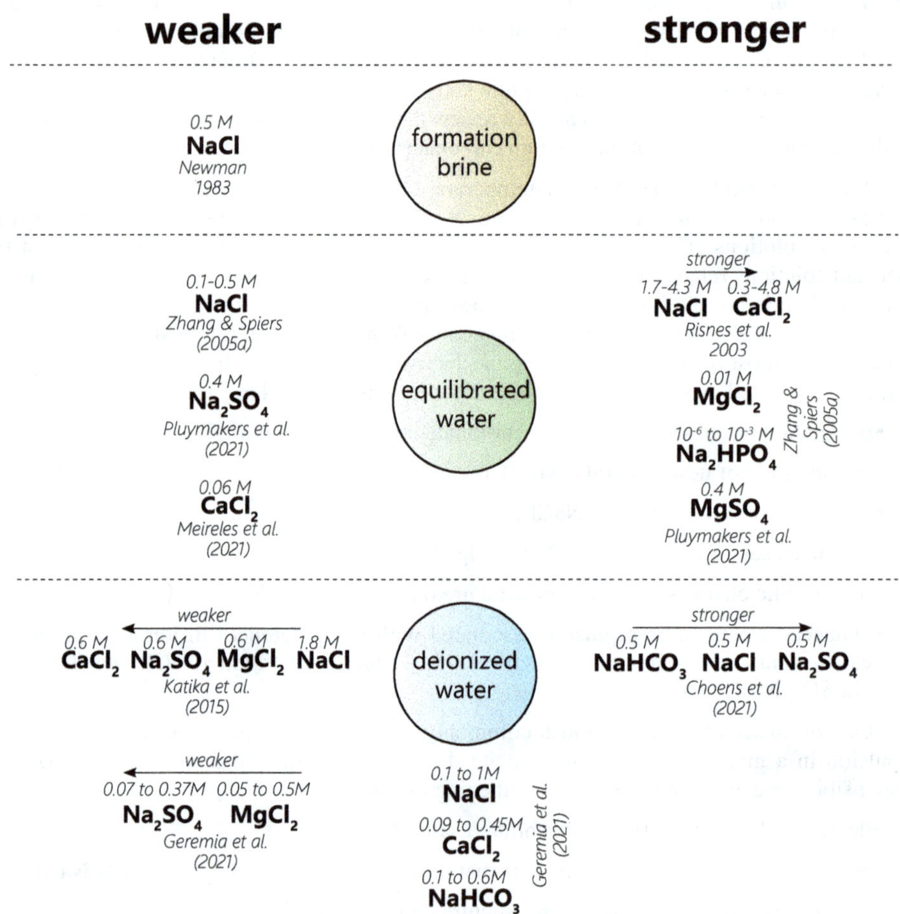

Figure 5. Summary of weakening (salts on the **left**) or strengthening (salts on the **right**) effects of salt solutions on calcium carbonate with respect to formation brine, $CaCO_3$-equilibrated water, or deionized water. Salts aligned with the reference solution (only for deionized water) show no significant strengthening or weakening effect. Data is based on the room temperature studies listed in Table 1.

Sodium salts. The most frequently used salts in studies reported in Table 1 were sodium salts: NaCl, Na_2SO_4, and $NaHCO_3$. NaCl has been reported to weaken the calcium carbonate samples by: repulsive electrostatic forces (Katika et al. 2015; Nermoen et al. 2018; Geremia et al. 2021), enhanced calcite dissolution and pressure solution (Newman 1983; Zhang and Spiers 2005a), and enhanced subcritical cracking of calcite crystals (Liteanu and Spiers 2009). The strengthening induced by NaCl was in turn attributed to: arrested pressure solution at low salt concentrations (Liteanu and Spiers 2009; Diao et al. 2020), arrested subcritical cracking (Choens et al. 2021), and decreased water activity at high NaCl concentration (Risnes et al. 2005). Na_2SO_4 was found toaffect the rock strength by modulating electrostatic repulsive forces with both the weakening (Katika et al. 2015; Nermoen et al. 2018; Pluymakers et al. 2021) and strengthening (Geremia et al. 2021) observed and was reported to strengthen the calcium carbonate by arrested subcritical cracking (Choens et al. 2021). Most of the above studies explored how NaCl and Na_2SO_4 salts affect the compaction of salt solution-saturated calcium carbonate samples. Experiments on pre-compacted calcite and crushed limestone powders have shown that $CaCO_3$-equilibrated NaCl salt solution at concentrations between 0.1–0.5 M cause significant increase in compaction of these samples relative to non-saline solutions due to enhanced pressure solution processes (Zhang and Spiers 2005a). Similar experiments in the presence of subcritical CO_2 performed at 80 °C indicate that NaCl salt reduces compaction of calcite aggregates at concentrations of 1M and below because of reduced intergranular pressure solution with respect to CO_2-free NaCl solutions, while high concentration of these salts (2–3 M) increase compaction rates via enhanced subcritical cracking (Liteanu and Spiers 2009). Experiments on compacted granular calcite assemblages by Choens et al. (2021) show that 0.5 M NaCl, Na_2SO_4 and $NaHCO_3$ greatly reduced deformation and microfracturing density with respect to deionized water. These observations were attributed to formation of calcium-anion complexes at fracture tips during consolidation, which limit cracking due to the reduction of active surface sites available for Ca–CO_3 bond hydrolysis (Ilgen et al. 2018). A considerable set of results comes from experiments on porous chalks. Early experiments by Newman (1983) have shown a significant compaction of chalk saturated with 0.5 M NaCl solution in comparison to oil-saturated samples, which was attributed mainly to chalk dissolution in the presence of salt. Risnes et al. (2003) has demonstrated that the tensile strength of chalk is increased in the presence of concentrated NaCl (1.7 to 4.3 M) salt equilibrated with the rock relative to chalk-equilibrated water without the added ions and attributed this effect to decreased water activity in the presence of concentrated salts. Katika et al. (2015) showed that chalk weakens in the presence of 1.8 M NaCl because of the enhanced repulsion between grains with respect to deionized water, however this weakening is more pronounced in 0.6 M Na_2SO_4 because of the presence of divalent ions. Similarly, enhanced electrostatic repulsion acting between chalk grains in less concentrated NaCl (0.657 M) has been reported by Nermoen et al. (2018) resulting in samples weaker than the oil-saturated ones and in 0.4 M Na_2SO_4 by Pluymakers et al. (2021) with respect to $CaCO_3$-equilibrated water. However, Geremia et al. (2021) have reported chalk weakening in 0.1 to 1 M NaCl and in 0.1 to 0.6 M $NaHCO_3$ comparable to that observed in deionized water and chalk strengthening in 0.07 to 0.37 M Na_2SO_4, attributed to the differences in the magnitude of ion-specific repulsion between chalk grains induced by these two salts.

Calcium salts. Another frequently used salt was $CaCl_2$. In the studies reported in Table 1, $CaCl_2$ was mainly causing weakening of calcium carbonate samples. The associated mechanisms were repulsive surface forces acting between calcite grains in the presence of $CaCl_2$ (Katika et al. 2015; Geremia et al. 2021; Meireles et al. 2021) and accelerated pressure solution (Diao et al. 2020). One study reported strengthening due to decreased water activity in the presence of concentrated $CaCl_2$ (Risnes et al. 2003). While Katika et al. (2015) reports very significant chalk weakening due to strong electrostatic repulsion activated in the presence of divalent Ca^{2+} (0.6 M $CaCl_2$) with respect to deionized water, Geremia et al. (2021) show comparable chalk weakening in $CaCl_2$ solutions (0.09 to 0.45 M) and deionized water.

Magnesium salts. The remaining studies investigated the effect of magnesium salts on the chalk strength. $MgCl_2$ was found to strengthen calcium carbonate samples by surface passivation by adsorbed Mg^{2+}, which prevented calcite dissolution (Zhang and Spiers 2005a), contributing to weaker repulsion between calcite grains (Geremia et al. 2021) and reduced intergranular pressure solution at low magnesium salt concentration (Liteanu and Spiers 2009). In turn $MgCl_2$ contributed to weakening by the following mechanisms: enhanced electrostatic repulsion (Katika et al. 2015; Nermoen et al. 2018), precipitation of Mg-minerals (Madland et al. 2011), and enhanced subcritical cracking at high magnesium salt concentration (Liteanu and Spiers 2009). Strengthening was also reported for $MgSO_4$ salt and attributed to weakened repulsion between calcite grains in limestone relative to $CaCO_3$-saturared water. Due to the variety of reported mechanisms for Mg^{2+}, it is interesting to look at these works in more detail. Mechanical testing performed on pre-compacted calcite and limestone powders showed that the addition of 0.01 M of $MgCl_2$ (but also 0.0001–0.001 M Na_2HPO_4) to $CaCO_3$-equilibrated water decreased compaction relative to $CaCO_3$-equilibrated water without ions. These observations have been attributed to passivation of calcite surfaces in the presence of these ions, which adsorb to calcite and prevents its dissolution (Zhang and Spiers 2005a,b) and similar results were observed at 150 °C (Zhang et al. 2011). Experiments in the presence of subcritical CO_2 performed at 80 °C indicate that $MgCl_2$ salt reduced compaction of calcite aggregates at concentrations of 1M and below because of reduced intergranular pressure solution with respect to CO_2-free NaCl solutions, while high concentration of these salts (2–3 M) increase compaction rates via enhanced subcritical cracking (Liteanu and Spiers 2009). Long term exposure (200 days) of limestone to $CaCO_3$-equilibrated 0.4 M $MgSO_4$ solutions significantly strengthened the porous limestone, which has been attributed to ion-induced alteration of surface charges on mineral grains in the absence of significant limestone dissolution (Pluymakers et al. 2021). Some of the works listed in Table 1 studied the mechanical behavior of chalk in the presence of Mg^{2+} at elevated temperatures, which are more typical for reservoir conditions. Korsnes et al. (2008) have demonstrated that chalk saturated with seawater becomes weaker with the increasing temperature, while the opposite trend has been shown for deionized water. These findings have been mainly attributed to the adsorption of Mg^{2+} onto calcite surfaces from seawater, which affected interactions between calcite grains. It has been also shown that $MgCl_2$ (0.1 to 0.22 M) and seawater decreases the mechanical strength of chalk in reference to NaCl (0.66 M) and distilled water as tested by Madland et al. (2011) at 130 °C. This weakening effect of Mg^{2+} has been attributed to chalk dissolution and precipitation of Mg-bearing minerals. The weakening of chalk due to precipitation of magnesium minerals have been confirmed in long term experiments at elevated temperatures by Minde et al. (2020).

Important insight on the effect of salts on the deformation of calcite-bearing rocks in the presence of water and dissolved salt ions comes from experiments on subcritical crack growth, i.e. crack growth at stresses below the failure stress (Atkinson and Meredith 1987), in calcite and carbonate rocks, summarized in Table 2. Dunning et al. (1994) have performed experiments in oxalic and hydrochloric acids at pH 2 and have shown that both acids enhance subcritical calcite cracking with respect to water. Since calcite is highly soluble in HCl and poorly soluble in oxalic acid, the authors concluded that dissolution cannot be the main mechanism behind the observed weakening. Moreover, they have shown that Na_2CO_3 and NaOH at pH 11.4 have opposite effects, where sodium carbonate causes enhanced subcritical cracking and sodium hydroxide strengthens calcite with respect to tests in water. Experiments by Røyne et al. (2011) have demonstrated that water decreases surface energy of calcite and lowers the threshold energy release rate, which is required for a crack to propagate relative to glycol. It has been later found that NaCl salt enhances subcritical cracking at low and moderate concentrations (< 0.8 M) and arrest cracking at high concentrations, which was attributed to electrolyte-concentration sensitive surface forces operating at the crack tip (Rostom et al. 2013). In a follow-up work, the effect of ionic strength was further investigated and it has been shown that Na_2SO_4 has an opposite effect than NaCl salt: Na_2SO_4 arrests subcritical cracking of calcite at low concentrations (0.004 M) and enhances it at higher concentrations (0.4 M) with respect to pure water (Bergsaker et al. 2016). This work has also indicated that $MgCl_2$ and $MgSO_4$ do not have any significant effect on subcritical calcite cracking. Ilgen et al. (2018) have used dilute hydrochloric, sulfuric, and oxalic acids and have shown that pH in the range 2 to 6.5 is not correlated with the rates of subcritical fracture growth in calcite. Instead, the Ca-anion complexation in the different tested acids was proposed to control the strength of calcite subjected to fracture. The same phenomenon has been also shown to control the strength of calcite assemblages in the presence of $NaHCO_3$, Na_2SO_4, NaCl, $Na_3C_6H_5O_7$, with the lowest microfracture density in the presence of sulphate and chloride anions (Choens et al. 2021). Some of these experimental results could be well predicted by the surface complexation-based model by Zeng et al. (2020), which links surface potential of calcite to energy release rate during subcritical fracture propagation. As such, the main physicochemical mechanisms affecting the subcritical crack propagation in calcite immersed in aqueous salt solutions according to studies in Table 2 include: nanoscale surface forces acting between calcite surfacyes at the crack tip, changes in surface energy of the calcite due to adsorbed ionic species, and formation of Ca–anion surface complexes. In these studies, calcite dissolution has been ruled out as a factor promoting subcritical cracking.

Table 2. Effect of salts on subcritical calcite cracking. C is concentration. Symbol * means that the salt solution was presaturated with $CaCO_3$ powder.

Authors	Salt	pH	C (M)	Effect	Dominant mechanism
Dunning et al. (1994)	HCl	2	0.01	weakening	changes in calcite strength were not caused by dissolution
	$C_2H_2O_4$		0.05		
	Na_2CO_3	11.4	0.1	weakening	
	NaOH		0.005	strengthening	
Rostom et al. (2013)	NaCl	-	0.01–4	weakening effect at low and moderate concentrations (< 0.8 M)	electrolyte-concentration sensitive surface forces operating at the crack tip
	NH_4Cl*		0.7	moderate strengthening effect	
Bergsaker et al. (2016)	Na_2SO_4*	8.24–8.5	0.004, 0.4	strengthening at 0.004 M and weakening at 0.4 M	ion specific changes in surface energy that cannot be explained by changes in pH, zeta potential, or ionic strength
	$MgSO_4$*		0.003, 0.3	no significant effect, but promote crack healing	
	$MgCl_2$*		0.004, 0.4		
	CH_3COOH*	5.5–7.5	–	no effect of pH	
Ilgen et al. (2018)	HCl	2.1–5.3		no effect of pH on fracture propagation rate, but the rate depends on the type of anion: $HCO_3^- > Cl^- > SO_4^{2-}$	favourability of Ca-anion complex formation
	H_2SO_4	1.4–4.5	–		
	$H_2C_2O_2$	3.9–5.3			
Choens et al. (2021)	$NaHCO_3$	6.68–8.28		most pronounced strengthening in the presence of sodium sulphate and sodium chloride with respect to deionized water	strong Ca-anion complexes arrest subcritical fracturing
	Na_2SO_4		0.5		
	NaCl				
	$Na_3C_6H_5O_7$				

How calcite grains interact with each other in calcium carbonate rocks may be affected by the presence of organic matter and foreign mineral particles adsorbed on carbonate surfaces (Okhrimenko et al. 2013; Lakshtanov et al. 2018). Preserved remnants of calcite biominerals formed by organisms are commonly present in chalk, suggesting that surface-adsorbed organic material protects them from recrystallization in water (Hassenkam et al. 2009). Such surfaces are heterogenous, with randomly distributed hydrophilic and hydrophobic areas (Pedersen et al. 2016; Matthiesen et al. 2017; Hassenkam et al. 2009). Detailed reviews on how organic molecules modify behavior of calcite surfaces exposed to water have been recently published (Ganor et al. 2009; Al-Busaidi et al. 2019; Claesson et al. 2024). Experiments by Sachdeva et al. (2019) confirm that such changes in chalk wettability caused by the presence of adsorbed organic material affect the mechanical response of chalk in the presence of $MgCl_2$ salt solutions. A differing magnitude of water-weakening is also related to the presence of varying amounts of mineral impurities in calcium carbonate rocks (Amour et al. 2021).

EXPERIMENTAL MEASUREMENTS OF INTERACTIONS BETWEEN CALCITE SURFACES

We previously discussed, which mechanisms sensitive to electrolyte composition may explain deformation processes in fluid-saturated calcium carbonate rocks subject to stress. In this section, we will review experimental measurements of surface forces acting between calcite surfaces in aqueous electrolyte solutions, both the direct distance-resolved studies with atomic force microscopy (AFM) and surface force apparatus (SFA) and the indirect studies performed with other experimental methods. As previously outlined, the total interaction between two charged mineral surfaces at separation distances larger than several nanometers, can be in many cases sufficiently estimated by the Derjaguin–Landau–Verwey–Overbeek (DLVO) theory of colloidal stability, which considers the joint action of attractive van der Waals (vdW) and repulsive electrical double layer (EDL) surface forces (Derjaguin and Landau 1941; Verwey and Overbeek 1955; Israelachvili 2015; Butt and Kappl 2018). However, recent experiments highlight the importance of short-range hydration forces acting between hydrophilic calcite surfaces at distances typically below 5 nm (Røyne et al. 2015; Diao and Espinosa-Marzal 2016, 2019; Guo and Kovscek 2019; Fu et al. 2021). We will mainly focus on these two types of interactions and on how are they affected by the ionic composition of the salt solutions used in these measurements. Figure 6 schematically shows the main possible contributions to forces measured between similarly charged mineral surfaces such as calcite.

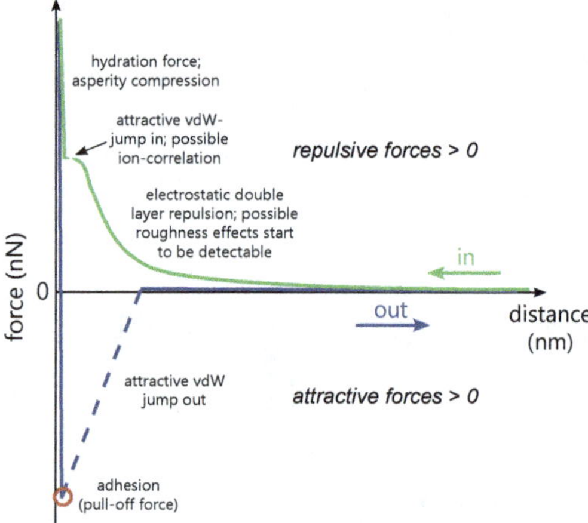

Figure 6. Schematic representation of nanoscale surface forces measured between two similarly charged mineral surfaces in an aqueous solution, measured as a function of a separation distance. Main and possible contributions to the measured forces were marked in the repulsive and attractive regions. A typical range for the presented distance (x axis) between surfaces is tens of nanometers.

Table 3. Summary of experimental measurements of distance resolved forces in a symmetrical surface configuration between two calcite surfaces in various aqueous solutions. Symbol * means that a given solution was presaturated with respect to calcite.

Authors	Surfaces	Method	Solutions	Main observations	Interpretation
Pourchet et al. (2013)	single crystal, cleaved (104) calcite	AFM	12 mM $Ca(OH)_2$ + $NaNO_3$ (IS = 0.12 M) 12 mM $Ca(OH)_2$ + 12 mM Na_2SO_4 + $NaNO_3$ (IS = 0.12 M)	stronger adhesion in the presence of sulphate ions	sulphate adsorption lowers the surface potential of calcite and screens electrical double layer repulsion
Røyne et al. (2015)	single crystal, cleaved (104) calcite	AFM	mixtures* of deionized water and ethylene glycol with water volume fractions of 0, 0.25, 0.5, 0.75, and 1.0	adhesion increases with the increasing ethylene glycol concentration	water reduces adhesion because of repulsive hydration forces acting between hydrophilic calcite surfaces
Javadi and Røyne (2018)	single crystal, cleaved (104) calcite	AFM	NaCl*, 0–1.2 M	adhesion increases with the increasing NaCl concentration, especially above 0.1 M	higher salt concentration leads to progressively weaker secondary hydration and increasing attraction due to instantaneous ion-ion correlation
Dziadkowiec et al. (2018)	polycrystalline calcite films	SFA	$CaCO_3$-saturated water	only repulsive forces	repulsive forces are related to hydration forces and to the increasing roughness of polycrystalline calcite surfaces immersed in calcite-saturated water
Dziadkowiec et al. (2019)	polycrystalline calcite films	SFA	NaCl*, $CaCl_2$*, and $MgCl_2$* with ionic strength of 0.01, 0.1, and 1 M	only repulsive forces in all solutions with the lowest magnitude for $MgCl_2$	ion-specific nucleation in confinement generates long-range repulsion between rough calcite surfaces
Dziadkowiec et al. (2021)	single crystal, cleaved (104) calcite	AFM	calcite-saturated water with pH 9–12, $CaCl_2$*, 0–0.7 M	adhesion in the presence of Ca^{2+} is lower than for NaCl and increases slightly with the increasing $CaCl_2$ concentration; in calcite-saturated water adhesion is somewhat correlated with the dissolved Ca^{2+} content, with purely repulsive forces measured at pH 12 (lowest Ca^{2+})	repulsive hydration effects associated with the outer-sphere electrostatic binding of strongly hydrated Ca^{2+} cations are responsible for weak adhesion in $CaCl_2$ solutions
Li et al. (2023)	polycrystalline aggregate of calcite against single crystal (100) calcite substrate	AFM	1–100 mM NaCl* at pH from 5.5 to 9.5; 1–2 mM $MgCl_2$* at pH 6.5 and 9.5	In NaCl, adhesion decreases with the increasing pH (apart from pH 5.5, which shows the strongest repulsion); there is no major change in adhesion with the increasing NaCl concentration from 1 to 100 mM; higher Mg^{2+} concentration decreases adhesion at pH 6.5, but increases adhesion at pH 9.5	ion adsorption changes surface potential and screens EDL repulsion

Direct force measurements between two calcite surfaces

Table 3 summarizes recent experiments, in which distance-resolved surface forces were measured in a symmetrical surface configuration between two calcite surfaces immersed in water or aqueous salt solutions. Because of a challenging nature of brittle calcite surfaces for surface force measurements and specific requirements of experimental techniques, there are currently only several such studies. The measurements were performed mainly with AFM and SFA. The symmetrical calcite–calcite surface configuration in AFM is typically achieved in AFM by modifying the AFM cantilever with a small single crystal cleaved calcite (Pourchet et al. 2013; Røyne et al. 2015; Javadi and Røyne 2018; Dziadkowiec et al. 2021) or a polycrystalline aggregate of calcite particles (Li et al. 2023), which are glued to an AFM cantilever. In SFA, the measurements are performed between two extended polycrystalline calcite surfaces deposited as a thin, continuous film using the atomic layer deposition technique (Dziadkowiec et al. 2018, 2019). The greatest benefit of this surface configuration is a chemical symmetry of the system since the presence of another surface with different chemical composition modifies the measured interaction forces to significant but often difficult to quantify extent. The drawbacks of involve poor experimental resolution and uncontrolled geometry of the contact, which makes it difficult to precisely relate the measured force to disjoining pressure. In addition, in such setup, the roughness and electrostatic effects may overlap and are often difficult to distinguish, while hydration effects present at the smallest surface separations are hard to resolve. As such, the main parameters that can be accessed in these measurements are ion-specific adhesion (attractive force measured on retraction prior to adhesive jump-out event) and the magnitude and range of repulsive forces (Fig. 6).

Forces between two calcite surfaces in water. Most of the available studies in calcite–calcite system measured forces between two freshly cleaved calcite surfaces, with one of them, much smaller in size, attached to a tipless AFM cantilever. Measurements by Røyne et al. (2015) revealed that strong repulsion act between two calcite surfaces in calcite-saturated water, with repulsive pressure significantly exceeding the range of typical electrical double layer repulsion for calcite in water, which was attributed to repulsive hydration force and supported by the presence of adhesive forces in ethylene glycol. Adhesion and frequency of adhesive force runs was correlated with the increasing ethylene glycol concentration in water-glycol mixtures. In addition to repulsive forces measured on approach, low pull-off adhesion force have been detected in some calcite–calcite pairs in this system, and its occurrence was attributed to varying roughness of calcite–calcite contacts (Røyne et al. 2015; Javadi and Røyne 2018; Dziadkowiec et al. 2021). This pull-off adhesion was not evidently correlated with the pH and thus dissolved Ca^{2+} concentration in calcite saturated water, however frequency of adhesive force runs was the highest for calcite-saturated water with the highest Ca^{2+} content and the runs were purely repulsive at pH 12, in which the Ca^{2+} concentration was the lowest (Dziadkowiec et al. 2021). This was interpreted in terms of potential determining properties of Ca^{2+} towards calcite surface, with the highest magnitude of zeta potential measured for the lowest Ca^{2+} concentration, reproducible in various limestone and calcite setups (Foxall et al. 1979; Heberling et al. 2011; Pourchet et al. 2013; Alroudhan et al. 2016; Song et al. 2017).

Forces between two calcite surfaces in salt solutions. Other works in this system investigated forces in the presence of electrolytes, with some of the data summarized in Figure 7. Pourchet et al. (2013) measured forces in high pH electrolytes and demonstrated higher adhesion in the presence of sulphate ions, which was correlated with sulphate ions lowering the zeta potential of calcite. Thus, sulphate-induced changes in adhesion were attributed to screened electrical double layer repulsion, however, only one sulphate concentration was tested in the force measurements and no systematic effect of sulphate on calcite–calcite interactions was studied. Later works by Javadi and Røyne (2018) and Dziadkowiec et al. (2021) studied adhesive forces between two calcite surfaces as a function

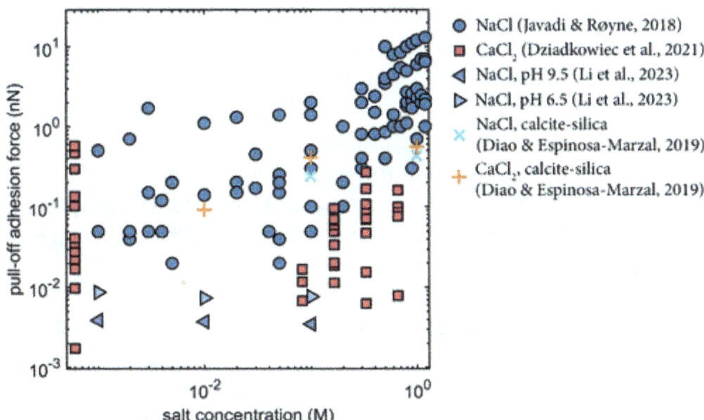

Figure 7. Comparison of reported literature data on pull-off adhesion force between two calcite surfaces in electrolyte solutions. Data for calcite–silica surfaces is plotted for comparison. In the dataset by Diao and Espinosa-Marzal (2019) only median values are plotted. In the dataset by Li et al. (2023) only mean values are plotted and are expressed here in nN, assuming equal radius of curvature of calcite particles of 5 μm. Thus, the magnitude of the adhesion force might by misestimated, while the trends remain accurate.

of NaCl and $CaCl_2$ electrolyte concentration. For NaCl, the load-dependent pull-off adhesion force generally increased with the increasing NaCl concentration, especially at concentrations above 100 mN. This increasing adhesion has been attributed to weaker cation hydration at higher salt concentration and to the increasing importance of ion correlation forces (Javadi and Røyne 2018). Adhesion pull-off forces measured in $CaCl_2$ solutions were relatively weak and up to three orders of magnitude lower than in NaCl at the same ionic strength. Since the DLVO modelling based on the zeta potential values reported by Heberling et al. (2011) predicted attractive DLVO forces for calcite surfaces for the tested $CaCl_2$ ionic strengths in these experiments, the weak adhesion was attributed to repulsive non-DLVO hydration forces due to adsorption of hydrated Ca^{2+}. Interestingly, Diao and Espinosa-Marzal (2019) measured opposite trend in adhesion for calcite–silica surfaces, where the pull-off adhesion force was higher in $CaCl_2$ than in NaCl. Li et al. (2023) measured forces between two calcite surfaces in NaCl and $MgCl_2$ solutions with varying pH. This work reports that adhesion between two calcite surfaces decreases as a function of pH from 6.5 to 9.5, apart from pH 5.5 for which the lowest adhesion was measured. Adhesion was weakly modulated by increasing NaCl concentration from 1 to 100 mM both at pH 6.5 and 9.5, which is generally in agreement with previous measurements of Javadi and Røyne (2018), where significant increase in adhesion was observed only at NaCl concentrations higher than 100 mM. $MgCl_2$ showed a more complex effect on adhesion with pull-off forces decreasing as a function of $MgCl_2$ concentration at pH 6.5 and increasing at pH 9.5, although only low concentrations of 1 and 2 mM were probed. In the work by Li et al. (2023), a polycrystalline aggregate of calcite particles was attached to an AFM tip (instead of a small cleaved calcite fragment in earlier works) and a better experimental resolution was achieved. As such, the authors could conclude that the measured forces and the trends in adhesion could be well interpreted with the DLVO forces for all conditions, with some discrepancies at distances below 3 nm, at which surface roughness and hydration forces modify the DLVO interactions. This study was however performed with (100) calcite surfaces, which may differ in surface charge and surface reactivity with respect to typical (104) equilibrium calcite faces.

Additional insight into calcite–calcite comes from the surface forces apparatus experiments (SFA), in which forces were measured between two polycrystalline calcite films in calcite-saturated water and in NaCl, $CaCl_2$, and $MgCl_2$ salts (Dziadkowiec et al. 2018, 2019) These experiments show that small calcite crystals bound by (104) faces in the film show higher reactivity than cleaved (104) single crystal calcite surfaces and are prone to roughening, even in calcite-saturated water. The authors interpret that both the range and the magnitude of repulsive forces are positively correlated with the increasing surface roughness. In addition, in the confined SFA settings, reprecipitation events are possible because of the reduced transport in a confined water film. During these events, calcite reprecipitation was correlated with the appearance long-range repulsion and linked to the nucleation of amorphous calcium carbonate (ACC). The onset of the ACC nucleation in these experiments depended on cation type, with $CaCl_2$ and NaCl promoting the fastest reprecipitation and $MgCl_2$ inhibiting it significantly (Dziadkowiec et al. 2019).

Direct force measurements in asymmetrical surface configurations

A significant number of AFM studies investigated forces as a function of surface separation between calcite and another surface such as silica, gold, and functionalized or bare AFM tips. These are summarized in Table 4. In these AFM setups, the geometry of the AFM probe is well defined, which generally provides much better experimental resolution than the experiments in a calcite–calcite surface configuration. The interactions in asymmetrical setup might be however modified by the presence of a chemically different interface, which alters ionic populations in overlapping double layers during force-distance measurements to a varied extent, especially if mineral surfaces have a significantly different point of zero charge in a given electrolyte solution.

A substantial amount of systematic data has been acquired in a calcite–silica AFM setup in $CaCl_2$, NaCl, $MgCl_2$, and $NiCl_2$ solutions, which provide molecular-scale insight into confined hydration layers on calcite (Diao and Espinosa-Marzal 2016, 2019; Fu et al. 2021). These high-resolution experiments detect EDL forces at surface separations larger than 5 nm and oscillatory, ion-specific hydration repulsive force acting at surface separations below 5 nm. The experimentally measured repulsive hydration forces are typically much higher in magnitude (up to two orders of magnitude) than electrostatic repulsive EDL forces acting in a given electrolyte solution (Diao and Espinosa-Marzal 2016, 2018, 2019; Fu et al. 2021). In the presence of $CaCl_2$, these hydration layers were composed of hydrated outer-sphere Ca^{2+} and partially dehydrated inner-sphere Ca^{2+}, with their relative proportions depending on the electrolyte concentration and surface charge of calcite. The proportion of inner-sphere Ca^{2+} increased with the increasing confinement. The magnitude and range of repulsive hydration forces depended on salt concentration with the strongest repulsion measured in 1 mM Ca^{2+} solution. Hydration repulsion was screened at 10 and 100 mM, with the minimum applied work needed to squeeze out hydrated Ca^{2+} at 10 mM $CaCl_2$. Measurements in other salt solutions confirm that the structure of the hydration layer on calcite is salt concentration-dependent but also reveal its strong ionic specificity. In NaCl, a much smaller number of molecular water layers was detected than in $CaCl_2$, with dehydrated Na^+ adsorbed closer to the calcite surface (Diao and Espinosa-Marzal 2019). A smaller number of hydration layers in comparison with Ca^{2+} was also detected in the presence of Ni^{2+} cations, which, similarly to Na^+, are located closer to the calcite surface and can screen the surface charge of calcite at lower concentrations than Ca^{2+} (Fu et al. 2021). Measurements performed in $MgCl_2$ with multiple hydration layers detected at generally the largest surface separations in comparison to other cations, indicated that Mg^{2+} ions may significantly disturb the ordering of hydration layers on calcite and are not efficient in screening of the surface charge of calcite (Fu et al. 2021). These works conclude that counterions with larger hydration enthalpy, which require more energy to lose their hydration shells fully or in part, are located further away from the calcite surface (Diao and Espinosa-Marzal 2016, 2019; Fu et al. 2021). Significant pull-off adhesive force due to van der Waals attraction and possible ion correlation forces is detected at and above 10 mM concentrations for $CaCl_2$ and at and above 100 mM for NaCl (Diao and Espinosa-Marzal 2019).

Ion-specific repulsive hydration forces were also detected between calcite and gold-coated silica by Guo and Kovscek (2019). Although water layering was not resolved in these measurements, there is a clear relationship between the decay length of short-range repulsive hydration forces and the type of cation present in electrolyte solutions, with Na^+ decreasing and

Table 4. Summary of experimental measurements of distance resolved forces in an asymmetrical surface configuration with one calcite surface various aqueous solutions. Symbol * means that a given solution was presaturated with respect to calcite.

Authors	Surfaces	Method	Solutions	Main observations	Interpretation
Diao and Espinosa-Marzal (2016)	colloidal silica probe against single crystal, cleaved (104) calcite	AFM	0 - 100 mM $CaCl_2$*	oscillatory repulsive force detected at surface separations < 3 nm, with range and magnitude dependent on ionic concentration	hydration force due to adsorption of outer- and inner-sphere Ca^{2+} ions
Hidema et al. (2016)	calcite particle against glass, copper, silica and silane-coated surfaces	AFM	$CaCO_3$-saturated water	adhesion increases with decreasing hydrophilicity of surfaces	water adsorbed on calcite and on hydrophilic surfaces prevents strong adhesion
Eichmann and Burnham (2017)	bare silicon tip or -COOH functionalized tip against calcite surface	AFM	seawater, brine, and Ca-enriched seawater	adhesion is systematically lower in the presence of Ca^{2+} both for bare tips and -COOH functionalized tips	Ca^{2+} adsorption onto surfaces enhances repulsion as it stabilizes thicker hydration layers
Generosi et al. (2017)	$-CH_3$ functionalized AFM tip against cleaved (104) calcite	AFM	NaCl* and NaCl* with added $MgSO_4$, $MgCl_2$, and Na_2SO_4; all solutions with IS = 0.5 M	$MgSO_4$ decreases adhesion force between hydrophobic tip and calcite surface	wettability changes to more hydrophilic calcite surface induced by Mg^{2+} adsorption

Table 4 (cont´d). Summary of experimental measurements of distance resolved forces in an asymmetrical surface configuration with one calcite surface various aqueous solutions. Symbol * means that a given solution was presaturated with respect to calcite.

Authors	Surfaces	Method	Solutions	Main observations	Interpretation
Diao and Espinosa-Marzal (2019)	silica AFM cantilever against cleaved (104) calcite	AFM	NaCl* and $CaCl_2$* at 0–1 M	pull-off adhesion force is higher for $CaCl_2$ than for NaCl at a given concentration and increases with the increasing concentration; friction force is weakened by Na^+ more than by Ca^{2+} at low applied stresses; at high applied stresses Na^+ increases friction force	ion-specific adhesion is the outcome of possible ion correlation forces and bridging of negatively charged surfaces by cations; friction force depends on the interfacial structure and the influence of cations on the friction force is altered when pressure solution is activated at high applied stresses
Guo and Kovscek (2019)	gold-coated silica against cleaved (104) calcite surface	AFM	$MgCl_2$*, $MgSO_4$*, Na_2SO_4*, all at 1 to 150 nM; NaCl at 1 to 500 mM*; brine*	attractive force is measured only in NaCl and Na_2SO_4 at higher salt concentrations; Na^+ reduces the decay length of short-range repulsion while Mg^{2+} increases it	Na^+ ions disturb water structure on calcite surface while Mg^{2+} ions are adsorbed at larger distances from the calcite surface, thus Na^+ ions reduce the range and magnitude of hydration forces and Mg^{2+} ions make calcite more hydrophilic
Fu et al. (2021)	colloidal silica microsphere or silica tip against cleaved (104) calcite	AFM	$CaCO_3$-saturated water and $NiCl_2$*, $CaCl_2$*, $MgCl_2$* at 0.1 to 10 mM	at surface separations > 5 nm, ion-specific and concentration-dependent forces can be well described by DLVO modelling with surface charge screening ability decreasing in order $Ni^{2+} > Ca^{2+} > Mg^{2+}$; at lower surface separations, ion-specific oscillatory hydration force is detected,	structure of hydration layers indicates that Ni^{2+} is adsorbed closer to the surface than Ca^{2+} and Mg^{2+}; Mg^{2+} may disrupt the structure of hydration layers on calcite; at the point of charge neutralization for a given ion, the structure of hydration layers undergoes a significant change

Mg^{2+} increasing the decay length of short-range hydration forces. Thus, the authors interpret that Na^+ ions can disturb the interfacial water structure and adsorb closer to the calcite surface while Mg^{2+} are located further away from the calcite surface, making calcite more hydrophilic, which corresponds to thicker hydration layers in the presence of Mg^{2+}. Attractive forces were only detected in NaCl and Na_2SO_4 solutions at concentrations of 150 mM and attributed to van der Walls attraction and to ion correlation, which makes the attraction stronger in Na_2SO_4 because of the presence of multivalent sulphate anions. Only repulsive interactions were measured in $MgCl_2$ and $MgSO_4$ at concentrations of 150 mM and below (Guo and Kovscek 2019). More hydrophilic calcite surface in the presence of Mg^{2+} has been also suggested by Generosi et al. (2017) who measured forces between a hydrophobic $-CH_3$ terminated AFM tip and calcite. This work showed reproducibly lower adhesion in the presence of $MgSO_4$ than for NaCl, which was attributed to salt-induced changes in the structure of calcite's hydration layers and thus wettability. A similar effect of stabilizing thicker hydration layers as for Mg^{2+} has been reported by Eichmann and Burnham (2017) for Ca^{2+}, with calcium ions contributing to stronger repulsion between –COOH-functionalized or bare silicon AFM tips and calcite.

Summarized effects of main salt ions on surface forces between calcite surfaces

In experimental measurements, two calcite surfaces do not show attraction nor strong adhesion when immersed in water, a phenomenon related to hydrophilic nature of these surfaces, decrease of their interfacial energy with respect to surfaces in air, and to the presence of repulsive hydration forces, which counteract the attractive vdW forces (Røyne et al. 2015; Diao and Espinosa-Marzal 2016; Guo and Kovscek 2019). Ions in electrolyte solutions can modify the forces acting between calcite surfaces, leading to more adhesive or more repulsive contacts between calcite surfaces in salt solutions than in water (Javadi and Røyne 2018; Dziadkowiec et al. 2021). Based on the findings summarized in Tables 3 and 4, while the surface forces acting above the surface separations of a few nanometers can be usually well described by the electrical double layer forces, the forces acting at lower surface separations, typically below 3 nm and the adhesion behavior seems to be dominated by ion-specific effects such as hydration forces (Diao and Espinosa-Marzal 2016, 2018, 2019; Javadi and Røyne 2018; Guo and Kovscek 2019; Diao et al. 2020; Dziadkowiec et al. 2021; Fu et al. 2021). Here, we summarize the main findings on surface forces measured in electrolyte solutions both in symmetrical and asymmetrical surface configurations, grouped by the tested salt solution composition.

Sodium salts. Sodium salts have been used in several studies both between two calcite surfaces and in asymmetrical surface configuration as summarized in Figure 8. These works indicate in a fair agreement that high sodium salt concentration (> 100 mM) promotes more adhesive or attractive interactions between calcite surfaces and contributes to reduced range and magnitude of short-range (< 3 nm) hydration forces. At higher surface separations (> 3 nm) forces can be described by EDL theory. In particular:

- Sodium salts significantly influence adhesion both in symmetrical (calcite–calcite) and asymmetrical surface configurations (calcite–silica, calcite–gold), especially at concentrations higher than 100 mM. Above this value, adhesion or attraction increases with the increasing salt concentration (Javadi and Røyne 2018; Diao and Espinosa-Marzal 2019; Guo and Kovscek 2019). Below 100 mM adhesion is low (Javadi and Røyne 2018), not detected (Diao and Espinosa-Marzal 2019), or depends weakly on sodium salt concentration (Li et al. 2023). Increasing adhesion could be explained by weaker hydration repulsion at higher sodium salts concentration (Javadi and Røyne 2018; Guo and Kovscek 2019) and increasing importance of ion correlation forces (Javadi and Røyne 2018; Diao and Espinosa-Marzal 2019);
- Sodium salts affect the magnitude and range of hydration forces resolved at low surface separations in asymmetrical surface configurations. Hydration forces display either oscillatory character, with much lower number of hydration layers resolved in NaCl than

in $CaCO_3$-saturated water (Diao and Espinosa-Marzal 2019), or show purely repulsive character with the range dependent on salt concentration and anion type (Guo and Kovscek 2019). These findings were attributed to Na^+ ions altering the interfacial water structure on calcite.

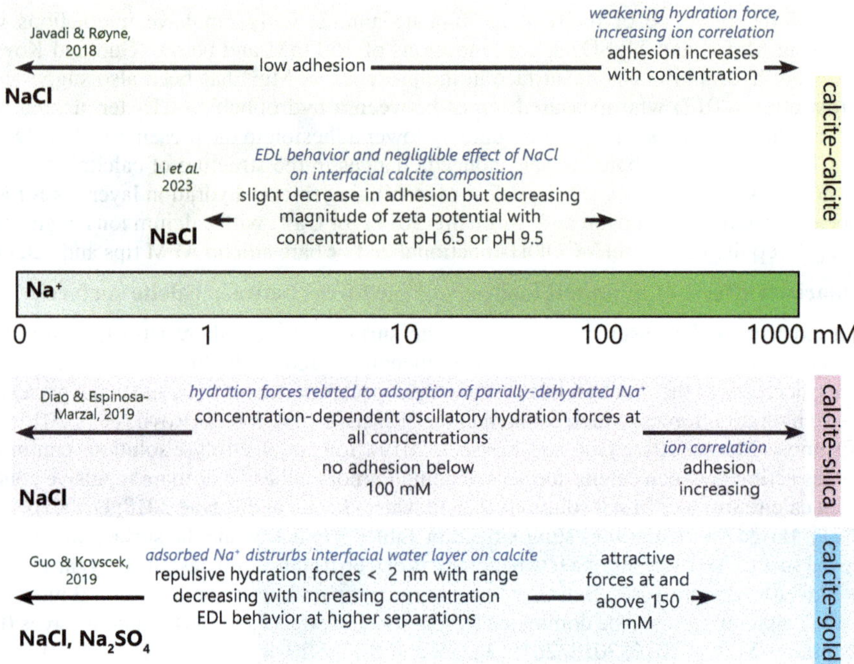

Figure 8. Summarized effects of sodium salts on forces between calcite surfaces. The **green bar** shows sodium salt concentration.

Calcium salts. $CaCl_2$ salt was the only calcium salt used to measured forces between two calcite surfaces (Dziadkowiec et al. 2021) or between calcite and colloidal silica (Diao and Espinosa-Marzal 2016; Fu et al. 2021) or calcite and bare silicon AFM tip (Diao and Espinosa-Marzal 2019). In these works, it was demonstrated that calcium chloride affects both the measured adhesion and hydration forces, as summarized in Figure 9. It was particularly shown that:

- Adhesion forces measured between two calcite surfaces are significantly weaker than those measured in NaCl and do not show significant increase with the increasing $CaCl_2$ salt concentration (Dziadkowiec et al. 2021), which was attributed to stronger repulsive hydration associated with the adsorbed Ca^{2+}. In asymmetrical calcite–silica system, the adhesion was however higher than in NaCl solutions and significantly increased with the increasing salt concentration, which was attributed to attractive ion correlation forces (Diao and Espinosa-Marzal 2019);
- Oscillatory concentration-dependent hydration forces are also detected in $CaCl_2$ solutions, with their range dependent on salt concentration. These forces are generally stronger in magnitude and with larger range than in NaCl solutions at the same concentration, which was attributed to Ca^{2+} ions being more hydrated than Na^+ and residing mainly on top of calcite hydration layers (Diao and Espinosa-Marzal 2016; Fu et al. 2021)

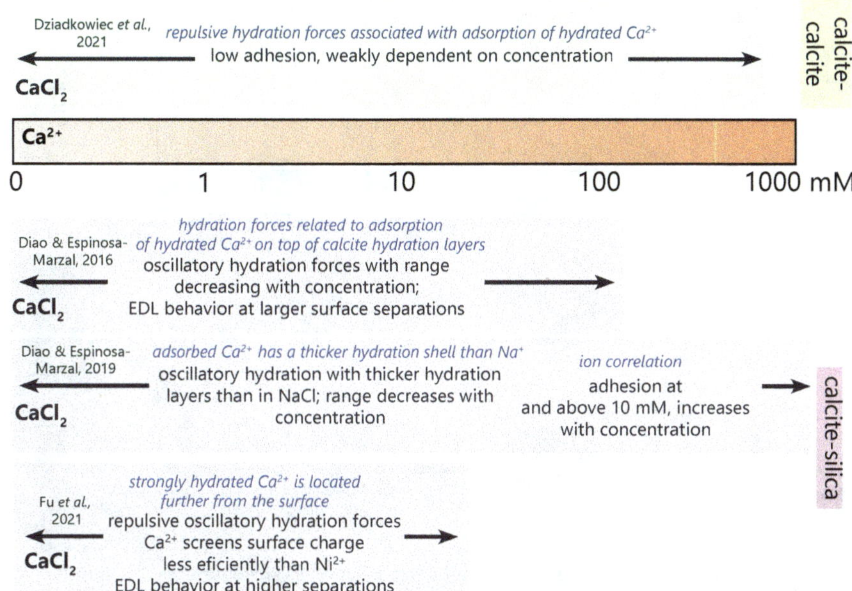

Figure 9. Summarized effects of calcium salts on forces between calcite surfaces. The **orange bar** shows sodium salt concentration.

Magnesium salts. There are only a few experimental studies of surface forces between calcite surfaces in the presence of magnesium salts. These works indicate that magnesium salts lower the adhesion between calcite surfaces and contribute to strong hydration forces, detected at low surface separations. These studies are summarized in Figure 10. Most importantly:

- Even low concentrations of magnesium salts (< 2 mM) lower adhesion between calcite surfaces with respect to sodium salts at pH 6.5, with the opposite effect at pH 9.5. These findings were attributed to high sensitivity of calcite zeta potential to Mg^{2+} concentration (Li et al. 2023). Mg^{2+} seems also to prevent calcite surfaces from recrystallization, which happens much faster in NaCl and $CaCl_2$ solutions (Dziadkowiec et al. 2019);
- Repulsive hydration forces detected at surface separations below a few nm increase in strength and magnitude with the increasing magnesium salt concentration (Guo and Kovscek 2019) as detected between calcite and gold surfaces. In calcite–silica setup, the hydration forces showed oscillatory character, but the resolved layering indicated that Mg^{2+} is the most efficient cation to disturb hydration layers on calcite in comparison with Ca^{2+} and Ni^{2+} (Fu et al. 2021)

INTERACTIONS OF SALT IONS WITH CALCIUM CARBONATE SURFACES

Information relevant to ion-specific attractive and repulsive interactions between calcite surfaces can also be accessed by many different theoretical and experimental methods other than the direct distance-force measurements. Some examples are given in Table 5.

Rheological behavior of calcium carbonate

Aggregation and rheological behavior of calcite particles has been shown to depend on the types of ions present in the interstitial solution. Nyström et al. (2001) showed that increasing NaCl concentration from 0 to 0.5 M led to stronger calcite aggregation, while for $CaCl_2$,

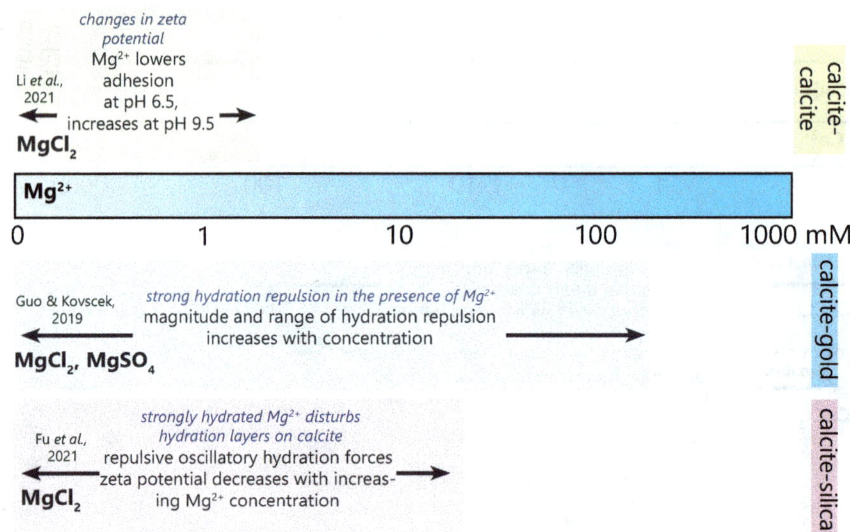

Figure 10. Summarized effects of magnesium salts on forces between calcite surfaces. The **blue bar** shows magnesium salt concentration.

$BaCl_2$, and $LaCl_3$ salts, aggregation behavior depended on the salt concentration in a more complex way, with a strongest aggregation correlated roughly with the lowest zeta potential and thus the magnitude of the EDL repulsion. For Ca^{2+} and Ba^{2+} the minimum in yield stress and thus the strongest aggregation occurred at 10 mM, while for Li^{3+} at 0.5 mM. These results are generally in a good agreement with rheological behavior of calcite pastes reported by Liberto et al. (2019) with Na^+ enhancing interparticle attraction and Ca^{2+} making the calcite paste less rigid due to enhanced repulsion between particles. The elastic modulus measured for calcite paste was the lowest (highest interparticle repulsion) at conditions for which the EDL repulsive energy barrier in $Ca(OH)_2$ solution was the highest. The magnitude of ion-specific interparticle repulsion has been also estimated from experiments on salt solutions-induced porosity changes in aggregated calcite powders by Meireles et al. (2020). $CaCl_2$ has been shown to yield the highest porosity in calcite powders in comparison with $MgCl_2$ and NaCl, while Na_2SO_4 yielded powders with the lowest porosity. In all these works, the interactions between calcite particles could be well predicted using the DLVO theory.

Zeta potential of calcium carbonate

As such, it is vital to understand the effect of various ions on the zeta potential of calcite and limestone. Measurements of zeta potential can serve to estimate the magnitude of the repulsive EDL forces in a given electrolyte solution and provide information how various salt ions affect the surface charge of calcite. Zeta potential refers to the electrical potential at the shear plane, which is the region of the diffuse part of the electrical double layer, where excess counter- and co-ions are no longer strongly bound to the charged mineral surface but become mobile (Hunter 2013; Bard et al. 2022). There are numerous works, which studied zeta potential of various calcium carbonate samples as a function of electrolyte concentration and compositions. A systematic review of these studies is out of scope here and we only summarize main findings; see for example comprehensive reviews by Al Mahrouqi et al. (2017) and Bonto et al. (2022). It has been established that principal potential determining ions for calcite, which control the development of surface charge, are the calcite lattice ions Ca^{2+} and CO_3^{2-}, but also Mg^{2+}, which can become incorporated into the calcite lattice (Al Mahrouqi et al. 2017). The magnitude of zeta

Table 5. Summary of experimental measurements and theoretical studies that provide additional insight on interactions between calcite surfaces. Please note that this list is not exhaustive. Symbol * means that a given solution was presaturated with respect to calcite.

Author	Material	Method	Solutions	Main Observations	Interpretation
Nyström et al. (2001)	calcite dispersions	zeta potential and rheological measurements	NaCl (0–0.8 M), $CaCl_2$ (0–0.12 M), $BaCl_2$ (0–0.2 M), and $LaCl_3$ (0–0.01 M)	adding salts influences the aggregation behavior of calcite, which correlates with changes in calcite zeta potential; multivalent ions are more efficient to reduce the zeta potential of calcite	aggregation of calcite is sensitive to electrostatic repulsion between calcite particles
Ricci et al. (2013)	(104) calcite surface	AFM	NaCl*, RbCl*, $CaCl_2$* at concentrations of 1–100 mM	adsorbed Na^+ and Rb^+ cations can be well visualized on calcite surface, while location of Ca^{2+} is less evident	hydrated Ca^{2+} with larger hydration shells are located further from the calcite surface than smaller hydrated cations, such as Na^+ and Rb^+
Levenson and Emmanuel (2017)	limestone samples	AFM	NaCl, 0–500 mM	grain detachment in flow experiments is observed, however the frequency of grain detachment shows no correlation with NaCl salinity	repulsive forces in water are responsible for grain detachment but sample heterogenous character can mask subtle differences in salinity dependence on detachment events
Al Mahrouqi et al. (2017)	limestone samples	zeta potential measurements using streaming potential method (SPM)	NaCl* (0.05–4 M), and NaCl (0.5 or 2 M) spiked with varying concentrations of $CaCl_2$*, $MgCl_2$*, or Na_2SO_4*	pH does not directly affect the zeta potential and instead pH-dependent concentration of Ca^{2+} is important; SO_4^{2-} affect zeta potential mainly by formation of complexes with dissolved Ca^{2+}	Ca^{2+}, Mg^{2+} are potential determining ions for calcite, while SO_4^{2-} affects zeta potential mainly by moderating equilibrium Ca^{2+} concentration
Diao and Espinosa-Marzal (2018)	single crystal calcite	AFM friction force	$CaCO_3$-saturated water and $CaCl_2$* at 1 to 1 M	friction force is lower in all salt solutions than for dry contact; the influence of salt on friction force depends on the applied load and ion concentration; dependence of concentration on friction force varies in each applied load regime	applied load affects the main friction mechanism with viscous shear of lubricating water film at low stresses, thermally activated slip and decrease of mobility of confined solution at intermediate stresses, and pressure-solution facilitated slip at the highest stresses
Alotaibi et al. (2018)	calcite or limestone	electrophoretic zeta potential measurements	NaCl (0.1M), $CaCl_2$ (0.05 M), $MgCl_2$ (0.06 M), Na_2SO_4 (0.04 M)	NaCl and Na_2SO_4 yield calcite similarly negatively charged as in deionized water. For limestone, zeta potential is less negative in sodium salts than in water. $CaCl_2$ and $MgCl_2$ reversed the zeta potential of both calcite and limestone particles towards the positive values.	ion-specific changes in zeta potential can be linked to various mechanisms of ion adsorption at calcite/brine interfaces

Table 5 (cont´d). Summary of experimental measurements and theoretical studies that provide additional insight on interactions between calcite surfaces. Please note that this list is not exhaustive. Symbol * means that a given solution was presaturated with respect to calcite.

Author	Material	Method	Solutions	Main Observations	Interpretation
Kirch et al. (2018)	calcite	molecular dynamics simulations	NaCl, $CaCl_2$	Na^+ and Ca^{2+} ions adsorbed onto calcite surfaces are located at different distances from the calcite surfaces; Na^+ ions are adsorbed closer to the calcite surface, while Ca^{2+} are located further away	binding of cations onto calcite is affected by the strong ordering of water molecules onto calcite and the resulting spatially alternated electrostatic potential and the hydration properties of the cations
Liberto et al. (2019)	nanocalcite powder	rheological measurements	$Ca(OH)_2$ at 3–50 mM and NaOH at 94 mM with 3 mM of $Ca(OH)_2$	minimum of shear modulus is observed at Ca^{2+} concentration corresponding to strongest electrostatic repulsion between calcite surfaces; rigid paste is obtained in the presence of sodium hydroxide	DLVO electrostatic forces can explain the rheological behavior of calcite particles in the presence of Ca^{2+} and Na^+ ions
Meireles et al. (2020)	calcite powder obtained from chalk	porosity changes measured with nuclear magnetic resonance (NMR) relaxometry	$CaCl_2$, $MgCl_2$, Na_2SO4, NaCl; at ionic strengths from 1.8 mM to 1.8 M	porosity of calcite powder columns is ion-specific, with the highest porosity in general for brines containing Ca^{2+} and the lowest for brines with SO_4^{2-}	differences in porosity reflects disjoining pressure acting between calcite grains, with the strongest electrostatic repulsion in the presence of Ca^{2+} and the weakest in the presence of SO_4^{2-}
Ali et al. (2020)	(104) calcite surface	molecular dynamics simulations	1–3 M NaCl, 2 M KCl, 2 M $MgCl_2$	K^+ adsorbs further from the surface than Na^+, divalent Mg^{2+} maintains strong hydration shell, is located the furthest from calcite surface and cause limited changes to surface distribution of water	addition of salt disturbs well-organized structure of oxygen atoms in the first and second hydration layers on calcite, which affects surface density distribution of water, its hydrogen bond network and its residence times on calcite surface
Tetteh et al. (2020)	crushed limestone	electrophoretic zeta potential measurements	NaCl*, KCl*, $MgCl_2$*, $CaCl_2$* and Na_2SO_4* at ionic strength of 0.04 M	zeta potential is negative for all tested salts at 20 °C and becomes less negative in the following order: NaCl < KCl < Na_2SO_4 < $CaCl_2$ < $MgCl_2$	Na^+ and K^+ are inert cations that do not alter surface charge of calcite but affect the thickness of EDL, while Ca^{2+}, Mg^{2+}, and SO_4^{2-} cause surface charge alteration due to their strong adsorption onto calcite surface
Fu et al. (2021)	calcite single crystals	SFA and AFM friction force measurements	$CaCO_3$-saturated water (SFA); $NiCl_2$*, $CaCl_2$*, $MgCl_2$* at 0–100 mM (AFM)	pressure-induced dissolution is related to frictional weakening; friction response is modified by the presence of ions: Mg^{2+} reduces friction and friction force does not decrease with increasing load, Ca^{2+} promotes pressure solution-facilitated slip and Ni^{2+} has a complex-concentration dependent behavior	freshly dissolved, hydrated ions present in the confined solution allow water molecules retained in hydration shells to remain more lubricious than pure water; when re-precipitation of material dissolved due to applied pressure happens, friction force increases

potential in the presence of calcium ions generally increases with the increasing concentration of Ca^{2+} and zeta potential quickly becomes positive due to charge reversal of calcite at relatively low mM Ca^{2+} concentrations (Thompson and Pownall 1989; Pierre et al. 1990; Huang et al. 1991; Cicerone et al. 1992; Alroudhan et al. 2016), however, some more complex dependence of zeta potential on Ca^{2+} concentration has been also reported (Nyström et al. 2001). Mg^{2+} exhibits similar behavior as Ca^{2+} (Zhang et al. 2007; Chen et al. 2014; Alroudhan et al. 2016), while zeta potential drops with the addition of CO_3^{2-} (Douglas and Walker 1950; Smallwood 1977; Moulin and Roques 2003). The effect of sulphate is less clear, and it is suggested that sulphate may affect the surface charge of calcite only indirectly, by controlling the concentration of dissolved Ca^{2+} while other works support strong adsorption of sulphate onto calcite (Alroudhan et al. 2016; Al Mahrouqi et al. 2017; Bonto et al. 2022). Contrary to other mineral surfaces, surface charge of calcite is not strongly dependent on pH. The observed changes of zeta potential in response to pH can be well explained by varying equilibrium concentration of potential determining ions at a given solution pH (Al Mahrouqi et al. 2017). Other ions are generally considered as indifferent, which means that they do not alter the surface charge of calcite (Al Mahrouqi et al. 2017; Tetteh et al. 2020). Instead, they affect zeta potential mainly by controlling the thickness of the electrical double layer, which affects the magnitude but not the polarity of zeta potential. Most recent works suggest however that adsorption of Na^+ and Cl^- may modify the surface charging behavior of calcite (Heberling et al. 2021). Interestingly, Al Mahrouqi et al. (2017) showed variation of zeta potential among apparently identical samples of natural carbonates, suggesting that the type and amount of mineral impurities in carbonate matrix can cause significant variations in the measured zeta potential with respect to carbonates composed purely of calcite, typically making it lower. This reflect the fact that although the changes of zeta potential with the changing electrolyte concentration are quite consistent for various synthetic and natural calcium carbonate samples, both the absolute magnitude of zeta potential and the concentration of Ca^{2+} at which the point of zero charge is measured vary significantly. For example, by comparing data from several zeta potential studies, one may note that the point of zero charge lies between 0.1 and 10 mM of dissolved Ca^{2+} for pure calcite but at much higher concentrations, up to 100 mM of Ca^{2+}, for limestone (Foxall et al. 1979; Huang et al. 1991; Cicerone et al. 1992; Heberling et al. 2011; Pourchet et al. 2013; Alroudhan et al. 2016; Song et al. 2017; Dziadkowiec et al. 2021). This implies that minute changes in dissolved Ca^{2+} concentration can affect the magnitude of the electrical double layer repulsion between calcite surfaces in pure calcite systems, while this not may be the case for natural rock samples.

Calcium carbonate/electrolyte interface

Finally, since the forces acting between calcite surfaces in electrolyte solutions are governed by the composition of calcite–electrolyte interfaces, it is vital to take a brief look at the ion-specific effects on the composition of calcite/electrolyte interfaces. Experimental and modelling studies of the isolated calcite–water interfaces suggest an unconventional interfacial structure of strongly hydrophilic calcite where water is directly adsorbed to the calcite surface (Stipp 1999; Fenter et al. 2000; Bohr et al. 2010). Experiments by Stipp and Hochella Jr (1991) first revealed the presence of hydrolysis species H^+ and OH^- terminating bulk calcite surface in surface-sensitive spectroscopic and diffraction experiments, which adsorb to under-coordinated Ca and O surface atoms. Numerous later works provided detailed descriptions of calcite–water interfaces, where two layers of water are associatively adsorbed directly onto calcite bulk termination (Söngen et al. 2021). Although the details of the described water structure on cleaved (104) calcite surface vary in terms of bonding strength and location over the surface atoms, density, distance from the surface, the extent of ordering, and surface residence times for water molecules (Kerisit et al. 2003; Geissbühler et al. 2004; Perry IV et al. 2007; Heberling et al. 2011; Fukuma et al. 2015; De La Pierre et al. 2016; Söngen et al. 2018; Mokkath 2022; Dickbreder et al. 2023; Heggemann et al. 2024), it is established that calcite–water interface in air and in water is understood at a very detailed level; see also recent review article by Claesson et al. (2024). As such, since the water is strongly adsorbed onto calcite (Brekke-Svaland and Bresme 2018), significant short-range repulsive hydration forces are expected to act between two calcite surfaces and overcome the van der Waals attraction at small surface separations. The presence of salt ions adsorbed onto calcite can however modify the structure of hydration layers on calcite, leading to ion-specific effects on hydration repulsion (Zarzycki 2023). Adsorbed salt ions display ion-specific adsorption behavior and may adsorb on top or within the calcite surface water layers. Despite a quite significant number of experimental and theoretical studies that explored the effect of ions on the structure of calcite hydration layer, there is no full, detailed agreement on how various ions modify the hydration layer on calcite yet (Stipp 1999; Kerisit and Parker 2004; Imada et al. 2013; Ricci et al. 2013; Araki et al. 2014; Heberling et al. 2014, 2021; Hofmann et al. 2016; Kirch et al. 2018; Santos et al. 2019; Ali et al. 2020; Badizad et al. 2020; Brugman et al. 2020; John and Kühnle 2022). The general agreement is however that more hydrated divalent cations are located further away from the calcite surface (Ricci et al. 2013; Kirch et al. 2018; Ali et al. 2020; Badizad et al. 2020; Brugman et al. 2020), often on top of the two strongly adsorbed water layers on calcite, while monovalent cations can partially lose their hydration shell and enter these hydration water layers (Ricci et al. 2013; Heberling et al. 2014; Hofmann et al. 2016; Santos et al. 2019; Ali et al. 2020; Zarzycki 2023). Some works report however that monovalent cations, such as Rb^+ or K^+, do not alter the structure of water layers on calcite (Imada et al. 2013; John and Kühnle 2022). In addition, the extension of ordered water structure has been reported for Mg^{2+} (Araki et al. 2014) in disagreement with other works (Brugman et al. 2020). The ion-specific effects have been also reported for anions (Hofmann et al. 2016).

CONCLUSIONS

We have summarized available studies on the effect of various salt solutions on the mechanical strength of natural and synthetic calcium carbonate samples and on the forces acting between calcite surfaces. Our goal was to address the relationship between the surface forces acting between calcite surfaces and the mechanical strength of calcium carbonate rocks. The overview on the effect of ions on interactions involving calcite surfaces lets us briefly identify main, recurrent cation-specific effects, despite quite limited data on surface forces acting between calcite surfaces: Na^+ in sodium salts promotes adhesive interactions between calcite surfaces, especially at concentrations above 100 mM. Ca^{2+} weakens adhesion between calcite surfaces with respect to NaCl, while varying effects were reported for asymmetrical surface configurations with one calcite surface: both weakening or strengthening effects on adhesion were shown. More complex effects of Ca^{2+} concentration on the strength of calcite contacts were reported, with the strongest adhesive or attractive interactions at the point of zero charge at which Ca^{2+} screens the surface charge of calcite. Mg^{2+} weakens attractive or adhesive interactions in the calcite system although more limited data with only narrow range of concentrations is available. Detailed data on ion-specific hydration forces agree well with these trends, with Ca^{2+} and Mg^{2+} contributing to longer ranged repulsive hydration forces than monovalent Na^+ and divalent Ni^{2+}. On the other hand, data on rock deformation do not show such clear trends with Na^+ salts having both strengthening or weakening effects on calcium carbonate rocks or assemblages. Ca^{2+} salts are more frequently reported to have a weakening effect on calcium carbonate rocks, although less data is available. Interestingly, Mg^{2+} salts, which promote repulsive forces in calcite systems, have more often been shown to have a strengthening effect on rocks. It is important to highlight here that data on the effects of salt solutions on both the mechanical strength on calcium carbonate rocks and on the surface forces acting between calcite surfaces is limited and were obtained in various experimental setups, following varying methodologies. As such, it is currently not possible to conclude to what extent the ion-specific surface forces can control the mechanical strength of calcium carbonate rocks immersed in aqueous solutions. Effects of ion-specific precipitation–dissolution reactions in carbonate systems, presence of organic and inorganic impurities, temperature effects, and other factors have to be taken into consideration as well.

We can however propose possible directions to improve the understanding of the relationship between ion-specific surface forces and the mechanical strength of calcium carbonate rocks in various salt solutions. As the experimental data on both is not systematic, and obtaining such data is technically difficult and time consuming, it could be beneficial to provide a better theoretical framework to understand the relationship between salt solution composition and surface forces acting between calcite surfaces to serve as a guide for experimental development. Currently, changes in cohesion in mechanical testing of calcium carbonate rocks are most often attributed to electrostatic and DLVO interactions. However, in the conditions of mechanical testing and when subjecting rocks to high pressures, it is possible that the majority of wetted grain contacts are separated by water films that are only a few nanometers thick. In such cases, short-range forces associated with structuring of hydrated cations in the overlapping double layers, such as hydration forces, are also necessary to consider. If relevant, hydration forces associated with calcite surfaces may provide a higher-magnitude repulsion (more positive disjoining pressure) than solely electrical double layer repulsive forces, which are typically weaker in magnitude but act at larger surface separations. It is also not clear to what extent attractive ion correlation forces, which become significant at high electrolyte concentrations, contribute to the total interaction forces between calcite surfaces. One reason that hydration and ion-correlation forces are rarely considered as possible causes of changes in mechanical strength of brine-saturated calcium carbonate rocks is that there is currently no straightforward theoretical framework to assess the contribution of these interaction forces. Surface complexation modelling, which is a tool to model the surface charge of mineral surfaces in the presence of electrolytes, could not only be used to predict the surface charge of calcite more correctly and thus provide a better description of DLVO forces but also provide a better understanding of adsorption of ions at calcite–water interfaces. This could provide valuable information on how to account for the other interaction forces, such as hydration and ion correlation forces.

ACKNOWLEDGMENTS

JD acknowledges funding from The Research Council of Norway, grant nr 344828. Small parts of this review have been adapted from the introduction to doctoral thesis of Joanna Dziadkowiec, titled 'Interactions between Confined Calcite Surfaces in Aqueous Solutions: A surface forces apparatus study', which can be accessed at the University of Oslo depository: http://urn.nb.no/URN:NBN:no-70262. We thank two anonymous Reviewers for their helpful suggestions for improving this manuscript.

REFERENCES

Abd El-Aal AK, Zakhera M, Khalifa MA, Qadri ST, Nabawy BS (2021) Carbonate strength classification based on depositional textures and fossil content of the Lower Eocene Drunka Formation, Assiut Area, central Egypt. J Petrol Sci Eng 207:109061

Abu-Mahfouz IS, Iakusheva R, Finkbeiner T, Cartwright J, Vahrenkamp V (2023) Rock mechanical properties of immature, organic-rich source rocks and their relationships to rock composition and lithofacies. Petrol Geosci 29:petgeo2022-2021

Ajalloeian R, Mansouri H, Baradaran E (2017) Some carbonate rock texture effects on mechanical behavior, based on Koohrang tunnel data, Iran. Bull Eng Geol Environ 76:295–307

Al-Busaidi IK, Al-Maamari RS, Karimi M, Naser J (2019) Effect of different polar organic compounds on wettability of calcite surfaces. J Petrol Sci Eng 180:569–583

Al Mahrouqi D, Vinogradov J, Jackson MD (2017) Zeta potential of artificial and natural calcite in aqueous solution. Adv Colloid Interface Sci 240:60–76

Alam MM, Borre MK, Fabricius IL, Hedegaard K, Røgen B, Hossain Z, Krogsbøll AS (2010) Biot's coefficient as an indicator of strength and porosity reduction: Calcareous sediments from Kerguelen Plateau. J Petrol Sci Eng 70:282–297

Alcantar N, Israelachvili J, Boles J (2003) Forces and ionic transport between mica surfaces: Implications for pressure solution. Geochim Cosmochim Acta 67:1289–1304

Ali A, Le TT B, Striolo A, Cole DR (2020) Salt effects on the structure and dynamics of interfacial water on calcite probed by equilibrium molecular dynamics simulations. J Phys Chem C 124:24822–24836

Alotaibi MB, Cha D, Alsofi AM, Yousef AA (2018) Dynamic interactions of inorganic species at carbonate/brine interfaces: An electrokinetic study. Colloids Surf A 550:222–235

Alroudhan A, Vinogradov J, Jackson M (2016) Zeta potential of intact natural limestone: Impact of potential-determining ions Ca, Mg and SO_4. Colloids Surf A 493:83–98

Alzabeebee S, Mohammed DA, Alshkane YM (2022) Experimental study and soft computing modeling of the unconfined compressive strength of limestone rocks considering dry and saturation conditions. Rock Mech Rock Eng 55:5535–5554

Amour F, Christensen H, Hajiabadi M, Nick H (2021) Effects of porosity and water saturation on the yield surface of Upper Cretaceous reservoir chalks from the Danish North Sea. J Geophys Res: Solid Earth 126:e2020JB020608

Andriani G, Walsh N (2007) The effects of wetting and drying, and marine salt crystallization on calcarenite rocks used as building material in historic monuments. Geol Soc London Spec Publ 271:179–188

Angheluta L, Mathiesen J, Aharonov E (2012) Compaction of porous rock by dissolution on discrete stylolites: A one-dimensional model. JGeophys Res: Solid Earth 117:B08203

Araki Y, Tsukamoto K, Takagi R, Miyashita T, Oyabu N, Kobayashi K, Yamada H (2014) Direct observation of the influence of additives on calcite hydration by frequency modulation atomic force microscopy. Crystal Growth Des 14:6254–6260

Atkinson BK, Meredith PG (1987) The theory of subcritical crack growth with applications to minerals and rocks. Fracture Mech Rock 2:111–166

Badizad MH, Koleini MM, Greenwell HC, Ayatollahi S, Ghazanfari MH, Mohammadi M (2020) Ion-specific interactions at calcite–brine interfaces: a nano-scale study of the surface charge development and preferential binding of polar hydrocarbons. Phys Chem Chem Phys 22:27999–28011

Baechle GT, Eberli GP, Weger RJ, Massaferro JL (2009) Changes in dynamic shear moduli of carbonate rocks with fluid substitution. Geophysics 74:E135–E147

Bard AJ, Faulkner LR, White HS (2022) Electrochemical Methods: Fundamentals and Applications. John Wiley and Sons

Baud P, Rolland A, Heap M, Xu T, Nicolé M, Ferrand T, Reuschlé T, Toussaint R, Conil N (2016) Impact of stylolites on the mechanical strength of limestone. Tectonophysics 690:4–20

Baykasoğlu A, Güllü H, Çanakçı H, Özbakır L (2008) Prediction of compressive and tensile strength of limestone via genetic programming. Expert Syst Appl 35:111–123

Bell FG (2005) Rock properties and their assessment. *In*: Selley, RC, Cocks RM, Plimer IR (eds). Encyclopedia of Geology. Elsevier, p 566–580

Bergsaker AS, Røyne A, Ougier-Simonin A, Aubry J, Renard F (2016) The effect of fluid composition, salinity, and acidity on subcritical crack growth in calcite crystals. J Geophys Res: Solid Earth 121:1631–1651

Bergström L (1997) Hamaker constants of inorganic materials. Adv Colloid Interface Sci 70:125–169

Boggs S (2009) Petrology of Sedimentary Rocks. Cambridge University Press

Bohr J, Wogelius RA, Morris PM, Stipp SL (2010) Thickness and structure of the water film deposited from vapour on calcite surfaces. Geochim Cosmochim Acta 74:5985–5999

Bonto M, Eftekhari AA, Nick HM (2022) Electrokinetic behavior of artificial and natural calcites: A review of experimental measurements and surface complexation models. Adv Colloid Interface Sci 301:102600

Brantut N, Heap MJ, Baud P, Meredith PG (2014) Mechanisms of time-dependent deformation in porous limestone. J Geophys Res: Solid Earth 119:5444–5463

Brekke-Svaland G, Bresme F (2018) Interactions between hydrated calcium carbonate surfaces at nanoconfinement conditions. J Phys Chem C 122:7321–7330

Brignoli M, Santarelli F, Papamichos E (1995) Capillary effects in sedimentary rocks: application to reservoir water-flooding. ARMAUS Rock Mechanics/Geomechanics Symposium, p ARMA-95-0619

Briševac Z, Špoljarić D, Gulam V (2014) Estimation of uniaxial compressive strength based on regression tree models. Rudarsko-geološko-naftni zbornik 29:39–47

Briševac Z, Maričić A, Kujundžić T, Hrženjak P (2023) Saturation influence on reduction of compressive strength for carbonate dimension stone in Croatia. Minerals 13:1364

Brugman SJ, Raiteri P, Accordini P, Megens F, Gale JD, Vlieg E (2020) Calcite (104) surface–electrolyte structure: A 3D comparison of surface X-ray diffraction and simulations. J Phys Chem C 124:18564–18575

Budd DA (2002) The relative roles of compaction and early cementation in the destruction of permeability in carbonate grainstones: a case study from the Paleogene of west-central Florida, USA. J Sediment Res 72:116–128

Butt H-J, Kappl M (2018) Surface and Interfacial Forces. John Wiley and Sons

Chen J, Chough SK, Chun SS, Han Z (2009) Limestone pseudoconglomerates in the Late Cambrian Gushan and Chaomidian Formations (Shandong Province, China): soft-sediment deformation induced by storm-wave loading. Sedimentology 56:1174–1195

Chen L, Zhang G, Wang L, Wu W, Ge J (2014) Zeta potential of limestone in a large range of salinity. Colloids Surf A 450:1–8

Cheshomi A, Sheshde EA (2013) Determination of uniaxial compressive strength of microcrystalline limestone using single particles load test. J Petrol Sci Eng 111:121–126

Choens R, Wilson J, Ilgen A (2021) Strengthening of calcite assemblages through chemical complexation reactions. Geophys Res Lett 48:e2021GL094316

Chuhan FA, Kjeldstad A, Bjørlykke K, Høeg K (2003) Experimental compression of loose sands: relevance to porosity reduction during burial in sedimentary basins. Can Geotech J 40:995–1011

Ciantia MO, Castellanza R, Crosta GB, Hueckel T (2015a) Effects of mineral suspension and dissolution on strength and compressibility of soft carbonate rocks. Eng Geol 184:1–18

Ciantia MO, Castellanza R, Di Prisco C (2015b) Experimental study on the water-induced weakening of calcarenites. Rock Mech Rock Eng 48:441–461

Cicerone DS, Regazzoni AE, Blesa MA (1992) Electrokinetic properties of the calcite/water interface in the presence of magnesium and organic matter. J Colloid Interface Sci 154:423–433

Claesson PM, Wojas NA, Corkery R, Dedinaite A, Schoelkopf J, Tyrode E (2024) The dynamic nature of natural and fatty acid modified calcite surfaces. Phys Chem Chem Phys 26:2780–2805

Clark AC, Vanorio T (2016) The rock physics and geochemistry of carbonates exposed to reactive brines. J Geophys Res: Solid Earth 121:1497–1513

Croizé D, Ehrenberg SN, Bjørlykke K, Renard F, Jahren J (2010) Petrophysical properties of bioclastic platform carbonates: implications for porosity controls during burial. Marine Petrol Geol 27:1765–1774

Croizé D, Renard F, Gratier J-P (2013) Compaction and porosity reduction in carbonates: A review of observations, theory, and experiments. Adv Geophys 54:181–238

Daraei A, Zare S (2018) Determination of critical saturation degree in rocks based on maximum loss of uniaxial compression strength and deformation modulus. Geomech Geophysi Geo-Energy Geo-Res 4:343–353

Davarpanah M, Ahmadi M, Török, Á., Vásárhelyi B (2019) Investigation of the mechanical properties of dry, saturated and frozen highly porous limestone. ISRMCongress, p ISRM-14CONGRESS-2019-2280

Davarpanah SM, Sharghi M, Tarifard A, Török Á, Vásárhelyi B (2022) Studies on the mechanical properties of dry, saturated, and frozen marls using destructive and non-destructive laboratory approaches. Iran J Sci Technol Trans Civil Eng 46:1311–1328

De La Pierre M, Raiteri P, Gale JD (2016) Structure and dynamics of water at step edges on the calcite {1014} surface. Crystal Growth Des 16:5907–5914

Deng H, Fitts JP, Crandall D, McIntyre D, Peters CA (2015) Alterations of fractures in carbonate rocks by CO_2-acidified brines. Environ Sci Technol 49:10226–10234

Derjaguin B, Landau L (1941) Theory of the stability of strongly charged lyophobic sols and the adhesion of strongly charged particles in solution of electrolytes. Acta Physicochim URSS 14:39

Descamps F, Faÿ-Gomord O, Vandycke S, Schroeder C, Swennen R, Tshibangu J-P (2017) Relationships between geomechanical properties and lithotypes in NW European chalks. Geol Soc London Spec Publ 458:227–244

Diao Y, Espinosa-Marzal RM (2016) Molecular insight into the nanoconfined calcite–solution interface. PNAS 113:12047–12052

Diao Y, Espinosa-Marzal RM (2018) The role of water in fault lubrication. Nat Commun 9:2309

Diao Y, Espinosa-Marzal RM (2019) Effect of fluid chemistry on the interfacial composition, adhesion, and frictional response of calcite single crystals—implications for injection-induced seismicity. J Geophys Res: Solid Earth 124:5607–5628

Diao Y, Li A, Espinosa-Marzal RM (2020) Ion specific effects on the pressure solution of calcite single crystals. Geochim Cosmochim Acta 280:116–129

Dickbreder T, Lautner D, Köhler A, Klausfering L, Bechstein R, Kühnle A (2023) How water desorbs from calcite. Phys Chem Chem Phys 25:12694–12701

Donaldson Jr SH, Røyne A, Kristiansen K, Rapp MV, Das S, Gebbie MA, Lee DW, Stock P, Valtiner M, Israelachvili J (2015) Developing a general interaction potential for hydrophobic and hydrophilic interactions. Langmuir 31:2051–2064

Douglas H, Walker R (1950) The electrokinetic behaviour of Iceland Spar against aqueous electrolyte solutions. Trans Faraday Soc 46:559–568

Dunham RJ (1962) Classification of carbonate rocks according to depositional textures. *In:* Classification of Carbonate Rocks—A Symposium. WE Ham (ed) AAPG Memoir. American Association of Petroleum Geologists, p 108–121

Dunning J, Douglas B, Miller M, McDonald S (1994) The role of the chemical environment in frictional deformation: stress corrosion cracking and comminution. Pure Appl Geophys 143:151–178

Dziadkowiec J, Javadi S, Bratvold JE, Nilsen O, Røyne A (2018) Surface Forces Apparatus measurements of interactions between rough and reactive calcite surfaces. Langmuir 34:7248–7263

Dziadkowiec J, Zareeipolgardani B, Dysthe DK, Røyne A (2019) Nucleation in confinement generates long-range repulsion between rough calcite surfaces. Sci Rep 9:8948

Dziadkowiec J, Ban M, Javadi S, Jamtveit B, Røyne A (2021) Ca^{2+} ions decrease adhesion between two (104) calcite surfaces as probed by atomic force microscopy. ACS Earth Space Chem 5:2827–2838

Ebhardt G (1968) Recent Developments in Carbonate Sedimentology in Central Europe. Springer, p 58–65

Eichmann SL, Burnham NA (2017) Calcium-mediated adhesion of nanomaterials in reservoir fluids. Sci Rep 7:11613

Embry AF, Klovan JE (1971) A Late Devonian reef tract on northeastern Banks Island, NWT. Bull Can Petrol Geol 19:730–781

Eom N, Parsons DF, Craig VS (2017) Roughness in surface force measurements: Extension of DLVO theory to describe the forces between hafnia surfaces. J Phys Chem B 121:6442–6453

Ercikdi B, Karaman K, Cihangir F, Yılmaz T, Aliyazıcıoğlu Ş, Kesimal A (2016) Core size effect on the dry and saturated ultrasonic pulse velocity of limestone samples. Ultrasonics 72:143–149

Eslami J, Grgic D, Hoxha D (2010) Estimation of the damage of a porous limestone from continuous (P-and S-) wave velocity measurements under uniaxial loading and different hydrous conditions. Geophys J Int 183:1362–1375

Fabricius IL (2003) How burial diagenesis of chalk sediments controls sonic velocity and porosity. AAPG Bull 87:1755–1778

Fabricius IL (2007) Chalk: composition, diagenesis and physical properties. Bull Geol Soc Denmark 55:97–128

Fenter P, Geissbühler P, DiMasi E, Srajer G, Sorensen L, Sturchio N (2000) Surface speciation of calcite observed in situ by high-resolution X-ray reflectivity. Geochim Cosmochim Acta 64:1221–1228

Folk RL (1962) Spectral subdivision of limestone types. Classification of Carbonate Rocks. *In:* Classification of Carbonate Rocks—A Symposium. WE Ham (ed) AAPG Memoir. American Association of Petroleum Geologists, p 62–84

Foxall T, Peterson GC, Rendall HM, Smith AL (1979) Charge determination at calcium salt/aqueous solution interface. J Chem Soc Faraday Trans 1 75:1034–1039

Fruth L, Orme G, Donath F (1966) Experimental compaction effects in carbonate sediments. J Sediment Res 36:747–754

Fu B, Diao Y, Espinosa-Marzal RM (2021) Nanoscale insight into the relation between pressure solution of calcite and interfacial friction. J Colloid Interface Sci 601:254–264

Fukuma T, Reischl B, Kobayashi N, Spijker P, Canova FF, Miyazawa K, Foster AS (2015) Mechanism of atomic force microscopy imaging of three-dimensional hydration structures at a solid–liquid interface. Phys Rev B 92:155412

Gandin A, Capezzuoli E (2014) Travertine: distinctive depositional fabrics of carbonates from thermal spring systems. Sedimentology 61:264–290

Ganor J, Reznik IJ, Rosenberg YO (2009) Organics in water-rock interactions. Rev Mineral Geochem 70:259–369

Geissbühler P, Fenter P, DiMasi E, Srajer G, Sorensen L, Sturchio N (2004) Three-dimensional structure of the calcite–water interface by surface X-ray scattering. Surf Sci 573:191–203

Generosi J, Ceccato M, Andersson M, Hassenkam T, Dobberschütz S, Bovet N, Stipp S (2017) Calcite wettability in the presence of dissolved Mg^{2+} and SO_4^{2-}. Energy Fuels 31:1005–1014

Geremia D, David C, Ismail R, El Haitami A (2021) An integrated study of water weakening and fluid rock interaction processes in porous rocks: linking mechanical behavior to surface properties. Appl Sci 11:11437

Grabbe A, Horn RG (1993) Double-layer and hydration forces measured between silica sheets subjected to various surface treatments. J Colloid Interface Sci 157:375–383

Gratier J-P, Renard F, Labaume P (1999) How pressure solution creep and fracturing processes interact in the upper crust to make it behave in both a brittle and viscous manner. J Struct Geol 21:1189–1197

Gray C, Stiles PJ (2018) Nonlinear electrostatics: the Poisson–Boltzmann equation. Eur J Phys 39:053002

Griffith AA (1921) VI. The phenomena of rupture and flow in solids. Philos Trans R Soc London Ser A 221:163–198

Gundersen E, Renard F, Dysthe DK, Bjørlykke K, Jamtveit B (2002) Coupling between pressure solution creep and diffusive mass transport in porous rocks. J Geophys Res: Solid Earth 107:ECV 19-11-ECV 19–19

Guo H, Kovscek AR (2019) Investigation of the effects of ions on short-range non-DLVO forces at the calcite/brine interface and implications for low salinity oil-recovery processes. J Colloid Interface Sci 552:295–311

Hallsworth CR, Knox R (1999) BGS rock classification scheme. Volume 3, classification of sediments and sedimentary rocks.

Harris PM, Kendall CG SC, Lerche I (1985) Carbonate cementation—a brief review. The Society of Economic Paleontologists and Mineralogists; Carbonate Cements SP36.

Hassenkam T, Skovbjerg LL, Stipp SL S (2009) Probing the intrinsically oil-wet surfaces of pores in North Sea chalk at subpore resolution. PNAS 106:6071–6076

Heberling F, Trainor TP, Lützenkirchen J, Eng P, Denecke MA, Bosbach D (2011) Structure and reactivity of the calcite–water interface. J Colloid Interface Sci 354:843–857

Heberling F, Bosbach D, Eckhardt J-D, Fischer U, Glowacky J, Haist M, Kramar U, Loos S, Müller HS, Neumann T (2014) Reactivity of the calcite–water-interface, from molecular scale processes to geochemical engineering. Appl Geochem 45:158–190

Heberling F, Klačić T, Raiteri P, Gale JD, Eng PJ, Stubbs JE, Gil-Díaz T, Begović T, Lützenkirchen J (2021) Structure and surface complexation at the calcite (104)–water interface. Environ Sci Technol 55:12403–12413

Hebib R, Belhai D, Alloul B (2017) Estimation of uniaxial compressive strength of North Algeria sedimentary rocks using density, porosity, and Schmidt hardness. Arab J Geosci 10:1–13

Heggemann J, Aeschlimann S, Dickbreder T, Ranawat YS, Bechstein R, Kühnle A, Foster AS, Rahe P (2024) Water adsorption lifts the (2 × 1) reconstruction of calcite (104) Phys Chem Chem Phys 26:21365–21369

Heggheim T, Madland M, Risnes R, Austad T (2005) A chemical induced enhanced weakening of chalk by seawater. J Petrol Sci Eng 46:171–184

Heidari M, Mohseni H, Jalali SH (2018) Prediction of uniaxial compressive strength of some sedimentary rocks by fuzzy and regression models. Geotech Geol Eng 36:401–412

Hellmann R, Renders PJ, Gratier J-P, Guiguet R (2002) Water–rock interactions, ore deposits, and environmental geochemistry: A tribute to David A Crerar 7:153–178

Henry JP, Paquet J, Tancrez J (1977) Experimental study of crack propagation in calcite rocks. Int J Rock Mech Mining Sci Geomech Abstr, p 85–91. Elsevier

Heuberger MP, Zachariah Z, Spencer ND, Espinosa-Marzal R (2017) Collective dehydration of ions in nano-pores. Phys Chem Chem Phys 19:13462–13468

Hidema R, Toyoda T, Suzuki H, Komoda Y, Shibata Y (2016) Adhesive behavior of a calcium carbonate particle to solid walls having different hydrophilic characteristics. Int J Heat Mass Transfer 92:603–609

Hiorth A, Cathles L, Madland M (2010) The impact of pore water chemistry on carbonate surface charge and oil wettability. Transp Porous Media 85:1–21

Hofmann S, Voïtchovsky K, Spijker P, Schmidt M, Stumpf T (2016) Visualising the molecular alteration of the calcite (104)–water interface by sodium nitrate. Sci Rep 6:21576

Huang YC, Fowkes FM, Lloyd TB, Sanders ND (1991) Adsorption of calcium ions from calcium chloride solutions onto calcium carbonate particles. Langmuir 7:1742–1748

Hunter RJ (2013) Zeta Potential in Colloid Science: Principles and Applications. (Vol. 2): Academic Press.

Ilgen AG, Mook W, Tigges AB, Choens R, Artyushkova K, Jungjohann K (2018) Chemical controls on the propagation rate of fracture in calcite. Sci Rep 8:16465

Imada H, Kimura K, Onishi H (2013) Water and 2-propanol structured on calcite (104) probed by frequency-modulation atomic force microscopy. Langmuir 29:10744–10751

Israelachvili J (2015) Intermolecular and surface forces. (3rd Edition ed.): Academic press.

Jalali SH, Heidari M, Mohseni H (2017) Comparison of models for estimating uniaxial compressive strength of some sedimentary rocks from Qom Formation. Environ Earth Sci 76:1–15

Javadi S, Røyne A (2018) Adhesive forces between two cleaved calcite surfaces in NaCl solutions: The importance of ionic strength and normal loading. J Colloid Interface Sci 532:605–613

John S, Kühnle A (2022) Hydration structure at the calcite–water (10.4) interface in the presence of rubidium chloride. Langmuir 38:11691–11698

Kasim M, Shakoor A (1996) An investigation of the relationship between uniaxial compressive strength and degradation for selected rock types. Eng Geol 44:213–227

Katika K, Addassi M, Alam MM, Fabricius IL (2015) The effect of divalent ions on the elasticity and pore collapse of chalk evaluated from compressional wave velocity and low-field Nuclear Magnetic Resonance (NMR). J Petrol Sci Eng 136:88–99

Kerisit S, Parker SC (2004) Free energy of adsorption of water and metal ions on the {1014} calcite surface. J Am Chem Soc 126:10152–10161

Kerisit S, Parker SC, Harding JH (2003) Atomistic simulation of the dissociative adsorption of water on calcite surfaces. J Phys Chem B 107:7676–7682

Kibikas WM, Choens RC, Bauer SJ, Shalev E, Lyakhovsky V (2023) Water-weakening and time-dependent deformation of organic-rich chalks. Rock Mech Rock Eng 56:8041–8059

Kirch A, Mutisya SM, Sánchez VM, De Almeida JM, Miranda CR (2018) Fresh molecular look at calcite–brine nanoconfined interfaces. J Phys Chem C 122:6117–6127

Kjellander R, Marcelja S, Pashley R, Quirk J (1988) Double-layer ion correlation forces restrict calcium-clay swelling. J Phys Chem 92:6489–6492

Korsnes RI, Madland MV, Austad T, Haver S, Røsland G (2008) The effects of temperature on the water weakening of chalk by seawater. J Petrol Sci Eng 60:183–193

Kuznetsov V (2023) Trend and phasing of carbonate accumulation in the history of the Earth. Doklady Earth Sci 510:449–452

Labuz JF, Zang A (2014) Mohr–Coulomb failure criterion. The ISRM Suggested Methods for Rock Characterization, Testing and Monitoring: 2007–2014, Springer, p 227–231

Lakshtanov L, Okhrimenko D, Karaseva O, Stipp S (2018) Limits on calcite and chalk recrystallization. Crystal Growth Des 18:4536–4543

Leighton M, Pendexter C (1962) Carbonate rock types. *In:* Classification of Carbonate Rocks—A Symposium. WE Ham (ed) AAPG Memoir. American Association of Petroleum Geologists, p 33–61

Levenson Y, Emmanuel S (2017) Repulsion between calcite crystals and grain detachment during water–rock interaction. Geochem Perspect Lett 3:133–141

Li H, Zhong Z, Liu X, Sheng Y, Yang D (2018) Micro-damage evolution and macro-mechanical property degradation of limestone due to chemical effects. Int J Rock Mech Mining Sci 110:257–265

Li A, Chang J, Shui T, Liu Q, Zhang H, Zeng H (2023) Probing interaction forces associated with calcite scaling in aqueous solutions by atomic force microscopy. J Colloid Interface Sci 633:764–774

Liberto T, Barentin C, Colombani J, Costa A, Gardini D, Bellotto M, Le Merrer M (2019) Simple ions control the elasticity of calcite gels via interparticle forces. J Colloid Interface Sci 553:280–288

Lifshitz E (1955) The theory of molecular attractive forces between solid bodies. J Exp Theor Phys USSR 29:83–94

Lin K, Yu T, Ji W, Li B, Wu Z, Liu X, Li C, Yang Z (2023) Carbonate rocks as natural buffers: Exploring their environmental impact on heavy metals in sulfide deposits. Environ Pollut 336:122506

Lisabeth HP, Zhu W (2015) Effect of temperature and pore fluid on the strength of porous limestone. J Geophys Res: Solid Earth 120:6191–6208

Liteanu E, Spiers CJ (2009) Influence of pore fluid salt content on compaction creep of calcite aggregates in the presence of supercritical CO_2. Chem Geol 265:134–147

Liu L, Liu H, Stuedlein AW, Evans TM, Xiao Y (2019) Strength, stiffness, and microstructure characteristics of biocemented calcareous sand. Can Geotech J 56:1502–1513

Lokier SW, Al Junaibi M (2016) The petrographic description of carbonate facies: are we all speaking the same language? Sedimentology 63:1843–1885

Lomboy G, Sundararajan S, Wang K, Subramaniam S (2011) A test method for determining adhesion forces and Hamaker constants of cementitious materials using atomic force microscopy. Cement Concrete Res 41:1157–1166

Madhubabu N, Singh P, Kainthola A, Mahanta B, Tripathy A, Singh T (2016) Prediction of compressive strength and elastic modulus of carbonate rocks. Measurement 88:202–213

Madland M, Hiorth A, Omdal E, Megawati M, Hildebrand-Habel T, Korsnes R, Evje S, Cathles L (2011) Chemical alterations induced by rock–fluid interactions when injecting brines in high porosity chalks. Transp Porous Media 87:679–702

Malkin A (2012) Regularities and mechanisms of the Rehbinder's effect. Colloid J 74:223–238

Matter JM, Kelemen PB (2009) Permanent storage of carbon dioxide in geological reservoirs by mineral carbonation. Nat Geosci 2:837–841

Matthiesen J, Hassenkam T, Bovet N, Dalby K, Stipp S (2017) Adsorbed organic material and its control on wettability. Energy Fuels 31:55–64

Maugis D (1985) Subcritical crack growth, surface energy, fracture toughness, stick-slip and embrittlement. J Mater Sci 20:3041–3073

Mazidi SM, Haftani M, Bohloli B, Cheshomi A (2012) Measurement of uniaxial compressive strength of rocks using reconstructed cores from rock cuttings. J Petrol Sci Eng 86:39–43

Megawati M, Hiorth A, Madland M (2013) The impact of surface charge on the mechanical behavior of high-porosity chalk. Rock Mech Rock Eng 46:1073–1090

Meireles LT P, Storebø EM, Lykke Fabricius I (2020) Effect of electrostatic forces on the porosity of saturated mineral powder samples and implications for chalk strength. Geophysics 85:MR37–MR50

Meireles LT, Storebø EM, Welch MJ, Fabricius IL (2021) Water weakening of soft and stiff outcrop chalk induced by electrical double layer disjoining pressure. Int J Rock Mech Mining Sci 141:104700

Michalske TA, Freiman SW (1982) A molecular interpretation of stress corrosion in silica. Nature 295:511–512

Minde MW, Zimmermann U, Madland MV, Korsnes RI, Schulz B, Gilbricht S (2020) Mineral replacement in long-term flooded porous carbonate rocks. Geochim Cosmochim Acta 268:485–508

Mircescu CV, Har N, Tămaş T (2022) Microfacies, physical and mechanical properties of carbonate rocks from the Apuseni Mountains, Romania: implication for delineating potential ornamental limestone extraction areas. Carbonates Evaporites 37:27

Mitchell RH (2005) Carbonatites and carbonatites and carbonatites. Can Mineral 43:2049–2068

Mokkath JH (2022) Water–calcite (104) surface interactions using first-principles simulations. J Phys Chem Solids 161:110394

Mortimore R, Stone K, Lawrence J, Duperret A (2004) Chalk physical properties and cliff instability. Geol Soc London, Eng Geol Spec Publ 20:75–88

Moulin P, Roques H (2003) Zeta potential measurement of calcium carbonate. J Colloid Interface Sci 261:115–126

Nara Y, Kashiwaya K, Nishida Y, Ii T (2017) Influence of surrounding environment on subcritical crack growth in marble. Tectonophysics 706:116–128

Nermoen A, Korsnes RI, Storm EV, Stødle T, Madland MV, Fabricius IL (2018) Incorporating electrostatic effects into the effective stress relation—Insights from chalk experiments. Geophysics 83:MR123–MR135

Newman GH (1983) The effect of water chemistry on the laboratory compression and permeability characteristics of some North Sea chalks. J Petrol Technol 35:976–980

Nicolas A, Fortin J, Regnet J, Dimanov A, Guéguen Y (2016) Brittle and semi-brittle behaviours of a carbonate rock: influence of water and temperature. Geophys J Int 206:438–456

Nur A, Byerlee JD (1971) An exact effective stress law for elastic deformation of rock with fluids. J Geophys Res 76:6414–6419

Nyström R, Lindén M, Rosenholm JB (2001) The influence of Na^+, Ca^{2+}, Ba^{2+}, and La^{3+} on the ζ potential and the yield stress of calcite dispersions. J Colloid Interface Sci 242:259–263

Okhrimenko DV, Dalby KN, Stipp SL (2013) Adsorption properties of chalk: Contributions from calcite and clays. Procedia Earth Planet Sci 7:632–635

Orowan E (1944) The fatigue of glass under stress. Nature 154:341–343

Palchik V, Hatzor YH (2004) The influence of porosity on tensile and compressive strength of porous chalks. Rock Mech Rock Eng 37:331–341

Papamichos E, Brignoli M, Santarelli F (1997) An experimental and theoretical study of a partially saturated collapsible rock. Mech Cohesive-frictional Mater: 2:251–278

Parsegian V, Zemb T (2011) Hydration forces: Observations, explanations, expectations, questions. Curr Opinion Colloid Interface Sci 16:618–624

Parsons DF, Ninham BW (2010) Charge reversal of surfaces in divalent electrolytes: the role of ionic dispersion interactions. Langmuir 26:6430–6436

Parsons DF, Walsh RB, Craig VS (2014) Surface forces: Surface roughness in theory and experiment. J Chem Phys 140:164701

Pashley R (1981) Hydration forces between mica surfaces in aqueous electrolyte solutions. J Colloid Interface Sci 80:153–162

Pashley R, Israelachvili J (1984) DLVO and hydration forces between mica surfaces in Mg^{2+}, Ca^{2+}, Sr^{2+}, and Ba^{2+} chloride solutions. J Colloid Interface Sci 97:446–455

Pedersen N, Hassenkam T, Ceccato M, Dalby K, Mogensen K, Stipp S (2016) Low salinity effect at pore scale: probing wettability changes in Middle East limestone. Energy Fuels 30:3768–3775

Peng J, Cai M (2019) A cohesion loss model for determining residual strength of intact rocks. Int J Rock Mech Mining Sci 119:131–139

Perry IVT D, Cygan RT, Mitchell R (2007) Molecular models of a hydrated calcite mineral surface. Geochim Cosmochim Acta 71:5876–5887

Pierre A, Lamarche J, Mercier R, Foissy A, Persello J (1990) Calcium as potential determining ion in aqueous calcite suspensions. J Dispersion Sci Technol 11:611–635

Pluymakers A, Ougier-Simonin A, Barnhoorn A (2021) Ion-species in pore fluids with opposite effects on limestone fracturing. Geomech Energy Environ 26:100233

Pomar L, Hallock P (2008) Carbonate factories: a conundrum in sedimentary geology. Earth Sci Rev 87:134–169

Pourchet S, Pochard I, Brunel F, Perrey D (2013) Chemistry of the calcite/water interface: Influence of sulfate ions and consequences in terms of cohesion forces. Cement Concrete Res 52:22–30

Prakash A, Holyoke IIIC W, Kelemen PB, Kirby SH, Kronenberg AK, Lamb WM (2023) Carbonates and intermediate-depth seismicity: Stable and unstable shear in altered subducting plates and overlying mantle. PNAS 120:e2219076120

Rajabzadeh M, Moosavinasab Z, Rakhshandehroo G (2012) Effects of rock classes and porosity on the relation between uniaxial compressive strength and some rock properties for carbonate rocks. Rock Mech Rock Eng 45:113–122

Regnet J, David C, Robion P, Menéndez B (2019) Microstructures and physical properties in carbonate rocks: A comprehensive review. Mar Petrol Geol 103:366–376

Rehbinder P, Lichtman V (1957) Effect of surface active media on strain and rupture in solids. Proc Second Int Congress Surf Activity. Buttersworths, p 563–580

Renard F, Ortoleva P, Gratier JP (1997) Pressure solution in sandstones: influence of clays and dependence on temperature and stress. Tectonophysics 280:257–266

Ricci M, Spijker P, Stellacci F, Molinari J-F, Voïtchovsky K (2013) Direct visualization of single ions in the Stern layer of calcite. Langmuir 29:2207–2216

Risnes R, Flaageng O (1999) Mechanical properties of chalk with emphasis on chalk-fluid interactions and micromechanical aspects. Oil Gas Sci Technol 54:751–758

Risnes R, Haghighi H, Korsnes R, Natvik O (2003) Chalk–fluid interactions with glycol and brines. Tectonophysics 370:213–226

Risnes R, Madland M, Hole M, Kwabiah N (2005) Water weakening of chalk—Mechanical effects of water–glycol mixtures. J Petrol Sci Eng 48:21–36

Rolland A, Toussaint R, Baud P, Conil N, Landrein P (2014) Morphological analysis of stylolites for paleostress estimation in limestones. Int J Rock Mech Mining Sci 67:212–225

Rostom F, Røyne A, Dysthe DK, Renard F (2013) Effect of fluid salinity on subcritical crack propagation in calcite. Tectonophysics 583:68–75

Rozhko AY (2020) Effects of pore fluids on quasi-static shear modulus caused by pore-scale interfacial phenomena. Geophys Prospect 68:631–656

Rutter EH (1974) The influence of temperature, strain rate and interstitial water in the experimental deformation of calcite rocks. Tectonophysics 22:311–334

Rutter E (1976) A discussion on natural strain and geological structure-the kinetics of rock deformation by pressure solution. Philos Trans R Soc London Ser A 283:203–219

Røyne A, Bisschop J, Dysthe DK (2011) Experimental investigation of surface energy and subcritical crack growth in calcite. J Geophys Res: Solid Earth 116:B04204

Røyne A, Dalby KN, Hassenkam T (2015) Repulsive hydration forces between calcite surfaces and their effect on the brittle strength of calcite-bearing rocks. Geophys Res Lett 42:4786–4794

Sachdeva JS, Nermoen A, Korsnes RI, Madland MV (2019) Impact of initial wettability and injection brine chemistry on mechanical behaviour of kansas chalk. Transp Porous Media 128:755–795

Sadeghi E, Nikudel MR, Khamehchiyan M, Kavussi A (2022) Estimation of unconfined compressive strength (UCS) of carbonate rocks by index mechanical tests and specimen size properties: central Alborz Zone of Iran. Rock Mech Rock Eng, 1–21

Santos MS, Castier M, Economou IG (2019) Molecular dynamics simulation of electrolyte solutions confined by calcite mesopores. Fluid Phase Equilibria 487:24–32

Scholle PA, Halley RB (1989) Burial diagenesis: out of sight, out of mind! Carbonate Sediment Petrol 4:135–160

Shepherd M (2009) Carbonate reservoirs. Oil field production geology: AAPG Memoir 91, p. 301-309.

Skinner AJ, LaFemina JP, Jansen HJ (1994) Structure and bonding of calcite: a theoretical study. Am Mineral 79:205–214

Smallwood P (1977) Some aspects of the surface chemistry of calcite and aragonite Part I: An electrokinetic study. Colloid Polymer Sci 255:881–886

Song J, Zeng Y, Wang L, Duan X, Puerto M, Chapman WG, Biswal SL, Hirasaki GJ (2017) Surface complexation modeling of calcite zeta potential measurements in brines with mixed potential determining ions (Ca^{2+}, CO_3^{2-}, Mg^{2+}, SO_4^{2-}) for characterizing carbonate wettability. J Colloid Interface Sci 506:169–179

Song Z, Cheng Y, Tian X, Wang J, Yang T (2020) Mechanical properties of limestone from Maixi tunnel under hydro-mechanical coupling. Arab J Geosci 13:1–13

Sopacı E, Akgün H (2015) Geotechnical assessment and engineering classification of the Antalya tufa rock, southern Turkey. Eng Geol 197:211–224

Stipp S (1999) Toward a conceptual model of the calcite surface: hydration, hydrolysis, and surface potential. Geochim Cosmochim Acta 63:3121–3131

Stipp SL, Hochella Jr MF (1991) Structure and bonding environments at the calcite surface as observed with X-ray photoelectron spectroscopy (XPS) and low energy electron diffraction (LEED) Geochim Cosmochim Acta 55:1723–1736

Stipp S, Gutmannsbauer W, Lehmann T (1996) The dynamic nature of calcite surfaces in air. Am Mineral 81:1–8

Söngen H, Reischl B, Miyata K, Bechstein R, Raiteri P, Rohl AL, Gale JD, Fukuma T, Kühnle A (2018) Resolving point defects in the hydration structure of calcite (10.4) with three-dimensional atomic force microscopy. Phys Rev Lett 120:116101

Söngen H, Schlegel SJ, Morais Jaques Y, Tracey J, Hosseinpour S, Hwang D, Bechstein R, Bonn M, Foster AS, Kühnle A (2021) Water orientation at the calcite-water interface. JPhys Chem letters 12:7605–7611

Talesnick M, Hatzor YH, Tsesarsky M (2001) The elastic deformability and strength of a high porosity, anisotropic chalk. Int J Rock Mech Mining Sci 38:543–555

Tegethoff FW, Rohleder J, Kroker E (2001) Calcium Carbonate: From the Cretaceous Period into the 21st Century. Springer Science and Business Media

Tetteh JT, Alimoradi S, Brady PV, Ghahfarokhi RB (2020) Electrokinetics at calcite-rich limestone surface: Understanding the role of ions in modified salinity waterflooding. J Mol Liquids 297:111868

Thompson DW, Pownall PG (1989) Surface electrical properties of calcite. J Colloid Interface Sci 131:74–82

Toussaint R, Aharonov E, Koehn D, Gratier J-P, Ebner M, Baud P, Rolland A, Renard F (2018) Stylolites: A review. J Struct Geol 114:163–195

Török, Á., Vásárhelyi B (2010) The influence of fabric and water content on selected rock mechanical parameters of travertine, examples from Hungary. Eng Geol 115:237–245

Vajdova V, Baud P, Wong TF (2004) Compaction, dilatancy, and failure in porous carbonate rocks. J Geophys Res: Solid Earth 109

Van Stappen JF, De Kock T, De Schutter G, Cnudde V (2019) Uniaxial compressive strength measurements of limestone plugs and cores: a size comparison and X-ray CT study. Bull Eng Geol Environ 78:5301–5310

Vásárhelyi B (2005) Technical note statistical analysis of the influence of water content on the strength of the Miocene limestone. Rock Mech Rock Eng 38:69–76

Verwey E, Overbeek JT G (1955) Theory of the stability of lyophobic colloids. J Colloid Sci 10:224–225

Wang H-L, Yin Z-Y (2021) Unconfined compressive strength of bio-cemented sand: state-of-the-art review and MEP-MC-based model development. J Cleaner Prod 315:128205

Wennerström H, Jönsson B, Linse P (1982) The cell model for polyelectrolyte systems. Exact statistical mechanical relations, Monte Carlo simulations, and the Poisson–Boltzmann approximation. J Chem Phys 76:4665–4670

Weyl PK (1959) Pressure solution and the force of crystallization: a phenomenological theory. J Geophys Res 64:2001–2025

Wilson JL (2012) Carbonate Facies in Geologic History. Springer Science and Business Media

Wolthers M, Charlet L, Van Cappellen P (2008) The surface chemistry of divalent metal carbonate minerals; a critical assessment of surface charge and potential data using the charge distribution multi-site ion complexation model. Am J Sci 308:905–941

Wong LN Y, Maruvanchery V, Liu G (2016) Water effects on rock strength and stiffness degradation. Acta Geotechnica 11:713–737

Wright V (1992) A revised classification of limestones. Sediment Geol 76:177–185

You M (2011) Strength and damage of marble in ductile failure. J Rock Mech Geotech Eng 3:161–166

Zarzycki P (2023) Distance-dependent dielectric constant at the calcite/electrolyte interface: Implication for surface complexation modeling. J Colloid Interface Sci 645:752–764

Zeng L, Chen Y, Lu Y, Hossain MM, Saeedi A, Xie Q (2020) Role of brine composition on rock surface energy and its implications for subcritical crack growth in calcite. J Mol Liquids 303:112638

Zhang X, Spiers C (2005a) Compaction of granular calcite by pressure solution at room temperature and effects of pore fluid chemistry. Int J Rock Mech Mining Sci 42:950–960

Zhang X, Spiers CJ (2005b) Effects of phosphate ions on intergranular pressure solution in calcite: An experimental study. Geochim Cosmochim Acta 69:5681–5691

Zhang P, Tweheyo MT, Austad T (2007) Wettability alteration and improved oil recovery by spontaneous imbibition of seawater into chalk: Impact of the potential determining ions Ca^{2+}, Mg^{2+}, and SO_4^{2-}. Colloids Surf A 301:199–208

Zhang X, Spiers CJ, Peach CJ (2010) Compaction creep of wet granular calcite by pressure solution at 28 °C to 150 °C. J Geophys Res: Solid Earth 115:B09217

Zhang X, Spiers C, Peach C (2011) Effects of pore fluid flow and chemistry on compaction creep of calcite by pressure solution at 150 C. Geofluids 11:108–122

Reviews in Mineralogy & Geochemistry
Vol. 91A pp. 295–336, 2025

Measurements of the Electrostatic Potential at the Mineral/Electrolyte Interface

Tin Klačić[1], Jozefina Katić[2], Davor Kovačević[1], Danijel Namjesnik[1], Ahmed Abdelmonem[3], Tajana Begović[1,*]

*[1]Department of Chemistry, Faculty of Science, University of Zagreb
Horvatovac 102A, HR-10000 Zagreb, Croatia*
*tklacic@chem.pmf.hr, davork@chem.pmf.hr, dnamjesnik.chem@pmf.hr, tajana.chem@pmf.hr**
**corresponding author*

*[2]Department of Electrochemistry, Faculty of Chemical Engineering and Technology,
University of Zagreb, Marulićev trg 19, HR-10000 Zagreb, Croatia*
jkatic@fkit.unizg.hr

*[3]Institute of Meteorology and Climate Research, Karlsruhe Institute of Technology (KIT),
76344, Eggenstein-Leopoldshafen, Germany*
ahmed.abdelmonem@kit.edu

INTRODUCTION

The electrostatic potential at mineral/aqueous electrolyte interfaces contains contributions that arise from charged solid surfaces, water dipoles, dissolved hydrated ions and other electrostatic contributions. It depends on the composition of the solution, the concentration and type of dissolved ions, the type of solvent and the reactions of these species with the mineral surface. In general, the electrostatic potential decreases with distance from the surface. In the electrical interfacial layer several planes with specific electrostatic potentials are defined. By measuring these interfacial potentials (such as inner surface potential, diffuse layer potential or electrokinetic potential), information is obtained about electrostatic interactions, surface reactions, and adsorption of ions, all of which have an impact on the modeling and application of mineral/aqueous electrolyte interfaces. The interfacial potentials can be determined utilizing different experimental methods such as electrochemically based methods, optical or spectroscopic methods, atomic force microscopy, or determined indirectly from the mobility of charged species in an electric field. The interpretation of these results depends on the applied models of electrical interfacial layer and corresponding surface reactions.

In this article, after a short introduction on the relationship of interfacial potentials with the processes within the electrical interfacial layer and thermodynamics of surface reactions, different experimental methods of the determination of interfacial potentials have been described. In the last part of this article, the potentials at the interface of silica/aqueous electrolyte solution, obtained by different experimental methods, are presented and compared.

INTERFACIAL POTENTIALS WITHIN THE ELECTRICAL INTERFACIAL LAYER

The mineral surface upon contact with aqueous electrolyte environments undergoes surface reactions and redistribution of the local charges (water molecules/quadrupoles and dissolved ions) that form electrical interfacial layer (EIL) with relatively complex interfacial structure (McNaught and Wilkinson 2019; Lin et al. 2020). EIL models are widely used to describe the structure of

1529-6466/25/0091A-0009$05.00 (print)
1943-2666/25/0091A-0009$05.00 (online)
http://dx.doi.org/10.2138/rmg.2025.91A.09

the mineral/aqueous electrolyte solution interface and to define and quantify the appropriate parameters such as surface charge density, interfacial potentials, and capacity of the layers etc.

In this review, we will describe and analyze several very different experimental methods used for determination of electrostatic potential, i.e., interfacial potentials at different distances from the solid surfaces to understand and distinguish the meaning of different specific measured potentials. Later on, we will give an overview of the measured surface potential values obtained for silica/aqueous electrolyte solution. Finally, we will discuss the advantages and compatibility of the experimental techniques used and the comparability of results.

In general, upon contact of mineral solid phase with the electrolyte solution an electrochemical interface (an electrified interface) is formed characterized by potential differences, charge densities, electric currents and electrostatic potentials. The electrostatic external (Volta) potential is defined as the work required to bring a unit point charge from the infinity to the point just outside of the surface of the solid phase, while the electrostatic internal (Galvani) potential, is defined as the work to bring a unit point charge from infinity to a point inside the solid phase (Schmickler 1996; Sato 1998) The difference between internal (Galvani) and external (Volta) potential is caused by the inhomogeneous charge distribution at the surface. Detailed description is given by Trasatti and Parsons (1986). The electric field producing this potential is wholly concentrated in the solid phase's surface layer, where an electrical interfacial layer is formed due to the non-uniform charge distribution. The value of the electrostatic potential depends on the structure and chemical properties of the solid phase surface and varies with the surface charge (Bagotsky 2005).

The process of formation of the interfacial layer is controlled by the structure of the solid surface and chemical and physical interactions of surface atoms with adjacent water molecules (Rehl et al. 2022) and dissolved ions. When metal oxide is exposed to the aqueous water solution, the surface oxygen atoms react with water molecules forming the amphoteric hydroxyl sites $\equiv MOH^z$ with charge number z, surface sites $\equiv MOH_2^{z+1}$ with the charge number $z+1$, and the surface sites $\equiv MO^{z-1}$ with the charge number $z-1$. These surface sites may undergo the following surface protonation reactions:

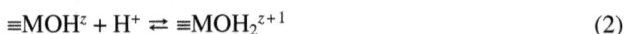

$$\equiv MO^{z-1}\ H^+ \rightleftarrows \equiv MOH^z \tag{1}$$

$$\equiv MOH^z + H^+ \rightleftarrows \equiv MOH_2^{z+1} \tag{2}$$

Charge numbers of surface groups are related to the coordination of metal atoms in metal oxides (Van Riemsdijk et al. 1986). Reactions of protonation and deprotonation result in pH-dependent surface charge and potential. These equilibrium reactions with corresponding equilibrium constants depend on the concentration of active surface sites, pH, type and the concentration of counterions and on the electrical surface charge (Kallay et al. 2011). The protonation and deprotonation reactions depend also on the degree of protonation of neighbouring surface groups (Brown 2009). The formation of charged surface sites on the mineral surface affects the distribution of ions near the interface in order to compensate the surface charge.

To develop a reliable and accurate description of the electrical interfacial layer, it is necessary to define the surface reactions and the structure of the electrical interfacial layer. The surface reactions are commonly described by using one of the Surface Complexation Models (SCM) (Lützenkirchen 2006). The structure of a thin liquid interfacial layer at a solid/liquid interface, possessing the opposite charge to the solid can be described by assuming one of the several existing models of the electrical interfacial layer, and one of them, is the general model (Gouy–Chapman–Stern model; Kallay et al. 2010) presented in Figure 1.

The general model of the electrical interfacial layer is composed of a compact inner part (also called the Stern layer) with fixed surface sites ($\equiv MOH^z$, $\equiv MOH_2^{z+1}$ and $\equiv MO^{z-1}$) and

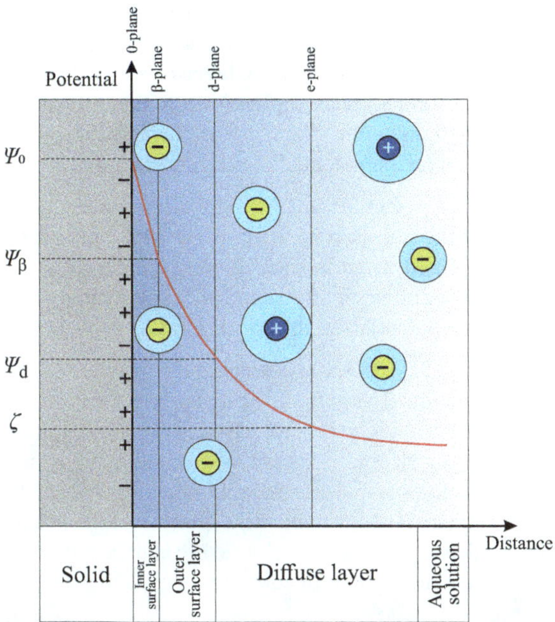

Figure 1. General model of the electrical interfacial layer at the solid/liquid interface. The **red line** presents the change of the electrostatic potential with distance from the solid. The **vertical lines** mark the position of 0-plane, β-plane, d-plane and e-plane.

adsorbed surface ions, and a diffuse part containing mobile solvent molecules and ions. Due to the presence of those two layers, the electrical interfacial layer is often called the double layer (DL), or by introducing the additional layers the triple layer (TL), the four-layer (FL) etc. According to the general model several different planes are defined within the EIL:

(i) The 0-plane divides the charged solid from the liquid phase and represents the plane of closest approach of nonhydrated ions to the surface. At the solid surfaces, active sites react with potential determining ions, which for most mineral surfaces, not only metal oxides (Miller et al. 2004; Wolthers et al. 2008; Heberling et al. 2011), are the hydrogen and hydroxide ions.

(ii) The β-plane divides the compact inner and outer part of the Helmholtz layer and represents the closest approach of the hydrated ions. At the β-plane non-specifically associated counterions are located reducing the effective surface charge.

(iii) The d-plane is the onset of the diffuse part of EIL. The d-plane and the 0-plane are separated by an unknown distance. In the double layer model, the β-plane and d-plane are located at the same distance from the solid surface, while in the triple layer model the β-plane and d-plane are separated.

(iv) The e-plane (electrokinetic slip plane; shear plane) is located within the diffuse layer dividing the mobile part of EIL from the stagnant part.

The above-mentioned surface planes are characterized by different types of interfacial potentials and define the regions/layers within the EIL with various properties (orientation of water dipoles and structure of interfacial water, transport properties, thermodynamic properties etc.). The values of interfacial potential and surface charge densities are controlled by the surface reactions of the fixed surface groups with dissolved ions as well as association of counterions and adsorption of other dissolved species from the bulk of the solution. Moreover, electrostatic potential affects the state of active surface charge, associated counterions, water molecules and ion distribution within the EIL. A layer of water dipoles and ions screens the inner surface potential at the solid surface (0-plane) while the diffuse potential is screened by ions in the diffuse layer. The magnitude of the diffuse layer potential is lower than the inner surface potential, but higher than the potential at the electrokinetic e-plane (Fig. 1).

Therefore, the interfacial potentials are fundamental parameters that describe the state of the interface and explain the transport of the particles, interactions between them, surface reactions, adsorption of ions and macromolecules. When measuring or specifying the value of interfacial potentials, it should be clearly stated which potential is meant, and how is the measured value related to the potentials in the corresponding model of the EIL.

Within the EIL the value of electrostatic potential decreases with distance from the mineral surface to the bulk of the solution. The bulk of the solution is the reference point characterized by the zero value of the electrostatic potential. There is also an electrostatic potential inside the solid mineral. The internal electric potential within the solid is assumed to be constant and related to the electrostatic potential at the solid surface exposed to the liquid medium; i.e., the inner surface potential. In this chapter, the inner surface potential is marked as Ψ_0. The inner surface potential depends on the crystal structure and the type and arrangement of atoms on the surface as well as of the interactions of surface groups with potential determining ions.

The inner surface charge density (σ_0) is partially compensated by association of charged surface groups with oppositely charged counterions. These associated counterions are located at the β-plane. The layer between these two planes is called the inner surface layer, inner Helmholtz layer or Stern layer. The inner surface layer acts like a capacitor. In the inner surface layer, we assume a linear drop in potential due to constant capacitance C_1 (expressed per surface area):

$$C_1 = \frac{\sigma_0}{\Psi_0 - \Psi_\beta} = \frac{\varepsilon_0 \varepsilon_{in}}{d} \tag{3}$$

where d is a thickness of the inner layer, and ε_0 is the electric permittivity of vacuum. In the inner surface layer, relative permittivity of water (dielectric constant) ε_{in} is smaller than the relative permittivity of bulk water (Teschke et al. 2001; Wander and Clark 2008; Boamah et al. 2019).

The region farther away from the β-plane, in which non-specifically adsorbed ions are accumulated and distributed by the influence of the charged surface and thermal motion, is called the diffuse layer of capacitance C_d. Non-specifically adsorbed ions inside the diffuse layer are mobile. The potential at the onset of diffuse layer (Ψ_d) affects the ion distribution within the diffuse layer. Within the diffuse layer the electrokinetic e-plane (slip plane), is located dividing the mobile from the stagnant liquid at the interface and is characterized by the electrokinetic ζ-potential, sometimes called the effective surface potential. It can be assumed that $\Psi_d = \zeta$, if the electrokinetic slip plane is located at the onset of the diffuse layer. Potential drop within diffuse layer is defined by the classical Poisson–Boltzmann equation (Gray and Stiles 2018). The surface charge density of the diffuse layer is $\sigma_d = -\sigma_s = -(\sigma_0 + \sigma_\beta)$, and is equal in magnitude but opposite in sign to the net (effective) surface charge density σ_s.

In the interpretation of the experimental data, one often uses various simplifications of the general model. For example, since the slip plane lies close to the onset of the diffuse layer, the diffuse layer potential may be identified with the zeta potential:

$$|\Psi_0| > |\Psi_\beta| > |\Psi_d| \approx |\zeta| \tag{4}$$

Basic Stern Model (BSM) (James and Parks 1982) does not consider the potential drop between β- and d-planes but takes into account that the slip plane is shifted from the onset of the diffuse layer (by $l_e > 0$) so that

$$|\Psi_0| > |\Psi_\beta| \approx |\Psi_d| > |\zeta| \tag{5}$$

The electroneutrality of solid mineral surface is described by different condition and composition of the aqueous solution. For minerals it is usually the pH value at which the surface potential or charge is equal to zero. The point of zero charge (pH_{pzc}) is the pH at which the surface charge density at 0-plane is zero ($\sigma_0 = 0$). The point of zero potential (pH_{pzp}) is the pH at which the inner surface potential is zero ($\Psi_0 = 0$). The isoelectric point (pH_{iep}) is the pH at which the electrokinetic potential is zero ($\zeta = 0$). These three values are equal in the case of neglected or symmetric counterion adsorption (Kallay et al. 2007a).

TYPES OF INTERFACIAL POTENTIALS

Potentials at certain distances from the surface plane are related to the measured or calculated values of a real system of a solid surface in contact with an aqueous electrolyte solution. All interfacial potentials are defined with respect to the potential in bulk solution. In literature, the term "surface potential" refers to different interfacial potentials. For this purpose, the meaning and significance of individual interfacial potentials is explained in the following chapters.

Inner surface potential

The inner surface potential—the electrostatic potential at the onset of the inner layer, Ψ_0 (Kallay et al. 2010), is assigned to the 0-plane where the charged surface groups are located. The inner surface potential affecting the state of active surface sites ($\equiv MOH^z$, $\equiv MOH_2^{z+1}$ and $\equiv MO^{z-1}$) corresponds to the 0-plane in which the surface sites are located. Due to the inner surface potential, electrostatic interactions between surface sites and potential determining ions occur, the surface sites activity coefficient ($\gamma_{\equiv S}$) changes,

$$\mathrm{R}T \ln \gamma_{\equiv S} = z_{\equiv S} \cdot \mathrm{F} \cdot \Psi_0 \tag{6}$$

where $z_{\equiv S}$ is charge number, T is thermodynamic temperature, R is the gas constant and F is the Faraday constant. Activity coefficients as well as the inner surface potential affect equilibrium reactions (1) and (2) with corresponding thermodynamic equilibrium constants:

$$K_1^\circ = \exp(\Psi_0 \mathrm{F}/\mathrm{R}T) \cdot \frac{\{\equiv \mathrm{MOH}^z\}}{\{\equiv \mathrm{MO}^{z-1}\} \cdot a_{\mathrm{H}^+}} \tag{7}$$

$$K_2^\circ = \exp(\Psi_0 \mathrm{F}/\mathrm{R}T) \cdot \frac{\{\equiv \mathrm{MOH_2}^{z+1}\}}{\{\equiv \mathrm{MOH}^z\} \cdot a_{\mathrm{H}^+}} \tag{8}$$

where curly braces denote surface concentration (the amount or number of moles per surface area). According to Equations (7) and (8), the inner surface potential at the 0-plane is related to the thermodynamic equilibrium constants, pH and the ratio of the surface concentrations of positively and negatively charged surface sites:

$$\Psi_0 = \frac{\mathrm{R}T \ln 10}{2\mathrm{F}} \log(K_1^\circ K_2^\circ) - \frac{\mathrm{R}T \ln 10}{\mathrm{F}} \mathrm{pH} - \frac{\mathrm{R}T \ln 10}{2\mathrm{F}} \log\left(\frac{\{\equiv \mathrm{MOH}_2^{z+1}\}}{\{\equiv \mathrm{MO}^{z-1}\}}\right) \tag{9}$$

where the product of thermodynamic equilibrium constants is related to the electroneutrality point pH_{eln} (sometimes called the pristine point of zero charge):

$$pH_{eln} = \frac{1}{2}\log(K_1^\circ K_2^\circ) \tag{10}$$

Electroneutrality point where all three zero charge points coincide ($pH_{pzp}(\Psi_0=0) = pH_{pzc}(\sigma_0=0) = pH_{iep}(\zeta=0)$) is determined by thermodynamic equilibrium constants of surface reactions (Pyman et al. 1979). In such a case the measurements of isoelectric point or the point of zero charge enable the evaluation of inner surface potentials from the measured open-circuit potentials of single crystal electrode, gate potential of ion sensitive field effect transistor, flat band potential obtained from C–V diagrams. This will be explained in the following chapters that describe the methods of interfacial potential measurement.

By inserting the Equation (10) into the Equation (9), the following relationship is obtained:

$$\Psi_0 = \frac{RT\ln 10}{F}(pH_{eln} - pH) - \frac{RT\ln 10}{2F}\log\left(\frac{\{\equiv MOH_2^{z+1}\}}{\{\equiv MO^{z-1}\}}\right) \tag{11}$$

The second term in Equation (11) is Nernstian potential, while the first term determines the deviation from the Nernstian potential. If the first term in Equation (11) is constant or zero, which happens if the surface concentrations of active sites are very high or the values of the thermodynamic equilibrium constants K_1° and K_2° are significant (Preočanin and Kallay 2013), the function Ψ_0(pH) is linear with the slope equal to the Nernstian slope or lower:

$$\Psi_0 = -\alpha\frac{RT\ln 10}{F}(pH - pH_{eln}) \tag{12}$$

where α represents the deviation of $d\Psi_0/dpH$ from the Nernstian slope ($\alpha = -Fd\Psi_0/dpH\,/RT\ln 10$). Accordingly, inner surface potential measurements and the deviation from the Nernstian behaviour may be used to obtain the ratio of surface concentrations of negative and positive sites, and thus the evaluation of interfacial equilibrium parameters.

Diffuse layer potential

The diffuse layer potential at the onset of the diffuse layer, Ψ_d, is the potential at the boundary between the diffuse and the non-diffuse parts of the electrical interfacial layer (Delgado et al. 2007). According to the Gouy–Chapman theory (Gouy 1910; Chapman 1913), or more precisely the Gouy–Chapman–Herzfeld theory as suggested by Ivanov (2023), the potential at the onset of the diffuse layer is related to the surface charge density of the diffuse layer σ_d by

$$\sigma_d = \sqrt{8\varepsilon_r\varepsilon_0 RTc}\,\sinh\left(-\frac{zF\Psi_d}{2RT}\right) \tag{13}$$

where ε_r and ε_0 are the relative permittivity of the medium and permittivity of vacuum, R and F are the molar gas constant and Faraday constant, T is the temperature, c is the concentration of electrolyte and z is the valence of ions that constitute the electrolyte. It should be pointed out that Equation (13) is only valid for a z:z valent electrolyte whereas for the case of a mixture of dissolved salts Grahame derived a more general expression (Grahame 1947).

$$\sigma_d = \left\{2\varepsilon_r\varepsilon_0 RT\sum_{i=1}^{N} c_i \cdot \left[e^{-\left(\frac{z_i F\Psi_d}{RT}\right)} - 1\right]\right\}^{1/2} \tag{14}$$

In this equation, c_i and z_i represent the concentration and valence of the i-th ion in the bulk of a solution and all other physical quantities were mentioned earlier. Measurement of surface forces by the atomic force microscope (AFM) enables the determination of the surface charge and surface potential on the onset of the diffuse layer (Butt et al. 2005).

Electrokinetic ζ-potential

The potential at the slip plane, i.e., electrokinetic potential ζ, is the potential at the boundary between the hydrodynamically mobile and immobile fluid. Electrokinetic phenomena involve fluid motion due to the electrical properties of interfaces under steady-state and isothermal conditions (Delgado et al. 2007). The value of the ζ-potential is equal to or lower in magnitude than the diffuse layer potential. The difference between those two interfacial potentials is a function of the concentration and type of ions in the solution; i.e., ionic strength:

$$I_c = \frac{1}{2}\sum_i c_i z_i^2 \tag{15}$$

At low electrolyte concentration, the decrease of the potential as a function of distance is small and $\zeta \approx \Psi_d$.

The electrokinetic potential can be determined by means of various electrokinetic methods mainly based on electrokinetic phenomena such as electrophoresis, streaming potential (current) measurements, electroosmosis, electrosedimentation, etc. (Delgado et al. 2007).

The potential drop

For the interpretation of experimental data, sometimes a certain value of the potential is not specified, but rather noted as the potential drop across the whole electrical interfacial layer. The potential drop across the solid–liquid interface Ψ_{surf} (Brown et al. 2016b; Ma and Geiger 2021) is the sum of the potential drop across the inner Helmholtz (Stern) layer (Ψ^{drop}_{Stern}) and the potential of the diffuse layer (ζ): $\Psi_{surf} = \Psi^{drop}_{Stern} + \zeta$; as the potential drop across the inner Helmholtz (Stern) layer is equal to the difference between potentials at the 0-plane and d-plane (assuming that the β-plane and d-plane are located at the same distance) and electrokinetic potential is equal to the diffuse layer potential ($\Psi_d = \zeta$):

$$\Psi_{surf} \approx \Psi_0 - \Psi_d + \Psi_d \approx \Psi_0 \tag{16}$$

For the colloid particles the potential drop across the solid–liquid interface can be determined by in situ X-ray photoelectron spectroscopy, as will be shown later in this paper (Brown et al. 2016b; Ma and Geiger 2021).

METHODS FOR THE MEASUREMENTS OF INTERFACIAL POTENTIALS

The measurement of the interfacial potentials is extremely important for the refinement of models that describe the interfacial layer, but also for understanding the processes that take place in the interfacial area and between charged particles. The value of specific surface potential was attempted to be comprehended and determined by different experimental techniques (Gonella et al. 2021). Moreover, the value of interfacial potentials was tried to be calculated using computational methods such as classical molecular dynamics (CMD) simulations (Předota et al. 2016; Ulman et al. 2019; Zarzycki 2023), grand canonical Monte Carlo simulations (Zarzycki and Rosso 2010), "ab initio" molecular dynamics (AIMD) (Lowe et al. 2018), and recently the implantation of machine learning to MD simulations (Hellström et al. 2019; Quaranta et al. 2019; Calegari Andrade and Selloni 2020; Eckhoff and Behler 2021).

A number of attempts have been employed to measure the interfacial potentials (Fig. 2). Generally, these experimental methods can be classified as:

(i) Electrochemically based methods which include Ion Sensitive Field Effect Transistors (ISFET), Capacity-voltage (C–V) measurements with electrolyte oxide semiconductor devices (EOS), Open-Circuit Potential (OCP) measurements with single crystal electrodes (SCrE);

(ii) Nonlinear optical methods (Second Harmonic Generation (SHG) and Sum Frequency Generation (SFG));

(iii) Spectroscopic techniques such as X-ray photoelectron spectroscopy (XPS);

(iv) Atomic force microscopy;

(v) Electrokinetic methods (such as streaming potential measurement for flat surfaces and electrophoresis for particles).

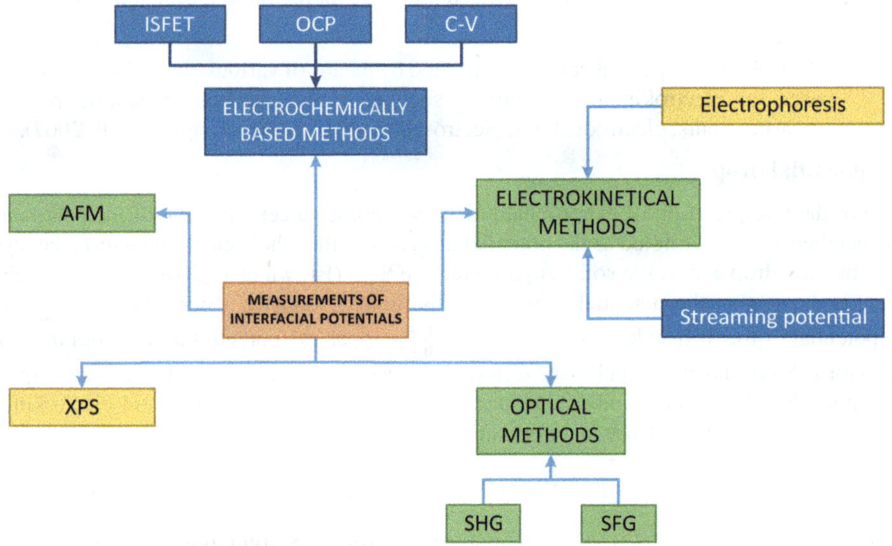

Figure 2. Interfacial potential measurement techniques for flat surfaces (**blue and green**) and particles (**yellow and green**).

While the experimentally obtained values may often not be directly comparable, they can provide a reasonably accurate representation of the state within the interfacial layer. Sometimes we use different techniques to measure the potential at different distances from the surface of a solid. It is important to distinguish the different electrostatic potentials within the electrical interfacial layer.

When minerals are exposed to moist air or aqueous solutions, ageing, surface transformation or dissolution take place, changing the structure and chemical composition of the mineral surface. By measuring the interfacial potentials, useful information about the rate or mechanisms of these processes can be obtained. It may be expected that these structural changes will affect the inner surface potential more than the electrokinetic potential.

Electrochemically based methods for determination of the inner surface potential

Over the decades, several electrochemical methods for determining the inner surface potential have been developed. These methods imply the measurement of some electrical response related to the inner surface potential of the investigated material, such as open-circuit potential (E_{OCP}), flat band potential (V_{FB}) derived from capacitance–voltage measurements, or gate potential (V_g). The working electrodes have been prepared by deposition of colloid particles on the metal wire (Penners et al. 1986), deposition of ice on the metal wire (Kallay and Čakara 2000), or by thermal and electrochemical oxidation of metallic plate (Avena et al. 1993). Also, for the construction of the working electrodes, the metal oxide single crystals (SCr) (Kallay et al. 2005), ion-sensitive field-effect transistors (ISFET) (Bergveld 1972; Bousse et al. 1992; Bergveld et al. 1995; van Hal et al. 1995; Liao et al. 1999; Lubbers and Schober 2009) and electrolyte-oxide-semiconductor (EOS) system (Siu and Cobbold 1979; Van den Vlekkert et al. 1988) have been used. Figure 3 shows the basic principles of inner surface potential measurements using the mentioned methods.

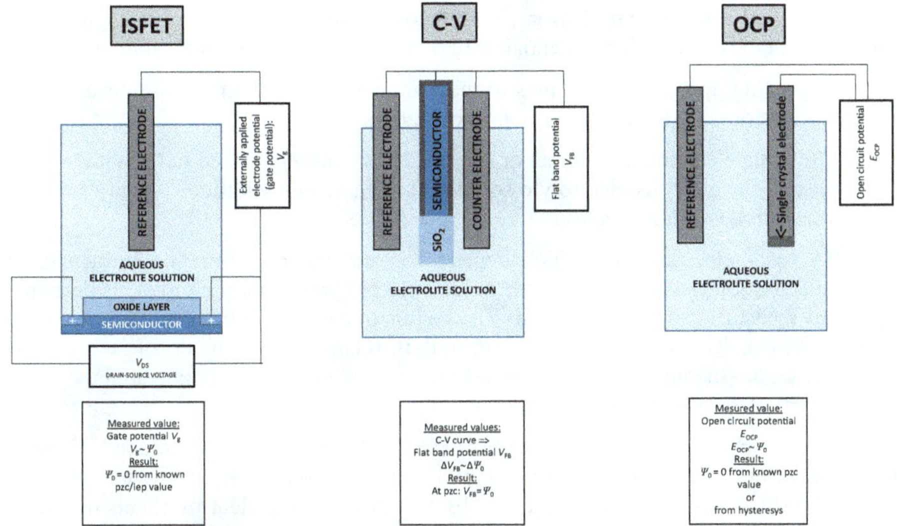

Figure 3. Electrochemically based methods for determination of inner surface potential: ISFET – ion-sensitive field-effect transistors; capacity–voltage (C–V) method with electrolyte/oxide/semiconductor probe (EOS) or electrolyte/insulator/semiconductor probe (EIS); OCP – open-circuit potential measurements.

The ion-sensitive field-effect transistors. The ion-sensitive field-effect transistor (ISFET) is a type of field-effect transistor that is specifically designed to measure ion concentrations in a solution. ISFETs measurement is based on the principle that the gate voltage of the transistor is affected by the concentration of ions in the surrounding solution. The structure of the ISFET device is closely related to the metal-oxide–semiconductor field-effect transistor (MOSFET) which is based on the modulation of current in the semiconductor by an electric field. An ISFET is a MOSFET with the gate connection separated from the SiO_2 in the form of a reference electrode inserted in the aqueous solution which is in contact with the gate oxide (Bergveld 2003). The first ISFET was introduced by Bergveld in 1970 (Bergveld 1970). He omitted the metal gate from a MOSFET and exposed the oxide layer to an electrolyte solution. Since then, ISFET sensors have been developed for the determination of pH, the concentration of other dissolved ions (Lubbers and Schober 2009) and as a biosensor (Nguyen et al. 2019; Cao et al. 2023). Bergveld (Bergveld 1970) originally prepared the SiO_2 layer on Si plate by thermal oxidation.

Lately the ISFET devices were prepared with other materials, e.g., Si_3N_4 (Matsuo and Wise 1974), Al_2O_3 (Abe et al. 1979) and Ta_2O_5 (Matsuo and Esashi 1981). ISFET probe consists of three regions (Fig. 3): the semiconductor (most often silicon), the hydrated oxide layer and the aqueous solution. Between the mentioned three phases, there are two interfaces: hydrated oxide/aqueous solution interface and semiconductor/ hydrated oxide interface. The reference electrode acts as a gate with a gate voltage (V_g). A thin layer of hydrated metal oxide on the semiconductor acts as a dielectric between the semiconductor and the reference electrode. Due to te hydration of the oxide layer the properties of the oxide are changed and the ISFET can operate without an externally applied gate voltage. Measurements are based on the field effect over the insulating oxide layer that is coupled to the surface through an oxide layer. Due to an internal interaction between two corresponding interfaces, the field at the semiconductor/ hydrated oxide interface depends on the activity of the ions within the hydrated oxide/aqueous solution interface. The measured gate voltage of the ISFET probe is related to the inner surface potential of hydrated metal oxide and is sensitive to the concentration of ions in aqueous solution environment. The inner surface potential obtained by an ISFET is pH-dependent (pH is determined by the buffer capacity of the oxide surface; van Hal et al. 1996; Bergveld 2003). The chemically sensitive layer of the ISFET interacts with H^+ ions, leading to a change in the surface charge density and consequently in the inner surface potential. ISFET measurement involves the following steps:

1. Applying appropriate voltages to the source and drain terminals to ensure that the transistor is in the desired region of operation.
2. Reference electrode (calomel or Ag/AgCl electrode) is placed in the same solution. pH of the aqueous electrolyte solution is adjusted by addition of acid or base and measured by glass electrode.
3. A local electric field, which depends on the ion concentration influencing the electrical interfacial layer at the oxide / electrolyte solution interface is determined by measuring the gate potential (V_g) as a function of the pH of the aqueous electrolyte solution. The measured gate potential is related to the inner surface potential. To obtain the inner surface potential from the measured ISFET potential the value of point of zero potential should be known.

From the response times and sensitivity of ISFET electrodes it was concluded that surface reactions determine their response (Bousse 1982). The pH sensitivity of the hydrated oxide layer in an ISFET device can be explained by two possible physical mechanisms (Siu and Cobbold 1979):

(i) Surface reactions, Equations (1) and (2), between the electrolyte and oxide result in formation of positively or negatively charged surface groups (Fung et al. 1986), which subsequently react with the supporting electrolyte ions. The inner surface potential at the aqueous electrolyte solution/oxide interface influences the space charge in the semiconductor by field-effect coupling. This mechanism does not involve the transport of ions between the two phases. All interactions are electrostatic.

(ii) Slow diffusion of H^+ ions and counterions from the electrolyte through the semiconductor/oxide interface. Although the diffusion of species must occur to some extent, this process cannot be considered as the primary response mechanism (Siu and Cobbold 1979; de Rooij and Bergveld 1980).

The ISFET probes exhibit a sub-Nernstian response (Bergveld 2003) which could be explained by the site-dissociation model (Bousse et al. 1983; van Hal et al. 1996). The deviation from the Nernstian behavior reflects the surface buffer capacity and interfacial layer capacitance. ISFETs are used as a pH sensor (Pathak et al. 2024), but the response is often not linear and deviates from the Nernstian potential, especially for unreactive surfaces such as SiO_2 (Bousse et al. 1983).

According to Lubbers and Schober (2009) the glass electrode and the ISFET probe have different charge balance mechanisms. The glass balances the pH-dependent surface charge by mobile ions within its glass matrix. ISFET, on the other hand, has no capability to balance the surface charge and is thus depended on ionic strength (Eqn. 15) through the formation of an electrical interfacial layer. Deviation from the Nernstian potential and pH dependency of ISFET can be explained by the site binding model (Fung et al. 1986), the protonation and deprotonation of active surface sites, but also the presence of counterions. This is particularly noticeable at extreme pH values or in solutions with high ion concentration. The sensitivity of the ISFETs is also dependent on the thermodynamic equilibrium constants of surface protonation and deprotonation (Eqns. 7 and 8) and the total density of surface sites. The high ΔpK values ($\Delta pK = pK_2 - pK_1$) and the low total surface sites density cause the non-linearity and slope of the Ψ_0(pH) function lower than Nernstian, which is consistent with Equation (12). The ΔpK value is a dominant factor for pH response (Liao et al. 1999).

Capacitance–voltage measurements. Electrolyte-oxide-semiconductor systems (EOS) are a simpler configuration of ISFET (Siu and Cobbold 1979; Bousse 1982). EOS consist of three phases: electrolyte (E), oxide (O) and silica (S) and two interfaces between them: E/O and O/S. Schematic presentation of the EOS measuring cell is:

$$\text{Cu}\,|\,\text{Si}\,|\,\text{SiO}_2\,|\,\text{electrolyte}|\text{reference electrode}\,|\,\text{Pt}$$

More complex types of measuring cells exist, introducing additional phases and liquid junctions between electrolytes. The oxide layer on silicon can be deposited by thermal oxidation (de Rooij and Bergveld 1980) or electrochemical oxidation (Liao et al. 1999). The basic principles are shown in Figure 3.

Since the inner surface potential cannot be measured directly, the most valuable information has been obtained from capacity measurements. The capacitance–voltage (C–V) measurements at different pH-values of the aqueous electrolyte solution are performed by electrochemical impedance spectroscopy. Measurements are performed at low applied *ac* potential and perturbating frequencies and the current passing through the EOS interface as well as the phase difference between the applied potential and the detected current are measured. The total interfacial capacitance values (C_{interf}), as a function of applied potential, at the angular frequency of the voltage perturbation are equal to $C = (Q \cdot R_{\text{el}}^{1-n})^{\frac{1}{n}}$ considering the constant phase element (CPE) used instead of classic capacitor (Hsu and Mansfeld 2001; Harrington and Devine 2008; Orazem et al. 2013). Since non-homogeneity of the system (inhomogeneous current flow, capacitance dispersion, etc.) arising from surface roughness or surface heterogeneity is often present, the CPE is employed. The CPE is characterized with the Q as the frequency-independent CPE parameter related to the interfacial capacitance [$\Omega^{-1}\text{cm}^{-2}\text{s}^{n}$] and n as the CPE exponent representing a level of non-ideal capacitor behavior (non-dimensional unit) (Hsu and Mansfeld 2001; Orazem et al. 2013). In the above term, R_{el} represents electrolyte solution resistance [Ω cm^2].

The minimum of the C–V curve corresponds to the flat band potential (V_{FB}). Flat band potential refers to the potential at which the Fermi level of the O/S interface is equal to the intrinsic Fermi level of the semiconductor (Memming 2015). Under these conditions, the potential of oxide/semiconductor interface is constant so that the measured changes can be attributed entirely to the reactions of the electrolyte–semiconductor (E/O) interface.

To obtain V_{FB} value, the Mott–Schottky (M-S) analysis is performed considering the frequency dispersion due to the inhomogeneous current flow (Harrington and Devine 2008; Katić et al. 2016). Additionally, it should be considered that the total interfacial capacitance, C_{interf} includes contribution from capacitance of the inner Helmholtz layer, C_{H} in series with the capacitance of the semiconductor space-charge layer, C_{sc}. Since C_{H} is often larger than C_{sc}, the total C_{interf} can be equalized with C_{sc} and follows the Mott–Schottky relationship, which for n-type material is (Di Quarto et al. 2018; Hankin et al. 2019):

$$\frac{1}{C_{SC}^2} = \frac{2}{\varepsilon\varepsilon_0 e N_D}\left(V - V_{FB} - \frac{k_B T}{e}\right) \tag{17}$$

where V is the applied potential, V_{FB} is the flat band potential both expressed with a respect to the reference electrode in the electrochemical scale, ε_0 is the vacuum permittivity, ε is relative permittivity, e is the electron charge, N_D is the donor concentration, T is the temperature and k_B is the Boltzmann constant. From the C_{SC}^{-2} vs. V plot, the flat-band potential value can be determined as the intercept on the potential axis of the linear part of the Mott–Schottky plot (Molto et al. 2020; Patel et al. 2022; Poulain 2023).

The flat band potential is used as a characteristic potential of individual semiconductor electrodes in the same way as the potential of zero charge is used for metal electrodes. At the flat band potential, the charge of the space-charge layer is zero but the interfacial charge, the sum of surface states charge, inner Helmholtz layer charge and charge due to the surface adsorption is not zero.

The potential of the interfacial layer at the flat band potential depends on the concentration of potential-determining ions in the electrolyte solution. For most semiconductor electrodes in aqueous solutions, the interfacial potential across the layer is determined by the dissociation of surface hydroxyl groups; hence, the flat band potential is given as a linear function of pH (Sato 1998).

The flat-band potential is related to the inner surface potential. The flat-band potential of the EOS structure is defined as the value of the applied potential where the silicon inner surface potential is equal to zero (Bousse 1982). For direct evaluation of inner surface potential from the flat-band potential, other parameters such as the reference electrode potential and effective insulator charge should be known (Bousse et al. 1983). Therefore, only the shift, not absolute value, in flat-band potential is related to the shift in the inner surface potential; i.e., relative inner surface potential:

$$\Delta V_{FB} = -\Delta\Psi_0 \tag{18}$$

Knowing the value of zero inner surface potential (point of zero potential, which is close to point of zero charge and isoelectric point in absence of counterion association) enables the evaluation of absolute values of inner surface potential. The measured potential is again a consequence of surface reactions (Eqns. 1 and 2), and the shape of the Ψ_0(pH) function depends on the thermodynamic equilibrium constants (Eqn. 11). Linear dependences for TiO_2 and SnO_2 surfaces (Gerischer 1989), and nonlinear dependence for SiO_2–Si (Siu and Cobbold 1979) were measured.

Open-circuit potential measurements (OCP). Another type of electrochemically based method includes direct measurement of the potential difference of an open-circuit in which the measuring electrode and the reference electrode in aqueous electrolyte solution are located. The principles of experimental set-up are schematically shown in Figure 3. The measurement of open-circuit potential of a measuring electrode with respect to the reference electrode under conditions when the current tends to zero, using a high-resistance voltmeter (i.e., pH-meter) enables the evaluation of inner surface potential (Ψ_0). The measurement is simple and requires:

(1) A reference electrode (Ag/AgCl electrode is most often used).

(2) A calibrated pH electrode for the simultaneous measurement of pH-value of the aqueous electrolyte solution.

(3) A high-resistance voltmeter (pH-meter) for the measurement of the open-circuit potential (i.e., the overall electromotive force of the cell) during the addition of acid, base or salt solution.

(4) The evaluation of the inner surface potential from the measured open-circuit potential.

Since the measured open-circuit potential corresponds to the sum of all potential difference in the measuring cell and only potential difference between mineral and aqueous electrolyte solution depends on composition of the solution (pH and ionic strength), inner surface potential can be calculated knowing the pH at which $\Psi_0 = 0$; i.e., pH_{pzp}. The point of zero potential can be identified with the point of zero charge or the isoelectric point, or evaluated from the cyclic titration, which often exhibits a hysteresis loop (Preočanin et al. 2006; Zarzycki et al. 2011). The titration path is divided into a region where surface protonation predominantly takes place (acidimetric titration – addition of acid) and a region where protons are released from the surface (alkalimetric titration – addition of base). This approach takes advantage that the hysteresis area is directly proportional to the work done on/by the system during a titration cycle and enables the determination of the point of zero potential. In the absence of hysteresis and counterion association, another method based on a common intersection point of measured electrode potential functions at different ionic strengths (Zarzycki and Preočanin 2012) was proposed.

As a measuring electrode, several types of the probes have been used:

(i) Thin layer of metal oxide on metal plate. The first electrodes for the inner surface potential measurements of TiO_2 were prepared by thermal and electrochemical oxidation of metallic titanium (Avena et al. 1993). The preparation of the oxide layer on the metal plate is the same as in the preparation of the ISFET probes. A schematic representation of the cell used for measuring the open-circuit potential of Ti/TiO_2 electrode is (Avena et al. 1993):

$$\mathrm{Pt} \,|\, \mathrm{Ti} \,|\, \mathrm{TiO_2} \,|\, \mathrm{KNO_3(aq)} \,|\, \text{reference electrode} \,|\, \mathrm{Pt}$$

The open-circuit potential, measured by voltmeter, corresponding to the sum of all potential differences across all the interfaces, on the schematic representation is marked with |. The only interface which is pH-dependent is TiO_2 | KNO_3(aq) interface. Ti/TiO_2 electrodes showed a linear response in the range from pH = 5 to pH = 10 with slopes lower than Nernstian; precisely –(39 ± 5) mV (Avena et al. 1993).

(ii) Polycrystalline metal oxide coated wire. Penners et al. (1986) prepared the hematite electrode by coating roughened Pt plates with a thin layer of monodisperse colloid particles by extensive application of colloid suspension and drying after removal of the excess of suspension. The electrode potential of the prepared hematite electrode was measured with respect to the reference electrode. The response was linear and almost Nernstian between pH = 5 and pH = 10. However, due to the porosity of the oxide layer, the solution could have direct contact with the metal wire so that such an electrode behaves as an electrode of the second kind and the potential is predominantly determined by the redox equilibrium and influenced by the solubility of the oxide. The measured slope of the measured potential–pH function was found to be linear with the Nernstian slope. Such a result is in agreement with a surface complexation model involving a sufficiently high number of reactive surface sites.

(iii) Ice electrode. A compact ice layer was formed on the platinum wire and the measured potential was a result of reactions on ice/aqueous electrolyte solution interface (Kallay and Čakara 2000):

$$\mathrm{Pt} \,|\, \text{ice layer } (-2\,^\circ\mathrm{C}) \,|\, \text{aqueous electrolyte solution } (0\,^\circ\mathrm{C}) \,|\, \text{reference electrode}$$

The slope of the Ψ_0(pH) function was found to be non-linear and lower than the Nernstian potential. The results were explained based on the surface complexation model assuming an amphoteric nature of ice surface OH groups.

(iv) Single crystal electrode. An electrode made of a single crystal in a Plexiglas holder in which electrical contact is achieved firstly by using mercury, or in later variants conductive paint/glue (Fig. 4).

Cu(s)|conductive glue | single crystal | aqueous electrolyte solution | reference electrode

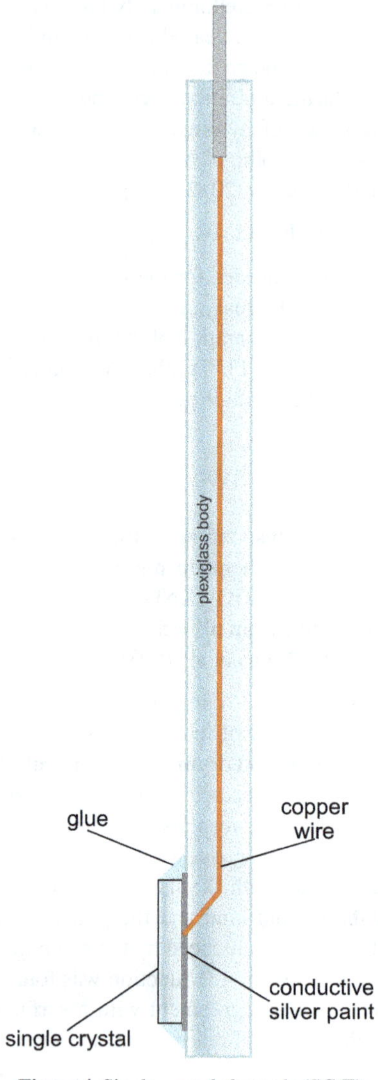

Figure 4. Single crystal electrode (SCrE).

The single crystals electrodes of hematite (Preočanin et al. 2006; Kallay et al. 2007b; Kallay and Preočanin 2008; Lützenkirchen et al. 2013, 2015; Shimizu and Boily 2014), TiO_2 (Kallay et al. 2007a, 2014; Kovačević et al. 2010; Preočanin et al. 2010a; Klačić et al. 2021; Jukić et al. 2023), SiO_2 (Brkljača et al. 2018b), Al_2O_3 (Abdelmonem et al. 2022), silver halides (Kallay et al. 2008, 2014; Preočanin et al. 2009, 2010b, 2011; Selmani et al. 2014), pyrite (Preočanin et al. 2007), ceria CeO_2 (Namjesnik et al. 2016; Klačić et al. 2019), fluorite CaF_2 (Preočanin et al. 2017b; Klačić et al. 2019), and calcite $CaCO_3$ (Heberling et al. 2021) were made and tested.

The open-circuit potential is measured with respect to the reference electrode using a pH-meter (i.e., high resistance voltmeter). The measured open-circuit potential is the sum of all potential differences in the measuring circuit. The only one potential difference that depends on the solution composition is the potential at the single crystal/ aqueous electrolyte solution interface. Thus, this potential difference can be identified with the inner surface potential at the 0-plane. The pH dependency of the inner surface potential is a consequence of the distribution of hydronium and hydroxide ions between the inner layer and the bulk of the solution. The measured SCrE potential is affected by the surface reactions (Eqns. 1 and 2) which include potential determining ions, but also counterions and adsorbed ions near the surface. The measured SCrE potentials do not provide directly the information on the extent to which interfacial water molecules affect the surface potential. The values of the metal oxide inner surface potential can be calculated from the measured open-circuit potential if the point of zero potential is known

$$\Psi_0 = E_{\text{OCP}} - E_{\text{pzp}} \tag{19}$$

where the constant term denoted as E_{pzp} represents sum of all potential jumps that are independent of the activities of potential determining ions. In the absence of specific ion adsorption or in the case of symmetrical association of counterions, the point of zero potential corresponds to the point of zero charge or isoelectric point. To solve the problem of determination of the point of zero potential several procedures have been proposed. One of them exploits the appearance of hysteresis of electrode potentials response during the acid base titration (Preočanin et al.

2006), next one uses a common intersection point of measured electrode potential functions at different ionic strengths (Zarzycki and Preočanin 2012). The simple method of the evaluation of the point of zero potential of metal oxide surfaces can be used if the inner surface potential function Ψ_0(pH) exhibits a saddle-like shape (Preočanin and Kallay 2013).

Non-linear optical spectroscopy

Second-order nonlinear optical techniques have been used as sensitive techniques for the characterization of mineral/aqueous electrolyte solution interfaces (Backus et al. 2021). Nonlinear optical spectroscopy relies on the nonlinear response of molecules under the influence of strong electromagnetic field, and in the case of minerals it is influenced by the charge of the mineral surface. The increase in the second harmonic field depends on pH; i.e., on the magnitude of the surface charge and potential (Ong et al. 1992). Solid material in non-linear optic spectroscopy could be single crystal (Hopkins et al. 2005; Lützenkirchen et al. 2018; Boily et al. 2019; Ma and Geiger 2021), or colloid particulate substance in a solution (Marchioro et al. 2019; Bischoff et al. 2020, 2021, 2022). Sum-frequency generation (SFG) and second-harmonic generation (SHG) spectroscopy are powerful nonlinear spectroscopic tools for probing molecules at surfaces and interfaces. In either of these techniques, the measured signal arises from the second-order susceptibility $\chi^{(2)}$ which represents an average response of all molecular responses. At a very high surface charge densities of mineral, in very acidic and basic region, far from the electroneutrality point, the large increase of SH signal is observed, and third-order nonlinear susceptibility can be measured $\chi^{(3)}$ (Ong et al. 1992). The $\chi^{(2)}$ and $\chi^{(3)}$ contributions originate in the inner surface layer and diffuse layer, respectively (Darlington et al. 2017). The second-order susceptibility $\chi^{(2)}$ depends on polarization of molecules and ions in the inner surface layer, while the effective third-order susceptibility $\chi^{(3)}$ describes the effect of the preferred dipolar orientation of more distant species; i.e., located in diffuse layer. It is important to note that the $\chi^{(3)}$ mentioned here is the effective third-order susceptibility and not real third-order optical susceptibility, which describes three-photon interactions as in triple-frequency generation. Only the component of $\chi^{(3)}$ that is perpendicular to the surface represents the average second-order hyperpolarizability, $\beta^{(2)}$, which is the interaction of two photons leading to second-harmonic frequency. $\beta^{(2)}$ is zero in centrosymmetric environment and non-zero at the interface in the presence of local electric field from the surface (Chu et al. 2023).

In SFG spectroscopy, two pulsed laser beams, typically an infrared (IR) and a visible (Vis) are temporally and spatially overlapped at an interface between two isotropic media. Two beams reflected from an interface induce a second-order polarization that oscillates at the sum of the two incident frequencies. A resulting signal at the sum-frequency of the two incident frequencies is generated in the direction governed by the conservation of momentum and under broken centrosymmetry selection rule, which inherently occurs at the interface and therefore makes this technique superior to linear spectroscopic methods, e.g., linear vibrational spectroscopy. When the surface is charged, that charge will align water molecules near the surface, further breaking the symmetry (Gonella et al. 2021). The resonant vibration modes of the molecules at the interface can be probed by tuning the IR frequency (Morita 2018).

SHG is a special case of SFG in which only one incident frequency is used, and a signal at twice the incident frequency is generated. SHG is non-resonant second-order optical process and hence provides no chemical information about the system, however geometrical information about the molecular orientations is still possible. The advantage of SHG is that the experiment is not complicated and the signal to noise ratio is relatively high compared to SFG and hence data acquisition is faster. In both systems, the generated signal is polarization dependent. Different polarization combinations for the incident and the generated beams allow for probing different tensor elements of the optical susceptibility of the molecules at the interface (Cyran et al. 2019). The relative intensities under different polarization combinations reflect the molecular orientations and the degree of ordering. After proper correction of the Fresnel factors, theoretical assumptions and data fitting, structural information about the molecular orientation can be obtained. The sign of the fitted nonlinear optical susceptibility indicates the absolute orientation of molecules with respect to the surface. Time-resolved experiments that provide dynamic information about the system are also possible. Scattering SHG and SFG can be used to investigate interactions at the surface of nano- and microparticles (Roke et al. 2004).

The three most common geometries that have been used in SFG at liquid–solid interface experiments for single crystal minerals are shown in Figure 5. The choice of the optimum geometry is defined by the linear optical properties of the interface, the nonlinearity of the SFG process, and the complexity of the setup. For example, if one of the bulk media at one side of the interface is opaque with respect to one or both incident beams, the interface can be accessed from the other side, and the same for the generated beam. In case of probing an interface between an aquatic solution next to a mineral in the OH vibrational region (3000 to 4000 cm^{-1}), it is convenient to have the incident beams reach the surface from the mineral side provided that this mineral is transparent for IR in this range of frequencies (e.g., silica, alumina, calcium floride, etc.). When using reflection geometry, the depth probed perpendicular to the interface in the second medium is constrained by the evanescent field, which relies on the incident angle. As a result, the contribution of the bulk SFG is restricted to a region extremely close to the interface. If the refractive indices of the adjacent media are similar, the reflected signal becomes minimal, making the transmission geometry a more favorable choice. In such a case, the signal can be generated along a distance that is of the order of fundamental wavelengths (Dalstein et al. 2017). Total internal reflection (TIR) is the best geometry for mineral samples of refractive indices significantly higher than that of bulk liquid and can be obtained in prism like single crystal. Due to the enhanced Fresnel factors, the signal in TIR can be increased above an order of magnitude in case of having the incident beams hitting the interface at the critical angle of TIR (Williams et al. 2000). However, working at the critical angle makes the signal extremely sensitive to variations in the refractive indices of the adjacent media, and hence proper Fresnel factors correction of the signal is mandatory.

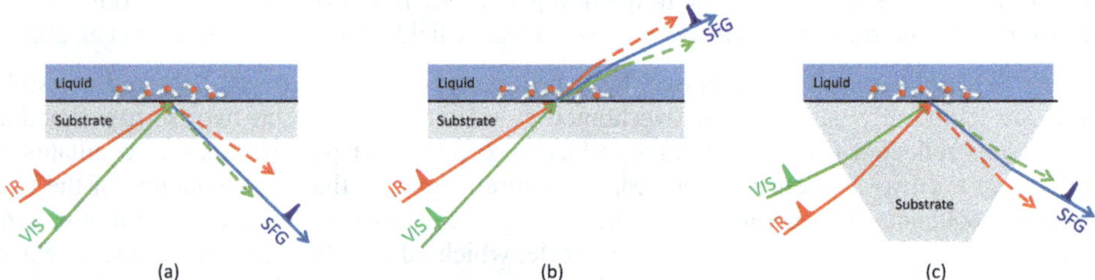

Figure 5. Schematic representation of three common SFG experimental geometries used in liquid–solid interface studies. **(a)** Reflection, **(b)** transmission, and **(c)** total internal reflection.

In SHG experiments, the second-harmonic electric field ($E_{2\omega}$) is directly proportional to the inner surface potential Ψ_0 (Ong et al. 1992). The second-order susceptibility $\chi^{(2)}$ contribution to the second-harmonic electric field can be evaluated by measuring under electroneutral surface condition; i.e., at pH_{iep} or pH_{pzc}:

$$E_{2\omega} - E_{2\omega}(\Psi_0 = 0) \sim \Psi_0 \tag{20}$$

which enables the absolute measurements of the inner surface potential. However, knowing the electroneutrality point is necessary.

The SHG measured signal originates from the instantaneous response of the electrons bound to the interfacial species to the incident optical fields. The photons generated at the interface interfere with the photons generated in the diffuse layer. The SHG amplitude and phase enable the determination of the total interfacial potential, which contains the surface charge and dipole, contributions to the potential at the interface (Chang et al. 2020):

$$\Psi_{total}(0) = \Psi_{el}(0) + \Psi_{dipol}(0) \tag{21}$$

The recent research of Ma and Geiger (2021) shows that separating the second- and third-order contributions of the SHG signal enables estimating interfacial structure and interfacial potential. From SHG amplitude and phase the total interfacial potential drop across the mineral/water interface, which contains the electrostatic and dipole contributions, could be calculated.

Cai et al. (2021) provided a more accurate model for the relationship of inner surface potential (Ψ_0) and nonlinear optical response. Moreover, performing an off-resonance phase-resolved SFG experiments enables obtaining the inner surface potential data with corresponding phase information, without performing calibration.

SHG can detect the orientation of water molecules even within the diffuse layer due to H-bonding and electrostatic interactions with the mineral surface as well as ion hydration (Gonella et al. 2021; Rehl and Gibbs 2021).

Polarimetric angle-resolved second harmonic scattering (AR-SHS; Marchioro et al. 2019) is used for the determination of the average surface potential of colloidal particles in solution. AR-SHS assumes exponential decay of the electrostatic potential near the interface. The nonresonantly scattered second harmonic (SH) light that is emitted from the particle interface and the electrical interfacial layer provides information for the determination of the interfacial potentials and orientation of water molecules at the interface.

X-ray photoelectron spectroscopy

X-ray photoelectron spectroscopy (XPS) technique operates based on photoelectric effect, i.e., monochromatic X-ray irradiation (hν) induces the emission of photoelectrons from the material being analysed. By measuring the kinetic energy, E_{kin}, of electrons that leave the surface without inelastic collisions, their binding energy E_B (BE) can be obtained: BE = E_{kin} − hν (Greczynski and Hultman 2022). Albeit typically operating under ultrahigh vacuum conditions, recently XPS measurements with liquid samples were also performed employing liquid microjet in combination with synchrotron radiation (Brown et al. 2013a,c) This approach enabled *in situ* XPS characterization of aqueous nanoparticle systems and direct determination of solid/liquid interface parameters including inner surface potential.

The BE of oxide nanoparticles in liquid suspension (obtained by XPS from liquid microjet) is sensitive to solution pH with observed shift to higher BE at pH < pH_{pzc} and to lower BE at pH > pH_{pzc} (Brown et al. 2013a). As defined before, at pH_{pzc} the concentrations of deprotonated and protonated surface mineral surface groups are equal (the surface is uncharged). Hence, the BE shift for an oxide nanoparticle is a result of a charge in the inner surface potential, Ψ_0 (Brown et al. 2016a):

$$BE^{salt} = BE^{pzc} + \Psi_0^{salt} e \tag{22}$$

with binding energy of electrons for charged surfaces (BE^{salt}) shifting relative to binding energy of uncharged surfaces (BE^{pzc}) in the same electrolyte. The change in inner surface potential values due to the solution pH variation can be considered having as having an analogous effect to applying an external DC potential to a mineral substrate (Brown et al. 2013a).

To determine the absolute value of inner surface potential the system should be near the electroneutrality region. However, due to the aggregation of uncharged particles the measurement in that region is difficult. Electroneutrality thus can be obtained by performing the measurements at different pH and extrapolation the results to pH_{pzc} (Brown et al. 2016a). This experimental technique is used to determine the inner surface potential of silica nanoparticles in highly concentrated solutions (Brown et al. 2013b, 2016a; Uzundal et al. 2019). The pH-dependent $\Delta\Psi_0$ measured by *in situ* XPS can be compared to the potential drop through the electrical interfacial layer (Butt et al. 2003).

Atomic force microscopy

The atomic force microscope was invented in the mid-eighties by Binnig, Quate, and Gerber (Binnig et al. 1986). Their invention was a major step forward in surface science because it enabled real topographical representations of surfaces in three dimensions with nanometre-scale resolution. The operating principle of the AFM is shown in Figure 6. The atomic force microscope probes the sample surface using a sharp tip mounted to the end of a spring cantilever. The tip is typically several micrometres long and with a radius of curvature of only a few nanometres. The AFM measures the force between the tip and the sample surface by reading the changes in the reflection of the laser beam due to the displacement of the cantilever caused by the interaction with the sample. An image of the relief of the sample surface is obtained with the help of a piezoelectric element that moves the sample under the tip or the tip over the sample surface with an accuracy of a few Å (Eaton and West 2010).

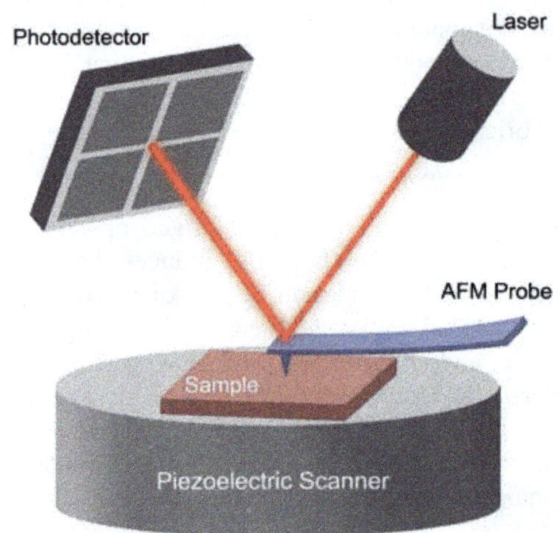

Figure 6. Basic principle of operation of an atomic force microscope.

Not long after the invention of the AFM, Ducker et al. (1991) on the one hand and Butt (1991a) on the other realized that by replacing the sharp AFM tip with a single spherical colloidal particle, the force between the particle and the sample can be very precisely described by the Derjaguin–Landau–Verwey–Overbeek (DLVO) theory (Derjaguin and Landau 1941; Verwey and Overbeek 1948). This discovery established a new subtype of AFM measurement called colloidal probe technique.

In this technique, the force between the colloidal particle and the sample surface is measured in liquid media as a function of the distance between them. The total force between the particle and surface is given by the contribution of the electrostatic (F_{el}) and van der Waals (F_{vdW}) interactions

$$F = F_{el} + F_{vdW} \tag{23}$$

The van der Waals force between a sphere of radius r and a flat surface is described by

$$F_{\mathrm{vdW}} = -\frac{r\mathrm{A_H}}{6x^2} \tag{24}$$

where x is a separation between them, and $\mathrm{A_H}$ is the Hamaker constant that depends on the chemical nature of the particle, surface, and medium (Israelachvili 2011). The electrostatic force between a sphere and a flat surface can be expressed by an equation developed by Butt (Butt 1991a,b; Butt et al. 2005) based on Parsegian and Gingell theory (Parsegian and Gingell 1972)

$$F = \frac{2\pi l_{\mathrm{D}} r}{\varepsilon_r \varepsilon_0}\left[\left(\sigma_{\mathrm{T}}{}^2 + \sigma_{\mathrm{S}}{}^2\right)\cdot \mathrm{e}^{-(2x/l_{\mathrm{D}})} + 2\sigma_{\mathrm{T}}\sigma_{\mathrm{S}} \cdot \mathrm{e}^{-(x/l_{\mathrm{D}})}\right] \tag{25}$$

Here, σ_{T} and σ_{S} are the surface charge densities of the tip and sample surface, ε_r and ε_0 are the relative permittivity of the medium and permittivity of vacuum, x is the distance between tip and sample surface, r is the radius of a spherical tip, and l_{D} is the Debye length. The Debye length represents the average thickness of the ionic atmosphere near a charged surface, and its magnitude depends solely on the properties of the solution

$$l_{\mathrm{D}} = \sqrt{\frac{\varepsilon_r \varepsilon_0 \mathrm{R}T}{2\mathrm{F}^2 I_c}} \tag{26}$$

where T is the temperature of the solution, I_c is the ionic strength of the solution (Eqn. 15), and R and F are the molar gas constant and Faraday constant. It should be pointed out that in the derivation of Equation (25) it was assumed that the surface charges of the tip and sample do not depend on the distance between them. In addition to the requirement that the surface charge densities are constant, there are three more limitations connected to the Equation (25). First, the surface potentials of the tip and sample should be lower than 50 mV. Second, the radius of the spherical particle should be much larger than the Debye length ($r \gg l_{\mathrm{D}}$), and third, the distance between the tip and the sample surface should be larger or equal to the Debye length ($x \geq l_{\mathrm{D}}$). A sum of Equations (24) and (25) gives one of the formulas for the total force between the particle and the sample surface that is used in AFM measurements for surface charge and surface potential determination

$$F = \frac{2\pi l_{\mathrm{D}} r}{\varepsilon_r \varepsilon_0}\left[\left(\sigma_{\mathrm{T}}^2 + \sigma_{\mathrm{S}}^2\right)\cdot \mathrm{e}^{-(2x/l_{\mathrm{D}})} + 2\sigma_{\mathrm{T}}\sigma_{\mathrm{S}} \cdot \mathrm{e}^{-(x/l_{\mathrm{D}})}\right] - \frac{rA_{\mathrm{H}}}{6x^2} \tag{27}$$

How exactly are the surface charge density and interfacial potentials determined with an AFM? First, the force between the colloidal probe and the sample is measured at certain conditions (e.g., pH, ionic strength, temperature, etc.) as the probe is approaching the surface of the sample. Then, the measured force–distance curve is fitted by the model (for example by Eqn. 27). In principle, there are two possibilities for the fitting procedure. The first case is when the colloidal probe and the sample are of the same material. Therefore, the tip and sample surface bear the same surface charge densities ($\sigma_{\mathrm{T}} = \sigma_{\mathrm{S}}$) and Equation (27) thus becomes

$$F = \frac{4\pi l_{\mathrm{D}} r \sigma_{\mathrm{S}}^2}{\varepsilon_r \varepsilon_0}\left[\mathrm{e}^{-(2x/l_{\mathrm{D}})} + \mathrm{e}^{-(x/l_{\mathrm{D}})}\right] - \frac{r\mathrm{A_H}}{6x^2} \tag{28}$$

The surface charge density of the sample can be explicitly determined from this equation by fitting the experimental data if the other parameters are known. The radius of the colloidal AFM tip can be independently measured microscopically, and the Debye length can be calculated according to Equation (26) for a given ionic strength, type of solvent, and temperature. The permittivity of the solvent and the Hamaker constant are usually known in the literature.

However, it is much more common in practice that the AFM probe and the sample are not made of the same material. In such cases, there is not one unique pair of surface charge densities that will fit the experimental force curves. For that reason, it is necessary to use the surface charge density of the tip from the independent measurements as a guide in fitting the experimental force–distance curves with a theory. As one example, the force–distance curves between a silica particle and a hematite flat surface are shown in Figure 7. The curves were generated according to Equation (27) for the cases of different pH values.

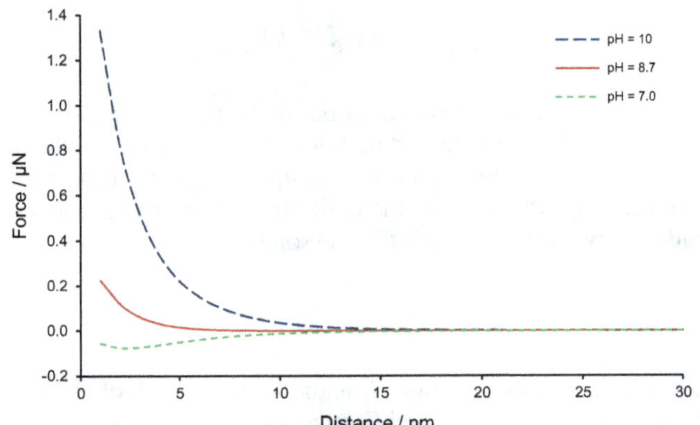

Figure 7. Force between a silica spherical particle of 2.5 µm radius and a hematite flat surface versus the distance between them. The curves were generated by Equation 27 for pH values of 10.0 ($\sigma_T = -0.13$ C/m^2 and $\sigma_S = -0.05$ C/m^2), 8.7 ($\sigma_T = -0.08$ C/m^2 and $\sigma_S = 0.0$ C/m^2), and 7.0 ($\sigma_T = -0.04$ C/m^2 and $\sigma_S = 0.06$ C/m^2) with 10 mM 1:1 background electrolyte at 25.0 °C. The surface charge densities of silica and hematite were chosen according to the experimental results reported in the literature (Kędra et al. 2021; Goel and Lützenkirchen 2022). The Hamaker constant for the silica–water–hematite system was set to be 1×10^{-20} J (Faure et al. 2011).

These surfaces have different point of zero charge. Silica has a pH_{pzc} of around 3.5 (Goel and Lützenkirchen 2022), while the pH_{pzc} of hematite is around 8.7 (Kędra et al. 2021). For pH values above the pH_{pzc} of hematite (e.g., pH = 10.0), where both materials are negatively charged, an electrostatic repulsion is observed (see Fig. 7). The repulsion decreases as the pH decreases, and at pH_{pzc} of hematite long-range electrostatic force diminishes, resulting in an overall repulsion between the two surfaces at close distances due to the overlapping of EILs. For pH values between the pH_{pzc} of the silica and the hematite (e.g., pH = 7.0), an electrostatic attraction between the oppositely charged surfaces can be observed. In this example, it was shown how the pH value of the solution affects the trend of the force–distance curve. By knowing the surface charge density of the colloidal particle at each pH, it is possible to determine the surface charge density of a sample for each of the curves based on Equation (27).

Whether the surface charge density is determined for the tip and sample made of the same or different materials, the potential is commonly calculated from the experimentally determined surface charge density by the Grahame equation (Eqn. 14). Although it is not entirely clear to which plane in the electrical interfacial layer the potential calculated by Equation (14) could be attributed, it is often compared to the zeta potential (Veeramasuneni et al. 1996; Hartley et al. 1997; Larson et al. 1997; Assemi et al. 2006; Herzberg et al. 2020). However, it is more likely that potential measured with AFM (Ψ_{AFM}) corresponds to the potential at the beginning of the diffuse layer, i.e., on the d-plane. Therefore, one would expect that Ψ_{AFM} is higher in magnitude, or eventually the same as ζ-potential. In most cases, this is true, but there are exceptions (Ruiz-Cabello et al. 2013). It seems that the exact definition of the potential obtained by AFM measurements still remains an open question. In the end, it should be mentioned that theoretical models for the force between a sharp conical AFM tip and a flat surface have also been developed (Butt 1991b; Drelich et al. 2008; Yin and Drelich 2008). As well as the colloidal probe technique, these models can be used to determine the sample surface charge density and potential. One of the main advantages of these models is the possibility of mapping the charge density and potential in the vicinity of the surface, which is extremely important for heterogeneous samples. However, determining surface charge densities by sharp tip-substrate models has less precision than the colloidal technique.

Methods for measuring electrokinetic potential

Electrokinetic (zeta, ζ) potential is the potential that characterizes the slip plane (Fig. 1). The state of the system in which the electrokinetic potential is equal to zero is called the isoelectric point (iep), and it is expressed as the corresponding pH-value (pH_{iep}). The absolute value of the electrokinetic potential increases as the pH is more different from pH_{iep}. The isoelectric point is sometimes mistakenly identified with the point of zero charge (pzc) (Sposito 1998), which represents the state of the system where the surface charge is equal to zero and is also expressed as a pH value (pH_{pzc}). When only potential-determining ions (p.d.i.) are bound in the system, pH_{iep} is equal to pH_{pzc}, while in the case of adsorption of ions different from p.d.i., the so-called specific adsorption, pH_{iep} and pH_{pzc} differ. In the case of anion adsorption, pH_{iep} moves towards lower pH values, and pH_{pzc} towards higher ones, while the opposite is true for cation adsorption. The question is often raised whether the adsorption of counterions should also be considered as specific adsorption. Many authors (Parks and Bruyn 1962; Blok and Bruyn 1970; Breeuwsma and Lyklema 1973), especially in the case of 1:1 electrolyte, do not consider counterion adsorption as specific adsorption. Davis et al. (1978) call such adsorption, since there is no shift in pH_{iep} or pH_{pzc}, symmetric specific adsorption. Sprycha (1984) showed, that independently of the concentration of the electrolyte, in the case of some electrolytes (NaCl, KCl, CsCl), no shift in pH_{pzc} or pH_{iep} occurs, while with some (LiCl) it does. Johnson, Scales and Healy (Johnson et al. 1999) also showed that, in the case of aluminum oxide, the presence of sodium, potassium and cesium nitrate, as well as potassium bromide, chloride and iodide, does not affect the shift of the isoelectric point, unlike lithium nitrate whose presence, in the range concentration 0.01 to 1 mol dm^{-3}, leads to a change in pH_{iep}.

The electrokinetic potential is not measured directly but is calculated from theoretical models based on the results obtained by measuring the electrical mobility of particles, which is based on different electrokinetic phenomena (Delgado et al. 2007). The technique for measuring the mobility of particles is chosen depending on the size of the particles and their electrical properties. Electrophoresis is most often used for small particles of negligible conductivity, while streaming potential, electroosmosis or sedimentation potential are used for larger ones.

As already emphasized, the most used methods for determining particle mobility are those based on electrophoresis. Electrophoresis is the movement of dispersed particles in an electric field. The charged particles together with the compact layer move towards the oppositely charged electrode, while the outer diffuse layer (outside the slip plane) remains stationary. The electrophoretic mobility of a particle, μ, is defined as,

$$\mu = \frac{v}{E} \tag{29}$$

where v is the electrophoretic velocity of the particle relative to the medium, and E is the electric field strength. There are several methods for determining electrophoretic mobility. One of them is the direct measurement of the velocity of individual particles in an electric field of known strength under a microscope (probably the most widespread method until the 1980s), and the second is the measurement of the velocity of movement of the boundary of the suspension in the electric field in the U-tube. Finally, a method has been developed that measures the velocity of particles using the Doppler effect, and this is the basis of the electrophoretic light scattering (ELS) measurements. The method is based on the analysis of the (laser) light scattered by moving particles (Uzgiris 1972; Ware and Flygare 1972; Malher et al. 1982a,b; Delgado et al. 2005). The dependence of the Doppler frequency shift on the electrophoretic mobility is shown by the expression

$$\nu_d = \frac{\mu E n}{\lambda} \sin \vartheta \tag{30}$$

where ν_d is the Doppler shift, n is the refractive index of the medium, λ is the wavelength of the incident light, and ϑ is the viewing angle. A term that relates electrophoretic mobility and electrokinetic potential

$$\mu = \frac{\varepsilon_r \varepsilon_0 \zeta}{\eta} \tag{31}$$

where η is the viscosity, ε_r the relative permittivity of the solvent, and ε_0 permittivity of vacuum, was first proposed by Marian Smoluchowski, and therefore this expression is known as the "Smoluchowski equation". It is obvious that the velocity of the particles depends on the charge, i.e., on the electrokinetic potential, which is proportional to the electrophoretic mobility. Equation (31) is valid when the term κr is sufficiently large with $\kappa = 1/l_D$ being the reciprocal Debye length and r the particle radius. In the case when $\kappa r < 1$, the Hückel–Onsager equation applies:

$$\mu = \frac{2}{3} \frac{\varepsilon_r \varepsilon_0 \zeta}{\eta} \tag{32}$$

For the transition range between low and high κr, Henry's formula can be applied

$$\mu = f(\kappa r) \frac{2\varepsilon_r \varepsilon_0 \zeta}{3\eta} \tag{33}$$

where $f(\kappa r)$ is the Henry coefficient. The function f varies from 1.0, for low values of κr, to 3/2 as κr approaches infinity.

Using the Smoluchowski theory one can in principle derive zeta potential from the electrophoretic mobility with the implicit assumption that the particles are perfectly rigid. However, in the regime of soft particles, it is necessary to apply corrections, i.e., models which account for porous nature of such particles. In that respect, Ohshima model (Ohshima 2007, 2012) should be mentioned. In that model electrophoretic mobility shows dependence on several additional parameters such as charge density of the polyelectrolyte layer and the value $1/\lambda$ known as the electrophoretic softness (Cametti 2011; Brkljača et al. 2018a).

Another commonly used method for the determination of the electrokinetic potential is based on the measurement of the streaming potential. The streaming potential is the electrical potential that is created, at zero electric current, by the flow of liquid under a pressure gradient through a capillary, plug, diaphragm, or membrane. If the surface of the chamber is charged, an electric double layer is created, and there is a local increase in the concentration of counterions and a local decrease in the concentration of coions. The electric double layer that is created is of the order of magnitude of the Debye length, which depends on the ionic strength. By applying a mechanical force, there is a charge movement in the double layer, which results in an electric current, which is also called a streaming current, and the resulting potential difference between the two reference electrodes at the beginning and at the end of the chamber is called the streaming potential. Such a streaming potential changes at different pressures and can be directly related to the ζ-potential, which is described by the following equation:

$$\zeta = \frac{\Delta U}{\Delta p} \cdot \frac{\eta \gamma}{\varepsilon_r \varepsilon_0} \tag{34}$$

In this equation, ΔU denotes the streaming potential, Δp the change in pressure, and the symbol γ in this case denotes the electrical conductivity of the solution.

A CASE: INTERFACIAL POTENTIALS AT THE SiO_2/AQUEOUS ELECTROLYTE INTERFACE

In previous chapters, we described different experimental methods used for determination of the interfacial potentials. Unfortunately, the interfacial potentials cannot be measured directly. They are obtained indirectly by measuring the physical properties that depend on them. Numerous studies have dealt with the effect of solution composition (type of solvent, pH, background electrolyte, ionic strength, etc.) on solid/aqueous electrolyte interfaces. To compare the data obtained by different experimental techniques and interpret the results, it is necessary to find one system that has been tested by different experimental methods. Although it would be better to examine the same sample by different techniques, it is not possible for several reasons. Established laboratories use a limited number of specific techniques and additionally, specific experimental techniques may require the use of different types of samples i.e., powders, suspensions, or flat surfaces. It is well known that the properties of the samples, especially colloid particles, differ and depend on the purity, the method of preparation, the porosity of the solid, the shape, the orientation of the crystal faces exposed to the aqueous solution, the composition of the aqueous solution, and in general the experimental conditions and protocols. For this purpose, we searched the available literature and found published data of interfacial potential values obtained for the silicon dioxide / aqueous electrolyte solution interface. Silica is an often-studied system, present in the environment and important for technological processes. Entire books (Iler 1979; Bergna 1994) and reviews (Bañuelos et al. 2023) are devoted to chemistry of silica and their properties. Silicon dioxide samples exist in variety of forms, such as crystalline quartz, or amorphous fused silica. The amorphous silica sample consists of small particles or porous aggregates (Iler 1979). Silica surface can be hydrophilic, with silanol groups (Si–OH) exposed at the surface, or hydrophobic with siloxane groups (Si–O–Si groups) exposed at the surface. The transition from one type of surface to another is reversible and occurs at high temperature by releasing the water molecules from the surface. By exposure of silica surface to water environment the reaction with water continues which leads to formation of silica gel (Van Roosmalen and Mol 1978; Iler 1979).

The silica/aqueous solution interface is pH-dependent due to the reaction with potential determining ions; i.e., H^+ and OH^- ions. The reactions of protonation and deprotonation of active surface groups regulate the properties of the electrical interfacial layer. The extent of the surface protonation and deprotonation depends on pH, the composition of the aqueous electrolyte solution but also on the structure of the silica surface. The crystallographic structures of silica samples are three-dimensional arrays of linked tetrahedrons, each consisting of a silicon atom coordinated by four oxygen atoms (Brown et al. 2014). The surface concentration of silanol groups on the silica is express as silanol number, the number of ≡SiOH groups per square nanometres. Zhdanov and Kiselev (1957) determined, by a deuterium exchange method, the silanol number of dehydrated but fully hydroxylated amorphous silica. They found out that the silanol number is 4.6 nm^{-2} and independent of origin of samples and structural characteristics of the silica. The number of ≡SiOH groups decreases with increasing temperature when the sample is heated under vacuum.

In aqueous solution, the silica surface is charged due to the deprotonation of surface silanol groups and back reaction of protonation of negatively charged $\equiv SiO^-$ groups:

$$\equiv SiOH \rightleftarrows \equiv SiO^- + H^+(aq);\ pK \quad (35)$$

A broad range of pK values of surface protonation (Eqn. 35) has been reported. Schindler and Kamber (1968) determined the intrinsic acidity constant pK = 6.8 of surface silanol groups by coulometric titration of silica gel. Sahai and Sverjensky (1997) predicted pK values for different types of silica particles from experimental titration data. The predicted pK value was 7.7. Sonnefeld (1995) estimated acidity equilibrium constant for silica particles in alkali chloride solution. He found that the value of apparent acidity constant depends on type of the background electrolyte (pK values from 6.44 for lithium chloride to 6.06 for cesium chloride). The alkali sequence is reflected in different values of the Stern Layer thickness, which is correlated with the radii of solvated ions.

The point of zero charge and the isoelectric point of silica are found to be in the range between pH = 2 and pH = 4 (Kosmulski 2023). The silica surface is negatively charged in the broad pH region, which prevents the particles from aggregation and keeps the suspension stable. Surface charge density of silica particles in different background electrolyte solutions was determined by potentiometric acid–base titrations (Tadros and Lyklema 1968; Sonnefeld 1995; Karlsson et al. 2001; Dove and Craven 2005; Goel and Lützenkirchen 2022).

Surface charge density of silica particles differs in the various electrolyte solutions containing the same concentrations of cations. For alkali cations the trend of $Cs^+ > K^+ > Na^+ > Li^+$ in the development of surface charge on silica was observed (Tadros and Lyklema 1968; Dove and Craven 2005) and Karlsson et al. (2001) report the correlation of surface charge density with the hydration size of the ion. While monovalent alkali cations show lyotropic behaviour, the divalent alkali earth cations exhibit a reversed lyotropic trend promoting the negative surface charge of silica in the order $Ba^{2+} < Sr^{2+} < Ca^{2+} < Mg^{2+}$(Dove and Craven 2005). The increasing of surface charge density of silica with decreasing cation size was confirmed by second harmonic generation measurements (Boamah et al. 2019). Sum-frequency generation techniques reveal the role of surface water structure and hydrogen-bonding to ion-specific effects (Rehl and Gibbs 2021). Due to the increasing of salt concentration, the restructuring of water takes place, leading to a decrease in order at the surface.

In Table 1 we have selected several published results obtained for different types of flat silica surfaces and for the SiO_2 particles. It is challenging to compare all these data collected for different types of silica surface, different methods and at different experimental conditions. For most of these data some physical properties related to the potential were measured, and many values of different interfacial potentials were calculated using various assumptions. In most cases, the measured signal is relative, and the absolute value is obtained by knowing some reference point, for example the point of zero potential. It is often assumed that the value of the point of zero potential is equal to the isoelectric point or the point of zero charge, which may not be true for the asymmetric association of anions and cations, or at high ionic strengths. In the following chapters we have compared these selected results.

Measurements by single crystal electrodes and OCP technique

The principle of determination of the inner surface potential by means of single crystal electrodes, measuring the open-circuit potential, is exactly the same as that of pH-measurement with a glass electrode. Both methods use the same set-up, pH-meter as a measuring device, a reference electrode and a working electrode (glass electrode or SCr electrode). The results of OCP measurements for SCr quartz (0001), SCr fused silica and glass electrode, with respect to the reference electrode are presented in Figure 8. Under equilibrium conditions (without applied external voltage and when electrical current decreases towards zero) the internal electrical potential within the solid phase is assumed to be constant. The resistance of the

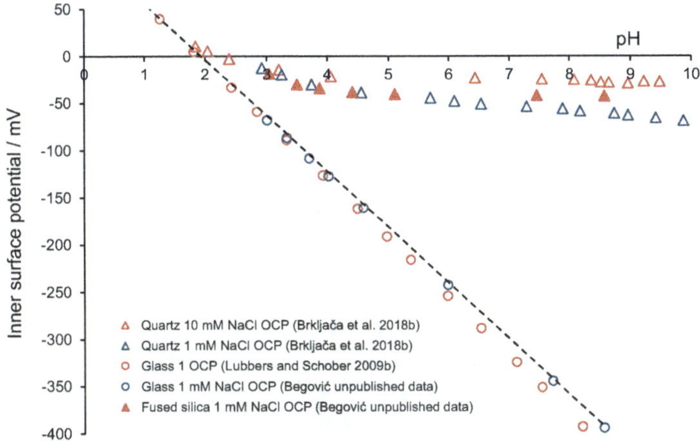

Figure 8. Inner surface potential of various SiO_2 flat surfaces: glass (**red circles**) (Lubbers and Schober 2009) (**blue circles**) (Begović unpublished data); fused silica in 1 mM NaCl (**solid red triangles**) (Begović unpublished data) and quartz (0001) plane in 1 mM (**open red triangles**) and 10 mM (**open blue triangles**) NaCl (Brkljača et al. 2018b) evaluated from OCP signal at 25 °C. Line(- - -): Ψ_0(pH) function with Nernstian slope.

quartz and fused silica single crystals are comparable to the resistance of glass used for the glass bubble of the glass electrode (Kallay et al. 2014; Preočanin et al. 2017a). The measured values of inner surface potential for the quartz (0001) crystal plane (Brkljača et al. 2018b) are significantly lower than the data measured for the glass with the glass electrode.

The first thing that can be noticed in Figure 8 is that the inner surface potential of the glass differs the most. The inner surface potential of glass was determined by measuring the open-circuit potential of the glass electrode with respect to the reference electrode. The glass electrode shows a linear dependence of pH following the Nernst equation. According to the Nernst equation, the slope of Ψ_0(pH) function depends on temperature ($-\alpha RT\ln 10/F$). It is expected that the slope of the glass electrode is close to the Nernstian slope $\alpha \approx 1$ (see data from Lubbers and Schober 2009), but often the experimentally determined slope is lower (unpublished data Begović), which is acceptable if $\alpha \geq 0.9$. The reason for the lower value of measured slope could be the contamination or drying up the hydrated gel of glass surfaces. The good pH-sensitivity is typical for a glass electrode and is used for common pH measurement.

Unlike for a glass electrode, the slopes of the Ψ_0(pH) function for quartz (0001) and fused silica are lower than the Nernstian slope. The reason for differences lies in the structure of the SiO_2. Glass consists of a disordered network of silicon and oxygen atoms with small amount of metal cations. In the contact with water or aqueous solution a glass surface can be considered as a layer of the partially ionized oxygen atoms; i.e., the amphoteric ≡SiOH surface groups (Oldham and Myland 1994). The highly hydrated glass surface forms a porous gel layer, highly sensitive to the concentration of hydrogen ions due to the extremely high number of ≡SiOH groups. The inner surface potential of quartz depends on the specific arrangement of silicon and oxygen atoms at the quartz (0001) surface, the formation of silanol groups and their degree of protonation. Deviation from the Nernstian slope yields information on the ratio of surface concentrations of (more) positively to (more) negatively charged surface groups, Equation (11). The ratio of free (more) positive and free (more) negative surface groups, causes a reduction of the slope and deviation from linearity. The surface concentration (the number per surface area) of silanol groups at quartz is significantly lower than the surface concentration (the number per surface area) of silanol groups in the gel phase at glass, which explains the difference between the response of a glass and a quartz electrode. The measured reduction of inner surface potential

Table 1. Selected interfacial potential measurements of different silica flat surfaces and particles.

Type of SiO_2	Type of measurement		Experimental conditions	Reference/figure
Si/SiO_2 wafer (thermal oxidation silicon)	Electrochemically based methods	C–V (EIS)	$1 < pH < 5$ 1000 mmol dm^{-3} $NaNO_3$ 100 mmol dm^{-3} $NaNO_3$ 10 mmol dm^{-3} Na NO_3	Bousse et al. (1983) Figure 11
Si/SiO_2 wafer (pyrogenic silicon layer)	Electrochemically based methods	C–V (EOS)	$1 < pH < 10$	Siu and Cobbold (1979) Figure 11
Si/SiO_2 wafer (thermal oxidation silicon)	Electrochemically based methods	ISFET C–V (EOS)	$3 < pH < 10$	de Rooij and Bergveld (1980) Figure 11
Si/SiO_2 wafer	Electrochemically based methods	ISFET	$2 < pH < 12$ 100 mmol dm^{-3} tetrabutylammonium chloride solution	Bergveld 2003) Figure 11
Si 20 nm SiO_2 layer 100 nm Si_3N_4 layer	Electrochemically based methods	ISFET	$3 < pH < 9$ 100 mmol dm^{-3} NaCl and added buffer	Bousse et al. (1992) Figure 11
Si/SiO_2 wafer (thermal oxidation of silicon)	Electrochemically based methods	EOS	$1 < pH < 11$ 100 mmol dm^{-3} NaCl and added buffer	Diot et al. (1985) Figure 11
glass	Electrochemically based methods	OCP	$2 < pH < 14$	Lubbers and Schober 2009) Figure 8
glass	Electrochemically based methods	OCP	$3 < pH < 11$ 1 mmol dm^{-3} NaCl	Begović Unpublished data Figure 8
quartz (0001)	Electrochemically based methods	OCP	$2 < pH < 10$ 10 mmol dm^{-3} NaCl 1 mmol dm^{-3} NaCl	Brkljača et al. (2018b) Figure 8
fused silica	Electrochemically based methods	OCP	$3 < pH < 11$ 1 mmol dm^{-3} NaCl	Begović Unpublished data Figure 8
fused silica	Optical	SHS	$2 < pH < 11$ 500 mmol dm^{-3} NaCl	Ong et al. (1992) Figure 12
SiO_2 300 nm particles	Optical	SHS	pH = 5.7, 10 and 11 addition of NaOH	Marchioro et al. (2019) Figure 12

Table 1 (cont´d). Selected interfacial potential measurements of different silica flat surfaces and particles.

Type of SiO_2	Type of measurement		Experimental conditions	Reference/figure
quartz (0001)	Electrokinetic	Streaming potential	2 < pH <10 1 mmol dm^{-3} NaCl	Brkljača et al. (2018b) Figure 9
glass			3 < pH < 10 adjusted by adding standard dilute solutions of either HCl or NaOH into pure water	Gu and Li (2000) Figure 9
Si/SiO_2 (20 nm layer) and Si/Si_3N_4 (100 nm layer)			3 < pH < 9 100 mmol dm^{-3} NaCl and added buffer	Bousse et al. (1992) Figure 9
fused silica			3 < pH < 9 1 mmol dm^{-3} $NaNO_3$	Hartley et al. (1997) Figures 9, 10
silanated glass			2 < pH < 9 10 mmol dm^{-3} KCl 1 mmol dm^{-3} KCl 0.1 mmol dm^{-3} KCl	Lützenkirchen et al. (2010a) Figure 9
SiO_2 300 nm particles		Electrophoresis	pH = 5.7, 10 and 11 addition of NaOH	Marchioro et al. (2019) Figure 12
SiO_2 200 nm Geltech			3 < pH < 10 unknown electrolyte	Nagel et al. (2017) Figure 9
SiO_2 colloid particles 4–5 µm			4 < pH < 9 0.1 mmol dm^{-3} KNO_3 1 mmol dm^{-3} KNO_3	Larson et al. (1997) Figure 10
SiO_2 9 nm particles	XPS		3.5 < pH <10 50 mmol dm^{-3} NaCl	Brown et al. (2016b) Figure 11
SiO_2 colloid particles 4–5 µm	AFM		4 < pH < 9 0.1 mmol dm^{-3} KNO_3 1 mmol dm^{-3} KNO_3	Larson et al. (1997) Figure 10
fused silica flats and SiO_2 colloid particles 4-6 µm	AFM		4 < pH < 9 0.1 mmol dm^{-3} KNO_3 1 mmol dm^{-3} $NaNO_3$	Hartley et al. (1997) Figure 10

of (0001) quartz due to increasing of ionic strength (Fig. 8) is a consequence of screening of the surface by ions and reduction of electrostatic interactions. Interestingly, the inner surface potential of the amorphous, fused silica, is similar to the inner surface potential for a well-defined quartz (0001) surface. The structure of the fused silica is like the structure of the glass with disordered network of silicon and oxygen atoms (Oldham and Myland 1994). However, the surface of fused silica has not been hydrated prior to the measurement, and it could be assumed that the inner surface potential is also determined by a limited number of charged surface groups like in the case of the well-defined quartz (0001) surface.

An additional benefit of determining the inner surface potential using single crystal electrode and OCP technique is the possibility of determination of the inner surface potential of an individual crystal plane (Kallay et al. 2014). The surface of the real crystalline particles consists of different crystal planes which in contact with aqueous environment exhibit different surface properties. Planes of the same particle are electrically connected through the bulk of crystal. The value of common inner surface potential lies between values of separated individual planes and one plane may dominate. These types of measurements open possibilities for linking surface properties of individual crystal planes (flat surfaces) with average surface properties of colloid and nanoparticles.

Electrokinetic and AFM measurements

Streaming potential and streaming current are electrokinetic phenomena caused by the flow of the aqueous electrolyte solution through a microchannel with electrically charged solid walls. The electrical interfacial layer at the solid–liquid interface results in a net electrical current, which is measured by electrodes, located at the ends of the microchannel. The measured value of streaming potential difference or streaming current is related to the ζ-potential. The ζ-potential of quartz (0001) plane (Brkljača et al. 2018a), fused silica (Hartley et al. 1997), Si/SiO_2 wafer (Bousse et al. 1992), glass (Gu and Li 2000), and silanated glass (Lützenkirchen et al. 2010b) as a function of pH is presented in Figure 9. Two important observations emerge from the analysis of these results. First, the zeta potential of all studied SiO_2 surfaces is negative over almost all investigated pH-ranges. The isoelectric points were found to be in expected region, between pH = 2 and pH = 4 (Kosmulski 2023). Second, electrokinetic potentials measured for the different flat surfaces of silica in 1 mM NaCl, KCl, or $NaNO_3$ aqueous solution are comparable. The only exception is obtained for the Si/SiO_2 wafer. In the case of the Si/SiO_2 wafer, absolute values of ζ-potential are significantly lower, but in those measurements the pH was adjusted and kept constant using a buffer. In buffer solution, the ions concentration is high, which increases the ionic strength of the solution. Moreover, ions can adsorb at the silica surface which affects the state within the interfacial layer.

The fact that the zeta potential for different types of flat silica surfaces (Fig. 9) differs less than the inner surface potential of corresponding silica surfaces (Fig. 8) indicates that ions within the EIL have a different impact on those two potentials. While for the inner surface potential, surface reactions and concentrations of charged surface groups ($\equiv SiO^-$ and $\equiv SiOH$) play a decisive role, the electrokinetic potential is affected by the state and ions in the diffuse part of the EIL. In addition, it can be observed that the absolute values of the electrokinetic potentials (Fig. 9) are higher than the absolute value of the inner surface potential measured by means of OCP methods (Fig. 8). Therefore, the expected decrease of the absolute values of the potential due to the increase in distance from the surface; i.e., $|\zeta| < |\Psi_0|$, was not observed. This finding may raise doubts about the reliability of the applied experimental techniques but also raises a question about the procedures used for the evaluation of the inner surface or zeta potential from the corresponding measured signal. The inner surface potential, i.e., the potential at the 0-plane, can be influenced by adsorbed counterions. The measured values of the inner surface potential provide essential information whether the species are bound on a metal oxide interface as partially dehydrated and directly bounded inner sphere complexes or

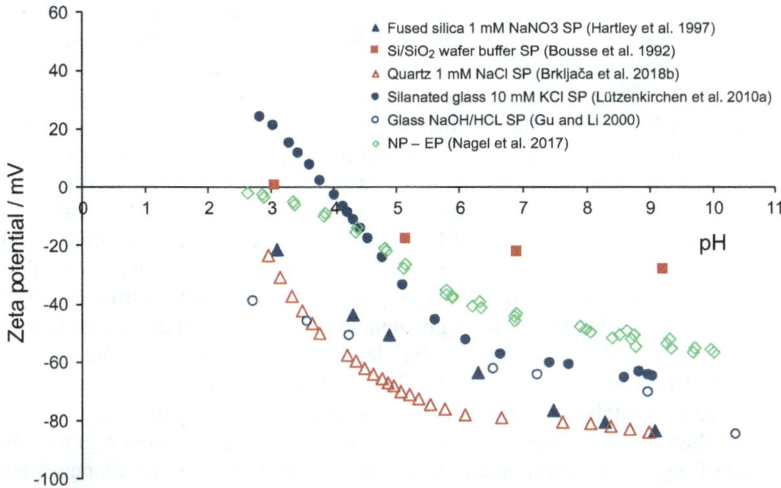

Figure 9. Electrokinetic potential of various SiO_2 flat surfaces: glass (**open blue circles**; Gu and Li 2000); silanated glass (**filled blue circles**; Lützenkirchen et al. 2010a); fused silica in 1 mM $NaNO_3$ (**blue triangles**; Hartley et al. 1997), quartz (0001) plane in 1 mM NaCl (**red triangles**; Brkljača et al. 2018b), and Si/SiO_2 wafer (**filled red squares**; Bousse et al. 1992) evaluated streaming potential measurements at 25 °C. The **open green diamonds** represent the electrokinetic potential of SiO_2 particles (Nagel et al. 2017).

via electrostatic forces as hydrated outer sphere surface complexes. The inner surface potential could be sensitive to the orientation of water dipoles within the electrical interfacial layer (Brkljača et al. 2018a; Gonella et al. 2021). On the other hand, the electrokinetic surface potential provides information on the sign of the net surface charge, the isoelectric point, and the adsorption of charged species within the entire electrical interfacial layer. The significance of the ζ-potential lies in the fact that its value is close to the potential at the onset of the diffuse layer. Generally, the measured electrokinetic potential is not sensitive to the impact of water dipoles on the ion distribution within the EIL (Bonthuis et al. 2011). In the comprehensive review, Banuelos et al. (2023) presented and analyzed the simultaneous measurements of different interfacial potentials of SiO_2 and other metal oxide surfaces, obtained from different experimental techniques. While some surfaces show expected behavior with measured electrokinetic potential being lower in magnitude than inner surface potential, some surfaces show the opposite trend which cannot be explained by classical EIL models.

Besides different types of SiO_2 samples, it is also interesting to compare the electrokinetic potentials of flat SiO_2 surfaces and SiO_2 nanoparticles. Electrokinetic zeta potentials for silica particles (Nagel 2017) are generally lower than the electrokinetic potential obtained for flat surfaces (see Fig. 9). The increase in the absolute value of the electrophoretic mobility and corresponding electrokinetic potential with increasing particle size has already been determined experimentally (Kędra-Królik et al. 2017). The surface charge density of nanoparticles is pH-dependent first of all due to protolytic reactions (Abbas et al. 2008; Holmberg et al. 2013; Barisik et al. 2014). In addition, it can vary with particle size due to the ratio of the electrical double layer thickness and particle diameter.

But unlike the electrokinetic potential, for certain pH and ionic strength, the magnitude of surface charge decreases with particle size and reaches a constant value at critical particle size. Knowing the electrokinetic potential and the slip plane separation distance, it is possible to calculate the potential Ψ_d. As a rule, the electrokinetic potential is lower than the potential at the onset of the diffuse layer, and their mutual dependence is derived from the Gouy–Chapman theory. In discussing the relationship between Ψ_d and zeta potential, it should first be emphasized that some authors (Hunter and Wright 1971; Sprycha 1984; Sprycha and Szczypa 1984) assume that the distance between the onset of the diffuse layer and the slip plane is so small that the potential difference between the two abovementioned potentials can be ignored ($\zeta = \Psi_d$). As the potential obtained from the AFM experiments is often related to the potential on the onset of the diffuse layer, it is useful to compare the results of the AFM and electrokinetic measurements in justification of the Ψ_d and ζ-potential equalization. Figure 10 shows one such comparison for symmetric (silica particle/silica plate system) and asymmetric (silica particle/alumina plate system) AFM experiments. In both cases, the agreement between the electrophoresis results and the colloidal probe AFM results is very good. From these findings, one can conclude that the diffuse layer potentials derived from fitting the AFM forces-separation curves with DLVO theory are equivalent to ζ-potentials and that the distance between the d-plane and the slip plane could be negligible. On the other hand, various authors (Healy and White 1978; Chow and Takamura 1988; Torres et al. 1988; Hesleitner et al. 1991) found that this distance is enough large that neglecting it could cause a significant error, especially in systems with high ionic strength and high interfacial potential. Moreover, in the case where the adsorption of organic acids on metal oxides was considered using simultaneous adsorption and electrokinetic measurements (Kovačević et al. 1998), the best results were obtained for the assumed slip plane separation distance of (15 ± 5) Å. From the results published so far, it is therefore possible to conclude that the distance between the onset of the diffuse layer and the slip plane shouldn't be ignored and that it is between 5 and 20 Å.

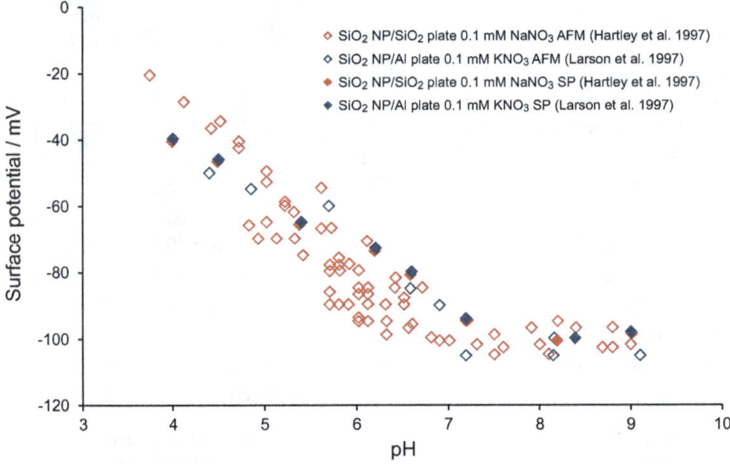

Figure 10. Comparison of the AFM-measured diffuse layer potentials (**open red and blue diamonds**) and the independently measured electrokinetic potentials (**filled red and blue diamonds**) for the silica particles as a function of pH in 0.1 mmol dm^{-3} $NaNO_3$ (Hartley et al. 1997) and 0.1 mmol dm^{-3} KNO_3 (Larson et al. 1997) solution. The diffuse layer potentials were derived from fitting the AFM force-separation curves with DLVO theory for silica particle/silica plate system (Hartley et al. 1997) and silica particle/alumina plate system (Larson et al. 1997).

Measurements by electrochemically based techniques and XPS

Figure 11 shows the results of the inner surface potentials for the Si/SiO_2 system measured by electrochemical methods. It is considered that these methods determine the surface potential on the 0-plane, that is, the inner surface potential. Measurement with ISFET sensors is a more direct method in which the gate potential, which is related to the inner surface potential, is measured. In C–V method, to determine the flat band potential, which is related to the inner surface potential, it is necessary to perform impedance measurements, evaluate the capacitance using appropriate approximation and produce the capacitance–voltage curves. In both above mentioned methods external voltage during measurements is applied, in ISFET devices the direct current (DC), and in C–V method alternating current (AC). Note that in the OCP method with single crystal electrodes no external voltage is applied, but the equilibrium voltage is measured when the current flowing through the electrodes tends to zero. In general, the values of the inner surface potentials measured with the OCP method are lower than the inner surface potentials determined by the ISFET and C–V methods. The difference can occur precisely because of the applied external field. The external electric field affects the state in the surface layer of SiO_2 and thus the state in the interfacial layer. Additional measurements at an applied potential equal to the OCP potential would give different values of the inner surface potential, closer to the values determined in the OCP measurements.

An additional reason for the difference in results may be the type of SiO_2 surface that was investigated. While single crystals (quartz or fused silica) were used in OCP measurements, in other electrochemical measurements (ISFET and C–V), a layer of SiO_2 was formed on a silicon wafer. The surface of the SiO_2 film is more like the surface of particles, has a higher specific surface area, and higher surface roughness.

Furthermore, all presented inner surface potentials in Figure 11 measured by the ISFET probe and the C–V method are comparable except for the results from Bousse et al. (1992). As for corresponding electrokinetic potential for the same particles these results were measured in a solution in which buffer was added to keep the pH constant. Buffers affect the state in the interfacial layer, they can bind to the surface and change the surface potential.

Brown et al. (2016a) utilized the X-ray photoelectron spectroscopy from a liquid microjet to measure the interfacial potential of 9 nm colloidal silica nanoparticles dispersed in different aqueous electrolyte solutions. Data are shown in Figure 11, together with the inner surface potential values obtained by electrochemically based techniques. Authors emphasized that

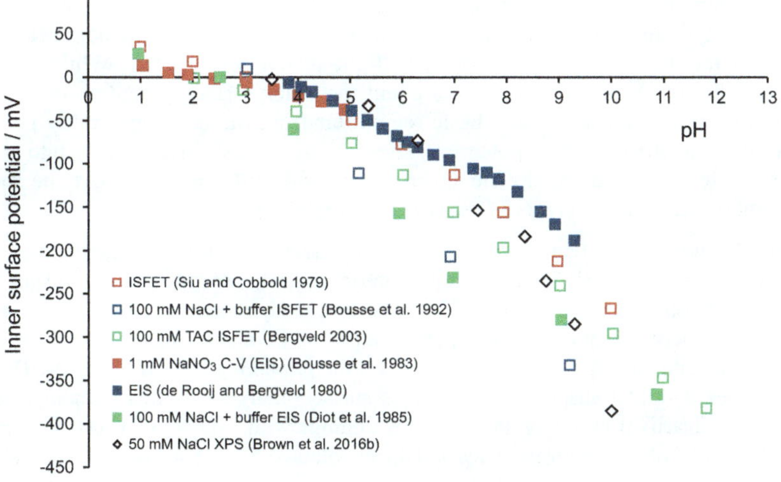

Figure 11. Inner surface potential of Si/SiO_2 wafer determined by the electrochemically based technique: ISFET (**open squares**), C–V (**filled squares**) and SiO_2 nanoparticles measured by XPS (**open diamond**) at 25 °C.

XPS results for silica nanoparticles (Brown et al. 2016a) are consistent with results obtained for flat silica surfaces by means of electrochemically based methods (ISFET, C–V and EOS), as shown in Figure 11, even though these two methods use different samples, flat SiO_2 on Si wafer and silica nanoparticles, respectively. We already highlighted that the inner surface potentials obtained by electrochemically based methods are much higher than the surface potentials evaluated by means of the OCP method (equilibrium conditions when the current flowing tends to zero). This means that interfacial potential determined by XPS is significantly higher than potentials determined by OCP method. The interfacial potential values obtained by XPS measurements cannot be directly compared to the electrokinetic potential values (Fig. 8) since the ζ-potential values do not represent true surface potential values but instead the values of potential at the slip plane. Uncertainty in the location of the slip plane with respect to the particles' interface makes extrapolating Ψ_0 from ζ-potential difficult (Butt et al. 2003). As can be compared, the $|\zeta\text{-potential}|$ values are lower than $|\Psi_0|$ values due to the charge screening by the electrolyte ions in the EIL (Brown et al. 2013a).

On the other hand, the surface potential values obtained by XPS method and electrochemically based methods are comparable (Fig. 11). The change in photoelectron energies and consequently inner surface potential values due to the solution pH variation during XPS measurements with microjet liquid can be considered having analogous effect as applying an external DC potential to a silica substrate (Brown et al. 2013a). This causes energy levels shift arising from the (de)acceleration of the outgoing photoelectron due to the electric field created by the external potential at the sample interface by establishing a surface potential relative to the reference state of the grounded analyzer (Ulgut and Suzer 2003). The photoelectron energies are altered shifting energy levels downward (with positive potential slowing down the outgoing photoelectron, effectively reducing its kinetic energy and increasing the apparent binding energy) or upward (with a negative applied potential) (Brown et al. 2016a). The obtained inner surface potential values are comparable since XPS measurements follow the effect of the external applied potential influence as well as the electrochemically based methods (ISFET, C–V EOS).

Measurements by non-linear optical methods

As mentioned earlier, SHG and SFG signals depend on the second-order nonlinear susceptibility $\chi^{(2)}$ and the third-order nonlinear susceptibility $\chi^{(3)}$ which describe how surfaces and interfaces respond to nonlinear light–matter interactions. While the second-order susceptibility $\chi^{(2)}$ gives information about the inner surface potential, the third-order susceptibility $\chi^{(3)}$ in very acidic and basic region, far from the electroneutrality point, gives information about diffuse layer potential. These methods also give insight into the water structure within the electrical double layer and provide information about the molecular arrangement of the first few layers of water molecules at the surface.

Figure 12 shows the potentials extracted from the SHG at SiO_2 fused silica / water interface for a planar surface (Ong et al. 1992) and polarimetric Angle-Resolved SHG for 300 nm colloidal particles in solution (Marchioro et al. 2019). The first study proposes a two-site model at the planar surface, where one of the silanol sites has a pK_a of 4.5 and occupies 19% of the sites and the other silanol site has a pK_a of 8.5 and occupies the remaining 81% of the sites. The second study examines colloidal suspension, computes surface charge densities without assuming a specific charge distribution model and validates findings with MD simulations. The difference between the results obtained for particles and those obtained for flat surfaces is noticeable.

The two sets of three points measured by Marchioro et al. (2019) represent the inner surface potentials and electrokinetic potential, respectively, of the same silica particles. Both potentials are negative, electrokinetic potentials almost pH independent, and measured values of inner surface potential vary as a function of pH. At pH 6 and 10, $\zeta \approx \Psi_0$, $\chi^{(2)}$is negative,

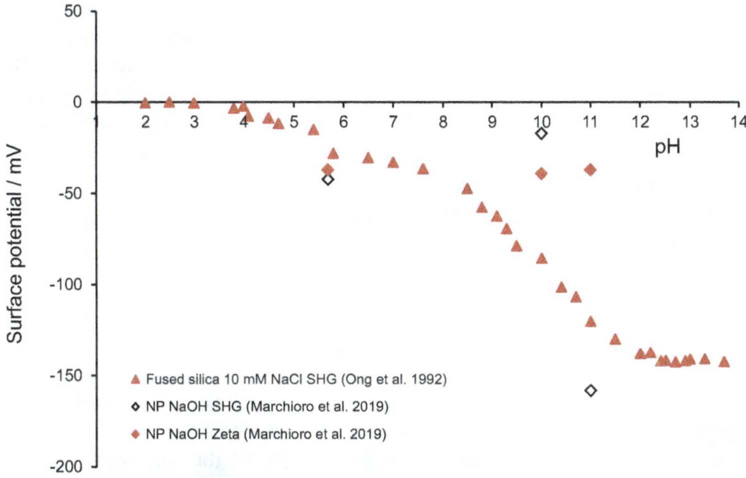

Figure 12. Interfacial potential versus pH in the bulk solution for a planar silica surface (**red triangles**) calculated from SHG measurements (Ong et al. 1992) and 300 nm colloidal silica particle suspension (Marchioro et al. 2019) calculated from Angle-Resolved SHG measurements (**open diamonds**) and streaming potential measurements (**filled diamonds**)

corresponding to water hydrogen atoms oriented away from the surface, and oxygen atoms facing the surface. At pH 11, $\zeta \gg \Psi_0$, $\chi^{(2)}$ is positive, corresponding to water oxygen atoms oriented away from the surface, and hydrogen atoms facing the surface.

The values of inner surface potentials evaluated for the planar surface (Ong et al. 1992) are the most comparable to the results obtained by measuring OCP with single-crystal electrodes of the quartz (0001) plane (Brkljača et al. 2018b) and fused silica (Begović unpublished data), Figure 8. It can be concluded that this type of measurement, like the OCP method, provides data on the part of the electrical interfacial layer at the solid silica surface in contact with an aqueous electrolyte solution; i.e., on the inner surface potential. More recent measurements with evaluated surface potential values, for silica, have unfortunately not been found.

An interesting study was recently published by Parshotam et al. (2022). They analyzed the fused silica surface by utilizing vibrational sum frequency generation spectroscopy (VSFG) in combination with the maximum entropy method of isotopically diluted water and electrokinetic measurements and elucidated the spectral response of vibrational coupling in different parts of the electric double layer. From the inner layer spectra, at pH = 2, the presence of intermolecular coupling, was related to the double hydrogen bonds between the interfacial water molecules.

Recently, a set-up for simultaneous measurement of SHG and streaming potential (SP) at mineral-water interfaces was presented (Lützenkirchen et al. 2018). The set-up directly yields SHG data and streaming potential for the same mineral–water interface. The optical geometry was applied in the same configuration of TIR-SFG experiments where two incident beams, however with the same frequency in case of SHG, and one generated beam propagate through the sample crystal to allow for better separation between the background accompanying the fundamental beam and the signal from the probed interface. Unlike the case where the beams propagate in the liquid phase (Fitts et al. 2005) and hence the SHG data mimic the zeta potential vs. pH (Lützenkirchen et al. 2010b), the propagation of the beams through the crystal produce data that are clearly reminiscent of the expected inner surface potential vs pH curve (Lützenkirchen et al. 2010b; Preočanin et al. 2017a). It was found that the zeta potential of either hydrophobic surface or hydrophilic surface is proportional to the respective second harmonic generation signal in the vicinity of the isoelectric point. The proportionality occurred over the full pH range when SSP (S-polarized SHG output, S-polarized beam 1 input, and P-polarized beam 2 input) polarization combination was used, while diverged from the linear to plateau when PPP (P-polarized SHG output, P-polarized beam 1 input, and P-polarized beam 2 input) polarization combination was used. The authors concluded that the SSP signal relates to zeta potentials, while the PPP might rather represent inner surface potential, since typical inner surface potential would further increase with decreasing pH. However, it was not very clear if the anomalous PPP behavior was due to sensitivity of the corresponding surface nonlinear susceptibility tensor, $\chi^{(2)}$, to the fluctuations in the beam incident angle. Ultimately, the authors proposed that more effort be made to quantify aspects of the different components of the nonlinear susceptibility tensor.

CONCLUSIONS

When a mineral is exposed to aqueous solutions, charged surface groups and an electrical interfacial layer are formed. One of the quantities that describes the electrical interfacial layer is the electrostatic potential, an electrical property of the interface that arises due to the interaction between species in the liquid phase and atoms at the mineral solid surface. It is beyond doubt one of the fundamental parameters which describes colloidal systems and interfaces and due to its importance, it has been the focus of many extensive studies.

We combined and compared the most used techniques for determination of interfacial potentials. None of them is a direct method. Indirectly, some physical property is measured that is related to the interfacial potentials. The EIL is a complex physical system; it consists of two different phases and their interface with an inhomogeneous distribution of substances and properties. The electrostatic potential depends on the distance from the surface; therefore, the evaluated values of potential reflect the principle of the certain experimental method. Each of the techniques described has advantages and disadvantages. Some are simpler, some involve several steps and various approximations. For most of the techniques, a reference point should be assumed, and that is usually the point of zero potential. For the analysis of the results, it is important to consider which system is analyzed, whether it is a flat surface or dispersed particles, what is the particle size, what is the composition of the solution (pH and ionic strength), whether the surface is contaminated, how the surface was pretreated before measurements and so on. However, the availability of various experimental techniques enables the determination of interfacial potentials of various samples, from well-defined flat surfaces to surfaces of colloids and nanoparticles.

- OCP measures the electrode potential between a SCrE and a reference electrode under equilibrium condition and zero current. The measured values reflect the state at the 0-plane which is mostly affected by surface concentrations of charged surface groups and their surface reactions with potential determining ions. The effect of ionic strength is high due to the shading of solid surface with counterions and inner sphere complexes. The influence of outer sphere complexes on the inner surface potential is significantly lower.
- During electrochemically based measurements, the C–V and IFSET probes, the external voltage is applied which can affect the state in the EIL and the value of apparent inner surface potential. The obtained inner surface potentials are a linear function of pH and close to the Nernstian potential. The effect of ionic strength on measured inner surface potential is small. In data processing the capacitance correction of C_{interf} for the C_{H} component, use of a constant phase element, CPE and use of appropriate electric equivalent circuit for impedance data modelling, and the frequency dispersion should be taken into consideration.

- The values of inner surface potential obtained from XPS measurements are related to the binding energy of electrons at the solid surface and they are close to potentials obtained by electrochemical methods. It is not clear whether the strong X-rays affect the state within the EIL and consequently the resulting values of inner surface potential. Despite the great sensitivity of this method the disadvantage is the limited availability of synchrotron facilities.
- Electrokinetic zeta potential is strongly affected by the presence of ions within the diffuse layer and gives information about the total surface charge density of the EIL. It is close to the diffuse layer potential. For the evaluation of electrokinetic potential from the measured electrical mobility of particles different models could be assumed.
- AFM provides information about the interaction force between the AFM tip and the sample surface. To obtain surface charge density and interfacial potential, this force as a function of distance is modeled by DLVO theory. In some cases, to extract the interfacial potential of the sample from force–distance measurements, the surface charge density of the AFM tip needs to be known. One of the main advantages of AFM over other methods is the possibility of mapping potentials at the surface of a heterogenic sample. The interfacial potential obtained from the AFM experiments is often related to the potential on the onset of the diffuse layer.
- The SHG signal is sensitive to the surface, temperature and solution composition changes. It may not represent the entire electrical double layer, and the corresponding interfacial potential (inner surface potential or diffuse layer potential) depends on the order of optical susceptibility.

ACKNOWLEDGMENTS

This work has been supported by Croatian Science Foundation under the POLYMIN2 project (IP-2020-02-9571)

REFERENCES

Abdelmonem A, Zhang Y, Braunschweig B, Glikman D, Rumpel A, Peukert W, Begović T, Liu X, Lützenkirchen J (2022) Adsorption of CTAB on sapphire-c at high pH: Surface and zeta potential measurements combined with sum-frequency and second-harmonic generation. Langmuir 38:3380–3391, https://doi.org/10.1021/acs.langmuir.1c03069

Abe H, Esashi M, Matsuo T (1979) ISFET's using inorganic gate thin films. IEEE Trans Electron Devices 26:1939–1944, https://doi.org/10.1109/t-ed.1979.19799

Assemi S, Nalaskowski J, Miller JD, Johnson WP (2006) Isoelectric point of fluorite by direct force measurements using atomic force microscopy. Langmuir 22:1403–1405, https://doi.org/10.1021/la052806o

Avena MJ, Cámara OR, De Pauli CP (1993) Open circuit potential measurements with Ti/TiO_2 electrodes. Colloids and Surfaces 69:217–228, https://doi.org/10.1016/0166-6622(93)80003-x

Backus EHG, Schaefer J, Bonn M (2021) Probing the mineral–water interface with nonlinear optical spectroscopy. Angew Chemie Int Ed 60:10482–10501, https://doi.org/10.1002/anie.202003085

Bagotsky VS (2005) Fundamentals of Electrochemistry. Wiley,

Bañuelos JL, Borguet E, Brown GE, Cygan RT, DeYoreo JJ, Dove PM, Gaigeot M-P, Geiger FM, Gibbs JM, Grassian VH, Ilgen AG, Jun Y-S, Kabengi N, Katz L, Kubicki JD, Lützenkirchen J, Putnis C V., Remsing RC, Rosso KM, Rother G, Sulpizi M, Villalobos M, Zhang H (2023) Oxide– and silicate–water interfaces and their roles in technology and the environment. Chem Rev 123:6413–6544, https://doi.org/10.1021/acs.chemrev.2c00130

Bergna HE (1994) Colloid chemistry of silica. Adv Chem 16:1–47, https://doi.org/10.1021/ba-1994-0234.ch001

Bergveld P (1970) Development of an ion-sensitive solid-state device for neurophysiological measurements. IEEE Trans Biomed Eng BME-17:70–71, https://doi.org/10.1109/tbme.1970.4502688

Bergveld P (1972) Development, operation, and application of the ion-sensitive field-effect transistor as a tool for electrophysiology. IEEE Trans Biomed Eng BME-19:342–351, https://doi.org/10.1109/tbme.1972.324137

Bergveld P (2003) Thirty years of ISFETOLOGY. Sensors Actuators B Chem 88:1–20, https://doi.org/10.1016/s0925-4005(02)00301-5

Bergveld P, van Hal REG, Eijkel JCT (1995) The remarkable similarity between the acid-base properties of ISFETs and proteins and the consequences for the design of ISFET biosensors. Biosens Bioelectron 10:405–414, https://doi.org/10.1016/0956-5663(95)96887-5

Binnig G, Quate CF, Gerber C (1986) Atomic force microscope. Phys Rev Lett 56:930–933, https://doi.org/10.1103/physrevlett.56.930

Bischoff M, Biriukov D, Předota M, Roke S, Marchioro A (2020) Surface potential and interfacial water order at the amorphous TiO_2 nanoparticle/aqueous interface. J Phys Chem C 124:10961–10974, https://doi.org/10.1021/acs.jpcc.0c01158

Bischoff M, Biriukov D, Předota M, Marchioro A (2021) Second harmonic scattering reveals ion-specific effects at the SiO_2 and TiO_2 nanoparticle/aqueous interface. J Phys Chem C 125:25261–25274, https://doi.org/10.1021/acs.jpcc.1c07191

Bischoff M, Kim NY, Joo JB, Marchioro A (2022) Water orientation at the anatase TiO_2 nanoparticle interface: A probe of surface p*K*a values. J Phys Chem Lett 13:8677–8683, https://doi.org/10.1021/acs.jpclett.2c02453

Blok L, Bruyn PLD (1970) The ionic double layer at the interface. J Colloid Interface Sci 32:518–526, https://doi.org/10.1016/0021-9797(70)90141-4

Boamah MD, Ohno PE, Lozier E, Van Ardenne J, Geiger FM (2019) Specifics about specific ion adsorption from heterodyne-detected second harmonic generation. J Phys Chem B 123:5848–5856, https://doi.org/10.1021/acs.jpcb.9b04425

Boily J-F, Fu L, Tuladhar A, Lu Z, Legg BA, Wang ZM, Wang H (2019) Hydrogen bonding and molecular orientations across thin water films on sapphire. J Colloid Interface Sci 555:810–817, https://doi.org/10.1016/j.jcis.2019.08.028

Bonthuis DJ, Gekle S, Netz RR (2011) Dielectric profile of interfacial water and its effect on double-layer capacitance. Phys Rev Lett 107:166102, https://doi.org/10.1103/physrevlett.107.166102

Bousse L (1982) Single electrode potentials related to flat-band voltage measurements on EOS and MOS structures. J Chem Phys 76:5128–5133, https://doi.org/10.1063/1.442812

Bousse L, De Rooij NF, Bergveld P (1983) The influence of counter-ion adsorption on the ψ_0/pH characteristics of insulator surfaces. Surf Sci 135:479–496, https://doi.org/10.1016/0039-6028(83)90237-6

Bousse LJ, Mostarshed S, Hafeman D (1992) Combined measurement of surface potential and zeta potential at insulator/electrolyte interfaces. Sensors Actuators B Chem 10:67–71, https://doi.org/10.1016/0925-4005(92)80013-n

Breeuwsma A, Lyklema J (1973) Physical and chemical adsorption of ions in the electrical double layer on hematite (α-Fe_2O_3). J Colloid Interface Sci 43:437–448, https://doi.org/10.1016/0021-9797(73)90389-5

Brkljača Z, Lešić N, Bertović K, Dražić G, Bohinc K, Kovačević D (2018a) Polyelectrolyte-coated cerium oxide nanoparticles: insights into adsorption process. J Phys Chem C 122:27323–27330, https://doi.org/10.1021/acs.jpcc.8b07115

Brkljača Z, Namjesnik D, Lützenkirchen J, Předota M, Preočanin T (2018b) Quartz/aqueous electrolyte solution interface: Molecular dynamic simulation and interfacial potential measurements. J Phys Chem C 122:24025–24036, https://doi.org/10.1021/acs.jpcc.8b04035

Brown ID (2009) Recent developments in the methods and applications of the bond valence model. Chem Rev 109:6858–6919, https://doi.org/10.1021/cr900053k

Brown MA, Beloqui Redondo A, Sterrer M, Winter B, Pacchioni G, Abbas Z, Van Bokhoven JA (2013a) measure of surface potential at the aqueous–oxide nanoparticle interface by XPS from a liquid microjet. Nano Lett 13:5403–5407, https://doi.org/10.1021/nl402957y

Brown MA, Duyckaerts N, Redondo AB, Jordan I, Nolting F, Kleibert A, Ammann M, Wörner HJ, van Bokhoven JA, Abbas Z (2013b) Effect of surface charge density on the affinity of oxide nanoparticles for the vapor–water interface. Langmuir 29:5023–5029, https://doi.org/10.1021/la4005054

Brown MA, Jordan I, Beloqui Redondo A, Kleibert A, Wörner HJ, van Bokhoven JA (2013c) In situ photoelectron spectroscopy at the liquid/nanoparticle interface. Surf Sci 610:1–6, https://doi.org/10.1016/j.susc.2013.01.012

Brown MA, Arrigoni M, Héroguel F, Beloqui Redondo A, Giordano L, van Bokhoven JA, Pacchioni G (2014) pH dependent electronic and geometric structures at the water–silica nanoparticle interface. J Phys Chem C 118:29007–29016, https://doi.org/10.1021/jp502262f

Brown MA, Abbas Z, Kleibert A, Green RG, Goel A, May S, Squires TM (2016a) Determination of surface potential and electrical double-layer structure at the aqueous electrolyte–nanoparticle interface. Phys Rev X 6:11007, https://doi.org/10.1103/physrevx.6.011007/figures/8/medium

Brown MA, Goel A, Abbas Z (2016b) Effect of electrolyte concentration on the stern layer thickness at a charged interface. Angew Chemie Int Ed 55:3790–3794, https://doi.org/10.1002/anie.201512025

Butt H-J (1991a) Measuring electrostatic, van der Waals, and hydration forces in electrolyte solutions with an atomic force microscope. Biophys J 60:1438–1444, https://doi.org/10.1016/s0006-3495(91)82180-4

Butt H-J (1991b) Electrostatic interaction in atomic force microscopy. Biophys J 60:777–785, https://doi.org/10.1016/s0006-3495(91)82112-9

Butt H, Graf K, Kappl M (2003) Physics and Chemistry of Interfaces. Wiley-VCH Verlag GmbH & Co. KGaA, https://doi.org/10.1002/3527602313

Butt H-J, Cappella B, Kappl M (2005) Force measurements with the atomic force microscope: Technique, interpretation and applications. Surf Sci Rep 59:1–152, https://doi.org/10.1016/j.surfrep.2005.08.003

Cai C, Azam MS, Hore DK (2021) Determining the surface potential of charged aqueous interfaces using nonlinear optical methods. J Phys Chem C 125:25307–25315, https://doi.org/10.1021/acs.jpcc.1c07761

Calegari Andrade MF, Selloni A (2020) Structure of disordered TiO_2 phases from ab initio based deep neural network simulations. Phys Rev Mater 4:113803, https://doi.org/10.1103/physrevmaterials.4.113803

Cametti C (2011) Dielectric properties of soft-particles in aqueous solutions. Soft Matter 7:5494–5506, https://doi.org/10.1039/c0sm01150a

Cao S, Sun P, Xiao G, Tang Q, Sun X, Zhao H, Zhao S, Lu H, Yue Z (2023) ISFET-based sensors for (bio)chemical applications: A review. Electrochem Sci Adv 3:, https://doi.org/10.1002/elsa.202100207

Chang H, Ohno PE, Liu Y, Lozier EH, Dalchand N, Geiger FM (2020) Direct Measurement of Charge Reversal on Lipid Bilayers Using Heterodyne-Detected Second Harmonic Generation Spectroscopy. J Phys Chem B 124:641–649, https://doi.org/10.1021/acs.jpcb.9b09341

Chapman DL (1913) LI. A contribution to the theory of electrocapillarity. London, Edinburgh, Dublin Philos Mag J Sci 25:475–481, https://doi.org/10.1080/14786440408634187

Chow RS, Takamura K (1988) Effects of surface roughness (hairiness) of latex particles on their electrokinetic potentials. J Colloid Interface Sci 125:226–236, https://doi.org/10.1016/0021-9797(88)90071-9

Chu B, Marchioro A, Roke S (2023) Size dependence of second-harmonic scattering from nanoparticles: Disentangling surface and electrostatic contributions. J Chem Phys 158:, https://doi.org/10.1063/5.0135157/2881500

Cyran JD, Donovan MA, Vollmer D, Siro Brigiano F, Pezzotti S, Galimberti DR, Gaigeot M-P, Bonn M, Backus EHG (2019) Molecular hydrophobicity at a macroscopically hydrophilic surface. Proc Natl Acad Sci 116:1520–1525, https://doi.org/10.1073/pnas.1819000116

Dalstein L, Potapova E, Tyrode E (2017) The elusive silica/water interface: isolated silanols under water as revealed by vibrational sum frequency spectroscopy. Phys Chem Chem Phys 19:10343–10349, https://doi.org/10.1039/c7cp01507k

Darlington AM, Jarisz TA, DeWalt-Kerian EL, Roy S, Kim S, Azam MS, Hore DK, Gibbs JM (2017) Separating the pH-dependent behavior of water in the Stern and diffuse layers with varying salt concentration. J Phys Chem C 121:20229–20241, https://doi.org/10.1021/acs.jpcc.7b03522

Davis JA, James RO, Leckie JO (1978) Surface ionization and complexation at the oxide/water interface: I. Computation of electrical double layer properties in simple electrolytes. J Colloid Interface Sci 63:480–499, https://doi.org/10.1016/s0021-9797(78)80009-5

Delgado A V, González-Caballero F, Hunter RJ, Koopal LK, Lyklema J (2005) Measurement and Interpretation of Electrokinetic Phenomena (IUPAC Technical Report). Pure Appl Chem 77:1753–1805, https://doi.org/10.1351/pac200577101753

Delgado AV, González-Caballero F, Hunter RJ, Koopal LK, Lyklema J (2007) Measurement and interpretation of electrokinetic phenomena. J Colloid Interface Sci 309:194–224, https://doi.org/10.1016/j.jcis.2006.12.075

Derjaguin B, Landau L (1941) Theory of the stability of strongly charged lyophobic sols and of the adhesion of strongly charged particles in solutions of electrolytes. Acta Physicochim URSS 14:633, https://doi.org/10.1016/0079-6816(93)90013-l

Diot JL, Joseph J, Martin JR, Clechet P (1985) pH dependence of the Si/SiO_2 interface state density for EOS systems: Quasi-static and AC conductance methods. J Electroanal Chem Interfacial Electrochem 193:75–88, https://doi.org/10.1016/0022-0728(85)85053-1

Dove PM, Craven CM (2005) Surface charge density on silica in alkali and alkaline earth chloride electrolyte solutions. Geochim Cosmochim Acta 69:4963–4970, https://doi.org/10.1016/j.gca.2005.05.006

Drelich J, Long J, Yeung A (2008) Determining surface potential of the bitumen-water interface at nanoscale resolution using atomic force microscopy. Can J Chem Eng 85:625–634, https://doi.org/10.1002/cjce.5450850509

Ducker WA, Senden TJ, Pashley RM (1991) Direct measurement of colloidal forces using an atomic force microscope. Nature 353:239–241, https://doi.org/10.1038/353239a0

Eaton P, West P (2010) Atomic Force Microscopy. Oxford University Press Inc., New York

Eckhoff M, Behler J (2021) Insights into lithium manganese oxide–water interfaces using machine learning potentials. J Chem Phys 155:, https://doi.org/10.1063/5.0073449

Faure B, Salazar-Alvarez G, Bergström L (2011) Hamaker constants of iron oxide nanoparticles. Langmuir 27:8659–8664, https://doi.org/10.1021/la201387d

Fitts JP, Shang X, Flynn GW, Heinz TF, Eisenthal KB (2005) Electrostatic surface charge at aqueous/α-Al_2O_3 single-crystal interfaces as probed by optical second-harmonic generation. J Phys Chem B 109:7981–7986, https://doi.org/10.1021/jp040297d

Fung CD, Cheung PW, Ko WH (1986) Generalized theory of an electrolyte-insulator-semiconductor field-effect transistor. IEEE Trans Electron Devices ED-33:8–18, https://doi.org/10.1109/t-ed.1986.22429

Gerischer H (1989) Neglected problems in the pH dependence of the flatband potential of semiconducting oxides and semiconductors covered with oxide layers. Electrochim Acta 34:1005–1009, https://doi.org/10.1016/0013-4686(89)87133-6

Goel A, Lützenkirchen J (2022) relevance of colloid inherent salt estimated by surface complexation modeling of surface charge densities for different silica colloids. Colloids Interfaces 6:23, https://doi.org/10.3390/colloids6020023/s1

Gonella G, Backus EHG, Nagata Y, Bonthuis DJ, Loche P, Schlaich A, Netz RR, Kühnle A, McCrum IT, Koper MTM, Wolf M, Winter B, Meijer G, Campen RK, Bonn M (2021) Water at charged interfaces. Nat Rev Chem 2021 57 5:466–485, https://doi.org/10.1038/s41570-021-00293-2

Gouy M (1910) Sur la constitution de la charge électrique à la surface d'un électrolyte. J Phys Théorique Appliquée 9:457–468, https://doi.org/10.1051/jphystap:019100090045700

Grahame DC (1947) The electrical double layer and the theory of electrocapillarity. Chem Rev 41:441, https://doi.org/10.1021/cr60130a002

Gray CG, Stiles PJ (2018) Nonlinear electrostatics. The Poisson–Boltzmann equation. Eur J Phys 39:053002, https://doi.org/10.1088/1361-6404/aaca5a

Greczynski G, Hultman L (2022) A step-by-step guide to perform X-ray photoelectron spectroscopy. J Appl Phys 132:, https://doi.org/10.1063/5.0086359

Gu Y, Li D (2000) The ζ-potential of glass surface in contact with aqueous solutions. J Colloid Interface Sci 226:328–339, https://doi.org/10.1006/jcis.2000.6827

van Hal REG, Eijkel JCT, Bergveld P (1995) A novel description of ISFET sensitivity with the buffer capacity and double-layer capacitance as key parameters. Sensors Actuators B Chem 24:201–205, https://doi.org/10.1016/0925-4005(95)85043-0

van Hal REG, Eijkel JCT, Bergveld P (1996) A general model to describe the electrostatic potential at electrolyte oxide interfaces. Adv Colloid Interface Sci 69:31–62, https://doi.org/10.1016/s0001-8686(96)00307-7

Hankin A, Bedoya-Lora FE, Alexander JC, Regoutz A, Kelsall GH (2019) Flat band potential determination: avoiding the pitfalls. J Mater Chem A 7:26162–26176, https://doi.org/10.1039/c9ta09569a

Harrington SP, Devine TM (2008) Analysis of electrodes displaying frequency dispersion in Mott–Schottky tests. J Electrochem Soc 155:C381, https://doi.org/10.1149/1.2929819

Hartley PG, Larson I, Scales PJ (1997) Electrokinetic and direct force measurements between silica and mica surfaces in dilute electrolyte solutions. Langmuir 13:2207–2214, https://doi.org/10.1021/la960997c

Healy TW, White LR (1978) Ionizable surface group models of aqueous interfaces. Adv Colloid Interface Sci 9:303–345, https://doi.org/10.1016/0001-8686(78)85002-7

Heberling F, Trainor TP, Lützenkirchen J, Eng P, Denecke MA, Bosbach D (2011) Structure and reactivity of the calcite–water interface. J Colloid Interface Sci 354:843–857, https://doi.org/10.1016/j.jcis.2010.10.047

Heberling F, Klačić T, Raiteri P, Gale JD, Eng PJ, Stubbs JE, Gil-Díaz T, Begović T, Lützenkirchen J (2021) Structure and surface complexation at the calcite(104)–water interface. Environ Sci Technol 55:12403–12413, https://doi.org/10.1021/acs.est.1c03578

Hellström M, Quaranta V, Behler J (2019) One-dimensional vs. two-dimensional proton transport processes at solid–liquid zinc-oxide–water interfaces. Chem Sci 10:1232–1243, https://doi.org/10.1039/c8sc03033b

Herzberg M, Dobberschütz S, Okhrimenko D, Bovet NE, Andersson MP, Stipp SLS, Hassenkam T (2020) Comparison of atomic force microscopy and zeta potential derived surface charge density. Europhys Lett 130:36001, https://doi.org/10.1209/0295-5075/130/36001

Hesleitner P, Kallay N, Matijevic E (1991) Adsorption at solid/liquid interfaces. 6. The effect of methanol and ethanol on the ionic equilibria at the hematite/water interface. Langmuir 7:178–184, https://doi.org/10.1021/la00049a032

Hopkins AJ, McFearin CL, Richmond GL (2005) Investigations of the solid–aqueous interface with vibrational sum-frequency spectroscopy. Curr Opin Solid State Mater Sci 9:19–27, https://doi.org/10.1016/j.cossms.2006.04.001

Hsu CH, Mansfeld F (2001) Technical Note: Concerning the conversion of the constant phase element parameter Ψ_0 into a capacitance. Corrosion 57:747–748, https://doi.org/10.5006/1.3280607

Hunter RJ, Wright HJL (1971) The dependence of electrokinetic potential on concentration of electrolyte. J Colloid Interface Sci 37:564–580, https://doi.org/10.1016/0021-9797(71)90334-1

Iler RK (1979) The Chemistry of Silica : Solubility, Polymerization, Colloid and Surface Properties, and Biochemistry. John Wiley & Sons

Israelachvili JN (2011) Intermolecular and Surface Forces. Elsevier,

Ivanov VD (2023) Who should be credited for the Gouy–Chapman model? J Electrochem Soc 170:106507, https://doi.org/10.1149/1945-7111/ad041f

James RO, Parks GA (1982) Characterization of aqueous colloids by their electrical double-layer and intrinsic surface chemical properties. *In:* Matijević E (eds) Surface and Colloid Science. Springer US, Boston, MA, p 119–216

Johnson SB, Scales PJ, Healy TW (1999) The binding of monovalent electrolyte ions on α-alumina. I. Electroacoustic studies at high electrolyte concentrations. Langmuir 15:2836–2843, https://doi.org/10.1021/la980875f

Jukić J, Juračić T, Josić E, Namjesnik D, Begović T (2023) Effects of polyion adsorption on surface properties of TiO_2. Adsorption, https://doi.org/10.1007/s10450-023-00397-9

Kallay N, Čakara D (2000) Reversible charging of the ice-water interface: I. Measurement of the surface potential. J Colloid Interface Sci 232:81–85, https://doi.org/10.1006/jcis.2000.7193

Kallay N, Preočanin T (2008) Measurement of the surface potential of individual crystal planes of hematite. J Colloid Interface Sci 318:290–295, https://doi.org/10.1016/j.jcis.2007.09.090

Kallay N, Dojnović Z, Čop A (2005) Surface potential at the hematite–water interface. J Colloid Interface Sci 286:610–614, https://doi.org/10.1016/j.jcis.2005.01.032

Kallay N, Preočanin T, Ivšić T (2007a) Determination of surface potentials from the electrode potentials of a single-crystal electrode. J Colloid Interface Sci 309:21–27, https://doi.org/10.1016/j.jcis.2006.10.075

Kallay N, Preočanin T, Marković J, Kovačević D (2007b) Adsorption of organic acids on metal oxides: Application of the surface potential measurements. Colloids Surfaces A Physicochem Eng Asp 306:40–48, https://doi.org/10.1016/j.colsurfa.2006.11.001

Kallay N, Preočanin T, Šupljika F (2008) Measurement of surface potential at silver chloride aqueous interface with single-crystal AgCl electrode. J Colloid Interface Sci 327:384–387, https://doi.org/10.1016/j.jcis.2008.08.041

Kallay N, Preočanin T, Kovačević D, Lützenkirchen J, Chibowski E (2010) Electrostatic potentials at solid/liquid interfaces. Croat Chem Acta 83:357–370

Kallay N, Preočanin T, Kovačević D, Lützenkirchen J, Villalobos M (2011) Thermodynamics of the reactions at solid/liquid interfaces. Croat Chem Acta 84:1–10, https://doi.org/10.5562/cca1864

Kallay N, Preočanin T, Sapunar M, Namjesnik D (2014) Common surface potential of two different crystal planes. Surf Innov 2:142–150, https://doi.org/10.1680/si.13.00029

Karlsson M, Craven C, Dove PM, Casey WH (2001) Surface charge concentrations on Silica in different 1.0 M metal-chloride background electrolytes and implication for dissolution rates. Aquat Geochem 7:13–32, https://doi.org/10.1023/a:1011377400253/metrics

Katić J, Metikoš-Huković M, Šarić I, Petravić M (2016) Semiconducting properties of the oxide films formed on tin: Capacitive and XPS Studies. J Electrochem Soc 163:C221–C227, https://doi.org/10.1149/2.0961605jes

Kędra-Królik K, Rosso KM, Zarzycki P (2017) Probing size-dependent electrokinetics of hematite aggregates. J Colloid Interface Sci 488:218–224, https://doi.org/10.1016/j.jcis.2016.11.004

Kędra K, Łazarczyk M, Begović T, Namjesnik D, Lament K, Piasecki W, Zarzycki P (2021) Electrochemical perspective on hematite–malonate interactions. Colloids Interfaces 5:47, https://doi.org/10.3390/colloids5040047

Klačić T, Sadžak A, Jukić J, Preočanin T, Kovačević D (2019) Surface potential study of ceria/poly(sodium 4-styrenesulfonate) aqueous solution interface. Colloids Surfaces A Physicochem Eng Asp 570:32–38, https://doi.org/10.1016/j.colsurfa.2019.03.002

Klačić T, Katić J, Namjesnik D, Jukić J, Kovačević D, Begović T (2021) Adsorption of polyions on flat TiO_2 surface. Minerals 11:1164, https://doi.org/10.3390/min11111164

Kosmulski M (2023) The pH dependent surface charging and points of zero charge. X. Update. Adv Colloid Interface Sci 319:102973, https://doi.org/10.1016/j.cis.2023.102973

Kovačević D, Kallay N, Antol I, Pohlmeier A, Lewandowski H, Narres HD (1998) The use of electrokinetic potential in the interpretation of adsorption phenomena:: Adsorption of salicylic acid on hematite. Colloids Surfaces A Physicochem Eng Asp 140:261–267, https://doi.org/10.1016/s0927-7757(97)00283-5

Kovačević D, Mazur D, Preočanin T, Kallay N (2010) Electrical interfacial layer at TiO_2/poly(4-styrene sulfonate) aqueous interface. Adsorption 16:405–412, https://doi.org/10.1007/s10450-010-9234-1

Larson I, Drummond CJ, Chan DYC, Grieser F (1997) Direct force measurements between silica and alumina. Langmuir 13:2109–2112, https://doi.org/10.1021/la960684h

Liao H-KK, Chi L-LL, Chou J-CC, Chung W-YY, Sun T-PP, Hsiung S-KK (1999) Study on pH_{pzc} and surface potential of tin oxide gate ISFET. Mater Chem Phys 59:6–11, https://doi.org/10.1016/s0254-0584(99)00033-4

Lin S, Xu L, Chi Wang A, Wang ZL (2020) Quantifying electron-transfer in liquid–solid contact electrification and the formation of electric double-layer. Nat Commun 2020 111 11:1–8, https://doi.org/10.1038/s41467-019-14278-9

Lowe BM, Skylaris C-K, Green NG, Shibuta Y, Sakata T (2018) Calculation of surface potentials at the silica–water interface using molecular dynamics: Challenges and opportunities. Jpn J Appl Phys 57:04FM02, https://doi.org/10.7567/jjap.57.04fm02

Lubbers B, Schober A (2009) Comparing the ISFET to the glass electrode: advantages, challenges and similarities. Chem Analityczna 54:1121–1148

Lützenkirchen J (ed) (2006) Surface Complexation Modelling. Academic Press, Amsterdam, London

Lützenkirchen J, Richter C, Brandenstein F (2010a) Some data and simple models for the silanated glass–electrolyte interface. Adsorption 16:249–258, https://doi.org/10.1007/s10450-010-9228-z

Lützenkirchen J, Zimmermann R, Preočanin T, Filby A, Kupcik T, Küttner D, Abdelmonem A, Schild D, Rabung T, Plaschke M, Brandenstein F, Werner C, Geckeis H (2010b) An attempt to explain bimodal behaviour of the sapphire c-plane electrolyte interface. Adv Colloid Interface Sci 157:61–74, https://doi.org/10.1016/j.cis.2010.03.003

Lützenkirchen J, Preočanin T, Stipić F, Heberling F, Rosenqvist J, Kallay N (2013) Surface potential at the hematite (001) crystal plane in aqueous environments and the effects of prolonged aging in water. Geochim Cosmochim Acta 120:479–486, https://doi.org/10.1016/j.gca.2013.06.042

Lützenkirchen J, Heberling F, Šupljika F, Preočanin T, Kallay N, Johann F, Weisser L, Eng PJ (2015) Structure–charge relationship—The case of hematite (001). Faraday Discuss 180:55–79, https://doi.org/10.1039/c4fd00260a

Lützenkirchen J, Scharnweber T, Ho T, Striolo A, Sulpizi M, Abdelmonem A (2018) A set-up for simultaneous measurement of second harmonic generation and streaming potential and some test applications. J Colloid Interface Sci 529:294–305, https://doi.org/10.1016/j.jcis.2018.06.017

Ma E, Geiger FM (2021) Divalent ion specific outcomes on Stern layer structure and total surface potential at the silica:water interface. J Phys Chem A 125:10079–10088, https://doi.org/10.1021/acs.jpca.1c08143

Malher E, Martin D, Duvivier C (1982a) No Title. Stud Biophys 90:33

Malher E, Martin D, Duvivier C, Volochine B, Stoltz JF (1982b) New device for determination of cell electrophoretic mobility using Doppler velocimetry. Biorheology 19:647–654, https://doi.org/10.3233/bir-1982-19506

Marchioro A, Bischoff M, Lütgebaucks C, Biriukov D, Předota M, Roke S (2019) Surface characterization of colloidal silica nanoparticles by second harmonic scattering: quantifying the surface potential and interfacial water order. J Phys Chem C 123:20393–20404, https://doi.org/10.1021/acs.jpcc.9b05482

Matsuo T, Esashi M (1981) Methods of ISFET fabrication. Sensors Actuators 1:77–96, https://doi.org/10.1016/0250-6874(81)80006-6

Matsuo T, Wise KD (1974) An integrated field-effect electrode for biopotential recording. IEEE Trans Biomed Engr 21:485–487

McNaught AD, Wilkinson A (2019) The IUPAC Compendium of Chemical Terminology. International Union of Pure and Applied Chemistry (IUPAC), Research Triangle Park, NC

Memming R (2015) Semiconductor Electrochemistry. Wiley-VCH Verlag GmbH

Miller JD, Fa K, Calara JV, Paruchuri VK (2004) The surface charge of fluorite in the absence of surface carbonation. Colloids Surfaces A Physicochem Eng Asp 238:91–97, https://doi.org/10.1016/j.colsurfa.2004.02.030

Molto C, Etcheberry A, Grand PP, Goncalves AM (2020) Study of photo-oxidized n-type textured silicon surface through electrochemical impedance spectroscopy. J Electrochem Soc 167:146505, https://doi.org/10.1149/1945-7111/abc0a6

Morita A (2018) Theory of Sum Frequency Generation Spectroscopy. Springer Nature Singapore

Nagel J, Kroschwald F, Bellmann C, Schwarz S, Janke A, Heinrich G (2017) Immobilisation of different surface-modified silica nanoparticles on polymer surfaces via melt processing. Colloids Surfaces A Physicochem Eng Asp 532:208–212, https://doi.org/10.1016/j.colsurfa.2017.05.016

Namjesnik D, Mutka S, Iveković D, Gajović A, Willinger M, Preočanin T (2016) Application of the surface potential data to elucidate interfacial equilibrium at ceria/aqueous electrolyte interface. Adsorption 22:825–837, https://doi.org/10.1007/s10450-016-9785-x

Nguyen HH, Lee SH, Lee UJ, Fermin CD, Kim M (2019) Immobilized enzymes in biosensor applications. Materials (Basel) 12:121, https://doi.org/10.3390/ma12010121

Ohshima H (2007) Electrokinetics of soft particles. Colloid Polym Sci 285:1411–1421, https://doi.org/10.1007/s00396-007-1740-7

Ohshima H (2012) Electrical phenomena in a suspension of soft particles. Soft Matter 8:3511, https://doi.org/10.1039/c2sm07160f

Oldham KB, Myland JC (1994) Fundamentals of electrochemical science. Academic Press

Ong S, Zhao X, Eisenthal KB (1992) Polarization of water molecules at a charged interface: second harmonic studies of the silica/water interface. Chem Phys Lett 191:327–335, https://doi.org/10.1016/0009-2614(92)85309-x

Orazem ME, Frateur I, Tribollet B, Vivier V, Marcelin S, Pébère N, Bunge AL, White EA, Riemer DP, Musiani M (2013) Dielectric properties of materials showing Constant-Phase-Element (CPE) impedance response. J Electrochem Soc 160:C215–C225, https://doi.org/10.1149/2.033306jes

Parks GA, Bruyn PL (1962) The zero point of charge of oxides. J Phys Chem 66:967–973, https://doi.org/10.1021/j100812a002

Parsegian VA, Gingell D (1972) On the electrostatic interaction across a salt solution between two bodies bearing unequal charges. Biophys J 12:1192–1204, https://doi.org/10.1016/s0006-3495(72)86155-1

Parshotam S, Rehl B, Busse F, Brown A, Gibbs JM (2022) Influence of the hydrogen-bonding environment on vibrational coupling in the electrical double layer at the silica/aqueous interface. J Phys Chem C 126:21734–21744, https://doi.org/10.1021/acs.jpcc.2c06412

Patel MY, Mortelliti MJ, Dempsey JL (2022) A compendium and meta-analysis of flatband potentials for TiO_2, ZnO, and SnO_2 semiconductors in aqueous media. Chem Phys Rev 3:, https://doi.org/10.1063/5.0063170

Pathak Y, Mishra P, Sharma M, Solanki S, Agarwal VV, Chaujar R, Malhotra BD (2024) Experimental circuit design and TCAD analysis of ion sensitive field effect transistor (ISFET) for pH sensing. Mater Sci Eng B 299:116951, https://doi.org/10.1016/j.mseb.2023.116951

Penners NHG, Koopal LKK, Lyklema J (1986) Interfacial electrochemistry of haematite (α-Fe_2O_3): homodisperse and heterodisperse sols. Colloids and Surfaces 21:457–468, https://doi.org/10.1016/0166-6622(86)80109-3

Poulain R (2023) How flat is the flatband potential? J Mater Chem A 11:17787–17796, https://doi.org/10.1039/d3ta03621a

Předota M, Machesky ML, Wesolowski DJ (2016) Molecular origins of the zeta potential. Langmuir 32:10189–10198, https://doi.org/10.1021/acs.langmuir.6b02493

Preočanin T, Kallay N (2013) Evaluation of surface potential from single crystal electrode potential. Adsorption 19:259–267, https://doi.org/10.1007/s10450-012-9448-5

Preočanin T, Čop A, Kallay N (2006) Surface potential of hematite in aqueous electrolyte solution: Hysteresis and equilibration at the interface. J Colloid Interface Sci 299:772–776, https://doi.org/10.1016/j.jcis.2006.02.013

Preočanin T, Tuksar M, Kallay N (2007) Mechanism of charging of the pyrite/aqueous interface as deduced from the surface potential measurements. Appl Surf Sci 253:5797–5801, https://doi.org/10.1016/j.apsusc.2006.12.090

Preočanin T, Šupljika F, Kallay N (2009) Evaluation of interfacial equilibrium constants from surface potential data: Silver chloride aqueous interface. J Colloid Interface Sci 337:501–507, https://doi.org/10.1016/j.jcis.2009.05.051

Preočanin T, Selmani A, Mazur D, Kallay N (2010a) The effect of electrolytes on the surface potential at the rutile/ aqueous interface. Appl Surf Sci 256:5412–5415, https://doi.org/10.1016/j.apsusc.2009.12.121

Preočanin T, Šupljika F, Kallay N (2010b) Charging of silver bromide aqueous interface: Evaluation of interfacial equilibrium constants from surface potential data. J Colloid Interface Sci 346:222–225, https://doi.org/10.1016/j.jcis.2010.02.053

Preočanin T, Šupljika F, Kallay N (2011) Charging of silver bromide aqueous interface: Evaluation of enthalpy and entropy of interfacial reactions from surface potential data. J Colloid Interface Sci 354:318–321, https://doi.org/10.1016/j.jcis.2010.09.077

Preočanin T, Namjesnik D, Brown MA, Lützenkirchen J (2017a) The relationship between inner surface potential and electrokinetic potential from an experimental and theoretical point of view. Environ Chem 14:295, https://doi.org/10.1071/en16216

Preočanin T, Namjesnik D, Klačić T, Šutalo P (2017b) The effects on the response of metal oxide and fluorite single crystal electrodes and the equilibration process in the interfacial region. Croat Chem Acta 90:333–344, https://doi.org/10.5562/cca3206

Pyman MAF, Bowden JW, Posner AM (1979) The movement of titration curves in the presence of specific adsorption. Soil Res 17:191–195

Quaranta V, Behler J, Hellström M (2019) Structure and dynamics of the liquid–water/zinc-oxide interface from machine learning potential simulations. J Phys Chem C 123:1293–1304, https://doi.org/10.1021/acs.jpcc.8b10781

Di Quarto F, Di Franco F, Santamaria M, La Mantia F (2018) Differential capacitance measurements on passive films. *In:* Encyclopedia of Interfacial Chemistry. Elsevier, p 75–92

Rehl B, Gibbs JM (2021) Role of ions on the surface-bound water structure at the silica/water interface: Identifying the spectral signature of stability. J Phys Chem Lett 12:2854–2864, https://doi.org/10.1021/acs.jpclett.0c03565

Rehl B, Ma E, Parshotam S, DeWalt-Kerian EL, Liu T, Geiger FM, Gibbs JM (2022) Water structure in the electrical double layer and the contributions to the total interfacial potential at different surface charge densities. J Am Chem Soc 144:16338–16349, https://doi.org/10.1021/jacs.2c01830

Van Riemsdijk WH, Bolt GH, Koopal LK, Blaakmeer J (1986) Electrolyte adsorption on heterogeneous surfaces: adsorption models. J Colloid Interface Sci 109:219–228, https://doi.org/10.1016/0021-9797(86)90296-1

Roke S, Bonn M, Petukhov A V. (2004) Nonlinear optical scattering: The concept of effective susceptibility. Phys Rev B 70:115106, https://doi.org/10.1103/physrevb.70.115106

de Rooij NF, Bergveld P (1980) The influence of the pH on the electrolyte–SiO_2–Si system studied by ion-sensitive fet measurements and quasi-static C–V measurements. Thin Solid Films 71:327–331, https://doi.org/10.1016/0040-6090(80)90167-4

Van Roosmalen AJ, Mol JC (1978) An infrared study of the silica gel surface. 1. Dry silica gel. J Phys Chem 82:2748–2751, https://doi.org/10.1021/j100514a026

Ruiz-Cabello FJM, Maroni P, Borkovec M (2013) Direct measurements of forces between different charged colloidal particles and their prediction by the theory of Derjaguin, Landau, Verwey, and Overbeek (DLVO). J Chem Phys 138:234705, https://doi.org/10.1063/1.4810901

Sahai N, Sverjensky DA (1997) Evaluation of internally consistent parameters for the triple-layer model by the systematic analysis of oxide surface titration data. Geochim Cosmochim Acta 61:2801–2826, https://doi.org/10.1016/s0016-7037(97)00128-2

Sato N (1998) Electrochemistry at Metal and Semiconductor Electrodes. Elsevier

Schindler P, Kamber HR (1968) Die Acidität von Silanolgruppen. Vorläufige Mitteillung. Helv Chim Acta 51:1781–1786, https://doi.org/10.1002/hlca.19680510738

Schmickler W (1996) Interfacial Electrochemistry. Oxford University Press

Selmani A, Lützenkirchen J, Kallay N, Preočanin T (2014) Surface and zeta-potentials of silver halide single crystals: pH-dependence in comparison to particle systems. J Phys Condens Matter 26:244104, https://doi.org/10.1088/0953-8984/26/24/244104

Shimizu K, Boily J-F (2014) Electrochemical properties and relaxation times of the hematite/water interface. Langmuir 30:9591–9598, https://doi.org/10.1021/la501669a

Siu WM, Cobbold RSC (1979) Basic properties of the electrolyte–SiO_2–Si system: Physical and theoretical aspects. IEEE Trans Electron Devices 26:1805–1815, https://doi.org/10.1109/t-ed.1979.19690

Sonnefeld J (1995) Surface charge density on spherical silica particles in aqueous alkali chloride solutions. Colloid Polym Sci 273:932–938, https://doi.org/10.1007/bf00660370

Sposito G (1998) On Points of Zero Charge. Environ Sci Technol 32:2815–2819, https://doi.org/10.1021/es9802347

Sprycha R (1984) Surface charge and adsorption of background electrolyte ions at anatase/electrolyte interface. J Colloid Interface Sci 102:173–185, https://doi.org/10.1016/0021-9797(84)90211-x

Sprycha R, Szczypa J (1984) Estimation of surface ionization constants from electrokinetic data. J Colloid Interface Sci 102:288–291, https://doi.org/10.1016/0021-9797(84)90221-2

Tadros TF, Lyklema J (1968) Adsorption of potential-determining ions at the silica–aqueous electrolyte interface and the role of some cations. J Electroanal Chem Interfacial Electrochem 17:267–275, https://doi.org/10.1016/s0022-0728(68)80206-2

Teschke O, Ceotto G, de Souza EF (2001) Interfacial water dielectric–permittivity-profile measurements using atomic force microscopy. Phys Rev E 64:011605, https://doi.org/10.1103/physreve.64.011605

Torres R, Kallay N, Matijevic E (1988) Adsorption at solid/solution interfaces. V: Surface complexation of iminodiacetic acid on hematite. Langmuir 4:706–710

Trasatti S, Parsons R (1986) Interphases in systems of conducting phases (Recommendations 1985). Pure Appl Chem 58:437–454, https://doi.org/10.1351/pac198658030437

Ulgut B, Suzer S (2003) XPS studies of SiO_2/Si system under external bias. J Phys Chem B 107:2939–2943, https://doi.org/10.1021/jp022003z

Ulman K, Poli E, Seriani N, Piccinin S, Gebauer R (2019) Understanding the electrochemical double layer at the hematite/water interface: A first principles molecular dynamics study. J Chem Phys 150: 041707, https://doi.org/10.1063/1.5047930

Uzgiris EE (1972) Electrophoresis of particles and biological cells measured by the Doppler shift of scattered laser light. Opt Commun 6:55–57, https://doi.org/10.1016/0030-4018(72)90247-7

Uzundal CB, Sahin O, Gokturk PA, Wu H, Mugele F, Ulgut B, Suzer S (2019) X-ray photoelectron spectroscopy with electrical modulation can be used to probe electrical properties of liquids and their interfaces at different stages. Langmuir 35:16989–16999, https://doi.org/10.1021/acs.langmuir.9b03134

Veeramasuneni S, Yalamanchili MR, Miller JD (1996) Measurement of interaction forces between silica and α-alumina by atomic force microscopy. J Colloid Interface Sci 184:594–600, https://doi.org/10.1006/jcis.1996.0656

Verwey EJW, Overbeek JTG (1948) Theory of the Stability of Lyophobic Colloids. Elsevier Publishing Inc., New York,

Van den Vlekkert HH, Kloeck B, Kloeck D, Prongue D, Berthoud J, Hu B, De Rooij NF, Gilli E, De Crousaz PH (1988) A pH–ISFET and an integrated pH–pressure sensor with back-side contacts. Sensors Actuators 14:165–176, https://doi.org/10.1016/0250-6874(88)80063-5

Wander MCF, Clark AE (2008) Structural and dielectric properties of quartz–water interfaces. J Phys Chem C 112:19986–19994, https://doi.org/10.1021/jp803642c

Ware B., Flygare W. (1972) Light scattering in mixtures of BSA, BSA dimers, and fibrinogen under the influence of electric fields. J Colloid Interface Sci 39:670–675, https://doi.org/10.1016/0021-9797(72)90075-6

Williams CT, Yang Y, Bain CD (2000) Total internal reflection sum-frequency spectroscopy: a strategy for studying molecular adsorption on metal surfaces. Langmuir 16:2343–2350, https://doi.org/10.1021/la9910091

Wolthers M, Charlet L, Van Cappellen P (2008) The surface chemistry of divalent metal carbonate minerals; a critical assessment of surface charge and potential data using the charge distribution multi-site ion complexation model. Am J Sci 308:905–941, https://doi.org/10.2475/08.2008.02

Yin X, Drelich J (2008) Surface charge microscopy: Novel technique for mapping charge-mosaic surfaces in electrolyte solutions. Langmuir 24:8013–8020, https://doi.org/10.1021/la801269z

Zarzycki P (2023) Distance-dependent dielectric constant at the calcite/electrolyte interface: Implication for surface complexation modeling. J Colloid Interface Sci 645:752–764, https://doi.org/10.1016/j.jcis.2023.04.169

Zarzycki P, Preočanin T (2012) Point of zero potential of single-crystal electrode/inert electrolyte interface. J Colloid Interface Sci 370:139–143, https://doi.org/10.1016/j.jcis.2011.12.068

Zarzycki P, Rosso KM (2010) Nonlinear response of the surface electrostatic potential formed at metal oxide/electrolyte interfaces. A Monte Carlo simulation study. J Colloid Interface Sci 341:143–152, https://doi.org/10.1016/j.jcis.2009.09.002

Zarzycki P, Chatman S, Preočanin T, Rosso KM (2011) Electrostatic potential of specific mineral faces. Langmuir 27:7986–7990, https://doi.org/10.1021/la201369g

Zhdanov SP, Kiselev AW (1957) No Title. Zhur Fiz Khim 31:2213

Reviews in Mineralogy & Geochemistry
Vol. 91A pp. 337–352, 2025

Surface Complexation Reactions in Oxide Nanopores

Anastasia G. Ilgen

Geochemistry Department
Sandia National Laboratories
Albuquerque, NM 87123, U.S.A.

agilgen@sandia.gov

INTRODUCTION

Nanoconfined chemical systems are abundant in terrestrial environments. In this chapter, we define an interfacial system as *nanoconfined* when it measures less than 1000 nm in at least one dimension. Considering the effects of nanoconfinement on surface reactivities is crucial for accurately predicting the fate and transport of chemical species in the environment, where nanoscale pores and thin films on surfaces are prevalent (Fig. 1). Various nanoconfined geometries can exist in nature, including pores/cages (3-D confinement), channels (2-D confinement) and slits (1-D confinement). Because the influence of geometry has yet to be clearly quantified, in this chapter we refer to all discussed systems as *nanopores*, regardless of whether they are characterized by 1-D, 2-D, or 3-D confinement. For our purposes here, we consider surface complexation reactions in all these systems as nanoconfined. A few selected examples in Figure 1 illustrate typical cases where nanoconfinement effects on chemistry can be observed. Future research should focus on enhancing SCMs for nanoporous systems, and the author hopes this chapter will inspire such efforts.

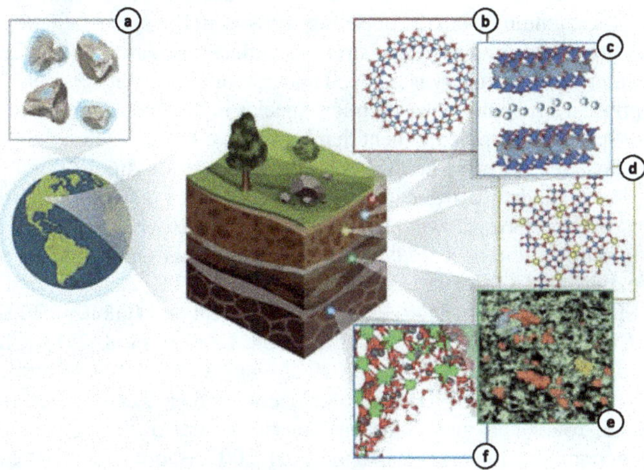

Figure 1. Nanoconfined solid–water interfaces are abundant in the environment and define the fate and transport of chemical species: **(a)** atmospheric dust with thin water films; **(b)** nanochannels in tubular minerals, such as imogolite (1-D confinement); **(c)** nanoslits in clay minerals (2-D confinement); **(d)** nanocages in zeolite minerals (3-D confinement); **(e)** nanopores in sedimentary rocks, such as shale (Ilgen et al. 2018); **(f)** nanocracks in minerals; molecular rendering of a single crack tip is shown for calcite (Choens et al. 2021). Mineral structures obtained from the American Mineralogist Crystal structure database and visualized with the Vesta 3.3.9 program.

1529-6466/25/0091A-0010$05.00 (print)
1943-2666/25/0091A-0010$05.00 (online)
http://dx.doi.org/10.2138/rmg.2025.91A.10

First, we consider why geometric constraints influence interfacial chemistry in nanopores filled with aqueous solutions. All chemical reactions, including surface complexation reactions, are described by their free energy change (ΔG). The ΔG of surface complexation reactions (ΔG_{ads}) can be strongly influenced by confinement because it can alter the structure and properties of water and the charge distributions within the interfacial region. It is important to understand the nanoconfinement effects on all free energy contributions for surface complexation reactions. Let us examine the energetic contributions to the overall free energy of adsorption, which can be expressed as (James and Healy, 1972):

$$\Delta G_{ads} = \Delta G_{coul} - \Delta G_{hydr} + \Delta G_{chem} \quad (1)$$

where ΔG_{coul} represents electrostatic interactions (can be favorable or unfavorable for adsorption, depending on the charges of surface sites and adsorbing species); ΔG_{hydr} is the hydration free energy (always unfavorable for adsorption; however, the solvation free energy can change under nanoconfinement); and ΔG_{chem} is the free energy of chemical bond formation between the adsorbing ion and the surface site (favorable for adsorption). As we will show later in the chapter, macroscopic adsorption trends and spectroscopic data on surface complexes for oxide and silicate nanopores indicates that ΔG_{hydr} is the key property that can be predictive of the extent to which nanoconfinement affects surface complexation reactions. The magnitude of ΔG_{hydr} change in nanopores is defined by the properties of nanoconfined aqueous solutions, which are dependent on the chemistry of the confining surfaces (e.g., such as surface charge and surface wettability related to the solid–water interfacial tension).

As shown in the other chapters, all amorphous and crystalline oxide surfaces (and all other mineral surfaces) submerged in aqueous solutions possess surface charge due to (de)protonation and/or (de)hydroxylation reactions. The surface charge causes ions to arrange near surface–water interfaces in a somewhat predictable fashion known as the electric double-layer (EDL) (Fig. 2a). In dilute aqueous solutions (~ few mM ionic strength), the extent of the EDL can reach several nanometers; in pure water at pH 7, where the concentrations of [H^+] and [OH^-] ions are on the order of 10^{-7} M, the EDL extends to ~1000 nanometers (Debye 1923; Ilgen et al. 2023b). Therefore, the EDLs extending from the opposing surfaces may overlap in nanopores, depending on the nanopore dimensions and the ionic strength of aqueous solution (Fig. 2b). The effect of EDL overlap on interfacial chemistry is not well understood; however, conceptually such overlap is expected to impose a structure on the chemical species in the affected regions, including water molecules, as they possess a dipole moment that is sensitive to the electric field within the EDL. Experimentally, the strength of H-bonds—the key measure of dipole–dipole interactions in water—has been shown to decrease due to nanoconfinement in 4 nm and 7 nm silica pores. (Ilgen et al. 2023c). Researchers have shown various properties of nanoconfined H_2O deviate from its bulk counterpart: nanoconfinement leads to a decrease in H_2O density, solid–water interfacial tension, and freezing point (Takei et al. 2000; Koga et al. 2001; Senapati and Chandra 2001; Levinger 2002; Striolo et al. 2005; Marti et al. 2006; Hirunsit and Balbuena 2007; Alexiadis and Kassinos 2008; Cicero et al. 2008; Rasaiah et al. 2008; Le Caër et al. 2011; Chakraborty et al. 2017; Ruiz Pestana et al. 2018; Knight et al. 2019; Zaragoza et al. 2019; Motevaselian and Aluru 2020). The changes in the physical properties of nanoconfined H_2O can be related to the alteration in the translational, rotational, and vibrational motions due to non-negligible surface–water interactions in nanoconfined domains (Tsukahara et al. 2007: Baum et al. 2019; Senanayake et al. 2021). Nanoconfinement can also lead to an increase in the apparent viscosity of H_2O (Sansom and Biggin 2001; Mattia and Calabrò 2012; Ortiz-Young et al. 2013). These nanoconfinement effects on H_2O structure and the structure of aqueous solutions in the interfacial regions can impact the inter-ionic distances, and therefore, free energies of nanoconfined species (Brubach et al. 2001; Musat et al. 2008; Israelachvili 2011; Baum et al. 2019; Knight et al. 2019, 2020). Furthermore, nanoconfinement effects on solution structure and water dynamics can impact the likelihood of molecular collisions during surface complexation reactions.

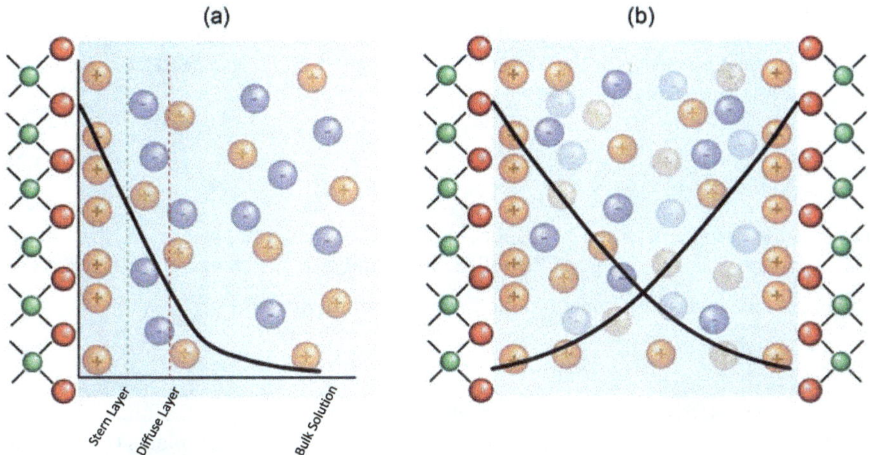

Figure 2. Electric Double Layer (EDL) **(a)** structure; and **(b)** possible EDL overlap in nanopores.

Secondly, it is important to note that researchers in fields outside of geoscience have also observed unpredictable reactivities of nanoporous solids and nanoparticle surfaces while studying catalysis (Na and Somorjai 2015; Shen et al. 2022), water purification (Argyris et al. 2010; Epsztein et al. 2020), material synthesis (Kobayashi et al. 2016; Shen et al. 2022), flow in nanochannels (Lee et al. 2012; Kou et al. 2015; Zhang et al. 2020), and other areas. Here, we focus primary on the geoscience-relevant solids, where nanoconfined surfaces exhibit unexpected reactivities, with an emphasis on surface complexation reactions and the properties of reactive systems pertinent for accurate SCMs. In the other chapters, the specific parameters necessary for constructing accurate SCMs are addressed in detail. Here, we will illustrate how these parameters can be impacted by the dimensions and geometry of the oxide-water domain where surface complexation reactions take place. Specifically, we will discuss how nanoconfinement of interfacial reactions can affect solution structures, H-bond strengths, and ion solvation energetics, and therefore influence surface complexation reactions (Wang 2014; Knight et al. 2018, 2020; Nelson et al. 2018; Ilgen et al. 2021). In this chapter, we present current findings on oxides and aluminosilicates, including zeolites, clay minerals, and mica minerals, while excluding soft confining materials such as reverse micelles, which are more relevant to research on biological systems than to geochemistry.

SOLUTION STRUCTURE AND DIELECTRIC PROPERTIES IN NANOPORES

Triple layer model of solution structures at interfaces

The triple layer model (TLM) refers to a representation of the EDL where three interfacial regions and associated parameters are considered (Fig. 3). Currently, TLM is one of the most detailed approaches for constructing SCMs because it includes both strong (chemisorption) and weak (physisorption) interactions with the surfaces, as well as the properties of the ion diffuse layer. The downside of using the TLM is having to include multiple parameters that should be known or well-approximated for an accurate SCM. The TLM consists of an α-plane or 0-plane, which coincides with the solid's surface; a β-plane which terminates the plane where counter-ions are tightly bound at charged surfaces (inner Stern layer); and a d-plane, which cuts through the center of the diffuse layer near the surfaces (Sun and Li 2021).

Parameters of interest are shown in Figure 3. First, the individual layer capacitance values per surface area(C_1, C_2 in $\mu F\ cm^{-2}$) need to be included. Related to the capacitance values, the individual surface potentials at the 0-plane, β-plane, and d-plane also need to be included. The surface potentials are usually expressed as follows (Sun and Li 2021):

$$\psi_0 - \psi_\beta = \sigma_0 / C_1 \tag{2}$$

$$\psi_\beta - \psi_d = \sigma_d / C_2 \tag{3}$$

where ψ is the surface potential corresponding to one of the three planes, C is the capacitance of the plane, and σ_d is the charge of the diffuse layer and σ_0 is the charge of the 0-plane (Fig. 3).

Capacitance is directly related to the relative permittivity (ε) via:

$$C = \frac{\varepsilon_r \varepsilon_0 A}{d} \tag{4}$$

where ε_r is the relative permittivity of a phase (also referred to as the dielectric constant, unitless), ε_0 is the permittivity of vacuum (in F/m), A is the area of the plates (in m^2), and d is the distance between the plates in meters (or planes in the TLM representation).

Other parameters that are necessary to include in an SCM are related to surface reactivities and surface complexation reactions. These specifically include equilibrium constants for surface (de)protonation and (de)hydroxylation reactions. The equilibrium constants for surface complexation reactions involving neutral and charged aqueous species are obtained by fitting a SCM to experimental data for adsorption isotherms and adsorption edges (pH-dependent adsorption). In the following sections, we will detail how dielectric properties, capacitance, and surface potentials may be affected by nanoconfinement, followed by a discussion on how equilibrium chemistry is impacted.

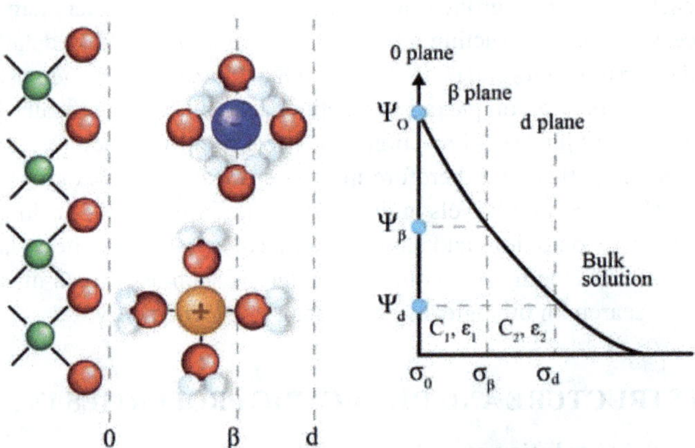

Figure 3. The triple layer model and associated physico-chemical variables that define this model. Symbols are: ψ_0, ψ_β, ψ_d are the surface potential values corresponding to the three planes (0, β and d planes), C are the capacitance values for the layers between planes, ε_1, ε_2, are relative permittivity values for the corresponding layers, and σ_d is the charge at the diffuse layer plane and σ_0 is the charge at the 0-plane.

Dielectric properties, capacitance, and surface potentials in nanopores

In SCMs that include a description of the Stern layer, at least one static ε value is used (Davis and Kent 1990). As described above, in SCMs that utilize the TLM framework, two ε values are included. There are significant uncertainties associated with determining the "dielectric constant" of aqueous solutions when considering unconfined solid surfaces and the conditions become even more complicated for nanoconfined systems. Due to EDL overlap and its effect on solution structures, ε values depend on (1) the nanoconfinement characteristics—e.g., pore size/geometry and chemistry of the solid surface; and (2) the solution composition and the ion-specific effects (Leung 2023). Furthermore, the dielectric characteristics of solutions depend on the distance from a solid surface, and therefore are best represented as a spatially variant "dielectric profile," rather than a constant value (Ilgen et al. 2023b; Leung 2023). Here, we only consider the perpendicular tensor component ($\varepsilon_\perp$) of the overall dielectric profile in nanopores since it has been measured in a laboratory. The $\varepsilon_\perp$ was quantified for 2-D slit pores in between muscovite mica sheets (Teschke et al. 2000), in boron nitride (Fumagalli et al. 2018), and in silica channels (Morikawa et al. 2015). These measurements show that the $\varepsilon_\perp$ value decreases to ~2 (compared to a value of 78 for bulk H_2O) in these systems (Fumagalli et al. 2018; Teschke et al. 2000). The decrease in the dielectric response in nanopores is not surprising, because numerous previous studies have shown that at solid–water interfaces H_2O dipoles align within the EDL, leading to a decrease in the dielectric response (Brovchenko and Oleinikova 2008). What is surprising and still requires further research is understanding why ε decreases have been measured for even very large nanochannels (<1000 nm) (Morikawa et al. 2015); the long-range impact on ε cannot be explained by the van der Waals forces or the Debye length of most electrolytes (Debye 1923).

While nanoconfined pure water is expected to behave like a classical dielectric material with the electric field attenuated by a factor of $1/\varepsilon_\perp(\mathrm{r})$ at distances r from the surface (Jeanmairet et al. 2019), the solution properties change dramatically when ions are present, as such solutions conduct electric charge. Some researchers postulate that the low dielectric response of H_2O at solid surfaces enhances ion-pairing (Matyushov 2021). However, the role of the $\varepsilon_\perp$ component of the dielectric profile on ion–ion interactions is still debated, with some findings indicating that in 2-D pores ion-ion interactions at the same distance from the surface depend on the geometric mean of the perpendicular ($\varepsilon_\perp$) and parallel ($\varepsilon_\parallel$) components (Alvarez et al. 2021; Matyushov 2021).

NANOCONFINEMENT EFFECTS ON ADSORPTION COMPLEXES

Speciation of adsorbed cations in nanopores

To construct an accurate SCM, the products of surface complexation reactions must be known. Nanoconfinement has been shown to alter the abundances of surface-bound species and even promote new surface species (such as polynuclear complexes, not observed in the unconfined counterparts) (Knight et al. 2020; Ilgen et al. 2021). The reaction steps during adsorption are shown in Figure 4, and each of these steps can be impacted by nanoconfinement. A consistent observation is that inner-sphere cation adsorption is enhanced in nanopores. Different studies on oxides and silicates demonstrate the enhancement of inner-sphere complexation (chemisorption), which has been reported for very small (< 1 nm) nanopores (Ferreira and Schulthess 2011; Schulthess et al. 2011; Ferreira et al. 2012a,b, 2013), as well as for larger nanopores with diameters on the order of 4–7 nm (Knight et al. 2020; Ilgen et al. 2021).

Studies on zeolites with pore diameters of ~0.3–0.5 nm report that the adsorption of Na^+, Ca^{2+}, Ni^{2+}, Cu^{2+}, and Mn^{2+} does not follow the same trends as those observed for unconfined silicate surfaces. Using batch adsorption measurements, flow microcalorimetry, nuclear magnetic resonance (NMR), and electron paramagnetic resonance (EPR) spectroscopy measurements Ferreira, Schulthess and others show that in progressively smaller zeolite pores, monovalent Na^+ cations are selected over the divalent Ca^{2+} and Ni^{2+} (Ferreira and Schulthess 2011; Schulthess et al. 2011; Ferreira et al. 2012a,b, 2013). They attribute this trend to the fact that Na^+ has a lower free energy of hydration (ΔG_{hydr}), and the small diameter of zeolite pores requires a dehydration step for a cation to fit in a nanopore. In the case of zeolites, cations

Ca^{2+} and Ni^{2+} with more negative ΔG_{hydr} cannot overcome the energetic barrier of dehydrating to enter a nanopore. These studies also highlight the role that pore diameter plays in whether inner- or outer-spere surface complexes will form. In 0.74 nm pores, Na^+ and Ca^{2+} adsorb as outer-sphere complexes, while in 0.5–0.6 nm pores Na^+ forms inner-sphere-, and Ca^{2+} forms outer-sphere complexes (Ferreira et al. 2013). These observations were summarized as a general principle that weakly hydrated ions are preferentially adsorbed in nanopores that are smaller than the hydrated ion diameter, therefore, the ratio of the solvated ion diameter to the nanopore diameter defines the "strength" of the ion–surface interaction (Ferreira and Schulthess 2011; Schulthess et al. 2011; Ferreira et al. 2013). In a similar vein, ion exclusion in nanopores has been observed in separation science (Shefer et al. 2021), and one proposed explanation for this exclusion is the increase in the energy barriers for solute desolvation under nanoconfinement (Bowen and Welfoot 2002). The energy barrier (ΔW) in Joules, calculated from the Born model, is dictated by the ion charge (z_i), the hydrodynamic radius of a solute i (a_s, m), and the difference between the dielectric constants of the fluids in nanopores (ε_p) compared to the bulk water (ε_b):

$$\Delta W_i = \frac{z_i^2 e^2}{8\pi\varepsilon_0 a_s}\left(\frac{1}{\varepsilon_p} - \frac{1}{\varepsilon_b}\right) \tag{5}$$

where e is the elementary charge and ε_0 is the permittivity of vacuum. With decreasing pore diameter, ε_p decreases, and therefore $1/\varepsilon_p$ increases, thus raising the energy barrier for ions to enter a nanopore. The phenomenon of salt rejection in nanoporous membranes is proposed to be due to the increase in the energy barrier for ion diffusion into the pores with decreasing pore diameter (Bowen and Welfoot 2002). Based on Equation (5), ion charge amplifies this effect, with the energy barrier increasing more steeply for ions with higher valence.

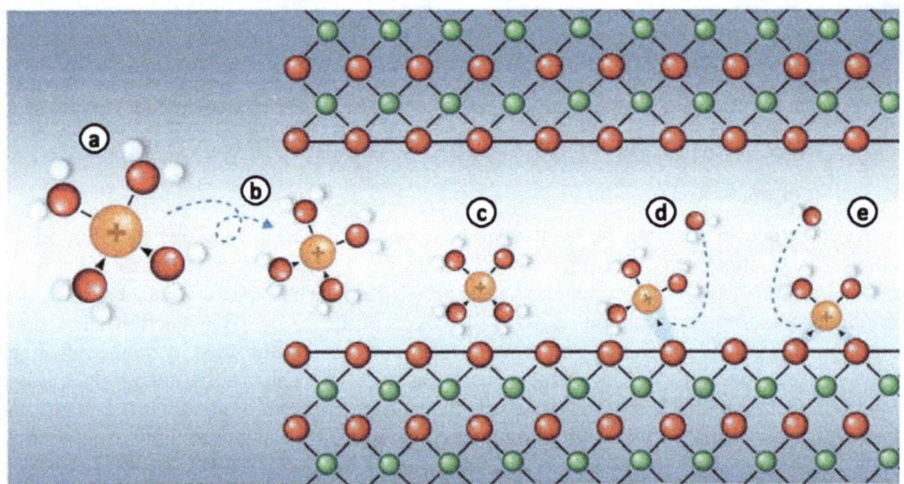

Figure 4. Surface complexation reaction steps and structures in nanopores: **(a)** Hydrated ion enters a nanopore; **(b)** hydrated ion diffuses through a nanopore; **(c)** hydrated ion adsorbs on nanopore surface as outer-sphere complex (physisorption); **(d)** hydrated ion loses one H_2O molecule from its hydration shell and forms an inner-sphere mono-dentate complex (chemisorption); **(e)** hydrated ion loses second H_2O molecule from its hydration shell and forms an inner-sphere bi-dentate complex (chemisorption).

To continue our discussion on the role of ion hydration energy during adsorption in nanopores, let us consider a series of cations with gradual changes in hydration energy, while the charge of the cation remains the same: lanthanides (Ln^{3+}). Ilgen et al. (2021) explored surface complexation reactions within 4 nm and 7 nm SiO_2 nanopores and compared them to similar reactions on non-porous SiO_2 surfaces. Importantly, Si–OH (silanol) surface group densities were similar between these substrates (~2 Si–OH groups per nm^2), which indirectly implies that reactive surface moieties, surface pH_{PZC} and other surface characteristics were similar. Therefore, the experimental design included controlling pore diameters and ΔG_{hydr} (and ionic radius) of the adsorbing lanthanide cations. The ΔG_{hydr} for the selected Ln^{3+} cations (Nd^{3+}, Tb^{3+}, and Lu^{3+}) range from –3280 to –3515 kJ mol^{-1} (D'Angelo and Spezia 2012). To quantify the local coordination environment of adsorbed Ln^{3+} cations on SiO_2 surfaces, X-ray absorption fine structure (XAFS) spectroscopy was used (Ilgen et al. 2021). XAFS is an element-specific technique that allows for precise determination of local coordination geometry of adsorbed species. It accurately quantifies the bond lengths within the first and second coordination shells around the element of interest. Ilgen et al. reported that with nanoconfinement two changes occur: (1) inner-sphere complex formation is promoted, and (2) Ln^{3+} cations begin to form polymeric surface species. The proposed explanation is that with decreasing dielectric response of water in nanopores, the ΔG_{hydr} values of Ln^{3+}cations in nanopores also decrease, making it easier (less energetically demanding) to dehydrate any given Ln^{3+} cation, despite the increase in the energy barrier to ion's entry into the nanopore discussed above. To form an inner-sphere complex, partial dehydration of Ln^{3+} is required (Ilgen et al. 2021). The same study also shows that across the Ln^{3+} series, nanoconfinement effects on adsorption were greater for the weaker-hydrated Ln^{3+}cations (those with less negative ΔG_{hydr} values: Nd^{3+}, Eu^{3+}, Tb^{3+}), than for stronger-hydrated Ln^{3+}cations (Tm^{3+}, Lu^{3+}). The hydrated radii of the examined Ln^{3+} cations ranged from 2.32 Å (Nd^{3+}) to 1.96 Å (Lu^{3+}) (D'Angelo and Spezia 2012), which is approximately 20–30 times smaller than the SiO_2 nanopore diameters of 40 and 70 Å (4 and 7 nm), therefore, steric effects are likely not the primary driving force in the observed adsorption behavior. This study concludes that the free energy of adsorption, and specifically the ΔG_{hydr} contribution to the overall ΔG_{hydr} (Eqn. 1), is the main influence on surface complexation in nanopores (Ilgen et al. 2021). Electrostatic effects can become important for lanthanides at pH >7, because Ln^{3+} cations are Brønsted acids capable of hydrolyzing water; however, pK_a values for the first hydrolysis product for Ln^{3+} cations in this study are >7 (Shiery et al. 2021). Therefore, in the adsorption experiments on SiO_2 surfaces, Ln^{3+} cations are in the aqua-ion form $[Ln{\cdot}(H_2O)_8]^{3+}$ while SiO_2 surfaces are negatively charged at the pH 6 used in these experiments.

A study of Ln^{3+} adsorption on confined and unconfined Al_2O_3 uncovered more intriguing details about the impact of nanoconfinement on surface complexation reactions involving lanthanides (Ilgen et al. 2023a). In agreement with the study on SiO_2 nanopores, ΔG_{hydr} values were used to predict that weaker-hydrated Nd^{3+} would be impacted by nanoconfinement more than the stronger-hydrated Lu^{3+}. This prediction was supported because Lu^{3+} formed inner-sphere complexes (chemisorption) on both confined and unconfined surfaces, and Nd^{3+} formed outer-sphere (physisorption) on unconfined, and inner-sphere on nanoconfined surface (Fig. 5).

To explain these observations for Al_2O_3 surfaces, let us consider the energetic contributions to the overall ΔG_{ads} as shown in Equation (1). In the case of Al_2O_3 surfaces, ΔG_{coul} was unfavorable for adsorption because all Al_2O_3 surfaces are positively charged at pH 6.0. The $pH_{PZC,}$ where surface has zero charge, is 8.7 for γ-Al_2O_3 with 2 nm pores (Wang et al. 2002), 10.9 for porous Al_2O_3 with 1.3 nm pores (Sun et al. 2011), and 9.7 for α-Al_2O_3 (corundum) (Sun et al. 2011). This means that the coulombic interactions involving Ln^{3+} and Al_2O_3 surfaces are unfavorable, and when Lu^{3+} and Nd^{3+} are compared, is more unfavorable for Nd^{3+} because of its larger solvation shell and less effective charge screening compared to the tighter-packed Lu^{3+} hydration shell. The chemical free energy change (ΔG_{chem}) can be qualitatively assessed by examining the metal–oxygen bond lengths. Based on the XAFS analysis, the Lu–O bond length is ~ 0.2 Å shorter compared to Nd–O bond length, therefore, ΔG_{chem} for Lu^{3+} is more favorable for adsorption than for the analogous Nd^{3+} complexes. If ΔG_{hydr} was the primary contributor to ΔG_{ads} we would expect to see that Nd^{3+}, instead of Lu^{3+},

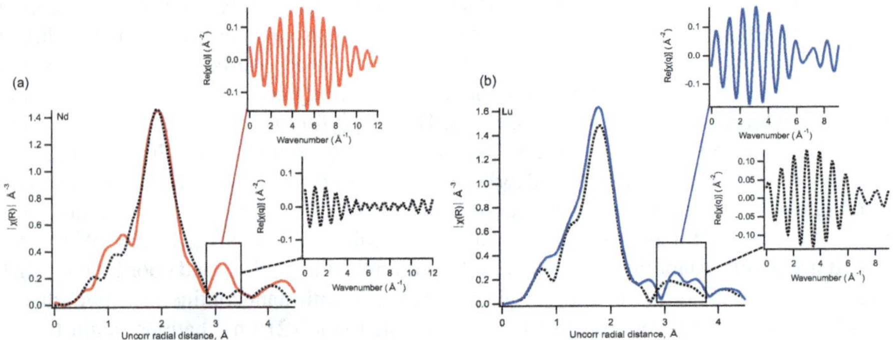

Figure 5. Fourier transform of the X-ray absorption fine structure (XAFS) spectra for (**a**) Nd and (**b**) Lu adsorbed on corundum (**dashed lines**) and on nanoconfined Al_2O_3 surfaces (**solid lines**). The insets in each (a) and (b) panels show back-transformed spectra of the isolated 2nd shell. Nd has a 2nd shell when adsorbed on the nanoconfined Al_2O_3 surfaces that corresponds to Nd–Al backscattering due to inner-sphere complexation, while no 2nd shell is present for Nd adsorbed on corundum (outer-sphere complexes). Lu forms inner-sphere complexes on both corundum and nanoconfined Al_2O_3. [Reproduced from Ilgen et al. (2023a) CC-BY 4.0]

would be more prone to forming inner-sphere complexes, because 1–2 water molecules need to be removed from the 1st shell prior to adsorption and it is easier to do so for an ion with less negative ΔG_{hydr}. However, solvation free energy is *also* impacted by nanoconfinement, and the tabulated values measured for bulk solutions may no longer apply. Ilgen et al. proposed that once Nd^{3+} enters a nanopore its ΔG_{hydr} becomes less negative due to nanoconfinement and lower dielectric response of nanopore solutions, which allows Nd^{3+} to shed 1–2 H_2O molecules prior to adsorption at a lower energetic cost. Ilgen et al. concluded that the ΔG_{chem} term for Lu^{3+} must compensate for the unfavorable ΔG_{hydr} contribution in both nanoconfined and unconfined systems. Because Nd^{3+} has less effective charge screening and less favorable ΔG_{chem}, compared to Lu^{3+}, it forms outer-sphere complexes on unconfined surfaces.

At the time of writing, researchers are still working to understand how nanoconfinement influences each of the energetic contributions to the overall adsorption free energy. We anticipate that all three energy contributions discussed are impacted: (1) solvation free energies become less negative in nanopores when compared to the ΔG_{hydr} values measured in bulk solutions (Ilgen et al. 2021); (2) electrostatic interactions and ΔG_{coul} are likely amplified in nanopores because of the decreased charge screening due to decrease in H-bond strength (Ilgen et al. 2023c); and (3) ΔG_{chem} is likely impacted due to subtle changes in the configuration of surface reactive sites resulting from pore surface curvature and other effects discussed below.

Other studies have addressed the adsorption of Cu^{2+} and Fe^{2+}/Fe^{3+} cations within 4 and 7 nm SiO_2 nanopores using experiments (Knight et al. 2018, 2020) and predicted surface reactivities in model SiO_2 pores using classical molecular dynamics (CMD) simulations (Greathouse et al. 2021). These cations (Cu^{2+} and Fe^{2+}/Fe^{3+}) adsorb as inner-sphere complexes, and similar to Ln^{3+} cations, Cu^{2+} also tends to form polymeric surface species in nanopores (Knight et al. 2020; Greathouse et al. 2021). The CMD simulations predict an increased fraction of inner-sphere complexes for Cu^{2+} and Fe^{2+}/Fe^{3+} when the pore diameter is decreased (Knight et al. 2020; Greathouse et al. 2021). These simulations also show that the primary driver for Fe^{2+}/Fe^{3+} adsorption on SiO_2 surfaces is surface charge, with greater negative charge resulting in a higher number of surface complexes (Greathouse et al. 2021).

No inner-sphere enhancement was observed during Zn^{2+} adsorption in larger SiO_2 nanopores with diameters of ~10 to 330 nm in controlled pore glass (Nelson et al. 2018). Using XAFS spectroscopy, Nelson *et al.* show that for all pore diameters, Zn^{2+} forms inner-sphere mono-dentate complexes with SiO_2 surfaces. The zinc cation has variable coordination; it can have either tetrahedral or octahedral coordination in its first shell. In this case, for confined and unconfined SiO_2 surfaces, Zn^{2+} exhibited a larger fraction of tetrahedral coordination with increasing surface coverage (Nelson et al. 2018). One distinction in smaller nanopores (~10 nm diameter) was that the change from octahedral to tetrahedral Zn^{2+} surface complexes occurred at a lower surface coverage. The increase in the tetrahedral-to-octahedral ratio of surface complexes with decreasing pore diameter was attributed to (1) a decrease in the ΔG_{hydr}, which would favor fewer H_2O molecules in the 1st coordination shell of Zn^{2+}; and/or (*2*) an increase in the available surface Si–OH sites with decreasing pore diameter (Nelson et al. 2018). Furthermore, when Zn^{2+} forms a mono-dentate complex, it promotes de-protonation of two surface sites: the Si–OH site where Zn^{2+} adsorbs and a neighboring site (Nelson et al. 2018). Nelson et al. did not find any pore-size dependency in the macroscopic adsorption trends. However, Ilgen et al. (2021) and Knight et al. (2018, 2020) documented macroscopic pore-size dependent adsorption for Cu^{2+} and Ln^{3+}. These studies showed that cations with lower (less negative) ΔG_{hydr} (Cu^{2+}, Nd^{3+}, Eu^{3+}, Tb^{3+}) have higher surface coverages at the same equilibrium concentration in 4 nm pores compared to 7 nm pores. The non-porous SiO_2 had the highest surface coverage likely due to the diffusion barrier (ΔW_i for ions in nanopores discussed above). Pore-size dependency gradually diminished for Ln^{3+} cations with more negative ΔG_{hydr} (Tm^{3+}, Lu^{3+}) (Ilgen et al. 2021).

Molecular-scale information about ion coordination environments in nanopores, as discussed above, is extremely limited. A few *macroscopic* adsorption studies attempted to infer surface complex geometries in such systems. Adsorption of Cd^{2+} on non-porous SiO_2 was compared to its adsorption on SiO_2 with 3.8–6.4 nm pores: the surface-area-normalized surface coverage is higher for non-porous SiO_2, but once it is normalized by the Si–OH surface site density, Cd^{2+} adsorption is higher in SiO_2 nanopores (Prelot et al. 2011). To explain this observation, Prelot *et al.* proposed that Cd^{2+} forms inner-sphere complexes by coordinating to two Si–OH sites (bi-dentate); therefore Si–OH surface site density is the main chemical driver for Cd^{2+} adsorption (Prelot et al. 2011). For Sr^{2+}, adsorption on SiO_2 with 5-6 nm pores follows a Langmuir isotherm model, and exhibits decreased adsorption with increasing ionic strength, likely indicating that a mixture of inner- and outer-sphere surface complexes form on these materials (Zhang et al. 2015). Additionally, microscopic adsorption was studied for U(VI) species (here U(VI) refers to all aqueous species) on non-porous α-Al_2O_3 (corundum) and Al_2O_3 with ~1.3 nm pores (Sun et al. 2011). When U(VI) adsorbs within Al_2O_3 nanopores, the reaction is pH-dependent while it is independent of ionic strength, likely indicating inner-sphere complexation. On corundum U(VI) adsorption is dependent on both pH and ionic strength, likely indicative of outer-sphere complexation (Sun et al. 2011). Furthermore, U(VI) adsorption in nanopores was shown to be irreversible, while it is fully reversible for corundum (Sun et al. 2011). This irreversibility was also supported by another study, where the reduction of U(VI) adsorbed in Al_2O_3 nanopores to U(IV) was impossible; however, it proceeded easily when U(VI) was adsorbed on unconfined corundum surfaces (Jung et al. 2012).

These findings illustrate that the exact surface complex geometry depends on the typical variables associated with surface and solution chemistry but is also dictated by the geometric constraints of reactive nanopores. Very few studies have directly probed speciation in nanopores, both in nanoconfined solutions and on nanoconfined surfaces. Due to the complicated chemical nature of these systems, predicting surface speciation of ions adsorbed in nanopores is challenging, making it exceedingly difficult to design robust SCMs for nanoconfined surfaces.

NANOCONFINEMENT EFFECTS ON EQUILIBRIUM CONSTANTS

Solvation of nanoconfined aqueous species

As demonstrated in the discussion above, ΔG_{hydr} is one of the primary contributors to adsorption free energy that can influence the extent to which nanoconfinement affects surface complexation reactions (Ilgen et al. 2021, 2023a). We also discussed that ΔG_{hydr} is likely lower (less negative) in nanopores, compared to the tabulated solution values, because of the nanoconfinement effects on the structure and dynamics of nanoconfined H_2O. For example, H-bonding interactions can be selectively weakened or strengthened due to nanoconfinement, with consequences for solvated species in nanopores (Ilgen et al. 2023c). Assuming the lower dielectric response of nanoconfined fluids discussed earlier, the ΔG_{hydr} for a nanoconfined species is predicted to be less negative compared to an unconfined counterpart (Ilgen et al. 2023b; Nelson et al. 2018):

$$\Delta G_{\mathrm{hydr}} = \frac{z_i^2 e^2}{r_i}\left(\frac{1}{\varepsilon_{\mathrm{p}}} - \frac{1}{\varepsilon_{\mathrm{b}}}\right) \tag{6}$$

where z_i is ion charge, r_i is the radius of the solvated ion (including H_2O molecules), ε_p is the dielectric constant of fluid in the nanopore, ε_b is the dielectric constant in bulk solution, and e is the elementary charge. Based on Equation (6), if ΔG_{hydr} were to become *less* negative, the Gibbs free energy of adsorption ΔG_{ads} would then become *more* negative, thus shifting the adsorption equilibrium constant K_{ads} towards the products (adsorbed species).

Nanoconfinement effects on adsorption equilibrium constants

The effect of nanoconfinement on adsorption equilibrium constants K_{ads} can be quantified through either direct measurement of reactant and product concentrations or *via* measuring or calculating the free energy of adsorption reaction and/or its main contributions, such as the enthalpy of adsorption ΔH_{ads}.

Equilibrium constants are calculated using:

$$\ln K_{\mathrm{ads}} = -\frac{\Delta G_{\mathrm{ads}}}{\mathrm{R}T} \tag{7}$$

where R is the universal gas constant, and T is temperature.

The ΔH_{ads} can be measured using immersion calorimetry or flow microcalorimetry. With either technique, the heat of reaction is measured in a batch or a flow reactor (Allen et al. 2017; Ferreira et al. 2013; Ilgen et al. 2021; Knight et al. 2020; Wu and Navrotsky 2013). These calorimetry techniques measure the total reaction enthalpy, which arises from different reaction steps (Fig. 4), including ion (de)hydration, physisorption, and chemisorption processes. Flow microcalorimetry studies on Cu^{2+} and Ln^{3+} cation adsorption, in combination with the structural XAFS analysis described above show that the reaction is *endothermic* for unconfined surfaces and strongly *exothermic* in SiO_2 and Al_2O_3 nanopores (Fig. 6) (Knight et al. 2020; Ilgen et al. 2021, 2023a). Ilgen et al. proposed that the ΔH_{ads} values switch from endothermic to exothermic in nanopores due to two phenomena: (1) a decrease in the cations' ΔG_{hydr} due to a decrease in the dielectric response of nanoconfined water (Morikawa et al. 2015; Fumagalli et al. 2018; Motevaselian and Aluru 2020), which means the energetic cost of removing 1–2 H_2O molecules from the hydration shell is lowered (Ilgen et al. 2021); and (2) adsorbing cations (Cu^{2+}, Nd^{3+}, Tb^{3+}, and Lu^{3+}) forming dimers and other polymeric species in nanopores (Knight et al. 2020; Ilgen et al. 2021; Leung and Greathouse 2022).

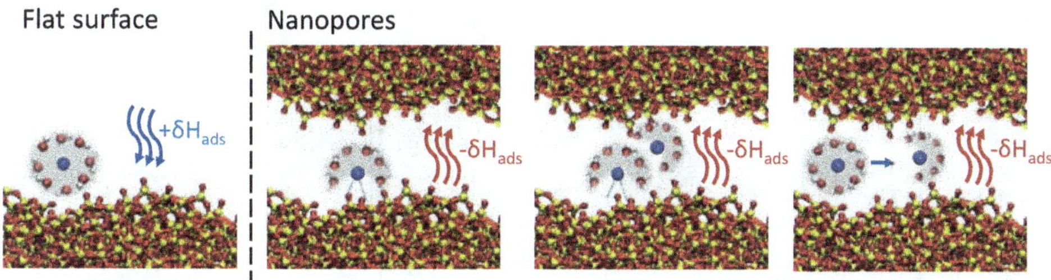

Figure 6. Adsorption of lanthanide cations onto unconfined and confined SiO_2 surfaces. Surface complexation is endothermic on unconfined surfaces and becomes exothermic when surfaces are nanoconfined. More outer-sphere complexes are formed on unconfined surfaces, and more inner-sphere complexes are formed in nanopores. Additionally, polymeric surface species are observed in nanopores. [Reprinted with permission from Bañuelos et al. (2023). Copyright 2023 American Chemical Society.]

Immersion calorimetry—where material is immersed in a solution and the heat is measured—was used to assess small molecule interactions with SiO_2 nanopores. The "small molecules" in this case included water, ethanol, triethylamine, sodium chloride, and sodium (bi)carbonate solutions. Negative (exothermic) ΔH_{ads} values were observed for water, ethanol, triethylamine, sodium chloride, and sodium (bi)carbonate solutions (Wu and Navrotsky 2013). Wu and Navrotsky (2013) show that ΔH_{ads} becomes progressively more negative with decreasing pore diameter and with increasing Si–OH surface site densities.

In zeolite pores, the adsorption of Na^+ and Ca^{2+} produces no heat in 0.7 nm pores, while in 0.5–0.6 nm pores, Na^+ adsorption is exothermic, and Ca^{2+} adsorption produces no heat (Ferreira et al. 2013).

While most of calorimetry studies on cation adsorption in SiO_2 nanopores indicate *exothermic* adsorption; (e.g., Na^+–Wu and Navrotsky 2013; Cu^{2+} and Ln^{3+}–Knight et al. 2020; Ilgen et al. 2021), Cd^{2+} adsorption is *endothermic* for both non-porous SiO_2 and SiO_2 with 3.8-6.4 nm pore diameters (Prelot et al. 2011). ΔH_{ads} for Cd^{2+} also depends on the pore diameter, it decreases (becomes more positive) as the pore diameter increases (Prelot et al. 2011). Apparent ΔG_{ads} values become more negative as pore diameter increases, indicating that adsorption becomes more favorable, therefore K_{ads} values increase with increasing pore diameter. Prelot et al. attribute their observations to the potential curvature effects in nanopores, where decreasing pore diameter and increasing curvature reduce the average distance between the available Si–OH surface sites, which can affect how favorable the adsorption reaction is (entropic effects). The most positive ΔH_{ads} values (*endothermic*) were recorded for non-porous SiO_2 (Prelot et al. 2011). This study further highlights that ΔH_{ads} and ΔG_{ads} change due to nanoconfinement, and existing theories cannot predict them based on pore chemistry and pore dimensions alone. Ion-specific effects must be considered.

The adsorption equilibrium constants K_{ads} calculated based on the measured concentrations of Ln^{3+} cations in solutions and Ln adsorbed on non-porous and porous SiO_2 show a dependency on nanoconfinement. For non-porous SiO_2, K_{ads} values vary linearly as a function of lanthanide mass and surface loading; however, in SiO_2 nanopores, they deviate from linearity (Fig. 7) (Ilgen et al. 2021). Importantly, the shifts from the linear distribution of K_{ads} constants are element-specific and are more pronounced for the heavier lanthanides with more negative ΔG_{hydr}. For Al_2O_3 with pore diameters ranging from 2.5 to 20 nm, it was observed that K_{ads} values for Ni^{2+} decreased with decreasing pore diameter (Baca et al. 2008).

Because nanoconfinement-driven shifts in K_{ads} values are ion-specific, it is extremely difficult to predict how equilibrium constants will change under nanoconfinement. Therefore, it is also challenging to construct robust SCMs that take these effects into account.

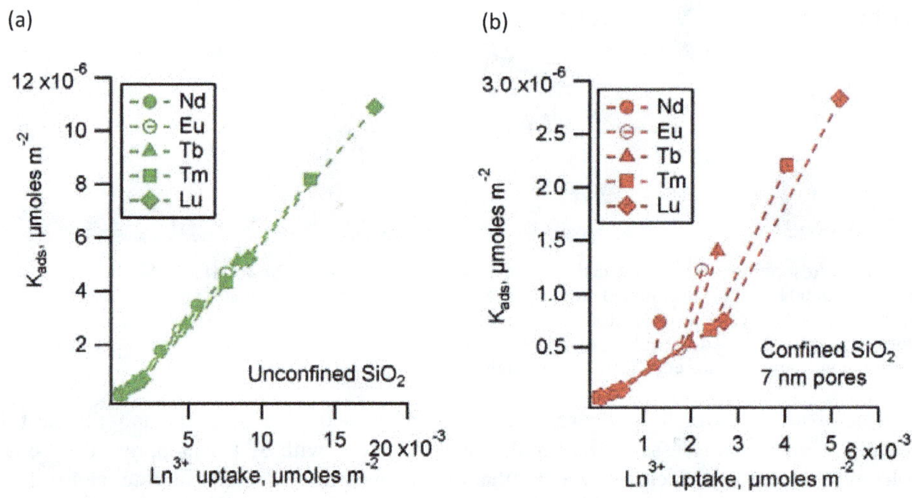

Figure 7. Adsorption equilibrium constants calculated as ratio of products to reactants in batch adsorption studies for **(a)** lanthanide Ln^{3+} adsorption on non-porous SiO_2 and **(b)** porous SiO_2. [Reproduced from Ilgen et al. (2021). United States Government works].

Equilibrium constants for surface (de)protonation and (de)hydroxylation reactions

At the time of writing, there are almost no studies that quantify the shifts in the pK_a values due to confinement for surface functional groups, such as Si–OH, Al–OH, Fe–OH, Mn–OH, or others relevant to surface complexation on oxide and silicate surfaces. The few existing studies used acid-base titrations to measure pH_{PZC} and pK_a values (note that these macroscopic measurements can be associated with large uncertainties because they cannot distinguish the contributions from nanopores *vs.* external surfaces of the particles). The microscopically-observed pH_{PZC} were similar for 2-nm porous and non-porous γ-Al_2O_3; however, there were measurable differences in the surface de-protonation constants pK_a for porous Al_2O_3 ($pK_{a1} = 9.0$ and $pK_{a2} = 10.3$) and non-porous Al_2O_3 ($pK_{a1} = 7.7$ and $pK_{a2} = 11.0$) (Wang et al. 2002). It is important to note that ΔpK_a calculated as the difference between the pK_{a2} and pK_{a1} values is 1.3 pH units for porous, and 3.3 pH units for non-porous Al_2O_3. Also, for porous Al_2O_3 the surface charge density was ~2 times higher than that for non-porous Al_2O_3 (Wang et al. 2002). A similar narrowing of the pH range (ΔpK_a) due to nanoconfinement was reported by Sun et al. for Al_2O_3 with ~1.3 nm pores, which had pK_a values of $pK_{a1} = 10.8$ and $pK_{a2} = 11.0$, while for non-porous corundum the values were $pK_{a1} = 8.9$ and $pK_{a2} = 10.9$; the reported pH_{PZC} values were 10.9 for porous Al_2O_3, and 9.7 for corundum (Sun et al. 2011). In agreement with Wang et al. (2002), Baca et al. (2008) also show minor pH_{PZC} shifts for Al_2O_3 with pore diameter from about 20 to 50 nm. Contrary to the reports of Wang et al. (2002) and Sun et al. (2001), Baca et al. (2008) show that the ΔpK_a values become larger as pore diameters in Al_2O_3 are decreased.

Batch titrations and SCM were used to determine surface charge densities of SiO_2 materials with varying pore diameters. Murota and Saito (2022) show that the surface charge densities decrease with decreasing pore diameters in SiO_2 (pore diameter range from 3.4 to 12.4 nm); this trend was attributed to the non-negligible curvature effects in nanopores and overlapping EDLs. However, for Al_2O_3 nanopores, Baca et al. (2008) conclude that the surface charge is roughly the same for pore diameters ranging from about 20 nm to 50 nm.

Case study: SCM for cation adsorption in silica nanopores

One example study where an SCM was developed to model Cs^+ and Sr^{2+} adsorption in SiO_2 nanopores highlights the difficulty of constructing such models (Murota and Saito 2022; Murota et al. 2023). Murota et al. (2023) developed an SCM conceptually represented as cylinders with fixed diameters and infinite lengths and incorporating a radial one-dimensional potential. This model was used to fit experimentally determined surface charge densities of porous SiO_2 and Cs^+ and Sr^{2+} adsorption edges. The model includes the description of the Stern layer (β-plane), with fitted capacitance values for this layer to capture the changes in capacitance values for pores with different diameters. The model also incorporates the Poisson–Boltzmann (PB) equation written using cylindrical coordinates allowing for the inclusion of overlapping EDLs (Murota and Saito 2022; Murota et al. 2023). Despite the elegant construction of this model that seemingly accounts for important phenomena in nanopores, it fails to accurately capture Cs^+ and Sr^{2+} adsorption data, especially when ionic strength of the background electrolyte is low, and the surface charge is not screened effectively (Murota et al. 2023). Future work should include important phenomena discussed in this chapter to develop representative SCMs that can capture experimental data and predict adsorption behaviors in nanopores.

CONCLUSIONS

Determining surface properties, solution structures, and surface complexation reactions in nanopores is an exciting field of research. As we show in this chapter, some universal trends are beginning to emerge. For example, nanoconfinement can make the free energy of adsorption reactions more favorable due to decreased dehydration costs in nanopores—especially those that are less than 10 nm in diameter—when ionic strength is low and EDLs from the opposing surfaces overlap. However, surface curvature and other geometric constraints may alter this observed trend if, for example, entropic contributions become significant. The main chemical properties that control surface complexation reactions in nanopores are (Ilgen et al. 2023b): (1) surface chemistry; (2) pore diameter and pore topology that define solution structure within the overlapping EDLs; and (3) the hydration free energy of the adsorbing ion, ΔG_{hydr} (Ferreira and Schulthess 2011; Ilgen et al. 2021; Schulthess et al. 2011). Current models, including SCMs, do not capture the macroscopic adsorption data due to a limited understanding of how these three main classes of chemical variables determine the pathways, products, and energetics of surface complexation reactions in nanopores (Rodriguez et al. 2009; Argyris et al. 2010; Jiang and Ward 2014; Patra et al. 2014; Li et al. 2017; Knight et al. 2018, 2019, 2020; Nelson et al. 2018; Greathouse et al. 2021; Ilgen et al. 2021, 2023a,b,c). These challenges need to be addressed to enable the development of new approaches for SCMs for nanopores that achieve the same accuracy as SCMs developed for unconfined surfaces.

ACKNOWLEDGEMENTS

This material is based upon work supported by the U.S. Department of Energy, Office of Science, Office of Basic Energy Sciences, Chemical Sciences, Geosciences, and Biosciences Division under Field Work Proposal Number 23-015452 at Sandia National Laboratories. This article has been authored by an employee of National Technology & Engineering Solutions of Sandia, LLC under Contract No. DE-NA0003525 with the U.S. Department of Energy (DOE). The employee owns all right, title and interest in and to the article and is solely responsible for its contents. The United States Government retains and the publisher, by accepting the article for publication, acknowledges that the United States Government retains a non-exclusive, paid-up, irrevocable, world-wide license to publish or reproduce the published form of this article or allow others to do so, for United States Government purposes. The DOE will provide public access to these results of federally sponsored research in accordance with the DOE Public Access Plan https://www.energy.gov/downloads/doe-public-access-plan. This paper describes objective technical results and analysis. Any subjective views or opinions that might be expressed in the paper do not necessarily represent the views of the U.S. Department of Energy or the United States Government.

REFERENCES

Alexiadis A, Kassinos S (2008) Molecular simulation of water in carbon nanotubes. Chem Rev 108:5014–5034

Allen N, Machesky ML, Wesolowski DJ, Kabengi N (2017) Calorimetric study of alkali and alkaline-earth cation adsorption and exchange at the quartz-solution interface. J Colloid Interface Sci 504:538–548

Alvarez F, Arbe A, Colmenero J (2021) Unraveling the coherent dynamic structure factor of liquid water at the mesoscale by molecular dynamics simulations. J Chem Phys 155:244509

Argyris D, Cole DR, Striolo A (2010) Ion-specific effects under confinement: the role of interfacial water. ACS Nano 4:2035–2042

Baca M, Carrier X, Blanchard J (2008) Confinement in nanopores at the oxide/water interface: Modification of alumina adsorption properties. Chemistry 14:6142–6148

Bañuelos JL, Borguet E, Brown Jr GE, Cygan RT, DeYoreo JJ, Dove PM, Gaigeot M-P, Geiger FM, Gibbs JM, Grassian VH (2023) Oxide– and silicate–water interfaces and their roles in technology and the environment. Chem Rev 123:106413–106544

Baum M, Rieutord F, Juranyi F, Rey C, Rébiscoul D (2019) Dynamical, structural properties of water in silica nanoconfinement: impact of pore size, ion nature, and electrolyte concentration. Langmuir 35:10780–10794

Bowen WR, Welfoot JS (2002) Modelling the performance of membrane nanofiltration—critical assessment, model development. Chem Eng Sci 57:1121–1137

Brovchenko I, Oleinikova A (2008) Interfacial and Confined Water. Elsevier.

Brubach J-B, Mermet A, Filabozzi A, Gerschel A, Lairez D, Krafft M, Roy P (2001) Dependence of water dynamics upon confinement size. J Phys Chem B 105:430–435

Chakraborty S, Kumar H, Dasgupta C, Maiti PK (2017) Confined water: structure, dynamics, thermodynamics. Acc Chem Res 50:2139–2146

Choens R, Wilson J, Ilgen A (2021) Strengthening of calcite assemblages through chemical complexation reactions. Geophys Res Lett 48:e2021GL094316

Cicero G, Grossman JC, Schwegler E, Gygi F, Galli G (2008) Water confined in nanotubes, between graphene sheets: A first principle study. J Am Chem Soc 130:1871–1878

D'Angelo P, Spezia R (2012) Hydration of lanthanoids (III), actinoids (III): an experimental/theoretical saga. Chemistry 18:11162–11178

Davis JA, Kent DB (1990) Surface complexation modeling in aqueous geochemistry. Rev Mineral Geochem 23:177–260

Debye P (1923) The theory of electrolytes. I. Freezing point depression and related phenomena. Phys J 24: 305–324

Epsztein R, DuChanois RM, Ritt CL, Noy A, Elimelech M (2020) Towards single-species selectivity of membranes with subnanometre pores. Nat Nanotechnol 15:426–436

Ferreira D, Schulthess C (2011) The nanopore inner sphere enhancement effect on cation adsorption: Sodium, potassium, calcium. Soil Sci Soc Am J 75:389–396

Ferreira DR, Schulthess CP, Amonette JE, Walter ED (2012a) An electron paramagnetic resonance spectroscopy investigation of the retention mechanisms of Mn, Cu in the nanopore channels of three zeolite minerals. Clays Clay Miner 60:588–598

Ferreira DR, Schulthess CP, Giotto MV (2012b) An investigation of strong sodium retention mechanisms in nanopore environments using nuclear magnetic resonance spectroscopy. Environ Sci Technol 46:300–306

Ferreira D, Schulthess C, Kabengi N (2013) Calorimetric evidence in support of the nanopore inner sphere enhancement theory on cation adsorption. Soil Sci Soc Am J 77:94–99

Fumagalli L, Esfandiar A, Fabregas R, Hu S, Ares P, Janardanan A, Yang Q, Radha B, Taniguchi T, Watanabe K (2018) Anomalously low dielectric constant of confined water. Science 360:1339–1342

Greathouse JA, Duncan TJ, Ilgen AG, Harvey JA, Criscenti LJ, Knight AW (2021) Effects of nanoconfinement and surface charge on iron adsorption on mesoporous silica. Environ Sci Nano 8:1992–2005

Hirunsit P, Balbuena PB (2007) Effects of confinement on water structure, dynamics: a molecular simulation study. J Phys Chem C 111:1709–1715

Ilgen AG, Aman M, Espinoza DN, Rodriguez MA, Griego J, Dewers TA, Feldman JD, Stewart T, Choens RC, Wilson J (2018) Shale–brine–CO_2 interactions and the long-term stability of carbonate-rich shale caprock. Int J Greenhouse Gas Control 78:244–253

Ilgen AG, Kabengi N, Leung K, Ilani-Kashkouli P, Knight AW, Loera L (2021) Defining silica–water interfacial chemistry under nanoconfinement using lanthanides. Environ Sci Nano 8:432–443

Ilgen AG, Kabengi N, Smith JG, Sanchez KM (2023a) Ion solvation as a predictor of lanthanide adsorption structures, energetics in alumina nanopores. Commun Chem 6:172

Ilgen AG, Leung K, Criscenti LJ, Greathouse JA (2023b) Adsorption at nanoconfined solid–water interfaces. Annu Rev Phys Chem 74:169–191

Ilgen AG, Senanayake HS, Thompson WH, Greathouse JA (2023c) Structure and energetics of hydrogen bonding networks in dilute HOD/H_2O solutions confined in silica nanopores. Environ Sci: Nano 10:3025–3038

Israelachvili JN (2011) Intermolecular and Surface Forces. Academic Press

James RO, Healy TW (1972) Adsorption of hydrolyzable metal ions at the oxide–water interface. IIIA Thermodynamic model of adsorption. J Colloid Interface Sci 40:65–81

Jeanmairet G, Rotenberg B, Borgis D, Salanne M (2019) Study of a water-graphene capacitor with molecular density functional theory. J Chem Phys 151:124111

Jiang Q, Ward MD (2014) Crystallization under nanoscale confinement. Chem Soc Rev 43:2066–2079

Jung HB, Boyanov MI, Konishi H, Sun Y, Mishra B, Kemner KM, Roden EE, Xu H (2012) Redox behavior of uranium at the nanoporous aluminum oxide–water interface: Implications for uranium remediation. Environ Sci Technol 46:7301–7309

Knight AW, Tigges A, Ilgen A (2018) Adsorption of copper on mesoporous silica: The effect of nano-scale confinement. Geochem Trans 19:13

Knight AW, Kalugin NG, Coker E, Ilgen AG (2019) Water properties under nano-scale confinement. Sci Rep 9:8246

Knight AW, Ilani-Kashkouli P, Harvey JA, Greathouse JA, Ho TA, Kabengi N, Ilgen AG (2020) Interfacial reactions of Cu (ii) adsorption, hydrolysis driven by nano-scale confinement. Environ Sci Nano 7:68–80

Kobayashi Y, Horie Y, Honjo K, Uemura T, Kitagawa S (2016) The controlled synthesis of polyglucose in one-dimensional coordination nanochannels. Chem Commun 52:5156–5159

Koga K, Gao G, Tanaka H, Zeng XC (2001) Formation of ordered ice nanotubes inside carbon nanotubes. Nature 412:802–805

Kou J, Yao J, Lu H, Zhang B, Li A, Sun Z, Zhang J, Fang Y, Wu F, Fan J (2015) Electromanipulating water flow in nanochannels. Angew Chem Int Ed 54:2351–2355

Le Caër S, Pin S, Esnouf S, Raffy Q, Renault JP, Brubach J-B, Creff G, Roy P (2011) A trapped water network in nanoporous material: the role of interfaces. Phys Chem Chem Phys 13:17658–17666

Lee KP, Leese H, Mattia D (2012) Water flow enhancement in hydrophilic nanochannels. Nanoscale 4:2621–2627

Leung K (2023) Finding infinities in nanoconfined geothermal electrolyte static dielectric properties, implications on ion adsorption/pairing. Nano Lett 23:8868–8874

Leung K, Greathouse JA (2022) Ab initio molecular dynamics free energy study of enhanced copper (II) dimerization on mineral surfaces. Commun Chem 5:76

Levinger NE (2002) Water in confinement. Science 298:1722–1723

Li L, Kohler F, Røyne A, Dysthe DK (2017) Growth of calcite in confinement. Crystals 7:361

Marti J, Nagy G, Guardia E, Gordillo M (2006) Molecular dynamics simulation of liquid water confined inside graphite channels: dielectric, dynamical properties. J Phys Chem B 110:23987–23994

Mattia D, Calabrò F (2012) Explaining high flow rate of water in carbon nanotubes via solid–liquid molecular interactions. Microfluidics, Nanofluidics 13:125–130

Matyushov DV (2021) Dielectric susceptibility of water in the interface. J Phys Chem B 125:8282–8293

Morikawa K, Kazoe Y, Mawatari K, Tsukahara T, Kitamori T (2015) Dielectric constant of liquids confined in the extended nanospace measured by a streaming potential method. Anal Chem 87:1475–1479

Motevaselian MH, Aluru NR (2020) Universal reduction in dielectric response of confined fluids. ACS Nano 14:12761–12770

Murota K, Saito T (2022) Pore size effects on surface charges, interfacial electrostatics of mesoporous silicas. Phys Chem Chem Phys 24:18073–18082

Murota K, Takahashi Y, Saito T (2023) Adsorption of cesium, strontium on mesoporous silicas. Phys Chem Chem Phys 25:16135–16147

Musat R, Renault JP, Candelaresi M, Palmer DJ, Le Caër S, Righini R, Pommeret S (2008) Finite size effects on hydrogen bonds in confined water. Angew Chem Int Ed 47:8033–8035

Na K, Somorjai GA (2015) Hierarchically nanoporous zeolites, their heterogeneous catalysis: current status and future perspectives. Catalysis Lett 145:193–213

Nelson J, Bargar JR, Wasylenki L, Brown Jr GE, Maher K (2018) Effects of nano-confinement on Zn (II) adsorption to nanoporous silica. Geochim Cosmochim Acta 240:80–97

Ortiz-Young D, Chiu H-C, Kim S, Voïtchovsky K, Riedo E (2013) The interplay between apparent viscosity, wettability in nanoconfined water. Nat Commun 4:1–6

Patra S, Pandey AK, Sarkar SK, Goswami A (2014) Wonderful nanoconfinement effect on redox reaction equilibrium. RSC Adv 4:33366–33369

Prelot B, Lantenois S, Chorro C, Charbonnel M-C, Zajac J, Douillard JM (2011) Effect of nanoscale pore space confinement on cadmium adsorption from aqueous solution onto ordered mesoporous silica: a combined adsorption, flow calorimetry study. J Phys Chem C 115:19686–19695

Rasaiah JC, Garde S, Hummer G (2008) Water in nonpolar confinement: From nanotubes to proteins and beyond. Annu Rev Phys Chem 59:713–740

Rodriguez J, Elola MD, Laria D (2009) Polar mixtures under nanoconfinement. J Phys Chem B 113:12744–12749

Ruiz Pestana L, Felberg LE, Head-Gordon T (2018) Coexistence of multilayered phases of confined water: The importance of flexible confining surfaces. ACS Nano 12:448–454

Sansom MS, Biggin PC (2001) Water at the nanoscale. Nature 414:157–159

Schulthess C, Taylor R, Ferreira D (2011) The nanopore inner sphere enhancement effect on cation adsorption: Sodium, nickel. Soil Sci Soc Am J 75:378–388

Senanayake HS, Greathouse JA, Ilgen AG, Thompson WH (2021) Simulations of the IR, Raman spectra of water confined in amorphous silica slit pores. J Chem Phys 154:104503

Senapati S, Chandra A (2001) Dielectric constant of water confined in a nanocavity. J Phys Chem B 105:5106–5109

Shefer I, Peer-Haim O, Leifman O, Epsztein R (2021) Enthalpic and entropic selectivity of water and small ions in polyamide membranes. Environ Sci Technol 55:14863–14875

Shen Y, Wang X, Lei J, Wang S, Hou Y, Hou X (2022) Catalytic confinement effects in nanochannels: from biological synthesis to chemical engineering. Nanoscale Adv 4:1517–1526

Shiery RC, Cooper KA, Cantu DC (2021) Computational prediction of all lanthanide aqua ion acidity constants. Inorg Chem 60:10257–10266

Striolo A, Chialvo A, Gubbins K, Cummings P (2005) Water in carbon nanotubes: Adsorption isotherms, thermodynamic properties from molecular simulation. J Chem Phys 122:234712

Sun Y, Li Y (2021) Application of surface complexation modeling on adsorption of uranium at water–solid interface: A review. Environ Pollut 278:116861

Sun Y, Yang S, Sheng G, Guo Z, Tan X, Xu J, Wang X (2011) Comparison of U (VI) removal from contaminated groundwater by nanoporous alumina and non-nanoporous alumina. Sep Purif Technol 83:196–203

Takei T, Mukasa K, Kofuji M, Fuji M, Watanabe T, Chikazawa M, Kanazawa T (2000) Changes in density, surface tension of water in silica pores. Colloid Polymer Sci 278:475–480

Teschke O, Ceotto G, De Souza E (2000) Interfacial aqueous solutions dielectric constant measurements using atomic force microscopy. Chem Phys Lett 326:328–334

Tsukahara T, Hibara A, Ikeda Y, Kitamori T (2007) NMR study of water molecules confined in extended nanospaces. Angew Chem 119:1199–1202

Wang Y (2014) Nanogeochemistry: nanostructures, emergent properties and their control on geochemical reactions and mass transfers. Chem Geol 378:1–23

Wang Y, Bryan C, Xu H, Pohl P, Yang Y, Brinker CJ (2002) Interface chemistry of nanostructured materials: Ion adsorption on mesoporous alumina. J Colloid Interface Sci 254:23–30

Wu D, Navrotsky A (2013) Small molecule–silica interactions in porous silica structures. Geochim Cosmochim 109:38–50

Zaragoza A, González MA, Joly L, López-Montero I, Canales M, Benavides A, Valeriani C (2019) Molecular dynamics study of nanoconfined TIP4P/2005 water: how confinement, temperature affect diffusion and viscosity. Phys Chem Chem Phys 21:13653–13667

Zhang N, Liu S, Jiang L, Luo M, Chi C, Ma J (2015) Adsorption of strontium from aqueous solution by silica mesoporous SBA-15. J Radioanal Nucl Chem 303:1671–1677

Zhang T, Javadpour F, Li X, Wu K, Li J, Yin Y (2020) Mesoscopic method to study water flow in nanochannels with different wettability. Phys Rev E 102:013306

Reviews in Mineralogy & Geochemistry
Vol. 91A pp. 353–382, 2025

11

Transport and Surface Complexation in Subsurface Flow-through Systems

Massimo Rolle[1,2]

[1]Institute of Applied Geosciences, Technical University of Darmstadt, Darmstadt, Germany
[2]Department of Environmental and Resource Engineering, Technical University of Denmark, Lyngby, Denmark

massimo.rolle@tu-darmstadt.de

Lucien Stolze

Earth and Environmental Sciences Area, Lawrence Berkeley National Laboratory, Berkeley, CA 94720, USA

lstolze@lbl.gov

Jacopo Cogorno

Department of Environmental and Resource Engineering, Technical University of Denmark, Lyngby, Denmark

jacogo@dtu.dk

Muhammad Muniruzzaman[1,2]

[1]Water Management Solutions, Geological Survey of Finland, Espoo, Finland
[2]Institute of Geosciences, University of Bonn, Bonn, Germany

md.muniruzzaman@gtk.fi

INTRODUCTION

In subsurface environments the cycling of nutrients and the fate and transport of contaminants is controlled by the coupling of physical flow and transport processes with biogeochemical reactions (Steefel et al. 2005; Essaid et al. 2015; Rolle and Le Borgne 2019). Surface complexation reactions occurring at the mineral–water interface in porous media are of key importance in a number of subsurface flow-through systems both under natural conditions and in engineering applications (Davis and Kent 1990; Appelo and Postma 2005; Sposito 2008; Bañuelos et al. 2023). Notable examples include the transport of pollutants in contaminated groundwater flow systems, where the chemical and electrostatic interactions between dissolved contaminants and different subsurface minerals may control the contaminant migration. Other important examples of flow-through systems in which the reaction at solid surfaces significantly impact pore water quality are the interface zones between different environmental compartments. For instance, surface/solution interactions play an important role for the composition of pore water infiltrating in soils and unsaturated zones, in determining the impact on water quality of hyporheic exchange between groundwater and surface water, and at the mixing zone between fresh and salt water in coastal aquifers. Taking into account surface complexation reactions is also important in subsurface engineering applications in which subsurface flow and transport processes are manipulated by human interventions. For instance, different approaches have been proposed for in situ remediation of contaminated

1529-6466/25/0091A-0011$05.00 (print)
1943-2666/25/0091A-0011$05.00 (online)
http://dx.doi.org/10.2138/rmg.2025.91A.11

groundwater to prevent the spreading and migration of contaminants by controlling the groundwater flow and/or by inducing or enhancing chemical and biological reactions. Other subsurface engineering fields of growing interest are river bank filtration systems and managed aquifer recharge (MAR). In MAR applications, the subsurface serves to store water for future use and typically involve the infiltration or pumping of surface water in subsurface formations. In such context, reactions at mineral–water interfaces are of interest and often of concern since they can control the release and transport of trace elements and pollutants.

Figure 1 provides a schematic illustration of important examples of subsurface processes and applications in which surface complexation reactions at the mineral/water interface are of key interest.

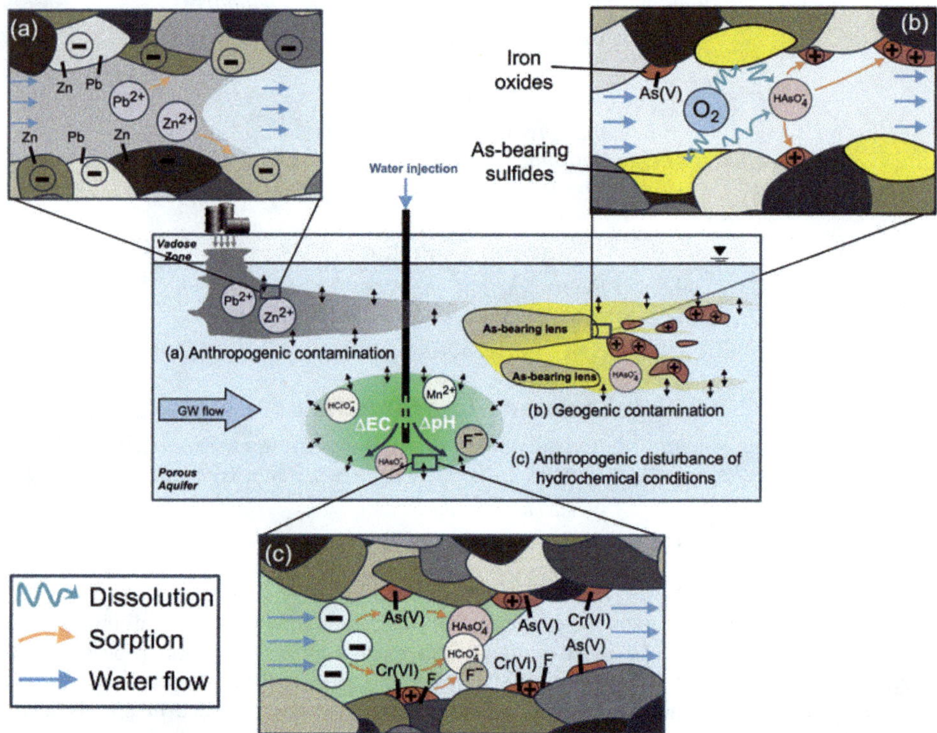

Figure 1. Schematic illustration of the environmental relevance of surface complexation reactions in subsurface flow-through systems: **(a)** plume of heavy metals from a contamination point source, **(b)** arsenic release from geogenic sources and interactions with iron minerals, **(c)** geochemical drivers of trace elements mobilization and transport in MAR applications.

In this chapter, we address the coupling of transport and surface complexation reactions in subsurface flow-through systems. In the first part, we discuss the typical scales at which these processes are investigated and we illustrate some relevant examples at the laboratory and field scales, as well as the reactive transport modeling tools that have been developed to quantitatively describe these systems. In the second part, we take a closer look at the coupling of transport and surface reactions. In particular, we focus on aspects that have recently advanced the understanding of transport processes in saturated porous media. Such aspects include the effect of incomplete mixing, due to the formation of solute concentration gradients within pore channels at high groundwater flow velocities, the impact of dimensionality and the effects of considering transport processes in multiple dimensions (i.e., 2D and fully 3D porous media), the impact of Coulombic interactions on the displacement of charge solutes, and the importance of heterogeneity. The latter is ubiquitous in subsurface formations at all scales and encompasses both physical properties (i.e., hydraulic conductivity, porosity and tortuosity) and chemical and electrostatic properties (i.e., mineral composition, surface charge). We focus on an example of surface complexation reactions on a mineral assemblage, including a quartz and a metal oxide surface, and we illustrate the behavior of such reactive system in different flow-through systems involving the non-trivial coupling between the physical mass transfer processes and the surface reactions.

SCALES FOR TRANSPORT AND SURFACE COMPLEXATION

The study of subsurface processes is performed at different scales. The key scales that are typically considered in the investigation of surface complexation reactions in flow-through systems are the laboratory and the field scales.

Laboratory scale

Experimental investigations at the laboratory scale have been typically performed in batch and column setups. These systems are complementary and provide the opportunity of exploring different aspects of surface/solution interactions. Batch experiments have been fundamental to illuminate the mechanisms of surface complexation reactions and many important works have been based on these experimental systems (e.g., Schindler et al. 1976; Dzombak and Morel 1990; Hiemstra and van Riemsdijk 1996; Davis et al. 1998; Dixit and Hering 2003). Although this approach probably remains the most widely used, the possibility to investigate surface/solution interactions in flow-through column setups has attracted the attention of a large number of researchers. Column systems offer several advantages to investigate surface complexation in subsurface porous media, such as: (i) higher solid to solution ratios representing more closely natural soils and sediments; (ii) immobile porous media with less abrasion and disturbance of mineral surfaces; (iii) water flows through the porous media, characteristic of natural subsurface environments and avoiding accumulation of reaction products; (iv) the possibility to integrate in a simple setup the effects of hydrodynamics, variable hydrochemistry, mass transfer limitations and surface reactions. Therefore, column experiments have become a fundamental tool to investigate the fate and transport of a wide number of environmental pollutants and major ions. Column experiments have been performed to investigate transport and surface complexations of key environmental contaminants such as uranium (Kohler et al. 1996; Barnett et al. 2000), arsenic (Darland and Inskeep 1997a,b; Radu et al. 2005), chromium (Jardine et al. 1999), molybdenum (Sun and Selim 2019), fluoride (Meeussen et al. 1996), phosphate (Stollenwerk 1996), heavy metals (Papini et al. 1999; Delholme et al. 2004; Hanna et al. 2009), rare earth elements (Iqbal et al. 2023), pH fronts (Scheidegger et al. 1994, McNeece and Hesse 2017; McNeece et al. 2018). Such column experiments have been performed in a wide variety of porous media, including quartz sand, granular media coated and/or mixed with specific mineral phases (e.g., iron, aluminum and manganese oxides, clay minerals, carbonates), as well as natural soils and sediments.

In a study on the mobility of heavy metals, Hanna et al. (2009) investigated sorption of lead and zinc in roadside soils. These soils can be heavily polluted and behave as a source of contaminants potentially affecting other environmental compartments. The authors performed column experiments in which the elution and transport of Pb and Zn were studied in presence of acetic acid and EDTA, at pH 5 and 7, respectively. Figure 2a shows the measured breakthrough curves of major cations and heavy metals at the outlet of the column for the case of transport with acetic acid. These conditions were selected to mimic flushing by infiltrating water containing organic acids used as additives in de-icing salts. The breakthrough of heavy metals was delayed and their peak concentrations were observed after approximately two pore volumes (PVs), whereas the major cations, released by exchange with protons, eluted earlier from the columns (1 PV). The measured Pb and Zn concentrations also show long tailing patterns due to the desorption from the soil surfaces.

Another interesting example is the study of Fakhreddine et al. (2015), who investigated the impact of water composition on arsenic desorption from sediments of a shallow aquifer in Orange County (USA). Their column experiments mimicked the injection of treated recharge water in MAR operations and showed different extents of arsenic elution from the sediments in the case of unamended recharge water compared to the scenarios in which the recharge water was enhanced in calcium and magnesium content through its amendments with lime, dolomitic lime and gypsum (Fig. 2b). The authors reported that, in presence of divalent cations, arsenic elution decreased due to the electrostatic interactions at the surface of clay minerals. The proposed mechanism of surface/solution interaction was the formation of cation-bridging complexes changing the charge of the clay and resulting in surface complexation and reduced elution of arsenate species.

Column experiments have also been used to study transport in subsurface porous media in presence of mass transfer limitations (e.g., Brusseau et al. 1989; Darland and Inskeep 1997a; Fesch et al. 1998). The occurrence of fast and transient groundwater flow, the heterogeneity in the sediment texture and in the pore space, and the presence of preferential flow paths such as macropores and fractures can lead to mass-transfer limited displacement of solutes from the bulk pore water to reactive surface sites. Significant impact of mass transfer limitations on macroscopic transport was observed during the study of uranium migration in sediments from the Hanford site (e.g., Qafoku et al. 2005; Liu et al. 2008). For instance, the work of Liu et al. (2008) investigated desorption of U(VI) from aquifer sediments packed in columns of different dimensions with fine and coarse sediments, respectively. The imposed water flow was dynamic, with intermittent and stop flow events, and resulted in a kinetic desorption of uranium from the sediments (Fig. 2c). Kinetic effects were observed in the breakthrough curves of the eluted U(VI), both from the fine, sieved sediments in the small column and from the coarse, field-textured sediments packed in the larger column. In the latter setup, mass transfer limitation effects were more pronounced and were observed not only for U(VI) but also for different conservative tracers injected in the setup.

Other interesting aspects that have emerged in column transport experiments are the effects of surface reactions on the breakthrough time and characteristic shapes (e.g., self-sharpening and/or rarefaction/diffusive waves) of eluted fronts of charged species such as metals, protons and major ions. The interpretation of observed spatial profiles and breakthrough curves can illuminate the mechanisms of surface/water interactions controlling solute transport in granular porous media (e.g, Bürgisser et al. 1994; McNeece and Hesse 2016). For instance, the

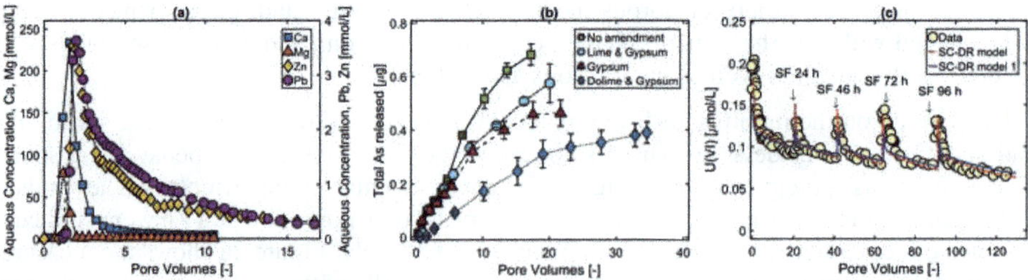

Figure 2. Examples of transport and surface complexation in column setups, focusing on: **(a)** heavy metals in soils (modified from Hanna et al. 2009); **(b)** arsenic release from sediments at a MAR site (modified from Fakhreddine et al. 2015); **(c)** uranium displacement in dynamic flow experiments (modified from Liu et al. 2008). **Symbols** refer to measurements at the column outlets and **lines** in panel (c) are model results.

interactions of hydrodynamics with surface chemistry reactions can lead to fast pH and metal pulses observed in goethite-coated glass beads columns (Prigiobbe et al. 2012; Prigiobbe and Bryant 2014). As will be discussed below, aspects that have not been covered extensively in well-controlled laboratory experiments are the study of transport and surface complexation in multidimensional setups, the effects of fast flow and incomplete mixing in permeable granular media, as well as the impact of Coulombic interactions between charged species in the pore water (i.e., multicomponent ionic transport).

Field scale

Field work at contaminated sites shed light on the importance of microscopic surface/solution interactions on the macroscopic fate and transport of contaminants of primary concern. Extensive work focusing on transport and surface complexation reactions in aquifer systems has been done at several contaminated sites such as Cape Code, USA (Kent at al. 1994; Stollenwerk et al. 1996, 1998; Davis et al. 2000), Naturita, USA (Davis et al. 2004; Curtis et al. 2006), Grindsted, Denmark (Kjoller et al. 2004), and Hanford, USA (Zachara et al. 2016). The main focus has been on reactive transport of trace metals and metalloids (e.g., uranium, zinc, nickel, molybdenum, chromium, selenium etc.), as well as on changes in hydrochemistry and transport of other charged species (e.g., phosphate, protons, fluoride, etc.).

As an example of advances from field investigation in contaminated aquifers, the Cape Code research site highlighted important aspects of transport and surface complexation in sandy aquifers. The site was equipped with a dense and high-resolution network of observation wells and thorough hydrogeological and hydrochemical investigation was performed (Le Blanc et al. 1991; Davis et al. 2000). The conditions were particularly suitable to study reactive transport under variable geochemical conditions since a sewage plume was released in the shallow permeable aquifer thus perturbing the original conditions. Beside the sewage contamination, the site was also used to conduct well-controlled tracer experiments. Different contaminants were injected in the sand and gravel aquifer, typically in mixtures with a tracer and the spatio-temporal distributions of the different solute plumes were monitored in the high-resolution network of observation points. Figure 3 shows the outcomes of a tracer experiment in which nickel, complexed with the organic ligand EDTA, and bromide were released and their displacement was monitored in successive sampling and measurement campaigns (Hess et al. 2002).

The tracer studies at Cape Cod have illuminated the distinct transport behavior of different contaminants, undergoing surface reactions with the aquifer solids, compared to the distribution of the bromide tracer. The field experiments also provided unique datasets for the development and testing of reactive transport modeling incorporating surface complexation formalisms. Detailed geochemical characterization and laboratory experiments performed with aquifer sediments and under conditions mimicking those observed in the sand and gravel aquifer were useful to constrain important properties such as the quantity and composition of the surface coatings of the solid grains (Coston et al. 1995; Fuller et al. 1996; Zhang et al. 2011).

Investigation of geogenic contamination also highlighted the key role of solid/solution interactions in subsurface flow-through systems. A notable example is the one of arsenic, whose release and transport are strongly affected by complexation reactions on mineral surfaces. In particular, arsenic has a high affinity for iron (hydr)oxide surface sites that provide large surface areas and charged surface sites. Many factors can control sorption of arsenic to these minerals, including the oxidation state of As, the pH and pore water composition, the concentration of competing ligands, and the mineralogy of the Fe-oxides (e.g., Dixit and Hering 2003; Stachowicz et al. 2006, 2008). Evidence from field studies has shown that pH variations and competitive sorption with other ions in groundwater are important mechanisms, besides the dissolution of iron (hydr)oxides, of arsenic release from aquifer sediments and responsible for the increase of arsenic concentrations in groundwater (e.g., Biswas et al. 2014; Stolze et al. 2019a). In a forced

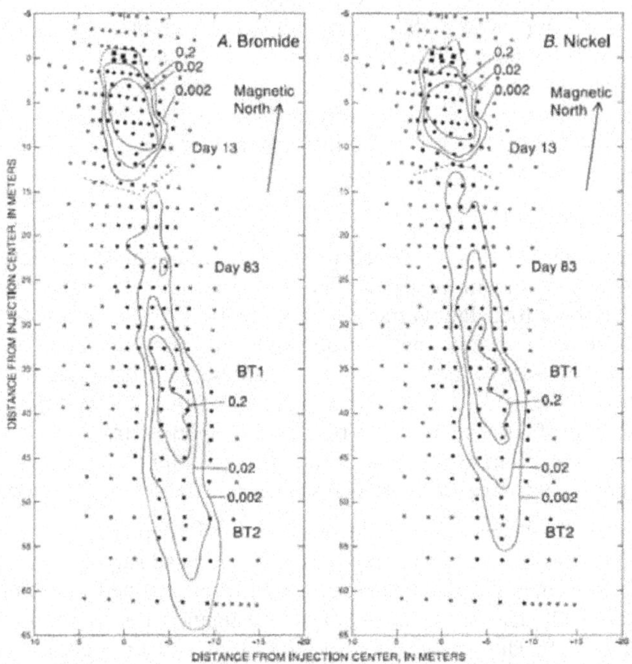

Figure 3. Plan view of the Cape Code sand and gravel aquifer with the symbols representing the injection wells (**solid squares**) and the multilevel monitoring wells (**circles**). Isocontour lines represent the concentrations of bromide **(A)** and nickel **(B)** at 13 and 83 days from the injection [Used by permission of Wiley, from Hess et al. (2002), Water Resources Research, Vol. 38, 1161, Fig. 2].

gradient field experiment in a flood-plain aquifer adjacent to a channel tributary to the Red River (Vietnam), Jessen et al. (2012) observed As(III) desorption from the sediment upon infiltration of the channel water, which perturbed the water chemistry and induced a response from the mineral surface. The measurements of arsenic and multiple solutes, including competing sorption ligands, allowed the authors to compare the performance of different surface complexation models to describe multicomponent adsorption under field conditions.

Surface complexation reactions can play an important role also in engineered field applications in subsurface flow-through systems such as managed aquifer recharge (MAR), which aims to balance water disparities by storing in the subsurface excess water volumes during wet periods and making them available for dry seasons. MAR applications are based on different infiltration (e.g., infiltration basins and riverbank filtration) and injection methods (Fakhreddine et al. 2021). Typically, in MAR operations waters with different composition such as the recharge water and the native groundwater are mixed and interact with the sediments. This can be beneficial since mixing and sorption processes can dilute or remove contaminants and improve the quality of the recharge water, but can also lead to the release of geogenic contaminants (e.g., As, F, Mn, Cr, U, Ni, Sb, etc.) via dissolution and desorption. Transport and surface complexation reactions are key processes controlling the mobility and fate of the contaminants. For instance, studies at MAR injection sites, where the release of arsenic was of primary concern (Wallis et al. 2011; Rathi et al. 2017; Fakhreddine et al. 2020), have highlighted that sorption/desorption onto/from iron hydr(oxide) is one of the key factors controlling the concentration of As in groundwater.

Reactive transport modeling

Reactive transport modeling is an essential tool to quantitatively describe water flow, solute transport and reactive processes in porous media (Steefel et al. 2005; Li et al. 2017). A number of codes have been developed to simulate reactive transport and complex biogeochemical reaction networks in the subsurface, including surface complexation reactions (Steefel et al. 2015). Examples of these advanced reactive transport simulators include PHREEQC (Parkhurst and Appelo 2013), PHAST (Parkhurst et al. 2010), CrunchFlow (Steefel and Lasaga 1994), HBGC123D (Gwo et al. 2001), PFLOTRAN (Lichtner et al. 2013), OpenGeoSys (Kolditz et al. 2012), MIN3P (Mayer et al. 2001), PHT3D (Prommer and Post 2010), HYDRUS/HPx (Simunek and van Genuchten 2008), ORCHESTRA (Meussen 2003). In recent years, the release of modules such as IPhreeqc (Charlton and Parkhurst 2011) and PhreeqcRM (Parkhurst and Wiessmeier 2015) has facilitated the coupling of advanced multidimensional and multiphysics simulators with the widely used geochemical code PHREEQC (e.g., Nardi et al. 2014; Muniruzzaman and Rolle 2016; Jara et al. 2017; Sprocati et al. 2019, 2023; Ahmadi et al. 2022a). Surface complexation reactions can be incorporated and solved in the framework of modern reactive transport simulators. Surface complexation models based on different descriptions of chemical and electrostatic interactions at the surface/solution interface, such as the constant capacitance model, the diffuse layer model and the triple layer model (e.g., Davis and Kent 1990; Appelo and Postma 1999; Koretsky 2000; Goldberg et al. 2007), have been implemented within the framework of reactive transport simulators. As described by Davis et al. (1998), applications of surface complexation models for mineral assemblage can be divided into two main categories: the CA approach, considering the interactions of the solute species with distinct minerals composing the assemblage, and the GC approach, considering sorption equilibria with generic surface sites (Dzombak et al. 2025, this volume). Both of these approaches have been successfully implemented in reactive transport simulators of subsurface flow-through systems.

A large number of experimental studies have benefited from the unique capabilities of reactive transport simulators to integrate water flow, solute transport, aqueous speciation and surface complexation reactions and, thus, to allow the quantitative interpretation of breakthrough curves and spatial profiles obtained in column experiments. In the following, we provide three examples of reactive transport modeling applied to laboratory datasets from column setups:

- As an example of successful application of the CA approach, Meeussen et al. (1996) simulated reactive transport of fluoride in a column setup with goethite-coated sand by explicitly considering the goethite and silica surfaces. The mechanistic description of transport and an electrostatic surface complexation model describing the reactions at the two surfaces allowed capturing very well not only the transport of fluoride but also important hydrochemistry dynamics such as the propagation of acidity fronts resulting from the surface reactions of protons in the goethite–silica sand column and from the change in ionic strength between injected and resident solutions.
- Parkhurst et al. (2003) simulated column experiments investigating sorption/desorption of phosphorous in sediments collected at the Cape Code field site. They employed an electrostatic surface complexation model based on the GC approach, describing protonation, deprotonation and phosphate binding reactions occurring at a generic surface site. In addition to the SCM, exchange reactions were defined for the major cations. The model captured the breakthrough curves of phosphorus and pH in column experiments under variable hydrochemistry conditions switching between the average composition of treated sewage effluent and the one of uncontaminated groundwater.
- Mass transfer limitations and their effects on surface complexation reactions have been described with reactive transport models based on different approaches. The experiments of mass transfer limited uranium transport in the columns

packed with sediments from the Hanford site described above (Fig. 2c) allowed to develop and successfully test chemical versus physical non-equilibrium models. The former approach describes diffusion-limited surface complexation reactions with a distribution of first-order kinetics of surface complexation reactions (Qafoku et al. 2005, Liu et al. 2008). In the physical non-equilibrium approach, surface complexation reactions are considered instantaneous and the mass transfer limitations are described as kinetic exchange between mobile and immobile pore water. Dual-domain (i.e., one mobile and one immobile continuum) and/or multiple continua have been used to represent mass exchange between the advective region of the porous medium and the surface complexation sites in intragrains, aggregates and coating domains (Fesch et al. 1998; Liu et al. 2008; Greskowiak et al. 2011).

Reactive transport simulations are precious tools also at the field scale. In fact, during solute and contaminant transport in aquifer systems, many physical, geochemical, electrostatic and biological processes occur simultaneously and reactive transport modeling offers a unique framework to integrate these processes. Integrating surface complexation mechanisms in field-scale reactive transport simulators is important to understand and predict the migration of contaminants of concern taking into account spatially and temporally variable hydrochemical conditions and subsurface heterogeneity, which are common conditions in field-scale applications. Many studies have demonstrated the need and added value of multispecies reactive transport models implementing surface complexation reactions and allowing to advance and improve the description of subsurface flow-through systems compared to simplified formulations based on empirical distribution coefficients and sorption isotherms (e.g., Freundlich and Langmuir models). For instance, at the Cape Code field mentioned above, reactive transport modeling was successfully applied to describe the pH change due to the release of the sewage plume and its impact on transport of different contaminants such as zinc (Kent et al. 2000, 2007), molybdenum (Stollenwerk 1996), and phosphate (Stollenwerk 1998, Kent et al. 2007). Reactive transport models implementing surface complexation reactions have also been applied to describe systems impacted by transient surface-groundwater interactions, such as those observed at the Hanford site and affecting uranium transport (Greskowiak et al. 2011; Ma et al. 2014), as well as the river–groundwater exchange important for arsenic transport documented at a field site in Vietnam (Wallis et al. 2020). Furthermore, the model-based description of transport and surface complexation reactions is important in engineering applications such as remediation of contaminated sites and managed aquifer recharge (e.g., Masi et al. 2017; Rathi et al. 2017).

Reactive transport modeling is also essential for upscaling and predicting the large-scale impact of surface complexation reactions in heterogeneous aquifer systems. Only a few studies have used reactive transport models to investigate the coupling of transport and surface reactions in multi-dimensional domains with spatially variable distributions of hydraulic properties and minerals, the latter controlling the extent of surface complexation. For instance, Wang et al. (2018) performed reactive transport simulations of Cr(VI) transport and surface complexation in 2D synthetic scenarios characterized by spatial physical and geochemical heterogeneity and investigated the effects of connectivity and spatial correlation lengths on the surface complexation capacity of the system.

A CLOSER LOOK AT SURFACE COMPLEXATION IN FLOW-THROUGH SYSTEMS: COUPLING OF PHYSICAL AND CHEMICAL PROCESSES

In this section we analyze in detail the coupling of physical processes and surface reactions in flow-through systems. We present an illustrative example of surface complexation reactions at mineral surface/solution interface and we explore different aspects of coupling this reactive system in flow-through porous media with different processes and conditions, including flow, transport, and electrostatic interactions, as well as system dimensionality. We then discuss the impact of physical, chemical and electrostatic heterogeneity on reactive transport in a two-dimensional domain representative of an aquifer cross section.

Reactive transport system

The considered reactive system focused on the interactions between natural sand and protons and major ions in the pore water solution. Although surface complexation reactions in the subsurface have been mostly studied for iron and aluminum oxides (e.g., Davis et al. 1978; Dzombak and Morel 1986, 1987; Meeussen et al. 1996; Appelo et al. 2002; Dixit and Hering 2003; Catalano et al. 2008; Hiemstra 2018), silicate minerals are the most abundant materials in aquifer systems and their importance and non-trivial surface behavior have been investigated in a number of recent studies (e.g., Sulpizi et al. 2012; Pfeiffer-Laplaud et al. 2015; McNeece and Hesse 2017; Garcia et al. 2019; Neumann et al. 2021). Furthermore, quartz grains are typically coated by a complex mineral assemblage of iron, aluminum and other oxides that, despite representing a minor fraction of the solid, impacts the overall surface chemistry behavior and the interactions with the ions in pore water. The study of the reactive system presented in this section is relevant to understand subsurface systems where the propagation of acidic fronts is of concern (Kjoller et al. 2004) and also for the migration of trace elements since pH and major ions control the surface charge behavior of minerals, which, in turn, can greatly impact the dynamics of trace elements displacement.

The conceptual model for the reactive transport system is illustrated in Figure 4. The figure highlights the interactions between the dissolved ions and the mineral surfaces on the sand grains (i.e., quartz and metal oxide minerals). Sodium ions can bind as outer-sphere complexes to the deprotonated quartz surface sites. The released protons form an acidic front that is transported downgradient by the flowing pore water but their transport is not conservative. The H^+ ions interact with the surface of the metal oxides, which becomes more positively charged. The natural sand surface was described using a "hybrid" approach, inspired both by the component additive model (Davis et al. 1998; i.e., considering two distinct surfaces: quartz and metal oxide) and by the general composite approach (Davis et al. 1998; i.e., considering a generic metal oxide to represent the natural coating of quartz sand). The generic metal oxide surface was attributed a surface charge behavior representative of iron and aluminum oxides (PZC in the range 6.7–10) and significantly different from quartz (PZC in the range 1.8–3).

Figures 4b,c present schematic illustrations of the surface complexation models used for the quartz and metal oxide surfaces, as well as the computed surface charge behavior with varying pH at different ionic strength of the pore water solution (0.1–100 mM NaBr). The adsorption processes on the quartz surface were simulated by applying the Basic Stern model (BS), whereas the triple layer model (TLM) was adopted to describe sorption on the metal oxides. These models account for the impact of electrostatic potential and have been widely used in previous studies on quartz (McNeece and Hesse 2016, 2017; McNeece et al. 2018; García et al. 2019) and oxide surfaces (Koretsky 2000; Villalobos and Leckie 2001; Peacock and Sherman 2004). They allow capturing the impact of major ions on surface protonation/deprotonation reactions over a wide range of ionic strength by distinguishing inner- and outer-sphere surface complexes (Hayes and Leckie 1987; Hayes et al. 1990). The surface properties and reactions considered in the two selected surface complexation models are summarized in Table 1.

Calibration was performed using the particle swarm optimization (PSO) method implemented in the MATLAB® environment (e.g., Rawson et al. 2016; Stolze et al. 2019b), with parallelization of the simulations to constrain the parameterization on the entire set of experimental observations.

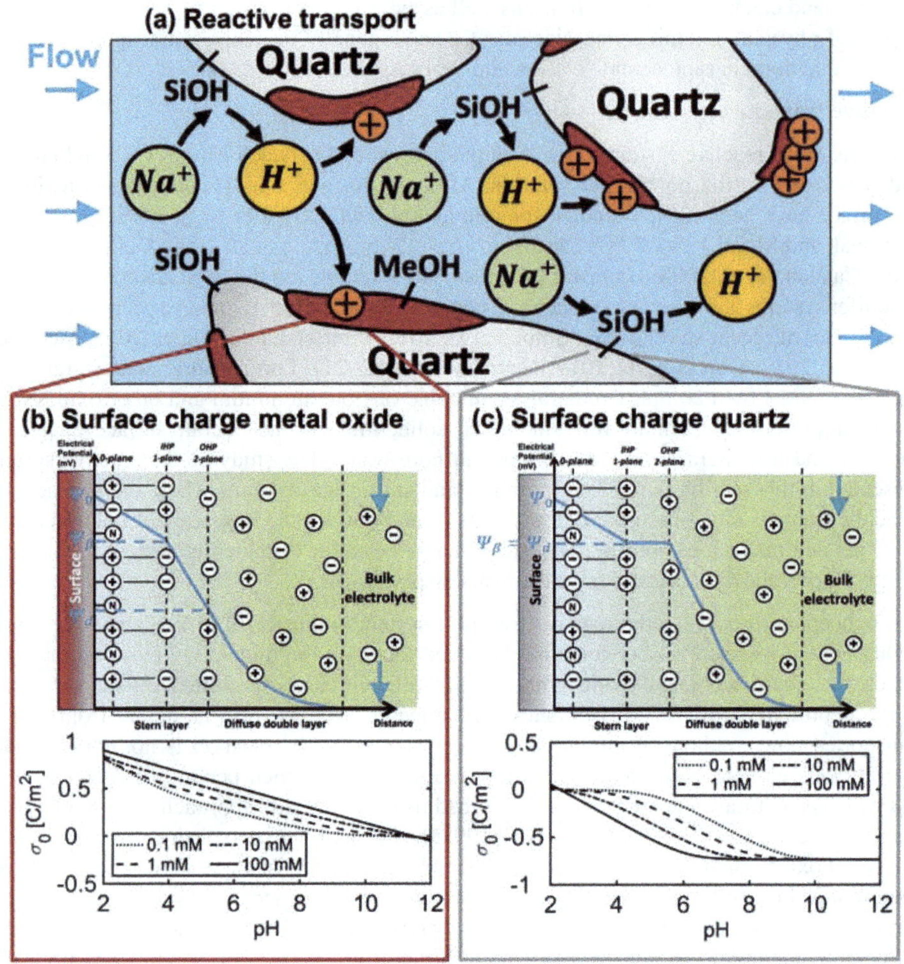

Figure 4. Conceptual model of the reactive transport system **(a)** with illustration of interactions at the mineral surfaces and computed surface charge with varying pH and ionic strength for metal oxide **(b)** and quartz **(c)**.

The governing reactive transport equations for the 1D and 2D flow-through setups discussed in the following are expressed as:

$$\theta\frac{\partial C_i}{\partial t}+(1-\theta)\rho_s\frac{\partial S_i}{\partial t}=\nabla\cdot\left(-\theta\mathbf{v}C_i+\left(\sum_{j=1}^{N}\theta\left[\mathbf{D}_{ij}\left(1+\frac{\partial\ln\gamma_j}{\partial\ln C_j}\right)\nabla C_j\right]\right)\right)-\sum_{i=1}^{N}\upsilon_i R_i \qquad (1)$$

where C_i is the concentration of a solute species, S_i is the concentration of a surface species, θ is the porosity, ρ_s is the density of the solid grains, $\mathbf{D}_{ij}$ is the cross-coupled dispersion tensor, γ_i is the activity coefficient of a solute species, R_i is the reaction rate and υ_i represents the stoichiometric coefficient. These general mass balance equations describe multicomponent ionic transport and surface complexation reactions in saturated porous media. Coulombic interactions between charged species in the pore water are included considering a Nernst–Planck based formulation of the diffusive/dispersive fluxes (Rolle et al. 2013a, 2018; Rasouli et al. 2015; Tournassat and Steefel 2015):

Table 1. Surface properties, reactions and optimized parameters for the sand porous medium.

Quartz (Basic Stern)			
Surface area A_{quartz} [m^2/g sand]	0.129		
Site density N_{quartz} [sites/nm^2]	4.6		
Capacitance C_{quartz} [F/m^2]	8.55		
pH_{PZC}, quartz [–]	1.80		
Reactions, affinity constants and charge allocation (Δz_0 and Δz_1)			
	log *K*	**Δz_0**	**Δz_1**
$SiOH + H^+ \rightleftharpoons SiOH_2^+$	0.90	1	0
$SiOH \rightleftharpoons SiO^- + H^+$	–2.7	–1	0
$SiO^- + Na^+ \rightleftharpoons SiONa$	0.76	0	1
$SiOH_2^+ + Cl^- \rightleftharpoons SiOH_2Cl$	–1.90	0	–1
$SiOH_2^+ + Br^- \rightleftharpoons SiOH_2Br$	2.15	0	–1
Metal oxide (TLM)			
Mass m_{oxide} [μg/g sand]	12		
Surface area A_{oxide} [m^2/g oxide]	968		
Site density N_{oxide} [sites/nm^2]	8.70		
Capacitance inner–Stern layer C_1, oxide [F/m^2]	1.90		
Capacitance inner–Stern layer C_2, oxide [F/m^2]	0.2		
pH_{PZC}, oxide [–]	9.03		
Reactions, affinity constants and charge allocation (Δz_0 and Δz_1)			
	log *K*	**Δz_0**	**Δz_1**
$MeOH + H^+ \rightleftharpoons MeOH_2^+$	9.31	1	0
$MeOH \rightleftharpoons MeO^- + H^+$	–8.74	–1	0
$MeO^- + Na^+ \rightleftharpoons MeONa$	–0.22	0	1
$MeOH_2^+ + Cl^- \rightleftharpoons MeOH_2Cl$	2.20	0	–1
$MeOH_2^+ + Br^- \rightleftharpoons MeOH_2Br$	1.46	0	–1
$MeOH + H^+ + CO_3^{2-} \rightleftharpoons MeOCO_2^- + H_2O$	17.40	–0.13	–0.87

$$\mathbf{J}_i = -\mathbf{D}_i \nabla C_i - \mathbf{D}_i C_i \nabla \ln \gamma_i - \mathbf{D}_i \frac{z_i \mathrm{F}}{\mathrm{R}T} C_i \nabla \Phi \quad (2)$$

where $\mathbf{D}_i$ is the self-dispersion tensor of species i.

In a one-dimensional system the cross-coupled dispersion tensor, $\mathbf{D}_{ij}$, is represented by the longitudinal component, whereas in two-dimensional systems, oriented along the principal flow direction, it has two diagonal entries, $\mathbf{D}_{ij}^{L}$ and $\mathbf{D}_{ij}^{T}$, representing the longitudinal and transverse cross-coupled dispersion matrices (Muniruzzaman et al. 2014; Muniruzzaman and Rolle 2015):

$$\mathbf{D}_{ij}^{\mathrm{L}} = \delta_{ij} D_i^{\mathrm{L}} - \frac{z_i z_j D_i^{\mathrm{L}} D_j^{\mathrm{L}} C_i}{\sum_{j=1}^{N} \left(z_j^2 D_j^{\mathrm{L}} C_j \right)} \tag{3}$$

$$\mathbf{D}_{ij}^{\mathrm{T}} = \delta_{ij} D_i^{\mathrm{T}} - \frac{z_i z_j D_i^{\mathrm{T}} D_j^{\mathrm{T}} C_i}{\sum_{j=1}^{N} \left(z_j^2 D_j^{\mathrm{T}} C_j \right)} \tag{4}$$

where δ_{ij} is the Kronecker delta, which is equal to 1 when and equal to 0 when D_i^{L} and D_i^{T} are the longitudinal and transverse components of the hydrodynamic dispersion of species i, which were parameterized according to a linear (e.g., De Carvalho and Delgado 2005) and a nonlinear compound-specific formulation (Chiogna et al. 2010; Rolle et al. 2012), respectively:

$$D_i^{\mathrm{L}} = D_i^{\mathrm{P}} + 0.5vd \tag{5}$$

$$D_i^{\mathrm{T}} = D_i^{\mathrm{P}} + D_i^{\mathrm{aq}} \left(\frac{Pe_i^2}{Pe_i + 2 + 4\delta^2} \right)^{\beta} \tag{6}$$

where D_i^{P} is the pore diffusion coefficient, approximated as the product between θ and the self-diffusion coefficient of the different ionic species D_i^{aq}, Pe is the Péclet number, $Pe_i = vd / D_i^{\mathrm{aq}}$, with d being the characteristic grain size of the porous medium. δ is the ratio between the length of a pore channel and its hydraulic radius, and β is an empirical exponent accounting for the incomplete mixing in the pore channels. δ and β were set to 5.37 and 0.5, respectively, based on a compilation of flow-through experiments (Ye et al. 2015a).

The reactive transport models used to study transport and surface complexation under the different flow-through conditions described in the next sections, were implemented in the geochemical code PHREEQC (Parkhurst and Appelo 2013) coupled with MATLAB® by using the IPhreeqc module (Charlton and Parkhurst 2011) for the 1D systems. A similar approach was employed to simulate the 2D laboratory and field-scale systems, but in these cases PHREEQC was coupled to a multidimensional and multicomponent transport simulator (Muniruzzaman and Rolle 2019) by using the module PhreeqcRM (Parkhurst and Wiessmeier 2015).

pH front propagation in 1D columns

Proton propagation fronts were investigated in a series of 21 flow-through column experiments in different porous media, including two types of natural sand and quartz beads, and considering the injection of both salt and acidic solutions (Stolze et al. 2020). Here, we report on a subset of that study, limiting our attention to one porous medium (i.e, quartz sand from Dorsten, Germany) and on the salt (NaBr) injections performed at different concentrations.

Figure 5 illustrates the observations of dissolved species at the outlet of the column setups. pH was measured with an electrode in a flow-through vial and dilution effects were removed by applying the continuous stirred tank reactor equation (Wright and Kravaris 1989). Major cations and anions were determined by inductively-coupled-plasma optical emission spectroscopy and ion chromatography. Figures 5a-d show the proton breakthrough curves at the outlet of the columns. A sharp drop in pH is observed after 1 PV followed by a slow recovery towards the pH of the injected solution. The amount of protons released from the sand surface differs considerably between the experiments performed at different ionic strengths and increases at higher injected NaBr concentrations. Moreover, a maximum capacity of proton release is observed, as shown by the similar H^+ breakthrough curves in the case of injection of the 10 and 100 mM NaBr solutions. The impact of the surface reactions on solute displacement is also apparent from the

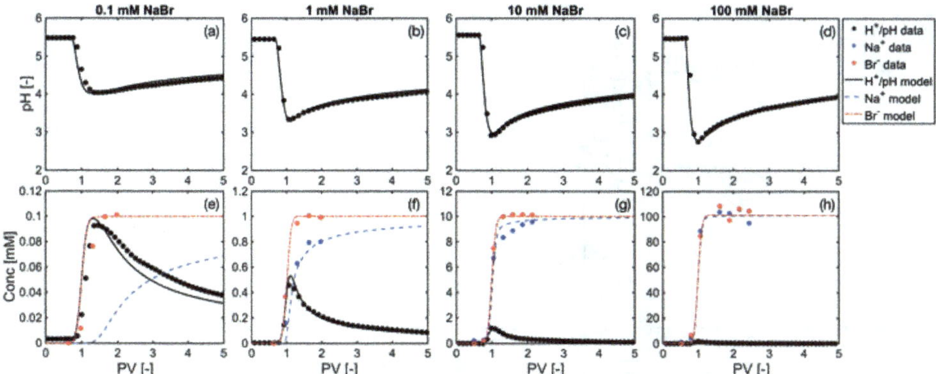

Figure 5. Measured (**symbols**) and simulated (**lines**) breakthrough of pH (**a–d**) and major ions (**e–h**) at the outlet of the columns after the continuous injection of different NaBr solutions. [Modified from Stolze et al. (2020)].

measured major ions breakthrough curves (Fig. 5e–h). While the transport behavior of the anion Br^- appears to be little affected by surface/solution interactions, the displacement of sodium is retarded. The retardation of Na^+ is particularly important at low ionic strengths, whereas its effect is less pronounced when higher salt concentrations are injected. Na^+ retardation is not visible for the highest value of NaBr injected concentration (100 mM, Fig. 5h) although in that experiment the release of protons is the highest (Fig. 5d). This behavior is due to the fact that, at high concentrations, the amount of Na^+ in solution largely exceeds the amount adsorbed.

The reactive transport simulations performed with the 1D transport and surface complexation model described above captured the experimental observations. The simulations reproduced the pH propagation fronts observed in the different porous media (Fig. 6 in Stolze et al. 2020) and showed the capability of the "hybrid" approach formulation considering both quartz and metal oxide surfaces to precisely simulate the distinct features of the observed pH fronts (Fig. 5a–d). The multicomponent reactive transport model also allowed describing the behavior of the other charged species in the column setup and the impact of ionic strength of the injected salt solution on their transport and retardation (Fig. 5e–h).

Furthermore, the model allowed exploring the surface behavior. Figure 6 displays simulated spatial profiles of different solute and surface species and their temporal dynamics. The profiles show the evolution of pH and protons (Fig. 6a–b), the change in surface composition (Fig. 6c–f) and the change of surface charge density of quartz (Fig. 6g–i) along the length of the column in the case of 100 mM NaBr injection. As the NaBr propagates through the porous medium, it contacts an increasing amount of quartz surface, thus releasing more protons along the flow path. The change in quartz surface composition indicates that the proton release is primarily due to the formation of sodium outer-sphere complexes and the concomitant decrease of SiOH and, to a lower extent, of surface species, compensating the positive charge brought to the surface by Na^+ (Fig. 6b-e). The effect of Na^+ on quartz deprotonation is also confirmed by the decrease of the charge density at the quartz 0-plane mirrored by the surface charge density increase at the surface 1-plane and, to a lesser extent, at the surface 2-plane (Fig. 6g–i).

Impact of flow velocity and incomplete mixing

In this section, we examine the same reactive transport system described above with the aim to investigate the effects of mass-transfer limitations on surface complexation reactions in granular porous media. In particular, we consider the impact of incomplete mixing induced by high pore water velocities. Under these conditions, incomplete mixing results in non-uniform solute distribution and concentration gradients within individual pore channels

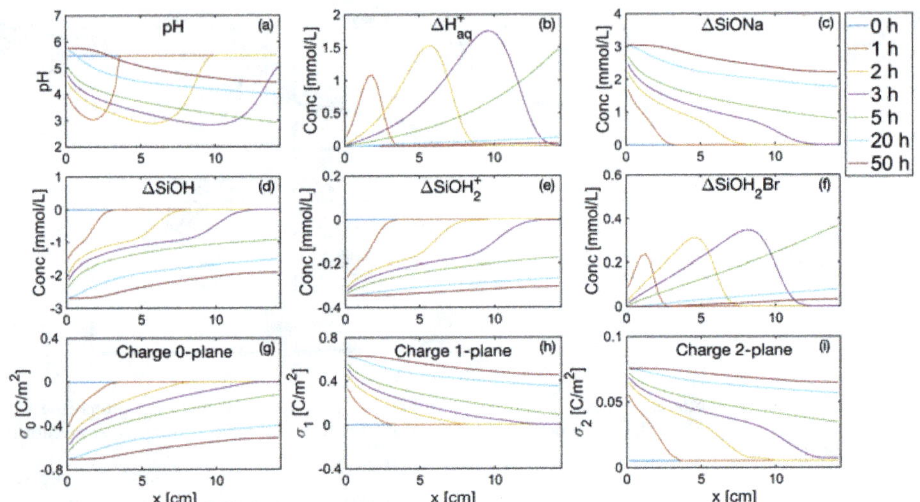

Figure 6. Simulated spatial profiles of pH **(a)**, protons released **(b)**, surface species **(c–f)** and surface charge **(g–i)** for the case of 100 mM NaBr injection [Used by permission of Elsevier, from Stolze et al. (2020), Geochimica et Cosmochimica Acta, Vol. 15, 132–149, Fig. 8].

(e.g., Li et al. 2008; Rolle et al. 2013b; Rolle and Kitanidis 2014; Jimenez-Martinez et al. 2015; Wienkenjohann et al. 2023). The effects of these pore-scale mass-transfer limitations on macroscopic transport and reactions at the mineral-solution interface have been recently investigated in a series of 36 column experiments considering sandy porous media with different grain sizes, different groundwater flow velocities and hydrochemical conditions (Stolze and Rolle 2022). The same column setup and sand described in the previous section were used. The focus was on the investigation of incomplete mixing and its impact on surface reactions and pore water quality through the continuous injection of salt solutions (NaBr at different ionic strengths) triggering changes in protonation of the sand surfaces. Figure 7 illustrates the experimental conditions, the incomplete solute mixing in pore channels, and the surface complexation and physical transport models used to simulate the flow-through experiments.

The experiments showed strong deprotonation of the sand surface and the release of H^+ in the pore water triggered by the contact of the sand grains with the injected NaBr solution. Distinct elution of protons was observed under the different tested conditions, highlighting the significant impact of the pore water flow velocity and of the porous media grain sizes on the surface reactions under flow-through conditions. Regarding the hydrochemical conditions, the highest concentration and amount of protons released in the column experiments were obtained when injecting more concentrated NaBr solutions. Differences with respect to the grain sizes were observed particularly at higher ionic strengths and the highest proton concentrations were observed for the porous medium with the smallest grain size. Finally, focusing on the flow-through conditions, the experimental results show that the applied flow rates strongly affect the concentration and the total amount of eluted protons. For all porous media and salt injections, higher pore water velocities decreased the release of protons from the reactions occurring at the sand surface.

Reactive transport simulations were performed to quantitatively interpret the experimental observations. The reactive transport model was based on the same surface complexation reactive module described above, considering the surfaces of quartz and a metal oxide with

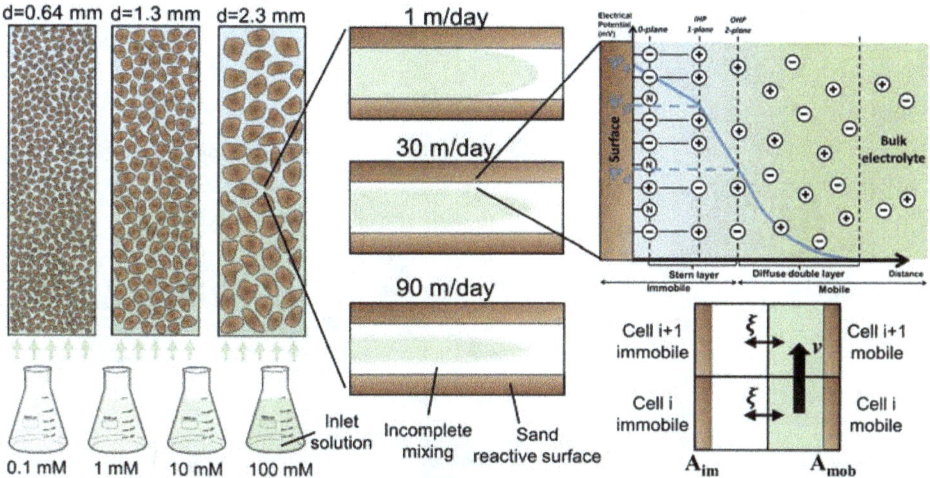

Figure 7. Setups with different grain sizes, electrolyte solutions and flow velocities to investigate the impact of pore-scale incomplete mixing on surface complexation reactions [CC-BY by Stolze and Rolle (2022), from Journal of Contaminant Hydrology, Vol. 246, 103965, Fig. 1].

a basic Stern model and a triple layer model, respectively. Transport was performed with the classic 1D advection-dispersion equation for the experiment at low flow velocity. To capture the physical non-equilibrium induced by high flow rates, a dual domain mass transfer (DDMT) formulation was adopted (e.g., Valocchi 1985; Liu et al. 2008; Muniruzzaman and Rolle 2017). The surface complexation parameters of the reactive transport models were calibrated using the experimental dataset at slow flow velocity (1 m/day) and assuming local physical equilibrium (i.e., no DDMT included in the transport description). The model allowed reproducing the H^+ breakthrough curves measured for the different grain diameters and injected solutions with a single set of surface complexation parameters, as well as the transport of the major ions. Na^+ was strongly retarded at low ionic strengths by the formation of outer-sphere complexes at the quartz surface, whereas Br^- showed little interaction with the surfaces of the considered porous media. These simulations revealed that the difference in proton release in the different porous media at the slow flow velocity primarily results from the distinct amounts of oxide coating in the sandy porous media considered. Such oxide coatings and their protonation behavior buffer the decrease in pH caused by the deprotonation of the quartz surface. The surface complexation reactive module developed and validated for the slow flow experiments (i.e., physical equilibrium conditions) was subsequently applied to the high velocity cases. In these experiments the fast flow induced physical non-equilibrium conditions causing incomplete mixing and mass transfer limitations in the pore channels. These conditions resulted in the decrease of emitted proton concentrations both at 30 and 90 m/day. The macroscopic DDMT model coupled to the tested surface complexation reactive module could reproduce the measured proton concentrations for the different grain sizes and ionic strengths tested in the experiments (Fig. 8). Conversely, a classic advection-dispersion formulation of the reactive transport model (dashed lines in Fig. 8) overestimated the H^+ concentrations measured at the outlet of the columns. This overestimation was particularly pronounced for the cases with larger grain sizes where pore-scale concentration gradients and incomplete mixing are more pronounced.

This example highlights the important interactions of physical flow and mass transfer processes with surface complexation reactions and the capability of a simple macroscopic formulation of reactive transport modeling to capture the key features and concentration patterns induced by microscopic mass transfer limitations and surface/solution interactions.

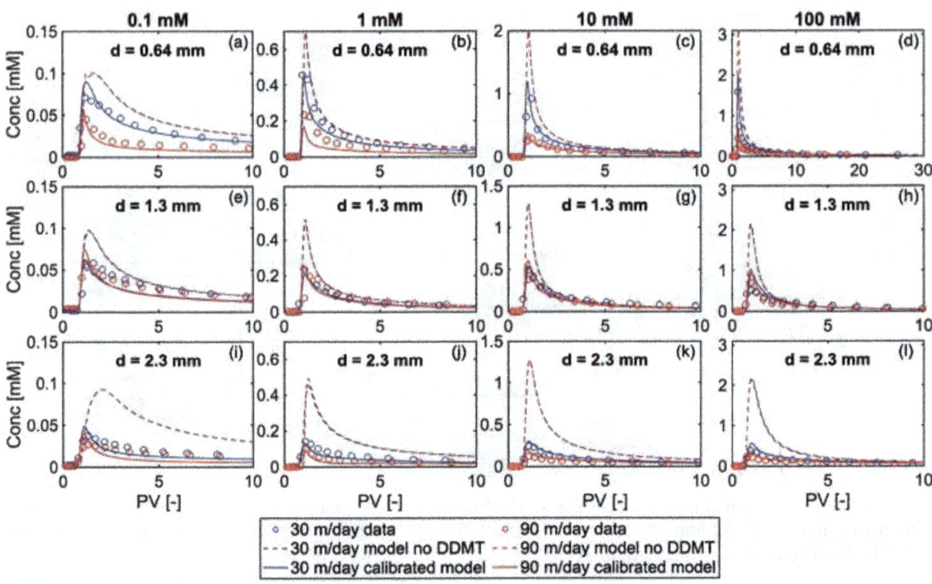

Figure 8. Measured (**symbols**) and simulated (**lines**) proton concentrations at the column outlet for the flow-through experiments performed at high flow velocity (30 and 90 m/day). The **solid lines** represent the outcomes of the dual-domain mass transfer model taking into account incomplete mixing, whereas the results of the physical equilibrium are shown by the **dashed lines** [CC-BY by Stolze and Rolle (2022) from Journal of Contaminant Hydrology, Vol. 246, 103965, Fig. 8].

Dimensionality effects

The reactive transport behavior of the sandy porous medium described above was investigated in another recent study (Cogorno et al. 2022) exploring other aspects of the coupling between physical mass transfer and surface/solution interactions. In that work, the focus was on the impact of system dimensionality and Coulombic interactions between charged species in pore water on surface complexation reactions. To this end, flow-through experiments were designed and performed in a 1D column and a 2D flow-through chamber with the same length and filled with the same sandy porous medium. The experiments were performed with equal flow velocity and inlet mass fluxes to allow a direct comparison of the results. Continuous injection of electrolyte solutions and acidic solutions were performed in both experimental setups. Figure 9a schematically illustrates the experimental setups, their geometry, inlet conditions and outlet measurements of pH and major ion concentrations.

Figure 9b illustrates the conceptual model for the case of continuous injection of the electrolyte solution (100 mM NaBr), highlighting the impact of the salt injection on the proton release from quartz and protonation of the metal oxide surfaces, as well as the Coulombic interactions between the ions in the pore water and on their lateral displacement and gradients at the plume fringe. As discussed above, in the experiments with NaBr injection in the sandy porous medium, sodium sorbs to the quartz surface and displaces protons. The released H^+ further interact with the metal oxide coatings present on the sand. The electrostatic interactions between the ions in the pore water affect the displacement of Br^-, which becomes progressively more coupled to H^+ at the plume fringe, due to the Na^+ retardation (i.e., formation of outer-sphere complexes at the quartz surfaces). Conversely, the electroneutrality of the pore water is dominated by the ions present in excess concentration (i.e., Na^+ and Br^-) in the core of the plume. The effects induced by system dimensionality were quantified by comparing the proton fluxes eluted at the outlet of the 1D column and of the 2D flow-through chamber. Figure 10a shows

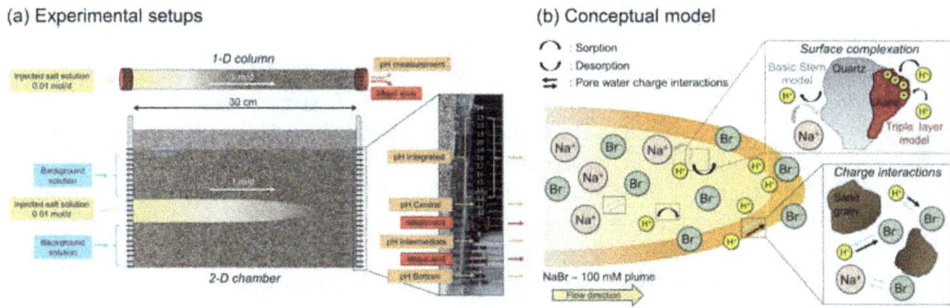

Figure 9. Schematic of the 1D column and the 2D flow-through chamber, with inset showing the sampling at the outlet of the 2D setup **(a)**. Conceptualization of charge interactions in the pore water and at the surface/solution interface during the injection of the 100 mM NaBr solution **(b)**. [Modified from Cogorno et al. (2022).]

the flux-weighted boundary-normalized integrated breakthrough curves of H^+ (i.e., normalized by the sum of the flux-weighted inflow concentrations of H^+ and Na^+) at the outlet of both experimental setups. The results are very different and show that the amount of protons released in the 2D chamber is considerably higher than in the 1D column. For instance, the peak proton concentration in the 2D system is almost three times higher than the one observed in the 1D column. Since the injected mass fluxes were identical in both setups, the increased impact of surface complexation reactions in the 2D chamber depends on the greater amount of surface area exposed to the injected Na^+, which displaces laterally due to the transverse dispersive fluxes in 2D. Modeling was performed by solving the governing multicomponent reactive transport equations in 1D and 2D (Eqs. 1-5) and considering the same component additive surface complexation description of the surface/solution reactions. The results show that the simulations capture the amount and concentration observed in both experimental setups and allowed quantifying the increase in the moles of protons released (+61%) in the 2D chamber.

Figure 10b shows the dynamics of the transverse pH fronts triggered by the continuous injection of the 100 mM NaBr solution in the 2D chamber. The vertical profiles of proton concentration initially increase to reach a maximum after ~7h from the injection of the NaBr solution and then diminish to reach steady-state conditions, at which the effects of surface complexation reactions and proton release from the surface vanish. After reaching the peak concentration, the pH fronts develop a vertically more spread shape with a plateau corresponding to the core of the plume where the surface of the quartz was saturated by the injected Na^+.

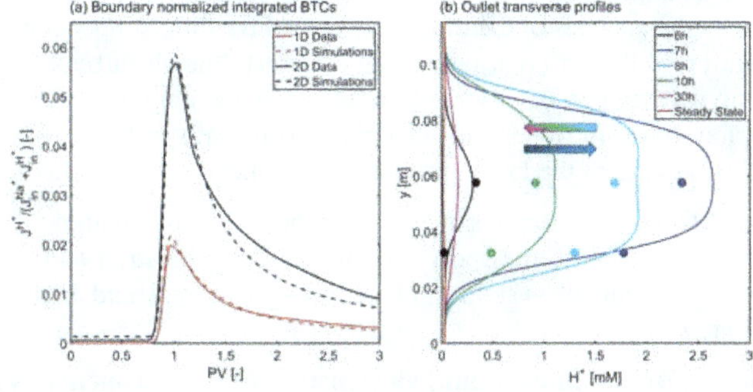

Figure 10. Measured (**solid lines**) and simulated (**dashed lines**) normalized proton fluxes at the outlet of the 1D and 2D setups **(a)**. Temporal evolution of H^+ transverse spatial profiles in the 2D setup: the symbols are the measurement at outlet ports, the lines are the simulations, and the arrows indicate the temporal development of the H^+ profiles induced by the continuous 100 mM NaBr injection **(b).** [Modified from Cogorno et al. (2022).]

The 2D multicomponent reactive transport model also allows mapping and visualizing the spatial distribution of the key dissolved and adsorbed species and surface properties in the 2D setup. Figure 11 shows the outcomes after 0.6 PV from the NaBr injection. The computed concentrations of the aqueous species are reported in the first row (Fig. 11a–c) and allow visualizing the H^+ plume induced by the surface reactions and the continuously injected plumes of Na^+ and Br^-. The quartz surface generating the acidic plume deprotonates (Fig. 11d) upon sodium outer-sphere surface complexes formation (Fig. 11e), thus retarding Na^+ displacement (Fig. 11g) and modifying the quartz surface charge density at the 0- and 1-planes (Fig. 11h and 11i). The multicomponent ionic transport features of the system are also apparent from the excess of proton and bromide concentrations (compared to sodium) at the plume fringe due to the Na^+ delay and the electroneutrality constraint in the pore water solution (Fig. 11j–l).

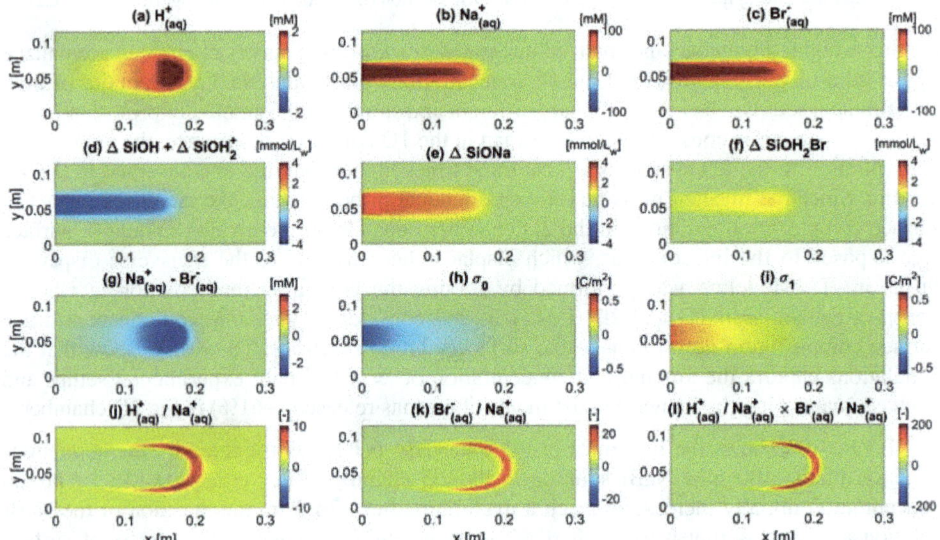

Figure 11. Simulated distribution of dissolved species **(a–c, g, j–l)**, surface species **(d–f)**, and surface charge **(h–i)** after 0.6 PV of continuous NaBr injection in the 2D setup [Used by permission, from Cogorno et al. (2022), Geochimica et Cosmochimica Acta, Vol. 318, 230-246, Fig. 9].

The impact of Coulombic interactions on the displacement of charged species in the 2D setup is illustrated by the analysis of the transverse dispersive flux components. Figure 12 shows the spatial distribution of the different terms in the Nernst–Planck fluxes (Eqn. 2) for H^+ after 0.6 PVs. Interesting patterns can be observed: for instance, the migration flux of H^+ has positive and negative contributions at the inner and outer plume fringe, respectively (Fig. 12c), depending on the electrostatic coupling of the lateral proton displacement with the other ions in solution.

The example of NaBr injection in sandy porous media discussed above reveals interesting effects of multicomponent ionic transport and surface complexation and reveals a different behavior of the same chemical system studied in experimental and modeling setups with different dimensionality.

Transport in physically, chemically and electrostatically heterogeneous media

In this last section, we explore transport and surface complexation in physically, chemically and electrostatically heterogeneous porous media at the field scale. We consider a two-dimensional domain (20 m × 2 m) representing a cross section of an aquifer in which

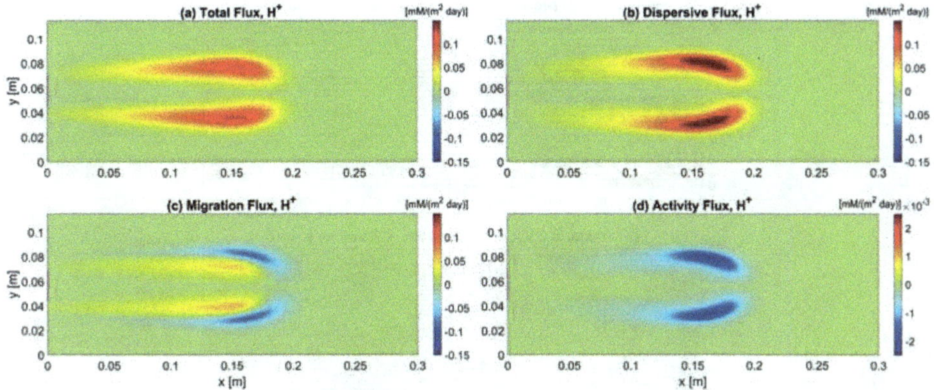

Figure 12. Maps of the different components of the Nernst–Planck fluxes for H^+ in the 2D setup after 0.6 PV [Modified from Cogorno et al. (2022)].

reactive inclusions, with the same properties and reactive surfaces (i.e., quartz and metal oxide, Table 1) considered in the experiments described above, are embedded in an inert matrix. Binary fields are generated with a stochastic approach (Dykaar and Kitanidis 1992) based on a Gaussian covariance model for an auxiliary variable and a thresholding process to obtain 20% of the domain coverage by the reactive inclusions (e.g., Lee et al. 2018; Muniruzzaman and Rolle 2023). We focus on three different scenarios with the following key features:

- Scenario 1: Physically homogeneous domain with the same hydraulic conductivity ($K = 6.14 \times 10^{-3}$ m/s) in the inert sandy matrix and in the reactive inclusions;
- Scenario 2: Physically, chemically and electrostatically heterogeneous domain with the same hydraulic conductivity as in the previous case for the reactive inclusions ($K = 6.14 \times 10^{-3}$ m/s) and high hydraulic conductivity ($K = 6.14 \times 10^{-2}$ m/s) assigned to the background inert matrix;
- Scenario 3: Physically, chemically and electrostatically heterogeneous domain with the same hydraulic conductivity as in the previous cases for the reactive inclusions ($K = 6.14 \times 10^{-3}$ m/s) and low hydraulic conductivity ($K = 6.14 \times 10^{-4}$ m/s) assigned to the background inert matrix.

Fixed head boundary conditions are applied at inlet and outlet boundaries to establish values of hydraulic gradient resulting in average flow velocities (~1 m/day) typical of sandy aquifers. The steady-state groundwater flow is solved in a rectangular grid ($\Delta x = 0.1$ m and $\Delta z = 0.01$ m) and the computed distribution of hydraulic head and stream functions allowed the construction of streamline-oriented grids (Cirpka et al. 1999) used for the solution of the multicomponent ionic transport equations (Eqn. 1). Figure 13 illustrates the three heterogeneous domains, with the spatially-distributed reactive inclusions and the computed streamlines.

In the physically homogeneous domain (Scenario 1), the computed streamlines show a regular pattern and the flow velocity is uniform in the whole domain. In the other two cases the streamline behavior is irregular due to the different hydraulic conductivities of the reactive inclusions causing focusing and defocusing of the groundwater flow (e.g., Rolle et al. 2009; Ye et al. 2015b) and resulting in spatially variable groundwater flow velocities. The considered reactive transport problem was similar to the cases investigated in the laboratory and discussed in the previous sections. A solution of 100 mM NaBr was continuously injected at the inlet boundary of the heterogeneous domains. The surface complexation reactions, occurring in the reactive inclusions, were simulated according to the same reactive module described above and

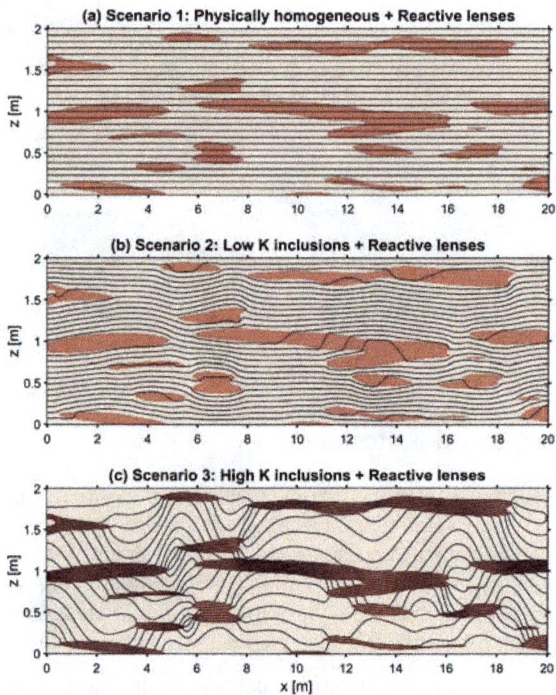

Figure 13. Heterogeneous field-scale setups. Scenario 1: Physically homogeneous with reactive inclusions **(a)**; Scenario 2: Low-permeability reactive inclusions **(b)**; Scenario 3: High-permeability reactive inclusions **(c)**.

the reactive properties summarized in Table 1. The interplay between the groundwater flow and the spatially distributed reactive zones resulted in different distribution of main chemical species within the heterogeneous domains. Figure 14 shows the spatial distribution of key aqueous and surface species for the different scenarios after flushing 0.4 pore volumes. In the physically homogeneous case (Scenario 1), the fronts of dissolved species are sharp and show that the ions have travelled until ~8 m. The H^+ distribution reflects the location of the reactive inclusions from where the protons are released (panel a1). The changes in the spatial distribution of surface species are indicative of the effects of surface complexation reactions in the reactive zones, showing a depletion of protonated sites (panel c1) and an increase of sodium surface complexes (panel d1).

In the case with low-permeability reactive inclusions (Scenario 2), the fronts of dissolved species are more irregular as a result of the superposition of the spatially distributed reactive zones and the heterogeneous groundwater flow. The latter cause a farther downgradient propagation of the transported ions compared to the physically homogeneous case due to the faster water flow in the permeable matrix surrounding the low-K zones. The reactive inclusions act as sources of protons, due to the surface reactions triggered by the salt injection, which are slowly released by back-diffusion into the permeable matrix (panel a2). Concentration gradients are also observed for the injected sodium that can slowly enter the low-permeability zones. Mass transfer limitations in this scenario are also noticeable in the spatial distribution of the surface species (panels c2 and d2). Finally, in Scenario 3, the high-permeability inclusions cause a much faster displacement of the solutes in the domain. The protons are completely released from the reactive zones on the left of the domain and a significant amount of H^+ has already reached the outlet boundary. The main Na^+ front appears to be located approximately in the middle of the domain, although the ions transported by flow focusing on the main high-K inclusions already reached the domain's outlet. The maps of the surface species also show that

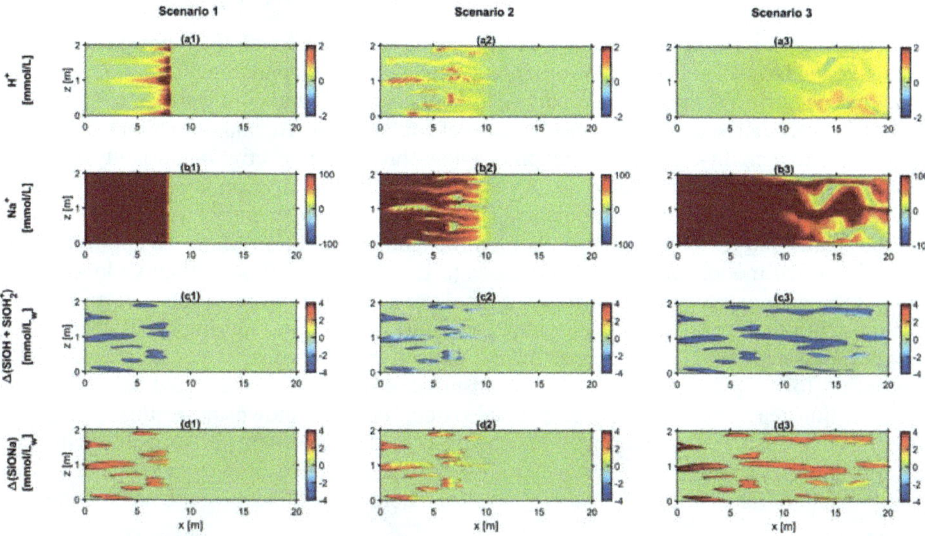

Figure 14. Computed distributions of aqueous and surface species in the heterogeneous field scale scenarios. Scenario 1: Physically homogeneous with reactive inclusions (**a1–d1**); Scenario 2: Low-permeability reactive inclusions (**a2–d2**); Scenario 3: High-permeability reactive inclusions (**a3–d3**).

surface complexation reactions have already occurred in most of the reactive inclusions present in the cross section (panels c3 and d3). The effect of the different heterogeneity considered in the simulations can be appreciated not only from the spatial distribution of ions and surface species at a given time in the domains but also from the breakthrough of solutes at the outlet.

Figure 15 shows flux-weighted breakthrough curves of pH at the outlet boundary. This integrated measure of proton concentrations also shows important differences between the considered scenarios. In the physically homogeneous scenario, the pH drop due to the deprotonation of the quartz surface is the most pronounced, reaching almost a value of pH = 2.5 at 1 PV, and the breakthrough has a very sharp front followed by a slow recovery. The low-permeability reactive inclusions (Scenario 2) result in an earlier breakthrough and also in higher pH values corresponding to the peak concentrations of protons eluted. Finally, in Scenario 3, the high-K inclusions cause a very early breakthrough and the strong flow focusing events also affect the integrated pH breakthrough curve (Fig. 15). The curve shows a faster recovery of the pore water pH, which tends to recover more rapidly towards the value of the salt solution injected at the inlet boundary.

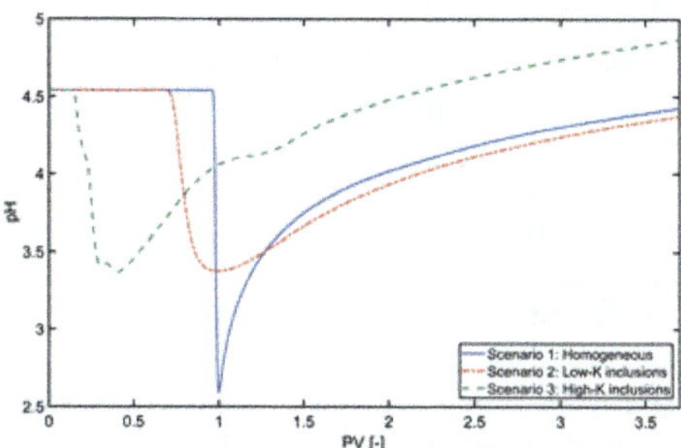

Figure 15. Flux-weighted pH breakthrough curves computed at the outlet of the heterogeneous domains.

CONCLUSIONS AND OUTLOOK

Subsurface environments are dynamic systems in which water flow, physical transport processes and biogeochemical reactions interact and determine contaminant fate and transport. Reactions occurring at the interface between mineral surfaces and pore water in porous media control the attenuation and release mechanisms of contaminants and major ions and, ultimately, the groundwater quality. Therefore, adopting a flow-through perspective on the study of surface complexation reactions is of pivotal importance for a quantitative understanding of natural and engineered subsurface environments. The inherent heterogeneity of subsurface systems and the spatial and temporal variability of hydrochemical conditions limit the applicability of (over)simplified models of sorption processes (e.g., Bethke and Brady 2002; Goldberg et al. 2007) and require rigorous approaches for the description of transport processes and surface reactions. In this work, we have provided examples highlighting the importance of surface complexation in flow-through porous media, as well as the typical scales and investigation approaches to study reactive transport in subsurface systems. For a selected case of surface complexation reactions on quartz and metal oxide, we have shown the results of coupling such reactive system, considering both chemical and electrostatic interactions, with flow and transport processes. The latter include fast flow velocities leading to incomplete mixing, dimensionality effects and the impact of physical, chemical and electrostatic heterogeneity.

Many aspects of surface complexation reactions in flow-through systems still need to be explored and/or require further investigation. In the following we outline some research directions that, in our view, are of great interest and may be promising avenues for future investigation.

There has been an evolution in the study of surface complexation mechanisms from selected contaminants and selected mineral surfaces towards considering more complex hydrochemical conditions including multicomponent adsorption and ligand competition reactions. An aspect that has not received broad attention yet is the study of such complex hydrochemical conditions in system of mineral aggregates in which the pore water composition can differently affect the charge and surface behavior of different minerals, which can also interact and change, in turn, the pore water chemistry. Another aspect that has been explored (e.g., Darland and Inskeep 1997a, Liu et al. 2008) but is not commonly investigated is the impact of varying hydrodynamic conditions, which in laboratory setups could be mimicked with stop flow events, abrupt changes, and cycles of different conditions. Testing such systems will allow representing more realistically dynamic natural environments and engineered systems and capturing key aspects of the coupling between mass transfer processes and surface reactions in flow-through systems.

Furthermore, despite decades of research, datasets on surface complexation are still limited to a few minerals and contaminants. The latter are most often heavy metals and metalloids; however, surface complexation approaches could be extended to other charged species whose fate and transport in the subsurface is of growing interest and concern such as rare earth elements (Neumann et al. 2023) and persistent and mobile organic compounds like PFAS, pharmaceuticals and pesticides (e.g., Luo et al. 2022; Cogorno and Rolle 2024a,b).

Transport in natural porous media is inherently multidimensional. Therefore, experiments in multidimensional setups will allow exploring the effects of flow and transport in 2D and fully 3D porous media on surface complexation reactions. Multidimensional setups will enable capturing important feedbacks and coupling effects between mass transfer processes and hydro- and surface chemistry that cannot be observed in classic 1D columns, such as the impact of lateral acidity and salinity gradients and the effects of electrostatically coupled displacement of ionic species on surface reactions. Multidimensional setups will also allow investigating surface complexation reactions in presence of complex patterns of water flow and solute transport induced by porous media heterogeneity and anisotropy (e.g., Chiogna et al. 2014, 2015; Ye et al. 2016), as well as chemical (e.g., Battistel et al. 2019, 2021; Salehikhoo et al. 2013; Stolze et al. 2022) and electrostatic (Muniruzzaman and Rolle 2021) heterogeneity.

The investigation of surface complexation both at laboratory and field scales has been mostly focused on saturated porous media. Future studies can explore multiphase systems, such as the vadose zone (e.g., Hopmans and van Genuchten 2005; Jimenez-Martinez et al. 2020; Stolze et al. 2024), unsaturated mine wastes (e.g., Acero et al. 2009; Muniruzzaman et al. 2020 and 2021; Muniruzzaman and Pedretti 2021; Seigneur et al. 2021) and evaporative and interface environments (e.g., Or et al. 2013; Haberer et al. 2015; Ahmadi et al. 2022b, 2024). In these systems, water chemistry and contaminant fate and transport can be controlled by two fluid phases and exchange of volatile components at the gas/water interface can affect reactions at the water/mineral interface. Surface reactions, in turn, could impact the distribution and exchange of species between the aqueous and gas phases in unsaturated porous media. Another important aspect that has received limited attention is the effect of temperature fluctuations on surface complexation reactions. Indeed, temperature can significantly impact the sorption and retention of contaminants and major ions by modifying the prevailing thermodynamic conditions, as well as the surface complexation parameters at the mineral–water interface (e.g., Lützenkirchen and Finck 2019; Machesky et al. 2023). These temperature dependent effects are of growing interest for several subsurface applications, such as hydrothermal systems and nuclear waste repository (e.g., Liu et al. 2015; Montavon et al. 2023).

Field studies on contaminant transport and surface complexation have also evolved in the last years. The investigation of groundwater plumes, which was instrumental to develop the understanding of surface/solution interactions in aquifer systems (e.g., Davis et al. 2000; Stollenwerk 1998; Curtis et al. 2006), has been accompanied by studies focusing on the interfaces between different environmental compartments (e.g., Zachara et al. 2016; Wallis et al. 2020). Besides transport and surface complexation reactions under natural flow conditions, engineered setups are also of interest. For instance, managed aquifer recharge systems, increasingly used to mitigate water scarcity, represent a unique opportunity for future understanding of the role of surface/solution interactions in aquifer systems. In fact, the controlled hydrodynamic and hydrochemical forcing of subsurface conditions, the extensive network of monitoring locations, and the multiple cycles of injection/extraction can illuminate spatial and temporal evolution of surface reactions under flow-through conditions.

Finally, the development of reactive transport simulators has greatly contributed to the quantitative understanding of surface complexation in flow-through systems. Envisioned directions of further modeling development include: (i) the coupling of pore scale and hybrid scale flow and transport models with surface complexation reactions (Maes and Geiger 2018), (ii) the extension and unification of databases and adopted formulations of surface reactions (Lützenkirchen et al. 2015), (iii) the improvement of the efficiency of multi-physics-chemistry codes and their capability to perform high-resolution simulations in multiple dimensions, (iv) the development of upscaling approaches and stochastic simulations to transfer detailed small scale process knowledge of transport and surface/solution reactions to larger field scales and to rigorously quantify uncertainty of model predictions, (v) the applications of data assimilation and inverse modeling techniques, often used in subsurface and contaminant hydrology, to image subsurface physical and chemical heterogeneity (Fakhreddine et al. 2016) and to investigate multi-solutes and multi-minerals transport and surface complexation in physically, chemically and electrostatically heterogeneous flow-through systems.

These (and many other) directions of future investigation can contribute to shed new light on the interplay of flow and transport processes with surface complexation reactions in flow-through porous media, thus highlighting and quantitatively capturing the critical role of these coupled reactive transport processes for the understanding, management and remediation of subsurface systems.

REFERENCES

Acero P, Ayora C, Carrera J, Saaltink MW, Olivella S (2009) Multiphase flow and reactive transport model in vadose tailings. Appl Geochem 24:1238–1250

Ahmadi N, Muniruzzaman M, Sprocati R, Heck K, Mosthaf K, Rolle M (2022a) Coupling soil/atmosphere interactions and geochemical processes: A multiphase and multicomponent reactive transport approach. Adv Water Resour 169:104303

Ahmadi N, Acocella M, Fries E, Mosthaf K, Rolle M (2022b) Oxygen propagation fronts in porous media under evaporative conditions at the soil/atmosphere interface: Lab-scale experiments and model-based interpretation. Water Resour Res 58: e2021WR031668

Ahmadi N, Muniruzzaman M, Cogorno J, Rolle M (2024) Oxidative dissolution of sulfide minerals in porous media under evaporative conditions: Multiphase experiments and process-based modeling. Water Resour Res 61: e2024WR037317

Appelo CAJ, Postma D (1999) A consistent model for surface complexation on birnessite (-MnO_2) and its application to a column experiment. Geochim Cosmochim Acta 63:3039–3048

Appelo CAJ, Postma D (2005) Geochemistry, Groundwater and Pollution, second ed., CRC Press, London

Appelo CAJ, van der Weiden MJJ, Torunassat C, Charlet L (2002) Surface complexation of ferrous iron and carbonate on ferrihydrite and the mobilization of arsenic. Environ Sci Technol 36:3096–3103

Bañuelos JL, Borguet E, Brown Jr GE, Cygan RT, DeYoreo JJ, Dove PM, Gaigeot MP, Geiger FM, Gibbs JM, Grassian VH, Ilgen AG (2023) Oxide– and silicate–water interfaces and their roles in technology and the environment. Chem Rev 123:6143–6544

Barnett MO, Jardine PM, Brooks SC, Selim HM (2000) Adsorption and transport of Uranium(VI) in subsurface media. Soil Sci Am J 64:908–917

Battistel M, Muniruzzaman M, Onses F, Lee J, Rolle M (2019) Reactive fronts in chemically heterogeneous porous media: experimental and modeling investigation of pyrite oxidation. Appl Geochem 100:77–89

Battistel M, Stolze L, Muniruzzaman M, Rolle M (2021) Arsenic release and transport during oxidative dissolution of spatially-distributed sulfide minerals. J Hazard Mater 409:1–14

Bethke CM, Brady PV (2002) How the K_d approach undermines ground water cleanup. Ground Water 38:435–443

Biswas A, Gustafsson JP, Neidhart H, Halder D, Kundu AM, Chatterjee D, Berner Z, Bhattacharya P (2014) Role of competing ions in the mobilization of arsenic in groundwater of Bengal Basin: Insight from surface complexation modeling. Water Res 55:30–39

Brusseau ML, Rao PSC, Jessup RE, Davidson JM (1989) Flow interruption: A method for investigating sorption nonequilibrium. J Contam Hydrol 4:223–240

Bürgisser C, Schidegger AM, Borkovec M, Sticher H (1994) Chromatographic charge density determination of materials with low surface area. Langmuir 10:855–860

Catalano JG, Park C, Fenter P, Zhang Z (2008) Simultaneous inner- and outer-sphere arsenate adsorption on corundum and hematite. Geochim Cosmochim Acta 72:1986–2004

Charlton SR, Parkhurst DL (2011) Modules based on the geochemical model PHREEQC for use in scripting and programming languages. Comput Geosci 37:1653–1663

Chiogna G, Eberhardt C, Cirpka OA, Grathwohl P, Rolle M (2010) Evidence of compound-dependent hydrodynamic and mechanical transverse dispersion by multitracer laboratory experiments. Environ Sci Technol 44:688–693

Chiogna G, Rolle M, Bellin A, Cirpka OA (2014) Helicity and flow topology in three-dimensional anisotropic porous media. Adv Water Resour 73:134–143

Chiogna G, Cirpka OA, Rolle M, Bellin A (2015) Helical flow in three-dimensional nonstationary anisotropic heterogeneous porous media. Water Resour Res 51:261–280

Cirpka OA, Frind EO, Helmig R (1999) Streamline-oriented grid generation for transport modelling in two-dimensional domains including wells. Adv Water Resour 22:697–710

Cogorno J, Rolle M (2024a) Impact of variable water chemistry on PFOS-goethite interactions: Experimental evidence and surface complexation modeling. Env Sci Technol 58:1731–1740

Cogorno J, Rolle M (2024b) Multicomponent and surface charge effects on PFOS sorption and transport in goethite-coated porous media under variable hydrochemical conditions. Env Sci Technol 58:13866–13878

Cogorno J, Stolze L, Muniruzzaman M, Rolle M (2022) Dimensionality effects on multicomponent ionic transport and surface complexation in porous media. Geochim Cosmochim Acta 318:230–246

Coston JA, Fuller CC, Davis JA (1995) Pb^{2+} and Zn^{2+} adsorption by a natural aluminium- and iron-bearing surface coating on an aquifer sand. Geochim Cosmochim Acta 59:3535–3547

Curtis GP, Davis JA, Naftz DL (2006) Simulation of reactive transport of uranium(VI) in groundwater with variable chemical conditions. Water Resour Res 42: W04404

Darland JE, Inskeep WP (1997a) Effects of pore water velocity on the transport of arsenate. Env Sci Technol 31:704–709

Darland JE, Inskeep WP (1997b) Effects of pH and phosphate competition on the transport of arsenate. J Environ Qual 26:1133–1139

Davis JA, Kent DB (1990) Surface complexation modeling in aqueous geochemistry. Rev Mineral Geochem 23:177–260

Davis J, James R, Leckie J (1978) Surface ionization and complexation at the oxide/water interface. I. Computation of electrical double layer properties in simple electrolytes. J. Colloid Interface Sci 63:480–499

Davis JA, Coston JA, Kent DB, Fuller CC (1998) Application of the surface complexation concept to complex mineral assemblages. Environ. Sci. Technol. 32(19): 2820–2828

Davis JA, Kent DB, Coston JA, Hess KM, Joye JL (2000) Multispecies reactive tracer test in an aquifer with spatially variable chemical conditions. Water Resour Res 36:119–134

Davis JA, Meece DE, Kohler M, Curtis GP (2004) Approaches to surface complexation modeling of Uranium(VI) adsorption on aquifer sediments. Geochim Cosmochim Acta 68:3621–3641

De Carvalho JRF, Delgado JMPQ (2005) Overall map and correlation of dispersion data for flow through granular packed beds. Chem Eng Sci 60:365–75

Delholme C, Hebrard-Labit C, Spadini L, Gaudet JP (2004) Experimental study and modeling of the transfer of zinc in a low reactive sand column in the presence of acetate. J Contam Hydrol 70:205–224

Dixit S, Hering J (2003) Comparison of arsenic(V) and arsenic(III) sorption onto iron oxide minerals: implications for arsenic mobility. Environ Sci Technol 37:4182–4188

Dykaar BB, Kitanidis PK (1992) Determination of the effective hydraulic conductivity for heterogeneous porous media using a numerical spectral approach: 1. Method. Water Resour Res 28:1155–1166

Dzombak D, Morel F (1986) Sorption of cadmium on hydrous ferric oxides at high sorbate/sorbent ratios: Equilibrium, kinetics and modeling. J Colloid Interf Sci 112:588–598

Dzombak D, Morel F(1987) Adsorption of inorganic pollutants in aquatic systems. J Hydraul Eng 111:430–475

Dzombak D, Morel F (1990) Surface Complexation Modeling—Hydrous Ferric Oxide. New York: Wiley

Dzombak DA, Allison JD, Lillys TP, Mills J (2025) Practical application of surface complexation models: Evolution, approaches, and examples. Rev Mineral Geochem 91A:413–456

Essaid HI, Bekins BA, Cozzarelli IM (2015) Organic contaminant transport and fate in the subsurface: evolution of knowledge and understanding. Water Resour Res 51:4861–4902

Fakhreddine S, Dittmar J, Phipps D, Dadakis J, Fendorf S (2015) Geochemical triggers of arsenic mobilization during aquifer managed recharge. Environ Sci Technol 49:7802–7809

Fakhreddine S, Lee J, Kitanidis PK, Fendorf S, Rolle M (2016) Imaging geochemical heterogeneities using inverse reactive transport modeling: An example relevant for characterizing arsenic mobilization and distribution. Adv Water Resour 88:186–197

Fakhreddine S, Prommer H, Gorelick S, Dadakis J, Fendorf S (2020) Controlling arsenic mobilization during managed aquifer recharge: The role of sediment heterogeneity. Environ Sci Technol 54:8728–8738

Fakhreddine S, Prommer H, Scanlon BR, Ying SC, Nicot JP (2021) Mobilization of arsenic and other naturally occurring contaminants during managed aquifer recharge: A critical review. Environ Sci Technol 55:2208–2223

Fesch C, Simon W, Haderlein SB, Reichert P, Schwartzenbach RP (1998) Nonlinear sorption and nonequilibrium solute transport in aggregated porous media: Experiments, process identification and modeling. J Contam Hydrol 31:373–407

Fuller CC, Davis JA, Coston JA, Dixon E (1996) Characterization of metal adsorption variability in a sand and gravel aquifer, Cape Cod, Massachussetts, U.S.A. J Contam Hydrol 22:165–187

García D, Lützenkirchen J, Petrov V, Siebentritt M, Schild D, Lefèvre G, Rabung T, Altmaier M, Kalmykov S, Duro L, Geckeis H (2019) Sorption of Eu(III) on quartz at high salt concentrations. Colloids Surf A 578:123610

Goldberg S, Criscenti LJ, Turner DR, Davis JA, Cantrell KJ (2007) Adsorption–desorption processes in subsurface reactive transport modeling. Vadose Zone J 6:407–435

Greskowiak J, Hay MB, Prommer H, Liu C, Post VEA, Ma R, Davis JA, Zheng C, Zachara JM (2011) Simulating adsorption of U(VI) under transient groundwater flow and hydrochemistry: Physical versus chemical nonequilibrium model. Water Resour Res 47: W08501

Gwo JP, D'Avezedo EF, Frenzel H, Mayes M, Yeh GT, Jardine PM, Salvage KM, Hoffman FM (2001) HBGC123D: a high-performance computer model of coupled hydrogeological and biogeochemical processes. Comput Geosci 27:1231–1242

Haberer CM, Muniruzzaman M, Grathwohl P, Rolle M (2015) Diffusive/Dispersive and reactive fronts in porous media: Fe (II)-oxidation at the unsaturated/saturated interface. Vadose Zone J 15:1–14

Hanna K, Lassabatere L, Bechet B (2009) Zinc and lead transfer in a contaminated roadside soil: Experimental study and modeling. J Hazard Mat 161:14991505

Hayes K, Leckie JO (1987) Modeling ionic-strength effects on cation adsorption at hydrous oxide–solution interfaces. J Colloid Interface Sci 115: 564–572

Hayes KF, Redden G, Wendell E, Leckie JO (1990) Surface complexation models: An evaluation of model parameter estimation using FITQL and oxide mineral titration data. J Colloid Interf Sci 142:448–469

Hess KM, Davis JA, Kent DB, Coston JA (2002) Multispecies reactive tracer test in an aquifer with spatially variable chemical conditions, Cape Cod, Massachusetts: Dispersive transport of bromide and nickel. Water Resour Res 38:1161

Hiemstra T (2018) Ferrihydrite interaction with silicate and competing oxyanions; Geometry and hydrogen bonding of surface species. Geochim Cosmochim Acta 238:453–576

Hiemstra T, van Riemsdijk W (1996) A surface structural approach to ion adsorption: the charge distribution (CD) model. J Colloid Interf Sci 179:488–508

Hopmans JW, van Genuchten MT (2005) Vadose zone—Hydrologic processes. *In:* Encyclopedia of Soils in the Environment 4:209–216

Iqbal M, Marsac R, Davranche M, Dia A, Hanna K (2023) A mechanistic surface complexation approach for the prediction of rare earth element reactive transport in quartz porous media. Chem Geol 634:121601

Jara D, de Dreuzy JR, Cochepin B (2017) TReacLab: An object-oriented implementation of non-intrusive splitting methods to couple independent transport and geochemical software. Comput Geosci 109:281–294

Jardine PM, Fendorf SE, Mayes MA, Larsen IL, Brooks SC, Bailey WB (1999) Fate and transport of hevaent chromium in undisturbed heterogeneous soil. Env Sci Technol 33:2939–2944

Jessen S, Postma D, Larsen F, Nhan PQ, Hoa LQ, Trang PTK, Long TV, Viet PH, Jakobsen R (2012) Surface complexation modeling of groundwater arsenic mobility: Results of a forced gradient experiment in a Red River flood plain aquifer, Vietnam. Geochim Cosmochim Acta 98:186–201

Jimenez-Martinez J, de Anna P, Tabuteau H, Turuban R, Le Borgne T, Méheust Y (2015) Pore-scale mechanisms for the enhancement of mixing in unsaturated porous media and implications for chemical reactions. Geophys Res Lett 42:5316–5324

Jimenez-Martinez J, Alcolea A, Straubhaar JA, Renard P (2020) Impact of phases distribution on mixing and reactions in unsaturated porous media. Adv Water Resour 144:103697

Kent DB, Davis JA, Anderson LCD, Rea BA, Waite TD (1994) Transport of chromium and selenium in the suboxic zone of a shallow aquifer: Influence of redox and adsorption reactions. Water Resour Res 30:1099–1114

Kent DB, Abrams RH, Davis JA, Coston JA, LeBlanc DR (2000) Modeling the influence of variable pH on the transport of zinc in a contaminated aquifer using semiempirical surface complexation models. Water Resour Res 36:3411–3425

Kent DB, Wilkie JA, Davis JA (2007) Modeling the movement of a pH perturbation and its impact on adsorbed zinc and phosphate in a wastewater-contaminated aquifer. Water Resour Res 43:1–17

Kjoller C, Postma D, Larsen F (2004) Groundwater acidification and the mobilization of trace metals in a sandy aquifer. Environ Sci Technol 38:2829–2835

Kohler M, Curtis GP, Kent DB, Davis JA (1996) Experimental investigation and modeling of uranium(VI) transport under variable chemical conditions. Water Resour Res 32:3539–3551

Kolditz O, Bauer S, Bilke L, Bottcher N, Delfs JO, Fischer T, Gorke UJ, Kalbacher T, Kosakowski G, McDermott CI, Park CH, Radu F, Rink K, Shao H, Shao HB, Sun F, Sun YY, Singh AK, Taron J, Walther M, Wang W, Watanabe N, Wu N, Xie M, Xu W, Zehner B (2012) OpenGeoSys: An open-source initiative for numerical simulation of thermo-hydro-mechanical/chemical (THM/C) processes in porous media. Environ. Earth Sci. 67:589–599

Koretsky C (2000) The significance of surface complexation reactions in hydrologic systems: A geochemist's perspective. J Hydrol 230:127–171

LeBlanc DR, Garabedian SP, Hess KM, Gelhar LW, Quadri RD, Stollenwerk KG, Wood WW (1991) Large-scale natural-gradient tracer test in sand and gravel, Cape Cod, Massachusetts, 1, Experimental design and observed tracer movement. Water Resour Res 27:895–910

Lee J, Rolle M, Kitanidis PK (2018) Longitudinal dispersion coefficients for numerical modeling of groundwater solute transport in heterogeneous formations. J Contam Hydrol 212:41–54

Li L, Steefel CI, Yang L (2008) Scale dependence of mineral dissolution rates within single pores and fractures. Geochim Cosmochim Acta 72:360–377

Li L, Maher K, Navarre-Sitchler A, Druhan J, Meile C, Lawrence C, Moore J, Perdrial J, Sullivan P, Thompson A, Jin L (2017) Expanding the role of reactive transport models in critical zone processes. Earth Sci Rev 165:280–301

Liu C, Zachara JM, Qafoku NP, Wang Z (2008) Scale-dependent desorption of uranium from contaminated subsurface sediments. Water Resour Rese 44:W08413

Lichtner PC, Hammond GE, Lu C, Karra S, Bisht G, Andre B, Mills RT, Kumar J (2013) PFLOTRAN User Manual: A Massively Parallel Reactive Flow and Transport Model for Describing Surface and Subsurface Processes

Liu C, Zachara JM, Qafoku NP, Wang Z (2008) Scale-dependent desorption of uranium from contaminated subsurface sediments. Water Resour Res 44:W08413

Liu X, Lu X, Cheng J, Sprik M, Wang R (2015) Temperature dependence of interfacial structures and acidity of clay edge surfaces. Geochim et Cosmochim Acta 160:91–99

Lützenkirchen J, Finck N (2019) Treatment of temperature dependence of interfacial speciation by speciation codes and temperature congruence of oxide surface charge. Appl Geochem 102:26–33

Lützenkirchen J, Marsac R, Kulik DA, Payne TE, Xue Z, Orsetti S, Haderlein SB (2015) Treatment of multi-dentate surface complexes and diffuse layer implementation in various speciation codes. Appl Geochem 55:128–137

Luo T, Pokharel R, Chen T, Boily JF, Hanna K (2022) Fate and transport of pharmaceuticals in iron and manganese binary oxide coated sand columns. Environ Sci Technol 57:214–221

Ma R, Liu C, Greskowiak J, Prommer H, Zachara J, Zheng C (2014) Influence of calcite on uranium(VI) reactive transport in the groundwater–river mixing zone. J Contam Hydrol 156:27–37

Machesky ML, Kubicki JD, Palmer DA, Wesolowski DJ (2023) Zn^{2+}, Co^{2+}, and Ni^{2+} adsorption at the rutile–water interface to 250 °C. ACS Earth Space Chem 7:1781–1790

Maes J, Geiger S (2018) Direct pore-scale reactive transport modelling of dynamic wettability changes induced by surface complexation. Adv Water Resour 111:6–19

Masi M, Ceccarini A, Iannelli R (2017) Multispecies reactive transport modelling of electrokinetic remediation of harbour sediments. J Hazard Mater 326:187–196

Mayer KU, Blowes DW, Frind EO (2001) Reactive transport modeling for the treatment of an in situ reactive barrier for the treatment of hexavalent chromium and trichloroethylene in groundwater. Water Resour Res 37:3091–3103

McNeece CJ, Hesse M (2016) Reactive transport of aqueous protons in porous media. Adv Water Resour 97:314–325

McNeece CJ, Hesse M (2017) Challenges in coupling acidity and salinity transport in porous media. Environ Sci Technol 51:11799–11808

McNeece CJ, Lützenkirchen J, Hesse M (2018) Chromatographic analysis of the acidity–salinity transport system. J Contam Hydrol 216:27–37

Meeussen JCL (2003) ORCHESTRA: An object-oriented framework for implementing chemical equilibrium models. Environ Sci Technol 37:1175–1182

Meeussen JCL, Scheidegger A, Hiemstra T, van Riemsdijk, Borkovec M (1996) Predicting multicomponent adsorption and transport of fluoride at variable pH in a goethite–silica sand system. Environ Sci Technol 30:481–488

Montavon G, Ribet S, Bailly C, Hassan Loni Y, Madé B, Grambow B (2023) U(VI) retention in compact Callovo-Oxfordian clay stone at temperature (20–80 °C); What is the applicability of adsorption models? Appl Clay Sci 244:107093

Muniruzzaman M, Pedretti D (2021) Mechanistic models supporting uncertainty quantification of water quality predictions in heterogeneous mining waste rocks: a review. Stoch Environ Res Risk Assess 35:985–1001

Muniruzzaman M, Rolle M (2016) Modeling multicomponent ionic transport in groundwater with IPhreeqc coupling: Electrostatic interactions and geochemical reactions in homogeneous and heterogeneous domains. Adv Water Resour 98:1–15

Muniruzzaman M, Rolle M (2017) Experimental investigation of the impact of compound-specific dispersion and electrostatic interactions on transient transport and solute breakthrough. Water Resour Res 53:1189–1209

Muniruzzaman M, Rolle M (2019) Multicomponent ionic transport modeling in physically and electrostatically heterogeneous porous media with PhreeqcRM coupling for geochemical reactions. Water Resour Res 55:11121–11143

Muniruzzaman M, Rolle M (2021) Impact of diffuse layer processes on contaminant forward and back diffusion in heterogeneous sandy–clayey domains. J Contam Hydrol 237:1–16 Muniruzzaman M, Rolle M (2015) Impact of multicomponent ionic transport on pH fronts propagation in saturated porous media. Water Resour Res 51:6739–6755

Muniruzzaman M, Rolle M (2023) Relevance of charge interactions for contaminant transport in heterogeneous formations: a stochastic analysis. Stoch Environ Res Risk Assess 37:4399–4416

Muniruzzaman M, Haberer CM, Grathwohl P, Rolle M (2014) Multicomponent ionic dispersion during transport of electrolytes in heterogeneous porous media: Experiments and model-based interpretation. Geochim Cosmochim Acta 141:656–669

Muniruzzaman M, Karlsson T, Ahmadi N, Rolle M (2020) Multiphase and multicomponent simulation of acid mine drainage in unsaturated mine waste: modeling approach, benchmarks and application examples. Appl Geochem 120:104677

Muniruzzaman M, Karlsson T, Ahmadi N, Kauppila P, Kauppila T, Rolle M (2021) Weathering of unsaturated waste rocks from Kevitsa and Hitura mines: Pilot-scale lysimeter experiments and reactive transport modeling. Appl Geochem 130:104984

Nardi A, Idiart A, Trinchero P, de Vries LM, Molinero J (2014) Interface COMSOL-PHREEQC (iCP) an efficient numerical framework for the solution of coupled multiphysics and geochemistry. Comput Geosci 69:10–21

Neumann J, Brinkmann H, Britz S, Lützenkirchen J, Bok F, Stockmann M, Brendler V, Stumpf T, Schmidt M (2021) A comprehensive study of the sorption mechanism and thermodynamics of f-element sorption onto K-feldspar. J Coll Interf Sci 591:490–499

Or D, Lehmann P, Shahraeeni E, Shokri N (2013) Advances in soil evaporation physics – A review. Vadose Zone J 12:1–16

Papini MP, Kahie YD, Troia B, Majone M (1999) Adsorption of lead at variable pH onto a natural porous medium: Modeling of batch and column experiments. Environ Sci Technol 33:44574464

Parkhurst DL, Appelo CAJ (2013) Description of input and examples for PHREEQC version 3—a computer program for speciation, batch-reaction, one-dimensional transport, and inverse geochemical calculations, U.S. Geological Survey Techniques and Methods, book 6, chap. A43, 497 p., available only at http://pubs.usgs.gov/tm/06/a43

Parkhurst DL, Weissmeier L (2015) PhreeqcRM: A reaction module for transport simulators based on the geochemical model PHREEQC. Adv Water Resour 83:176–189

Parkhurst DL, Stollenwerk KG, Colman JA (2003) Reactive-transport simulation of phosphorus in the sewage plume at the Massachusetts Military Reservation, Cape Cod, Massachusetts. USGS Water-Resources Investigations Report 2003–4017

Parkhurst DL, Kipp KL, Charlton SR (2010) PHAST Version 2— a program for simulating groundwater flow, solute transport, and multicomponent geochemical reactions: U.S. Geological Survey Techniques and Methods 6–A35, p. 235 Peacock CL, Sherman DM (2004) Copper(II) sorption onto goethite, hematite and lepidocrocite: A surface complexation model based on ab initio molecular geometries and EXAFS spectroscopy. Geochim Cosmochim Acta 68:2623–2637

Pfeiffer-Laplaud M, Costa D, Tielens F, Gaigeot MP, Sulpizi M (2015) Bimodal acidity at the amorphous silica–water interface. J Phys Chem C 119:27354–27362

Prigiobbe V, Bryant SL (2014) pH-dependent transport of metal cations in porous media. Environ Sci Technol 48:3752–3759

Prigiobbe V, Hesse MA, Bryant SL (2012) Fast strontium transport induced by hydrodynamic dispersion and pH-dependent sorption. Geophys Res Lett 39: L18401

Prommer H, Post VEA (2010) PHT3D, A Reactive Multicomponent Transport Model for Saturated Porous Media. User's Manual v2.10, http://www.pht3d.org

Qafoku NP, Zachara JM, Liu C, Gassman PL, Qafoku OS, Smith S (2005) Kinetic desorption and sorption of U(VI) during reactive transport in a contaminated Hanford sediment. Environ Sci Technol 39:3157–3165

Radu T, Subacz J, Phillippi J, Barnett MO (2005) Effects of dissolved carbonate on arsenic adsorption and mobility. Environ Sci Technol 39:7875–7882

Rasouli P, Steefel CI, Mayer UK, Rolle M (2015) Benchmarks for multicomponent diffusion and electrochemical migration. Computat Geosci 19:523–533

Rawson J, Prommer H, Siade AJ, Carr J, Berg M, Davis JA, Fendorf S (2016) Numerical modeling of arsenic mobility during reductive iron-mineral transformations. Environ Sci Technol 50:2459–2467

Rathi B, Siade AJ, Donn MJ, Helm L, Morris R, Davis JA, Berg M, Prommer H (2017) Multiscale characterization and quantification of arsenic mobilization and attenuation during injection of treated coal seam gas coproduced water into deep aquifers. Water Resour Res 53:10779–10801

Rolle M, Kitanidis PK (2014) Effects of compound-specific dilution on transient transport and solute breakthrough: A pore-scale analysis. Adv Water Resour 71:186–199

Rolle M, Le Borgne T (2019) Mixing and Reactive Fronts in the Subsurface. Rev Mineral Geochem 85:111–142

Rolle M, Eberhardt C, Chiogna G, Cirpka OA, Grathwohl P (2009) Enhancement of dilution and transverse reactive mixing in porous media: Experiments and model-based interpretation. J Contam Hydrol 110:130–142

Rolle M, Hochstetler DL, Chiogna G, Kitanidis PK, Grathwohl P (2012) Experimental investigation and pore-scale modeling interpretation of compound-specific transverse dispersion in porous media. Transport Porous Med 93:347–362

Rolle M, Muniruzzaman M, Haberer CM, Grathwohl P (2013a) Coulombic effects in advection-dominated transport of electrolytes in porous media: Multicomponent ionic dispersion. Geochim Cosmochim Acta 120:195–205

Rolle M, Chiogna G, Hochstetler DL, Kitanidis PK (2013b) On the importance of diffusion and compound-specific mixing for groundwater transport: An investigation from pore to field scale. J Contam Hydrol 153:51–68

Rolle M, Sprocati R, Masi M, Jin B, Muniruzzaman M (2018) Nernst–Planck based description of transport, Coulombic interactions and geochemical reactions in porous media: Modeling approach and benchmark experiments. Water Resour Res 54:3176–3195

Salehikhoo F, Li L, Brantley SL (2013) Magnesite dissolution rates at different spatial scales: The role of mineral spatial distribution and flow velocity. Geochim Cosmochim Acta 108:91–106

Scheidegger A, Bürgisser C, Borkovec M, Sticher H, Meeussen H, van Riemsdijk (1994) Convective transport of acid and bases in porous media. Water Resour Res 30:2937–2944

Schindler PW, Fürst B, Dick R, Wolf PU (1976) Ligand properties of surface silanol groups. 1. Surface complex formation with Fe^{3+}, Cu^{2+}, Cd^{2+} and Pb^{2+}. J Coll Interf Sci 55:469–475

Seigneur N, Vrines B, Beckie RD, Mayer KU (2021) Reactive transport modelling to investigate multi-scale waste rock weathering processes. J Contam Hydrol 236:103752

Simunek J, van Genuchten MT (2008) Modeling non-equilibrium flow and transport processes using HYDRUS. Vadose Zone J 7:782–797

Sposito G (2008) The chemistry of soils. Oxford University Press, Inc., New York, USA

Sprocati R, Masi M, Muniruzzaman M, Rolle M (2019) Modeling electrokinetic transport and biogeochemical reactions in porous media: A multidimensional Nernst–Planck–Poisson approach with PHREEQC coupling. Adv Water Resour 127:134–147

Sprocati R, Gallo A, Caspersen MB, Rolle M (2023) Impact of variable density on electrokinetic transport and mixing in porous media. Adv Water Resour 174:104422

Stachowicz M, Hiemstra T, van Riemsdijk W (2006) Surface speciation of As(III) and As(V) in relation to charge distribution. J Colloid Interf Sci 302:62–75

Stachowicz M, Hiemstra T, van Riemsdijk W (2008) Multi-competitive interactin of As(III) and As(V) oxyanions with Ca^{2+}, Mg^{2+}, and ions on goethite. J Colloid Interface Sci 320:400–414

Steefel CI, Lasaga AC (1994) A coupled model for transport of multiple chemical species and kinetic precipitation/dissolution reactions with application to reactive flow in single phase hydrothermal systems. Am J Sci 294:529–592

Steefel CI, DePaolo DJ, Lichtner PC (2005) Reactive transport modeling: An essential tool and a new research approach for the Earth sciences. Earth Planet Sci Lett 240:539–558

Steefel CI, Appelo CA, Arora B, Jacques D, Kalbacher T, Kolditz O, Lagneau V, Lichtner PC, Mayer KU, Meeussen JC, Molins S (2015) Reactive transport codes for subsurface environmental simulation. Comput Geosci 19:445–78

Stollenwerk KG (1996) Simulation of phosphate transport in sewage-contaminated groundwater, Cape Cod, Massachusetts. Appl Geochem 11:317–324

Stollenwerk KG (1998) Molybdate transport in a chemically complex aquifer: Field measurements compared with solute-transport model predictions. Water Resour Res 34:2727–2740

Stolze L, Rolle M (2022) Surface complexation reactions in sandy porous media: Effects of incomplete mixing and mass-transfer limitations in flow-through systems. J Contam Hydrol 246:103965

Stolze L, Zhang D, Guo H, Rolle M (2019a) Surface complexation modeling of arsenic mobilization from goethite: Interpretation of an in-situ experiment. Geochim Cosmochim Acta 248:274–288

Stolze L, Zhang D, Guo H, Rolle M (2019b) Model-based interpretation of groundwater arsenic mobility during reductive transformation of ferrihydrite. Environ Sci Technol 53:6845–6854

Stolze L, Wagner JB, Damsgaard C, Rolle M (2020) Impact of surface complexation and electrostatic interactions on pH front propagation in silica porous media. Geochim Cosmochim Acta 15:132–149

Stolze L, Battistel M, Rolle M (2022) Oxidative dissolution of arsenic-bearing sulfide minerals in groundwater: Impact of hydrochemical and hydrodynamic conditions on arsenic release and surface evolution. Environ Sci Technol 56:5049–5061

Stolze L, Arora B, Dwivedi D, Steefel CI, Bandai T, Wu Y, Nico P (2024) Climate forcing controls on carbon terrestrial fluxes during shale weathering. PNAS 121: e2400230121

Sulpizi M, Gaigeot M, Sprik M (2012) Silica-water interface: How the silanols determine the surface acidity and modulate the water properties. J Chem Theory Comput 8:1037–1047

Sun W, Selim HM (2019) Transport and retention of Molybdenum(VI) on iron oxide-coated sand: A modified multi reaction model. Appl Geochem 108:104387

Tournassat C, Steefel CI (2015) Ionic transport in nano-porous clays with consideration of electrostatic effects. Rev Mineral Geochem 80:287–329

Valocchi A (1985) Validity of the local equilibrium assumption for modeling sorbing solute transport through homogeneous soils. Water Resour Res 21:808–820

Villalobos M, Leckie JO (2001) Surface complexation modeling and FTIR study of carbonate adsorption to goethite. J Colloid Interf Sci 235:15–32

Wallis I, Prommer H, Pichler T, Post V, Norton SB, Annable MD, Simmons CT (2011) Process-based reactive transport model to quantify arsenic mobility during aquifer storage and recovery of potable water. Environ Sci Technol 45:6924–6831

Wallis I, Prommer H, Berg M, Siade AJ, Sun J, Kipfer R (2020) The river-groundwater interface as a hotspot for arsenic release. Nat Geosci 13:288–295

Wang L, Wen H, Li L (2018) Scale dependence of surface complexation capacity and rates in heterogeneous media. Sci Tot Environ 635:1547–1555

Wienkenjohann H, Jin B, Rolle M (2023) Diffusive-dispersive isotope fractionation of chlorinated ethenes in groundwater: The key role of incomplete mixing and its multi-scale effects. Water Resour Res 59: e2022WR034041

Wright RA, Kravaris C (1989) Nonlinear pH control in a CSTR. American Control Conf Proc, p 1540–1544

Ye Y, Chiogna G, Cirpka OA, Grathwohl P, Rolle M (2015a) Experimental investigation of compound-specific dilution of solute plumes in saturated porous media: 2-D vs. 3-D flow-through systems. J Contam Hydrol 172:33–47

Ye Y, Chiogna G, Cirpka OA, Grathwohl P, Rolle M (2015b) Enhancement of plume dilution in three-dimensional porous media by flow-focusing in high-permeability inclusions. Water Resour Res 51:5582–5602

Ye Y, Chiogna G, Cirpka OA, Grathwohl P, Rolle M (2016) Experimental investigation of transverse mixing in porous media under helical flow conditions. Phys Rev E 94:013113

Zachara JM, Chen X, Murray C, Hammond G (2016) River stage influences on uranium transport in a hydrologically dynamic groundwater surface water transition zone. Water Resour Res 52:1568–1590

Zhang S, Kent DB, Elbert DC, Shi Z, Davis JA, Veblen DR (2011) Mineralogy, morphology, and textural relationships in coatings on quartz grains in sediments in a quartz-sand aquifer. J Contam Hydrol 124:57–67

Reviews in Mineralogy & Geochemistry
Vol. 91A pp. 383–412, 2025

12

History, Algorithms, Model Uncertainty, and Common Pitfalls of Traditional SCM Fitting Procedures

Norbert Jordan

Helmholtz-Zentrum Dresden-Rossendorf e.V., Institute of Resource Ecology, Bautzner Landstraße 400, 01328 Dresden, Germany

n.jordan@hzdr.de

Frank Heberling

Institute for Nuclear Waste Disposal, Karlsruhe Institute of Technology, Hermann-von-Helmholtz-Platz 1, 76344 Eggenstein-Leopoldshafen, Germany

frank.heberling@kit.edu

Jeffrey Kelling

Helmholtz-Zentrum Dresden – Rossendorf e.V., Institute of Radiation Physics, Bautzner Landstraße 400, 01328 Dresden, Germany
Faculty of Natural Sciences, Chemnitz University of Technology, Straße der Nationen 62, Chemnitz 09111, Germany

j.kelling@hzdr.de

Johannes Lützenkirchen

Institute for Nuclear Waste Disposal, Karlsruhe Institute of Technology, Hermann-von-Helmholtz-Platz 1, 76344 Eggenstein-Leopoldshafen, Germany

johannes.luetzenkirchen@kit.edu

INTRODUCTION

High quality experimental data, realistic, and spectroscopically confirmed surface species with ascertained stoichiometries are state-of-the-art for deriving thermodynamic constants for surface complexation models (SCMs) of well-defined surfaces. As these models manage to capture more and more physical details, protonation and ion affinity constants approach towards reliable thermodynamic parameters. From the early 1960s where "trial and error" as well as graphical approaches were the only available options for model calibration, to the introduction of numerical fitting software as well as shell optimizers, the world of surface complexation modeling underwent spectacular developments, one of the latest being the introduction and application of artificial intelligence. Nevertheless, understanding of the optimization procedures, discussion of the adjustable parameters, their respective uncertainties and mutual parameter interdependencies, together with the awareness of common pitfalls are still required of model developers and users. In this chapter, we aim at briefly describing the history of SCM fitting procedures, sketch the most commonly applied parameter optimization algorithms, emphasize the importance of experimental uncertainties and related model uncertainties, and drive the attention of the readers to some pitfalls. Ultimately, a comparative fitting exercise of surface charge data from literature with different codes and shell optimizers is discussed.

1529-6466/25/0091A-0012$05.00 (print)
1943-2666/25/0091A-0012$05.00 (online)
http://dx.doi.org/10.2138/rmg.2024.91A.12

NUMERICAL APPROACHES TO SURFACE COMPLEXATION AND PARAMETER OPTIMIZATION

History

The history of numerical calculations of surface chemical equilibria in aqueous suspensions was triggered, as the concomitant experimental work, by the related but earlier research in aqueous solutions. It is obvious that the treatment of complexation reactions in aqueous solutions is so similar to that of interfacial chemical reactions, that the subsequent developments are closely related. This concerns the analogy between dissolved ligands that can complex, for example, metal ions and surface ligands (e.g., surface hydroxyls on oxidic surfaces) that may form surface complexes with dissolved metal ions in an adsorption reaction. In the mostly environmentally oriented communities dealing with contaminant uptake on mineral surfaces, the spectroscopic works involving various techniques have shown that inner-sphere surface complexes do exist. Such reactions need to be taken into account for surfaces and in solutions, in models capable of handling naturally relevant solution compositions with far higher complexity than simple electrolytes like NaCl. On the other extreme, in numerous publications on applied interfacial phenomena, the possibilities and advantages of surface complexation models are disregarded, and instead, uptake data are modelled with e.g., empirical Langmuir, Freundlich, Temkin or similar isotherms. The fit parameters of these empirical approaches are only valid under the conditions at which experimental data where acquired, i.e., such models do not allow the possibility to transfer the obtained adsorption parameters to e.g., a different pH value or to situations involving the presence of potentially complexing or specifically adsorbing ligands (Martínez et al. 2023), and are therefore outdated. The huge advantage of surface complexation models, as for aqueous solution equilibria, is embedded in the framework of stability constants and activity models plus the electrostatic terms that allow to build self-consistent databases that consider all (known/required) species in solution and on surfaces. The stability constants for all those species are obtained from fits to appropriate experimental data. While in the absence of specific knowledge about the structure and composition, the best fit and simplest models were chosen initially to represent experimental results, nowadays a plethora of methods is available to obtain information about structure and composition of mineral–water interfaces, and this information is used as input to design constraints to the chemical equilibrium models. This progress is amazing. It is expected that future progress, via artificial intelligence (AI) e.g., involving big-data approaches, will facilitate the design of self-consistent databases (Li and Zarzycki 2022). The numerical treatment of the stability constant determination for simultaneous equilibria in multi-component systems appears to be a case that is ideal for AI, since it is in principle a very systematic procedure. AI may permit to automatically update databases with new information, correct existing errors, to identify inconsistencies and gaps in the experimental information or parameter space.

As explained above, there is a logical correspondence between the treatment of aqueous solution and surface chemistry, which in both cases follow a step-wise procedures. At first in SCMs, one needs to create a model able to describe the surface charge data of the solid phase, and then a model to describe the binding of contaminants at the surface can be developed. The first step always involves acquisition of pH- or concentration-dependent properties of the individual components of a target system. For a pure solution example, in the case of europium complexation by oxalate, one would start by determining the hydrolysis of europium in the absence of oxalate and the acid-base properties of oxalate in the absence of europium in a chosen background electrolyte (ideally in a system that does not involve any specific ion interaction with europium or oxalate). In the case of a surface, for example europium adsorption to silica, one needs to know the acid base behavior of both europium containing solution (or aqueous chemistry in general) and the silica surface as well as the bulk silica solubility and europium solubility limits. Once the individual systems are sufficiently well-defined (either via own, relevant experiments or previously published studies), the combined systems can be targeted. All of this is done in the absence of carbon dioxide, which is forming complexes with europium and can in principle adsorb to surfaces and consequently modify their surface charge. The individual systems are ultimately defined by chemical reactions and concomitant stability constants (typically for room temperature). The problem, thus, lies in defining appropriate chemical reactions and determining their stability constants. The determination of stability constants for aqueous solutions was originally done using graphical methods due to the lack of computers (Rossotti and Rossotti 1955), and similarly, graphical methods were applied for surfaces (James et al. 1978). Previously, several books originating from symposia dealing with geochemical modelling were published, mainly dealing with simulation examples and available numerical tools (Jenne 1979; Melchior and Bassett 1990). In surface complexation modeling, numerical parameter estimation was initiated through the development of FITEQL (Westall and Morel 1977; Westall 1982a,b,c), but had begun much earlier for solution equilibrium systems.

Numerical approaches

The numerical treatment of chemical equilibria in solution started in the early 1960s. Such treatment was, as first mentioned in previous review articles on the topic, assigned to the application of "pit-mapping" (Dyrssen et al. 1961), an approach where after defining the system in terms of species only one stability constant is numerically adjusted, while keeping all others constant and pursuing this for all species and repeating until the sum of least squares is as low as possible. The application of computer programs appears to have been accessory to the earlier used graphical methods (Rossotti and Rossotti 1955; Sillén 1956). In this period, the program LETAGROP was written and further developed (Ingri and Sillén 1962; Sillén 1962). The visionary decision to pursue the use of what they called "High-Speed Computers" at the time can hardly be overrated. In a subsequent paper (Sillén 1964) the program LETAGROP VRID was described, a general minimization routine for chemical equilibria to be defined within the code. Noteworthy, as later for FITEQL (Westall and Morel 1977), the solution of the chemical equilibrium problem and the fitting procedure were combined in one specific code.

Another early development in the solution chemistry community was the family of GAUSS-G/SCOGS codes (Tobias and Yasuda 1963; Perrin and Sayce 1967). A comprehensive overview of the developments, including even at an early stage the treatment of spectroscopic data, can be found in the book edited by Leggett (Leggett 1985). The treatment of the equilibria in the above cited codes is a combination of mass law equations and mole balances (MLE) and computers were involved from the beginning for optimizing reaction constants (i.e., not only simulation).

In the early development, and most relevant to the surface complexation community was the development of the codes of the MINEQL family, initially REDEQL (Morel and Morgan 1972; McDuff and Morel 1975), then via MINEQL (Westall 1976; Westall et al. 1976), to MICROQL (Westall 1979a,b).[1] All these simulation tools building on the MLE formalism culminated in various versions of FITEQL (Westall and Morel 1977; Westall 1982a,c; Herbelin and Westall 1994, 1996, 1999). FITEQL is a coupled simulation/fitting tool which cleverly uses the Newton–Raphson method with analytical derivatives and can optimize "chemical" parts of the problem, such as free component concentrations and total component concentrations or stability constants.[2]

[1] Other developments resulted in codes like MINTEQ (Krupka and Morrey 1985) or Hydraql (Papelis et al. 1988).

[2] It is possible to standardize e.g., stock-solution concentrations of phosphate or arsenate by potentiometric titration and subsequently fitting total concentration and the system inherent equilibrium constants (i.e., deprotonation). This all allows to quantify the system under study in a very precise way and contribute to the determination of reliable stability constants.

Westall found a way to include the surface equilibria and the electrostatic effects in the MLE formalism (Westall 1980), and early on warned that the electrical double layer structure cannot be inferred from the experimental data usually available at the time (Westall and Hohl 1980). Later more modern variants like some FITEQL3 versions and finally FITEQL4 (involving "modern" windows) were developed (Herbelin and Westall 1994, 1996, 1999). Nowadays, a modified version (Gustafsson 2003) of FITEQL4 (which requires windows-XP environment or emulation) has replaced FITEQL or MINFIT (using a MINEQL based code as a simulation tool), and appears to be in use in rare cases (Xie et al. 2016) as a combined code that couples solution of the (forward) equilibrium calculations and (backward) parameter optimizations. In this respect, as discussed later, shell optimizers have certain advantages. Codes like ECOSAT (Keizer and Van Riemsdijk 1994) coupled to FIT (Kinniburgh 1993) as a shell optimizer require old computers, and are therefore bound to disappear, similar to other "homemade" codes like SOLGASWATER (Eriksson 1979), which was not only coupled to an excel-solver for optimizing parameters at Umeå University, but also transformed to a windows environment[3]. The time from the late seventies to the nineties saw strong development in this respect, and many of the codes had their own specific capabilities that were e.g., directed by the focus of the experimental research at the respective laboratories. Moreover, PHREEQC (Parkhurst and Appelo 1999) or Geochemist's workbench (Bethke and Yeakel 2009) have become increasingly popular, and include surface complexation modeling capabilities. It seems that FITEQL is still around and John Westall still has impact. FITEQL had the disadvantage that certain parameters, like capacitances, could not directly be optimized, and that coupling of different sources of experimental data in one fitting exercise was awkward. This was amended when shell optimizers such as PEST (Doherty 1994) or UCODE (Hill 1998; Poeter and Hill 1998, 1999) became available. Their use is not restricted to surface complexation modeling, and UCODE for example was developed for groundwater flow applications. Numerous couplings of these optimization codes of general purpose have since then been applied, and the number of available speciation codes is steadily increasing, including coupling to various general purpose fitting routines. As pointed out, some of the programs are losing ground just due to the fact that they have to be run on old computers or emulators, like ECOSAT/FIT (Kinniburgh 1993; Keizer and Van Riemsdijk 1994) or FITEQL version 4 (Herbelin and Westall 1999). To some extent they have also been replaced, e.g., ORCHESTRA (Meeussen 2003) has replaced ECOSAT in Wageningen. ORCHESTRA as an open-source code that is object oriented can be seen as a very general tool, and as for shell optimizers, the development of these codes appears to go away from lab-specific codes to implementations that can be further developed.

Law of mass action and Gibbs energy minimization

To give a brief insight into the formalism used by many speciation codes, we briefly summarize the mathematical approach to solve MLE for multiple interdependent equilibria according to the approach used in FITEQL (Westall 1980). As an example, we use the very simple system as in the comparative fitting exercise discussed in detail later, a silica surface in KCl solution, which reacts as function of pH. Mass balances are calculated for so called components (sometimes also termed "master species", e.g., in PHREEQC). In the exemplary system, these would be four components, the surface silanol groups, ≡SiOH (the "≡" sign denotes binding to the surface), and the solution species, H^+, K^+, and Cl^-. In many cases, protons are treated as special species with known activity. Thus, calculations are performed at fixed pH, and mass balancing is only required for ≡SiOH, K^+, and Cl^-. For components for which mass balancing is required, the total amount available in the system must be defined. In the model system, the components undergo two relevant reactions R_1 and R_2:

$$\equiv SiOH \rightleftharpoons \equiv SiO^- + H^+ \quad (R1)$$

[3] We have decided to cite web-links to allow the readers an easy access to more detailed information. https://www.winsgw.se/

$$\equiv SiOH + K^+ \rightleftharpoons \equiv SiO^- \cdots K^+ + H^+ \quad \text{(R2)}$$

Thus, there are six possible product species in the system: H^+, K^+, Cl^-, $\equiv SiOH$, $\equiv SiO^-$, and $\equiv SiO^- \cdots K^+$. A Basic-Stern type electrostatic model is used in this example, where the $\equiv SiO^-$ and $\equiv SiO^- \cdots K^+$ species create a negative charge in the surface plane (0-plane), and the $\equiv SiO^- \cdots K^+$ species creates a positive charge in the plane separating the Stern layer from the diffuse layer (β-plane). It needs to be specified how all possible product species may form from the components, and an equilibrium constant for each of these reactions needs to be provided. This leads to the typical matrix notation which is demonstrated in Table 1 for the given example.

The first column in Table 1 contains the activities of the product species, listed in a vector in logarithmic form, $(\mathbf{C}_{\log} + \mathbf{G}_{\mathrm{Clog}})$, where $\mathbf{G}_{\mathrm{Clog}}$ would be a vector containing the logarithmic aqueous activity coefficients for each product species. The following columns, from 2 to 7, resemble the A-matrix of stoichiometric coefficients after multiplication with the vector $(\mathbf{X}_{\log} + \mathbf{G}_{\mathrm{Xlog}})$. $(\mathbf{X}_{\log} + \mathbf{G}_{\mathrm{Xlog}})$ contains the component species activities. $\mathbf{X}_{\log}$ are the adjustable concentrations of component species, optimized to find the equilibrium composition of the system. It is important to note the two additional components in columns 6 and 7, used to adjust the electrostatic (interfacial) potentials, in this case in two planes 0 and β: ψ_0 and ψ_β. The last column contains the logarithmic equilibrium constants. The first four lines in the Table 1 are identity reactions of the component species. They are required to make sure the component species are included in the mass balance.

One might also discuss here the calculation of activities, which is in principle necessary whenever stability constants are used from databases, where they are referenced to infinite dilution. The calculation of activity coefficients is typically restricted to aqueous species and component species. Care must be taken as to whether, e.g., molar or molal concentrations are used. Also, in many if not most cases, the concentrations are not divided by the corresponding standard state concentrations, which would make the number dimensionless (as would be required, e.g., to take the logarithm). Inherently, since the standard state concentrations have values of 1 ($mol \cdot kg^{-1}$ or $mol \cdot L^{-1}$), this does not affect numerical values, but is formally incorrect. The use of molalities has various advantages, including work at high temperatures, where molarities change for constant molalities, high salt concentrations, where the activity of water decreases, and also with respect to the standard state molality, which makes changes in temperature more traceable. The above involve simple conversions, but they need to be done self-consistently. At high salt concentrations and with variable temperatures, the molar concentration scale becomes awkward.

For the surface species in electrostatic SCMs, the interaction coefficients are in almost all cases restricted to the electrostatic terms. These become huge and therefore other contributions to non-ideality are neglected (the more so since they cannot be experimentally determined).

The treatment of activities and electrostatics in speciation codes requires the definition of specific components or separate procedures. In a MINEQL type algorithm, the total concentration of surface charge can be computed from the surface speciation by counting charges in separate surface planes. With equations for charge/potential relationships the potentials can be obtained and inserted in the mass law equations. For the calculation of activities this can be done in a similar way, by calculating the ionic strength from the aqueous solution equilibrium composition. The ionic strength can then be used via the Davies equation for example to calculate activity coefficients, which can be applied in the mass law equation to correct the equilibrium constant. Both cases require additional relations compared to a pure equilibrium solution calculation, and this and the necessary adjustment of derivatives are straightforward. These details are mentioned here, but in the following we will for the sake of simplicity equate activities with concentrations.

Table 1. Full set of chemical equations, describing all reactions for the exemplary chemical system described in the text. "la" denotes the decadic logarithm of species activities. Surface species typically have activity coefficients =1 and activity is equal to concentration. Note that the definition of the activity coefficients of surface species can vary among the codes.

Product species	Components, stoichiometries coefficients, Δz and Boltzmann factors					Eqn. const.
$\mathbf{C}_{log} + \mathbf{G}_{Clog}$			$\mathbf{A}\cdot(\mathbf{X}_{log} + \mathbf{G}_{Xlog})$			$-\mathbf{K}_{log}$
	H^+	K^+	≡SiOH	$\frac{-2.303F}{RT}\psi_0$	$\frac{-2.303F}{RT}\psi_\beta$	
$la(H^+)$	$+1\ la(H^+)$	+0	+0	+0	+0	−0
$la(K^+)$	+0	$+1\ la(K^+)$	+0	+0	+0	−0
$la(\equiv SiOH)$	+0	+0	$+1\ la(\equiv SiOH)$	+0	+0	−0
$la(\equiv SiO^-)$	$-1\ la(H^+)$	+0	$+1\ la(\equiv SiOH)$	$-1\frac{-2.303F}{RT}\psi_0$	+0	$-\log_{10} K_1$
$la(\equiv SiO^- \cdots K^+)$	$-1\ la(H^+)$	$+1\ la(K^+)$	$+1\ la(\equiv SiOH)$	$-1\frac{-2.303F}{RT}\psi_0$	$+1\frac{-2.303F}{RT}\psi_\beta$	$-\log_{10} K_2$

With this formalism all MLE, including the electrostatic corrections for the surface species can be calculated as follows (Eqn. 1):

$$\mathbf{C}_{log} + \mathbf{G}_{Clog} = \mathbf{A}\cdot(\mathbf{X}_{log} + \mathbf{G}_{Xlog}) - \mathbf{K}_{log} \tag{1}$$

The transform of the **A**-matrix may now be used to calculate the mass balance of the chemical system (Eqn. 2):

$$\mathbf{T} = \mathbf{A}^{\mathrm{T}}\cdot\mathbf{C} \tag{2}$$

Note that in this equation the non-logarithmic form of **C** without activity coefficients is used. **T** contains the total amounts of all component species. The last two components in **T** and **X** deserve special attention. In **T** they contain the charge transferred into the 0- and β-plane by the surface reactions. In **X** they contain the so-called Boltzmann-factors for the 0- and β-plane, including ψ_0 and ψ_β, as shown in Table 1. The speciation code needs to find **X** such that **T** matches the predefined total amounts of all components in the chemical system for which a mole balance is included, in order to define the equilibrium speciation, and simultaneously it needs to adjust the 0- and β-plane potential to satisfy the electrostatic balance equations. Very robust procedures exist to solve such equilibrium speciation problems using a Newton–Raphson routine (Westall 1980) or similar (cf. next section) approaches.

In Table 1, the sum to the right in each row corresponds to a product on the linear scale and yields the species concentration. The sum of the concentrations (including the multiplicator, so for a stoichiometric coefficient of zero, there is no contribution) for each column downwards yields the total component concentration.

For the aqueous component species, this is, e.g., an equation $K^+ = K^+$ with a stability of unity ($\log_{10} K = 0$). So, to the right, this is equation does not help solve the equilibrium problem, but downwards in includes the concentration of the aqueous K^+ ion in the mole balance for potassium. For the electrostatic components, no separate entry in the mass law equation is necessary. The free value (i.e., the corresponding component species "concentration" of the electrostatic component, i.e., the potential, which is applied in the mass law equations) is calculated from a separate set of equation (the charge/potential relationship). This is similar for the activity corrections. As shown by Westal (1982c), this can be elegantly done by distinguishing coefficients for the mass law equations and the mole balances, which is implemented in codes starting with FITEQL2 and has various additional advantages.

The other, less frequently applied approach, involves calculation of the Gibbs energy of the system, which should be at a minimum for equilibrium conditions. The Gibbs Energy Minimization (GEMs) approach is for example applied in the GEMs software developed at the Paul Scherrer Institute in Switzerland.[4] Kulik and coworkers developed powerful tools in this context including surface complexation modeling (Kulik and Peretyashko 1998; Wagner et al. 2012; Kulik et al. 2013), and including fitting routines (Miron et al. 2015). The GEMs software is based on the mass balance of the entire system (Kulik et al. 2004; Akula 2020). Independent components (ICs) refer to elements and electrical charges (comparable to components or master species in the example above), whereas dependent components (DCs) refer to species (comparable to product species in the example above; Kulik et al. 2004; Akula 2020). Thermodynamic phases with more than one DC can be defined (Kulik et al. 2004; Akula 2020). During GEMs runs, the activities and concentrations of the DCs are calculated separately with the corresponding reference states and activity coefficients (Kulik et al. 2004; Akula 2020). The Interior Point Algorithm (IPM) of GEMs calculates the speciation vector x containing the molar quantity of DCs as well as a vector $\mathbf{u}$ including the chemical potential of the ICs (Kulik et al. 2004; Akula 2020).

In GEMs, the Gibbs free energy of the system (Eqn. 3) is given by:

$$\mathbf{G}(\mathbf{X}) = \sum_k \sum_j X_j \nu_j,\ j \in L_k,\ k \in \phi \tag{3}$$

where ν_j is the chemical potential of the j-th DC and L_k a subset of DC in the k-th phase (Kulik et al. 2004; Akula 2020). ν_j is represented (Eqn. 4) by:

$$\nu_j = \frac{G^{\circ}_{j,T}}{RT} + \ln C_j + \ln \gamma_j + C_F + \text{constant},\ j \in L_k \tag{4}$$

where $G^{\circ}_{j,T}$ refers to the standard molar Gibbs free energy (from the used thermodynamic database), $C_j = f(X_j)$ the concentration, γ_j the activity coefficient of the j-th DC, and C_F the Coulombic parameter applied for charged surface species (Kulik et al. 2004; Akula 2020). The *constant* converts from the practical to the rational (mole fraction) standard-state concentration scale (Kulik et al. 2004). The non-linear minimization IPM algorithm computes the vector $\mathbf{X}$ and dual vector $\mathbf{u}$ (Eqn. 5) by checking the Karpov–Kuhn–Tucker (KKT) conditions (Kulik et al. 2004; Akula 2020):

$$\nu_j - \sum_i a_{ij} u_i \geq 0,\ i \in N \tag{5}$$

A species at equilibrium (Eqn. 6) is thus described by (Kulik et al. 2004; Akula 2020)

$$\sum_i a_{ij} u_i = \frac{G^{\circ}_{j,T}}{RT} + \ln C_j + \ln \gamma_j + C_F + \text{constant} \tag{6}$$

This equation is then used to compute activities, saturation indices, and activity functions (pH, p_e, E_h) (Akula 2020). Contrary to law of mass action algorithms, GEM is particularly adapted for solving phase equilibrium problems for complex heterogeneous systems including several non-ideal multicomponent phases (Miron et al. 2015). More detailed explanations can be found elsewhere[4] (Kulik et al. 2004, 2013; Miron et al. 2015). CemGEMS, a web application for academics and industrials for the thermodynamic modeling of cementitious materials enabling to address the composition, hydration, leaching of cements at temperatures ranging from 0–99 °C and pressures going from 1 to 100 bar, was recently developed (Kulik et al. 2021).

[4] https://gems.web.psi.ch/

Parameterization of SCM

All these approaches require balances (mass conservation), which is typically easily solved for aqueous solutions, since ligand or metal concentrations can be analytically determined. It is more of a problem for surface ligands, not only due to the general heterogeneity of real surfaces, but also due to the potential presence of various surface functional groups on one single crystal plane. The concentrations/densities of these reactive sites can be determined from the structure for idealized surfaces with known termination, but such idealized terminations may not be realistic, unless one deals with single crystals or similar well-defined samples. Currently, techniques that allow determining many details for particles are becoming available (Livi et al. 2017, 2023). But even then, idiomorphic particles are required, which may not always be a realistic representation of environmental samples. Separate approaches can be found in Boily (2025, this volume) and Hiemstra and Lützenkirchen (2025, this volume).

The development of a surface complexation model typically involves the definition of the surface chemical model (which implies that all aqueous reactions are adequately defined) and the evaluation of the unknown parameters (typically stability constants for chosen species, capacitance values, charge distribution factors, and in some cases total concentrations, i.e., site densities, of surface functional groups). The stoichiometry of surface complexes and the concomitant distributions of charge may also be treated as a priori unknowns. The goal is to limit the number of adjustable parameters by making use of independent information from (surface-) spectroscopy, surface diffraction, high resolution atomic force microscopy or atomistic modeling like classical molecular dynamics simulations or quantum chemistry. At the same time a maximum of experimental data points should be used to calibrate the model. The aim is then to find the parameters that best describe the experimental results, which is done by inverse modeling. The numerical approaches shortly discussed in the following section compare experimental values with simulation results. The adjustable parameters are varied to minimize an "objective function", which is typically a weighted sum of squared residuals (WSOSR). The squared residuals are squared differences between observations and simulations. The squares will yield positive numbers and are required to avoid that large positive deviations and large negative deviations of the model from the observation will just cancel in the sum of differences. Sometimes, the WSOSR is divided by the degrees of freedom, i.e., a number that is defined by the difference between the number of data points and the number of optimized parameters. Many flavors of objective functions are in use, as will become clear in the following sections where we shortly describe some of the algorithms for inverse modeling used in the context of surface complexation modeling.

DESCRIPTION OF THE ALGORITHMS USED IN FITTING PROCEDURES

Newton–Raphson

The Newton–Raphson method, developed by Isaac Newton and Joseph Raphson, has entailed the development of the most commonly used iterative root-finding algorithm for non-linear equations (Deuflhard 2012; Dedieu 2015; Schlegel 2016; i.e., the root corresponds to $f(x) = 0$). In 1669, Newton was dealing with polynomials only and in 1690 Raphson introduced the notion of derivative as well as the general form of the method (Polyak 2007). Let us begin with a single variable equation $f(x)$. The algorithm uses an initial single guess of the root (open method), here denoted x_0. The true root r is defined as $r = x_0 + h$ (Deuflhard 2012; Schlegel 2016),[5] where the perturbation h measures how distant the initial guess x_0 is from the true value of the root r (Deuflhard 2012; Schlegel 2016). The Taylor series expansion (Eqn. 7) of

[5] Weisstein EW "Newton's Method." From MathWorld--A Wolfram Web Resource. https://mathworld.wolfram.com/NewtonsMethod.html

the function $f(x)$ for $r = x_0 + h$ (Press et al. 1992) is given by:

$$f(r) = f(x_0 + h) \approx \sum_{n=0}^{\infty} \frac{1}{n!} f^{(n)}(x_0) h^n \approx f(x_0) + f'(x_0) h + \frac{1}{2} f''(x_0) h^2 + \ldots \qquad (7)$$

The linear approximation of the function $f(x)$ is given by the first-order Taylor polynomial, whereas the second-order Taylor polynomial correspond to the quadratic approximation, again for univariate functions.[6]

If h is sufficiently small, the root of the linear tangent approximates the root position (as it corresponds to the 1st order Taylor expansion[7] (Press et al. 1992), i.e., after truncation after the first order terms results). The root (Eqn. 8) can thus be expressed as (Deuflhard 2012; Schlegel 2016):[5]

$$0 = f(r) = f(x_0 + h) \approx f(x_0) + f'(x_0) h \qquad (8)$$

with h (Eqn. 9) approximately being:

$$h \approx -\frac{f(x_0)}{f'(x_0)} \qquad (9)$$

Inserting into $r = x_0 + h$ (Eqn. 10), we get:

$$r = x_0 + h \approx x_0 - \frac{f(x_0)}{f'(x_0)} \qquad (10)$$

the new approximated root x_1 (estimate of r) is thus expressed as (Eqn. 11):

$$x_1 = x_0 - \frac{f(x_0)}{f'(x_0)} \qquad (11)$$

From x_1 the new estimate x_2 (Eqn. 12) becomes:

$$x_2 = x_1 - \frac{f(x_1)}{f'(x_1)} \qquad (12)$$

The iteration process of the Newton–Raphson algorithm for root finding (Press et al. 1992) is consequently expressed as (Eqn. 13):

$$x_{n+1} = x_n - \frac{f(x_n)}{f'(x_n)} = x_n - \left(f'(x_n)\right)^{-1} f(x_n) \qquad (13)$$

In that way, several improved approximations of the root are successively generated until a solution with the desired accuracy (stopping criterion) has been obtained. Note, that the Newton–Raphson method converges quadratically as approaching the root (i.e., with a convergence of order 2 (Press et al. 1992; Dedieu 2015),[5] i.e., the number of significant figures is approximately doubled after each step (Press et al. 1992)), and is thus very efficient in finding a minimum. An illustration of the tangent step of the Newton–Raphson algorithm is presented in Figure 1. After selection of the initial guess x_0, a tangent line is drawn through the point $(x_0, f(x_0))$. The intersection between the x-axis and the tangent line provides the next approximate root, i.e., x_1 (Press et al. 1992).

Since this is an iterative algorithm, convergence heavily depends on the selection of the initial guess (Casella and Bachmann 2021). In case $f'(x)$ is close to zero or the initial guess far from the true solution, the Newton–Raphson algorithm may fail to converge (Schlegel 2016;

[6] Çapar Y (2020) Taylor Expansion. https://yasincapar.com/taylor-expansion/

[7] Çapar Y (2020) The Newton-Raphson Method https://yasincapar.com/the-newton-raphson-method/

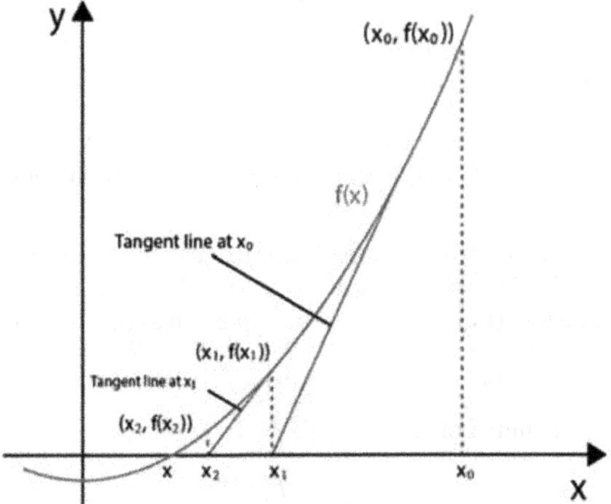

Figure 1. Illustration of the tangent step of the Newton–Raphson algorithm. Reproduced with permission from "Newton's Method Calculator". Allmath.com. Accessed 07.01.2025. https://www.allmath.com/newtons-method-calculator.php.

Casella and Bachmann 2021)[8] or converges only slowly or even converges to a different root. Slow convergence may also occur close to local maxima or minima leading to infinite oscillation (Press et al. 1992)[8]. This indicates that appropriate initial guesses (as typically specified by the user) can be quite important. Given a twice differentiable function (of one variable), the Newton–Raphson algorithm can also be used for minimization purposes (Sia 2018)[9], i.e., finding the root of $f'(x) = 0$. The second-order Taylor expansion of $f(x)$ around $x_0 + h$ (Sia 2018)[9] is obtained as (Eqn. 14):

$$f(x_0+h) \approx f(x_0) + f'(x_0)h + \frac{1}{2}f''(x_0)h^2 \tag{14}$$

x_{n+1} is defined in order to minimize this quadratic approximation in h, considering that at the minimum the derivative $f'(x_0 + h) = 0$ (Sia 2018).[9] Thus (Eqn. 15),

$$0 = \frac{\partial}{\partial h}\left(f(x_0) + f'(x_0)h + \frac{1}{2}f''(x_0)h^2 \right) = f'(x_0) + f''(x_0)h \tag{15}$$

the minimum being achieved (Eqn. 16) for

$$h = -\frac{f'(x_0)}{f''(x_0)} \tag{16}$$

Note that this consists in fact in finding the root of the derivative of quadratic approximation (second-order Taylor) by applying the first-order Taylor expansion.

In this case, the iteration process of the Newton–Raphson algorithm to find a minimum can consequently be expressed as (Eqn. 17):

$$x_{n+1} = x_n + h = x_n - \frac{f'(x_n)}{f''(x_n)} = x_n - \left(f''(x_n)\right)^{-1} f'(x_n) \tag{17}$$

where $f'(x_n)$ and $f''(x_n)$ correspond to the gradient (gradient, i.e., first derivative) and the second

[8] Newton–Raphson method algorithm and flowchart. https://www.codewithc.com/newton-raphson-method-algorithm-flowchart/

[9] Newton's method in optimization. https://en.wikipedia.org/wiki/Newton%27s_method_in_optimization

derivative of the function $f(x)$, respectively (Sia 2018).[9] Note that the second derivative must be positive, ensured by the convexity of the function in case of a minimum. A negative second derivative value would correspond to a maximum of the concave function $f(x)$.

The univariable Newton–Raphson algorithm can be extended to multidimensional problems (Eqn. 18) with n equations and n unknows and arises for root finding:[10]

$$x_{n+1} = x_n - \left(\mathbf{J}f\left(x_n\right)\right)^{-1} f\left(x_n\right) \tag{18}$$

and for minimization purposes (Eqn. 19):[9]

$$x_{n+1} = x_n - \left(\nabla^2 f\left(x_n\right)\right)^{-1} \nabla f\left(x_n\right) = x_n - \left(\mathbf{H}f\left(x_n\right)\right)^{-1} \nabla f\left(x_n\right) \tag{19}$$

with ∇f being the gradient (vector matrix of first derivatives of dimension $n \times 1$), $\mathbf{J}$ being the Jacobian matrix (dimension $n \times n$) comprising all first-order partial derivatives, and $\mathbf{H}$ being the Hessian matrix (dimension $n \times n$) of all second-order partial derivatives (which must be positive definite to give a minimum). Note, that the term $-\left(\mathbf{H}f\left(x_n\right)\right)^{-1} \nabla f\left(x_n\right)$ is also called the Newton direction[9]. One major known drawback is the cost of calculating and storing the inverse of the Hessian matrix at each iteration[9]. If the Hessian matrix is not-invertible (also called singular or degenerate, with determinant equal to zero), the solution can also diverge.[9] Overall, the risk to reach a local minimum rather than a global minimum again strongly depends on the choice of the initial guess(es).

Levenberg–Marquardt

The Levenberg–Marquardt (LM) algorithm also solves non-linear least squares equations and can be seen as a combination of the Gauss–Newton algorithm and the gradient descent method (Lösler and Hennes 2008)[11]. The Gauss–Newton method can be considered as an extension of the Newton–Raphson algorithm towards multivariate functions and applied to find the roots of the first derivatives, i.e., minima, of a convex objective function. Contrary to the Newton–Raphson method, the Gauss–Newton algorithm does not require calculating the second-order derivatives in form of the Hessian matrix (Gratton et al. 2007). In the Gauss–Newton method, the Hessian matrix is approximated by $\left(\mathbf{H}f\left(x_n\right)\right) \approx \left(\mathbf{J}f\left(x_n\right)\right)^{\mathrm{T}} \left(\mathbf{J}f\left(x_n\right)\right)$ (Gargiani et al. 2020).

The LM algorithm is also sometimes described as a trust-region modification of the Gauss–Newton algorithm (Ranganathan 2004). One of the common minimizing techniques is the gradient descent approach, which updates parameters in the "downhill" direction, i.e., the opposite direction of the gradient of the objective function (Gavin 2022). Assuming that the objective function is roughly quadratic in the parameters close to the ideal solution, the Gauss–Newton approach minimizes the sum of square residuals objective function (Gavin 2022).

The LM algorithm as published by Kenneth Levenberg (Levenberg 1944) and further developed by Marquardt (1963) relies on an iterative approach, with the need of an initial guess, provided by the user.[11]

Marquardt (1963) described this so called damped Gauss–Newton method (Lösler and Hennes 2008; Eqn. 20):

$$\left(\mathbf{A}_{\beta_k}^{\mathrm{T}} \mathbf{A}_{\beta_k} + \mu \mathbf{I}\right) x = -\mathbf{A}_{\beta_k}^{\mathrm{T}} \mathbf{w}_{\beta_k} \tag{20}$$

Here, $\mathbf{A}_{\beta_k}$ is the Jacobian matrix of the function under investigation with the parameters β_k of the kth iteration (i.e., the matrix of the first derivates of the objective function with respect to the adjustable parameters), I the identity matrix, $\mathbf{w}_{\beta_k}$ a vector containing the function residuals

[10] Newton's method. https://en.wikipedia.org/wiki/Newton%27s_method

[11] Statistics How To Levenberg–Marquardt Algorithm (Damped Least Squares): Definition. https://www.statisticshowto.com/levenberg–Marquardt-algorithm/

(Lösler and Hennes 2008), whereas μ represents the non-negative damping parameter and the vector of increments x the damped Newton step (Lösler and Hennes 2008). Thus, the iteration process of the Levenberg–Marquardt algorithm can be expressed as (Eqn. 21):

$$\beta_{k+1} = \beta_k - \left(\nabla^2 \mathbf{w}_k + \mu \mathbf{I}\right)^{-1} \nabla \mathbf{w}_k \tag{21}$$

Note, that a modified version introducing the use of the diagonal matrix $\mathbf{A}_{\beta_k}^{\mathrm{T}} \mathbf{A}_{\beta_k}$ (Fletcher 1971) can also be found in descriptions of the LM algorithm (Eqn. 22):

$$\left(\mathbf{A}_{\beta_k}^{\mathrm{T}} \mathbf{A}_{\beta_k} + \mu \,\mathbf{diag}\left(\mathbf{A}_{\beta_k}^{\mathrm{T}} \mathbf{A}_{\beta_k}\right)\right) x = -\mathbf{A}_{\beta_k}^{\mathrm{T}} \mathbf{w}_{\beta_k} \tag{22}$$

After each iteration, the Levenberg–Marquardt algorithm is updating the solution via a combination of the gradient descent method and the Gauss–Newton algorithm (Gavin 2022)[11].

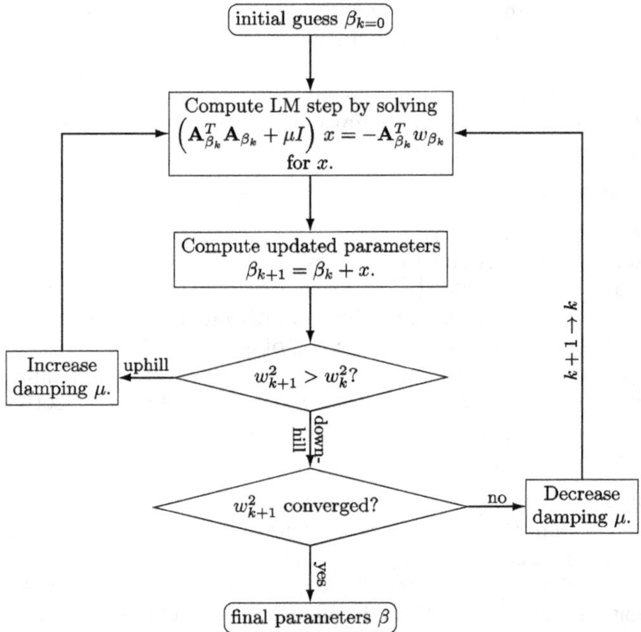

Figure 2. A flow chart of Levenberg–Marquardt algorithm (note that there can be several variants of how the damping parameter is adjusted, but these details are outside of the scope of this work).

In this context μ, the non-negative damping parameter, determines each iterative update (Gavin 2022)[11]. When μ is small, the update is closer to the Gauss–Newton method and the term $\nabla^2 \mathbf{w}_k$ dominates, whereas the gradient descent direction has more weight for large μ (Lösler and Hennes 2008; Gavin 2022)[11], with the component $\mu\mathbf{I}$ prevailing. At the beginning of the optimization, μ is set to a large value to allow the first updates to be small steps in the steepest-descent direction (Gavin 2022), and μ is further increased if a worse approximation occurs in the course of the optimization (Gavin 2022). A flow chart summarizing the Levenberg–Marquardt algorithm is presented in Figure 2.

The Levenberg–Marquardt algorithm converges faster than the simple Gauss–Newton or gradient descent methods, i.e., when the individual approaches are applied independently.[11] Even if the initial guess is far from the true solution, an optimal solution can still be found.[11]

Nevertheless, the Levenberg–Marquardt algorithm can get lost in parameter space for flat functions (Transtrum and Sethna 2012) and remains a local minimization method. Note that popular shell optimizers such as UCODE (Poeter and Hill 1998) are based on a modified version of the LM algorithm.

The Downhill Simplex method

The Downhill Simplex method is a direct search method for non-linear optimization problems, like the fitting of surface complexation model parameters. It is sometimes termed Nelder–Mead Simplex method reconciling the developers (Nelder and Mead 1965). Eponymous for the method is the form created by $n+1$ points in an n-dimensional space, called a simplex. Simple examples of simplices would be a triangle in 2D or a tetrahedron in 3D.

For an optimization problem with n parameters the simplex is an entity of $n+1$ different parameter sets, $x_1 \ldots x_{n+1}$. Associated with each parameter set is a value of the objective function $f(x)$, i.e., a goodness of fit (GoF) parameter like (Eqn. 23):

$$f(x) = \chi^2 = \frac{1}{d-n}\sum_{i=1}^{d}\left(\frac{y_{i,\text{exp}} - y_{i,\text{model}}}{y_{i,\text{err}}}\right)^2 \tag{23}$$

with d data points, n parameters, and y_{exp} and y_{err} the experimental data-values and their associated experimental uncertainty, respectively, and y_{model} model prediction.

The central scheme of the method is that (1) the points are ordered such that $f(x_1) < \ldots < f(x_n) < f(x_{n+1})$. Second (2), the centroid of the simplex, x_{avg}, is calculated, as the arithmetic mean of all points in the simplex except the one with the highest χ^2 (the worst fit, x_{n+1}). Then (3), the worst parameter set is "reflected" over the centroid to the other side of the simplex: $x_r = x_{\text{avg}} + a\,(x_{\text{avg}} - x_{n+1})$. If this process yields a new optimum ($f(x_r) < f(x_1)$) the new point is "expanded" further, to check if the fit gets even better: $x_e = x_{\text{avg}} + \gamma\,(x_r - x_{\text{avg}})$. Then a new simplex is formed, including x_e or x_r instead of x_{n+1} and the process restarts (back to (1)).

If the reflected point remains the worst fit, the simplex is "contracted". Two options for the contraction are available: (1) if $f(x_r) < f(x_{n+1})$: $x_c = x_{\text{avg}} + \rho\,(x_r - x_{\text{avg}})$, (2) if $f(x_r) \geq f(x_{n+1})$: $x_c = x_{\text{avg}} + \rho\,(x_{n+1} - x_{\text{avg}})$. If contraction leads to an improvement of the worst fitting parameter-set, the process restarts (back to (1)) with a simplex including x_c instead of x_{n+1}, otherwise the simplex is "shrunk" or "compressed", i.e., all points are moved towards the best fitting point: $x_i = x_1 + \sigma(x_i - x_1)$.

In the above, a, γ, ρ, and σ, are the simplex parameters or the reflection-, expansion-, contraction- and shrink coefficient, respectively. Standard values are: $a = 1$, $\gamma = 2$, $\rho = ½$, and $\sigma = ½$. If the method is visualized in 2D, the iterative movement of the points in parameter space gives the impression of the simplex crawling towards the minimum, which gave the method the name "Amoeba Method".

The search terminates when some convergence criterion is met. This may for example be the case for the standard deviation of all $f(x)$ falling below a certain limit, which is usually the case, when all points have moved very close together near a minimal objective function value. The choice of a suitable convergence limit may be critical to get good results for very flat objective functions.

We will not further detail the method here, as there are very good descriptions and demonstration examples available on the internet[12].

The Downhill Simplex method is the main optimization engine in a coupling between PHREEQC (Parkhurst and Appelo 2013) and Python software[13] developed (among other

[12] Nelder–Mead method. https://en.wikipedia.org/wiki/Nelder–Mead_method

[13] https://github.com/FHe/P3R

possibilities) for the adjustment of surface complexation model parameters (Heberling et al. 2021; Machesky et al. 2023). The Downhill Simplex method stands out by its robustness. Compared to Newton–Raphson or Levenberg–Marquardt type methods it is less efficient (in terms of iterations needed to find the next minimum), but also less prone to fall into local minima. It is better suited to fit many parameters at a time. Insensitive parameters are more likely to be left untouched instead of being shifted all the way to the edge of parameter space. On the other hand, methods like genetic algorithms can even handle more adjustable parameters and are even better suited to ensure finding a global minimum, however, at the cost of being even less efficient. Thus, among the mentioned optimization methods the Downhill Simplex method may be considered a good compromise concerning success rate, robustness, and efficiency.

A BRIEF SURVEY OF PARAMETER STATISTICS AND MODEL UNCERTAINTY

The discussion about model related uncertainties necessarily starts with the discussion of the inherent uncertainties of the data used for model calibration. In this context, it is highly desirable to (i) obtain high quality data and (ii) know the associated experimental errors.

Data uncertainty

The foundation of any statistical assessment of a model or its parameters in terms of goodness of fit or parameter uncertainty, are reliable estimates for the error bars associated with each experimental data point. These are often represented by the standard deviation of a normal distributed expected value, or a multiple of it. In the best case, error bars are derived from multiple independent experiments, so they represent analytical uncertainties (uncertainties related with analytical methods like measurements of element concentrations, potentiometric measurements, or zeta potential measurements) as well as procedural uncertainties (uncertainties related with experimental procedures like pipetting, pH adjustment, the composition of experimental solutions, or differences between aliquots of a powder). In cases where such an approach would cause disproportionate efforts, error bars often represent only a selected analytical uncertainty, or a number of uncertainties, combined via an error propagation procedure. The error in pH-measurements (particularly in particle suspensions) is also a source of experimental uncertainty, but hardly ever reported. If the pH values are re-adjusted in the experiments, they may be quite precise with little error, but adjusting pH values always involves a perturbation to the equilibria, which adjust themselves. It may be advisable to let the pH adjust itself instead. Repeated pH-measurements on one sample may then yield an idea of the uncertainty. Determining an additional value after solid liquid separation might supply additional pieces of information.

Goodness of fit

The goodness of fit describes how well a model represents certain data. It is usually estimated from some kind of weighted sum of squared residuals calculation. A popular statistical parameter used for this purpose is χ^2 (Eqn. 23). In addition to the squared difference between data ($y_{i,\mathrm{exp}}$) and model ($y_{i,\mathrm{model}}$) it includes a weighting of these differences by the value uncertainty ($y_{i,\mathrm{err}}$). The sum of the squared weighted differences is divided by the degrees of freedom ($d-n$) in the fit, with d data points and n parameters. In general, a smaller χ^2 represents a better fit. For an optimal model fit, χ^2 should be around one, meaning that on average differences between data and model are of the same size as data uncertainties (error bars). χ^2 allows comparing how well different modeling approaches represent a certain data set. Moreover, it can be used to take decisions on model complexity. For example, if one model results in a χ^2 close to one, but a second more complex model, including more parameters, would decrease χ^2 significantly below one, then the first model should be preferred because the second model likely fits some details in the data which are within the margin of the error bars. Another example: when the complexity of a model is increased (more parameters are included) the degrees of freedom $(d-n)$ decrease in principle increasing χ^2. A guideline may be that only if the improvement in model fit overcompensates this effect, the more complex model may be justified, i.e., it will finally lead to a decrease in χ^2. All these guidelines for model development strongly depend on the uncertainties assigned to the data points.

Parameter uncertainty

The next important question during model development is how to assign uncertainties (standard deviations) to the model parameters. This usually involves an evaluation of the partial derivatives of the objective function (e.g., χ^2) with respect to each adjustable parameter at a minimum of the objective function, i.e., around the best fit. The procedure described here is used in the P^{3}R optimization software (Heberling et al. 2021; Machesky et al. 2023). It is adopted from the USGS inverse modeling software UCODE (Poeter and Hill 1998), and evaluates parameter variances and covariances on the basis of "scaled sensitivities" of each adjustable parameter (p) on each data point. Scaled sensitivities are the partial derivatives of all model values with respect to a certain parameter ($\partial y_{i,\text{model}} / \partial p_j \cdot p_j$) multiplied by the parameter value and scaled with the value uncertainty ($y_{i,\text{err}}$). In the mentioned code, scaled sensitivities for all parameters are collected in a $(d \times n)$ matrix $\mathbf{X}$ (Eqn. 24), with indices $i = 1 \ldots d$, and $j = 1 \ldots n$:

$$x_{i,j} = \frac{\dfrac{\partial y_{i,\text{model}}}{\partial p_j} \cdot p_j}{y_{i,\text{err}}} \tag{24}$$

The partial derivatives are usually numerically estimated e.g., by a centered differencing scheme. Scaled sensitivities themselves are useful quantities during model development and optimization. They provide insight into how much and in which direction a certain parameter will influence the model fit at each data point, i.e., how sensitive data points are, and help constraining a certain parameter. Insensitive parameters, having no or hardly any effect on the model fit can be identified by low scaled sensitivities, and could be excluded from the list of adjustable parameters in subsequent optimization runs.

The step from scaled sensitivities to the covariance matrix, $\mathbf{V}$, uses an additional matrix of data weights, $\mathbf{W}$, a $(d \times d)$ matrix where $w_{i,i} = (1/y_{i,\text{err}})^2$ and all other entries are zero. The two matrices can be combined by (Eqn. 25):

$$\mathbf{V} = \chi^2 \left(\mathbf{X} \cdot \mathbf{W} \cdot \mathbf{X}^{\mathrm{T}}\right)^{-1} \tag{25}$$

$\mathbf{V}$ is an $(n \times n)$ matrix, and the standard deviations of parameters p_j (Eqn. 26) are given by:

$$\sigma\left(p_j\right) = \sqrt{v_{j,j}} \tag{26}$$

Correlations between two parameters p_i and p_j, can be calculated as (Eqn. 27):

$$\operatorname{Corr}\left(p_i, p_j\right) = \frac{v_{i,j}}{\sqrt{v_{j,j}}\sqrt{v_{i,i}}} \tag{27}$$

Strongly correlated parameters are quite common in surface complexation models. They should not be adjusted simultaneously in an automated optimization routine. In such cases it is the responsibility of the model developer to gather further information, to constrain one of the correlated parameters, such that the other parameter(s) may be adjusted independently. Sometimes, if the initial guess of a parameter is far off, its optimum value cannot be found by a given optimization routine, nevertheless, the parameter may still be relevant. Therefore, it is always useful to vary initial guesses.

Deriving parameter uncertainties from local sensitivity implies a linear approximation of the fitting procedure. For non-linear models, especially in cases where the choice of initial guesses can affect the final fit parameters (multiple local minima), a more general approach to error propagation should be considered. A robust, albeit computationally expensive one is the Monte Carlo (MC) error propagation method (Anderson 1976; Papadopoulos and Yeung 2001). In this technique, uncertainties on the data are taken as statistical uncertainties usually assumed to follow a normal distribution. Hence, a large number of random variants y^j of the data y are drawn from, e.g., a normal distribution with mean y_i and a standard deviation σ_i is obtained, corresponding to the given error $y_{i,\text{err}}$ as $y^j_{i,\text{MC}} = N\left(y_i, \sigma_i\right)$. For each random dataset y^j, the fitting procedure is repeated to obtain corresponding parameters estimates. The final parameters estimates are then computed as the mean over all MC samples, with the corresponding standard deviation providing a statistical error estimate.

Model uncertainties

In order to assess the predictive power of a model, for example for taking a decision in how far a model may provide sufficiently reliable output for a certain application, it is crucial to be able to assess the uncertainty of model simulations. In practice, this is often done using a Monte Carlo approach. A number of parameter sets can be created from normally distributed random values either (1) based on the respective best fit parameter value error estimates taken as standard deviations of individual normal distributions or (2) based on a covariance matrix (i.e., considering one multivariate normal distribution, including all parameter values, variances, and covariances). Numerous parameter sets resulting from either approach will then be used to perform forward model simulations. Finally, the average and the standard deviation of the model output are calculated for each data point (where the average should equal the best fit model), and can be assigned to the model as an uncertainty range.

Model parameters are commonly reported with their value and standard deviation. Thus, for most cases where the model uncertainty is calculated independently of a model development, parameter covariances are unknown, and only option (1) among the procedures described above is feasible.

Figure 3 exemplifies how this lack of information may lead to a tremendous overestimation of the model uncertainty. The example in Figure 3 represents a preliminary snapshot from a surface complexation model development for calcite (Heberling et al. 2021). Two datasets (surface and zeta potentials of calcite) both as a function of pH are shown (the same data sets in the upper and lower rows, note that different scales on the y-axes were chosen). In the upper graphs, model uncertainty is calculated excluding information about parameter correlation, leading to large uncertainty ranges, which might even suggest that the model has no predictive power. In the lower graphs, covariances are included, and model uncertainty decreases strongly. The graphs on the right try to rationalize this effect, based on a 2D example. Irrelevant portions of parameter space are probed during the selection of random parameter sets, when information about parameter correlation is omitted (i.e., no correlation is considered) even for the same parameter values and standard deviations, and this may cause unrealistically large model uncertainty ranges, because any parameter value combination is allowed, with most of the parameter combinations being unrealistic.

As a consequence, we emphasize the importance to report covariances or correlations between parameters, besides parameter values and standard deviations, or even better to provide the raw data and the model input files, so the best fit parameters and the covariance matrix can be reassessed easily, and potentially modified if additional data suggest to do so.

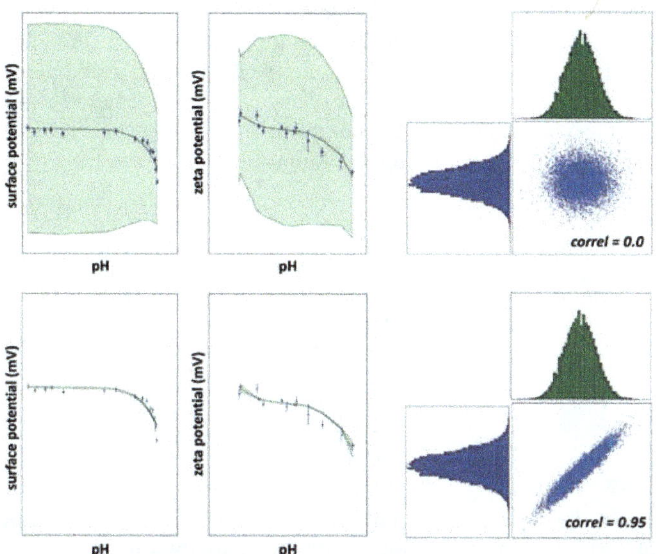

Figure 3. Exemplary model uncertainty calculations for a preliminary model development for two datasets of calcite surface- and zeta potential as function of pH (Heberling et al. 2021). The model uncertainty calculations in the upper row are based on individual parameter standard deviations for each parameter, which is equivalent to setting all covariances to zero. The calculation in the lower row is based on parameter sets collected from a multivariate normal distribution considering the full covariance matrix. The difference between the two approaches is illustrated on the right for a simple 2D parameter space. Neglecting parameter correlations leads to irrelevant regions of parameter space being probed during the random choice of parameter sets for the uncertainty calculation, and this may in cases as shown here lead to an enormous overestimation of model uncertainty.

COMMON ISSUES IN SURFACE COMPLEXATION MODELING

One unresolved issue in surface complexation modeling is the handling of multi-dentate surface complexes (Benjamin 2002; Wang and Giammar 2013; Lützenkirchen et al. 2015). Based on spectroscopic approaches such as IR and EXAFS and with support of atomistic simulations, the formation of monodentate, bidentate, tridentate, and up to tetradentate surfaces species has been revealed (Spadini et al. 1994; Zhang et al. 2004; Sherman et al. 2008; Jordan et al. 2014, 2018; Huittinen et al. 2022). This represents a binding of the sorbing ion to one, two, three, and four surface oxygens, respectively. The problem with simulating the surface complex geometries arises from the fact that different numerical formalisms have been implemented to deal with such species in commonly used geochemical speciation codes. As a consequence, published constants cannot necessarily be used with any code, when a model involves a multidentate surface complex. In case of formation of multi-dentate species, the surface complexation reaction may be expressed as:

$$d \equiv SOH^{z} + M^{x} + wH_2O \rightleftharpoons (\equiv SO)_d M(OH)_w^{x+dz-w} + (d+w)\,H^{+} \qquad \text{(Formalism 1)}$$

or

$$(\equiv SOH^{z})_d + M^{x} + wH_2O \rightleftharpoons (\equiv SO)_d M(OH)_w^{x+dz-w} + (d+w)\,H^{+} \qquad \text{(Formalism 2)}$$

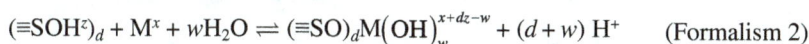

where ≡SOH stands for a surface site, z and x denote charges, and d and w correspond to stoichiometric coefficients. Westall compared formalism 2 to an example from solution for $d = 2$ (Westall 1982c). In this reasoning aqueous complexes between one oxalate (Ox^{2-}) with e.g., a divalent metal (Me^{2+}) to form MeOx in bidentate fashion would never involve the oxalate concentration squared in a mass law equation. On a surface, the functional groups involved in bidentate surface complex formation are as fixed as on a single oxalate (Morel and Hering 1993). In case two independent oxalates were involved in complex formation (i.e., $\mathrm{Me}(\mathrm{Ox})_2^{2-}$), the stoichiometric coefficient for oxalate would be 2 in the mass law equation. For low coverage, the concentration of the sites available for bidentate binding will be equal to the total amount of sites. At higher coverage, more serious problems arise.

For both formalisms, the mass balance is identical, whereas the application of the law of mass action will lead to differences if $d > 1$:

$$K_{\mathrm{M}}^{\mathrm{int},d} = \frac{\left[(\equiv\mathrm{SO})_d\mathrm{M}(\mathrm{OH})_w^{x+dz-w}\right]\left[\mathrm{H}^+\right]^{d+w}}{\left[\equiv\mathrm{SOH}^z\right]^d\left[\mathrm{M}^x\right]} \qquad \text{(Formalism 1)}$$

$$K_{\mathrm{M}}^{\mathrm{int}} = \frac{\left[(\equiv\mathrm{SO})_d\mathrm{M}(\mathrm{OH})_w^{x+dz-w}\right]\left[\mathrm{H}^+\right]^{d+w}}{\left[(\equiv\mathrm{SOH}^z)_d\right]\left[\mathrm{M}^x\right]} \qquad \text{(Formalism 2)}$$

where $K_{\mathrm{M}}^{\mathrm{int}}$ represents the intrinsic formation constant. It becomes obvious that the ratio between concentration of surface species is dependent on the denticity d for formalism 1, but not for formalism 2. Also, for the latter, the solid concentration would not impact the stability constant (Lützenkirchen et al. 2015). How this difference in the two formalisms is directly reflected in the speciation codes is also of interest. Formalism 2 can only be handled in codes which allow distinguishing between the mass action law equation and the mass balance coefficients. This depends on the respective code implementations, and can become difficult to trace.

Whereas the surface complexation constant of monodentate surface species is insensitive to the formalism, it has been shown that the surface concentration scale significantly impacts the numerical values related to multi-dentate sorption species (Lützenkirchen et al. 2015). As pointed out above, problems with site occupancy arise at higher coverage, when the solute can in principle not use all sites for bidentate/multidentate binding.

In a comparison, the two formalisms with the ECOSAT (version 4.8) (Keizer and Van Riemsdijk 1998), FITEQL (version 2.0) (Westall 1982c), PHREEQC (version 2.18.00) (Parkhurst and Appelo 1999), and Visual MINTEQ (version 3.1) (Gustafsson 2013) were used (Lützenkirchen et al. 2015). Parameters from Peacock and Sherman derived by means of EXAFS studies the formation of bidentate ($(\equiv\mathrm{FeOH})_2\mathrm{Cu(OH)}_2$) and tridentate ($(\equiv\mathrm{Fe_3O(OH)_2})$ $\mathrm{Cu_2(OH)_3}$) surface complexes upon Cu(II) interaction with goethite (Peacock and Sherman 2004). Using the FITEQL code (formalism 1) with the diffuse layer model (DLM), Peacock and Sherman obtained the following intrinsic surface complexation constants:

$$\equiv\mathrm{SOH} + \mathrm{H}^+ \rightleftharpoons \equiv\mathrm{SOH_2}^+ \quad \log_{10}K_{\mathrm{H1}} = 6.78 \qquad \text{(R3)}$$

$$\equiv\mathrm{SOH} \rightleftharpoons \equiv\mathrm{SO}^- + \mathrm{H}^+ \quad \log_{10}K_{\mathrm{H2}} = -10.10 \qquad \text{(R4)}$$

$$2\equiv\mathrm{SOH} + \mathrm{Cu}^{2+} + 2\,\mathrm{H_2O} \rightleftharpoons \equiv(\mathrm{SOH})_2\mathrm{Cu(OH)}_2 + 2\,\mathrm{H}^+ \quad \log_{10}K_{\mathrm{Cu}}^{\mathrm{int},2} = -3.10 \qquad \text{(R5)}$$

$$3\equiv\mathrm{SOH} + 2\,\mathrm{Cu}^{2+} + 3\,\mathrm{H_2O} \rightleftharpoons \equiv(\mathrm{SOH})_2\mathrm{SOCu_2(OH)_3} + 4\,\mathrm{H}^+ \quad \log_{10}K_{\mathrm{2Cu}}^{\mathrm{int},3} = -5.25 \qquad \text{(R6)}$$

Using Visual MINTEQ and PHREEQC to reproduce (forward modeling) the model simulations of Peacock and Sherman performed with FITEQL (Peacock and Sherman 2004) without any prior corrections resulted in huge differences as can be seen in Figure 4 (Lützenkirchen et al. 2015).

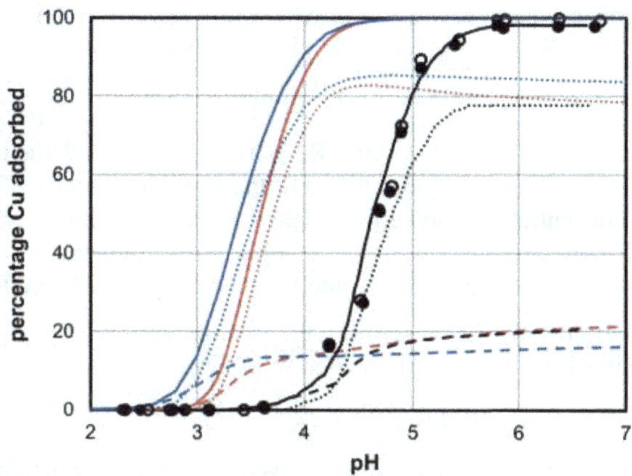

Figure 4. Copper pH adsorption edge for the Cu(II)–goethite system with bi- and tri-dentate species. Black circles are experimental data (Peacock and Sherman 2004). **Full red and blue lines** are simulations involving the DLM using Visual MINTEQ and PHREEQC, respectively, without any conversion of FITEQL intrinsic adsorption constants. **Dashed lines** are bidentate contributions (---) and tridentate contributions (····), respectively. The **black lines** were digitized from the original paper. The original model could not be exactly reproduced by the authors, as discussed elsewhere (Lützenkirchen et al. 2015). Adjusted from Lützenkirchen et al. (2015).

One option to overcome this issue is to generate general equations which will ease the transfer of the stability constants among the different concentration scales (Lützenkirchen et al. 2015). Earlier the consequences of a mishandling in the formulation of the mass action expressions for bidentate surface species in detail were discussed (Wang and Giammar 2013), namely:

- the limitation of the 1.0 mol·L^{-1} standard state concept and
- the relationship between the quantity of available multidentate sites vs. that of monodentate sites.

The authors recommended a few necessary details to be included in publications involving surface complexation modeling:

- clear mention of the mass action expression
- the value of exponents of the activity of available monodentate sites (1 or n) in the mass law expression of multidentate species
- information on the standard state (i.e., the numerical scale used for the activity of surface species)
- the basis of molarity, mole fraction, or coverage fraction used by the employed geochemical speciation code

The discussion can even be pushed further. Indeed, Benjamin (2002) discussed the impact of bidentate sorption on the concentration of available bidentate binding sites and on the consequent unavailability of unoccupied sites surrounding each adsorbed molecule, for which a Monte-Carlo model had been developed.

Furthermore, the way the formalism of the diffuse layer of electrostatic models is handled in different speciation codes is not always consistent. Indeed, using codes such as Visual MINTEQ, MINEQL, and FITEQL for asymmetrical electrolytes or for mixtures of symmetrical and asymmetrical electrolytes in Lützenkirchen et al. (2015) led to inconsistencies since they use the Gouy–Chapman equation which is limited to symmetrical electrolytes. This is not an issue in PHREEQC and ECOSAT which both apply generic equations for arbitrary electrolytes (Lützenkirchen et al. 2015).

The Gouy–Chapman equation (Eqn. 28), which relates the potential at the onset of the diffuse layer (ψ_d) to the diffuse layer charge (σ_d) is described as follows:

$$\sigma_d = -(8000\varepsilon RTI)^{1/2}\sinh(zF\psi_d/2RT) \tag{28}$$

with R being the universal molar gas constant (8.31451 J·K^{-1}·mol^{-1}), T the temperature (in K), I the ionic strength (in mol·L^{-1}), ε the permittivity of the medium (which changes as function of temperature and salt concentration), and z the valence of the symmetrical electrolyte (e.g., 1:1, 2:2, etc.). This Gouy–Chapman equation cannot be applied to asymmetrical electrolytes (e.g., 2:1, 3:1, etc.) or mixtures of symmetrical and asymmetrical electrolytes (Lützenkirchen et al. 2015).

The general equation (Eqn. 29) describing the relation between the charge of the diffuse layer and the potential, for instance implemented in ECOSAT, is given by:

$$\sigma_d = \pm(2000\varepsilon RT)^{1/2}\,\{\Sigma[i]\,(\exp(-z_iF\psi_d/RT) - 1)\}^{1/2} \tag{29}$$

where $[i]$ represents the concentration of the dissolved species i and z_i its charge. This general equation involves all aqueous species that are present in the solution for a given pH and ionic strength.

When the model aiming at describing the electrostatic interface layer contains a Stern layer with electrolyte binding (i.e., involving small diffuse layer potentials), using the wrong equation (Gouy–Chapman vs. general) is expected to have a minor impact since the surface potential is strongly screened by the Stern-layer, and the diffuse layer makes only a small contribution to the overall electroneutrality condition. However, the effect is considered to be at maximum when using the diffuse layer model. Implementation of the general equation in the speciation codes is strongly suggested, as it is the easiest option to avoid any issues. In the meantime, the treatment in Visual MINTEQ has been generalized in more recent versions of the codes (J.P. Gustafsson, personal communication).

Another issue in the models is the role of particle geometry. For very small particles, surface curvature starts to play a role (see chapter on ferrihydrite by Hiemstra et al. in this book). Numerical calculations for the diffuse layer potential and other quantities have been available since 1961 for spherical particles (Loeb et al. 1961), and approximate analytical solutions have been derived later (Ohshima et al. 1982). Most of the results suggest that spheres of the same material with smaller diameter exhibit a higher charge (Lützenkirchen 2002; Abbas et al. 2008) and that the geometry plays a role only for very small particles with particle sizes below about 10 nm (Abbas et al. 2008), but such a limit depends also on ionic strength and the absolute basic charge density (Barisik et al. 2014). The calculations by Barisik et al. do not consider a Stern layer and was based on COMSOL. A surface complexation code that has all the required details implemented does not seem to be yet available.

A very important point in surface complexation modeling is internal consistency of parameters. This concerns first of all the aqueous speciation scheme. Surface complexation parameters for the adsorption of a target cation, say europium, will depend on the chosen aqueous speciation of europium. Mixing surface complexation parameters with a different set of e.g., aqueous europium hydrolysis parameters will lead to wrong results. Similar issues can arise when ion pairs in solution are included in a speciation code database e.g., Eu^{3+} ion pairs with nitrate, which had not been included in the adsorption model calibration, although nitrate was used as background electrolyte. The consistency issue extends to surface parameters, like acid base constants or pair formation constants. Thus, mixing different models requires careful testing of parameter consistency (Lützenkirchen 2001a) and addition of reactions like pair formation constants in a contaminant adsorption model may require a recalibration of the adsorption model (Lützenkirchen 2001b). The consistency issue becomes even more acute with the use of charge distribution, which makes it highly unlikely that parameters can be easily transferred from one model to another.

COMPARISON OF TRADITIONAL FITTING PROCEDURES

In this section, several fitting procedures will be compared and differences will be discussed on the basis of a comprehensive example where one series of data are fit with different couplings of geochemical speciation codes and parameter optimization routines. The experimental data chosen for the fitting exercise are for the surface charge of silica (AEROSIL OX 50) in KCl solutions (Pilgrimm and Sonntag 1980; Pilgrimm 1981). Comparison of the data by Pilgrimm and independent data by Sonnefeld et al. (2001) in 1 mmol·L^{-1}, 10 mmol·L^{-1}, and 100 mmol·L^{-1} KCl, as well as unpublished data of JL in 10 mmol·L^{-1} KCl shows that the experimental results can be extremely well reproduced, even if the amount of solid is not the same. In the calculation of the charge density in Figure 5, the individually determined values of the specific surface areas were used. The surface chemistry is as simple as possible, i.e., data can be described with a Bolt-type silica behavior with one deprotonation step plus surface ion-pair formation. The model for interfacial electrostatics was also chosen to be as simple as possible while keeping it physically reasonable, i.e., a Basic Stern model with one capacitance (denoted Cap), resulting overall in three adjustable parameters, was employed. The site density was fixed at 4.75 sites·nm^{-2} based on the paper by Zhuralev (2000), i.e., the average between the two values 4.6 and 4.9 sites·nm^{-2}. An inter-comparison of codes is summarized in Table 2 and Table 3. More precisely, here, we use the exactly same type of input data (with the same number of digits in surface charge density (-*n.nn* E-n) and pH values (*n.nn*), the same activity coefficient corrections (i.e., Davies equation, with $\log_{10}\gamma_1 = -0.11$, -0.05, and -0.02 for 100 mmol·L^{-1}, 10 mmol·L^{-1}, 1 mmol·L^{-1}, respectively) for all codes. The observations and fitting results were plotted with two digits on *n.nn* E-n format to determine R^2. The fits were performed with different weights ($1/\sigma^2$) of the experimental data, i.e., 2.5×10^5, 4×10^4, and 1, resulting from the use of different uncertainties, namely $\sigma = 0.002$ C·m^{-2}, 0.005 C·m^{-2}, and 1 C·m^{-2}, respectively. The two involved surface chemical reactions have been introduced earlier:

$$\equiv\mathrm{SiOH} \rightleftharpoons \equiv\mathrm{SiO^-} + \mathrm{H^+},\ \log_{10} K_\mathrm{H} \tag{R1}$$

$$\equiv\mathrm{SiO^-} + \mathrm{K^+} \rightleftharpoons \equiv\mathrm{SiO^-\cdots K^+},\ \log_{10} K_\mathrm{K} \tag{R7}$$

The $\log_{10} K_\mathrm{H}$ value for the bare deprotonation lies between −7.06 and −7.10 and remains very simillar for all cases. The most significant differences can be observed for $\log_{10} K_\mathrm{K}$ (−0.55 to −0.92) and the capacitance (1.43–1.85 F·m^{-2}). Nevertheless, the overall differences in GoF and R^2 remain small among the applied codes and shell optimizers. Obviously, the GoF values are dependent on the uncertainties used for the experimental data points. Nevertheless, except for the PHREEQC/Python case, the uncertainties of the optimized parameters remained almost constant when the FIT, PEST, and UCODE shell optimizers were used (data not shown), with different weights for the experimental data points. Values which make sense should then be derived with the help of dedicated software such as SIMLAB (Saltelli 2003; Stockmann et al. 2017).

The correlation coefficients between all optimized parameters are given in Table 4.

Except for the PHREEQC/UCODE and PHREEQC/Python code combination where ALL parameters exhibit similarly high correlation coefficients, the highest correlation among the optimized parameters occurred between the capacitance Cap and the ion-pair formation constant ($C_{\mathrm{cap},\log_{10}K_\mathrm{K}}$).

The best fitting surface charge models for all code combinations tested in this work are shown in Figure 5 for the fitting involving unit weights.

Overall, a satisfactory agreement between the experimental surface charge curves with all simulations was obtained for all combination of codes, despite a small but systematic deviation observed at 1 mmol·L^{-1}. Since all codes exhibited overall very similar results, no other plots are shown. It is interesting to note that for the different code combinations, the best

Table 2. Fitting results with unit weights. The goodness of fit indicators from the respective output files were the Sum of Squared Residuals (SSR) and WSOSR, i.e., the same indicator but with different names in the different fitting codes.

Code/code combination	$\log_{10} K_H$ (R_1)	$\log_{10} K_K$ (R_7)	Cap ($F \cdot m^{-2}$)	GoF indicator	GoF	R^2
ECOSAT/FIT	−7.09 ± 0.14	−0.77± 4.09	1.68 ± 3.17	SSR	1.27×10^{-4}	0.9874
FITEQL[m 1]/PEST	−7.06 ± 0.06	−0.55 ± 0.29	1.43 ± 0.25	WSOSR	1.23×10^{-4}	0.9878
FITEQL[m 1]/UCODE[2] (User 1)	−7.06 ± 0.06	−0.63 ± 0.97	1.50 ± 0.65	WSOSR	1.23×10^{-4}	0.9877
FITEQL[m 1]/UCODE[2] (User 2)	−7.08 ± 0.06	−0.72 ± 0.99	1.63 ± 0.72	WSOSR	1.22×10^{-4}	0.9878
PHREEQC[3]/UCODE[4]	−7.10 ± 0.14	−0.92 ± 3.43	1.85 ± 2.86	WSOSR	1.25×10^{-4}	0.9785
PHREEQC[5]/Python	−7.083 ± 0.02	−0.91 ± 3.9	1.777 ± 1.5	WSOSR	1.24×10^{-4}	0.9877

Notes:
[1]FITEQL[m] refers to a modified version adopted from FITEQL 2.0
[2]UCODE refers to the version 3.061
[3]PHREEQC refers to the batch version version phreeqc-3.7.3-15968-x64
[4]UCODE refers to the 2005 version
[5]PHREEQC refers to an IPhreeqC.dll compiled from phreeqc-3.7.3 and implemented via phreeqpy

fit parameters were not the same even for such a simple system, although the model results are essentially equal (Fig. 5), and the various differences are likely related to differences in rounding for example.

The reason for the differing best fit parameters providing equally good model is certainly the high correlation between at least two of the parameters as discussed above. In such cases it is possible that tiny details in the settings, like numbers of digits written

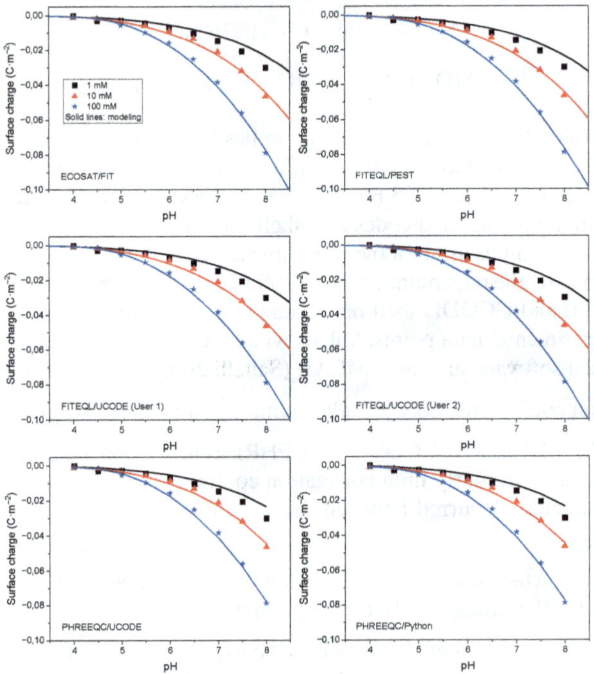

Figure 5. Surface charge of silica (AEROSIL OX 50) in different KCl concentrations (I = 1, 10 or 100 $mmol \cdot L^{-1}$). Experimental values (**symbols**) from the data of Pilgrimm (1981) and respective fits (**solid lines**) are shown.

Table 3. Fitting results for the PHREEQC/Python code combination using various data uncertainties. The GoF indicator from the respective output files was χ^2 as described above.

Code/Code combination (uncertainties)	$\log_{10} K_H$ (R1)	$\log_{10} K_K$ (R7)	Cap ($F \cdot m^{-2}$)	GoF (χ^2)	R^2
PHREEQC[1]/Python ($\sigma = 1\ C \cdot m^{-2}$)	−7.083 ± 0.02	−0.91 ± 3.9	1.777 ± 1.5	5.18×10^{-6}	0.9877
PHREEQC[1]/Python ($\sigma = 0.005\ C \cdot m^{-2}$)	−7.083 ± 0.001	−0.91 ± 0.02	1.777 ± 0.008	2.1×10^{-1}	0.9877
PHREEQC[1]/Python ($\sigma = 0.002\ C \cdot m^{-2}$)	−7.083 ± 0.001	−0.91 ± 0.01	1.777 ± 0.003	1.29×10^{0}	0.9877

Note:
1PHREEQC refers to an IPhreeqC.dll compiled from phreeqc-3.7.3 and implemented via phreeqpy

Table 4. Correlation coefficients between capacitance and ion-pair formation constant ($C_{cap, \log_{10} K_K}$), capacitance and protonation constant ($C_{cap, \log_{10} K_H}$), and protonation constant and ion-pair formation constant ($C_{\log_{10} K_H, \log_{10} K_K}$) (absolute values, all for unit weights). Note that with ECOSAT/FIT the electrolyte binding constant was fitted combined with the acidity constant, and the correlation coefficients (in brackets) are not directly comparable for this code combination.

Code combination	$C_{cap, \log_{10} K_K}$	$C_{cap, \log_{10} K_H}$	$C_{\log_{10} K_H, \log_{10} K_K}$
ECOSAT/FIT	(1.00)	0.93	(0.92)
FITEQL[m 1]/PEST	0.84	0.67	0.23
FITEQL[m 1]/UCODE[2] (User 1)	0.97	0.34	0.13
FITEQL[m 1]/UCODE[2] (User 2)	0.97	0.51	0.32
PHREEQC[3]/UCODE[4]	0.997	0.940	0.915
PHREEQC[5]/Python	0.997	0.940	0.915

Notes:
[1]FITEQL[m] refers to a modified version adapted from FITEQL 2.0
[2]UCODE refers to the version 3.061
[3]PHREEQC refers to the batch version version phreeqc-3.7.3-15968-x64
[4]UCODE refers to the 2005 version
[5]PHREEQC refers to an IPhreeqC.dll compiled from phreeqc-3.7.3 and implemented via phreeqpy

into output files, which are read by the optimization routine, or allowed maximum change in parameters or stop criteria (see the use of the same code combination by different users) are responsible for differences in the best fit parameters. In such ambiguous cases, it may be a good choice to explore parameter space by using a search algorithm rather than a least squares optimization routine (Zhao et al. 2015) and to report parameter space maps or ranges of parameter combinations that result in equally good fits, rather than best fit parameters. Here, such an approach was employed to elaborate the nature of the parameter correlations and to rationalize the various fitting results.

Figure 6 shows 2D parameter space maps for the three possible parameter combinations between $\log_{10} K_H$, $\log_{10} K_K$, and the capacitance.

The maps are based on the PHREEQC/Python combination employed here as one of the code couplings. The best fit χ^2 for these codes is 1.29. Color codes and iso-lines in Figures 6 and 7 refer to χ^2 variations with varying parameter combinations. The maps show between capacitance (Cap) and K^+ binding constant ($\log_{10} K_K$) a classical correlation where, e.g., the goodness of fit remains below 1.30 for $\log_{10} K_K$ varying between −0.7 and −0.95, if at the same time Cap varies between 1.65 and 1.80 $F \cdot m^{-2}$. For the other parameter combinations, the minima are better defined, which is also reflected in the lower correlation coefficients as shown in Table 4.

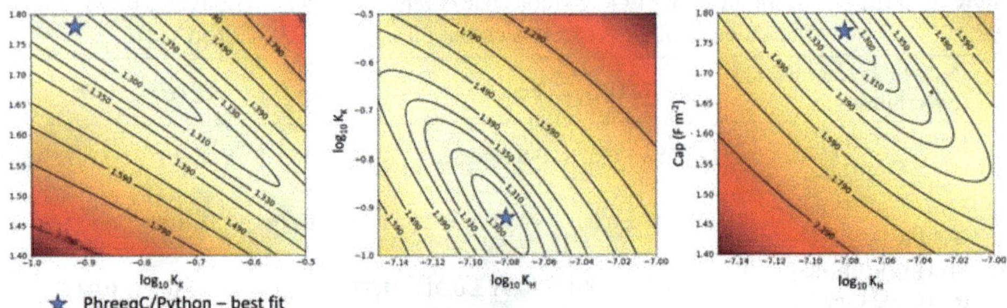

Figure 6. 2D parameter space maps. Parameter uncertainties are smallest, where the gradients in parameter space (χ^2) are high. The **blue asterisk** marks the best fit of the P^{3}R, PHREEQC/Python fitting tool, which was used to generate the parameter space maps. An uncertainty of $\sigma = 0.002$ C·m^{-2} was considered, thus referring to the last row in Table 3.

In Figure 7, the 2D parameter space map for the parameter combination ($\log_{10} K_K$, $\log_{10} K_H$) is calculated for the various best fit capacitances determined by the respective code combinations.

The maps in Figure 7 show how the minimal χ^2 region (according to the PHREEQC/Python combination) follows the respective best fit ($\log_{10} K_K$, $\log_{10} K_H$) combinations of the other codes as indicated by the various symbols in Figure 7. This demonstrates that there is a very flat region in parameter space, with χ^2 varying only from 1.29–1.32 (PHREEQC/Python) into which the best fit results of all code combinations fall.

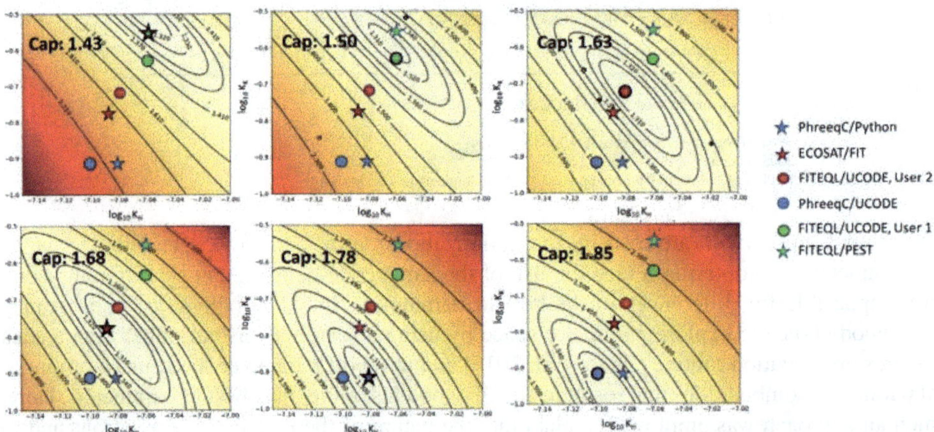

Figure 7. Parameter space ($\log_{10} K_H$, $\log_{10} K_K$) plane maps along the different capacitance values (Cap in F·m^{-2}) determined as best fit values by the various modeling code / fitting tool combinations. A very flat gradient in χ^2 exists along the best fit parameter sets. In the parameter space maps, always the symbols highlighted with a bold edge line correspond to the ones that match the capacitance (Cap) value of the map. The parameter space maps were all calculated with PHREEQC/Python.

CONCLUSIONS

The acquisition of high-quality data and the knowledge of their associated uncertainties are the basis of a successful parameterization of any surface complexation model. Despite all the mentioned developments in the codes and the fitting algorithms, together with the promising perspectives offered by artificial intelligence, the reliability and robustness of the experimental data still remain of major importance. Some details in SCMs that can lead to tremendous errors in simulations (such as the treatment of multidentate surface species) need to be carefully documented and recognized by model developers and users, respectively. We, furthermore, emphasize the need for internal consistency both between aqueous speciation schemes and surface complexation models, and within surface complexation models. This can be a challenge when introducing surface complexation databases in speciation codes, which contain their own aqueous speciation databases.

A comparison of code combinations, applied to a seemingly simple system, revealed pronounced differences in final values of adjustable parameters. These may arise from high parameter correlation, but also from settings in the codes or input files that are not necessarily compatible (such as convergence criteria). To obtain reasonable statistical quantities and reasonable parameter estimates and uncertainties, it is essential to involve reliable and/or reasonable experimental error estimates. These are ideally reported in the experimental parts of adsorption (and related solution) studies point by point. Reported parameters should ideally not only be accompanied by a measure of parameter uncertainty, but also by parameter covariances and/or correlations.

The most comprehensive available self-consistent databases for hydrous ferric oxide (Dzombak and Morel 1990) and gibbsite (Karamalidis and Dzombak 2010) include error margins of the stability constants and long sections discussing experimental data. This is what is needed more in the future, maybe also considering more sophisticated interfacial models. While these databases are comprehensive for single-solute systems, uncertainty will be unavoidable in multi-solute systems, where ternary complexes may occur. As in the case of aqueous solution, there is a wealth of systems to investigate before comprehensive databases will be ready to tackle the needs of environmental applications, and this statement extends drastically if temperatures different from room temperature shall be included.

Notwithstanding the above statements, the existing examples do show that it is possible to reduce uncertainty in dealing with adsorption phenomena by compiling available literature data and treating it within one consistent model concept. Picking one data source for a given system instead may be problematic. We emphasize the importance of maintaining the knowledge for selecting the best experimental data, which in turn requires that the decision takers know the experimental methods in detail. Finally, reliable surface complexation models allow a direct link between surface- and solution speciation, which makes them superior to previously common conditional Langmuir and Freundlich models or even constant-K_D approaches, that are still published in far too many papers, and are still commonly applied to real-world scenarios.

REFERENCES

Abbas Z, Labbez C, Nordholm S, Ahlberg E (2008) Size-dependent surface charging of nanoparticles. J Phys Chem C 112:5715–5723, https://doi.org/10.1021/jp709667u

Akula P (2020) Thermodynamic approach to computational modeling of chemically stabilized soils. Ph.D. Thesis, Texas A&M University

Anderson GM (1976) Error propagation by Monte-Carlo method in geochemical calculations. Geochim Cosmochim Acta 40:1533–1538, https://doi.org/10.1016/0016-7037(76)90092-2

Barisik M, Atalay S, Beskok A, Qian SZ (2014) Size dependent surface charge properties of silica nanoparticles. J Phys Chem C 118:1836–1842, https://doi.org/10.1021/jp410536n

Benjamin MM (2002) Modeling the mass-action expression for bidentate adsorption. Environ Sci Technol 36:307–313, https://doi.org/10.1021/es010936n

Bethke C, Yeakel S (2009) Geochemist's Workbench: Release 8.0 Reaction Modeling Guide; RockWare Incorporated: Golden, CO, USA

Boily J-F (2025) Molecular controls on complexation reactions and electrostatic potential development at mineral surfaces. Rev Mineral Geochem 91A:105–148

Casella F, Bachmann B (2021) On the choice of initial guesses for the Newton–Raphson algorithm. Applied Mathematics and Computation 398:125991, https://doi.org/10.48550/arXiv.1911.12433

Dedieu JP (2015) Newton–Raphson Method. *In*: Encyclopedia of Applied and Computational Mathematics. Engquist B (ed) Springer, Berlin, Heidelberg.

Deuflhard P (2012) A short history of Newton's Method. Documenta Mathematica. Extra Volume ISMP, p 25–30

Doherty J (1994) PEST: a unique computer program for model-independent parameter optimisation. *In:* Water Down Under 94: Groundwater/Surface Hydrology Common Interest Papers; Preprints of Paper, Barton, ACT: Institution of Engineers, Australia, p 551–554

Dyrssen D, Sillén LG, Ingri N (1961) "Pit-mapping" --- a General Approach for Computer Refining of Equilibrium Constants. Acta Chem Scand 15:694-696, https://doi.org/10.3891/acta.chem.scand.15-0694

Dzombak DA, Morel FMM (1990) Surface complexation modeling: Hydrous ferric oxide. Wiley, New York

Eriksson G (1979) Algorithm for the computation of aqueous multi-component, multiphase equilibria. Analytica Chimica Acta-Computer Techniques and Optimization 3:375–383, https://doi.org/10.1016/S0003-2670(01)85035-2

Fletcher R (1971) A Modified Marquardt Subroutine for Non-Linear Least Squares. UKAEA, Res. Group, report AERE-R-6799, Harwell, Berks, UK

Gargiani M, Zanelli A, Diehl M, Hutter F (2020) On the promise of the stochastic generalized Gauss–Newton method for training DNNs. arXiv:200602409

Gavin HP (2022) The Levenberg–Marquardt algorithm for nonlinear least squares curve-fitting problems. https://people.duke.edu/~hpgavin/ExperimentalSystems/lm.pdf

Gratton S, Lawless AS, Nichols NK (2007) Approximate Gauss–Newton methods for nonlinear least squares problems. SIAM J Optimization 18:106–132

Gustafsson JP (2003) Modelling molybdate and tungstate adsorption to ferrihydrite. Chem Geol 200:105–115, https://doi.org/10.1016/s0009-2541(03)00161-x

Gustafsson JP (2013) Visual MINTEQ, Version 3.1, KTH, Stockholm Sweden. https://vmintEqn.lwr.kth.se/

Heberling F, Klačić T, Raiteri P, Gale JD, Eng PJ, Stubbs JE, Gil-Díaz T, Begović T, Lützenkirchen J (2021) Structure and surface complexation at the calcite(104)–water interface. Environ Sci Technol 55:12403–12413, https://doi.org/10.1021/acs.est.1c03578

Herbelin AL, Westall JC (1994) FITEQL: A Computer Program for Determination of Chemical Equilibrium Constants from Experimental Data, Version 3.1. Report 94-01, Department of Chemistry, Oregon State University Corvallis, Oregon 97331, USA

Herbelin AL, Westall JC (1996) FITEQL: A Computer Program for Determination of Chemical Equilibrium Constants from Experimental Data, Version 3.2. Report 96-01, Department of Chemistry, Oregon State University Corvallis, Oregon 97331, USA

Herbelin AL, Westall JC (1999) FITEQL: A Computer Program for Determination of Chemical Equilibrium Constants from Experimental Data, Version 4.0. Report 99-01, Department of Chemistry, Oregon State University Corvallis, Oregon 97331, USA

Hiemstra T, Lützenkirchen J (2025) Development and modus operandi relating surface structure and ion complexation modeling for important metal (hydr)oxides. Rev Mineral Geochem 91A:13–84

Hill MC (1998) Methods and guidelines for effective model calibration; with application to UCODE, a computer code for universal inverse modeling, and MODFLOWP, a computer code for inverse modeling with MODFLOW. Water-Resources Investigations Report 98-4005, U.S. Geological Survey

Huittinen N, Virtanen S, Rossberg A, Eibl M, Lönnrot S, Polly R (2022) A combined extended X-ray absorption fine structure spectroscopy and density functional theory study of americium vs. yttrium adsorption on corundum (α-Al_2O_3). Minerals 12:18, https://doi.org/10.3390/min12111380

Ingri N, Sillén LG (1962) High-speed computers as a supplement to graphical methods. II. Some computer programs for studies of complex formation equilibria. Acta Chem Scand 16:173–191, https://doi.org/10.3891/acta.chem.scand.16-0173

James RO, Davis JA, Leckie JO (1978) Computer simulation of the conductometric and potentiometric titrations of the surface groups on ionizable latexes. J Colloid Interface Sci 65:331–344, https://doi.org/10.1016/0021-9797(78)90164-9

Jenne EA (ed) (1979) Chemical Modeling in Aqueous Systems. Speciation, Sorption, Solubility, and Kinetics. American Chemical Society

Jordan N, Ritter A, Scheinost AC, Weiss S, Schild D, Hübner R (2014) Selenium(IV) uptake by maghemite (γ-Fe_2O_3). Environ Sci Technol 48:1665–1674, https://doi.org/10.1021/es4045852

Jordan N, Franzen C, Lützenkirchen J, Foerstendorf H, Hering D, Weiss S, Heim K, Brendler V (2018) Adsorption of selenium(vi) onto nano transition alumina. Environ Sci Nano 5:1661–1669, https://doi.org/10.1039/c8en00293b

Karamalidis AK, Dzombak DA (2010) Surface complexation modeling: Gibbsite. Wiley. Hoboken

Keizer MG, Van Riemsdijk WH (1994) ECOSAT: A computer program for the calculation of speciation and transport in soil–water systems. Dept. of Soil Science and Plant Nutrition, Wageningen Agricultural University, Wageningen, the Netherlands

Keizer MG, Van Riemsdijk WH (1998) ECOSAT: A computer program for the calculation of speciation and transport in soil–water systems. Dept. of Soil Science and Plant Nutrition, Wageningen Agricultural University, Wageningen, the Netherlands

Kinniburgh DG (1993) FIT user guide. BGS Technical Report WD/93/23:40

Krupka KM, Morrey JR (1985) MINTEQ geochemical reaction code: status and applications. Jacobs GK, Whatley SK (Eds), Proceedings of the Conference on the Application of Geochemical Models to High-Level Nuclear Waste Repository Assessment, NUREG/CP-0062 (ORNL/TM-9585), Oak Ridge National Laboratory (1985), p 46–53

Kulik D, Berner U, Curti E (2004) Modeling chemical equilibrium partitioning with the GEMS-PSI code. PSI Scientific Report 2003 / Vol IV, Nuclear Energy and Safety. Villigen, Paul Scherrer Institut: p 109–122

Kulik DA, Peretyashko TS (1998) Comparison of sorption modelling by Law of mass action (FITEQL) and Gibbs energy minimisation (Selektor-A) codes. Mineral Mag 62:824-825

Kulik DA, Wagner T, Dmytrieva SV, Kosakowski G, Hingerl FF, Chudnenko KV, Berner UR (2013) GEM-Selektor geochemical modeling package: revised algorithm and GEMS3K numerical kernel for coupled simulation codes. Comput Geosci 17:1–24, https://doi.org/10.1007/s10596-012-9310-6

Kulik DA, Winnefeld F, Kulik A, Miron GD, Lothenbach B (2021) CemGEMS—An easy-to-use web application for thermodynamic modelling of cementitious materials. RILEM Tech Lett 6:36–52, https://doi.org/10.21809/rilemtechlett.2021.140

Leggett DJ (1985) The determination of formation constants. *In*: Computational Methods for the Determination of Formation Constants. Leggett DJ (ed) Springer US, Boston, MA, p 1–17

Levenberg K (1944) A method for the solution of certain non-linear problems in least squares. Q Appl Math 2:164–168, https://doi.org/10.1090/qam/10666

Li C, Zarzycki P (2022) A computational pipeline to generate a synthetic dataset of metal ion sorption to oxides for AI/ML exploration. Front Nucl Eng 1:977743, https://doi.org/10.3389/fnuen.2022.977743

Livi KJT, Villalobos M, Leary R, Varela M, Barnard J, Villacís-García M, Zanella R, Goodridge A, Midgley P (2017) Crystal face distributions and surface site densities of two synthetic goethites: Implications for adsorption capacities as a function of particle size. Langmuir 33:8924–8932, https://doi.org/10.1021/acs.langmuir.7b01814

Livi KJT, Villalobos M, Ramasse Q, Brydson R, Salazar-Rivera HS (2023) Surface site density of synthetic goethites and its relationship to atomic surface roughness and crystal size. Langmuir 39:556–562, https://doi.org/10.1021/acs.langmuir.2c02818

Loeb AL, Overbeek JTG, Wiersema PH (1961) The electrical double layer around a spherical colloid particle. Computation of the potential, charge density, and free energy of the electrical double layer around a spherical colloid particle. The MIT Press, Cambridge, Massachusetts

Logue BA, Smith RW, Westall JC (2004) U(VI) adsorption on natural iron-coated sands: comparison of approaches for modeling adsorption on heterogeneous environmental materials. Appl Geochem 19:1937–1951, https://doi.org/10.1016/j.apgeochem.2004.05.010

Lösler M, Hennes M (2008) An innovative mathematical solution for a time-efficient IVS reference point determination. http://www.gik.kit.edu/downloads/MC_029_FINAL.pdf

Lützenkirchen J (2001a) Evaluation of experimental procedures and discussion of two different modelling approaches with respect to long-term kinetics of metal cation sorption onto (hydr)oxide surfaces. Aquat Geochem 7:217–235, https://doi.org/10.1023/a:1012973630754

Lützenkirchen J (2001b) A discussion of the surface complexation modeling in the paper by Sarkar et al. (1999). Soil Sci Soc Am J 65:1348-1349, https://doi.org/10.2136/sssaj2001.6541348x

Lützenkirchen J (2002) Surface complexation models of adsorption: A critical survey in the context of experimental data *In*: Adsorption: Theory, Modeling, and Analysis (Surfactant Science Series, Volume 107). Tóth J (ed) Marcel Dekker, Inc., New-York, Basel, p 631–710

Lützenkirchen J, Marsac R, Kulik DA, Payne TE, Xue ZR, Orsetti S, Haderlein SB (2015) Treatment of multi-dentate surface complexes and diffuse layer implementation in various speciation codes. Appl Geochem 55:128-137, https://doi.org/10.1016/j.apgeochem.2014.07.006

Machesky ML, Ridley MK, Heberling F, Lützenkirchen J (2023) Proton uptake at the barite–aqueous solution interface: A combined potentiometric, electrophoretic mobility, and surface complexation modeling investigation. ACS Earth and Space Chemistry 7:1713–1726, https://doi.org/10.1021/acsearthspacechem.3c00109

Marquardt DW (1963) An algorithm for least squares estimation of nonlinear parameters. J Soc Indust Appl Math 11:431–441, https://doi.org/10.1137/0111030

Martínez RJ, Villalobos M, Loredo-Jasso AU, Cruz-Valladares AX, Mendoza-Flores A, Salazar-Rivera H, Cruz-Romero D (2023) Towards building a unified adsorption model for goethite based on direct measurements of crystal face compositions: I. Acidity behavior and As (V) adsorption. Geochim Cosmochim Acta 354:252–262, https://doi.org/10.1016/j.gca.2023.06.021

McDuff RE, Morel FMM (1975) Description and use of the chemical equilibrium program REDEQL2. WM Keck Laboratory of Environmental Engineering Science, California Institute of Technology

Meeussen JCL (2003) ORCHESTRA: An object-oriented framework for implementing chemical equilibrium models. Environ Sci Technol 37:1175–1182, https://doi.org/10.1021/es025597s

Melchior DC, Bassett RL (eds) (1990) Chemical Modeling of Aqueous Systems II. ACS Symposium Series, American Chemical Society

Miron GD, Kulik DA, Dmytrieva SV, Wagner T (2015) GEMSFITS: Code package for optimization of geochemical model parameters and inverse modeling. Appl Geochem 55:28–45, https://doi.org/10.1016/j.apgeochem.2014.10.013

Morel F, Morgan J (1972) Numerical method for computing equilibriums in aqueous chemical systems. Environ Sci Technol 6:58–67, https://doi.org/10.1021/es60060a006

Morel FMM, Hering JG (1993) Principles and Applications of Aquatic Chemistry. John Wiley & Sons

Nelder JA, Mead R (1965) A simplex-method for function minimization. Comput J 7:308–313, https://doi.org/10.1093/comjnl/7.4.308

Ohshima H, Healy TW, White LR (1982) Accurate analytic expressions for the surface charge density/surface potential relationship and double-layer potential distribution for a spherical colloidal particle. J Colloid Interface Sci 90:17–26, https://doi.org/10.1016/0021-9797(82)90393-9

Papadopoulos CE, Yeung H (2001) Uncertainty estimation and Monte Carlo simulation method. Flow Meas Instrum 12:291–298, https://doi.org/10.1016/s0955-5986(01)00015-2

Papelis C, Hayes KF, Leckie JO (1988) HYDRAQL: a program for the computation of chemical equilibrium composition of aqueous batch systems including surface-complexation modeling of ion adsorption at the oxide/solution interface. Technical Report 306, Department of Civil Engineering, Stanford University, Stanford, Calif., USA

Parkhurst DL, Appelo CAJ (1999) User's guide to PHREEQC (Version 2)—A computer program for speciation, batch-reaction, one-dimensional transport, and inverse geochemical calculations. U.S. Geological Survey Water-Resources Investigations Report 99-4259

Parkhurst DL, Appelo CAJ (2013) Description of input and examples for PHREEQC version 3—A computer program for speciation, batch-reaction, one-dimensional transport, and inverse geochemical calculations. U.S. Geological Survey Techniques and Methods, Book 6, Chapter 43

Peacock CL, Sherman DM (2004) Copper(II) sorption onto goethite, hematite and lepidocrocite: A surface complexation model based on ab initio molecular geometries and EXAFS spectroscopy. Geochim Cosmochim Acta 68:2623–2637, https://doi.org/10.1016/j.gca.2003.11.030

Perrin DD, Sayce IG (1967) Computer calculation of equilibrium concentrations in mixtures of metal ions and complexing species. Talanta 14:833–842, https://doi.org/10.1016/0039-9140(67)80105-x

Pilgrimm H (1981) Untersuchungen an der AEROSIL-Elektrolyt-Phasengrenze ohne Anwesenheit von Basenkationen. Colloid Polym Sci 259:1111–1115, https://doi.org/10.1007/BF01524898

Pilgrimm H, Sonntag H (1980) Die Struktur der elektrochemischen Doppelschicht an der AEROSIL-Elektrolyt-Phasengrenze. Z Phys Chemie 261:433–440, https://doi.org/10.1515/zpch-1980-26156

Poeter EP, Hill MC (1998) Documentation of UCODE, a computer code for universal inverse modeling. Water-Resources Investigations Resport 98-4080, U.S. Geological Survey

Poeter EP, Hill MC (1999) UCODE, a computer code for universal inverse modeling. Comput Geosci 25:457–462, https://doi.org/10.1016/S0098-3004(98)00149-6

Polyak BT (2007) Newton's method and its use in optimization. Eur J Oper Res 181:1086–1096, https://doi.org/ https://doi.org/10.1016/j.ejor.2005.06.076

Press WH, Teukolsky SA, Vetterling WT, Flannery BP (1992) Numerical Recipes in C, The Art of Scientific Computing, Second Edition. Cambridge University Press, Cambridge, New York, Port Chester, Melbourne, Sydney

Ranganathan A (2004) The Levenberg–Marquardt algorithm. Tutorial on LM Algorithm 11:101–110

Rossotti FJC, Rossotti HS (1955) Graphical Methods for Determining Equilihrium Constants.I. Systems of Mononuclear Complexes. Acta Chem Scand 9:1166–1176, https://doi.org/10.3891/acta.chem.scand.09-1166

Saltelli A (2003) SIMLAB 2.2 manual, simulation environment for uncertainty and sensitivity analysis. SIMLAB 2.2 Manual, JRC/ POLIS ScaRL, SIMLAB. I

Schlegel A (2016) Newton–Raphson method for root-finding. https://rpubs.com/aaronsc32/newton-raphson-method

Sherman DM, Peacock CL, Hubbard CG (2008) Surface complexation of U(VI) on goethite (α-FeOOH). Geochim Cosmochim Acta 72:298–310, https://doi.org/10.1016/j.gca.2007.10.023

Sia S (2018) Taylor Series approximation, newton's method and optimization. https://suzyahyah.github.io/calculus/optimization/2018/04/06/taylor-series-newtons-method.html

Sillén LG (1956) Some graphical methods for determining equilibrium constants. II. On "curve-fitting" methods for two-variable data. Acta Chem Scand 10:186–202, https://doi.org/10.3891/acta.chem.scand.10-0186

Sillén LG (1962) High-speed computers as a supplement to graphical methods. I. Functional behavior of the error square sum. Acta Chem Scand 16:159–172, https://doi.org/10.3891/acta.chem.scand.16-0159

Sillén LG (1964) High-speed computers as a supplement to graphical methods. III. Twist matrix methods for minimizing the error-square sum in problems with many unknown constants. Acta Chem Scand 18:1085–1098

Sonnefeld J, Löbbus M, Vogelsberger W (2001) Determination of electric double layer parameters for spherical silica particles under application of the triple layer model using surface charge density data and results of electrokinetic sonic amplitude measurements. Colloids Surf A 195:215-225, https://doi.org/https://doi.org/10.1016/S0927-7757(01)00845-7

Spadini L, Manceau A, Schindler PW, Charlet L (1994) Structure and stability of Cd^{2+} surface complexes on ferric oxides .1. Results from Exafs spectroscopy. J Colloid Interface Sci 168:73–86, https://doi.org/10.1006/jcis.1994.1395

Stockmann M, Schikora J, Becker DA, Flügge J, Noseck U, Brendler V (2017) Smart K_d-values, their uncertainties and sensitivities—Applying a new approach for realistic distribution coefficients in geochemical modeling of complex systems. Chemosphere 187:277–285, https://doi.org/10.1016/j.chemosphere.2017.08.115

Tobias RS, Yasuda M (1963) Computer analysis of stability constants in three-component systems with polynuclear complexes. Sov Phys Tech Phys 8:1307–1310

Transtrum MK, Sethna JP (2012) Improvements to the Levenberg–Marquardt algorithm for nonlinear least-squares minimization. arXiv preprint arXiv:12015885

Vulava VM, Kretzschmar R, Rusch U, Grolimund D, Westall JC, Borkovec M (2000) Cation competition in a natural subsurface material: Modelling of sorption equilibria. Environ Sci Technol 34:2149-2155, https://doi.org/10.1021/es990214k

Wagner T, Kulik DA, Hingerl FF, Dmytrieva SV (2012) GEM-Selektor geochemical modeling package: TSolMod library and data interface for multicomponent phase models. Can Mineral 50:1173–1195

Wang ZM, Giammar DE (2013) Mass action expressions for bidentate adsorption in surface complexation modeling: Theory and practice. Environ Sci Technol 47:3982–3996, https://doi.org/10.1021/es305180e

Westall J (1980) Chemical equilibrium including adsorption on charged surfaces. *In*: Particulates in Water. Kavanaugh MC, Leckie JO (eds). ACS, Washington, DC, p 33–44

Westall JC (1976) MINEQL: A computer program for the calculation of chemical equilibrium composition of aqueous systems. Massachusetts Institute of Technology, Water Quality Laboratory, Cambridge, Massachusetts

Westall JC (1979a) MICROQL. II. Computation of adsorption equilibria in BASIC. Technical Report, Swiss Federal Institute of Technology, EAWAG, Dübendorf, Switzerland

Westall JC (1979b) MICROQL. I. A chemical equilibrium program in BASIC. Technical Report, Swiss Federal Institute of Technology, EAWAG, Dübendorf, Switzerland

Westall JC (1982a) FITEQL. A computer program for determination of equilibrium constants from experimental data. Version 1.2. Report 82–01, Department of Chemistry, Oregon State University. Corvallis, Oregon 97331

Westall JC (1982b) FITEQL, a computer program for determination of chemical equilibrium constants from experimental data. Version 2.0. report 82–02. Department of Chemistry, Oregon State University. Corvallis, Oregon 97331

Westall JC (1982c) FITEQL: A Computer Program for Determination of Chemical Equilibrium Constants from Experimental Data, version 2.0. Report 82-02, Department of Chemistry, Oregon State University. Corvallis, Oregon 97331

Westall J, Hohl H (1980) A comparison of electrostatic models for the oxide/solution interface. Adv Colloid Interface Sci 12:265-294, https://doi.org/10.1016/0001-8686(80)80012-1

Westall JC, Morel FMM (1977) FITEQL: A general algorithm for the determination of metal–ligand complex stability constants from experimental data. *In*: Technical Note 19. Ralph M. Parsons Laboratory, Department of Civil Engineering, Massachusetts

Westall JC, Zachary JL, Morel FMM (1976) MINEQL: A General Algorithm for The Computation of Chemical Equilibrium in Aqueous Systems. Abstracts of Papers of the American Chemical Society 172:8

Xie XF, Giammar DE, Wang ZM (2016) MINFIT: A spreadsheet-based tool for parameter estimation in an equilibrium speciation software program. Environ Sci Technol 50:11112–11120, https://doi.org/10.1021/acs.est.6b03399

Zhang Z, Fenter P, Cheng L, Sturchio NC, Bedzyk MJ, Předota M, Bandura A, Kubicki JD, Lvov SN, Cummings PT, Chialvo AA (2004) Ion adsorption at the rutile–water interface: Linking molecular and macroscopic properties. Langmuir 20:4954–4969, https://doi.org/10.1021/la0353834

Zhao CL, Ebeling D, Siretanu I, van den Ende D, Mugele F (2015) Extracting local surface charges and charge regulation behavior from atomic force microscopy measurements at heterogeneous solid–electrolyte interfaces. Nanoscale 7:16298-16311, https://doi.org/10.1039/c5nr05261k

Zhuravlev LT (2000) The surface chemistry of amorphous silica. Zhuravlev model. Colloids Surf A 173:1–38, https://doi.org/10.1016/S0927-7757(00)00556-2

Reviews in Mineralogy & Geochemistry
Vol. 91A pp. 413–456, 2025

13

Practical Application of Surface Complexation Models: Evolution, Approaches, and Examples

David A. Dzombak

Carnegie Mellon University
Department of Civil and Environmental Engineering
5000 Forbes Avenue
Pittsburgh, PA 15213, USA

dzombak@cmu.edu

Jerry D. Allison

University of North Georgia
Department of Chemistry
3820 Mundy Mill Road
Oakwood, GA 30566, USA

jallison@ung.edu

Ted P. Lillys

RTI International
Center for Environmental Health, Risk, and Sustainability
3040 E Cornwallis Road
Research Triangle Park, NC 27709, USA

tlillys@rti.org

Jason Mills

United States Environmental Protection Agency
Economic and Risk Analysis Group
Office of Resource Conservation and Recovery
1301 Constitution Avenue NW
Washington, DC 20460, USA

Mills.Jason@epa.gov

INTRODUCTION

Following their introduction and development in the 1970s and 1980s, surface complexation models have been used for a diverse range of practical applications, from engineering analyses to interpretation of geochemical data to reactive transport modeling for inorganic contaminants in surface water and groundwater systems. The use of surface complexation models (SCMs) in practical applications has evolved and expanded steadily. As experience with practical application of SCMS has been gained, approaches for addressing

1529-6466/25/0091A-0013$05.00 (print)
1943-2666/25/0091A-0013$05.00 (online)
http://dx.doi.org/10.2138/rmg.2025.91A.13

the complexity of real systems have been developed and learned. Three general approaches have been used for practical applications of scms: the multisurface (or component additivity) approach, the single-surface/surrogate approach, and the generalized composite approach. The use of SCMs for practical applications has been facilitated by the development of standard parameter values and SCM databases for some sorbents of importance in soils, sediments, and aquifers, and by use of selective extractions to estimate the amounts of reactive sorbent surfaces available in complex systems.

The development of SCMs for the description of inorganic ion sorption on metal oxides emerged in the 1960s and evolved in the 1970s and 1980s. The history of their development has been well documented (Davis and Kent 1990; Dzombak and Morel 1990; Goldberg 1992; Jenne 1998; Ponthieu et al. 2006). Core to all surface complexation models is the use of reactions and mass action equations to describe the equilibrium sorption of ions at specific surface sites. Most surface complexation models also consider explicitly the contribution of electrostatic interactions to the energetics of the surface reactions, which for oxides are variable as the surface charge varies with the state of surface protonation and loading with other sorbing ions. Different molecular models of the interface and representations of surface charge conditions are used in surface complexation models, resulting in model-dependent equilibrium constants for the surface complexation reactions (Westall and Hohl 1980; Payne et al. 2006).

Surface complexation models have been deployed for a wide range of uses, from research on fundamental aspects of surface chemistry to diverse practical applications in engineering and science. Examples of uses in research include assistance with interpretation of spectroscopic data from studies of ion interactions with oxide surfaces (e.g., Hayes and Katz 1996; Wang et al. 2013), and with data from studies of electrolyte effects on colloids and their stability in aqueous suspension (e.g., Liang and Morgan 1990; Tombacz and Szekeres 2004). SCMs have also served as a foundation to advance theory for the structure of the oxide–water interface (e.g., Davis et al. 1978; Hiemstra and Van Riemsdijk 1996, 2006; Sverjensky and Sahai 1996). In such fundamental research applications, there is significant interest in the fidelity of SCMs with microscale mechanisms and processes. Examples of practical application of SCMs include use in interpreting data to understand geochemical systems (e.g., Smith et al. 1998), design of water treatment processes (e.g., Hering et al. 1996), and reactive transport modeling for metal ions in surface waters (e.g., Runkel et al. 1999) and subsurface systems (e.g., Kent et al. 2000). In practical applications, interests in SCMs focus on their central strengths, especially the ability to predict sorption for a range of solution conditions, and there is less concern with capturing the full mechanistic complexity of sorption processes with the models. Indeed, employing SCMs that engage more complex descriptions of the oxide–water interface can be a hindrance to practical application, as a larger number of parameters are typically involved with such models, increasing effort required on model parameterization and computational time, particularly when incorporated in transport models (Davis et al. 1998; Miller et al. 2010; Steefel 2019).

Practical application of SCMs began soon after their development and has evolved over time. Early on, the models were employed largely in a peripheral manner to help understand the behavior of real systems. The challenges of applying the models to natural systems or to treatment processes, with all of the complexity of these systems in terms of sorbent heterogeneity and solution conditions, were daunting to many. As experience with practical application of SCMs has been gained, however, approaches for addressing the complexity have been developed and learned, and the use of SCMs for practical application has expanded steadily.

The development of reaction databases for SCMs (Goldberg and Sposito 1984; Dzombak and Morel 1990; Tonkin et al. 2004; Mathur and Dzombak 2006; Karamalidis and Dzombak 2010) and of standard parameterization procedures (e.g., ISO 2012a,b,c) have facilitated the use of SCMs for practical applications.

Primary challenges for the application of SCMs relate to the complexity of sorbents in real systems and parameterization of SCMs for such systems. The sorbents in real systems typically consist of a mixture of mineral phases, sometimes including organic carbon materials. In contrast to this complexity, SCMs and their associated equilibrium constants have been developed based on data from well-defined laboratory systems with single phases. To apply SCMs to real systems one is confronted with the challenges of identifying the dominant sorbent phase(s) present, quantifying the amount(s) of the dominant sorbent phase(s), and quantifying the amount of accessible surface area of the dominant sorbent phase(s). In addition, the presence of particulate or sorbed organic carbon will usually require consideration of how to account for this material as a sorbent. Thanks to the work and creative insights of many in the decades since the development of SCMs, approaches to addressing these challenges have been developed and are the focus of this chapter.

The overall goal for the chapter is to provide an overview of the approaches and tools that have been developed for the practical application of SCMs. The chapter includes presentation of the three types of approaches that have been used in SCM practical applications, specific examples of the practical application of SCMs, and an outline of some research needs to advance practical application of SCMs.

SURFACE COMPLEXATION MODELS: DEVELOPMENT AND BASICS

Surface complexation models were originally developed to describe sorption of inorganic ions on the surfaces of hydrous metal oxides (Schindler and Kamber 1968; Stumm et al. 1970, 1976; Huang and Stumm 1973; Davis and Leckie 1978; Davis et al. 1978, 1980; Morel et al. 1981; Schindler 1981; Goldberg and Sposito 1984; Schindler and Stumm 1987). Most applications of surface complexation modeling are still focused on sorption to oxide/hydroxide surfaces and the surfaces of oxidic minerals such as clays, though the concept and approach has been extended to describe sorption of ions on other kinds of minerals including sulfides (Park and Huang 1989; Kornicker and Morse 1991; Ronngren et al. 1994) and carbonates (Comans and Middleburg 1987; Davis et al. 1987; Zachara et al. 1991; Brady et al. 1999).

The foundational concept for SCMs is that sorption of inorganic ions on mineral surfaces involves a combination of site-specific chemical bonding at reactive functional groups on the surface of the mineral coupled with non-site-specific electrostatic attraction (or repulsion) that depends on the charge on the surface. Extensive study of ion sorption on hydrous oxide surfaces, as cited above, has revealed that site-specific chemisorption is the dominant process involved for many inorganic ions.

Figure 1 presents a schematic illustration of the iron-oxide/water interface, with surface hydroxyl groups that result from interaction of surface iron and oxygen atoms with the H^+ and OH^- ions in the aqueous phase. In the case of hydrous metal oxides, such as the iron and aluminum oxides which are present in abundance in soils and sediments (Jenne 1977; Brady and Weil 2002; Sparks 2003; Sposito 2008), the surface hydroxyl groups comprise the reactive surface sites. These surface sites are considered in SCMs to be individual reactive entities, often designated by $\equiv$FeOH and $\equiv$AlOH for iron and aluminum oxide surfaces, respectively, as in Figure 1. The symbol $\equiv$ is used to represent bonds of the surface metal atom to other atoms in the solid lattice.

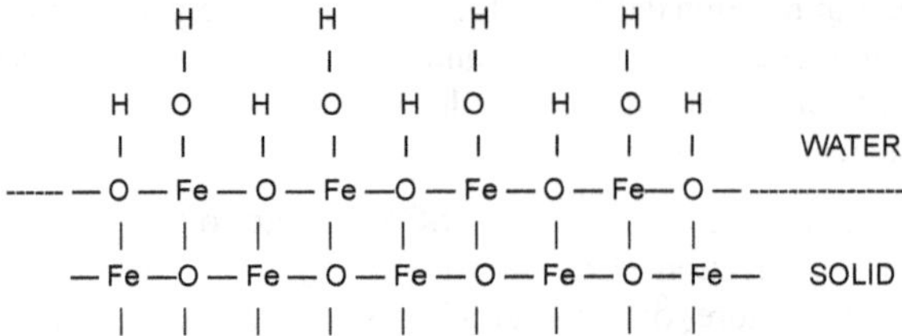

Figure 1. Schematic of the iron oxide/water interface.

Hydrous metal oxide and other mineral surfaces in aqueous systems typically exhibit a net surface charge due to uptake of ions on the surface (variable charge) and sometimes by structural atomic imbalances (permanent charge). The charge on oxide surfaces is derived predominantly by sorption of protons (H^+) and other ions on the surface and is thus variable. This charge can be positive or negative, depending on the types and amounts of ions sorbed from solution. Proton sorption and desorption from oxide surfaces, which is strongly pH dependent, is a primary determinant of oxide surface charge. Ions in solution will experience either attractive or repulsive electrostatic forces of interaction with a charged surface, depending on the ion charge relative to the net charge on the surface.

To account for the non-site-specific electrostatic energy of an ion interacting with the surface, most SCMs invoke a particular molecular model of the solid–water interface and use this model and Gouy–Chapman theory to develop a description of the electrostatic energy of interaction. One simple version of such an interfacial molecular model is the so-called Diffuse Layer or Two-Layer Model, illustrated in Figure 2. This model considers a single surface layer with a net charge determined by the presence of sorbed ions on the surface under particular solution conditions, and a "diffuse" layer of ions of opposite charge relative

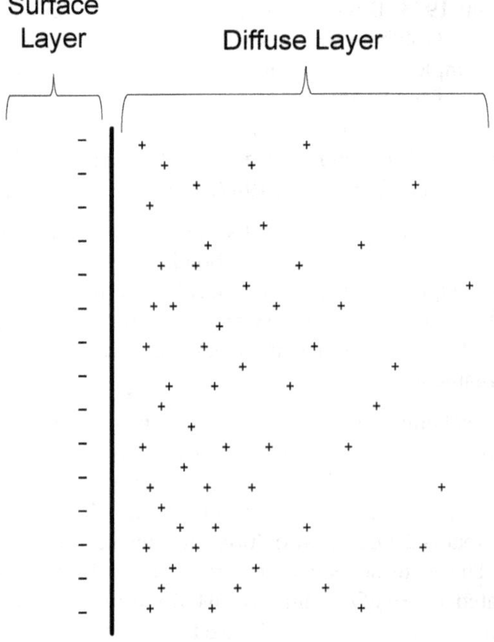

Figure 2. Charged oxide surface and diffuse layer of counterions in solution.

to the surface. The weakly sorbed "counterions" are held close to, but not in contact with, the surface by electrostatic forces of interaction. There are various other molecular models of the solid–water interface that have been employed in SCMs; these are covered amply elsewhere (e.g., Davis and Kent 1990; Goldberg 1992, 2013; Stumm 1992; Hiemstra and Van Riemsdijk 1996; Stumm and Morgan 1996; Langmuir 1997; Koretsky 2000). As demonstrated by Westall and Hohl (1980), ion sorption data can be fitted equally well with any of the various models employed for description of electrostatic interactions.

Surface complexation models are formulated by writing chemical reaction and mass action equations for the site-specific sorption of ions on surface sites in the same manner as for reactions of ions with other solution species. For oxide surfaces, these reactions encompass sorption and desorption of protons and other reactive cations and anions in solution. Table 1 presents example surface complexation reactions for sorption of protons, cations, and anions on iron oxide. Surface complexation reactions have been formulated to provide description of and consistency with experimental observations, including the release of protons with cation sorption and the uptake of protons with sorption of anions.

Table 1. Example reactions for surface complexation on iron oxide.

Acid–Base
$\equiv FeOH^o + H \rightleftharpoons \equiv FeOH_2^+$
$\equiv FeOH^o \rightleftharpoons \equiv FeO^- + H^+$
Cation Sorption
$\equiv FeOH^o + M^{2+} \rightleftharpoons \equiv FeOM^+ + H^+$
Anion Sorption
$\equiv FeOH^o + A^{2-} + H^+ \rightleftharpoons \equiv FeA^- + H_2O$

Notes:
Mass action equations: $K^{app} = K^{int} \times P = K^{int} \times f$(surface charge, T, I)
P = coulombic correction factor = $\exp(-\Delta ZF\Psi/RT)$
ΔZ = change in charge number of the surface species
F = Faraday Constant (96,485 coul·mol^{-1})
Ψ = surface potential (V = J·coul^{-1}); determined from surface charge and ionic strength, with use of a particular molecular model for the solid–water interface
R = Gas constant = 8.314 J·mol^{-1}·K^{-1}
T = temperature (K)
I = ionic strength (mol·L^{-1})

For equilibrium conditions, equilibrium constants (K) can be assigned to surface complexation reactions and reflect the energetics of reaction. Mass action expressions based on the reactions can be formed, and these reactions can be integrated into general chemical equilibrium models to incorporate sorption reactions and consider their influence on chemical speciation along with the other reactions (acid–base, oxidation–reduction, precipitation–dissolution, etc.) that determine the distribution of species in solution.

A distinguishing feature for surface complexation reactions relative to reactions among species in solution is the variable electrostatic energy of interaction associated with the variable surface charge determined by solution conditions (especially pH and ionic strength) and the extent of ion sorption under particular conditions. This means that equilibrium constants for surface complexation reactions, represented as K^{app} in Table 1, vary with surface charge condition. SCMs that include consideration of the electrostatic energy of interaction use the calculated surface charge and theory of electrostatics to calculate the electrostatic energy of interaction and a so-called coulombic correction factor, P (sometimes referred to as K^{coul}). This coulombic correction factor is employed to modify the so-called intrinsic equilibrium constant, K^{int}, which reflects the energy of interaction for the short-range, site-specific chemisorption reaction. Put another way, K^{app} reflects the observed overall energy of sorption, and this energy is viewed theoretically as the product of a short-range, chemical bonding component, corresponding to K^{int}, and a long-range, electrostatic interaction component, corresponding to P. Details and explanation of the formulation of surface complexation models, including with different molecular models of the solid–water interface, are provided elsewhere (e.g., Davis and Kent 1990; Dzombak and Morel 1990; Goldberg 1992, 2013; Stumm 1992; Morel and Hering 1993; Langmuir 1997; Koretsky 2000; Benjamin 2015; Brezonik and Arnold 2022).

The use of different molecular models for the solid–water interface results in different kinds of formulations of the surface charge–potential relationships, and thus different values of coulombic corrections for the chemical conditions of a particular data set. Thus, fitting of a set of sorption data with SCMs employing different descriptions of the interface and electrostatic energy of interaction results in different K^{int} values for the same surface complexation reaction (Westall and Hohl 1980; Payne et al. 2006).

The use of SCMs has been advanced by the development of thermodynamic databases for sorption of inorganic ions on some metal oxides of importance in natural aqueous systems and water treatment processes. Obtaining an internally consistent set of surface complexation reactions and associated K^{int} values requires the interpretation of experimental sorption data with the same SCM, and use of a consistent set of aqueous speciation reactions and equilibrium constants (Dzombak and Morel 1990; Lützenkirchen 2006). The simple Two-Layer (Diffuse Layer) Model has been employed to interpret compiled sorption data for Hydrous Ferric Oxide (Dzombak and Morel 1990), Hydrous Manganese Oxide (Tonkin et al. 2004), Goethite (Mathur and Dzombak 2006), and Gibbsite (Karamalidis and Dzombak 2010). Some of these databases have been incorporated in the general chemical equilibrium models VMINTEQ (Gustafsson 2020), MINEQL+ (ERS 2022), MINTEQA2 (USEPA 1991, 1998, 2024a), PHREEQC (Parkhurst and Appelo 2013), and others.

CHALLENGES FOR APPLICATION OF SURFACE COMPLEXATION MODELS IN SOIL, SEDIMENT AND AQUIFER SYSTEMS

Surface complexation models were developed with a focus on well-defined, pure-phase oxide/hydroxide minerals, and such minerals are important sorbents in natural systems. However, sorbent particles in near-surface soils, deeper soils and aquifer media, sediments at the bottom of water bodies, and suspended in water often comprise mixtures of oxide/hydroxide and other minerals. In addition to the common oxide/hydroxide minerals of aluminum, iron, manganese, and silicon, other sorbents that occur in natural systems include aluminosilicate clay minerals, carbonate minerals, and natural organic matter (NOM), either as particulate organic carbon or sorbed on the surfaces of minerals (Deutsch 1997; Jenne 1977; Sparks 2003). Some specific commonly occurring sorbent materials in natural systems are presented in Table 2. Separate, individual particles of these materials occur as well as agglomerates of these materials with each other. For example, as indicated in Table 2, iron and aluminum are often dissolved and re-precipitated as hydroxide coatings on other particles in natural systems. Also, in surface waters and near-surface soils and sediments where significant concentrations of NOM can be present, the NOM can sorb on mineral surfaces or exist as particulate matter, depending on solution conditions.

Considering the complexity of the mixture of sorbent particles in soils, sediments, and aquifer media, there are three primary challenges to applying surface complexation models to describe sorption of inorganic ions on the materials in these assemblages.

1. Identify the dominant mineral sorbent or sorbents in the mixture.
2. Determine the accessible surface areas of the dominant mineral sorbent(s).
3. If NOM is present, determine how to account for the effect of sorption on NOM.

Addressing each of these challenges explicitly is critical to parameterization of a surface complexation model for description and prediction of inorganic ion sorption in natural systems. Research on and experience with practical application of SCMs have shown that these challenges can be adequately addressed and useful systematic approaches for doing so have been developed.

Table 2. Some common components of soils, sediments, suspended sediments, and aquifer materials.

Oxides/Hydroxides
Gibbsite: $Al(OH)_3(s)$
Hydrous Ferric Oxide (Ferrihydrite): $Fe(OH)_3(am)$
Goethite: FeOOH(s)
Manganese Dioxide: $MnO_2(s)$
Quartz, am-Silica: $SiO_2(s)$, $SiO_2(am)$
Clays
Kaolinite: $Al_2Si_2O_5(OH)_4(s)$
Montmorillonite: $(Na,Ca)_{0.33}(Al,Mg)_2Si_4O_{10}(OH)_2{\cdot}nH_2O(s)$
Illite: $(K,H_3O)(Al,Mg,Fe)_2(Si,Al)_4O_{10}[(OH)_2,H_2O](s)$
Carbonates
Calcite: $CaCO_3(s)$
Dolomite: $CaMg(CO_3)_2(s)$
Siderite: $FeCO_3(s)$
Sulfates
Gypsum: $CaSO_4(s)$
Sulfides
Pyrite: $FeS_2(s)$
Natural Organic Material (NOM)
Humic acid
Fulvic acid
Agglomerates
NOM coated on $Al(OH)_3(s)$
$Al(OH)_3(s)$ coated on clay
$Fe(OH)_3(s)$ coated on $SiO_2(s)$

Identifying dominant sorbents

SCMs have been applied to describe and predict sorption of inorganic ions both in systems having a clearly dominant sorbent and in those with multiple sorbents present. Examples of applications in systems with a clearly dominant sorbent include $Al(OH)_3(s)$ in drinking water treatment (Dzombak and Morel 1987; Hering et al. 1996) and Hydrous Ferric Oxide in streams impacted by mine drainage (Smith et al. 1998). SCMs also have been applied to describe ion sorption in more complex systems with multiple sorbents. Examples include studies of sorption/desorption to/from ash materials (Meima and Comans 1998) and soils (Zachara et al. 1992; Weng et al. 2001; Dijkstra et al. 2004). In these cases, the ability of "multisurface" SCMs (discussed below) to provide accurate description, helpful insight, and reasonable predictions has been demonstrated. The SCM formulations in these applications necessarily involved selection of sorbents to include and approaches to parameterizing the models.

Iron and aluminum oxides/hydroxides, NOM, and to a lesser extent manganese oxides, metal sulfides, carbonates, and clays have been identified as the primary repositories of metals and other trace elements in soils and sediments (Jenne 1977). Due to their abundance and reactivity, iron oxides/hydroxides dominate inorganic ion sorption in many systems. Jenne (1998) noted (p. 30) that "the preponderance of evidence is that for transition metals such as Cd and Ni, iron oxides are generally the primary adsorbent in oxic environments unless unusual amounts of MnO_x, Al oxides, POC [particulate organic carbon], and rarely carbonates, etc., are present relative to the amount of oxidic Fe surfaces available." In developing a multisurface sorption model for use in assessing risk associated with a broad range of soil contaminants in Dutch soils, Dijkstra et al. (2009) included sorption on iron and aluminum oxides/hydroxides, clay, and particulate organic matter. Thus, while various components of soils and sediments can potentially be important sorbents under particular conditions, iron and aluminum oxides and NOM are most commonly the dominant sorbents in soil and sediment systems. For aquifer materials, where NOM is typically present only at very low levels, iron and aluminum oxides commonly dominate sorption when present, often as coatings on sand (e.g., Ryan and Gschwend 1992; Coston et al. 1995; Zhang et al. 2011).

With the dominance of iron and aluminum oxides/hydroxides as sorbents in many systems, some have used iron oxide/hydroxide as the sole dominant sorbent (e.g., Smith et al. 1998) or a generic Fe/Al oxide with mixed or fitted properties to represent the oxide/hydroxide sorbents in the system (e.g., Dijkstra et al. 2009). Adoption of either approach requires some explicit assumptions and decisions about model parameterization. These single-surface sorbent approaches are examined in more detail below.

Determining accessible surface areas

Whether one or multiple sorbents are considered in surface complexation modeling for real systems, a means of determining the amount of reactive surface is needed. While elemental or mineral composition data may be available for a soil, sediment, aquifer material, or suspended sediment, such data provide no information about what fraction of the element (e.g., iron, aluminum, manganese) or mineral is on the surface of a grain or particle and available for reaction with the solution phase in contact with the surface.

Selective chemical extractions have become widely accepted as a practical means of estimating the amount of iron (Fe) and aluminum (Al) oxides available for sorption reactions. The U.S. Department of Agriculture (USDA) has standard methods for several types of selective chemical extractions for soils (USDA 2022), and ISO standard methods also have been developed (ISO 2012a,b,c). As indicated in the introductory text for the ISO standard methods and in a wide range of geochemical studies (e.g., Meima and Comans 1998; Curtis et al. 2004; Dijkstra et al. 2004, 2009, 2018; Bonten et al. 2008; Groenenberg et al. 2012; Groenenberg and Lofts 2014), a driver of the frequent use of selective chemical extractions has been for parameterization of geochemical models. The limitations and operational nature of the selective extraction methods are recognized by the users and the research community (e.g., Fuller et al. 1996; Bacon and Davidson 2008; Groenenberg and Lofts 2014). As a practical matter, however, the methods are reproducible, have been demonstrated to be reasonably accurate through validation testing with model systems (e.g., Kostka and Luther 1994), are relatively inexpensive, and do not rely on modeling for interpretation.

Common selective chemical extraction methods for Fe and Al oxides. Methods for selective chemical extraction of the available, surface-reactive Fe and Al oxides from soils have been in use for at least a century. The methods presented by Kostka and Luther (1994), for example, are modifications of methods proposed and employed by others since the 1960s. The USDA NRCS Soil Survey Laboratory Manual (USDA 2022) cites references from 1953 and 1960 for the dithionite–citrate extraction method; the method of Mehra and Jackson (1960) is employed. Rennert et al. (2021) note that oxalate extraction has been used for approximately 100 years to extract elements from reactive oxide components of soils, citing the work of Tamm (1922) to extract Al, Fe, Si, and Ti with an ammonium oxalate/oxalic acid solution.

Table 3 provides a summary of commonly used selective chemical extraction methods for measurement of surface-reactive Fe and Al oxide phases in soils. Various modifications of these methods are employed, other selective extraction methods exist and are used, and there is no consensus on best methods to employ. Nevertheless, review of the literature indicates that these are the most commonly employed methods. The development of ISO methods for three of the extractants listed confirms their common acceptance and use.

Table 3. Summary of soil/sediment extraction methods and target Fe and Al mineral phases.

Extractant	Target mineral phase	Reference
Ascorbic acid	Amorphous iron oxide/hydroxide	Kostka and Luther (1994) ISO (2012a)
Hydroxylamine–hydrochloric acid (NH_2OH–HCl)	Amorphous iron oxide/hydroxide	Chao and Zhou (1983) Kostka and Luther (1994)
Ammonium oxalate/oxalic acid	Amorphous iron oxide/hydroxide*	Blakemore et al. (1987) Kostka and Luther (1994)
	Aluminum oxide/hydroxide	ISO (2012c)
Dithionite–citrate	Amorphous and crystalline iron oxide/hydroxides**	Kostka and Luther (1994) ISO (2012b)

Notes:

* Oxalate/Oxalic Acid extraction has also been used by numerous investigators to estimate content of amorphous Aluminum oxide/hydroxide content.

** Dithionite extraction captures amorphous and crystalline Iron oxide/hydroxide. The crystalline Iron oxide content is determined by subtracting the amorphous Iron oxide/hydroxide content as measured via extraction with ascorbic acid.

The preamble for the ISO selective extraction methods (2012a,b,c) justifies their selection as standard methods and as the best available methods to provide the amount of reactive Fe and Al oxides/hydroxides for use in geochemical modeling, which is used in risk assessments and other applications. The ISO methods are based on those developed at the University of Delaware for study of phase partitioning of metals in wetland sediments (Lord 1980; Ferdelman 1988). The methods are presented and validated in Kostka and Luther (1994).

Considering the validation results with the pure phase minerals, ascorbic acid appears to be fairly selective for amorphous iron oxyhydroxide. The oxalate–oxalic acid and hydroxylamine–HCl extractions appear to be only moderately selective for amorphous iron oxyhydroxide. Dithionite is intended to dissolve all iron minerals present, and consistently yields higher concentrations of Fe (and Al) than the ascorbic acid, oxalate–oxalic acid, and hydroxylamine–HCl extraction agents. Kostka and Luther (1994) are clear that the extractions provide operational parameters, not definitive delineation of phases. Concurrent extraction of aluminum oxide/hydroxide phases and iron oxide/hydroxide phases using the various extractants listed in Table 3 was not addressed by Kostka and Luther (1994), but concurrent iron and aluminum extractions using these methods have been reported by numerous others (e.g., EPRI 1986; Dijkstra et al. 2004, 2009; Groenenberg et al. 2012). The ISO (2012c) method for oxalate–oxalic acid extraction is specifically focused on aluminum oxides/hydroxides.

ISO standard methods for selective chemical extraction of Fe and Al oxides. The justification provided in the preamble for each of the ISO selective extraction methods for Fe and Al oxides/hydroxides is that geochemical models are needed for risk assessment, these models require certain input parameters, and experimental methods are needed to obtain the input parameters (ISO 2012a,b,c). The ISO selective extraction methods are justified as the best available methods to provide the amount of reactive Fe and Al oxides/hydroxides for use in geochemical modeling.

The stated aim of ISO 12782-1 (ISO 2012a) is to determine the content of reactive iron in the form of amorphous iron oxides/hydroxides in soils using ascorbic acid extraction. The procedure is based on the method of Ferdelman (1988) as described by Kostka and Luther (1994).

The stated aim of ISO 12782-2 (ISO 2012b) is to determine the content of crystalline iron oxides/hydroxides in soils using dithionite–citrate extraction. The procedure is based on the method of Lord (1988) and Canfield (1988) as described by Kostka and Luther (1994).

The stated aim of ISO 12782-3 (ISO 2012c) is to determine the content of amorphous aluminum oxides/hydroxides in soils using ammonium oxalate/oxalic acid extraction. The procedure is based on the method of Lord (1980).

Accounting for NOM

When natural organic matter is present in significant quantities in a sorbent medium, explicit accounting for NOM in modeling of ion binding is required for accurate description of sorption data. In a soil, sediment, or aquifer material, NOM can exist in the aqueous phase (dissolved organic matter, or DOM), as solid phase particulates (particulate organic matter, or POM), and sorbed on mineral surfaces. In soils, sorbed and particulate phase organic matter is often considered together as soil organic matter, or SOM.

The organic matter content of mineral soils in the top 1m of the Earth's surface ranges from about 1 to 6% by weight, with concentrations decreasing rapidly with depth (Brady and Weil 2002). The highest concentrations of organic matter in mineral soils are in the so-called *O*- and *A*- horizons of the soil, typically the top 20 cm (e.g., see examples in Figure 2.7 of Tipping 2002). SOM is often the dominant sorbing phase in the near-surface soil environment (Gustafsson et al. 2003). There are soils with higher organic matter contents (e.g., > 20%), such as peats and other histosols, but these are less common (Brady and Weil 2002).

The NOM content of deeper soils is generally less than 1% by weight and often much less than this amount. With increasing depth, the dominant form of organic matter shifts from the solid phase (sorbed and particulate) to the dissolved phase. Concentrations of DOM (or dissolved organic carbon, DOC) also decrease with depth, due to biodegradation and retention by sorption, and are generally quite low (< 1 mg/L) in aquifers (Thurman 1985; Pabich et al. 2001; Tipping 2002).

NOM in soils and sediments is dominated by humic substances, a complex, heterogeneous mixture of organic molecules resulting from degradation of plant, animal, and microbial matter (Thurman 1985; Brady and Weil 2002; Tipping 2002). Soluble NOM compounds are primarily polyelectrolytic organic acids ranging in molecular weight from 500 to 50,000 or greater. Fulvic acid (FA), soluble in both base and acid, comprises the molecules at the lower end of this range. Humic acid (HA), which is soluble in base but precipitates under acidic conditions, encompasses the larger compounds. Humin, the third component of humic substances in soils and sediments, is the fraction of organic matter that is not soluble in acid or base, and is considered unreactive with respect to inorganic ion binding. Humic substances comprise 60 to

80% of SOM (Brady and Weil 2002), and typically FA and HA are the dominant components of SOM (Tipping 2002). FA and HA also are the dominant components of aqueous phase DOM (Thurman 1985; Tipping 2002).

NOM can influence inorganic ion retention in soils, sediments, and aquifer materials through complexation in the dissolved phase and partitioning to sorbed or particulate organic matter in the solid phase. In general, the same models are used to describe ion interactions with NOM in both the dissolved- and solid-phase forms. These models, summarized below, require measurements for dissolved and solid phase organic matter (or organic carbon) content, measurements or assumptions about the fraction of organic matter that is humic acid (HA) and fulvic acid (FA), and assumptions about which fraction of the organic matter is unreactive with respect to ion binding (Dijkstra et al. 2009; Groenenberg et al. 2012). In the models for ion binding to organic matter commonly in use, all non-HA/FA components are usually considered to be unreactive (e.g., van Eynde et al. 2022).

The semi-mechanistic models in use for describing ion binding with dissolved and solid phase NOM consider reactions of ions with individual ligands associated with the HA and FA components of the NOM. The ligands, or more accurately ligand types, are specified to represent the reactivity with ions provided by the NOM, and not the actual ligands on the NOM molecules. Ligand types are represented either as a continuous distribution, with a continuous distribution of amounts of ligand types providing a continuous distribution of ion binding energies, or as a mixture of discrete ligands, each with a particular concentration and ion binding energy (Tipping 2002). For the continuous distribution approach, a normal or Gaussian distribution has been commonly employed (Tipping 2002). The two most commonly employed models for ion binding to NOM are the Non-Ideal Competitive Adsorption-Donnan (NICA-Donnan) Model (Kinniburgh et al. 1996, 1999; Koopal et al. 2005; van Riemsdijk et al. 2006), a type of continuous distribution model, and Model VII included in the Windermere Humic Acid Model (WHAM) (Tipping 1994, 1998), a type of discrete ligand model. Both of these models account for electrostatic effects and enable description and prediction of ion binding across a broad range of solution conditions. The models have been applied to describe ion sorption in particular soil systems (e.g., Gustafsson and van Schaik 2003; Bonten et al. 2008; Dijkstra et al. 2009; Groenenberg et al. 2010; van Eynde et al. 2022) with associated development of site-specific parameters from data fitting. In addition, both models have been applied to compilations of data for many soils and generic parameters have been developed from the simultaneous fitting of these data (Tipping 1998; Milne et al. 2001, 2003; Tipping et al. 2011). The generic model parameters have facilitated the use and practical application of both models.

This chapter is focused on the practical application of surface complexation models and does not delve into the details of the models employed for describing ion binding with NOM and their integration with SCMs. Nevertheless, consideration of ion binding to NOM is important in many systems, especially for near-surface soils. The lack of focus here is not intended to suggest that ion binding by NOM is rarely important, or that approaches for integrating NOM with SCMs are not adequately developed. Complete and definitively parameterized models with integrated consideration of ion binding on mineral surfaces and with solid-phase NOM have been developed and deployed for complex systems (e.g., Dijkstra et al. 2004; van Riemsdijk et al. 2006; de Vries and Groenenberg 2009; Groenenberg et al. 2017) and for regulatory purposes (de Vries et al. 2013; Dijkstra et al. 2018). Standard methods for parameterization have been developed for use of the NICA-Donnan Model, including standard methods for measurement and quantification of DOM and SOM (ISO 2012d,e).

APPROACHES FOR PRACTICAL APPLICATION OF SURFACE COMPLEXATION MODELS

Three general approaches have been employed for the application of surface complexation models to describe and predict ion sorption in soils, sediments, and aquifer media: multisurface models with component additivity; single-surface/surrogate models; and generalized composite models. These approaches are described here, and examples of how they have been applied are provided.

Multisurface approach

Multisurface SCMs assume additivity of independent sorbent action. Under this assumption, surface complexation reactions and parameters for each of several sorbents are included in an overall SCM, with the individual contribution of each sorbent to overall ion sorption considered in the model. So-called "component additivity" has been demonstrated to work reasonably well in both applied and more fundamental studies (Honeyman 1984; Payne et al. 2006; Bonten et al. 2008; Lund et al. 2008; Landry et al. 2009; Schaller et al. 2009; Reich et al. 2010; Groenenberg et al. 2012; Groenenberg and Lofts 2014). The work of Koretsky and colleagues (Lund et al. 2008; Landry et al. 2009; Schaller et al. 2009; Reich et al. 2010) with well-defined sorbent phase mixtures in laboratory systems has provided a level of validation for the multisurface or component additivity approach in such systems. Bradbury and Baeyens (1997, 2006, 2009a, 2009b, 2011) developed a type of component additivity model—a combined surface complexation and cation exchange model—to describe metal ion binding on the different types of surfaces of individual clay minerals. Davis et al. (1998) discuss limitations of the approach for accurate description of ion sorption with complex mineral assemblages in specific natural systems.

Identification of sorbents to include in multisurface SCMs relies on compositional analysis of the sorbent materials involved. Judgement is required about which sorbents to include and exclude, but compositional analysis data and knowledge about relative capabilities of sorbents guide sorbent selection. In general, multisurface SCMs have not included more than three sorbent surfaces.

An example of an application of a multisurface model is the analysis of metal leaching data from eight contaminated soil samples by Dijkstra et al. (2009). Data for the solid–water partitioning of 15 different metals from suspensions of contaminated soil samples adjusted to pH values between 2–12 for 24–48 hours were obtained in batch pH-static leaching experiments. The total concentrations of the metals in the soil samples "available" for sorption/desorption were estimated by extraction with 0.43 M HNO_3 (Dijkstra et al. 2004). The observed solid–water partitioning of the metals was explored with the assistance of chemical equilibrium modeling that included consideration of sorption on Fe and Al oxides/hydroxides, clay, and POM in the soil samples. Independent estimates of the amounts of the reactive solid phase components in the soils were obtained by measurements. The total amount of Fe oxides/hydroxides was estimated by extraction with dithionite–citrate, amorphous Fe hydroxide was estimated by extraction with ascorbic acid, and the amount of crystalline Fe oxide was estimated as the difference between the amount extracted with dithionite–citrate and ascorbic acid (see Table 3). The amount of amorphous Al hydroxide was estimated by extraction with oxalate/oxalic acid (see Table 3). The amount of clay in the samples was estimated as the grain size fraction less than 2 μm, as measured by a standard method involving fine particle settling from aqueous suspension. The total amount of reactive organic matter in the soils was estimated as the sum of extractable HA and FA amounts as measured with the procedure given in ISO (2012d). DOC was measured in the aqueous phase to enable consideration of partitioning of the metals to DOM.

With POM, Fe and Al oxides/hydroxides, and clay as the reactive solid phase components considered in the multisurface model of Dijkstra et al. (2009), three different submodels and various related assumptions were invoked to describe sorption on these different "surfaces." Sorption to POM was described with the NICA-Donnan Model (Kinniburgh et al. 1996, 1999). POM was considered to be entirely humic acid. The NICA-Donnan Model was also used to model partitioning to DOM, which was assumed to consist of 50% reactive humic substances. Both POM and DOM were represented using the generic humic acid model parameters developed by Milne et al. (2001, 2003). Sorption–desorption partitioning of ions with amorphous Fe hydroxide, crystalline Fe oxide, and amorphous Al hydroxide were considered separately with use of the Two Layer Surface Complexation Model, though the surface complexation reactions and constants for amorphous Fe hydroxide developed by Dzombak and Morel (1990) were used for all three surfaces due to the absence of similarly consistent databases for Al hydroxide and crystalline Fe oxide at the time the study was performed. Amorphous Al hydroxide was assumed to have the same properties and reactivity of amorphous Fe hydroxide. Crystalline iron oxide was assumed to have a lower surface area (and hence surface site concentrations) but the same reactivity as amorphous Fe hydroxide. Electrostatic, non-site-specific sorption of ions to permanent charge on clay surfaces was estimated using a Donnan model with an assumed charge density and Donnan volume, chosen based on average values for illitic clays.

Results of the leaching experiments and multisurface model predictions of Dijkstra et al. (2009) for Ni, Cu, Zn, Cd, and Pb, which all occur predominantly as divalent cations for the pH range studied, are presented in Figure 3. As indicated there, leached concentrations of these five elements varied by up to five orders of magnitude, with the lowest leachate concentrations, and hence the greatest retention in the solid phase, in the pH range 4–8. The distributions predicted with the multisurface model for each of the elements among the solid phase components for Soil V are shown in Figure 4. There it may be seen that POM was estimated to be the predominant reactive surface for Ni, Cu, and Cd, especially at lower pH values.

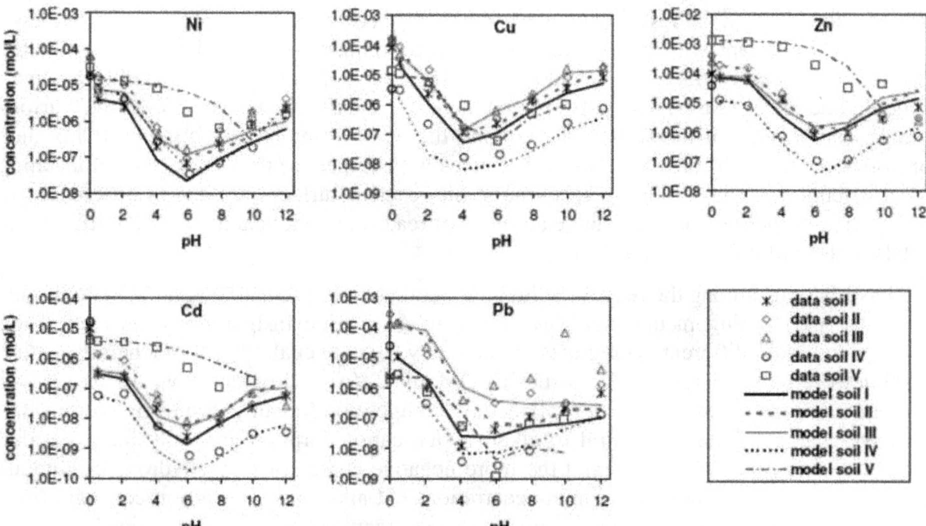

Figure 3. Equilibrium leached concentrations and multisurface model predictions for Ni, Cu, Zn, Cd, and Pb for Soils I–V. Total metal content and concentrations available for sorption processes (determined by extraction with 0.43 M HNO_3), converted to mol/L units, are shown at pH 0 and 0.5, respectively. [Used by permission of the American Chemical Society, from Dijkstra et al. (2009) Vol. 38, SI, Figure S2]

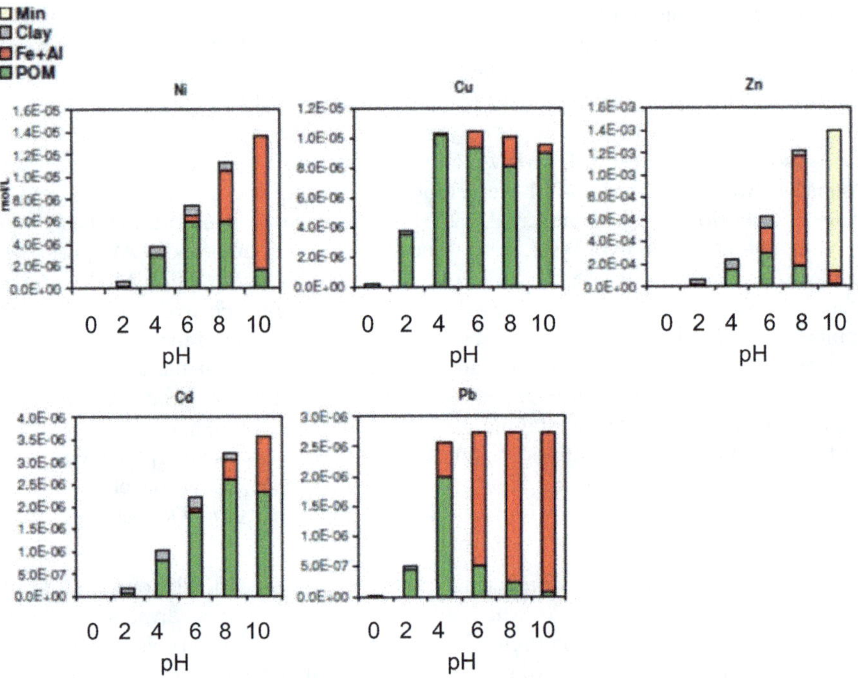

Figure 4. Multisurface model predictions for distribution of Ni, Cu, Zn, Cd, and Pb among the different surfaces considered in modeling leaching data for Soil V. POM = Particulate Organic Matter; Fe+Al = sum of amorphous Fe hydroxide, crystalline Fe oxide/hydroxide, and amorphous Al oxide/hydroxide; Clay = clay surfaces; Min = mineral precipitates. [Used by permission of the American Chemical Society, from Dijkstra et al. (2009) Vol. 38, SI, Figure S3]

Sorption on Fe and Al oxides, as estimated with surface complexation modeling, was assessed to be more important with increasing pH, and predominant for Zn and Pb. The estimated contribution of clay surfaces to overall sorption of the five metals was estimated to be small to negligible in all cases. Overall sorption and the relative contributions of the various sorbing components of the soils varied among the eight different soils studied. (Only data for Soil V are given in Fig. 4). These differences with respect to the pH-dependent leaching and distribution of the sorbing metals among the reactive surfaces are due to differences in soil-specific properties such as relative amounts of reactive surfaces and the availability of the metals in the particular soils (Dijkstra et al. 2009).

Distributions among the reactive solid phase components predicted with the multisurface model for the sorbing metals, such as in Figure 4, provide insights into the controlling mechanisms under different pH regimes. As noted by Dijkstra et al. (2009): "Generally, cation sorption to variably charged surfaces such as OM and Fe/Al-hydroxides is weaker at low pH than at neutral pH values, which is explained by competition for surface sites by protons and repulsive charge effects. At neutral to alkaline pH, cation sorption generally increases due to deprotonation of surface sites and the more negative charge of oxide/hydroxide minerals and OM, resulting in lower solution concentrations. At alkaline pH, solution concentrations increase again due to the formation of inorganic and organic complexes in solution."

The multisurface modeling approach requires decisions about which reactive surfaces to include, which models to use to describe ion interactions with those surfaces, and how to choose parameter values for the models selected. SCMs have an important role in such modeling to describe interactions with mineral surfaces, especially the Fe and Al oxide/hydroxide surfaces that are often important sorbents in soils, sediments, and aquifer media. The availability of internally consistent databases of "generic" reactions and parameter values for surface complexation modeling of ion sorption on Fe, Al, and Mn oxides (Dzombak and Morel 1990; Tonkin et al. 2004; Mathur and Dzombak 2006; Karamalidis and Dzombak 2010) and for modeling of ion binding on POM (Tipping 1998; Milne et al. 2001, 2003; Tipping et al. 2011), combined with development of standard methods for estimating the amounts of reactive surfaces of these phases (ISO 2012a,b,c,d), make it possible to implement the multisurface approach with little or no parameter fitting. This is a significant practical advantage for use of the multisurface approach, as has been emphasized by Dijkstra et al. (2009) and others.

Single-surface/surrogate approach

In contrast to the multisurface approach for implementation of SCMs, in which multiple sorbent surfaces are considered simultaneously, the single-surface or surrogate approach considers a single reactive surface in the solid phase. This approach has been used for a number of reasons, including focus on systems in which there is clearly one type of dominant sorbent phase, for computational simplicity, and in cases where there is limited information available about the composition of the sorbent. Computational simplicity is particularly of interest when a sorption model is being selected for integration in a reactive transport model. For implementation of the single-sorbent approach, a particular sorbent is selected and a surface complexation model is formulated with use of the sorbent as the dominant sorbing phase in the system. Hydrous Ferric Oxide (HFO) has often been selected for this purpose, due to its frequent presence in soil and aquifer materials, its dominance as a sorbing phase in natural systems (e.g., Jenne 1977), and the availability of an SCM database for HFO (Dzombak and Morel 1990).

It could be argued that selection of a single dominant sorbing phase is little different from using an empirical solid–water distribution (or partition) coefficient, in which the degree of sorption of an ion on a complex natural sorbent material is in proportion to the value of the distribution coefficient. Further, the inherent inaccuracy of representing a natural sorbent mixture with a single sorbent phase would seem to argue for use of a distribution coefficient for more accurate description of sorption. Use of an SCM, however, enables description of ion sorption across a range of solution conditions with one set of parameter values and allows for decrease in amount sorbed as the sorbent approaches saturation. Empirical distribution coefficients are relevant only to one particular solution condition and generally are applicable only to low sorbate-to-sorbent ratios. Description of ion binding to a natural sorbent under a range of conditions, such as different pH values and solute compositions, would require a large number of distribution coefficients. As demonstrated by the following examples, experience has shown that use of an SCM and the single-surface approach can be sufficiently accurate and provide useful insights for many systems.

Smith et al. (1998) conducted SCM modeling to assist with their studies of the partitioning of various metals between streambed and stream water in Colorado streams impacted by mine drainage. One of the investigations involved sampling of streambed sediment and stream water at two sites along St. Kevin Gulch, and performance of laboratory batch testing to assess equilibrium partitioning between streambed material and stream water for Cu, Cd, Ni, Pb, and Zn. The sediment and water samples were analyzed for elemental content and other characteristics. X-ray diffraction analysis of the sediment revealed that it consisted of a mixture of Fe oxide/hydroxide phases, specifically Schwertmannite ($Fe_8O_8(OH)_6SO_4$), Ferrihydrite (Hydrous Ferric Oxide), and Goethite (α-FeOOH). The equilibrium partitioning results as a function of pH for samples obtained at both sites are presented in Figure 5 (symbols). The amount of each metal sorbed is expressed as a percentage of the total metal in the sediment–water system. As evident in Figure 5, the relative sorption selectivity sequence as a function of pH was Pb > Cu > Zn > Ni ≈ Cd, similar to what is commonly observed for sorption on HFO.

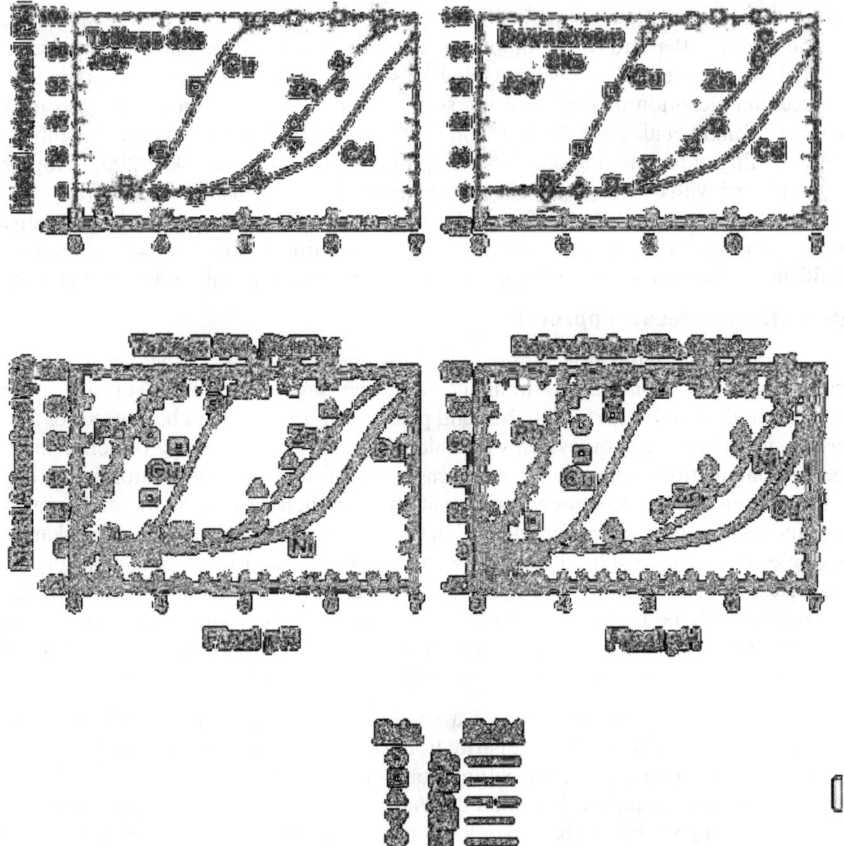

Figure 5. Comparison of experimental data (**symbols**) for Pb, Cu, Zn, Ni, and Cd equilibrium partitioning between streambed sediment and stream water from St. Kevin Gulch, Colorado with Two Layer Model simulations (**curves**) for sorption onto Hydrous Ferric Oxide. The concentration of the streambed sediment in the batch equilibration tests was 2.9 g/L. [Used by permission of Elsevier (Academic Press), from Smith et al. (1998), p. 530, Figure 2]

To assist with interpretation of the partitioning results of Figure 5 and the behavior of the metals in the St. Kevin Gulch stream system, Smith et al. (1998) assumed HFO to be the dominant sorbent in the system and employed the Two Layer SCM and HFO database of Dzombak and Morel (1990) in a general chemical equilibrium model. The modeling results are shown in Figure 5 (curves), where it may be seen that the model predictions were generally consistent with the data, with no adjustment of any parameters in the model. Smith et al. (1998) concluded that the assumption of HFO control of metal sorption in the sediment–water system adequately simulated observed sorption behavior, especially for Pb, Cu, and Zn. For Cd and Ni, they noted, as is evident in Figure 5, that the model consistently underestimated sorption of Cd and Ni to the sediment. Additional analysis led them to conclude that the model likely overpredicted the competition of Cd, Ni and the other metals with sulfate (SO_4^{2-}), which was abundant in the mine-drainage-impacted system.

Smith et al. (1998) also employed the Two Layer SCM and HFO database to interpret partitioning of As, Cd, Ni, Pb, and Zn between the aqueous phase and suspended particulate iron particles in mine drainage from 10 sites throughout Colorado. All sites "exhibited abundant hydrous iron oxide coatings in the mine drainage channel." Comparison of model predictions with measured concentrations of the target metals in the water and suspended particulate phases showed good agreement. The modeling was conducted with no adjustment of model parameters. Results confirmed that metal sorption on iron-rich particulates is an important control on metal solid–water partitioning and fate in the mine drainage waters studied.

The use of an SCM with a single, dominant reactive oxide surface, such as HFO, Goethite, or Gibbsite, has been employed in various practical applications (e.g., Dzombak 1988; Loux et al. 1989; Dzombak et al. 1992; Stollenwerk 1994, 1995; Hering et al. 1996; Meima and Comans 1998; Runkel et al. 1999; Boccelli et al. 2006; Goldberg 2014). In some cases SCM databases such as that for HFO (Dzombak and Morel 1990) with reactions and constants determined from fitting of data for pure phase oxide minerals have been used, in other cases the SCM database reactions and constants have been modified to fit data, and in some cases a generic reactive oxide surface has been assumed and reactions and constants have been determined from fitting system-specific data. The simple single-surface approach has been found useful for describing many systems. While the assumption of a single reactive surface sometimes is a simplistic representation of a complex system, the ability of an SCM to be easily integrated with a chemical equilibrium model for consideration of solution effects on sorbates and to capture the significant effect of pH on ion sorption are important benefits that have supported selection of the single-surface approach.

Generalized composite approach

Building on the concept of a generic reactive surface, Davis et al. (1998) proposed the generalized composite approach to apply surface complexation modeling for accurate description of sorption on natural sorbent material at a particular site. The generalized composite approach involves use of surface complexation reactions for a generic oxide surface site, with the stoichiometry of the reaction(s), the equilibrium constant(s) for the reaction(s), and the concentration of sites determined by fitting sorption data for the particular sorbent material. Davis et al. (1998) used the general composite approach rather than the multisurface approach because of the difficulty of identifying the reactive surfaces in complex assemblages of minerals and organic matter, and the challenge of applying electrostatic models developed for ideal surfaces to surfaces in complex natural mixtures of particles. The general composite approach requires characterization data for the sorbent material, at least specific surface area (BET surface area measurement is recommended by Davis et al.), and sorption data for the sorbates of interest on the particular sorbent material for the range of solution conditions of interest (especially pH). It is a more empirical SCM approach than the multisurface or single-surface SCM approaches that employ reactions and databases for pure phase sorbent materials, and potentially more accurate for a particular sorbent material and particular solution conditions. The same rationale could be used to justify a completely empirical approach such as use of K_d values and sorption isotherms for particular chemical conditions. Davis et al. (1998) emphasize, however, that use of the SCM approach, if in a more empirical manner, is preferable to such fully empirical approaches because the use of surface complexation reactions integrated in chemical equilibrium models allows for change in amount sorbed with changing geochemical conditions and sorbent loading.

Davis et al. (1998) conducted modeling of batch equilibration test data for sorption of Zn on well-characterized sand aquifer material from a much-studied shallow aquifer on Cape Cod, Massachusetts. Their study of the sand aquifer material grains revealed the presence of Fe and Al oxide coatings. The BET surface area of composite samples of the sand aquifer material was measured as 0.44 m^2/g. The density of reactive surface sites was assumed to be 3.84 $\mu mol \cdot sites/m^2$ based on analysis of estimated reactive site densities on a range of natural sorbent materials (Davis and Kent 1990). Batch sorption tests with the aquifer material were performed with use of artificial groundwater with a composition of the typical major ion composition of the groundwater at the site. Some of the sorption data obtained are presented in Figure 6, where fits of the data obtained with two different generalized composite SCM formulations are also provided.

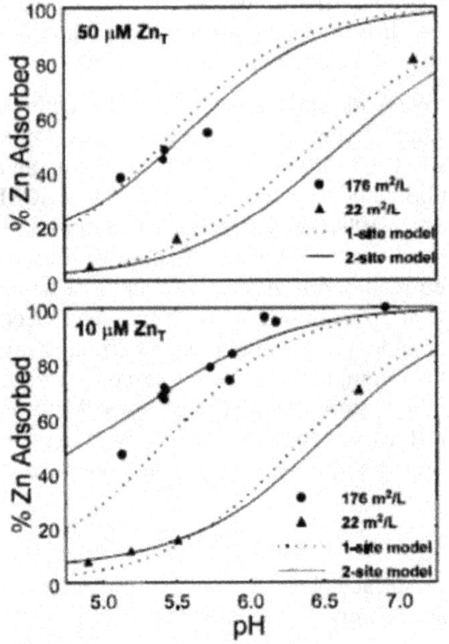

Figure 6. Zn^{2+} equilibrium sorption as a function of pH on Cape Cod aquifer material from batch tests conducted with two different Zn_T concentrations and two different total sorbent concentrations (expressed as m^2/L). **Dotted curves** represent best fits obtained with the 1-site nonelectrostatic, generalized composite SCM, and solid curves are best fits obtained with the 2-site nonelectrostatic, generalized composite SCM. [Used by permission of the American Chemical Society, from Davis et al. (1998) Vol. 32, p. 2825, Figure 4]

To fit the data of Figure 6 using the generalized composite SCM approach, Davis et al. (1998) investigated various surface complexation reaction formulations. The model fits shown in Figure 6 are for one-site and two-site generalized composite model formulations. The reaction employed for the one-site model was:

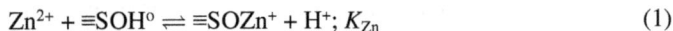

$$Zn^{2+} + \equiv SOH^o \rightleftharpoons \equiv SOZn^+ + H^+; K_{Zn} \tag{1}$$

where $\equiv SOH^o$ refers to a "structurally undefined, average functional group on the surface (assumed to be [an uncharged] amphoteric hydroxyl group)", and K_{Zn} is the apparent equilibrium constant for the surface complexation reaction. The reactions employed for the two-site model were:

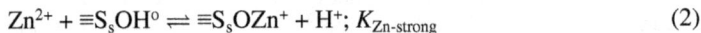

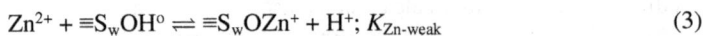

$$Zn^{2+} + \equiv S_sOH^o \rightleftharpoons \equiv S_sOZn^+ + H^+; K_{Zn\text{-strong}} \tag{2}$$

$$Zn^{2+} + \equiv S_wOH^o \rightleftharpoons \equiv S_wOZn^+ + H^+; K_{Zn\text{-weak}} \tag{3}$$

where $\equiv S_sOH^o$ and $\equiv S_wOH^o$ refer to strong and weak ion binding sites, respectively, and $K_{Zn\text{-strong}}$ and $K_{Zn\text{-weak}}$ are the apparent equilibrium constants for the reactions. The total site concentration used was the same for both the one-site and two-site models; the amounts of strong and weak sites for the two-site model were adjusted to provide the best fit of the data. For the relatively narrow pH range and the low Zn^{2+} sorption densities of the experimental data modeled, the concentrations of $\equiv SOH^o$ species were considered approximately constant and no reactions were included to describe the acid–base chemistry of the surface hydroxyl groups.

Davis et al. (1998) did not include in their generalized composite SCM any coulombic correction for electrostatic interactions. In this context, they refer to the generalized composite SCM as a "nonelectrostatic model," or NEM.

As indicated in Figure 6, both the one-site and two-site generalized composite models fit the experimental data well. The two-site model performed better for other data, not presented, for lower Zn_T concentrations. With use of the measured specific surface area for the solid phase and an assumed site density, the total site concentrations were determined and were not considered to be adjustable parameters by Davis et al. (1998). Thus, the number of fitting parameters for the one-site generalized composite model was one (K_{Zn}) and the number of fitting parameters for the two-site model was three: $K_{Zn\text{-strong}}$, $K_{Zn\text{-weak}}$, and the fraction of total sites that are strong binding sites.

Davis et al. (1998) also formulated and applied several versions of single-surface and multisurface (component additivity) models to describe the experimental data for sorption of Zn on the Cape Cod aquifer material. They considered single-surface models with iron oxides/hydroxides represented as HFO only ("Ferrihydrite DDL" model, using the reactions, constants and Diffuse Layer Model of Dzombak and Morel 1990), Goethite only ("Goethite" model, using a one-site NEM and selected literature data for Zn sorption on Goethite), and as a generic FeOH sorbent ("FeOH" model, using a one-site NEM and fitting of selected literature data for Zn sorption on Ferrihydrite). They also considered a single-surface generic AlOH sorbent ("AlOH" model, using a one-site NEM and fitting of selected literature data for sorption of Zn on poorly crystalline Alumina). Finally, Davis et al. considered a multisurface model consisting of a combination of the FeOH and AlOH sorbents. The model formulations are summarized in Table 4. The concentrations of reactive surface sites were estimated with use of aquifer-material extraction measurements and assumptions to convert the amount of extracted Fe or Al to site concentrations. Thus, none of the multisurface model formulations listed in Table 4 involved any fitting parameters. Note in Table 4 that the selective extraction methods employed to parameterize the various models were quite different, though a consistent assumption was employed in each case that 20% of the extracted Fe or Al represented the available ≡FeOH or ≡AlOH surface sites.

Table 4. Single-surface and multisurface model formulations applied by Davis et al. (1998) to describe Zn^{2+} sorption on Cape Cod aquifer material. See Figure 7.

Model label in Fig. 7	**Reactive surface estimate method**	**SCM applied**
Ferrihydrite DDL Model	30-min HH* extraction; yielded 0.9 μmol Fe/g	HFO reactions, constants, and Diffuse Layer Model from Dzombak and Morel (1990)
Goethite Model	24-hr DC* extraction; yielded 11 μmol Fe/g. SA = 50 m^2/g assumed	One-site NEM (Eqn. 1); K_{Zn} from fitting of selected literature data for Zn sorption on Goethite
FeOH Model	8-hr extraction with 4 M HCl at 100 °C; yielded 8 μmol Fe/g; assume 20% of extracted Fe present as ≡FeOH sites	One-site NEM (Eqn. 1); K_{Zn} from fitting of selected literature data for Zn sorption on Ferrihydrite
AlOH Model	8-hr extraction with 4 M HCl at 100 °C; yielded 17 μmol Al/g; assume 20% of extracted Al present as ≡AlOH sites	One-site NEM (Eqn. 1); K_{Zn} from fitting of selected literature data for Zn sorption on poorly crystalline Alumina
FeOH + AlOH Model	8-hr extraction with 4 M HCl at 100 °C; yielded 8 μmol Fe/g and 17 μmol Al/g; assume 20% of extracted Fe and Al present as ≡FeOH or ≡AlOH sites	Combination of FeOH and AlOH Models described above

Notes: * HH = hydroxylamine–HCL extraction; DC = dithionite–citrate extraction

The predictions of the various single-surface and multisurface models for one set of the data for Zn sorption on Cape Cod aquifer material are presented in Figure 7. As shown there, the predictions made with the "Ferrihydrite DDL" and "Goethite" SCMs significantly underestimated the observed Zn sorption. The consistent 5–10× differences between the predictions and the data for the pH range studied indicates, as noted by Davis et al. (1998), that the site concentrations employed in each model were insufficient to represent the amount of Zn sorption observed. This reflects on the efficiency of the extraction methods employed in collecting the surface-reactive Fe from the aquifer materials, and the assumption that only 20% of the extracted Fe is available as ≡FeOH surface sites. The predictions made with the "AlOH", "FeOH", and "FeOH+AlOH" models were closer to the observed data, with the "FeOH+AlOH" model providing a good fit of the data. While the model curves presented in Figure 7 show the results obtained with the single-surface and component additivity models constructed by Davis et al. and with the parameterization approaches they chose, general conclusions about relative performance of the various models are difficult because of the different selective extraction methods and hence surface site concentrations employed.

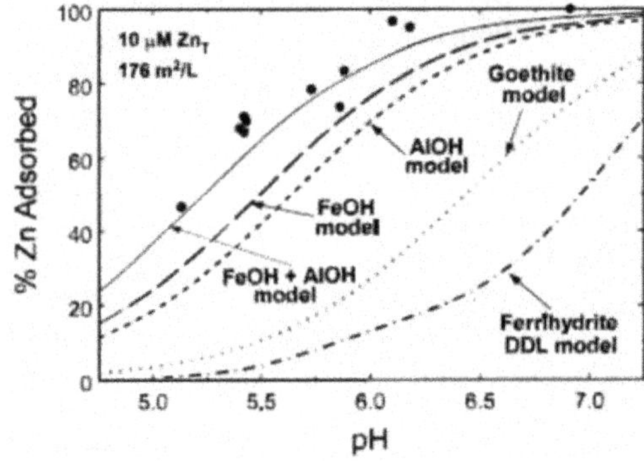

Figure 7. Zn^{2+} equilibrium sorption as a function of pH on Cape Cod aquifer material from batch tests conducted with Zn_T = 10 μM and total sorbent concentration = 176 m^2/L. **Curves** represent fits obtained with the various single-surface and multisurface SCMs listed in Table 4. [Used by permission of the American Chemical Society, from Davis et al. (1998) Vol. 32, p. 2826, Figure 6]

The generalized composite approach is a more empirical approach to surface complexation modeling that involves formulating generic surface reactions and determining the value of apparent equilibrium constants by fitting data for particular sorbents and solution conditions of interest. In a generalized composite SCM, like the multisurface and single-surface SCMs described previously, the amounts of reactive surface sites are estimated from measurements and with use of some needed assumptions. The more empirical nature of a generalized composite SCM is due to the reliance on fitting data for the particular system of interest in determining the surface complexation reactions and related equilibrium constants to include in the model. This could also be done with SCMs that include surface electrostatics and coulombic correction factors, though Davis et al. (1998) chose to omit consideration of electrostatics as a matter of rational consistency with the underlying tenet of the generalized composite approach that knowledge about natural sorbent surfaces is limited. While a generalized composite SCM may enable more accurate description and prediction of ion sorption in a particular system, implementation of it requires a significant amount of experimental data for the system of interest. This will be an impediment to its use in many practical applications, where such data will not be available due to time and budget constraints. For many applications, use of reactions and equilibrium constants determined for pure-phase sorbents, and the multisurface or single-surface SCM approach using databases for these sorbents, will be the only practical option. Further, for many practical applications, the inaccuracy in sorption estimation associated with use of SCM parameters determined for pure-phase sorbents as reflected in Figure 7 will be acceptable. Put another way, the multisurface models employed by Davis et al. (1998) to generate the sorption predictions shown in Figure 7 will be sufficiently accurate for many applications.

SURFACE COMPLEXATION MODELS AND DISTRIBUTION (K_d) COEFFICIENTS: COMPARISONS AND COMPLEMENTARITY

The equilibrium sorption-desorption partitioning of inorganic ions and other solutes between the solid and aqueous phase has long been described empirically with distribution coefficients, K_d (also referred to as partition coefficients, K_p) and isotherm equations (Deutsch 1997; Langmuir 1997; Jenne 1998; Benjamin 2015). The distribution coefficient is defined simply as:

$$K_d = C_s / C_{aq} \tag{4}$$

where C_s is the solid-phase concentration of the sorbing solute (the sorbate) and C_{aq} is the aqueous-phase concentration of the sorbate at equilibrium. Typical units for K_d are:

C_s: mg sorbate/kg sorbent (mass sorbate/mass sorbent)

C_{aq}: mg sorbate/L solution (mass sorbate/vol solution)

With these units, K_d thus has units of L/kg. If the volume of the aqueous solution is converted to mass of solution in kg through consideration of the density of water, then K_d would be apparently (or artificially) dimensionless. The empirical K_d descriptor of solid–water partitioning of sorbing solutes has been employed for accurate description of solute partitioning in particular systems with particular solution conditions and sorbent properties.

The simple linear relationship between the equilibrium aqueous and solid phase concentrations for a sorbing solute represented by K_d (Eqn. 4) also has the advantage of being readily incorporated in reactive transport models to account for sorption–desorption during solute transport in soils, sediments, and aquifer media (Freeze and Cherry 1979; Deutsch 1997; Jenne 1998; Zhu and Anderson 2002; USEPA 2003a,b,c; Stockmann et al. 2017). Most practical applications and regulatory assessments involving modeling of reactive transport of ions in soils, sediments and aquifer media employ K_d values to describe solid–water partitioning. A significant disadvantage of the use of K_d values, however, is that they apply only to the particular geochemical conditions for which have they have been determined experimentally. Relatedly, experimental measurements under a wide range of solution conditions, including changes in pH and the presence of competing sorbates, are needed to obtain a compilation of K_d values to employ across a range of conditions of interest (Dzombak 1988; Bethke 1996; Deutsch 1997; Bethke and Brady 2000; Zhu and Anderson 2002; Bradbury and Baeyens 2006, 2011). The significant limitations of employing single K_d values in reactive transport modeling, in which a parcel of water often encounters different solution and sorbent conditions along its transport pathway, has been well documented (e.g., see Bethke and Brady 2000; Zhu and Anderson 2002; Curtis et al. 2009; Steefel 2019).

The consideration of changing water chemistry along a transport flowpath in the subsurface is most comprehensively conducted by integrating the relevant equilibrium and kinetic chemistry equations with the equations for physical transport (e.g., Cederberg et al. 1985; Yeh and Tripathi 1989; Steefel and MacQuarrie 1996; Zhu and Anderson 2002). However, these "coupled reactive transport models" are relatively complex to formulate, require extensive knowledge of both the hydrogeology and geochemistry at the site of interest, and are relatively computationally expensive, especially for two- and three-dimensional simulations with higher levels of spatial and/or temporal resolution and/or long-duration temporal resolution (Zhu and Anderson 2002; Stockmann et al. 2017). For this reason, coupled reactive transport modeling has been used mostly in research and at extensively characterized sites (Zhu and Anderson 2002; Steefel 2019). The development and application of coupled reactive transport models in the long-term U.S. Geological Survey field research studies on transport of various dissolved metals in an aquifer on western Cape Cod, Massachusetts (Davis et al. 2000; Kent et al. 2000) and U(VI) transport in an aquifer near Naturita, Colorado (Curtis et al. 2004, 2006) are two excellent examples.

While most reactive transport models in practical use for contaminant fate and transport analyses at sites and for risk assessment use distribution coefficients to represent sorption, the limitations of a single K_d value to describe solid–water partitioning of sorbing ions across a spectrum of conditions encountered in subsurface transport has led to use of SCMs for calculation of K_d values. An approach that has been developed for use in risk assessments at contaminated sites (e.g., USEPA 2014 – Appendix H) and at sites of nuclear waste repositories (Stockmann et al. 2017) is to use SCMs in chemical equilibrium models to calculate a matrix of K_d values spanning a large range of potential geochemical conditions. As simulation of subsurface transport of an element of interest proceeds spatially and temporally, specified or calculated geochemical conditions for cells at particular locations in the transport domain and at particular times can be used as the basis for selection of the appropriate K_d value from the "lookup" matrix of K_d values. Such a "lookup" table or matrix with results of simulations of one model incorporated for use in another model is often referred to as a "reduced-order model." In preparing the matrix of K_d values with use of an SCM in a general chemical equilibrium model, decisions about dominant sorbents and parameterization still need to be made as discussed above and as is required for formulation of coupled transport models. The use of a reduced-order surface complexation model, or a "smart-K_d" model (to use the terminology of Stockmann et al. 2017), reduces the complexity of formulating and the computational burden of implementing an integrated SCM-transport model.

Stockmann et al. (2017) describe in detail the smart-K_d matrix approach and discuss its advantages and limitations. In addition to the computational efficiency of the approach in simulating sorption over a range of geochemical conditions, Stockmann et al. note and demonstrate how the development of the K_d-matrix facilitates sensitivity and uncertainty analyses. They also present a case study involving application of the smart-K_d approach to evaluate potential U(VI) migration in a long-term safety assessment for an aquifer at a nuclear waste repository site. The site has been extensively studied and characterized (e.g., Suter et al. 1998). Stockmann et al. discuss the decisions that they had to make about the species to include in the chemical equilibrium modeling, the ranges of concentrations to consider for those species, the sorbent surfaces to include and the properties of these surfaces, the SCM to employ, and other factors. They selected five parameters of the chemical system to vary in their analyses, which are reproduced here in Table 5 along with the concentration ranges considered and the type of probability distribution selected for each parameter. Using the chemical equilibrium model PHREEQC (Parkhurst and Appelo 2013), the Diffuse Layer (Two Layer) SCM, and the multisurface component additivity SCM approach, Stockmann et al. (2017) developed a smart-K_d matrix for U(VI) sorption under 10,000 different geochemical conditions. A histogram of the calculated K_d values, on a logarithmic scale, is provided in Figure 8. The distribution of calculated K_d values is approximately lognormal, with a skew

Table 5. Chemical parameters and concentration ranges employed by Stockmann et al. (2017) in smart-K_d matrix calculations for use in U(VI) reactive transport modeling in an aquifer at a nuclear waste repository site.

Parameter	Minimum conc*	Maximum conc*	PDF**
pH	6.4	8.2	uniform
Dissolved inorganic carbon	4.1×10^{-4}	5.3×10^{-3}	log-uniform
Ionic strength	9.4×10^{-4}	1.7×10^{-2}	log-uniform
$Ca_{T\text{-dissolved}}$	2.2×10^{-4}	3.1×10^{-3}	log-uniform
U(VI)	1.0×10^{-12}	1.0×10^{-6}	log-uniform

Notes:
* concentration units are mol/L, except for pH
** PDF = probability density function

to the left side of the distribution. The log K_d values for U(VI) sorption under the range of conditions considered ranged from –4.0 to 0.6, with a mean of –1.3. A benchmark constant-K_d value cited by Stockmann et al. for U(VI) sorption of log K_d = –2.7 is shown on Figure 8. They note that in safety assessments a conservative approach oriented toward worst-case scenarios is usually taken, which would correspond to greater attention to reactive transport modeling results obtained with lower log K_d values (i.e., the left side of the distribution in Fig. 8), which will show greater U(VI) mobility in groundwater with respect to sorption.

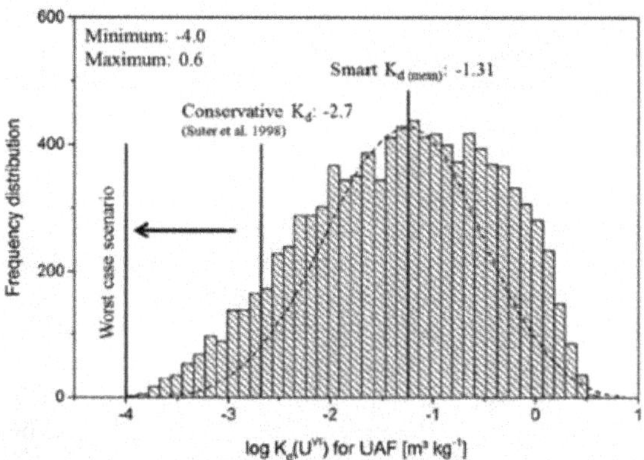

Figure 8. Histogram of 10,000 pre-calculated K_d values for U(VI) sorption on aquifer material (UAF = upper aquifer) at a potential nuclear waste repository site (Gorleben) in Germany. K_d in $m^3 \cdot kg^{-1}$, logarithmic scale. [Used by permission of Elsevier, from Stockmann et al. (2017) Vol. 187, p.283, Figure 4]

The reduced-order SCM smart-K_d example of Stockmann et al. (2017) was developed for a site with extensive geochemical and hydrogeological characterization data available, but such data availability is not typically the case in most practical applications. Especially with respect to chemical characterization of subsurface solids and groundwater, it is more commonly the case that limited data are available, requiring significant assumptions for formulation and parameterization of an SCM, and use of existing thermodynamic databases. There typically is no budget or time for development of site-specific data pertinent to parameterization of sorption models, such as sorption isotherms for a contaminant in contact with the soil/sediment materials under the specific groundwater conditions at the site(s) of interest. In the next section, we present a detailed example of formulation and implementation of a reduced-order SCM smart-K_d approach for reactive transport modeling of a contaminant conducted within the common data-constrained framework. The example SCM formulation and K_d matrix development is in the context of risk assessment for a class of contaminated sites with similar kinds of solid-phase source materials.

EXAMPLE SCM PRACTICAL APPLICATION: DEVELOPMENT OF CONTAMINANT K_d VALUES FOR COAL COMBUSTION RESIDUALS FOR USE IN PROBABILISTIC GROUNDWATER FATE AND TRANSPORT MODELING

This example application of a surface complexation model describes the development of concentration-dependent distribution coefficients, K_d, for use in modeling subsurface transport of metals and other contaminants leached from coal combustion residuals (CCR), part of a US Environmental Protection Agency (EPA) national-scale risk assessment of coal combustion wastes (USEPA 2014; see Appendix H).

Contaminant partitioning between sorbed and dissolved fractions was computed using the EPA equilibrium speciation model MINTEQA2 (USEPA 1991, 1998, 2024a) for 22 individual contaminants over a range of concentrations. Sorption isotherms were calculated for various settings of important geochemical parameters that strongly influence sorption. The complete set of isotherms indexed to the settings of these geochemical parameters were subsequently accessed during fate and transport modeling using the EPA Composite Model for Leachate Migration with Transformation Products (EPACMTP; USEPA 2003a,b,c). In the subsurface, metal and non-metal contaminants undergo reactions with ligands in the porewater and are sorbed onto solid-phase aquifer or soil matrix material. During contaminant transport, sorption to soil and aquifer material results in retardation of the contaminant front. Transport models such as EPACMTP incorporate the contaminant K_d into the overall retardation factor (the ratio of the average linear contaminant velocity to the average linear velocity of the groundwater). The larger the K_d value of a contaminant, the more its transport is retarded.

For an individual contaminant, K_d values in a soil or aquifer are dependent upon the contaminant concentration and various geochemical characteristics of the soil or aquifer and associated porewater. Geochemical parameters that have the greatest influence on the magnitude of K_d include the pH of the system and the nature and concentration of sorbents associated with the soil or aquifer matrix. In the subsurface beneath a disposal facility, the concentration of various leachate constituents may also influence K_d. This is especially true of some CCR leachates that may have high concentrations of calcium, sulfate, and other inorganic constituents. K_d values applicable to the broad range of possible geochemical conditions for specific contaminants are not available in the literature. It is for this reason that the EPA chose to use an equilibrium speciation model to estimate distribution coefficients for the CCR risk assessment (USEPA 2014). The use of a speciation model allows K_d values to be estimated for a range of total contaminant concentrations in various model systems designed to depict natural variability in those geochemical characteristics that most influence partitioning.

Description of MINTEQA2

Non-linear sorption isotherms for contaminants were produced using MINTEQA2 (version 4.02), a geochemical speciation model maintained and distributed by the EPA (USEPA 1991, 1998, 2024a). From input data consisting of total concentrations of chemical constituents, MINTEQA2 calculates the fraction of a contaminant that is dissolved, adsorbed, and precipitated at equilibrium. The total concentrations of major and minor ions, trace metals and other chemicals are specified in terms of key species known as components. MINTEQA2 includes an extensive database of solution species and solid phase species representing known reaction products of two or more of the components. When sorption reactions are included, the dimensionless K_d for a constituent can be calculated from the ratio of the constituent's equilibrium adsorbed concentration to its equilibrium dissolved concentration. The dimensionless K_d is converted to units of liters per kilogram (L/kg) by normalizing by the mass of soil with which one liter of porewater is equilibrated (the phase ratio). An isotherm is generated when the equilibrium distribution between adsorbed and dissolved fractions is estimated for a series of total contaminant concentrations.

Sorption of metal ions and other species can be included in general chemical equilibrium calculations by specifying surface complexation reactions developed for specific SCMs. MINTEQA2 provides several SCMs to describe sorption reactions, including the Triple-Layer, Constant Capacitance, and Diffuse Layer (or Two-Layer) Models. The MINTEQA2 Diffuse Layer Model conforms to that specified in Dzombak and Morel (1990). The version of MINTEQA2 used in this work provided a choice of two coherent databases in the context of the Diffuse Layer Model for surface complexation reactions on iron oxide: the Hydrous Ferric Oxide database (Dzombak and Morel 1990) and the Goethite database (Mathur and Dzombak 2006). The aqueous speciation reactions and equilibrium constants incorporated in MINTEQA2 are approximately the same as those used in the development of the SCM databases.

MINTEQA2 also includes the capability to consider ion binding to DOM. In the Gaussian distribution model for DOM provided in the version of MINTEQA2 employed, the DOM site binding affinities are assumed to exhibit a normal probability distribution (Perdue et al. 1984; Dobbs et. al. 1989; Allison and Perdue 1994). The acid–base and metal complexation reactions of DOM are each provided with a mean log equilibrium constant (log *K*) rather than the usual single value. The Gaussian DOM model is supplied with a database of reactions, including acid–base and major ion and trace metal reactions, each with its mean log *K* and standard deviation for depicting the Gaussian distribution (Susyeto et al. 1991).

Contaminants of interest

The contaminants for which K_d values were estimated using MINTEQA2 for the CCR risk assessment are listed in Table 6.

Several of these contaminants occur naturally in more than one oxidation state. The modeling described was restricted to the oxidation states most likely to occur in CCR-impacted subsurface systems or most likely to be mobile in groundwater. Table 6 lists the oxidation states modeled

Table 6. Contaminants for which distribution coefficients were computed in CCR risk assessment.

Contaminant	Oxidation state(s) modeled	Reason for modeling this oxidation state when other oxidation state(s) occur	MINTEQA2 component
Aluminum (Al)	+3		Al^{3+}
Antimony (Sb)	+5	Sorption reactions available for this state only	$Sb(OH)_6^-$
Arsenic (As)	+3, +5		H_3AsO_3, H_3AsO_4
Barium (Ba)	+2		Ba^{2+}
Beryllium (Be)	+2		Be^{2+}
Boron (B)	+3		H_3BO_3
Cadmium (Cd)	+2		Cd^{2+}
Chromium (Cr)	+3, +6		$Cr(OH)_2^+$, CrO_4^{2-}
Cobalt (Co)	+2		Co^{2+}
Copper (Cu)	+2	This state is the most environmentally relevant	Cu^{2+}
Fluorine (F)	−1		F^-
Iron (Fe)	+2	Fe(III) is likely to be precipitated	Fe^{2+}
Lead (Pb)	+2		Pb^{2+}
Manganese (Mn)	+2		Mn^{2+}
Mercury (Hg)	+2		$Hg(OH)_2$
Molybdenum (Mo)	+6	This state is the most environmentally relevant	MoO_4^{2-}
Nickel (Ni)	+2		Ni^{2+}
Selenium (Se)	+4, +6		$HSeO_3^-$, SeO_4^{2-}
Silver (Ag)	+1		Ag^+
Thallium (Tl)	+1	This state is the most environmentally relevant	Tl^{+1}
Vanadium (V)	+5	Most environmentally relevant and sorption reactions available for this state only	VO_2^+
Zinc (Zn)	+2		Zn^{2+}

for all contaminants. Sorption was modeled for two oxidation states each for arsenic, selenium, and chromium: As (III) and (V), Se (IV) and (VI), and chromium (III) and (VI). For antimony, molybdenum, thallium, copper, iron, and vanadium only one oxidation state was modeled although multiple oxidation states are known to occur naturally. For these metals, the choice of which state to model was dictated by practical aspects such as availability of sorption reactions and by subjective assessment of the appropriate oxidation state. The oxidation states modeled were antimony (V), molybdenum (VI), thallium (I), copper (II), iron (II), and vanadium (V).

Modeling approach

Expected variability in K_d for a particular contaminant at CCR disposal sites was accounted for in the MINTEQA2 modeling by including variability in important input parameters upon which K_d depends. The input parameters for which variability was incorporated include groundwater composition, pH, concentration of aquifer sorbents, and composition and concentration level of CCR leachate.

The chemistry of groundwater exerts an important influence on contaminant K_d values. For the purposes of the CCR risk assessment modeling, the influence of groundwater composition on contaminant sorption was assumed to be adequately represented by including two overall groundwater compositional types: carbonate and non-carbonate. The natural pH range for each of these groundwater types is restricted (about 7 to 9 for carbonate terrain waters and 4 to 10 for non-carbonate). As discussed later, CCR leachate ranges from acidic (pH<2) to highly alkaline (pH >12) and intrusion of leachate into the soil/aquifer system can impact the groundwater pH. For many metals, the magnitude of K_d is very much dependent on pH. This dependence was included in the estimated K_d values by equilibrating the leachate-impacted groundwater system at a series of pH values spanning the range of expected variability.

The influence of variability in sorption capacity of soil and aquifer materials on K_d estimates was included by equilibrating the groundwater systems with various concentrations of two commonly occurring natural adsorbents: "ferric oxide" (i.e., iron oxides/hydroxides) and particulate organic matter. Although other sorbents such as clay minerals, carbonate minerals, hydrous aluminum and manganese oxides, and silica may sorb metals in the subsurface, it was assumed that representation of ferric oxide and POM in the model is sufficient to provide a conservative (i.e., erring on the side of solute mobility) assessment of the sorption capacity of most natural groundwater systems.

CCR leachate can include elevated concentrations of inorganic constituents such as calcium, sulfate, sodium, potassium, and chloride (USEPA 2014). Elevated concentrations of iron and manganese also occur in certain types of leachate. Upon mixing with groundwater, elevated concentrations of leachate constituents may reduce sorption of metals due to competition for sorption sites or complexation with metals in solution. The influence of elevated concentrations of leachate constituents on K_d estimates was included in the modeling by adding leachate components at concentrations representative of CCR leachate (discussed below). In this way, the MINTEQA2 model systems included competitive sorption reactions, metal complexation, and elevated ionic strength effects from CCR leachate.

For each of the two groundwater types considered (carbonate and non-carbonate), the MINTEQA2 modeling to generate sorption isotherms was conducted separately for each contaminant. Each MINTEQA2 model execution to produce an isotherm was conducted for 44 different total contaminant concentrations and with a specific value of pH, concentration of ferric oxide sorbent, concentration of POM sorbent, and a specific setting of the leachate composition (or "richness"). Each resulting isotherm was labeled with the settings of these four master variables.

EPACMTP was subsequently run in Monte Carlo mode, with each execution selecting a value for each of the four master variables from a known or assumed national distribution of such values. EPACMTP then selected and used from the isotherm database generated with use of MINTEQA2 the previously computed isotherm most closely matching the values selected for the four master variables.

Characterization of model groundwater systems

The extent of contaminant sorption in a groundwater system is dependent upon the chemical characteristics of the porewater solution and the interactions among all solutes and the sorption sites present on the exposed surfaces of the soil and aquifer matrix material. The two groundwater compositional types considered—one representative of a carbonate-terrain system and one of a non-carbonate system—are correlated with the hydrogeologic environment parameter in EPACMTP which may take on one of 13 values, each indicative of a particular groundwater type (see Table 7). Issues of practicality (especially the number of model runs required) limited the number of groundwater types considered to just two. The broadest division that may be made of the 13 groundwater types in EPACMTP is carbonate and non-carbonate. The carbonate type corresponds to the "solution limestone" hydrogeologic environment setting in EPACMTP. The other 12 possible hydrogeologic settings in EPACMTP are represented by the non-carbonate groundwater type.

Table 7. Settings for hydrogeologic environment parameter in EPACMTP.

Hydrogeologic environment parameter	Environment (groundwater type) represented
1	Metamorphic and igneous (noncarbonate)
2	Bedded sedimentary rock (noncarbonate)
3	Till over sedimentary rock (noncarbonate)
4	Sand and gravel (noncarbonate)
5	Alluvial basins valleys and fans (noncarbonate)
6	River valleys and floodplains with overbank deposits (noncarbonate)
7	River valleys and floodplains without overbank deposit (noncarbonate)
8	Outwash (noncarbonate)
9	Till and till over outwash (noncarbonate)
10	Unconsolidated and semi-consolidated shallow aquifers (noncarbonate)
11	Coastal beaches (noncarbonate)
12	Solution limestone (carbonate)
13	Others (unclassified hydrogeologic environments) (noncarbonate)

Model groundwater compositions

For both groundwater types, representative charge-balanced groundwater chemistry compositions specified in terms of major ion concentrations and natural pH were selected from the literature. The carbonate system was represented by a well sample reported for a limestone aquifer (Freeze and Cherry 1979). This groundwater had a natural pH of 7.5 and was saturated with respect to calcite. The non-carbonate system was represented by a sample reported for an unconsolidated sand and gravel aquifer with a natural pH of 7.4 (White et al. 1963). An unconsolidated sand and gravel aquifer was selected to represent the non-carbonate compositional type because it is the most frequently occurring of the 12 non-carbonate hydrogeologic environments in EPACMTP. The compositions of both the carbonate and non-carbonate representative groundwater types are shown in Table 8. When EPACMTP was subsequently run in Monte Marlo mode, the hydrogeologic environment selected in the Monte Carlo simulation dictated the set of contaminant isotherms (carbonate or non-carbonate) that were accessed.

Table 8. Compositions of representative carbonate and non-carbonate groundwaters.

Constituent chemical	Concentrations (mg/L)	
	Carbonate groundwater	Non-carbonate groundwater
Ca	55	49
Mg	28	13
SO_4	20	27
HCO_3	265	384
Na	3.1	105
Cl	10	34
K	1.5	3.0
NO_3	—	7.8
F	—	0.3
SiO_2	—	21
pH	7.5	7.4
Temp	18 °C	14 °C
Other	Equilibrium with calcite	Equilibrium with quartz

Model sorbents

Ferric oxides/hydroxides and POM are among the most important sorbents in natural systems. The former may be present as crystalline minerals such as Hematite or Goethite or it may exist as poorly crystalline Hydrous Ferric Oxide dispersed in soil and aquifer media as discrete particles or as coatings on particles of other materials. The ferric oxide sorbent employed in the MINTEQA2 modeling was Goethite, a conservative choice with respect to contaminant transport because of its smaller surface area and lower surface site density compared to HFO.

For modeling metal ion sorption to POM, it was assumed that reactions on POM are analogous to those on DOM. Thus, the MINTEQA2 Gaussian DOM model was used to calculate contaminant binding with both DOM and POM.

The Goethite and POM databases of sorption reactions included in the MINTEQA2 modeling provided a means of including competition for sorption sites among major groundwater ions such as Ca^{2+}, Mg^{2+}, and SO_4^{2-} and the contaminant ions of interest. By including the acid–base reactions for these surfaces, the dependence of contaminant partitioning on pH was represented in the model results.

Details: Goethite sorbent. The concentrations of Goethite sorption sites used in the MINTEQA2 model runs were based on a measurement of ferric iron extracted from 30 soil samples obtained from five electric utility solid waste disposal sites across the U.S. (EPRI 1986). The iron extraction was conducted using the hydroxylamine hydrochloride method (EPRI 1986). This method of Fe extraction is intended to provide a measure of the ferric oxide/hydroxide present as mineral coatings and discrete particles and available for surface reactions with solutes in the associated pore water. The variability in ferric oxide sorbent concentration represented by the variability in extractable Fe from these samples was included in the modeling by selecting low, medium and high extractable Fe concentrations corresponding to the 17th, 50th and 83rd percentiles of the 30 sample measurements. The extractable Fe weight percentages used in the modeling are shown in Table 9.

Table 9. Concentration levels for Goethite sorbent.

Concentration level	Weight percent iron (extractable)	Goethite sorbent concentration (g/L)*
	Unsaturated zone	
Low	0.0182	1.325
Medium	0.0729	5.309
High	0.1190	8.667
	Saturated zone	
Low	0.0182	1.032
Medium	0.0729	4.136
High	0.1190	6.751

Note: * Calculated with use of the phase ratio (soil mass : solution volume), which was calculated as the mean bulk soil density divided by mean water content (mean porosity × mean water saturation). Using values of mean bulk soil density (1.6 kg/L), mean porosity (45%) and mean water saturation (77.7% for unsaturated zone, 100% for saturated zone) from EPACMTP, the phase ratio was determined to be 4.57 kg/L for the unsaturated zone and 3.56 kg/L for the saturated zone.

Although the same distribution of extractable Fe sorbent was used in modeling the saturated and unsaturated groundwater zones, the concentrations of sorbing sites corresponding to the low, medium, and high FeO_x (Goethite) settings in MINTEQA2 were different in the two zones because the assumed ratio of soil mass to solution volume (the phase ratio) was different. These phase ratio values (see Table 9) and the molar mass ratio of Goethite to Fe were used to convert the weight percent extractable Fe to the corresponding mass of Goethite for one liter of porewater solution (Table 9).

The specific surface area and site density used in the Diffuse Layer SCM were as prescribed by Mathur and Dzombak (2006) for Goethite. These values along with the molar concentration of ferric oxide sorbing sites are shown in Table 10. The complete database of Goethite sorption reactions used in MINTEQA2, including acid–base surface reactions and reactions for major ions, are provided in Appendix H of USEPA (2014).

Table 10. MINTEQA2 model parameters for Goethite sorbent.

Parameter	Model value
Specific surface area (m^2/g)	60
Site density (moles of sites per mole Fe)	0.018
Unsaturated zone: Site concentration (mol/L)	
Low	2.680×10^{-4}
Medium	1.074×10^{-3}
High	1.753×10^{-3}
Saturated zone: Site concentration (mol/L)	
Low	2.087×10^{-4}
Medium	8.365×10^{-4}
High	1.365×10^{-3}

Details: POM sorbent. The concentrations of the POM sorbent were obtained from organic matter distributions already present in EPACMTP for three soil types in the unsaturated zone: silty clay loam, sandy loam, and silty loam. Low, medium, and high POM content levels for the MINTEQA2 modeling in the unsaturated zone were established as the 7.5, 50, and 92.5 percentiles of the silty loam organic matter distribution. Low, medium, and high POM content levels were also established for the saturated zone as the 7.5, 50, and 92.5 percentile levels, respectively, of the sandy loam organic matter distribution. As was done for the Goethite sorbent, the concentration of POM included in the MINTEQA2 modeling was determined from the low, medium, and high content levels (expressed in weight percent POM) and the mass of soil for which pore space would accommodate one liter of porewater solution. Thus, phase ratios of 4.57 kg/L in the unsaturated zone and 3.56 kg/L in the saturated zone were used to compute the POM concentration for MINTEQA2 model runs.

A DOC distribution for the saturated zone was obtained from the EPA STORET database (USEPA 2024b). This distribution was based on 1343 groundwater samples and approximated by a log normal distribution with a median $\log_e$ DOC of 1.974 (corresponding to 7.2 mg C/L) and $\log_e$ standard deviation of 1.092. Assuming DOM is approximately 50 percent organic carbon, the DOC values were multiplied by two to approximate DOM concentrations. The MINTEQA2 modeling employed low, medium and high concentrations for DOM corresponding to the 7.5, 50.0, and 92.5 percentiles, respectively, of this approximated DOM distribution. An important point to note is that POM and DOM were treated as correlated variables in the MINTEQA2 modeling: the high DOM value was associated with the high POM value, the medium DOM with the medium POM, etc. Additional details are provided in Appendix H of USEPA (2014). The weight percent POM and concentrations (mg/L) of both POM and DOM are shown in Table 11 for all three concentration levels in both subsurface zones.

Table 11. POM and DOM concentration levels used in MINTEQA2 modeling.

	POM wt%	POM concentration (mg/L)	DOM concentration (mg/L)
		Unsaturated zone	
Low	0.034	1554	6.6
Medium	0.105	4799	20.4
High	0.325	14850	63.2
		Saturated zone	
Low	0.020	712	3
Medium	0.074	2630	14.4
High	0.275	9790	69.4

For both POM and DOM, a conservatively low site density of 1.2×10^{-6} moles of sites per mg organic matter was assumed, on the low end of carboxyl group content of humic and fulvic acids (e.g., Susetyo et al. 1990; Ritchie and Perdue 2003). The site concentrations for organic matter in both subsurface zones are listed in Table 12.

The Gaussian distribution model chosen to describe ion interactions with organic matter used a database of reactions developed by Susetyo et al. (1991) for the metals Al, Ba, Be, Cd, Cr(III), Cu, Fe(III), Mg, Ni, Pb, and Zn and for protons and various major ions binding with DOM. Sorption of metals onto POM was included in the model calculations by assuming that the reactions were identical to those for metal complexation with DOM. Details are provided in Appendix H of USEPA (2014).

Table 12. Site concentrations for POM and DOM components used in MINTEQA2 modeling.

	POM site concentration (mol/L)	DOM site concentration (mol/L)
	Unsaturated zone	
Low	1.865×10^{-3}	7.896×10^{-6}
Medium	5.758×10^{-3}	2.439×10^{-5}
High	1.782×10^{-2}	7.548×10^{-5}
	Saturated zone	
Low	8.544×10^{-4}	3.600×10^{-6}
Medium	3.161×10^{-3}	1.728×10^{-5}
High	1.175×10^{-2}	8.326×10^{-5}

Details: characterization of CCR leachate. Leachate from CCR facilities is known to span a broad pH range and to exhibit highly variable concentrations of inorganic ions (USEPA 2014). Based on a database of CCR leachate characteristics compiled for the CCR risk assessment (USEPA 2014), four representative leachate chemistries were selected for the MINTEQA2 modeling:

- Leachate 1 (L1)—Ash and Coal Waste. Characterized by low to mid pH (1.7–8.2), high SO_4, elevated Fe and Mn.
- Leachate 2 (L2)—Fluidized Bed Combustion (FBC) Waste. Characterized by very high pH (12–13), high Ca, high SO_4, some CO_3, may have high Na, K, Cl
- Leachate 3 (L3)—Flue Gas Desulfurization (FGD) Waste. Characterized by mid to high pH (8.7–11.3), high Ca, high SO_4, high Na, high Cl
- Leachate 4 (L4)—All Other Waste. This group has more variability in pH (3.9–12.3), Ca, and SO_4

The composition (all major ions) of each of the low, medium, and high samples for each leachate type was entered in the MINTEQA2 model in the runs used to estimate K_d. Details are provided in Appendix H of USEPA (2014).

MINTEQA2 modeling procedure

The MINTEQA2 modeling was conducted separately for each of the 22 contaminants (see Table 6). For each contaminant, the modeling was performed separately for each subsurface zone (unsaturated and saturated) for each of the two groundwaters (carbonate and non-carbonate). Thus, results were produced in four main categories: the carbonate groundwater unsaturated zone, the carbonate groundwater saturated zone, the non-carbonate groundwater unsaturated zone, and the non-carbonate groundwater saturated zone. Within each of these four categories, separate results were produced for each of the four leachate types. The same modeling procedure was followed for all contaminants, groundwaters, hydrologic zones, and leachate types. The modeling procedure consisted of two steps, each involving execution of the MINTEQA2 model. First, the sorbents were pre-equilibrated with the groundwater at the natural groundwater pH. This step was performed with groundwater at the ambient pH of the original groundwater for both carbonate and non-carbonate systems and before introduction of the contaminant and leachate. The purpose of the pre-equilibration step was to estimate the unknown sorbed species concentrations of major ions so that they could be included with the total concentrations for these constituents in subsequent model runs.

In the second step, leachate and the contaminant were added to the pre-equilibrated model systems and the new equilibrium distribution of contaminant among dissolved, sorbed, and precipitated phases was computed. In this step, the equilibrium pH was specified at a series of discrete values selected to span the range of expected pH values given the groundwater type and leachate type modeled. For each selected pH, a preliminary run was performed to estimate the ionic strength of the system being modeled. Subsequent model runs at 44 different total contaminant concentrations were constrained to this same ionic strength. Thus, each MINTEQA2 model execution to produce an isotherm for a particular contaminant was defined by the following selections of system variables:

- groundwater type (carbonate or non-carbonate)
- hydrologic zone (unsaturated or saturated)
- FeO_x sorbent concentration level (low, medium, high)
- POM/DOM concentration level (low, medium, high)
- equilibrium pH (range depends on groundwater type and leachate type; see Table 13)
- leachate type (Ash and Coal Waste, FBC Waste, FGD Waste, or Other Waste)
- leachate "richness" or ionic strength (low, medium, or high)

To establish the total concentrations of Ca, Mg, SO_4, Na, and other leachate constituents in the saturated zone, a constant dilution ratio of one part leachate to six parts groundwater was used for all modeling. For each system modeled (defined by the selections of system variables), the system was equilibrated at a series of 44 total concentrations of the contaminant spanning the range from 0.001 to 10,000 mg/L. The results were compiled as the equilibrium distribution of the contaminant among dissolved, sorbed, and precipitated fractions for each contaminant concentration.

Table 13. Range of equilibrium pH values used for waste leachate types in MINTEQA2 modeling.

	pH range	
Waste leachate type	**Carbonate groundwater**	**Non-carbonate groundwater**
Ash and coal waste	5.1–8.2	1.8–8.2
FBC waste	5.1–12.6	4.5–12.6
FGD waste	5.1–12.6	4.5–12.6
All other waste	5.1–12.6	4.5–12.6

Results

For each contaminant, the equilibrium dissolved, sorbed, and precipitated fractions computed by MINTEQA2, each indexed with the settings of all system variables, were compiled into a database for use in transport modeling with EPACMTP. The contaminant concentrations in the dissolved and sorbed fractions along with the corresponding total contaminant concentrations represent the non-linear sorption isotherm. With the consideration of 44 different total contaminant concentrations for MINTEQA2 simulations for each set of four master variable (FeO_x, POM, pH, and leachate richness) values, the related isotherm thus comprised 44 sets of sorbed and dissolved concentration results. The ratio of the equilibrium sorbed and dissolved concentrations recorded in the isotherm is the dimensionless K_d, which could be normalized by the appropriate phase ratio (4.57 kg/L and 3.56 kg/L for the unsaturated and saturated zones, respectively) to obtain the K_d in L/kg.

For each contaminant, the MINTEQA2 modeling resulted in 432 isotherms for the carbonate groundwater for the Ash and Coal Waste leachate type and 702 isotherms for each of the other leachate types. For the non-carbonate groundwater, 702 isotherms were produced for the Ash and Coal Waste leachate type and 756 for each of the other leachate types.

Example plots illustrating the influence of the master variables FeO_x, POM, and pH on the estimated K_d values are shown in Figures 9–11. Figure 9 shows K_d for As(V) versus total As(V) concentration for the carbonate groundwater saturated zone for the Other Waste leachate type at three levels of FeO_x sorbent concentration. The K_d versus total As(V) plots are for the medium level for leachate richness and the medium POM concentration level for pH 6.6. As expected, sorption is enhanced at the higher FeO_x concentrations resulting in larger K_d values.

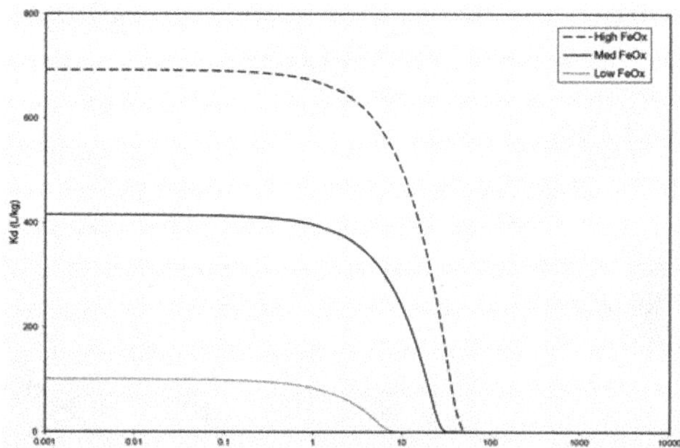

Figure 9. Estimated K_d versus total arsenic(V) concentration for three FeO_x sorbent concentrations levels (for "Other" waste type, carbonate groundwater, saturated zone, medium POM, medium leachate richness, pH 6.6). [Source: USEPA 2014]

The influence of POM on Pb sorption is shown in Figure 10. The impact of varying the POM concentration level differs among the various metals. The effect of varying the POM concentration level also depends on the pH. The variable impact of POM is due to two factors: the absence of organic matter reactions for anionic metals and the concurrent influence of DOM for those metals for which POM sorption reactions are included. In the MINTEQA2 modeling procedure used, increasing the POM sorbent concentration was always accompanied

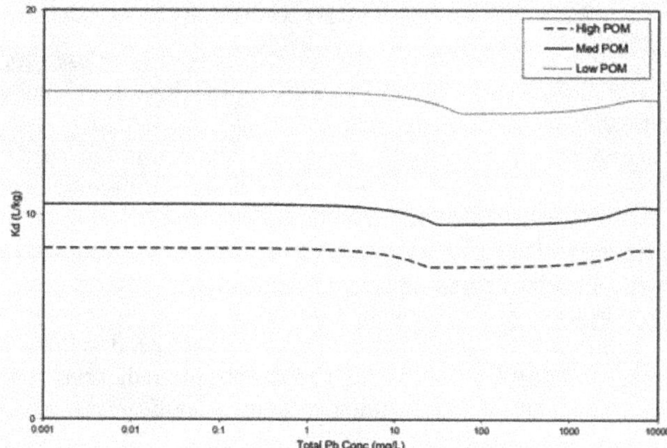

Figure 10. Estimated K_d versus total lead concentration for three POM/DOM concentrations levels (for non-carbonate groundwater, unsaturated zone, medium FeO_x, medium leachate richness, pH 8.2). [Source: USEPA 2014]

by a proportional increase in the DOM concentration. The overall impact on the amount of metal sorbed depends on the relative competition among all constituents in the systems for these two substances. The balance of this competition of POM and DOM shifts with pH because both POM and DOM undergo acid–base reactions. Figure 10 shows the result of varying the POM/DOM concentration level on Pb sorption for the non-carbonate groundwater unsaturated zone with medium FeO_x, medium leachate richness, and pH 8.2. In this example, the highest POM level results in the highest K_d values. However, this is not true for all cases—the outcome due to increasing the POM is dependent on the competition between the metal and major ions on all three competing substances (FeO_x, POM, and DOM).

Figure 11 depicts the K_d versus pH for Pb in a single system (non-carbonate groundwater, unsaturated zone, high FeO_x and POM sorbent levels, medium leachate richness level for Other Waste leachate type) at a single total Pb concentration (0.001 mg/L). K_d values in this system increase steadily from pH 4 to 7, then increase more significantly up to pH 9.8. After pH 9.8, K_d begins to decrease abruptly due to formation of Pb–OH solution species.

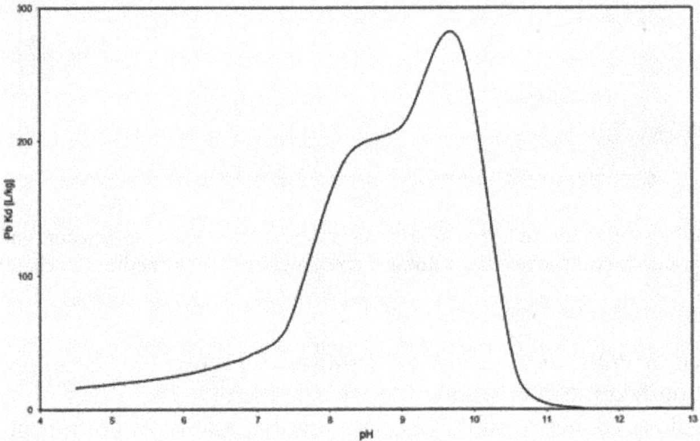

Figure 11. Estimated K_d versus pH for lead in a single system at a single total lead concentration of 0.001 mg/L (for "Other" waste type, non-carbonate groundwater, unsaturated zone, high FeO_x, high POM, medium leachate richness). [Source: USEPA 2014]

MOVING FORWARD: RESEARCH NEEDS

As discussed in the introduction to the chapter, SCMs have been used for many purposes, from fundamental surface science studies to various applications, including prediction of ion sorption in water treatment, contaminant fate and transport modeling, risk assessment, and others. For practical application of SCMs, the scientific foundation of SCMs provides important benefits including improved generality of ion solid–water partitioning predictions, greater insight into variables driving this phase partitioning, and more understanding of how ion sorption relates to system properties. However, for most practical applications inclusion in SCMs of full scientific complexity of surface properties, surface–surface interactions, and ion interactions with surfaces is not necessary or desirable. Consideration of more physical and chemical processes in SCMs brings in more parameters for which values need to be measured, estimated or assumed, and doesn't contribute to reducing uncertainty in predictions. The accuracy of ion sorption predictions with simple SCMs is sufficient for most applications, especially considering the uncertainty involved with predictions made with any SCM for a complex system.

Selection of any model for prediction of ion sorption in a practical application leads one to consider and make a decision about the appropriate balance of empiricism and scientific complexity, with consideration of time, budget, level of accuracy needed, and tools available.

These issues have been considered in some way by anyone who has ever applied an SCM to a complex system. Lützenkirchen (2006) discusses the challenge of balancing practicality and scientific complexity in surface complexation modeling in a direct and thought-provoking manner. As apparent from preceding discussions, most practical applications of SCMs have been in the context of research, but SCMs have slowly come into use in the broader community as somewhat standardized tools and approaches have been developed and adopted (e.g., de Vries et al. 2013; Dijkstra et al. 2018; USEPA 2014).

There are many research needs for advancement of surface complexation modeling, but there are different kinds of research needs for different uses of SCMs. Here the focus is on research needs for advancement of the practical application of surface complexation modeling.

Further develop SCM databases

The development of databases for modeling sorption of ions on oxides (e.g., Dzombak and Morel 1990; Tonkin et al. 2004; Mathur and Dzombak 2006; Karamalidis and Dzombak 2010) and organic matter (e.g., Tipping 1998; Milne et al. 2001, 2003; Tipping et al. 2011) have helped advance the practical application of SCMs over the past several decades. There is a need for updating of the existing databases to account for new data, to develop new databases for particular minerals such as other forms of iron and aluminum oxides, and to develop new databases for a wider selection of SCMs to supplement those already available for the Generalized Two-Layer Model. Jin et al. (2023) developed generic parameters for application of the CD-MUSIC Model to iron oxides/hydroxides, for example.

Further test multisurface SCMs

In a review of applications of multisurface (or component additivity or assemblage) SCMs to soils and sediments, Groenenberg et al. (2014) concluded that while the multisurface modeling approach has been tested and validated with synthetic systems and found to be fairly accurate, it has been less accurate for natural assemblages. Groenenberg and Lofts (2014) recommend a systematic study of the performance of the additivity approach with natural assemblages, with a focus on identifying and exploring discrepancies between experimental results and model predictions, and investigating specifically the improvements that are needed.

Perform uncertainty analyses

Groenenberg and Lofts (2014) and others have pointed to the need for uncertainty analyses for multisurface SCM model predictions. There are multiple factors that drive uncertainty in predictions made with multisurface SCMs, including the inherent model uncertainty (including the equilibrium assumption) as well as the uncertainty in selected dominant sorbents and the measured or estimated available reactive surface of those sorbents, and in model parameter values that can be measured or that are estimated or assumed. Some work on systematic analysis of uncertainty in SCM predictions has been conducted. Le and Hendriks (2014) investigated the effects of uncertainty in soil properties (pH, SOM, DOC, clay content, oxide content, reactive metal concentration) on sorption predictions for 17 metals made with a multisurface model. For most of the metals, the uncertainties in the sorption predictions were primarily associated with pH, SOM, and reactive metal concentration. Ciffroy and Benedetti (2018) evaluated uncertainty in multisurface model predictions of metal (Cd, Cu, Pb, Zn) sorption on particulates in surface waters. Sensitivity analyses to identify the main sources of uncertainty revealed that pH, SPM (suspended particulate matter), percentage of POC in SPM, and DOC were the most influential parameters. Another source of uncertainty in all uses of SCMs is the uncertainty in aqueous phase reactions and equilibrium constants which are critical to use of the models and usually assumed to be highly accurate as well as consistent with the aqueous reactions used in deriving the SCM databases. Uncertainties related to the aqueous phase reactions used may be unavoidable when surface reactions developed to describe data for well-characterized laboratory systems are applied in geochemical models of complex natural systems that may include reactions not accounted for in the derivation of the surface reactions. As Groenenberg and Lofts (2014) note, uncertainty analysis can help identify the most important sources of uncertainty in model predictions, which will help guide research to reduce uncertainty in SCM predictions.

Advance models for clays

While clays are minor contributors to ion sorption in many soils and sediments, in some systems they can be important, such as in engineered or natural clay barriers for waste containment. Various approaches have been employed to account for sorption on clays, including treating the sorbing surfaces as Al oxides or as Donnan ion exchangers. In the multisurface model of Dijkstra et al. (2009), clay surfaces are treated as Donnan ion exchangers to represent electrostatic, non-site-specific sorption of ions to permanent charge on the surfaces. Bradbury and Baeyens (2006) developed the 2-Site Protolysis Non-Electrostatic Surface Complexation and Cation Exchange (2SPNE SC/CE) Model to describe metal and radionuclide ion sorption on clays, with parameter values determined for montmorillonite (Bradbury and Baeyens 1997, 1999) and for illite (Bradbury and Baeyens 2009a,b). Goldberg (2014), Groenenberg and Lofts (2014) and others have called for the development of general models for clay surfaces present in soil/sediment mineral mixtures to advance the use of multisurface SCMs.

Advance models for NOM

There has been considerable advancement of models to describe ion binding to DOM and POM, as discussed earlier in the chapter. These models have been used to describe sorption on NOM for systems with significant organic matter content. Currently, sorption on NOM is modeled as sorption on one phase, with no distinction of whether the organic matter is sorbed on mineral surfaces or as separate POM phases. This approach may provide sufficient accuracy for most applications, but the potential for DOM to be a competing sorbate, especially for oxyanions, has been identified as a shortcoming (Goldberg 2014; Groenenberg and Lofts 2014). Alternative approaches for describing sorption on NOM in soils and sediments that contain NOM should be explored.

Further develop standard protocols for parameterizing models

The practical application of SCMs has been advanced considerably by development and use of both informal and formal standards and protocols for parameterizing SCMs. An example of the use of informal standards for parameters is the common use of certain specific surface area values for Hydrous Ferric Oxide and Goethite. In regard to formal standards, the development of the ISO standards for selective extractions of Al and Fe oxides/hydroxides (ISO 2012a,b,c) and for solid-phase NOM (2012d) provide examples of the kinds of standardization that will help advance practical applications of SCMs. All selective extraction approaches have limitations and associated uncertainty, but standard methods have many benefits for users including clarity about process and development of comparable data and model outputs that will outweigh the limitations in many situations. Confidence in use of the ISO standard selective extraction methods could be advanced by investigating the performance of the extraction methods on a set of soils with a range of properties and known elemental compositions.

Develop efficient and standard protocols for generalized composite SCM modeling

Generalized composite SCMs are of interest for accurate site-specific modeling. Applications to date have involved development of experimental site-specific ion partitioning data and fitting of these data to determine apparent equilibrium constants for specified reactions on generic surface sites and other model parameter values. There is a need for development of efficient approaches and standard protocols for implementation of the generalized composite approach at a site of interest for predictive modeling.

Develop standard protocols for reduced-order K_d models

Reduced-order models for prediction of K_d values spanning a wide range of conditions using SCMs are of significant interest for reactive transport modeling, as discussed earlier. A number of different methods for generating and using a matrix of calculated K_d values have been employed. A common protocol for use of SCMs to generate reduced-order K_d prediction models, oriented toward routine application at contaminated sites, would be helpful in advancing this modeling approach.

SUMMARY AND CONCLUSIONS

Since their introduction and development in the 1970s and 1980s, surface complexation models have been used for a diverse range of practical applications, including design and analysis of water treatment systems, reactive transport modeling for metals in surface water and groundwater systems, interpretation of geochemical data, and others. SCMs have been used to help understand and predict the performance of complex systems such as these because of the ability of SCMs to describe and predict sorption for a range of solution conditions and their compatibility with general chemical equilibrium models to account for solution chemistry. For practical applications of SCMs, there is less concern with capturing the full mechanistic complexity of sorption processes, and thus relatively simple SCMs such as the Constant Capacitance Model, the Two-Layer (or Diffuse Layer) Model, and various non-electrostatic models have been commonly employed. The use of SCMs for practical applications has been facilitated by the development of standard parameter values and SCM databases for some sorbents of importance in soils, sediments, and aquifers, e.g., Hydrous Ferric Oxide, Goethite, Gibbsite, and Hydrous Manganese Oxide, and by use of selective extractions to estimate the amounts of reactive sorbent surfaces available in complex systems.

The parameterization of SCMs for practical applications has been aided by the adoption and common use of selective extraction methods for measurement of surface-reactive Fe and Al oxide/hydroxide phases in soils, sediments, and aquifer media. Commonly employed methods include use of ascorbic acid for extraction of amorphous Fe oxides/hydroxides; dithionite–citrate for extraction of crystalline Fe oxides/hydroxides; and oxalate–oxalic acid for extraction of amorphous Al oxides/hydroxides. ISO standards have been developed for these methods and are contributing to their use. An ISO standard method has also been developed for extraction of NOM. Various modifications of the methods are employed, and other selective extraction methods are used. The common acceptance and use of selective extraction methods, despite their known limitations, is helping to advance practical application of SCMs.

Three general approaches have been used for practical applications of SCMs: the multisurface (or component additivity) approach, the single-surface/surrogate approach, and the generalized composite approach. Multisurface SCMs involve combination of SCM reactions and other parameters for each sorbent identified for a system, with the individual contribution of each sorbent to overall ion sorption considered in the model. The component additivity approach has been acceptably validated in laboratory studies with combinations of mineral sorbents. The multisurface modeling approach requires decisions about which reactive surfaces to include, which models to use to describe ion interactions with those surfaces, and how to choose parameter values for the models selected. In contrast to the multisurface approach for implementation of SCMs, in which multiple sorbent surfaces are considered, the single-surface or surrogate approach considers a single reactive surface in the solid-phase. This approach has been used for a number of reasons, including focus on systems in which there is clearly one type of dominant sorbent phase, for computational simplicity, and in cases where there is limited information available about the composition of the sorbent. A single-surface/surrogate SCM approach requires decisions about the dominant surface, the SCM model to employ, and parameter values for the model and sorbent selected. The generalized composite approach is a more empirical SCM approach that involves formulating generic surface reactions and determining the value of apparent surface complexation constants by fitting data for particular sorbents and solution conditions of interest. In a generalized composite SCM, the amounts of reactive surface sites are estimated from measurements and with use of some assumptions.

When natural organic matter is present in significant quantities in a soil, sediment, or aquifer material, accounting for NOM in modeling of ion binding is required for accurate description of sorption data. The semi-mechanistic models in use for describing ion binding with dissolved and solid phase NOM consider reactions of ions with ligand types associated with the humic and fulvic acid components of the NOM. The two most commonly employed models—the NICA-Donnan and the WHAM models—both account for electrostatic effects and enable description and prediction of ion binding across a broad range of solution conditions. In addition, both models have been applied to compilations of data for many soils and generic parameters have been developed from the fitting of these data. The generic model parameters have facilitated the use and practical application of both models. Models for ion binding with integrated consideration of surface complexation on mineral surfaces and ion binding on solid-phase NOM have been developed and deployed for complex systems.

There are various research needs for advancement of the practical application of surface complexation modeling. These include further development of SCM databases, further testing of the component additivity foundation of multisurface SCMs, advancement of models for ion binding on clays and NOM and their integration with SCMs, further development of standard methods and protocols for parameterizing SCM models, and development of standard protocols for generalized composite SCM modeling. In addition, with the interest in use of SCMs for K_d prediction, development of standard protocols for reduced-order K_d modeling is needed.

Practical applications of SCMs have increased steadily since the introduction of these models as research tools. Their utility for providing insights into and useful predictions of ion solid–water partitioning in complex soil, sediment, and aquifer media systems has been demonstrated. With continued experience and research, the utility and accuracy of SCMs in practical applications will grow in the years to come.

ACKNOWLEDGEMENTS

David Dzombak gratefully acknowledges the support of an Oak Ridge Institute for Science and Education Fellowship sponsored by the U.S. Environmental Protection Agency Office of Resource Conservation and Recovery. The views expressed in this article are those of the authors and do not necessarily represent the views or policies of the U.S. Environmental Protection Agency.

REFERENCES

Allison JD, Perdue EM (1994) Modeling metal–humic interactions with MINTEQA2. *In:* Humic Substances in the Global Environment and Implications on Human Health, Senesi N, Miano TM (eds), Elsevier, Amsterdam, p 927–942

Bacon JR, Davidson CM (2008) Is there a future for sequential chemical extraction? Analyst 133:25–46

Benjamin MM (2015) Water Chemistry, 2nd ed. Waveland Press, Long Grove IL, Chap 13

Bethke CM (1996) Geochemical Reaction Modeling. Oxford University Press, New York, Chap 8

Bethke CM, Brady PV (2000) How the K_d approach undermines ground water cleanup. Ground Water 38:435–443

Blakemore LC, Searle PL, Daly BK (1987) Methods for chemical analysis of soils. NZ Soil Bureau Scientific Report 80, NZ Soil Bureau, Lower Hutt, New Zealand

Boccelli DL, Small MJ, Dzombak DA (2006) Effects of water quality and model structure on arsenic removal simulation: An optimization study. Environ Eng Sci 23:835–850

Bonten LTC, Groenenberg JE, Weng, L, van Riemsdijk WH (2008) Use of speciation and complexation models to estimate heavy metal sorption in soils. Geoderma, 146:303–310

Bradbury MH, Baeyens B (1997) A mechanistic description of Ni and Zn sorption on Na-montmorillonite. Part II: modelling. J Contam Hydrol 27:223–248

Bradbury MH, Baeyens B (1999) Modeling the sorption of Zn and Ni and Zn on Ca-montmorillonite. Geochim Cosmochim Acta 63:325–336

Bradbury MH, Baeyens B (2006) A quasi-mechanistic non-electrostatic modelling approach to metal sorption on clay minerals. *In*: Surface Complexation Modeling. Lützenkirchen J (ed) Elsevier, Amsterdam, p 518–538

Bradbury MH, Baeyens B (2009a) Sorption modelling on illite Part I: Titration measurements and the sorption of Ni, Co, Eu, and Sn. Geochim Cosmochim Acta 73:990–1003

Bradbury MH, Baeyens B (2009b) Sorption modelling on illite Part II: Actinide sorption and linear free energy relationships. Geochim Cosmochim Acta 73:990–1003

Bradbury MH, Baeyens B (2011) Predictive sorption modelling of Ni(II), Co(II), Eu(III), Th(IV) and U(VI) on MX-80 bentonite and Opalinus clay: A bottom-up approach. Appl Clay Sci 52:27–33

Brady NC, Weil RR (2002) The Nature and Properties of Soils, 13th ed. Prentice Hall, Upper Saddle River NJ

Brady PV, Papenguth HW, Kelly JW (1999) Metal sorption to dolomite surfaces. Appl Geochem 14:569–579

Brezonik PL, Arnold WA (2022) Water Chemistry, 2nd ed. Oxford University Press, New York, Chap 12

Canfield DE (1988) Sulfate Reduction and Diagenesis of Iron in Anoxic Marine Sediments. PhD dissertation, Yale University

Cederberg GA, Street RL, Leckie JO (1985) A groundwater mass transport and equilibrium chemistry model for multicomponent systems. Water Resources Res 21:1095–1104

Chao TT, Zhou L (1983) Extraction techniques for selective dissolution of amorphous iron oxides from soils and sediments. Soil Sci Soc Am J 47:225–232

Ciffroy P, Benedetti M (2018) A comprehensive probabilistic approach for integrating natural variability and parametric uncertainty in the prediction of trace metals speciation in surface waters. Environ Poll 242:1087–1097

Comans RNJ, Middleburg JJ (1987) Sorption of trace metals on calcite: Applicability of the surface precipitation model. Geochim Cosmochim Acta 51:2587–2591

Coston JA, Fuller CC, Davis JA (1995) Pb^{2+} and Zn^{2+} adsorption by a natural aluminum- and iron-bearing surface coating on an aquifer sand. Geochim Cosmochim Acta 59:3535–3547

Curtis GP, Fox, P, Kohler, M, Davis JA (2004) Comparison of in situ uranium K_D values with a laboratory determined surface complexation model. Appl Geochem 19:1643–1653

Curtis GP, Davis JA, Naftz DL (2006) Simulation of reactive transport of uranium(VI) in groundwater with variable chemical conditions. Water Resources Res 42:W04404

Curtis GP, Kohler, M, Davis JA (2009) Comparing approaches for simulating the reactive transport of U(VI) in ground water. Mine Water Environ 28:84–93

Davis JA, Kent DB (1990) Surface complexation modeling in aqueous geochemistry. Rev Mineral 23:177–260

Davis JA, Leckie JO (1978) Surface ionization and complexation at the oxide/water interface: II. Surface properties of amorphous iron oxyhydroxide and adsorption of metal ions. J Colloid Interface Sci 67:90–107

Davis JA, Leckie JO (1980) Surface ionization and complexation at the oxide/water interface: III. Adsorption of anions. J Colloid Interface Sci 72:32–43

Davis JA, James RO, Leckie JO (1978) Surface ionization and complexation at the oxide/water interface: I. Computation of electrical double layer properties in simple electrolytes. J Colloid Interface Sci 63:480–499

Davis JA, Fuller CC, Cook AD (1987) A model for trace metal sorption processes at the calcite surface: Adsorption of Cd^{2+} and subsequent solid solution formation. Geochim Cosmochim Acta 51:1477–1490

Davis JA, Coston JA, Kent DB, Fuller CC (1998) Application of the surface complexation concept to complex mineral assemblages. Environ Sci Technol 32:2820–2828

Deutsch WJ (1997) Groundwater Geochemistry: Fundamentals and Applications to Contamination. Lewis Publishers, Boca Raton FL

de Vries W, Groenenberg JE (2009) Evaluation of approaches to calculate critical metal loads for forest ecosystems. Environ Poll 157:3422–3432

de Vries W, Groenenberg JE, Lofts S, Tipping E, Posch M (2013) Critical loads of heavy metals in soils. *In:* Heavy Metals in Soils: Trace Metals and Metalloids in Soils, Alloway BJ (ed), Springer, Dordrecht

Dijkstra JJ, Meeussen JCL, Comans RNJ (2004) Leaching of heavy metals from contaminated soils: an experimental and modeling study. Environ Sci Technol 38:4390–4395

Dijkstra JJ, Meeussen JCL, Comans RNJ (2009) Evaluation of a generic multisurface sorption model for inorganic soil contaminants. Environ Sci Technol 43:6196–6201

Dijkstra JJ, van Zomeren, A, Brand, E, Comans RNJ (2018) Site-specific aftercare completion criteria for sustainable landfilling in the Netherlands: geochemical modeling and sensitivity analysis. Waste Management 75:407–414

Dobbs JC, Susetyo, W, Carreira LA, Azarraga LV (1989) Competitive binding of protons and metal ions in humic substances by lanthanide ion probe spectroscopy. Anal Chem 61:1519–1524

Dzombak DA (1988) A general model for predicting sorption of inorganics on hydrous oxides. Proceedings of the Ground Water Geochemistry Conference, National Water Well Assoc., Denver, February, p 718–737

Dzombak DA, Morel FMM (1987) Adsorption of inorganic pollutants in aquatic systems. J Hydraulic Eng 113:430–475

Dzombak DA, Morel FMM (1990) Surface Complexation Modeling: Hydrous Ferric Oxide. Wiley-Interscience, New York

Dzombak, D, Morel, F, Mundt, J, Hirschberg, E, Huff G (1992) Modeling the leaching of metals from hazardous waste incinerator ash. Proceedings of the 1992 Incineration Conference, Albuquerque NM, May, p 501–506

EPRI (1986) Physi[c]ochemical Measurements of Soils at Solid-Waste Disposal Sites. EPRI Report EA-4417, prepared by Battelle Pacific Northwest Laboratories, Richland WA, Electric Power Research Institute, Palo Alto CA

ERS (2022) MINEQL+, Version 5.0. Environmental Research Software, Hallowell ME, available at https://mineql.com

Ferdelman DG (1988) The Distribution of Sulphur, Iron, Manganese, Copper, and Uranium in a Saltmarsh Sediment Core as Determined by a Sequential Extraction Method. MS dissertation, University of Delaware

Freeze RA, Cherry JA (1979) Groundwater. Prentice Hall, Englewood Cliffs NJ

Fuller CC, Davis JA, Coston JA, Dixon E (1996) Characterization of metal adsorption variability in a sand and gravel aquifer, Cape Cod, Massachusetts USA. J Contam Hydrol 22:165–187

Goldberg S (1992) Use of surface complexation models in soil chemical systems. Adv Agron 47:233–329

Goldberg S (2013) Surface complexation modeling. Reference Module in Earth Systems and Environmental Sciences. Elsevier

Goldberg S (2014) Application of surface complexation models to anion adsorption by natural materials. Environ Toxicol Chem 33:2172–2180

Goldberg S, Sposito G (1984) A chemical model of phosphate adsorption by soils: I. Reference oxide minerals. Soil Sci Soc Am J 48:772–778

Groenenberg JE, Lofts S (2014) The use of assemblage models to describe trace element partitioning, speciation, and fate: a review. Environ Toxicol Chem 33:2181–2196

Groenenberg JE, Koopmans GF, Comans RNJ (2010) Uncertainty analysis of nonideal competitive adsorption-Donnan model: Effects of dissolved organic matter variability on predicted metal speciation in soil solution. Environ Sci Technol 44:1340–1346

Groenenberg JE, Dijkstra JJ, Bonten LTC, de Vries, W, Comans RNJ (2012) Evaluation of the performance and limitations of empirical partition-relations and process based multisurface models to predict trace element solubility in soils. Environ Poll 166:98–107

Groenenberg JE, Romkens PFA, van Zomeran, A, Rodrigues SM, Comans RNJ (2017) Evaluation of the single dilute (0.43M) nitric acid extraction to determine geochemically reactive elements in soil. Environ Sci Technol 51:2246–2253

Gustafsson JP (2020) VisualMINTEQ, Version 3.1, available at: https://vminteq.com/

Gustafsson JP, van Schaik JWJ (2003) Cation binding in a mor layer: batch experiments and modeling. Eur J Soil Sci 54:295–310

Gustafsson JP, Pechova P, Berggren D (2003) Modeling metal binding to soils: The role of natural organic matter. Environ Sci Technol 37:2767–2774

Hayes KF, Katz LE (1996) Application of X-ray absorption spectroscopy for surface complexation modeling of metal ion sorption. *In*: The Physics and Chemistry of Mineral Surfaces. Brady P (ed) CRC Press, Boca Raton FL, p 147–223

Hering JG, Chen PY, Wilkie JA, Elimelech, M, Liang S (1996) Arsenic removal by ferric chloride. J Am Water Works Assoc 88:155–167

Hiemstra, T, van Riemsdijk WH (1996) A surface structural approach to ion adsorption: The charge distribution (CD) model. J Colloid Interface Sci 179:488–508

Hiemstra, T, van Riemsdijk WH (2006) On the relationship between charge distribution, surface hydration, and the structure of the interface of metal hydroxides. J Colloid Interface Sci 301:1–18

Honeyman BD (1984) Cation and anion adsorption at the oxide/solution interface in systems containing binary mixtures of adsorbents: An investigation of the concept of adsorptive additivity. PhD dissertation, Stanford University

Huang CP, Stumm W (1973) Specific adsorption of cations on γ-Al_2O_3. J Colloid Interface Sci 43:409–420

ISO (2012a) Soil Quality—Parameters for Geochemical Modelling of Leaching and Speciation of Constituents in Soils and Materials—Part 1: Extraction of Amorphous Iron Oxides and Hydroxides with Ascorbic Acid. ISO 12782–1:2012(E), International Standards Organization, Geneva, Switzerland

ISO (2012b) Soil Quality—Parameters for Geochemical Modelling of Leaching and Speciation of Constituents in Soils and Materials—Part 2: Extraction of Crystalline Iron Oxides and Hydroxides with Dithionite. ISO 12782–2:2012(E), International Standards Organization, Geneva, Switzerland

ISO (2012c) Soil Quality—Parameters for Geochemical Modelling of Leaching and Speciation of Constituents in Soils and Materials—Part 3: Extraction of Aluminium Oxides and Hydroxides with Ammonium Oxalate/Oxalic Acid. ISO 12782–3:2012(E), International Standards Organization, Geneva, Switzerland

ISO (2012d) Soil Quality—Parameters for Geochemical Modelling of Leaching and Speciation of Constituents in Soils and Materials—Part 4: Extraction of Humic Substances from Solid Samples. ISO 12782–4:2012(E), International Standards Organization, Geneva, Switzerland

ISO (2012e) Soil Quality—Parameters for Geochemical Modelling of Leaching and Speciation of Constituents in Soils and Materials—Part 5: Extraction of Humic Substances from Aqueous Samples. ISO 12782–5:2012(E), International Standards Organization, Geneva, Switzerland

Jenne EA (1977) Trace element sorption by sediments and soils: sites and processes. *In*: Molybdenum in the Environment, Vol. 2. Chappell WR (ed) Marcel Dekker, New York, p 425–553

Jenne EA (1998) Adsorption of metals by geomedia: Data analysis, modeling, controlling factors, and related issues. *In*: Adsorption of Metals by Geomedia. Jenne EA (ed) Academic Press, San Diego, p 1–73

Jin J, Liang Y, Wang M, Fang L, Xiong J, Hou J, Tan W, Koopal L (2023) Generic CD-MUSIC-eSGC model parameters to predict the surface reactivity of iron (hydr)oxides. Water Res 230:119534

Karamalidis AK, Dzombak DA (2010) Surface Complexation Modeling: Gibbsite. John Wiley& Sons, Hoboken, New Jersey

Kent DB, Abrams RH, Davis JA, LeBlanc DR (2000) Modeling the influence of variable pH on the transport of zinc in a contaminated aquifer using semiempirical surface complexation models. Water Resources Res 36:3411–3425

Kinniburgh DG, Milne CJ, Benedetti MF, Pinheiro JP, Filius J, Koopal LK, van Riemsdijk WH (1996) Metal ion binding by humic acid: Application of the NICA-Donnan model. Environ Sci Technol 30:1687–1698

Kinniburgh DG, van Riemsdijk WH, Koopal LK, Borkovec M, Benedetti MF, Avena MJ (1999) Ion binding to natural organic matter: Competition, heterogeneity, stoichiometry and thermodynamic consistency. Colloids Surf A 151:147–166

Koopal LK, Saito, T, Pinheiro JP, van Riemsdijk WH (2005) Ion binding to natural organic matter: General considerations and the NICA-Donnan model. Colloids Surf A 265:40–54

Koretsky CM (2000) The significance of surface complexation reactions in hydrologic systems: A geochemist's perspective. J Hydrol 230:127–171

Kornicker WA, Morse JW (1991) Interactions of divalent cations with the surface of pyrite. Geochim Cosmochim Acta 55:2159–2171

Kostka JE, Luther III GW (1994) Partitioning and speciation of solid phase iron in saltmarsh sediments. Geochim Cosmochim Acta 58:1701–1710

Landry CJ, Koretsky CM, Lund TJ, Schaller M, Das S (2009) Surface complexation modeling of Co(II) adsorption on mixtures of hydrous ferric oxide, quartz and kaolinite. Geochim Cosmochim Acta 73:3723–3737

Langmuir D (1997) Aqueous Environmental Geochemistry. Prentice-Hall, Upper Saddle River NJ, Chap 10

Le TTY, Hendriks AJ (2014) Uncertainties associated with lacking data for predictions of solid-solution partitioning of metals in soil. Sci Total Environ 490:44–49

Liang L, Morgan JJ (1990) Chemical aspects of iron oxide coagulation in water: laboratory studies and implications for natural systems. Aquat Sci 52:32–55

Lord CJ (1980) The Chemistry and Cycling of Iron, Manganese, and Sulphur in Salt Marsh Sediments. PhD dissertation, University of Delaware

Loux NT, Brown DS, Chafin CR, Allison JD, Hassan SM (1989) Chemical speciation and competitive cationic partitioning on a sandy aquifer material. Chem Speciation Bioavailability 1:111–125

Lund TJ, Koretsky CM, Landry CL, Schaller MS, Das S (2008) Surface complexation modeling of Cu(II) adsorption on mixtures of hydrous ferric oxide and kaolinite. Geochem Trans 9:9

Lützenkirchen J (2006) Preface. *In*: Surface Complexation Modeling. Lützenkirchen J (ed) Elsevier, Amsterdam, p v–x

Mathur SS, Dzombak DA (2006) Surface complexation modeling: Goethite. *In*: Surface Complexation Modeling. Lützenkirchen J (ed) Elsevier, Amsterdam, p 443–468

Mehra OP, Jackson ML (1960) Iron oxide removal from soils and clays by a dithionite–citrate system buffered with sodium bicarbonate. Clays Clay Minerals 7:317–327

Meima JA, Comans RNJ (1998) Application of surface complexation/precipitation modeling to contaminant leaching from weathered municipal solid waste incinerator bottom ash. Environ Sci Technol 32:688–693

Miller AW, Rodriguez DR, Honeyman BD (2010) Upscaling sorption/desorption processes in reactive transport models to describe metal/radionuclide transport: A critical review. Environ Sci Technol 44:7996–8007

Milne CJ, Kinniburgh DG, Tipping E (2001) Generic NICA-Donnan model parameters for proton binding by humic substances. Environ Sci Technol 35:2049–2059

Milne CJ, Kinniburgh DG, van Riemsdijk WH, Tipping E (2003) Generic NICA-Donnan model parameters for metal-ion binding by humic substances. Environ Sci Technol 37:958–971

Morel FMM, Yeasted JG, Westall JC (1981) Adsorption models: A mathematical analysis in the framework of general equilibrium calculations. *In*: Adsorption of Inorganics at Solid–Liquid Interfaces, Anderson MA, Rubin AJ (eds), Ann Arbor Science, Ann Arbor MI p 263–294

Morel FMM, Hering JG (1993) Principles and Applications of Aquatic Chemistry. Wiley-Interscience, New York, Chap 8

Pabich WJ, Valiela I, Hemond HF (2001) Relationship between DOC concentration and vadose zone thickness and depth below water table in groundwater of Cape Cod USA. Biogeochemistry 55:247–268

Park SW, Huang CP (1989) The adsorption characteristics of some heavy metal ions onto hydrous CdS(s) surface. J Colloid Interface Sci 128:245–257

Parkhurst DL, Appelo CAJ (2013), Description of input and examples for PHREEQC version 3—A computer program for speciation, batch-reaction, one-dimensional transport, and inverse geochemical calculations. *In*: US Geological Survey Techniques and Methods, book 6, chap. A43, 497 p., available at https://pubs.usgs.gov/tm/06/a43/

Payne TE, Davis JA, Ochs M, Olin M, Tweed CJ, Altmann S, Askarieh MM (2006) Comparative evaluation of surface complexation models for radionuclide sorption by diverse geologic materials. *In*: Surface Complexation Modeling. Lützenkirchen J (ed) Elsevier, Amsterdam, p 605–633

Perdue EM, Reuter JH, Parrish RS (1984) A statistical model of proton binding by humus. Geochim Cosmochim Acta 48:1257–1263

Ponthieu M, Juillot F, Hiemstra T, van Riemsdijk WH, Benedetti MF (2006) Metal ion binding to iron oxides. Geochim Cosmochim Acta 70:2679–2698

Reich,TJ, Das S, Koretsky CM, Lund TJ, Landry C (2010) Surface complexation modeling of Pb(II) adsorption on mixtures of hydrous ferric oxide, quartz and kaolinite. Chem Geol 275:262–271

Rennert T, Dietel J, Heilek S, Dohrmann R, and Mansfeldt T (2021) Assessing poorly crystalline and mineral–organic species by extracting Al, Fe, Mn, and Si using (citrate-) ascorbate and oxalate. Geoderma 397:115095

Ritchie JD, Perdue EM (2003) Proton-binding study of standard and reference fulvic acids, humic acids, and natural organic matter. Geochim Cosmochim Acta 67:85–96

Ronngren L, Sjoberg S, Sun ZX, Forsling W (1994) Surface reactions in aqueous metal sulfide systems. 5. The complexation of sulfide ions at the ZnS–H_2O and PbS–H_2O interfaces. J Colloid Interface Sci 162:227–235

Runkel RL, Kimball BA, McKnight DM, Bencala KE (1999) Reactive solute transport in streams: A surface complexation approach for trace metal sorption. Water Resources Res 35:3829–3840

Ryan JN, Gschwend PM (1992) Effect of iron diagenesis on the transport of colloidal clay in an unconfined sand aquifer. Geochim Cosmochim Acta 56:1507–1521

Schaller MS, Koretsky CM, Lund TJ, Landry CJ (2009) Surface complexation modeling of Cd(II) adsorption on mixtures of hydrous ferric oxide, quartz and kaolinite. J Colloid Interface Sci 339:302–309

Schindler PW (1981) Surface complexes at oxide–water interfaces. *In*: Adsorption of Inorganics at Solid–Liquid Interfaces, Anderson MA, Rubin AJ (eds), Ann Arbor Science, Ann Arbor MI p 1–49

Schindler PW, Kamber HR (1968) Die aciditat von silanolgruppen. Helv Chim Acta 51:1781–1786

Schindler PW, Stumm W (1987) The surface chemistry of oxides/hydroxides and oxide minerals. *In*: Aquatic Surface Chemistry, Stumm W (ed), Wiley-Interscience, p 83–110

Smith KS, Ranville JF, Plumlee GS, Macalady DL (1998) Predictive double-layer modeling of metal sorption in mine drainage systems. *In*: Adsorption of Metals by Geomedia. Jenne E (ed) Academic Press, San Diego, p 521–547

Sparks DA (2003) Environmental Soil Chemistry, 2nd ed. Academic Press, Amsterdam

Sposito G (2008) The Chemistry of Soils, 3rd ed. Oxford University Press, New York

Steefel CI (2019) Reactive transport at the crossroads. Rev Mineral Geochem 85:1–26

Steefel CI, MacQuarrie KTB (1996) Approaches to modeling of reactive transport in porous media. Rev Mineral 34:83–125

Stockmann M, Schikora J, Becker DA, Flugge J, Noseck U, Brendler V (2017) Smart K_d-values, their uncertainties and sensitivities—Applying a new approach for realistic distribution coefficients in geochemical modeling of complex systems. Chemosphere 187:277–285

Stollenwerk KG (1994) Geochemical interactions between constituents in acidic groundwater and alluvium in an aquifer near Globe, Arizona. Appl Geochem 9:353–369

Stollenwerk KG (1995) Modeling the effects of variable groundwater chemistry on adsorption of molybdate. Water Resources Res 31:347–357

Stumm W (1992) Chemistry of the Solid–Water Interface: Processes at the Mineral–Water and Particle–Water Interface in Natural Systems. Wiley-Interscience, New York

Stumm W, Morgan JJ (1996) Aquatic Chemistry: Chemical Equilibria and Rates in Natural Waters, 3rd ed. Wiley-Interscience, New York

Stumm W, Huang CP, Jenkins SR (1970) Specific chemical interactions affecting the stability of dispersed systems. Croat Chem Acta 42:223–244

Stumm W, Hohl H, Dalang F (1976) Interaction of metal ions with hydrous oxide surfaces. Croat Chem Acta 48:491–504

Susetyo W, Dobbs JC, Carreira LA (1990) Development of a statistical model for metal–humic interactions. Anal Chem 62:1215–1221

Susetyo W, Carreira LA, Azarraga LV, Grimm DM (1991) Fluorescence techniques for metal–humic interactions. Fresenius J Anal Chem 339:624–635

Suter D, Biehler D, Blaser P, Hollman A (1998) Derivation of a sorption data set for the Gorleben overburden. *In:* Proceedings DisTec 98, International Conference on Radioactive Waste Disposal, Hamburg, p 581–584

Sverjensky DA, Sahai N (1996) Theoretical prediction of single-site surface-protonation equilibrium constants for oxides and silicates in water. Geochim Cosmochim Acta 60:3773–3797

Tamm O (1922) Eine Methode zur Bestimmung des anorganischen Gelkomplexes im Boden [A method for determining the inorganic gel complex in soil]. Meddelanden fran Statens Skogsforsoksanstalt [Announcements from the Norwegian Forestry Research Institute] 19:385–404

Thurman EM (1985) Organic Geochemistry of Natural Waters. Kluwer Academic, Dordrecht

Tipping EA (1994) WHAM—A chemical equilibrium model and computer code for waters, sediments, and soils incorporating a discrete site/electrostatic model of ion-binding by humic substances. Comput Geosci 20:973–1023

Tipping EA (1998) Humic ion-binding Model VI: An improved description of the interactions of protons and metal ions with humic substances. Aquatic Geochemistry 4:3–48

Tipping EA (2002) Cation Binding by Humic Substances. Cambridge University Press, Cambridge UK

Tipping EA, Lofts S, Sonke JE (2011) Humic Ion-Binding Model VII: A revised parameterisation of cation-binding by humic substances. Environ Chem 8:225–235

Tombacz E, Szekeres M (2004) Colloidal behavior of aqueous montmorillonite suspensions: the specific role of pH in the presence of indifferent electrolytes. Appl Clay Sci 27:75–94

Tonkin JW, Balistrieri LS, Murray JW (2004) Modeling sorption of divalent metal cations on hydrous manganese oxide using the diffuse double layer model. Appl Geochem 19:29–53

USDA (2022) Kellogg Soil Survey Laboratory Methods Manual, Soil Survey Investigations Report No. 42, Version 6.0, Part 1: Current Methods. US Department of Agriculture, Natural Resources Conservation Service

USEPA (1991) MINTEQA2/PRODEFA2 A Geochemical Assessment Model for Environmental Systems: Version 3.0 User's Manual. EPA/600/3–91/021. US Environmental Protection Agency, Office of Research and Development, Athens, GA

USEPA (1998) MINTEQA2/PRODEFA2 A Geochemical Assessment Model for Environmental Systems: User Manual Supplement for Version 4.0. US Environmental Protection Agency, National Exposure Research Laboratory, Athens, GA. Available at: https://www.epa.gov/sites/default/files/documents/supple1.pdf

USEPA (2003a) EPA's Composite Model for Leachate Migration with Transformation Products (EPACMTP): Parameters/Data Background Document. US Environmental Protection Agency, Office of Solid Waste, Washington, DC. April. Available at: https://www.epa.gov/smm/epas-composite-model-leachate-migration-transformation-products-epacmtp

USEPA (2003b) EPACMTP Technical Background Document. US Environmental Protection Agency, Office of Solid Waste, Washington, DC. April. Available at: https://www.epa.gov/smm/epas-composite-model-leachate-migration-transformation-products-epacmtp

USEPA (2003c) Addendum to the EPACMTP Technical Background Document. US Environmental Protection Agency, Office of Solid Waste, Washington, DC. Available at: https://www.epa.gov/smm/epas-composite-model-leachate-migration-transformation-products-epacmtp

USEPA (2014) Final Human and Ecological Risk Assessment of Coal Combusion Residuals. RIN: R050-AE81, US Environmental Protection Agency, Office of Solid Waste and Emergency Response, Washington, DC, December. Available at: https://downloads.regulations.gov/EPA-HQ-OLEM-2019-0173-0008/content.pdf

USEPA (2024a) MINTEQA2 Equilibrium Speciation Model, version 4.03, released May 2006. US Environmental Protection Agency. Available at: https://www.epa.gov/ceam/minteqa2-equilibrium-speciation-model

USEPA (2024b) STORET: STOrage and RETreival water quality data repository. US Environmental Protection Agency. Available at: https://www3.epa.gov/storet/index-.html

van Eynde E, Groenenberg JE, Hoffland E, Comans RNJ (2022) Solid-solution partitioning of micronutrients Zn, Cu, and B in tropical soils: Mechanistic and empirical models. Geoderma 414:115773

van Riemsdijk WH, Koopal LK, Kinniburgh DG, Benedetti MF, Weng L (2006) Modeling the interactions between humics, ions, and mineral surfaces. Environ Sci Technol 40:7473–7480

Wang Z, Lee SW, Catalano JG, Lezama-Pacheco JS, Bargar JR, Tebo BM, Giammar DE (2013) Adsorption of Uranium(VI) to manganese oxides: X-ray absorption spectroscopy and surface complexation modeling. Environ Sci Technol 47:850–858

Weng L, Temminghoff EJM, van Riemsdijk WH (2001) Contribution of individual sorbents to the control of heavy metal activity in sandy soil. Environ Sci Technol 35:4436–4443

Westall JC, Hohl H (1980) A comparison of electrostatic models for the oxide/solution interface. Adv Colloid Interface Sci 12:265–294

White DE, Hem JD, Waring GA (1963) Data of Geochemistry. US Geological Survey Professional Paper 440-F, US Government Printing Office, Washington, DC

Yeh GT, Tripathi VS (1989) A critical evaluation of recent developments in hydrogeochemical transport models of reactive chemical components. Water Resources Res 25:93–108

Zachara JM, Cowan CE, Resch CT (1991) Sorption of divalent metals on calcite. Geochim Cosmochim Acta 55:1549–1562

Zachara JM, Smith SC, Resch CT, Cowan CE (1992) Cadmium sorption to soil separates containing layer silicates and iron and aluminum oxides. Soil Sci Soc Am J 56:1074–1084
Zhang S, Kent DB, Elbert DC, Shi Z, Davis JA, Veblen DR (2011) Mineralogy, morphology, and textural relationships in coatings on quartz grains in sediments in a quartz-sand aquifer. J Contam Hydrol 124:57–67
Zhu C, Anderson G (2002) Environmental Applications of Geochemical Modeling. Cambridge University Press, Cambridge, UK

www.ingramcontent.com/pod-product-compliance
Lightning Source LLC
Chambersburg PA
CBHW060425220326
41598CB00021BA/2298
9781946850164